OXFORD CLASSIC TEX
THE PHYSICAL SCIEN

M McMahon.
Liverpool 9/98

DYNAMICAL THEORY OF
CRYSTAL LATTICES

BY

MAX BORN

AND

KUN HUANG

CLARENDON PRESS · OXFORD

Oxford University Press, Great Clarendon Street, Oxford OX2 6DP

Oxford   New York

Athens   Auckland   Bangkok   Bogota   Buenos Aires   Calcutta
Cape Town   Chennai   Dar es Salaam   Delhi   Florence   Hong Kong   Istanbul
Karachi   Kuala Lumpur   Madrid   Melbourne   Mexico City   Mumbai
Nairobi   Paris   São Paolo   Singapore   Taipei   Tokyo   Toronto   Warsaw
and associated companies in
Berlin   Ibadan

Oxford is a trade mark of Oxford University Press

Published in the United States
by Oxford University Press Inc., New York

© Oxford University Press 1954

First published 1954
Reprinted in paperback 1969, 1985, 1988
First published in Oxford Classics series 1988

A catalogue record for this book is available from the British Library

Library of Congress Cataloging in Publication Data
(Data available)
ISBN 0 19 850369 5

Printed in Great Britain by Redwood Books Ltd, Trowbridge

PREFACE

THIS book was begun during the War, about fourteen years ago. My first book on crystals had appeared during the First World War, under the title *Dynamik der Kristallgitter* (Teubner, 1915). A few years later I was invited by Sommerfeld to write an article on the subject for the *Mathematical Encyclopaedia*. This appeared in volume v, p. 527, under the title 'Atomtheorie des festen Zustandes' and was published as a separate book (Teubner, 1923). This was two years before the discovery of quantum mechanics. A report on the situation up to the year 1933 has been published as an article in the *Handbuch der Physik* by Maria Göppert-Mayer and myself. There are several other articles in this *Handbuch* by K. F. Herzfeld, R. de W. Kronig, A. Smekal, H. G. Grimm, and H. Wolff, dealing with problems of lattice dynamics. Meanwhile several books on crystal theory have been published which take proper account of quantum mechanics. The most comprehensive is that by F. Seitz, *A Modern Theory of Solids* (McGraw-Hill, N.Y. and London, 1940); other books, for example that by N. F. Mott and R. W. Gurney, *Electronic Processes in Ionic Crystals* (Clarendon Press, Oxford, 1940), deal only with restricted sets of problems. Some special subjects, like the theory of specific heat, infra-red absorption, and Raman effect, have become standard chapters in general textbooks. A great number of single investigations have been published since my article in the *Encyclopaedia*.

The situation seemed to demand an attempt at a new, comprehensive presentation. But the subject has become much too large to be dealt with from all aspects. The contributions of my own school during the last few years have been mainly concerned with non-conducting materials. It seemed to be desirable to give a description of the methods and results in this field.

My plan was to start from the most general principles of quantum theory and to derive in a deductive way the structures and properties of crystals, as far as one could proceed.

I wrote a series of sections which now correspond to the backbone of the Chapters IV to VII of the present book. But with the end of the war my time was so occupied with other matters that I could not continue to write this book. It remained in my desk for several years until Dr. K. Huang, at that time an I.C.I. Fellow at Liverpool, came to work with me for the holidays. I gave him this manuscript and he was interested in the matter. He wrote several interesting papers on crystal theory, and I suggested that he should finish the book.

He accepted this and has succeeded in his task. However, the book has become rather different from my original plans. Dr. Huang, who is convinced that science's main purpose is its social usefulness, found my plan of an abstract, deductive presentation not to his taste. Therefore, he has written some introductory chapters of a more elementary character which should be easy to understand, and which lead slowly up to the general theory of the second half of the book. He has also rewritten my original text, generalizing it in many ways, and adding new sections.

Thus the final form and the wording of this book are essentially due to Dr. Huang. I have discussed the text with him and sometimes suggested alterations. He had to depart before it was quite finished and has sent me the remaining sections from China. I have checked the whole text and added a number of pages, footnotes, and some appendixes. The latter refer mainly to the historical aspect of the theory. Huang has often referred to newer events which he has witnessed, while I, being of an older generation, remember older developments. I have tried to amend this. But anyone interested in the sources should turn to my older books.

The book is not entirely a compilation of published results. The approach to the thermodynamics of lattices was sketched by me and worked out in considerable detail by Huang. His main contribution to this section is the extension of the theory of elasticity of lattices to finite strains. I think that the formulae given here, which represent the temperature-dependence of all parameters describing elastic, pyroelectric and piezoelectric properties of dielectrics, have never been given before. The formulae look rather long and complicated, but are really simple if one takes the trouble to study them. We have not discussed them in detail, and leave this to those who wish to apply them to special cases. Other sections which are entirely due to Huang are the phenomenological treatment of dispersion in the first part, and also its detailed atomistic treatment in the second part of the book, the quantum theory of width of the infra-red lines, and many other minor matters.

It may not be superfluous to mention some branches of crystal dynamics which are not in this book. There is first the theory of metals, which is a science by itself represented in several well-known textbooks (N. F. Mott and H. Jones, *The Theory of the Properties of Metals and*

Alloys, Oxford, Clarendon Press, 1936; A. H. Wilson, *The Theory of Metals*, Cambridge University Press, 2nd ed., 1953).

Then there is the wide field known under the name of 'Order-Disorder Theories', including the theory of alloys and of ferromagnetism. Here the lattice is regarded as a rigid frame; the problem consists in finding the statistical equilibrium distribution of particles, or of properties of particles, over the fixed lattice points. This has nothing to do with the dynamics of the lattice itself, and there are several accounts in the recent literature (F. C. Nix and W. Shockley, *Rev. Mod. Phys.* **10**, 1 (1938); J. H. Wannier, *Rev. Mod. Phys.* **17**, 50 (1945); L. D. Taschick and H. M. Jones, *Phys. Rev.* **91**, 1131 (1953); particularly dealing with ferromagnetism: P. R. Weiss, *Phys. Rev.* **74**, 1493 (1948)).

Other subjects omitted from the book are the theories of scattering of X-rays, electrons, and neutrons by crystal lattices. A great part of the extended literature on these subjects is concerned with the purely geometrical problem of determining lattice structures. But there are deep and important investigations on the propagation of rays of different kinds through crystal lattices which take account of the dynamical processes involved. They are all elaborations of the original work by P. P. Ewald on the dynamics of X-ray scattering. There exist fairly recent reports on scattering of X-rays and of electrons in two books by Max von Laue (*Röntgenstrahleninterferenzen*, Akad. Verlags-Ges., Becker u. Erler, Leipzig, 1941; *Materiewellen und ihre Interferenzen*, Akad. Verlags-Ges., Leipzig, 2nd ed., 1949). These books also contain sections on the interaction of lattice vibrations and the scattered particles. We have omitted these theories from our book only with regret, as they provide the most striking empirical evidence for the dynamics of lattice vibrations as treated here. But there are comprehensive accounts of these matters which made a repetition appear superfluous. (In the *Reports on Progress of Physics* are the following articles: K. Lonsdale (X-rays, experimental), **9**, 252 (1942); M. Born (X-rays, theoretical), **9**, 294 (1942); G. E. Bacon and K. Lonsdale (neutrons), **16**, 1 (1953). Further: R. D. Lowde (neutrons), *Proc. Roy. Soc.* A, **221** (1954).)

I have to say a word about the notation. It is essentially the same as that developed in my first book and consecutive papers, but adapted to British printing usage. Our main concern was economy with letters. The experimentalists dealing with crystal structures use three letters *h, k, l* for the Miller indices of lattice points: what a waste! If one were to follow this procedure, each letter would have to be used with many different meanings. Even with the greatest care, we could not completely avoid this. But we hope that we have succeeded in never using the same letter with two different meanings in any one formula. We have chosen the letters x, ξ for indicating points in real space, and y, η for points in reciprocal space. The relation between these spaces is expressed in terms of affine geometry, as is customary in general relativity. The three coordinates are always numbered 1, 2, 3 and indicated by small Greek letters used as subscripts.

The reason for my writing and signing this preface alone is not only the spatial separation from my collaborator, but my wish to make it clear that the book would never have been finished without his devoted and efficient labours. He has informed me that he is going to produce a Chinese edition.

I have to thank my former collaborator, Dr. Bhatia, for helping me in revising and checking the text and reading the proofs. Dr. J. M. Ziman, Oxford, and Dr. D. J. Hooton, Edinburgh, have given their assistance in the final corrections and proof-reading, and Dr. Hooton has produced the alphabetical index.

I am much indebted to Sir Ernest Oppenheimer and the firm Industrial Distributors, London, for financial help which made the production of this book possible.

The Clarendon Press, Oxford, has obliged me very much by following all my suggestions in preparing and printing the book.

M. B.

ACKNOWLEDGEMENTS

THE authors are grateful to the following for permission to use figures :

Professor E. A. Guggenheim for a figure from Fowler and Guggenheim, *Statistical Thermodynamics* (C.U.P.), Fig. 4.

Zeitschrift für Physik (Springer-Verlag), Figs. 1, 2, 3, 21.

Proceedings of the Royal Society, A, Figs. 8, 12, 18 *b*, 26.

Philosophical Transactions of the Royal Society, A, Figs. 10, 11, 15, 16, 17, 27.

Nature (Macmillan), Fig. 18 *a*.

Physical Review, Figs. 13, 14.

Proceedings of the Cambridge Philosophical Society, Figs. 24, 25.

Annalen der Physik (J. A. Barth), Fig. 5.

CONTENTS

PART I: ELEMENTARY THEORIES

PART I

ELEMENTARY THEORIES

I

ATOMIC FORCES

1. Theoretical considerations

CRYSTALLINE solids are distinguished from other states of matter by a periodic arrangement of the atoms; such a structure is called a crystal lattice. A precise description of the geometry of a lattice will be given later in § 22 of Chapter V. Essentially the regularity displayed by a crystal lattice is that of a three-dimensional mesh which divides space into identical parallelepipeds. Imagine a number of identical atoms placed at the intersections of such a mesh; then we have what is known as a *simple lattice* (or Bravais lattice). The interstitial parallelepipeds, which have atoms for corners, are referred to as the *elementary lattice cells*; in a simple lattice there is thus exactly one atom to each elementary cell. Now if the atoms are replaced by similarly oriented molecules, the result is a general lattice structure; clearly every cell contains as many atoms as there are in one molecule. The term molecule here describes the geometrical dispositions of the atoms and need not signify a real molecule (a group of atoms form a real molecule in a lattice only if they are more tightly bound to one another than to other atoms in the lattice).

In Appendix I, the familiar structure of the NaCl lattice is illustrated together with some other common lattice types. We note that despite the obvious cubic symmetry of the NaCl structure, the elementary cells have to be chosen as rhombohedra. The vectors a_1, a_2, a_3 shown in the figure are known as the *basic vectors*, which form the edges of the elementary cells. The sites of the Na^+ ions taken by themselves form a simple lattice; the NaCl structure is, on the other hand, a general lattice, since we can pair every Na^+ ion with one of its six neighbouring Cl^- and designate the pair a 'molecule' (of course no *real* molecules exist in this case, as every ion in the lattice is similarly related to all its six neighbours).

According to the underlying atomic forces, crystalline solids are roughly classified into the following four principal types:

(1) Ionic crystals,

(2) van der Waals crystals,

(3) Valency crystals,

(4) Metals.

A fundamental feature distinguishes types (1) and (2) from types (3) and (4). The distinction depends on the units with which the crystals are built. In a general way, one can describe the units as being *saturated* in the two former types, and *unsaturated* in the two latter. An atom (or ion) with all its electrons in closed shells (rare gas configurations) or a chemically saturated molecule is typical of a saturated unit; on the other hand, an atom that can readily form covalent bonds is an unsaturated unit. Speaking more physically, the electronic wave functions of unsaturated units, when brought together, are liable to be drastically altered, whereas the wave functions of saturated units are not so significantly affected. This difference can be traced to the fact that, in the unsaturated units, either the lowest electronic state is degenerate, or there are energy levels close to the ground state; both cases provide scope for electronic rearrangements under slight perturbations. The treatment of the crystals formed of saturated units is simpler; one can calculate the energy of interaction approximately by the quantum-mechanical perturbation theory. One important consequence is that the interaction can be considered as essentially operating between pairs, and the total energy in the lattice is the sum of the interactions between the units taken by pairs (two-body interaction). In fact, it is mainly in connexion with these crystals, namely, the ionic and van der Waals crystals, that the elementary theories are useful.

Ionic crystals. For example let us consider the alkali halides, which are the real crystals closest to the theoretical model for ionic crystals, and let us imagine building up their ionic lattices. The alkali atoms: Na, K, Rb, Cs (Li is not considered owing to the complications caused by its small size†) have one valence electron outside the complete shells; the halogen atoms: F, Cl, Br, I, on the other hand, just lack one electron to complete the outermost shell. From the atoms we form the free ions by transferring the odd valency electrons from the alkali atoms to the halogen atoms. The free ions have the stable rare gas configurations; the sequences, Na^+, K^+, Rb^+, Cs^+ and F^-, Cl^-, Br^-, I^- have the structures of the rare gas atoms, Ne, A, Kr, Xe. The extra stability achieved in the electronic configuration by forming the free ions from the atoms is not sufficient to offset the rise in electrostatic energy involved in separating the valence electrons from the alkali ions; this is clear from a comparison of the first ionization potentials of the alkali atoms with

† Cf. L. Pauling, *The Nature of the Chemical Bond* (2nd ed., Cornell, 1948), pp. 351–63.

the electron affinities of the halogen atoms given in Table 1. The electrostatic energy is, however, largely compensated when we next bring the ions together to form the lattice. Let us imagine the ions to be brought together in the following manner. We think of the ions as being initially arrayed in accordance with the desired lattice structure, but so far apart that their interaction is negligible. The ions are then brought together uniformly so that the same lattice structure is maintained throughout.

TABLE 1

	Na	K	Rb	Cs
Ionization potential in eV.	5·12	4·32	4·16	3·87

	F	Cl	Br	I
Electron affinity† in eV.	4·15	3·72	3·50	3·14

† See Tables 11 and 12.

We can crudely consider an ion as having a radius, beyond which the electron density is negligible. Then before the ions interpenetrate they interact like point charges $\pm e$. The electrostatic energy of a lattice of point charges cannot in general be calculated in an elementary way; the difficulty is due to the slow decrease of the Coulomb interaction with distance (long range force) which makes a direct summation procedure impossible. A general method of calculating this energy and a simpler method applicable to structures of high symmetry are described in Appendix II. Here it suffices to point out that the energy is inversely proportional to the linear dimensions of the lattice. Thus the energy per cell can be written quite generally for any ionic lattice in the alternative forms

$$-(ze)^2\frac{\alpha'}{r} = -(ze)^2\frac{\alpha''}{d}, \qquad (1.1)$$

where ze is the smallest ionic charge in the lattice, and r, d are respectively the nearest ion-ion distance in the lattice and the lattice constant, either of which gives a measure of the lattice dimensions. α' (or α''), which is a pure number known as *Madelung's constant*, depends only on the lattice structure and was first calculated by Madelung† with reference to real crystal lattices. A table of values for α' (and α'') for some common lattice types is quoted in Appendix II. Briefly, we shall refer to the above energy as *Madelung's energy*; it is the electrostatic energy between the ions in a lattice if they are considered as point charges.

† E. Madelung, *Phys. Zeit.* **19**, 524 (1918).

Madelung's energy decreases algebraically with decreasing lattice constant and tends to contract the lattice. Its effect on the lattice can thus be described as attractive. The attraction is unopposed until the neighbouring ions begin to interpenetrate, when other forces arise. Let us consider the effect of an overlap between ions. As we have explained, owing to the particular stability of saturated units, perturbation methods for energy calculation can be used. To a first approximation, the energy is given by the average value of the Hamiltonian calculated with the wave functions of the free ions (unperturbed wave functions). An exhaustive treatment of various properties of some alkali halide lattices has been given by Löwdin[†] on this basis. However, we shall follow a cruder method, based on the Thomas–Fermi–Dirac statistical method, due to Lenz, Gombás, and particularly Jensen;[‡] with this simpler method, it is easier to see the various factors which contribute to the resultant force.

In the statistical method, an electron density function $\rho(\mathbf{x})$ replaces the wave function; once the density function is known, the corresponding energy can be calculated. For two overlapping ions, Lenz and Jensen simply superpose the density functions of the free ions; in other words, the ions are assumed to be undeformed. This is the parallel to the wave mechanical first-order perturbation method, which does not allow for the effects of the distortions in the wave functions. In the Thomas–Fermi–Dirac method,[§] the following three energy terms are considered:

(i) The *classical Coulomb energy* of the average charge distribution, namely, $-e\rho(\mathbf{x})$ plus the nuclear charges.

(ii) The *zero-point kinetic energy* of the electrons as required by the exclusion principle, according to which a volume h^3 in phase space can accommodate no more than two electrons, h being Planck's constant, so that states of higher momenta will be involved with an increase in electron density ρ. The average kinetic energy turns out to be proportional to $\rho^{2/3}$ and the kinetic energy per unit volume is given by

$$\frac{3^{\frac{5}{3}}h^2}{40m}\left(\frac{1}{\pi}\right)^{\frac{2}{3}}\rho^{\frac{5}{3}}, \tag{1.2}$$

where m is the mass of the electron.

---

† P. Löwdin, *A Theoretical Investigation into some Properties of Ionic Crystals* (Uppsala, 1948). Similar calculations for NaCl were first carried out by R. Landshoff, *Zeit. f. Phys.* **102**, 201 (1936); *Phys. Rev.* **52**, 246 (1937).

‡ W. Lenz, *Zeit. f. Phys.* **77**, 713 (1932); P. Gombás, ibid. **121**, 523 (1943); H. Jensen, ibid. **77**, 722 (1932); ibid. **101**, 141; **101**, 164 (1936).

§ See also P. Gombás, *Theorie und Lösungsmethoden des Mehrteilchenproblems der Wellenmechanik* (Birkhauser, Basel, 1950); for general description and other references see Gombás's book.

(iii) The *exchange energy*, which is essentially a correction to (i). The Coulomb energy calculated classically is inaccurate in two ways. First, with the electrons represented as continuous clouds, (i) includes also the self-interaction of the charge cloud of an electron with itself. Secondly, the proper quantum-mechanical anti-symmetric wave function takes care of the fact that two electrons with parallel spins avoid close encounters; (i) takes no account of this effect. These effects can be taken into account by the addition of Dirac's exchange term:

$$-\frac{3^{\frac{4}{3}}e^2}{4}\left(\frac{1}{\pi}\right)^{\frac{1}{3}}\rho^{\frac{4}{3}} \qquad (1.3)$$

per unit volume, where e is the charge of the electron.

(Gombás† has further approximately corrected for the fact that electrons of opposed spins also avoid close encounters owing to their Coulomb repulsion. This correlational effect is roughly equivalent to raising the exchange term by a fraction. For our illustrative purpose, we shall consider the Thomas–Fermi–Dirac method without this further elaboration.)

The electron density function in the statistical method is determined by the condition that the corresponding energy is a minimum. Jensen‡ has calculated the density functions for all the alkali and halogen free ions. Using these density functions, we can readily calculate the total energies of the alkali halide lattices on the basis of the energy terms (i), (ii), and (iii), if the distortions of the ions are ignored. Before the ions overlap, obviously the energies (ii) and (iii) remain the same as for free ions, but the Coulomb energy (i) now includes the additional Coulomb energy between the ions. The latter is exactly the Madelung energy, which we have already discussed. When two ions overlap, we have to consider the following:

(a) Correction to the Madelung term so as to represent correctly the Coulomb energy (i). The term is attractive so long as neither nucleus has penetrated the other ion, for the electron cloud of either ion which has penetrated the other ion is now more strongly attracted to the other nucleus.

(b) Correction to the zero-point energy. Let ρ_1, ρ_2 respectively denote at any point in the overlap region the original densities of the free ions. The zero-point energy corresponding to the superposed density is given by

$$\frac{3^{\frac{5}{3}}h^2}{40m}\left(\frac{1}{\pi}\right)^{\frac{2}{3}}(\rho_1+\rho_2)^{\frac{5}{3}},$$

† Loc. cit., 1943.   ‡ Loc. cit., 1936.

whereas the same charge in the free ions originally has the kinetic
energy
$$\frac{3^{\frac{5}{2}}h^2}{40m}\left(\frac{1}{\pi}\right)^{\frac{2}{3}}(\rho_1^{\frac{5}{3}}+\rho_2^{\frac{5}{3}}).$$

Thus the correction term is

$$\frac{3^{\frac{5}{2}}h^2}{40m}\left(\frac{1}{\pi}\right)^{\frac{2}{3}}[(\rho_1+\rho_2)^{\frac{5}{3}}-\rho_1^{\frac{5}{3}}-\rho_2^{\frac{5}{3}}] \tag{1.4}$$

per unit volume. This term is positive and increases with the
extent of the overlap, leading thus to a repulsion between the ions.

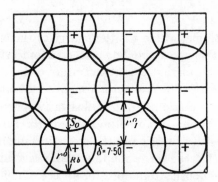

FIG. 1. Ions in cubic planes of RbI (Jensen).

(c) Following the same considerations as in (b), we get the following
correction to the exchange energy due to the overlap:

$$-\frac{3^{\frac{4}{3}}e^2}{4}\left(\frac{1}{\pi}\right)^{\frac{1}{3}}[(\rho_1+\rho_2)^{\frac{4}{3}}-\rho_1^{\frac{4}{3}}-\rho_2^{\frac{4}{3}}] \tag{1.5}$$

per unit volume. This expression is negative and decreases algebraically with increasing overlap. The corresponding force is thus
attractive.

Fig. 1 shows the ions in positions corresponding to the normal lattice
constant in a plane normal to an axis of the cube for a typical ionic
crystal (RbI); the radii indicated are those calculated for the free ions
by Jensen. Overlaps occur only between positive and negative ions
which are first neighbours and between negative ions which are second
neighbours. The curves in Fig. 2 represent the negative derivatives (with
respect to the lattice constant) of various energies as functions of the
lattice constant, namely, Madelung's term (VI), the correction terms (a),
(b), and (c) separately (I, III, II) for the first-neighbour interactions,
and the sum of the corrections (a), (b), and (c) for the second-neighbour

interactions between the negative ions (IV). Curve VII is the resultant
of all the distinct contributions.

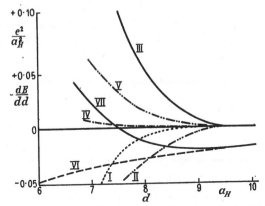

FIG. 2. Negative derivatives of various energies for RbI
(Jensen; a_{H} = hydrogen radius).

An energy term is attractive, if the corresponding curve in Fig. 2 lies
below the axis, and repulsive if the reverse is the case. At large values
of the lattice constant, only the attractive Madelung term is operative.
At small values of the lattice constant the resultant curve rises above the
axis, showing that here the repulsive force predominates. The point at
which the curve cuts the axis corresponds to equilibrium. We notice
that, of the various terms, only the kinetic energy (b) gives a repulsive
force. Pauli's principle is thus ultimately responsible for the bulk of a
lattice.

All the energies except Madelung's term have a very short range.
Curve V gives the sum of all the short range forces; it is seen to be always
repulsive. It is usual to consider these forces together as one single term,
which has been variously called the repulsive force, the exchange repulsive force, van der Waals force of the second kind, or the overlap force.
We shall use the last term, which seems the most appropriate. Owing
to the short range of the overlap force, the cohesive energy (energy
released on forming the lattice from the free ions) is mainly contributed
by Madelung's term. The total cohesive energy is numerically equal to
the area included between the resultant Curve VII and the axis; the
contribution due to the overlap force which is represented by the area
under the Curve V is clearly small, owing to restricted lateral extension
of the curve.

Though the wave-mechanical method must bear a close relation to
the above considerations, no detailed comparison of the individual energy

terms seems possible with the available data, for Pauli's principle requires a rearrangement in the free-ion wave functions, thus making it difficult to separate the various physical factors as we have done above. One result of using the wave-mechanical method has been particularly emphasized by Löwdin, namely, that part of the cohesive energy cannot be interpreted as due to interactions between particle-pairs. This part, according to Löwdin, can be of the order of 10 per cent. of the total cohesive energy!

van der Waals crystals. The Madelung term disappears in a crystal consisting of electrically neutral molecules. Electrostatic interactions are, however, not altogether absent, if the molecules are not spherically symmetric. According to electrostatics, a localized charge distribution, such as that found in a molecule, can be described in terms of its multipole moments. Thus if e_i denotes the element of charge at the position $\mathbf{x}^i$ (referred to a suitable centre in the molecule as origin), the *multipole moments* can be defined as a vector, and tensors of second and higher orders as follows:

$$m_\alpha = \sum_i e_i x_\alpha^i \quad \text{(dipole moments)}, \tag{1.6}$$

$$q_{\alpha\beta} = \sum_i e_i x_\alpha^i x_\beta^i \quad \text{(quadrupole moments)}.$$

Consider two neutral molecules 1, 2 with moments m_α^1, $q_{\alpha\beta}^1$ and m_α^2, $q_{\alpha\beta}^2$, defined relative to their respective centres, and let $\mathbf{R}$ be the vector joining the centres, from 1 to 2. The Coulomb energy between the two molecules can be expressed in terms of the moments as

$$- \sum_{\alpha\beta} m_\alpha^1 m_\beta^2 \left[\frac{\partial^2}{\partial y_\alpha \, \partial y_\beta} \frac{1}{|\mathbf{y}|} \right]_{\mathbf{y}\,=\,\mathbf{R}} -$$

$$-\tfrac{1}{2} \sum_{\alpha\beta\gamma} (m_\alpha^1 q_{\beta\gamma}^2 - m_\alpha^2 q_{\beta\gamma}^1) \left[\frac{\partial^3}{\partial y_\alpha \, \partial y_\beta \, \partial y_\gamma} \frac{1}{|\mathbf{y}|} \right]_{\mathbf{y}\,=\,\mathbf{R}} +$$

$$+\tfrac{1}{4} \sum_{\alpha\beta\gamma\lambda} q_{\alpha\beta}^1 q_{\gamma\lambda}^2 \left[\frac{\partial^4}{\partial y_\alpha \, \partial y_\beta \, \partial y_\gamma \, \partial y_\lambda} \frac{1}{|\mathbf{y}|} \right]_{\mathbf{y}\,=\,\mathbf{R}} - \dots. \tag{1.7}$$

This expression is obtained by straightforward expansion with respect to the coordinates of the charge elements of the two molecules. The convergence of the series depends clearly on the ratio of the size of the molecules to their separation $|\mathbf{R}|$. Different terms in the series can be described as interactions between the various multipoles of the two molecules. The dipole-dipole, dipole-quadrupole, quadrupole-quadrupole interactions, for instance, correspond to the terms of the second, third, and fourth orders respectively of the expansion.

In molecules where the charge distribution is continuous, the summations over the charge elements in (1.6) become integrations over charge densities. The corresponding interaction energy (1.7) will be referred to as the static Coulomb interaction; the reason will become clear presently.

The static multipole interactions never seem to be the dominant cohesive forces in crystals of neutral molecules. London† has shown that a force of a different nature, known as the *van der Waals force*, is usually responsible for the cohesion between neutral, saturated molecules. The van der Waals force is closely related to the multipole interactions. In fact, it is obtained if the multipole interactions are considered quantum-mechanically. Thus (1.7) should be regarded as a quantum-mechanical operator wherein e_i and $\mathbf{x}^i$ refer to the charge and the position vectors of the electrons. We can obtain the first-order perturbation in energy by forming the mean value of (1.7) with respect to the unperturbed wave function of the system. The latter is simply the product of the electronic wave functions (nuclei considered as fixed) of the unperturbed molecules 1 and 2. Thus it follows that the first-order energy is exactly the static Coulomb interaction which we have described above.

The van der Waals interaction is obtained by carrying the perturbation procedure on to the second order. Let us consider the dipole-dipole interaction. If we designate the electronic states of the unperturbed molecules respectively by i, j, and the excited states by i', j', the second-order energy is given by

$$E^{(2)} = \sideset{}{'}\sum_{i'} \sideset{}{'}\sum_{j'} \frac{\left| \sum_{\alpha\beta} \langle i|m_\alpha^1|i'\rangle \langle j|m_\beta^2|j'\rangle \left[\frac{\partial^2}{\partial y_\alpha\,\partial y_\beta} \frac{1}{|\mathbf{y}|} \right]_{\mathbf{y}=\mathbf{R}} \right|^2}{E_i + E_j - E_{i'} - E_{j'}}, \qquad (1.8)$$

where the primes exclude $i' = i$ and $j' = j$. E_i, E_j, $E_{i'}$, $E_{j'}$ are eigenvalues of the unperturbed molecules, and $\langle i|m_\alpha^1|i'\rangle$, $\langle j|m_\beta^2|j'\rangle$ the matrix elements of the dipoles between the states i, i' in molecule 1 and between the states j, j' in molecule 2. The energy $E^{(2)}$ depends in general not only on the separation between the molecules, but also on the orientation of the molecules with respect to the vector $\mathbf{R}$ joining them. We shall avoid this complication by averaging the expression over various orientations of the molecules. Remembering that the matrix elements transform as vector components when the molecules are rotated, we obtain in a straightforward way the average value of the second-order

† F. London, *Zeit. f. phys. Chem.* **11**, 222 (1930).

energy:

$$\overline{E^{(2)}} = -\frac{2B}{3R^6} \quad (R = |\mathbf{R}|),$$

$$B = -\sum_{i'}' \sum_{j'}' \frac{\left(\sum_{\alpha} |\langle i|m_{\alpha}^1|i'\rangle|^2\right)\left(\sum_{\beta} |\langle j|m_{\beta}^2|j'\rangle|^2\right)}{E_i + E_j - E_{i'} - E_{j'}}. \tag{1.9}$$

This term varies with the inverse sixth power of the molecular separation R. Moreover, if the molecules are in their ground states, all the denominators in (1.9) are negative; $\overline{E^{(2)}}$ is thus negative. Hence the corresponding van der Waals force is attractive.

In all but the simplest cases, the expression (1.9) is too complicated for actual calculations. Various crude approximations for the constant B have been proposed; their relative accuracies are difficult to gauge (cf. Margenau's review† on the subject). We shall describe a method due to London,‡ who correlates the van der Waals force with the polarizabilities of the molecules.

To obtain London's approximation we assume that the excited states i', j' with large matrix elements, and thus contributing essentially to the sum in (1.9), fall within relatively narrow energy ranges. We may then consider all the denominators in (1.9) as approximately the same, for all important terms, and we can write

$$B \simeq \frac{\left[\sum_{i'}' \left(\sum_{\alpha} |\langle i|m_{\alpha}^1|i'\rangle|^2\right)\right]\left[\sum_{j'}' \left(\sum_{\beta} |\langle j|m_{\beta}^2|j'\rangle|^2\right)\right]}{\Delta_1 + \Delta_2}, \tag{1.10}$$

where Δ_1, Δ_2 are roughly the excitation energies of the important groups of states i', j'.

This expression is closely related to the electric polarizabilities of the molecules. Consider, for instance, the molecule 1 placed in a constant electric field $\mathscr{E}$. Taking the interaction energy

$$-\sum_{\alpha, i} e_i \mathbf{x}_{\alpha}^i \mathscr{E}_{\alpha} = -\sum_{\alpha} m_{\alpha}^1 \mathscr{E}_{\alpha}$$

as a perturbation, we obtain the first-order perturbed wave function of the molecule as

$$\psi_i + \sum_{i'}' \frac{-\left(\sum_{\alpha} \langle i'|m_{\alpha}^1|i\rangle \mathscr{E}_{\alpha}\right)}{E_i - E_{i'}} \psi_{i'},$$

---

† H. Margenau, *Rev. Mod. Phys.* **11**, 1 (1939).
‡ F. London, *Zeit. f. Phys.* **63**, 245 (1930).

where the functions ψ_i, $\psi_{i'}$ are electronic wave functions of the unperturbed molecule. The corresponding mean value of the dipole moment is

$$\bar{m}_\alpha = m_\alpha^{(0)} + m_\alpha^{(1)} = \langle i|m_\alpha^1|i\rangle - 2\sum_{i'}{}' \frac{\langle i|m_\alpha^1|i'\rangle \sum_\beta \langle i'|m_\beta^1|i\rangle \mathscr{E}_\beta}{E_i - E_{i'}}$$

(here the matrix elements are assumed real); the first and second terms correspond respectively to the permanent dipole $m^{(0)}$ of the unperturbed molecule and the dipole $m^{(1)}$ induced by the field. Making the same assumption as before in arriving at (1.10), we can write the induced dipole components as

$$m_\alpha^{(1)} = \frac{2\sum_{i'}{}' \langle i|m_\alpha^1|i'\rangle \sum_\beta \langle i'|m_\beta^1|i\rangle \mathscr{E}_\beta}{\Delta_1}. \tag{1.11}$$

To avoid orientational effects, we consider the component of the induced moment in the direction of the field and average its value over various orientations of the molecule. The result is

$$\frac{2}{3}\frac{\sum_{i'}{}'\sum_\alpha |\langle i|m_\alpha^1|i'\rangle|^2}{\Delta_1}|\mathscr{E}| = \alpha_1|\mathscr{E}|. \tag{1.12}$$

The proportionality constant α_1 is the polarizability. After eliminating the sums over the matrix elements in (1.10) with the help of the polarizabilities, we obtain London's approximation for the van der Waals potential:

$$\frac{C}{R^6}, \qquad C = -\tfrac{2}{3}B = -\frac{3}{2}\left(\frac{\Delta_1\Delta_2}{\Delta_1+\Delta_2}\right)\alpha_1\alpha_2. \tag{1.13}$$

The matrix elements of the dipole are responsible for the absorption of radiation; the energies Δ_1, Δ_2 should thus be chosen in accordance with the strongest absorption frequencies of the molecules. When these frequencies are not known, the energies can be taken approximately as the ionization potentials of the corresponding molecules. The values of C for some similar molecules given in Table 2 are calculated by Margenau† with the above formula; the average discrepancy with values obtained by other approximations is about 20 per cent.

Similar considerations of the dipole-quadrupole and quadrupole-quadrupole interactions lead to additional van der Waals potentials

$$-\frac{C'}{R^8}, \qquad -\frac{C''}{R^{10}},$$

varying respectively with the inverse eighth and tenth powers of the

† Cf. F. London, ibid.

distance. Estimates of the constants C', C'' by Margenau† are also given in Table 2. The values are probably not reliable; however, they give an idea of the relative magnitudes of the various van der Waals potentials.

TABLE 2

Theoretical Values of van der Waals Potentials

Substance	C (erg cm.^{6})	C' (erg cm.^{8})	C'' (erg cm.^{10})	d cm.
H	$6·1 \times 10^{-60}$	$32·6 \times 10^{-76}$	$83·1 \times 10^{-92}$	$2·3 \times 10^{-8}$
He	1·23	1·89	1·65	1·24
Ne	4·67	6·9	5·3	1·48
A	55·4	120	136	1·47
Kr	107	275	370	1·6
Xe	233	710	1,120	1·74
H_2	11·4	31	45	1·65
N_2	57·2	120	130	1·45
O_2	39·8	96	120	1·55
CO_2	152	410	590	1·64
CH_4	112	310	440	1·66
NH_3	70	236	410	1·84
Cl_2	321	1,000	1,630	1·76
HCl	111	320	480	1·69
HBr	185	600	1,000	1·8
HI	370	1,360	2,700	1·92

In the same table, under the column marked d are given the distances at which the dipole-quadrupole potential equals the dipole-dipole potential. From these values, one expects that even for nearest neighbours, the dipole-quadrupole potential is only a relatively small fraction of the dipole-dipole potential. For molecules farther apart, all terms other than the dipole-dipole term become rapidly insignificant.

We have thus far considered only a pair of molecules. If we apply the perturbation theory similarly to a group of molecules, it follows directly‡ that the result is approximately equal to the sum of the interactions between the molecules taken in pairs. In other words, the van der Waals potentials are approximately additive.

The solidified rare gases are ideal examples of van der Waals crystals; the static Coulomb interactions are completely absent with the neutral and spherically symmetric atoms, and the attraction due to the van der Waals potential is the only force acting before the atoms overlap one another. The additional forces due to the overlap are exactly the same as in the case of the ionic crystals, consisting of the terms (a), (b), (c) and

† Loc. cit.
‡ F. London, loc. cit.; see Kun Huang and Avril Rhys, *Chin. J. Phys.* (1951), for an estimate of deviation from the additivity rule.

being repulsive in nature. The characteristic feature of the van der Waals crystals as a class is that they are formed of neutral, saturated molecules. As we have mentioned, such molecules are held together principally by the van der Waals forces. In Table 3 the van der Waals interaction energies calculated by London for a number of simple molecular crystals are compared with the experimentally determined cohesive energies.

TABLE 3

Cohesive Energies of van der Waals Crystals

Substance	Calculated van der Waals energy (k.cal./mol.)	Experimental cohesive energy (k.cal./mol.)
Ne	0·47	0·59
N$_2$	1·64	1·86
O$_2$	1·69	$\begin{cases} 1·89 \\ 2·06 \end{cases}$
A	2·08	2·03
CH$_4$	2·42	2·70
NO	2·89	4·29

The calculations are based on rather crude assumptions and the exact values are not significant, but both the order of magnitude and the correct relative magnitudes for different substances show clearly that the van der Waals forces are mainly responsible for the cohesion of these crystals.

It might seem surprising at first that the multipole static interactions for molecules with dipole moments are not more important. The dipole-dipole static interaction, for instance,

$$-\sum_{\alpha\beta} m_\alpha^1 m_\beta^2 \left(\frac{\partial^2}{\partial y_\alpha \partial y_\beta} \frac{1}{|\mathbf{y}|} \right)_{\mathbf{y}\,=\,\mathbf{R}},$$

varies generally with the inverse third power of the distance; its longer range would seem to favour a greater contribution to the cohesive energy. The explanation is that such interactions are strongly direction-dependent. Unlike the van der Waals interaction, the static interaction vanishes, when averaged over various orientations of the molecules. When the molecules are placed on a lattice they cannot all be favourably oriented with respect to one another.

To summarize, for crystals composed of saturated units, one can calculate the interaction energy by the perturbation theory. For large lattice constants, the first-order perturbation leads simply to the classical Coulomb interactions, and the second-order perturbation gives rise to the van der Waals attractions. If the units are electrically charged, the former predominate, consisting principally of the Madelung term (ionic crystals); if the units are electrically neutral, the latter predominate (van der Waals crystals). When the atoms (ions or molecules) begin to overlap, the first-order perturbation leads, by means of the exclusion principle, to a strong repulsion (overlap force); the expression for the van der Waals force, which is the result of the second-order perturbation, based on the multipole expansion, is less satisfactory for such closely situated atoms. However, the comparison of the quadrupole-dipole and dipole-dipole potentials indicates that the multipole expansion does not altogether fail and the dipole-dipole potential probably reproduces the correct order of magnitude.

About valency crystals and metals we shall make only such remarks as are relevant to our discussion.

In these solids, the electronic clouds of the constituent atoms undergo such radical rearrangements when the atoms approach one another that the electrons can no longer be attributed to individual atoms and the interactions cannot be considered even approximately as between pairs of atoms. In valency crystals, the electrons are shared between neighbouring atoms as valency bonds. The number of bonds that an atom can form is restricted; thus, whereas ionic and van der Waals crystals favour lattice structures which provide a large number of nearest neighbours, in valency crystals the maximum number of neighbours equals the number of valency bonds an atom can form. Moreover, the interactions of an atom with its different neighbours are interdependent. The most characteristic feature of the covalent bonds is their directional nature; an atom has a strong preference to maintain its neighbours in certain relatively fixed directions. One feature, however, the valency crystals have in common with the ionic and van der Waals crystals, namely, for all these crystals the lattice considered as a whole is saturated. That is, the originally unsaturated units of the valency crystals, once formed into the lattice, become saturated. As we shall see, this fact permits us to consider the crystal lattice as possessing an effective potential function for the movement of the nuclei. The existence of such a potential function provides the basis for the general mathematical treatment of crystal lattices to be given in Part II. From a formal point of view, the valency crystals differ from the ionic and van der Waals crystals only in the greater complexity in representing explicitly the potential function, a difference which does not in any way affect the general theory itself.

Metals form a class of solids so different from the others that we shall make no attempt to incorporate it in our treatment. Not only is the basic binding force in metals radically different from other solids, but many properties of particular interest in metals have no parallel in the other types of solids. It is interesting to contrast the metals with the valency crystals in this way: although both types are built of unsaturated units, whereas the valency crystal as a whole is saturated, the metallic lattice is not. In fact, in metals there is a practically continuous band of electronic energy levels contiguous to the ground state. All the characteristically metallic properties are the direct consequence of the unsaturated nature of the metallic lattice. The presence of the band of electronic levels precludes the possibility of considering the nuclear motions with the help of a potential function. It is of interest to note that, in the theory of metallic conduction, use is made of the current notion of lattice vibrations which is derived with the help of an effective potential function. In actual fact, the lattice vibrations do not form a closed system, for there is a continuous energy exchange between the electrons and the nuclear motion. It is possible that an explanation of the phenomenon of superconductivity may lie in a careful revision of this situation, as a recent theory put forward by Fröhlich[†] in fact suggests.

2. Ionic radii

In Table 4 are given the radii of the free alkali and halogen ions calculated by Jensen. Apart from giving a rough idea of the extensions of the ions, the values are not particularly significant, as the real wave-mechanical densities have no uniquely definable radii. However, with

TABLE 4

Radii of Free Ions and Additive Radii

(In atomic units)

	Na^+	K^+	Rb^+	Cs^+	F^-	Cl^-	Br^-	I^-
Free radii	2·85	3·25	3·65	3·85	5·10	5·35	5·50	5·65
Add. radii	2·08	2·30	2·53	2·65	4·06	4·41	4·64	4·85

these radii, Jensen has expressed the calculated nearest neighbour distance in the lattices at equilibrium in terms of a penetration distance defined by

$$s(A^+B^-) = r_f(A^+) + r_f(B^-) - r_0(A^+B^-), \qquad (2.1)$$

where $r_f(A^+)$, $r_f(B^-)$ are respectively the radii of the alkali and halogen ions concerned, and $r_0(A^+B^-)$ is the nearest neighbour distance. $s(A^+B^-)$ thus measures the mutual penetration of the nearest ion pairs. The

† H. Fröhlich, *Phys. Rev.* **19**, 845 (1950); *Proc. Roy. Soc.* A, **215**, 291 (1952).

values of $s(A^+B^-)$ are given in graphical form in Fig. 3 for the alkali
halides, where points on the same curve refer to lattices with the same
alkali ion.

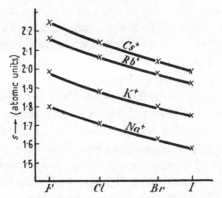

FIG. 3. Penetration distances $s(A^+B^-)$ for the alkali halides (Jensen).

From the curves, we can deduce a rule of great interest. We notice
that by vertical displacements the various curves can be brought practically into coincidence. Let us imagine such displacements effected and
denote by $-\delta(A^+)$ the displacement of the curve relating to an alkali ion
A^+. On the unified curve, to each halogen ion there corresponds a single
point; we denote the ordinate of the point for an ion B^- by $\delta(B^-)$. It
follows immediately that the penetration distances can be written as

$$s(A^+B^-) = \delta(A^+) + \delta(B^-). \qquad (2.2)$$

In place of the free ion radii, let us define new radii:

$$r(A^+) = r_f(A^+) - \delta(A^+); \qquad r(B^-) = r_f(B^-) - \delta(B^-). \qquad (2.3)$$

On combining (2.1) and (2.2), we find that

$$r_0(A^+B^-) = r(A^+) + r(B^-). \qquad (2.4)$$

Thus we have found a radius for every ion such that the nearest neighbour
distances in the alkali halides are given simply by the sums of the radii
of the ions concerned. We note, however, that the vertical displacements
$-\delta(A^+)$ are not uniquely defined. We can add an arbitrary constant to
all the $\delta(A^+)$ without affecting the above argument; the resulting positive-
ion radii would be raised by a constant and all negative-ion radii reduced
by the same constant. Under the heading 'additive radii' are given a
possible set of values for $r(A^+)$ and $r(B^-)$. The theoretical nearest neighbour distances calculated by Jensen are collected in Table 5. Given in

the same table for comparison are the corresponding sums of the additive radii as given in Table 4. The two sets of values are almost but not exactly equal because the curves in Fig. 3 can be brought into approximate but not exact coincidence.

TABLE 5

Nearest Neighbour Distances and Sums of Additive Radii

(In atomic units)

		Na$^+$	K$^+$	Rb$^+$	Cs$^+$
F$^-$	neighbour dist.	6·15	6·30	6·60	6·75
	$r(A^+)+r(B^-)$	6·14	6·36	6·59	6·71
Cl$^-$	neighbour dist.	6·50	6·70	6·95	7·10
	$r(A^+)+r(B^-)$	6·49	6·71	6·94	7·06
Br$^-$	neighbour dist.	6·75	6·95	7·20	7·35
	$r(A^+)+r(B^-)$	6·72	6·94	7·17	7·29
I$^-$	neighbour dist.	7·05	7·20	7·50	7·65
	$r(A^+)+r(B^-)$	6·93	7·15	7·38	7·50

Such additivity rules, first discovered by the crystallographers, hold approximately for a wide range of crystal lattices and are valuable in assisting crystallographic analyses. For a particular crystal, the suitable radii of the ions can be determined from the observed lattice constant, if the ratio of the radii is known. Assuming that the radii indicate the extensions of the electron orbits of the ions, Wasastjerna[†] has obtained the values of such ratios indirectly from the molar refractions of the ions, which should be roughly proportional to the volumes of the ions, and in this way deduced the radii for a number of monovalent and divalent ions. Taking the additivity rule as a criterion, Goldschmidt[‡] later obtained, by exhaustive analysis of empirical lattice constants, the suitable radii for a great number of ions. As the additivity rule, like the above analysis of Jensen's theoretical results, leaves the radii arbitrary to within a constant for each additive system (e.g. the alkali halides), Goldschmidt chooses his radii so that the values for F$^-$ and O^{--} agree with the values given by Wasastjerna. Such additive radii are usually known as the Goldschmidt radii; the Goldschmidt radii for some ions are collected in Table 6.

Similar radii have been deduced by Pauling[§] by a semi-theoretical method. Unlike the above analysis of Jensen's results, Pauling's work,

---

[†] J. A. Wasastjerna, *Soc. Sci. Fenn. Comm. Phys. Math.* **38**, 1 (1923).

[‡] V. M. Goldschmidt, *Skrifter det Norske Videnskaps Akademie* (1926), No. 2; (1927) No. 8.

[§] L. Pauling, *Nature of the Chemical Bond* (1948) (2nd ed.), pp. 343–7.

strictly speaking, does not demonstrate on a theoretical basis that the additivity rule should hold; rather he shows that in a theoretically plausible way certain radii can be deduced and the use of the additivity rule with these radii reproduces in fact the experimental lattice constant satisfactorily.

TABLE 6

Ionic Radii

(In angstrom units)

		Li^+	Be^{++}		
Pauling { univ.		0·60	0·44		
Pauling { crys.		0·60	0·31		
Goldschmidt		0·78	0·34		

	O^{--}	F^-	Na^+	Mg^{++}	Al^{3+}	Si^{4+}
Pauling { univ.	1·76	1·36	0·95	0·82	0·72	0·65
Pauling { crys.	1·40	1·36	0·95	0·65	0·50	0·41
Goldschmidt	1·32	1·33	0·98	0·78	0·57	0·39

	S^{--}	Cl^-	K^+	Ca^{++}	Sc^{3+}	Ti^{4+}
Pauling { univ.	2·19	1·81	1·33	1·18	1·06	0·96
Pauling { crys.	1·84	1·81	1·33	0·99	0·81	0·68
Goldschmidt	1·74	1·81	1·33	1·06	0·83	0·64

	Se^{--}	Br^-	Rb^+	Sr^{++}	Y^{3+}	Zr^{4+}
Pauling { univ.	2·32	1·95	1·48	1·32	1·20	1·09
Pauling { crys.	1·98	1·95	1·48	1·13	0·93	0·80
Goldschmidt	1·91	1·96	1·49	1·27	1·06	0·87

	Te^{--}	I^-	Cs^+	Ba^{++}	La^{3+}	Ce^{4+}
Pauling { univ.	2·50	2·16	1·69	1·53	1·39	1·27
Pauling { crys.	2·21	2·16	1·69	1·35	1·15	1·01
Goldschmidt	2·11	2·20	1·65	1·43	1·22	1·02

Following Pauling, let us consider the ions belonging to the same isoelectronic series (i.e. ions with the same number of electrons such as $\ldots, N^{---}, O^{--}, F^-, Ne, Na^+, Mg^{++}, Al^{+++}, \ldots$). If we consider the electrons in the ions as moving in a Coulomb field corresponding to some suitable effective charges, the outermost electrons in all these ions have the same total quantum number, and, moreover, the corresponding effective charges can be written as $Z - S$, where Z changes by one, as one passes from one ion to the next in the series and S, the screening constant, can be approximately determined in various ways. Roughly, the linear dimensions of the corresponding wave functions are inversely proportional to the effective charges $Z-S$ (strictly true for spherically symmetric wave functions). Pauling thus defines a set of radii inversely proportional to $Z-S$, and the proportionality constant is chosen so that the additivity rule reproduces the empirical lattice constant for the alkali halide lattice formed of the ions in the series (e.g. NaF for the Ne-like series). With the empirical lattice constants of NaF, KCl, RbBr, CsI (the last assumed to have the NaCl structure with a suitably chosen hypothetical value for the lattice constant), Pauling has determined in this way a set of radii of the ions with the structures of Ne, A, Kr, Xe. These he calls the *univalent radii*. As the equilibrium lattice constants depend on the attractive Madelung term as well as the overlap force, the univalent radii are not directly comparable with the Goldschmidt radii, owing to the steady increase of the Madelung term with the ionic charge as one passes along the isoelectronic series. Pauling interprets the univalent radii as the radii for which the additivity rule *would* hold, if the structure is of the NaCl type and the Madelung term were to have the value for univalent ions.

Radii comparable with the Goldschmidt radii (designated as *crystal radii* by Pauling) can be deduced from the univalent radii. One can imagine first the fictitious lattice which has the univalent Madelung term, and then find out how the lattice constant is modified, as the Madelung term is raised to the correct value. This depends obviously on the nature of the overlap force. The method used by Pauling to convert the univalent to the crystal radii will be described in the next section, where suitable expressions for the overlap force will be introduced. The values of Pauling's univalent and crystal radii are also collected in Table 6. The crystal radii are seen to be in general agreement with Goldschmidt's values.

In Table 7, the nearest neighbour distances in the alkali halides as determined experimentally are compared with the sums of Pauling's radii for the ions concerned.

3. Heuristic expressions for lattice energies

The principal interactions in ionic and van der Waals lattices, namely, the static Coulomb interaction, the van der Waals interaction, and the overlap force, are essentially two-body forces. If two lattice particles are spherically symmetric, their interaction energy is a function $\phi(r)$ of their distance apart (central force); this is the case with the simple ions, which have the rare-gas structures, and probably also with radicals, which are free to execute rotations (quantum-mechanical zero-point

TABLE 7

The Observed Nearest Neighbour Distances† and Sums of Pauling Radii

(In angstrom units)

		Li+	Na+	K+	Rb+	Cs+
F‑	radius sum	1·96	2·31	2·69	2·84	3·05
	obs. dist.	2·01	2·31	2·67	2·82	3·01
Cl‑	radius sum	2·41	2·76	3·14	3·29	..
	obs. dist.	2·57	2·81	3·14	3·29	..
Br‑	radius sum	2·55	2·90	3·28	3·43	..
	obs. dist.	2·75	2·98	3·29	3·43	..
I‑	radius sum	2·76	3·11	3·49	3·64	..
	obs. dist.	3·02	3·23	3·53	3·66	..

motion). The concept of additive two-body interactions is of course only approximate; the wave-mechanical calculations made by Löwdin, as we have mentioned, in fact already indicate that an appreciable fraction of the lattice cohesive energy cannot be represented in terms of two-body interactions. Moreover, in ionic crystals, an ion may be polarized by the resultant electric field due to other ions in the lattice. The effect of such polarizations on the lattice energy is not in the nature of a two-body interaction. Take the case of three particles 1, 2, 3: the interaction energy between 1 and 2 directly depends on the polarizations induced on these particles by 3. In many simple lattices, however, the polarization effects are largely suppressed by the high symmetry of the structure; although the electric field never vanishes over an entire ion, if an ion is sufficiently symmetrically situated in the lattice, the field vanishes at its centre and the polarization effect is small.‡ Only when such lattices are deformed in such a way as to destroy the symmetry will appreciable polarizations occur; later we shall have occasion to discuss occurrences of this nature.

The interaction potential $\phi(r)$ (assumed central) is the sum of an attractive part (Madelung and van der Waals terms) and a repulsive part (overlap term); as we have seen, no simple expression for the latter follows from the theoretical considerations. When it is not practicable to make a theoretical calculation of the overlap force, one usually has recourse to certain simple representations of the force, designed for a

† The values are not in agreement with the theoretical values in Table 5, showing that though Jensen's results reproduce the additivity rule, the absolute values are not accurate; a feature usual with approximate theoretical treatments. Jensen himself has considered the additivity rule by a different approach (H. Jensen, G. Meyer-Gossler, and H. Rohde, *Zeit. f. Phys.* **110**, 277 (1938)); the analysis given here is more in line with the energy calculations.

‡ T. Neugebauer and P. Gombás, ibid. **89**, 480 (1934).

heuristic purpose. Historically,[†] such representations in fact preceded detailed theoretical considerations such as described in § 1. This is not surprising, as the existence of the solid state is itself evidence for the presence of certain attractive forces; it is, moreover, clear that certain mechanisms must operate to set a limit to the packing of the atoms. In the earliest attempts to represent the forces, atoms were thus pictured as rigid spheres mutually attracted by a force varying as some inverse powers of the distance. This concept of the atoms is, however, not compatible with the fact that all solids are more or less compressible; it is thus useful only for crude considerations of the behaviour of gases and the geometrical configurations of atoms in molecules and solids. As more acceptable forms of the overlap energy, the two following expressions are in frequent use:

$$b/r^n, \tag{3.1 a}$$

$$\lambda e^{-r/\rho}. \tag{3.1 b}$$

The constants in these expressions are to be determined with the help of empirical data. Results of quantum-mechanical calculations favour the exponential form, whereas the inverse power form has the advantage of greater simplicity.

On the basis of (3.1 a), Lennard-Jones has studied atomic forces by investigating extensively the behaviour of gases.[‡] A hypothetical gas with non-interacting molecules obeys the equation of state for an *ideal gas*

$$p = \nu kT \quad \{k \text{ (the Boltzmann constant)} = 1 \cdot 3806 \text{ ergs/degree}\}, \tag{3.2}$$

where ν is the number of molecules per unit volume. The atomic forces can be investigated by the observed deviations of the real gases from the ideal. One can express the equation of state for a real gas in the following form:

$$p = \nu kT\{1 + \nu B'(T) + O(\nu^2) + ...\}. \tag{3.3}$$

The second term in the bracket represents the most important deviation from (3.2). It can be shown by statistical mechanics[§] that, if quantum effects are ignored (permissible for sufficiently high temperatures) the

---

[†] The pioneer work in this field is due to G. Mie, *Ann. d. Phys.* (4) **11**, 657 (1903); E. Grüneisen, ibid. **26**, 393 (1908); **39**, 257 (1912); M. Born and A. Landé, *Verh. d. D. Phys. Ges.* **20**, 210 (1918); M. Born, ibid. **21**, 13 (1919). A presentation of the development is given in an article of vol. v of *Encyclopädie der Mathematik* by M. Born, 'Atomtheorie des festen Zustandes', which also appeared as a book (Teubner, 1923).

[‡] See R. H. Fowler, *Statistical Mechanics* (1929), chap. x, where references to the original papers can be found.

[§] R. H. Fowler, *Statistical Mechanics* (2nd ed., Cambridge, 1936).

function of temperature $B'(T)$ is related to the interaction energy $\phi(r)$ by

$$B'(T) = 2\pi \int_0^\infty r^2(1 - e^{-\phi(r)/kT}) \, dr. \tag{3.4}$$

The integrand obviously vanishes if $\phi(r) = 0$; thus, loosely speaking, $B'(T)$ measures the volume surrounding a molecule where the molecular interaction energy is still appreciable compared with kT. The function $B'(T)$ is very simply related to the second virial coefficient, which is usually measured in experiments.

The simplest cases one can investigate are the rare gases, for which one can put

$$\phi(r) = -\frac{c}{r^6} + \frac{b}{r^n}, \tag{3.5}$$

where the first term represents the attractive van der Waals potential. The constants n, b, c must have such values that the function $B'(T)$ calculated from (3.4) agrees with that deduced from observed data. As the work of Lennard-Jones shows, although for given n the constants b and c can be determined with reasonable accuracy, the exponent n itself can be chosen with considerable latitude. In Table 8 are given several possible sets of values of n, b, c for Ne and A, as determined by Buckingham.[†]

TABLE 8

Overlap Forces and Lattice Properties of Ne *and* A

n	b (ergs cm.^{n})	c (ergs cm.^{6})	r_0 (10^{-8} cm.)	u ($eV/atom$)	lattice energy (cal./mol.)
Ne 9	$3 \cdot 50 \times 10^{-82}$	$1 \cdot 45 \times 10^{-59}$	$3 \cdot 15$	$0 \cdot 0224$	517
10	$7 \cdot 32 \times 10^{-90}$	$1 \cdot 14 \times 10^{-59}$	$3 \cdot 09$	$0 \cdot 0238$	549
12	$3 \cdot 55 \times 10^{-105}$	$8 \cdot 32 \times 10^{-58}$	$2 \cdot 99$	$0 \cdot 0264$	609
14	$1 \cdot 82 \times 10^{-120}$	$6 \cdot 78 \times 10^{-58}$	$2 \cdot 92$	$0 \cdot 0286$	660
exp.	. .	. .	$3 \cdot 20$	. .	590
A 9	$7 \cdot 68 \times 10^{-81}$	$1 \cdot 70 \times 10^{-58}$	$3 \cdot 88$	$0 \cdot 075$	1730
10	$2 \cdot 05 \times 10^{-88}$	$1 \cdot 37 \times 10^{-58}$	$3 \cdot 82$	$0 \cdot 0804$	1854
12	$1 \cdot 62 \times 10^{-103}$	$1 \cdot 03 \times 10^{-58}$	$3 \cdot 72$	$0 \cdot 0886$	2042
14	$1 \cdot 365 \times 10^{-118}$	$8 \cdot 67 \times 10^{-59}$	$3 \cdot 58$	$0 \cdot 121$	2782
exp.	. .	. .	$3 \cdot 80$	. .	2030

All the rare gases except helium solidify at sufficiently low temperatures (melting-points: 24° K. (Ne), 84° K. (A), 117° K. (Kr), 161° K. (Xe)) in the face-centred cubic structure (see Appendix I for description). Let us consider the properties of the crystals with the help of the forces determined from the gaseous data. We can introduce a systematic way of

† R. A. Buckingham, *Proc. Roy. Soc.* A, **168**, 264 (1938).

labelling lattice particles in a simple lattice. We notice that the structure itself provides a natural frame of coordinates (with, in general, oblique axes). If an arbitrary lattice point is chosen as the origin and the basic vectors $\mathbf{a}_1$, $\mathbf{a}_2$, $\mathbf{a}_3$ are used as units of length along the respective axes, the corresponding coordinates of the lattice particles are the whole numbers $l(l^1, l^2, l^3)$. The same numbers can also be used to label the elementary cells, if we associate with a lattice particle $l(l^1, l^2, l^3)$ the cell included in its positive quadrant. We shall refer to $l(l^1, l^2, l^3)$ as the lattice indices and to the cell $(0, 0, 0)$ as the zero-cell. As all atoms in a simple lattice are equivalent, the lattice energy u *per atom*, can be obtained from the interaction energy of the atom $(0, 0, 0)$ with all other atoms; thus, if $r(l)$ denotes the distance of atom l from the origin:

$$u = \tfrac{1}{2} \sum_{l}{}' \phi(r(l)) = \frac{1}{2} \sum_{l}{}' \left\{ -\frac{c}{r(l)^6} + \frac{b}{r(l)^n} \right\}, \tag{3.6}$$

where the prime over the summation sign excludes $l = 0$. The factor $\tfrac{1}{2}$ takes account of the fact that the interaction energy $\phi(r(l))$ is shared between the two atoms $(0, 0, 0)$ and (l^1, l^2, l^3). In the sum it is convenient to express $r(l)$ in terms of the nearest neighbour distance r as unit, hence we write

$$u(r) = -\frac{A}{r^6} + \frac{B}{r^n}, \quad A = -\frac{c}{2} \sum_{l}{}' \left(\frac{r}{r(l)} \right)^6, \quad B = \frac{b}{2} \sum_{l}{}' \left(\frac{r}{r(l)} \right)^n. \tag{3.7}$$

We notice that $[r/r(l)]$, and hence also the sums in A and B, are pure numbers completely determined by the structure; and they remain constant as the dimension of the lattice is varied by varying r. Efficient methods for evaluating such lattice sums have been described by Lennard-Jones and Ingham† who, in particular, have evaluated sums of the form

$$\sum_{l}{}' \left(\frac{r}{r(l)} \right)^n$$

for n up to thirty, for the simple cubic, body-centred cubic, and face-centred cubic structures, and also for a lattice of the NaCl type, where the sums are given separately for the interactions between a Na^+ site with all other Na^+ sites and between a Na^+ site and all Cl^- sites. With known b and c, the values of A and B can be found immediately with the help of their results.

For static equilibrium, the energy $u(r)$ must be a minimum, thus the

† J. E. Lennard-Jones and A. E. Ingham, *Proc. Roy. Soc.* A, **107**, 636 (1925).

equilibrium value of r is given by

$$r_0 = \left(\frac{nB}{6A}\right)^{1/(n-6)}. \tag{3.8}$$

Using this value in (3.7), we obtain for the lattice energy per atom:

$$u(r_0) = \frac{-A}{r_0^6}\left\{1 - \frac{6}{n}\right\}. \tag{3.9}$$

The values of r_0 and the cohesive energies $-u(r_0)$ (given in calories per mole) for the various sets of n, b, c permitted by the gaseous data are also given in Table 8.

The cohesive energy $-u(r_0)$ is not directly comparable with the sublimation energy at $0°$ K., for, according to quantum mechanics, the zero-point motion persists even at the absolute zero of temperature. The experimental values for the cohesive energy given in the table are the observed sublimation energies (extrapolated to $0°$ K.), to which the zero-point energies have been added. It is seen that by proper choice of the exponent n, fair agreement between the theoretical and experimental results can be secured.

In a face-centred cubic lattice each atom has twelve nearest neighbours. If in the overlap energy all but the contributions due to the nearest neighbours are ignored, we have

$$B = \frac{b}{2}\sum_l{}' \left(\frac{r}{r(l)}\right)^n \simeq 6b.$$

Comparing this with the rigorous values (Appendix III), we see that for $n = 8, 9, \ldots, 14$, the nearest neighbours account for $94, \ldots, 99 \cdot 5$ per cent. of the total overlap energy.

As gaseous data are not available for ions, in the case of ionic crystals the overlap force has to be determined from crystal data alone. In simple binary salts, the ions have equal and opposite charges $\pm ze$ (e.g. structures NaCl, CsCl, ZnS, ZnO); every ion is surrounded by M (coordination number) first neighbours of the opposite sign. The overlap energy between the positive and negative ions can be readily determined from experimental data, if we ignore the overlaps between all but the nearest neighbours. In this approximation, the use of the exponential form (3.1 b) is not more complicated than the inverse power formula; using the former, we can write the energy per cell as

$$u(r) = -\frac{A}{r} + Be^{-r/\rho}, \qquad A = (ze)^2\alpha', \qquad B = M\lambda_{+-}, \tag{3.10}$$

where the first term is the attractive Madelung energy, and ρ, λ_{+-} are the constants in the overlap potential between the two types of ions. As before, r denotes the distance between the nearest neighbours.

Remembering that the volume of the lattice cell v is proportional to the third power of r, we have

$$d \ln v = 3d \ln r \quad \text{or} \quad \frac{dv}{v} = \frac{3dr}{r}.$$

One finds easily for the pressure and the compressibility β:

$$p = -\frac{du}{dv} = \frac{1}{3v}\left\{ -\frac{A}{r} + B\left(\frac{r}{\rho}\right)e^{-r/\rho} \right\}, \tag{3.11}$$

$$\frac{1}{\beta} = -v\frac{dp}{dv} = \frac{1}{9v}\left\{ -\frac{A}{r} - B\left(\frac{r}{\rho}\right)e^{-r/\rho} + B\left(\frac{r}{\rho}\right)^2 e^{-r/\rho} \right\} + p. \tag{3.12}$$

For static equilibrium, $p = 0$, one has

$$\frac{A}{r_0} = B\left(\frac{r_0}{\rho}\right)e^{-r_0/\rho}. \tag{3.13}$$

Putting $r = r_0$ and $p = 0$ in (3.12) and eliminating B with (3.13), we find that

$$\frac{1}{\beta} = \frac{A}{9v_0 r_0}\left\{ -2 + \left(\frac{r_0}{\rho}\right) \right\}. \tag{3.14}$$

Putting $r = r_0$ in (3.10) and eliminating B with (3.13), we obtain for the lattice energy:

$$u(r_0) = -\frac{A}{r_0}\left\{ 1 - \left(\frac{\rho}{r_0}\right) \right\}. \tag{3.15}$$

In Table 9 are collected the observed nearest distances and compressibilities for the alkali halides. Using these values for r_0 and β we can determine ρ from (3.14) and then λ_{+-} and $-u(r_0)$ from (3.13) and (3.15). The calculated values are also given in the table. We note that the procedure is only approximate, as the empirical values for r_0 and ρ are room-temperature values and do not refer to the lattice in static equilibrium.

The theoretical cohesive energy $-u(r_0)$ is the energy necessary to disperse the lattice in static equilibrium into individual ions. The empirical values quoted in the table for comparison are obtained by a method which we shall describe at the end of this section. The table shows that there is a fair agreement between the theoretical and experimental values, the theoretical values being on the average 4 per cent. too low.

The two terms in (3.15) are due respectively to the Madelung term and the overlap energy. Table 9 shows that ρ/r_0 is of the order $1/10$;

TABLE 9

Overlap Energies and Cohesive Energies of the Alkali Halides

Sub-stance	r_0 (10⁻⁸ cm.)	β (10⁻¹²/barye)	r_0/ρ	ρ (10⁻⁸ cm.)	$\lambda_{+-}\, e^{-(r_0/\rho)}$ (10⁻¹³ ergs)	λ_{+-} (10⁻⁹ ergs)	$-u(r_0)$ (eV/cell)	Cohesive Energy (k.cal./mole)				
								Eq. (3.15)	Exp.	Error	Huggins	Error
LiF	2·010	1·17	8·25	0·244	4·05	1·05	11·00	253·5	··	··	243·6	··
LiCl	2·572	3·41	7·75	0·332	3·36	0·782	8·518	196·3	201·5	−5·2	200·2	−1·3
LiBr	2·745	4·31	7·88	0·348	3·10	0·821	8·000	184·4	191·5	−7·1	189·5	−2·0
LiI	3·000	6·01	8·02	0·374	2·79	0·849	7·337	169·1	180·0	−10·9	176·1	−3·9
NaF	2·310	2·11	8·03	0·288	3·62	1·11	9·541	219·9	··	··	215·4	··
NaCl	2·814	4·26	8·57	0·328	2·78	1·47	7·895	182·0	184·7	−2·7	183·5	−1·2
NaBr	2·981	5·08	8·95	0·333	2·52	1·94	7·494	172·7	175·9	−3·2	175·5	−0·4
NaI	3·231	7·07	8·90	0·363	2·33	1·71	6·914	159·3	166·3	−7·0	164·3	−2·0
KF	2·665	3·30	8·82	0·302	2·85	1·94	8·371	193·0	··	··	192·5	··
KCl	3·639	5·63	9·71	0·324	2·20	3·63	7·190	165·7	167·8	−2·1	167·9	−0·1
KBr	3·293	6·70	9·85	0·334	2·07	3·93	6·869	158·3	161·2	−2·9	161·3	−0·1
KI	3·526	8·54	10·10	0·349	1·89	4·60	6·429	148·2	152·8	−4·6	152·4	−0·4
RbF	2·815	4·1	8·85	0·318	2·70	1·88	7·929	182·8	··	··	183·0	··
RbCl	3·270	6·65	9·70	0·338	2·12	3·46	6·902	159·1	163·6	−4·5	162·0	−1·6
RbBr	3·427	7·94	9·77	0·350	2·00	3·51	6·592	151·9	158·0	−6·1	156·1	−1·9
RbI	3·663	9·57	10·40	0·352	1·76	5·80	6·208	143·1	149·7	−6·6	148·0	−1·7
CsF	3·004	4·25	10·55	0·284	2·12	8·09	7·579	174·7	··	··	175·7	··
CsCl	3·559	5·95	11·17	0·319	1·28	9·03	6·490	149·6	157·8	−8·2	153·1	−4·7
CsBr	3·713	7·6	11·15	0·332	1·23	8·54	6·221	143·4	152·3	−8·9	149·6	−2·7
CsI	3·95	8·57	11·65	0·340	1·11	12·67	5·868	135·2	145·4	−10·2	142·5	−3·2

therefore the value of the lattice energy is predominantly determined by the Madelung term. The above agreement thus confirms the essential correctness of the theoretical model for the ionic lattices.

Essentially the same results are obtained, if we use the inverse power for the overlap force. In place of (3.13)–(3.15) one has then

$$\frac{A}{r_0} = \frac{nB'}{r_0^n}, \tag{3.16}$$

$$\frac{1}{\beta} = \frac{A}{9v_0 r_0}(-1+n), \tag{3.17}$$

$$u(r_0) = -\frac{A}{r_0}\left(1 - \frac{1}{n}\right), \tag{3.18}$$

where $B' = Mb_{+-}$, b_{+-}, n being the constants in the potential between the two types of ions. Comparison of (3.14) and (3.17) shows that $n = r_0/\rho - 1$; thus the cohesive energy $-u(r_0)$ calculated from (3.18) will be about 1 per cent. lower than before, giving a slightly increased discrepancy as compared with the experimental results.

The conversion from Pauling's univalent radii to the crystal radii is conveniently effected with the help of (3.16), which may be written alternatively as

$$r_0 = \left(\frac{nB'}{A}\right)^{1/(n-1)}.$$

When the ionic charges are raised from $\pm e$ to $\pm ze$, A is multiplied by z^2. Thus the crystal radii can be obtained by multiplying the corresponding univalent radii by the factor $\{1/z\}^{2/(n-1)}$.

In the above calculation of the lattice energy, we have taken no account of the van der Waals potentials or the overlap potentials between other than nearest neighbours. Moreover, the experimental cohesive energies in the table refer to the lattice at the absolute zero of temperature, rather than the lattice in static equilibrium; thus the energy of the zero-point vibrations has to be subtracted from the theoretical cohesive energy for the latter to be comparable with the corresponding experimental value.

We notice that the cohesive energies for the ionic crystals are between a hundred and a thousand times higher than the rare gas crystals; accordingly the zero-point energy is, comparatively, very unimportant for the ionic crystals. It is estimated to be of the order of one k.cal. per mole for the alkali halides.[†] The consideration of the zero-point energies thus increases the average discrepancy between the experimental and theoretical cohesive energies by somewhat less than 1 per cent.

† M. L. Huggins, *J. Chem. Phys.* **5**, 143 (1937).

TABLE 10

van der Waals Interaction in Alkali Halide Lattices

Substance	LiF	LiCl	LiBr	LiI	NaF	NaCl	NaBr	NaI	KF	KCl
C (10^{-60} ergs cm.^{6})	18	113	183	363	46	180	271	482	167	452
D (10^{-76} ergs cm.^{8})	11	104	190	470	31	180	300	630	150	560
van der Waals energy (k.cal./mole)	4·5	6·3	7·0	8·2	4·9	5·9	6·3	6·8	7·6	7·6
Correction to energy (k.cal./mole) eqn. (3.19)	0·7	1·0	1·1	1·3	0·8	1·0	1·1	1·2	1·3	1·5

Substance	KBr	KI	RbF	RbCl	RbBr	RbI	CsF	CsCl	CsBr	CsI
C (10^{-60} ergs cm.^{6})	605	924	228	691	898	1330	495	1530	2070	2970
D (10^{-76} ergs cm.^{8})	800	1420	290	960	1340	2240	600	2600	3600	5800
van der Waals energy (k.cal./mole)	7·6	7·8	9·1	9·1	9·0	9·0	10·9	12·2	12·8	12·8
Correction to energy (k.cal./mole) eqn. (3.19)	1·6	1·6	1·6	1·8	1·8	2·0	2·5	3·0	3·2	3·4

By a careful analysis of optical data, Mayer† has estimated the van der
Waals potentials between the ions in the alkali halides. The energy per
cell due to van der Waals dipole-dipole and dipole-quadrupole interactions can be written respectively as $-C/r^6$, $-D/r^8$, where r, as before,
represents the nearest neighbour distance. The values of C, D estimated
by Mayer are collected in Table 10, where the corresponding contributions to the lattice cohesive energies are also given. We observe that the
van der Waals interactions increase with the sizes of the ions and contribute from four to ten k.cal. per mole to the cohesive energies of
the alkali halide lattices. These contributions must, however, not be
added directly to the earlier theoretical estimates of the cohesive energies.
In fact, if we include the terms $-C/r^6$, $-D/r^8$ in the energy expression
(3.10) and follow through the same developments as before, we shall
find that the ultimate energy expression (3.15) is increased by the further
terms (up to first-order terms in C and D):

$$-\frac{[(r_0/\rho)-6]^2+6}{(r_0/\rho)^2}\left(\frac{C}{r_0^6}\right)-\frac{[(r_0/\rho)-8]^2+8}{(r_0/\rho)^2}\left(\frac{D}{r_0^8}\right). \qquad (3.19)$$

The alterations in the estimates of the cohesive energies are thus very
much smaller than the actual contributions by the van der Waals interactions; for $(r_0/\rho) = 10$, the multiplicative factors in (3.19) are respectively 0·22 and 0·12. The reason is clear: the inclusion of the van der
Waals terms in (3.10) brings corresponding modifications to our estimates of the overlap potentials; the increase of the latter largely balances

† J. E. Mayer, *J. Chem. Phys.* **1**, 270 (1933).

the van der Waals contributions, as long as (r_0/ρ) is still comparable with the exponents of the van der Waals potentials, 6 and 8. The correction (3.19) is also given in Table 10; its value ranges from one to three k.cal. per mole.

Similar circumstances prevail, when we consider the overlap potentials between other than nearest neighbours. Thus if we include in the energy expression (3.10) a term $B' \exp\{-\sqrt{2}\,r/\rho\}$ ($\sqrt{2}\,r$ = second neighbour distance in a NaCl lattice) to represent the total energy contribution due to overlaps between second neighbours, we shall find that the ultimate energy expression (3.15) has the additional term (to the first order in the correction):

$$(3-2\sqrt{2})B' \exp\{\sqrt{2}\,r_0/\rho\} = 0\cdot17B' \exp\{\sqrt{2}\,r_0/\rho\}. \tag{3.20}$$

Later we shall see that the total overlap energies between second neighbours are in most cases less than one-fifth of the overlap energies between the nearest neighbours. We have already seen that the latter are of the order of one-tenth of the cohesive energies. Hence $B' \exp\{\sqrt{2}\,r_0/\rho\}$ is less than 2 per cent. of the cohesive energies and (3.20) will reduce the theoretical estimates of the cohesive energies by well under 1 per cent.

For given r_0 and β, the estimate of the cohesive energy based on the simple expression (3.10) should thus be quite accurate. Further elaborations by taking into account additional energy terms may modify appreciably the estimates of the overlap potentials, but will leave the estimate of the energy not substantially altered. This conclusion is true, so long as the rate of variation with r of the added energy terms is comparable with that of the overlap potentials. Substantial reductions in the average discrepancy between the theoretical and experimental values can only be sought in more accurate values of β and r_0. A decrease in β raises the value (r_0/ρ) (cf. (3.14)) and thus indirectly raises the cohesive energy (cf. (3.15)); a decrease in the value of r_0 similarly increases the cohesive energy (cf. (3.15)). The values of β and r_0 for the lattice in static equilibrium differ in fact from the room-temperature values in these directions. This point will be discussed in more detail in the next section.

The knowledge of β and r_0 alone is clearly not sufficient to determine the overlap potentials between the like ions which are never nearest neighbours. Using the known elastic constants of NaCl and KCl, Born and Brody† have determined partially the interactions between the like ions. They obtained the unlikely result that the corresponding

† M. Born and E. Brody, *Zeit. f. Phys.* **11**, 327 (1922).

potentials are attractive; this is probably due to the smallness of the second neighbour interactions and their estimates are likely to be grossly distorted by any inaccuracy in the method. The usual methods for discussing the overlap potentials in general depend on a systematic way of constructing the overlap potentials between the ions, and the idea of ionic radii is used as a guide for the method of construction. We shall describe briefly methods of this type, as variously developed by Lennard-Jones, Pauling, and by Born, Mayer, and Huggins.[†]

Lennard-Jones's work aimed at correlating the overlap potentials between ions with that between rare gas atoms; the latter, as we have seen, can be deduced from the observed second virial coefficients of the gases.

For a head-on collision between two rigid spheres, the nearest distance of approach between the centres is equal to the sum of the radii of the spheres. For atoms having a repulsive potential b/r^n, the corresponding distance is a function of the energy of collision W (i.e. total kinetic energy in the system of coordinates moving with the centre of mass of the particles), namely $(b/W)^{1/n}$. If the atoms are identical, we can ascribe to each a kinetic radius $\frac{1}{2}(b/W)^{1/n}$. Lennard-Jones's method is based on the following assumptions:

(i) a kinetic radius can be defined for every ion so that for a head-on collision between two ions, the nearest distance of approach is equal to the sum of the corresponding radii;

(ii) the kinetic radii of isoelectronic ions are proportional to the extensions of their respective electronic clouds, as a practical measure for which Lennard-Jones used the ionic radii deduced by Wasastjerna from molar refractions.

Consider the case of an ion i. Let the ratio of its Wasastjerna radius to that of its isoelectronic rare gas atom be ρ_i; it follows from assumption (ii) that its kinetic radius can be written as

$$\frac{\rho_i}{2}(b_i^0/W)^{1/n}, \qquad (3.21)$$

where b_i^0 refers to the constant b in the overlap potential between the rare gas atoms (index i indicating the rare gas atom in the isoelectronic series of i) so that $\frac{1}{2}(b_i^0/W)^{1/n}$ is the kinetic radius of the rare gas atom.

† J. E. Lennard-Jones, *Proc. Roy. Soc.* A, **106**, 452 (1924); **109**, 584 (1925); and see also R. H. Fowler, *Statistical Mechanics*, chap. x (Cambridge, 1929); L. Pauling, *The Nature of the Chemical Bond*, pp. 355–63 (Cornell, 1948); M. Born and J. E. Mayer, *Zeit. f. Phys.* **75**, 1 (1932); M. L. Huggins and J. E. Mayer, *J. Chem. Phys.* **1**, 643 (1933); M. L. Huggins, ibid. **5**, 143 (1937); **15**, 212 (1947).

At this stage, Lennard-Jones does not fix the exponent n, but considers b_i^0 as a function of n, which can be determined by gaseous data. For the interaction potential $b_{ij}/r^{n_{ij}}$ between two ions i and j, the exponent n_{ij} is fixed by interpolation from the exponents characteristic of their respective isoelectronic rare gas atoms ($n = 11, 9, 10, 11$ for Ne, A, Kr, Xe). The constant b_{ij} is then determined by assumption (i):

$$(b_{ij}/W)^{1/n_{ij}} = \left\{ \frac{\rho_i}{2}(b_i^0/W)^{1/n_{ij}} + \frac{\rho_j}{2}(b_j^0/W)^{1/n_{ij}} \right\},$$

or
$$b_{ij}^{1/n_{ij}} = \tfrac{1}{2}\{\rho_i(b_i^0)^{1/n_{ij}} + \rho_j(b_j^0)^{1/n_{ij}}\}, \qquad (3.22)$$

where the values of b_i^0 and b_j^0 corresponding to the exponent n_{ij} are to be used.

The above method is not free from arbitrariness, but the values of the lattice constants, cohesive energies, and compressibilities for the alkali-halides and the lattice constants of the binary salts of the bivalent ions Mg^{++}, Ca^{++}, Sr^{++}, Ba^{++} with O^{--}, S^{--}, Se^{--}, Te^{--} calculated by Lennard-Jones show satisfactory agreements with the observed values. The constants for the overlap potentials between a large number of ions obtained in this way have been tabulated by Lennard-Jones and Dent.[†] The original analysis of the rare gas data by Lennard-Jones was made before London put forward the theory of the van der Waals forces, and the attractive potentials were assumed to have the exponent 4. This affects somewhat the overlap potentials for the rare gases, upon which the force constants for the ions are based. Fowler[‡] has subsequently revised the values for the constants, by a method due originally to Lennard-Jones. Instead of using values from gas data, Fowler has taken all the exponents n_{ij} as 9 and determined the constants b_i^0 for the rare gas atoms indirectly by the use of the experimental values of the nearest neighbour distance in the lattices: NaF, KCl, RbBr, CsI. A table of the revised values can be found in his book on statistical mechanics.

For the alkali halides, Pauling has proposed the following expression for the overlap potentials:

$$\beta_{AB} B_0 \frac{(r_A + r_B)^{n-1}}{r^n} \qquad (n = 9), \qquad (3.23)$$

where, like Fowler, he adopts the uniform value 9 for the exponent. The factor β_{AB} originates from certain quantum-mechanical considerations, and depends only on the charges of the interacting ions, namely:

[†] J. E. Lennard-Jones and B. M. Dent, *Proc. Roy. Soc.* A, **112**, 230 (1926).
[‡] R. H. Fowler, *Statistical Mechanics* (2nd ed., Cambridge, 1936).

$\beta_{++} = 1 \cdot 25$, $\beta_{+-} = 1 \cdot 00$, $\beta_{--} = 0 \cdot 75$. With the following choice of the radii and the constant B_0

	Li⁺	Na⁺	K⁺	Rb⁺	Cs⁺	F⁻	Cl⁻	Br⁻	I⁻
$r(10^{-8}\,\text{cm})$	0·607	0·958	1·331	1·484	1·656	1·341	1·806	1·951	2·168

$$B_0 = 0 \cdot 02909 e^2$$

Pauling has shown that the calculated nearest neighbour distances for the seventeen alkali halides with the NaCl structure agree practically exactly with the highly accurate experimental values. This is particularly remarkable, if we observe that the deviations of the lithium salts from the additivity rule are of the order of up to 10 per cent. These deviations are due to the importance of the second neighbour repulsions in these salts, where the radii of the halide ions are between 2 and 3·5 times the radius of the lithium ion.

Using (3.23), we find that the overlap energy per cell due to nearest neighbours in a NaCl structure is

$$\frac{6 B_0 (r_+ + r_-)^8}{r^9},$$

and that due to second neighbours

$$\frac{6 B_0}{(\sqrt{2}\, r)^9} \{ 1 \cdot 25 (2 r_+)^8 + 0 \cdot 75 (2 r_-)^8 \}.$$

Dividing the latter by the former, we obtain the ratio

$$R = \frac{\text{overlap energy of second neighbours}}{\text{overlap energy of first neighbours}}$$

$$= \frac{1}{2^{4 \cdot 5}} \left\{ 1 \cdot 25 \left(\frac{2 \rho}{1 + \rho} \right)^8 + 0 \cdot 75 \left(\frac{2}{1 + \rho} \right)^8 \right\}, \quad (3.24)$$

which depends only on the radius ratio $\rho = r_+ / r_-$. The following values indicate this dependence:

$$\rho = 0 \cdot 3 \qquad 0 \cdot 4 \qquad 0 \cdot 5 \qquad 0 \cdot 6 \qquad 0 \cdot 8 \qquad 1 \cdot 0 \qquad 1 \cdot 2$$
$$R = 1 \cdot 06 \qquad 0 \cdot 58 \qquad 0 \cdot 33 \qquad 0 \cdot 20 \qquad 0 \cdot 10 \qquad 0 \cdot 09 \qquad 0 \cdot 14$$

Thus we see that for the salts of potassium, rubidium, and caesium, the contributions due to second neighbours are all less than 20 per cent. of that due to nearest neighbours. In the lithium salts, on the other hand, the contributions due to the first and second neighbours are comparable; for the extreme case of LiI, the respective contributions are approximately equal.

The most elaborate attempt at calculating crystal properties from suitably chosen interaction potentials is found in a series of works by Born, Mayer, Helmholz, and Huggins.† For the overlap potentials between the ions, Born and Mayer have proposed the expression

$$\beta_{AB}\, b e^{(r_A + r_B - r)/\rho}, \tag{3.25}$$

where β_{AB} is the same factor as in (3.23). The way in which the radii appear in the expression is suggested by the additivity rule. Consider for instance the approximate equation (3.13) (all but overlaps between nearest neighbours ignored!), where the constant B should now be replaced by $M\beta_{+-}\, b \exp[(r_+ + r_-)/\rho]$, i.e.

$$\frac{A}{r_0} = \left(\frac{r_0}{\rho}\right) M\beta_{+-}\, b e^{(r_+ + r_- - r_0)/\rho},$$

or, what is equivalent,

$$r_+ + r_- - r_0 = \rho\left\{\ln\frac{A}{b\beta_{+-} M\rho} - 2\ln\left(\frac{r_0}{\rho}\right)\right\}. \tag{3.26}$$

The right-hand side of the equation does not differ very appreciably for different alkali-halide lattices, for Table 9 shows that ρ will be only about one-tenth of the value of r. From LiF to RbI, r_0 changes by a factor 1·8 and the corresponding change of the right-hand side of (3.26) will be approximately 0·3 Å, which is about one-tenth of r_0. Hence b can, for instance, be chosen such that the right-hand side of (3.26) approximately vanishes for all cases:

$$r_+ + r_- \cong r_0;$$

the form (3.25) thus reproduces approximately the additivity rule. (The particular choice of b is, we notice, not actually necessary. The effect of introducing a different value of b on the potential (3.25) can be neutralized by a constant added to all the positive or all the negative ion radii in the formula.)

The constants in the potentials could, for instance, be assigned values so as to give the best fit for r_0 and β of all the alkali halide lattices in static equilibrium. However, as r_0 and β for the *static* lattices are not directly observable, Born and Mayer have used an alternative method based on an assumption due to Hildebrand,‡ namely, that at finite temperatures the energy of a lattice consists of two parts, one dependent only on its volume and the other only on temperature. It follows then by the use of thermodynamical relations that the first and second

† Loc. cit.
‡ J. H. Hildebrand, *Zeit. f. Phys.* **67**, 127 (1931); for a discussion of the accuracy of the method see K. Huang, *Phil. Mag.* **42**, 202 (1951).

derivatives of the lattice energy $u(r)$ can be expressed in terms of directly observable quantities:

$$\frac{du(r)}{dr} = \frac{3vT}{r\beta}\left(\frac{1}{V}\frac{\partial V}{\partial T}\right)_p, \tag{3.27}$$

$$\frac{d^2u(r)}{dr^2} = \frac{9v}{r^2\beta}\left\{1 + \frac{T}{\beta}\left[\left(\frac{\partial \beta}{\partial T}\right)_p + \frac{1}{V}\left(\frac{\partial V}{\partial T}\right)_p\frac{1}{\beta}\left(\frac{\partial \beta}{\partial p}\right)_T\right] + \frac{2}{3}\frac{T}{V}\left(\frac{\partial V}{\partial T}\right)_p\right\}, \tag{3.28}$$

where the nearest neighbour distance r, as well as all the quantities on the right, may refer to any arbitrary temperature. These relations, we notice, are the finite-temperature analogues of the earlier equations (3.11), (3.12) ($p = 0$), which determine the derivatives of $u(r)$ at the static equilibrium value r_0; (3.27), (3.28) reduce in fact to the earlier equations for $T = 0$. For the expression of $u(r)$, the above authors considered the Madelung term, the overlap potentials (3.25) between the nearest and second neighbours, and also the dipole-dipole, dipole-quadrupole van der Waals potentials. The constants in the overlap potentials are then determined so as to give the best fit for (3.27) and (3.28) (taken at room temperature) for all the alkali halide lattices. The values for the constants as determined by Huggins are:

	Li+	Na+	K+	Rb+	Cs+
$r_+ =$	0·570 Å	0·940	1·235	1·370	1·510

	F-	Cl-	Br-	I-
$r_- =$	1·050	1·435	1·560	1·750

$$b = 10^{-12}\text{ ergs}, \qquad \rho = 0{\cdot}3333 \times 10^{-8}\text{ cm}. \tag{3.29}$$

The corresponding calculated cohesive energies are also given in Table 9. The agreement with the experimental values is seen to be very satisfactory, especially as the experimental values are subject to possible errors of the order of a few kilo-calories per mole.

Let us return to discuss how the experimental values of the cohesive energies are obtained. The cohesive energy is the energy difference between the dispersed ions and the crystal lattice at the absolute zero of temperature. Formally we shall regard the dispersed ions as the gases of the ions at the absolute zero of temperature. We shall find the energy difference by considering a series of changes of states, which lead from the lattice at 0° K. eventually to the ion gases at 0° K. In following through the changes, it is more convenient to consider the changes in the heat function $H = E + PV$ rather than the energy E. Let us thus

consider the following changes and the corresponding changes in H.†
(In the following [ ], ( ) indicate the substances respectively in the
crystalline and gaseous states under unit atmospheric pressure. Un-
bracketed symbols refer to elements in their natural state under the
temperature specified.)

(i) $[A^+B^-]_{0°K.} \rightarrow [A^+B^-]_{298°K.}$, heating of the lattice from $0°$ K. to
$298°$ K. (room temperature). The corresponding ΔH is obtained by
integrating the heat capacity; the latter can be estimated in all cases.

(ii) $[A^+B^-]_{298°K.} \rightarrow [A]_{298°K.} + B_{298°K.}$, separation of the crystal into
the pure metal and the halide element in its natural state, which is in
most cases the diatomic gas. ΔH is the heat of formation of the salt.

(iii) $[A]_{298°K.} \rightarrow (A)_{298°K.}$, sublimation of the metal. The heat of
sublimation ΔH can be deduced indirectly from the vapour pressures
of the molten metal at high temperatures, the heat of fusion and the
heat capacities of the condensed phases up to high temperatures.

(iv) $B_{298°K.} \rightarrow (B)_{298°K.}$, transformation of the halide element from its
natural state to the monatomic gas. The corresponding ΔH is deducible
from the energy required for dissociation of the diatomic molecule.

(v) $(A)_{298°K.} \rightarrow (A)_{0°K.}$ and $(B)_{298°K.} \rightarrow (B)_{0°K.}$ Taking the gases
approximately as ideal gases, one has $\Delta H = -(5/2)RT$ per mole.

(vi) $(A)_{0°K.} \rightarrow (A^+)_{0°K.} + e$, removal of the valence electrons from the
alkali atoms. The corresponding energy change per atom is equal to
the ionization potential.

(vii) $e + (B)_{0°K.} \rightarrow (B^-)_{0°K.}$, the electrons being added to the halide
atoms to form negative ions. The energy per atom is $-E$, E being the
electron affinity; the latter has been determined for Cl, Br, and I by
Mayer and his collaborators.‡

As the heat function and energy are identical at $0°$ K., the sum of ΔH
in the above changes provides an experimental estimate of the cohesive
energy. The experimental cohesive energies for the alkali-halides and
the experimental data used are collected in Table 11.

When the value of the electron affinity is not known, the above
considerations can be used in reverse to determine the electron affinities

† Historically this method was first proposed and used by M. Born (*Verh. d. D. Phys.
Ges.* **21**, 679 (1919)) and represented in graphical form by F. Haber (ibid., p. 750); it is
now often quoted as the Born–Haber cycle.

For details, see J. E. Mayer and W. Helmholz, *Zeit. f. Phys.* **75**, 19 (1932); and
J. Sherman, *Chem. Rev.* **11**, 93 (1932). The latter two papers deal with ionic lattices
and introduce the Coulomb forces (Madelung's constant).

‡ K. J. McCallum and J. E. Mayer, *J. Chem. Phys.* **11**, 56 (1943); P. M. Doty and
J. E. Mayer, ibid. **12**, 323 (1944); P. P. Sutton and J. E. Mayer, ibid. **3**, 20 (1935).

TABLE 11

Experimental Values of the Cohesive Energies of the Alkali Halides

(All values refer to one mole of salt and in units of kilo-calories.)

	(i)	(ii)	(iii)	(iv)	(v)	(vi)	(vii)	Cohesive energy = (i)+(ii)+...+(vii)
LiF	1·4	145·2	38·3	31·8	−2·9	123·8	..	..
LiCl	2·0	97·4	38·3	28·8	−2·9	123·8	−85·8	201·6
LiBr	2·5	83·6	38·3	26·7	−2·9	123·8	−80·5	191·5
LiI	2·8	64·9	38·3	25·5	−2·9	123·8	−72·4	180·0
NaF	1·9	136·1	26·0	31·8	−2·9	117·9	..	..
NaCl	2·4	98·3	26·0	28·8	−2·9	117·9	−85·8	184·7
NaBr	2·5	86·2	26·0	26·7	−2·9	117·9	−80·5	175·9
NaI	2·8	69·4	26·0	25·5	−2·9	117·9	−72·4	166·3
KF	2·2	134·1	21·65	31·8	−2·9	99·5	..	..
KCl	2·5	104·1	21·65	28·8	−2·9	99·5	−85·8	167·8
KBr	2·7	94·0	21·65	26·7	−2·9	99·5	−80·5	161·2
KI	2·8	78·6	21·65	25·5	−2·9	99·5	−72·4	152·8
RbF	2·5	133·1	19·92	31·8	−2·9	96·0	..	..
RbCl	2·8	104·8	19·92	28·8	−2·9	96·0	−85·8	163·6
RbBr	2·9	95·9	19·92	26·7	−2·9	96·0	−80·5	158·0
RbI	3·0	80·6	19·92	25·5	−2·9	96·0	−72·4	149·7
CsF	2·7	131·8	19·11	31·8	−2·9	89·5	..	..
CsCl	2·8	106·3	19·11	28·8	−2·9	89·5	−85·8	157·8
CsBr	2·9	97·5	19·11	26·7	−2·9	89·5	−80·5	152·3
CsI	3·0	83·6	19·11	25·5	−2·9	89·5	−72·4	145·4

from the theoretical cohesive energies.[†] As different salts of the same electro-negative element (halide in the case of alkali halides) provide, in this way, each a distinct estimate of the electron affinity, the consistency between the different values serves as a check on the method. Take the case of fluorine. From Tables 9 and 11 we can immediately form the following values (in kilo-calories):

TABLE 12

Electron Affinity of Fluorine

Cohesive energy (Huggins)	243·6	215·4	192·5	183·0	175·7
Sum of (i)–(vi) (Table 11)	337·6	310·8	286·3	280·4	272·0
Electron affinity of F	94·0	95·4	93·8	97·4	96·3
Mean	95·4				

† This was done in the first publications by M. Born (*Verh. d. D. Phys. Ges.* **21**, 13, 679 (1919)) from which all these investigations started and which contains the first definition of electron affinity.

Sherman† has determined in this way the electron affinities of oxygen, sulphur, and selenium (to form divalent ions!) from the theoretical cohesive energies of the corresponding salts with the alkaline earth elements Mg, Ca, Sr, Ba. Using these values and the affinities of the halides, he has calculated the cohesive energies for a large number of compounds. The discrepancies with the experimental values have been taken by him as a measure of the degree of departure from an ideally ionic structure.

† Loc. cit.

LATTICE VIBRATIONS

4. Simple approximate treatment of thermodynamical behaviour

IN classical mechanics, it is well known that when a system of particles in stable equilibrium is disturbed, the system performs certain vibrations, such that every particle remains in the neighbourhood of its equilibrium position. For the description of the vibratory motions, a set of coordinates q_i can be introduced which are linear functions of the displacements of the particles and vary independently of one another, each as a sinusoidal function of time $A_i \sin(2\pi \nu_i t + \delta_i)$ (the frequency ν_i is determined by the nature of the forces, whereas A_i, δ_i are arbitrary constants). Such coordinates are known as the *normal coordinates* of the system, and the corresponding motions they describe as the *normal vibrations* (or *modes*). Their number must obviously be equal to the degrees of freedom of the system, namely, three times the number of particles. In many ways, the vibratory system is completely equivalent to a collection of independent simple harmonic oscillators with the respective frequencies ν_i. In particular, the possible quantum-mechanical energy levels are identical in the two cases. According to statistical mechanics, the energy levels of a system completely determine its thermodynamical functions; the Helmholtz free energy

$$F = E - TS \quad (E = \text{energy}; \ S = \text{entropy}) \quad (4.1)$$

is given quite generally by

$$F = -kT \ln Z, \quad (4.2)$$

where Z, the *partition function*, is the sum of the Boltzmann factors corresponding to all possible energy levels of the system:

$$Z = \sum_i e^{-\epsilon_i/kT}, \quad (4.3)$$

ϵ_i being the eigenvalues of the energy operator.

For a collection of independent oscillators, one can construct the free energies according to (4.2), (4.3) first for the oscillators separately and afterwards take the sum. The eigenvalues of an oscillator with frequency ν_i are

$$\tfrac{1}{2}h\nu_i, \ \tfrac{3}{2}h\nu_i, \ \tfrac{5}{2}h\nu_i, \ldots ;$$

the corresponding partition function and free energy are thus

$$Z_i = e^{-\frac{1}{2}h\nu_i/kT} \sum_{s=0}^{\infty} e^{-sh\nu_i/kT} = \frac{e^{-\frac{1}{2}h\nu_i/kT}}{1 - e^{-h\nu_i/kT}}, \quad (4.4)$$

$$F_i = \tfrac{1}{2}h\nu_i + kT \ln(1 - e^{-h\nu_i/kT}). \quad (4.5)$$

For a crystal lattice with the normal vibration frequencies ν_i, $\sum\limits_i F_i$ gives the part of its free energy due to the vibratory motions. The complete free energy includes in addition the energy U of the static lattice in which each lattice particle occupies its mean position; thus

$$F = U + \tfrac{1}{2} \sum_i h\nu_i + kT \sum_i \ln(1 - e^{-h\nu_i/kT}). \tag{4.6}$$

If a solid is deformed, the shifts in the mean positions of the particles alter the frequencies ν_i as well as the static energy U. At present we shall consider the isotropic change in volume V as the only form of deformation; thus in (4.6) both ν_i and U are to be considered as functions of V. From the free energy $F(V, T)$ as a function of V and T, all the thermodynamical behaviour concerning changes in volume and temperature can be deduced with the help of thermodynamical relations. The entropy is

$$S = -\left(\frac{\partial F}{\partial T}\right)_V, \tag{4.7}$$

hence the energy E is given by

$$E = F + TS = F - T\left(\frac{\partial F}{\partial T}\right)_V$$
$$= U + \tfrac{1}{2} \sum_i h\nu_i + kT \sum_i \frac{(h\nu_i/kT)}{e^{h\nu_i/kT} - 1}. \tag{4.8}$$

The heat capacity at constant volume is obtained by differentiating the energy with respect to temperature:

$$C_V = \left(\frac{\partial E}{\partial T}\right)_V = k \sum_i \frac{(h\nu_i/kT)^2 e^{h\nu_i/kT}}{(e^{h\nu_i/kT} - 1)^2}. \tag{4.9}$$

For $T = 0°$ K., the last term in (4.8) vanishes and the energy reduces to the static lattice energy U plus the zero-point energy $\tfrac{1}{2} \sum\limits_i h\nu_i$, to which reference has been made earlier.

For high temperatures such that $kT \gg h\nu_i$, we can expand the terms in the sums of (4.8) and (4.9) with respect to $(h\nu_i/kT)$ and obtain in this way

$$E \cong U + \tfrac{1}{2} \sum_i h\nu_i + 3kT \times (\text{number of particles}), \tag{4.10}$$

$$C_V \cong 3k \times (\text{number of particles}), \tag{4.11}$$

remembering that the total number of normal coordinates is equal to three times the number of particles. These relations express the classical energy partition law (the energy for each vibrational degree of freedom is kT) and the Dulong–Petit law of specific heat, which follows from it. These laws thus hold only when the temperature is so high that the vibratory modes are in such highly excited states that quantum effects become insignificant.

The equation of state, which connects the state parameters p, V, T, follows from the thermodynamical relation:

$$p = -\left(\frac{\partial F}{\partial V}\right)_T. \qquad (4.12)$$

Thus we find on substituting the expression (4.6) for the free energy in (4.12) that

$$p = -\frac{dU}{dV} - \sum_i \left(\frac{1}{2} + \frac{1}{e^{h\nu_i/kT}-1}\right)h\frac{d\nu_i}{dV}. \qquad (4.13)$$

Since ν_i and U are to be regarded as known functions of V, (4.13) provides the requisite relation between p, V, T. Often it is convenient to introduce, instead of the derivatives of the frequencies, the dimensionless quantities

$$\gamma_i = -\frac{d\ln\nu_i}{d\ln V} = -\frac{V}{\nu_i}\frac{d\nu_i}{dV}. \qquad (4.14)$$

In terms of γ_i, the equation of state becomes

$$p + \frac{dU}{dV} = \frac{1}{V}\sum_i \gamma_i\left\{\tfrac{1}{2}h\nu_i + \frac{h\nu_i}{e^{h\nu_i/kT}-1}\right\}. \qquad (4.15)$$

The actual normal vibrations in crystals are very complicated in nature, but for the purpose of constructing the thermodynamic functions some very crude approximations have proved successful.

The simplest possible approximation is usually referred to as the *Einstein model:*[†] it is assumed that all the vibrational frequencies ν_i are equal (strictly speaking, one could imagine this to be the case only if the lattice particles in a simple lattice were to vibrate independently of one another, each isotropically about its mean position). Let N be the number of cells per mole and n be the number of particles per cell. Then the energy per mole and the molar heat capacity can be written down directly from (4.8) and (4.9) by putting all the frequencies ν_i equal to ν_0:

$$E = N\left\{u(v) + 3nkT\left[\tfrac{1}{2}\xi + \frac{\xi}{e^\xi-1}\right]_{\xi=(\Theta_E/T)}\right\}, \qquad (4.16)$$

$$C_V = 3Nnk\left[\frac{\xi^2 e^\xi}{(e^\xi-1)^2}\right]_{\xi=(\Theta_E/T)}, \qquad (4.17)$$

where $u(v) = U/N$ is the static lattice energy per cell and

$$\Theta_E = \frac{h\nu_0}{k} \qquad (4.18)$$

† A. Einstein, *Ann. d. Phys.* [4], **22**, 180 (1907).

is a parameter having the dimension of temperature, called *Einstein's characteristic temperature*. In this model, the heat capacity as a function of temperature depends only on a single parameter, namely, Θ_E. We notice that Nn divided by the number of atoms per molecule is Avogadro's number $6\cdot022 \times 10^{23}$, the same for all substances. Thus if we measure the temperature for each substance in a suitably chosen unit $(\sim \Theta_E)$, the heat capacity divided by the number of atoms per molecule is the same function of the reduced temperature for all substances. We shall discuss how this theory compares with the experimental results later in conjunction with the results of the more elaborate Debye theory.

Speaking physically, the Einstein model appears very unrealistic for a monatomic substance, where all atoms are equivalent. For an atom is held to its equilibrium position only in virtue of its interaction with its surrounding atoms, and thus cannot possibly vibrate independently of the others. The situation becomes different, if one considers a molecular crystal. Consider, to start with, a diatomic gas, where the vibration frequency of the two atoms in the gaseous molecule is ν_0. In the gaseous phase, there is thus a part of the free energy exactly described by the Einstein model. In addition, there are of course contributions due to the translational and rotational motions of the molecules. When the molecules are condensed into a crystalline lattice, it may happen, as in the case of molecular crystals, that the molecules are held together by much weaker forces than that binding the atoms in the molecule. The atoms within each molecule will vibrate with respect to each other with much the same frequency as before, resulting in a contribution to the free energy of the Einstein type. The free translational and rotational motions of the molecules, on the other hand, lost on forming the lattice and become transformed into characteristic lattice vibrations; the corresponding contributions are not describable realistically in terms of the Einstein model. It is precisely the latter vibrations that are more adequately considered in the Debye model, which we shall presently discuss.

Before doing this, let us examine more closely how the vibration frequencies are best described for a macroscopic system. For a macroscopic specimen, the number of vibration frequencies is clearly enormous. What proves to be relevant is not so much the individual frequencies as the number of frequencies falling within any particular interval of frequency $(\nu, \nu+\Delta\nu)$. Let us denote this number by $N(\nu)\Delta\nu$. When the specimens are so large that the number of cells near the surface is only a negligible fraction of the total number N, $N(\nu)\Delta\nu$ becomes practically proportional to N. In a more precise way, this means that the limit

$$f(\nu)\Delta\nu = \lim_{N\to\infty} \frac{N(\nu)\Delta\nu}{N} \qquad (4.19)$$

exists; the function $f(\nu)$ defined by the limit will be called the *frequency distribution function*. It follows immediately from the consideration of the requisite degrees of freedom that $f(\nu)$ fulfils the normalization condition

$$\int_0^\infty f(\nu)\, d\nu = 3n. \qquad (4.20)$$

For a mole of the substance, we have then $N(\nu) = Nf(\nu)$, where N stands as previously for the number of cells in a mole. Using the frequency distribution function we can express (4.8), (4.9) in terms of integrals:

$$E = N\left\{u(\nu) + kT \int_0^\infty \left[\frac{1}{2} + \frac{1}{e^{h\nu/kT}-1}\right]\left(\frac{h\nu}{kT}\right)f(\nu)\, d\nu\right\}, \qquad (4.21)$$

$$C_V = Nk \int_0^\infty \frac{e^{h\nu/kT}}{(e^{h\nu/kT}-1)^2}\left(\frac{h\nu}{kT}\right)^2 f(\nu)\, d\nu. \qquad (4.22)$$

(In order to avoid complications by surface effects, when considering molar quantities one does not think of an actual specimen of one mole. Instead, one imagines very large specimens and divides the corresponding extensive quantities such as the energy and heat capacity, etc., by the number of moles contained. The molar quantities refer to the limits of the normalized values for infinitely large specimens. Thus the integral forms given in (4.21) and (4.22) are the precise expressions for the molar quantities E and C_V.)

In *Debye's model*,† one takes

$$\begin{aligned} f(\nu) &= C\nu^2 \quad \text{for} \quad \nu < \nu_m \\ &= 0 \quad \text{for} \quad \nu > \nu_m, \end{aligned} \qquad (4.23)$$

where ν_m, the maximum frequency, is a parameter characteristic of the substance. We shall return presently to discuss the physical significance of the above frequency distribution function; let us first discuss the corresponding expressions for the energy and heat capacity. The constant of proportionality C in (4.23) is not arbitrary; it follows immediately from the normalization condition for $f(\nu)$ that

$$C \int_0^{\nu_m} \nu^2\, d\nu = \frac{C\nu_m^3}{3} = 3n, \quad \text{or} \quad C = \frac{9n}{\nu_m^3}. \qquad (4.24)$$

† P. Debye, *Ann. d. Phys.* [4], **39**, 789 (1912).

Substituting (4.23) and (4.24) in (4.21) and (4.22), we obtain the energy per mole and the molar heat capacity:

$$E = N\left\{u(v) + 9nkT\left[\left(\frac{T}{\Theta_D}\right)^3 \int_0^{\Theta_D/T} \left(\frac{1}{2} + \frac{1}{e^\xi - 1}\right)\xi^3\, d\xi\right]\right\}, \qquad (4.25)$$

$$C_V = 9Nnk\left(\frac{T}{\Theta_D}\right)^3 \int_0^{\Theta_D/T} \frac{e^\xi}{(e^\xi - 1)^2}\xi^4\, d\xi, \qquad (4.26)$$

where

$$\Theta_D = \frac{h\nu_m}{k}. \qquad (4.27)$$

Θ_D is the characteristic parameter in this case and is usually known as the *Debye temperature*.

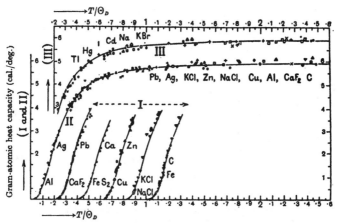

FIG. 4. Experimental and Debye's values for the heat capacity.

Also according to this theory, there is a universal curve for heat capacity versus temperature, if we divide the heat capacity for each substance by the number of atoms per molecule and express the temperature in the appropriate unit ($\sim \Theta_D$). Fig. 4 shows how well the experimental values can be fitted on to the Debye curve with suitably chosen values of Θ_D; the values used for Θ_D are given in Table 13. The curves I, II, III are identical Debye curves relatively shifted to avoid overcrowding of the experimental points. The agreement of the Debye curve and the experimental points is seen to be remarkably close. In Fig. 5 the Einstein and Debye values for the heat capacity are compared, where Θ_D is chosen equal to $0.75\Theta_E$. The two curves practically coincide except in the region of very low temperatures, where the Einstein values fall abruptly to values which are too low. The close agreement between the Debye and Einstein curves despite the radically different frequencies assumed, shows that the thermal properties are not sensitive to the frequency distributions except at very low temperatures. This is the main reason for the remarkable success of these simple theories.

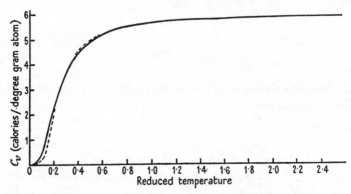

FIG. 5. Comparison of the Debye (——) and Einstein (- - - -) models.

TABLE 13
Debye Temperatures

Substance	Pb	Tl	Hg	I	Cd	Na	KBr	Ag	Ca
	88	96	97	106	168	172	177	215	226

Substance	KCl	Zn	NaCl	Cu	Al	Fe	CaF	FeS	C
	230	235	281	315	398	453	474	645	1860

To one not aware of the atomic constitution of solids, a solid would appear as an elastic continuum as conceived in the classical elasticity theory. Debye's frequency distribution is obtained in fact by regarding the crystalline lattice in this way. It is well known that all the motions taking place within an elastic continuum can be resolved into elastic waves of the form

$$\mathbf{u}(\mathbf{x}, t) = A\mathbf{n}\sin(2\pi\mathbf{y}.\mathbf{x} - 2\pi\nu t + \delta), \qquad (4.28)$$

$\mathbf{u}(\mathbf{x}, t)$ being the displacement of the medium at the point $\mathbf{x}$ and time t. (4.28) describes a plane wave of frequency ν with displacements in the direction of the unit vector $\mathbf{n}$ (the *polarization vector*), with wave normal parallel to $\mathbf{y}$ and wave-number equal to $|\mathbf{y}|$; $\mathbf{y}$ is known as the wave-number vector. A and δ are the usual arbitrary amplitude and phase associated with an oscillatory motion. For any given $\mathbf{y}$, there are three independent modes of elastic waves, with mutually perpendicular polarization vectors $\mathbf{n}_i$ ($i = 1, 2, 3$) and in general different frequencies ν_i. The corresponding phase velocities

$$c_i = \frac{\nu_i}{|\mathbf{y}|} \quad (i = 1, 2, 3) \tag{4.29}$$

and the polarization vectors $\mathbf{n}_i$ depend on the direction of $\mathbf{y}$ but not on its magnitude. We notice that owing to the presence of the arbitrary phase δ in (4.28), a similar cosine wave does not constitute any further independent mode of vibration; moreover, by keeping to a fixed sign for the νt term as we do here, we can count $\mathbf{y}$ and $-\mathbf{y}$ as two distinct vibrations, corresponding to identical progressive waves travelling opposite to one another.

(4.28) represents solutions of the elastic equations of motion and need not constitute the normal modes of vibration. The normal vibrations for a particular specimen are linear combinations of the above solutions (of the same frequency) which satisfy the boundary conditions on the surface (e.g. kept fixed or left free, etc.). When the specimen is sufficiently large, however, the distribution of the normal modes over frequency becomes practically independent of the shape of the specimen or the particular boundary conditions imposed.† For our purpose, it is simplest to take a large cubic specimen of volume V and impose the boundary condition that corresponding points on opposite faces of the cube must move in exactly the same way. The great advantage of this procedure is that we need not take combinations of the solutions (4.28); we can take as the normal modes those solutions which have the following wave-numbers:

$$\mathbf{y} = \frac{1}{\sqrt[3]{V}}(n_1, n_2, n_3) \qquad (n_1, n_2, n_3 = \text{integers}). \tag{4.30}$$

For such solutions obviously satisfy the requisite boundary condition. In particular, we note that if we represent these permitted values of $\mathbf{y}$ by points in the vector space of $\mathbf{y}$, the points are distributed with a uniform density (dimension = volume in the $\mathbf{y}$-space!) equal to V.

The above boundary condition is known as the *periodic boundary condition* first proposed by Born.‡ We can give the condition a different interpretation (cyclic condition) which makes its generalization to the

---

† See also W. Ledermann, *Proc. Roy. Soc.* A, **182**, 362 (1944). An elegant proof of this result has been recently given by R. Peierls (see Appendix IV).

‡ See M. Born, *Atomtheorie des festen Zustandes*, p. 557 (2nd ed., Teubner, 1923).

case of an atomic lattice straightforward. Thus, instead of a finite specimen, we imagine an infinitely extended medium. Since there is no boundary to reflect progressive waves, the solutions (4.28) all represent permissible modes of the medium. Clearly the modes cannot be enumerated, as **y** can take up a continuum of values. The condition (4.30) provides, so to speak, a procedure of sampling. Imagine a cubic volume V as before, but this time it is a portion of the infinitely extended medium. To this volume we ascribe those vibrations which repeat the vibration pattern within the cube periodically throughout the whole medium. As the geometrical configuration of the whole medium in such motions is completely determined by that in the cube, the number of such vibrations must evidently reproduce the dynamical degrees of freedom in the cube. Moreover, as can be easily verified, if the cube edges are increased by a factor s and hence the volume by s^3, the number of vibrations ascribed to the cube within any given frequency interval is raised by s^3. Thus by enlarging the cube indefinitely in this way, we should be taking in more and more of the vibrations in the medium, whereas the frequency distribution remains the same (a constant factor such as s^3 does not affect the distribution). It is thus plausible to regard the sampling as representative; and the above consideration of the degrees of freedom enables us to associate the vibrations with a finite volume. The vibrations thus ascribed to the cube V are evidently the same as before, i.e. those with the wave-numbers (4.30).

For an ideal isotropic solid, two vibrations ($i = 1, 2$) for a given **y** are transverse ($\mathbf{n}_1, \mathbf{n}_2 \perp \mathbf{y}$) and have the same frequency $\nu_1 = \nu_2 = \nu_t$; the remaining vibration ($i = 3$) is longitudinal ($\mathbf{n}_3 \| \mathbf{y}$) and has a distinct frequency ν_l. Owing to the isotropy, the phase velocities

$$c_1 = c_2 = \frac{\nu_t}{|\mathbf{y}|} = c_t, \qquad c_3 = \frac{\nu_l}{|\mathbf{y}|} = c_l \qquad (4.31)$$

are constants independent of the direction and the magnitude of **y**. The transverse vibrations with frequencies in the interval ν to $\nu + \Delta\nu$ thus occupy a spherical shell of the 'volume'

$$4\pi y^2 \Delta y = \frac{4\pi \nu^2}{c_t^3} \Delta\nu$$

in the **y**-space. The number of permitted **y** values in the shell is obtained by multiplying the volume by the density of their representative points, V:

$$\frac{4\pi V}{c_t^3} \nu^2 \Delta\nu.$$

As there are two transverse vibrations to each value of y, the number of transverse vibrations in the frequency interval ν to $\nu + \Delta\nu$ is thus

$$\frac{8\pi V}{c_t^3} \nu^2 \Delta\nu. \tag{4.32}$$

In exactly the same way we find the number of longitudinal vibrations in the same frequency interval to be

$$\frac{4\pi V}{c_l^3} \nu^2 \Delta\nu. \tag{4.33}$$

When we discuss the vibrations in terms of actual atomic lattices it will become clear that the wave-number y cannot be increased indefinitely; in fact all the distinct modes of vibration are included in a range of $|y|$ from zero to $1/a$, a being roughly the linear dimension of the lattice cell. That such a limit should exist is obvious, for otherwise the number of vibrations given by (4.32) and (4.33) would be infinite, rather than equal to the number of degrees of freedom of the atoms in V. One reasonable way to take account of this effect is to consider only the vibrations (4.32), (4.33) which lie within a sphere about the origin in the y-space; the radius of the sphere can then be determined so as to give the correct number of degrees of freedom. We notice that owing to the different velocities of the transverse and longitudinal waves, this implies that we cut off the transverse and longitudinal frequency spectra (4.32) and (4.33) at two different frequency maxima. Usually, however, one adopts the more straightforward way of cutting off both the longitudinal and transverse spectra at a common frequency ν_m. This procedure, we observe, leads directly to the frequency distribution function (4.23), which we have used. This latter method appears somewhat arbitrary, but as the difference in the two methods of cutting off the vibration spectra concerns only vibrations with very short wave-lengths for which the elastic treatment is in any case inadequate, whichever method is adopted is thus of no precise significance.

Adding (4.32) and (4.33) and comparing with $Nf(\nu)$, we find, introducing the volume per cell, $v = V/N$, that

$$f(\nu) = 4\pi v \left(\frac{2}{c_t^3} + \frac{1}{c_l^3}\right) \nu^2. \tag{4.34}$$

The constant of proportionality, as we have seen, is related to the cut-off frequency ν_m by (4.24), therefore

$$4\pi v \left(\frac{2}{c_t^3} + \frac{1}{c_l^3}\right) = \frac{9n}{\nu_m^3}, \tag{4.35}$$

which, we recall, follows from the normalization condition. The maximum frequency, and hence also the Debye temperature, can thus be calculated from the velocities of the elastic waves. As the crystalline solids are never isotropic, we can only check the above relation roughly by replacing $(2/c_t^3 + 1/c_l^3)$ by the following average over all directions of propagation (cf. § 6):

$$\frac{1}{4\pi} \int \left(\frac{1}{c_1^3} + \frac{1}{c_2^3} + \frac{1}{c_3^3} \right) d\omega, \tag{4.36}$$

where c_1, c_2, c_3 are the phase velocities for a given direction of propagation. The Debye temperatures calculated for a few compounds in this way are compared in the adjoining table with the values quoted earlier as determined by fitting the specific heat data.

TABLE 14

Comparison of Debye Temperatures determined from Elastic and Heat Data

	KCl	NaCl	CaF	FeS
Heat	230	281	474	645
Elastic	227	305	510	696

We shall return to a closer examination of Debye's theory when we have considered the lattice vibrations in more detail.

Using the Einstein and Debye frequency distributions, we readily obtain the corresponding free-energy expressions:

$$F = N\{u(v) + 3nkT[\tfrac{1}{2}\xi + \ln(1 - e^{-\xi})]_{\xi = \Theta_E/T}\} \quad \text{(Einstein model)}, \tag{4.37}$$

$$F = N\left\{u(v) + 9nkT\left[\left(\frac{T}{\Theta_D}\right) \int_0^{\Theta_D/T} [\tfrac{1}{2}\xi + \ln(1 - e^{-\xi})]\xi^2 \, d\xi\right]\right\} \quad \text{(Debye model)}. \tag{4.38}$$

We notice that in both expressions the vibrational contribution $F - Nu(v)$ is a function $\xi(V, T)$ of V, T of the following form:

$$\xi(V, T) = T\eta(T/\Theta), \tag{4.39}$$

where $\eta(T/\Theta)$ is a function of one argument T/Θ, Θ being a function of volume only. We easily verify that

$$\left(\frac{\partial \eta}{\partial \ln \Theta}\right)_T = -\frac{T\eta'}{\Theta} = -\left(\frac{\partial \eta}{\partial \ln T}\right)_V. \tag{4.40}$$

Using this relation we find that

$$\left(\frac{\partial \xi(V, T)}{\partial V}\right)_T = \frac{-\gamma}{V}\left(\frac{\partial \xi(T, V)}{\partial \ln \Theta}\right)_T = \frac{\gamma T}{V}\left\{\frac{\partial \eta}{\partial \ln T}\right\}_V, \qquad (4.41)$$

where we have written

$$\gamma = -\frac{d \ln \Theta}{d \ln V}. \qquad (4.42)$$

Expressing η in terms of ξ, $\eta = \frac{\xi}{T}$,

on the right-hand side of (4.41), we obtain the following relation, which holds generally for a function of the form (4.39):

$$\left(\frac{\partial \xi(V, T)}{\partial V}\right)_T = \frac{\gamma}{V}\left[T\left(\frac{\partial \xi(V, T)}{\partial T}\right)_V - \xi(V, T)\right]. \qquad (4.43)$$

Substituting the vibrational part of the free energy $F_{\mathrm{vib}} = F - Nu(v)$ into this relation, we find that

$$\left(\frac{\partial \{F - Nu(v)\}}{\partial V}\right)_T = \frac{\gamma}{V}\left\{T\left(\frac{\partial F}{\partial T}\right)_V - F + Nu(v)\right\}, \qquad (4.44)$$

where γ stands for the following expressions:

$$-\frac{d \ln \Theta_E}{d \ln V}, \qquad -\frac{d \ln \Theta_D}{d \ln V},$$

respectively, for the Einstein and Debye models. We notice that (see (4.1) and (4.7))

$$T\left(\frac{\partial F}{\partial T}\right)_V - F = -TS - F = -E;$$

hence (4.44) can be written as

$$-\left(\frac{\partial F}{\partial V}\right)_T + \frac{du(v)}{dv} = \frac{\gamma E_{\mathrm{vib}}}{V}, \qquad (4.45)$$

where $E_{\mathrm{vib}} = E - Nu(v)$ is the vibrational contribution to the energy. Introducing for $-(\partial F/\partial V)_T$ the pressure p (see (4.12)), we obtain finally the equation of state of Mie and Grüneisen:†

$$p + \frac{du}{dv} = \gamma \frac{E_{\mathrm{vib}}}{V}. \qquad (4.46)$$

This equation is probably of wider validity than either of the above special frequency distributions. The general equation of state (4.15) reduces to this form, if we assume all the γ_i to be equal. This is in fact the case for the actual vibrations in the simple linear case considered in the next section. Although the assumption of the equality of γ_i is not

† Cf. M. Born, *Atomtheorie des festen Zustandes* (2nd ed., Teubner, 1923), p. 652; E. Grüneisen, *Handb. der Phys.* 10, 22 (Springer, 1926).

generally true, it provides a very useful approximation for discussing the thermal properties.

The value of γ in (4.46) can of course depend on the volume, but if one makes a theoretical estimate of γ from the atomic forces, on the basis of either the Einstein or Debye model, its variation with volume is found to be slight. By treating γ as a constant (Grüneisen's constant), Grüneisen† has made extensive investigations as regards the various conclusions deducible from the equation of state and has obtained in most cases satisfactory experimental confirmations of the theoretical results. Following generally Grüneisen's method, we shall deduce certain relations giving the thermal variations of the volume and the compressibility; the results can then be used to calculate the volume and compressibility for the lattice in static equilibrium.

As we desire the volume and compressibility under normal pressures, which have negligible effects on solids, we can put $p = 0$ in the equation of state (4.46), obtaining

$$\frac{du}{dv} = \frac{\gamma E_{\text{vib}}}{V}. \tag{4.47}$$

Besides (4.47) a further relation is required to determine the thermal variation of the compressibility; it can be obtained as follows: Multiplying (4.46) by V and differentiating the equation with respect to V at constant temperature, we find that

$$p + V\left(\frac{\partial p}{\partial V}\right)_T + \frac{du}{dv} + v\frac{d^2u}{dv^2} = \gamma\left(\frac{\partial E_{\text{vib}}}{\partial V}\right)_T. \tag{4.48}$$

Since E_{vib} is a function of the form $\xi(V, T)$ (see (4.16), (4.25), (4.39)), we can apply the general relation (4.43), obtaining

$$\left(\frac{\partial E_{\text{vib}}}{\partial V}\right)_T = \frac{\gamma}{V}\left\{T\left(\frac{\partial E_{\text{vib}}}{\partial T}\right)_V - E_{\text{vib}}\right\} = \frac{\gamma}{V}\{TC_V - E_{\text{vib}}\}. \tag{4.49}$$

Substituting (4.47) in (4.48) and putting $p = 0$ (atmospheric pressure negligible), we find the requisite relation ($\beta = $ compressibility):

$$\frac{1}{\beta} = -v\left(\frac{\partial p}{\partial v}\right)_T = \frac{du}{dv} + v\frac{d^2u}{dv^2} - \frac{\gamma^2}{V}(TC_V - E_{\text{vib}}). \tag{4.50}$$

The significance of the two relations (4.47) and (4.50) for our purpose becomes more obvious if we express the static energy $u(v)$ more explicitly as a function of volume by a Taylor expansion with respect to $v-v_0$, where v_0 is the static equilibrium value satisfying the equilibrium condition

$$(du/dv)_{v_0} = 0. \tag{4.51}$$

† E. Grüneisen, loc. cit.

For ordinary temperatures $(v-v_0)/v_0$ is small $(< 1/30$ for the alkali halides); we shall thus retain only linear terms in $(v-v_0)/v_0$. Carrying out the expansion and remembering (4.51), we obtain the following approximate forms for (4.47) and (4.50):

$$\frac{v-v_0}{v_0} = \frac{\gamma E_{\mathrm{vib}}}{V}\beta_0, \tag{4.52}$$

$$\frac{1}{\beta} - \frac{1}{\beta_0} = \left(\frac{1}{\beta_0}\right)\left\{2 + \frac{v_0\left(\dfrac{d^3u}{dv^3}\right)_0}{\left(\dfrac{d^2u}{dv^2}\right)_0}\right\}\left(\frac{v-v_0}{v_0}\right) - \frac{\gamma^2}{V}(TC_V - E_{\mathrm{vib}}), \tag{4.53}$$

where β_0 denotes the compressibility for the lattice in static equilibrium:

$$\frac{1}{\beta_0} = v_0\left(\frac{d^2u}{dv^2}\right)_0. \tag{4.54}$$

Eliminating $(v-v_0)/v_0$ from (4.53) with the help of (4.52), we have

$$\frac{1}{\beta} - \frac{1}{\beta_0} = \left\{2 + \frac{v_0\left(\dfrac{d^3u}{dv^3}\right)_0}{\left(\dfrac{d^2u}{dv^2}\right)_0}\right\}\frac{\gamma E_{\mathrm{vib}}}{V} - \frac{\gamma^2}{V}(TC_V - E_{\mathrm{vib}}). \tag{4.55}$$

The relations (4.52) and (4.55) can be used directly to calculate the changes in volume and compressibility due to the thermal and zero-point vibrations, if the quantities on the right-hand side of the relations are known.

γ can easily be expressed in terms of directly measurable quantities. Let us differentiate the equation of state (4.46) with respect to T at constant volume. Since

$$\left(\frac{\partial E_{\mathrm{vib}}}{\partial T}\right)_V = C_V$$

we have

$$\left(\frac{\partial p}{\partial T}\right)_V = \frac{\gamma C_V}{V}. \tag{4.56}$$

Expressing $(\partial p/\partial T)_V$ in the form

$$\left(\frac{\partial p}{\partial T}\right)_V = \frac{\dfrac{1}{V}\left(\dfrac{\partial V}{\partial T}\right)_p}{-\dfrac{1}{V}\left(\dfrac{\partial V}{\partial p}\right)_T},$$

we can also write the above relation as

$$\gamma = \left(\frac{V}{C_V}\right)\frac{\dfrac{1}{V}\left(\dfrac{\partial V}{\partial T}\right)_p}{\dfrac{-1}{V}\left(\dfrac{\partial V}{\partial p}\right)_T}, \tag{4.57}$$

where $(1/V)(\partial V/\partial T)_p$ and $(-1/V)(\partial V/\partial p)_T$ are respectively the thermal expansion coefficient and the compressibility β. The values of γ for the alkali halides deduced with the above expression are given in Table 15; they are all in the neighbourhood of 1·5.

TABLE 15

The Grüneisen Constants γ, Nearest Neighbour Distances r_0, Compressibilities β_0 (for Lattice in Static Equilibrium), and the Revised Cohesive Energies of the Alkali Halides.

	γ	$r_0 (10^{-8}\,cm.)$	$\beta_0 (10^{-12}/barye)$	Cohesive energy (k.cal./mole) Calc. (3.15)	Exp.	Diff. (%)
LiF	1·99	1·983	1·09	257	..	..
LiCl	1·54	2·538	3·16	200	201·5	−1·5
NaF	1·57	2·283	1·97	223	..	..
NaCl	1·43	2·783	3·94	185	184·7	0·3
NaBr	1·55	2·946	4·63	176	175·9	0·1
NaI	1·59	3·189	6·40	162	166·3	−4·3
KF	1·48	2·637	3·06	196	..	..
KCl	1·34	3·108	5·20	168	167·8	−0·2
KBr	1·43	3·258	6·10	161	161·2	−0·2
KI	1·58	3·484	7·62	150·8	152·8	−2·0
RbF	1·28	2·788	3·82	185·2	..	..
RbCl	1·25	3·239	6·14	161·2	163·6	−2·4
RbBr	1·27	3·394	7·30	154·1	158·0	−3·9
RbI	1·50	3·622	8·62	145·6	149·7	−4·1
CsF	1·49	2·976	3·88	177·2	..	..
CsCl	1·97	3·520	5·18	152·4	157·8	−5·4
CsBr	1·93	3·665	6·10	146·4	152·3	−5·9
CsI	2·00	3·900	7·30	138·3	145·4	−7·1

To calculate $E_{\rm vib}$ requires a knowledge of the frequency distribution. Mayer and Helmholz[†] have given approximate distributions for the alkali halides, in some cases as pure Debye distributions, and in other cases, as superpositions of an Einstein and a Debye distribution. Their results are adopted in evaluating $E_{\rm vib}$ and $TC_V - E_{\rm vib}$ for the calculations carried out below.

There still remain the quantities depending on the static energy u,

[†] J. E. Mayer and L. Helmholz, *Zeit. f. Phys.* **75**, 19 (1932).

namely, the second and third derivatives at v_0. If we use the approximate form (3.10) for $u(v)$, we easily find with the help of the equilibrium condition (3.13) that

$$\frac{v_0\left(\dfrac{d^3u}{dv^3}\right)_0}{\left(\dfrac{d^2u}{dv^2}\right)_0} = -2 - \frac{1}{3}\left(\frac{r_0}{\rho}\right)\frac{\left(\dfrac{r_0}{\rho}\right)-3}{\left(\dfrac{r_0}{\rho}\right)-2}.$$

Hence we may write (4.55) in terms of (r_0/ρ) as

$$\frac{1}{\beta}-\frac{1}{\beta_0} = -\frac{1}{3}\left(\frac{r_0}{\rho}\right)\frac{\left(\dfrac{r_0}{\rho}\right)-3}{\left(\dfrac{r_0}{\rho}\right)-2}\frac{\gamma E_{\text{vib}}}{V}-\frac{\gamma^2}{V}(TC_V-E_{\text{vib}}). \tag{4.58}$$

A first approximation for β_0 can be immediately deduced from the room-temperature values of β, v with (4.58) by using on the right-hand side the approximate values for (r_0/ρ) given in Table 9 in the last section. From the values of β_0 so obtained and v_0 determined from (4.52), revised values for (r_0/ρ) can be calculated as in the last section with the help of (3.14). This process of successive approximations for determining (r_0/ρ), β_0, and v_0 with the equations (4.58), (4.52), and (3.14) can be carried on until no further changes occur in these quantities; the ultimate values are in fact already obtained in the second approximation. The values for β_0 and v_0 obtained in this way for the alkali halides are collected in Table 15. Revised values for the cohesive energies can now be calculated from (r_0/ρ) and v_0 as in the last section by the use of (3.15); the results for the alkali halides are also given in Table 15. If we compare these values with the values calculated in a similar way in § 3 (Table 9), where room-temperature values β, v have been used for β_0, v_0, we observe a significant improvement in the agreement with the empirical values in all cases. When we take into account the van der Waals interactions (see Table 10), the calculated values agree in most cases with the empirical values within the probable experimental error. There remains, however, a systematic increase in the deviation towards the heavier compounds. One may perhaps speculate that this is due either to an underestimate of the van der Waals interactions, or that with the more polarizable heavier ions, the van der Waals interactions are partly replaced by chemical forces of a more complex description.

The values of v_0 and β_0 at the absolute zero of temperature can be similarly deduced from (4.52) and (4.58) now that the values of (r_0/ρ)

are known. The values for $\{v(293°)-v(0°)\}/v_0$ as well as the values of $v(0°)$ and $\beta(0°)$ (the temperatures are indicated in the brackets) obtained in this way are given in Table 16. We have also given for comparison the values for $\beta(0°)$ obtained by Slater† by linear extrapolation from room temperature, and the values of $\{v(293°)-v(0°)\}/v_0$ as obtained by linear extrapolation, and as estimated by Hildebrand‡ from Henglein's§ low-temperature measurements. Hildebrand's values can be considered practically as empirical values, since the extrapolations from Henglein's values cover only a very small change in v; the calculated values are seen to be in good agreement with Hildebrand's values. The changes in β_0 and v_0 as estimated by linear extrapolations are exaggerated, as the temperature coefficients decrease to zero as the absolute zero of temperature is approached; the errors as shown by the calculated values are in fact quite considerable.

TABLE 16

Compressibilities and Volumes of the Alkali Halides at the Absolute Zero of Temperature

	$\beta(0° K.)$ in 10^{-12} cm.^{2} dyne^{-1}		Cell vol. $v(0° K.)$ in 10^{-24} cm.^{3}	$\dfrac{v(293° K.)-v(0° K.)}{v(0° K.)}$		
	Eqn. (4.58)	Linear extrapolation (Slater)	Eqn. (4.52)	Eqn. (4.52)	Hildebrand–Henglein	Linear extrapolation $\dfrac{1}{V}\left(\dfrac{\partial V}{\partial T}\right)_{293° K.}$
LiF	1·11	1·4	16·01	0·014	..	0·027
LiCl	3·20	2·7	33·31	0·021	..	0·036
NaF	1·99	..	24·23	0·017	..	0·026
NaCl	3·98	3·3	43·60	0·022	0·024	0·032
NaBr	4·70	3·9	51·66	0·025	0·027	0·035
NaI	6·45	..	65·38	0·031	0·030	0·040
KF	3·10	3·2	37·12	0·019	..	0·029
KCl	5·26	4·8	60·55	0·021	0·021	0·030
KBr	6·17	5·5	69·66	0·025	0·027	0·033
KI	7·75	7·0	85·15	0·029	0·030	0·037
RbF	3·86	..	43·71	0·020	..	0·028
RbCl	6·21	..	68·38	0·022	..	0·029
RbBr	7·35	6·5	78·55	0·024	..	0·031
RbI	8·70	7·6	95·46	0·029	..	0·035
CsF	3·94	..	53·07	0·021	..	0·028
CsCl	5·26	..	67·69	0·031	..	0·040
CsBr	6·17	..	76·22	0·033	..	0·041
CsI	7·35	..	91·57	0·035	..	0·043

† J. C. Slater, *Phys. Rev.* **23**, 488 (1924).
‡ J. H. Hildebrand, *Zeit. f. Phys.* **67**, 127 (1931).
§ F. A. Henglein, *Zeit. f. Elektrochem.* **31**, 424 (1925).

5. Vibrations of a diatomic chain

In this section we shall discuss lattice vibrations in terms of a simple model due originally to Born and Kármán.† The example is explicitly soluble and possesses many features common to lattice vibrations in general. The model consists of a chain of two different types of particles occupying alternate positions (Fig. 6). It is assumed that forces act only between neighbours; we shall denote the corresponding potential of

FIG. 6. Linear diatomic chain.

interaction by ϕ. Clearly, whenever the particles are evenly spaced, each particle is in equilibrium; the separation between neighbours for such a static configuration will be denoted by $s/2$ so that successive particles of the same kind are separated by s.

Let us regard the chain, for a moment, as an elastic string. The tension in the string, being equal simply to the attractive force between neighbours, is clearly given by

$$\phi'(s/2).$$

When the string is subject to a uniform extension, so that s becomes $s+\delta s$, the tension is increased by

$$\tfrac{1}{2}\delta s\phi''(s/2). \tag{5.1}$$

As the corresponding linear elastic strain (extension per unit length) is

$$\delta s/s, \tag{5.2}$$

the Young's modulus E of the string is obtained by dividing (5.1) by (5.2). Thus we have

$$E = \tfrac{1}{2}s\phi''(s/2). \tag{5.3}$$

The chain forms, so to speak, a linear lattice with two particles to each cell. Taking any arbitrary cell as reference, we can label different cells by an integral index l. In the following, we shall restrict the particles to move only along the length of the chain. When we discuss the lattice vibrations about any particular static configuration with a given spacing s, we shall denote the displacements of the two particles in cell l from the evenly spaced configuration respectively by u_l and u_l'. With the help of the potential function, the typical equations of motion for the two types of particles can be readily written down as:

$$\left.\begin{aligned} m\ddot{u}_l &= \phi'(u_l'-u_l+\tfrac{1}{2}s)-\phi'(u_l-u_{l-1}'+\tfrac{1}{2}s) \\ m'\ddot{u}_l' &= \phi'(u_{l+1}-u_l'+\tfrac{1}{2}s)-\phi'(u_l'-u_l+\tfrac{1}{2}s) \end{aligned}\right\}, \tag{5.4}$$

† M. Born and Th. von Kármán, *Phys. Zeit.* **13**, 297 (1912).

where m, m' are the masses of the particles. Making the assumption that the displacements are small, we can develop the potential functions in Taylor's series with respect to the displacements and neglect the second- and higher-order terms. Thus for small vibrations, (5.4) becomes

$$\left.\begin{array}{l} m\ddot{u}_l = \phi''(\tfrac{1}{2}s)\{(u_l'-u_l)-(u_l-u_{l-1}')\} \\ m'\ddot{u}_l' = \phi''(\tfrac{1}{2}s)\{(u_{l+1}-u_l')-(u_l'-u_l)\} \end{array}\right\}. \tag{5.5}$$

(5.5) expresses a system of simultaneous equations, infinite in number. Since the equations are linear and homogeneous, we can make use of complex solutions with the usual understanding. Let us seek to reduce the above equations by the following special substitution:

$$\left.\begin{array}{l} u_l = u \\ u_l' = u' \end{array}\right\} \times e^{2\pi i \eta l - i\omega t}. \tag{5.6}$$

One recognizes that these represent essentially progressive wave solutions, similar to the elastic waves discussed in the last section. ω is the circular frequency (2π times the frequency ν) and $|s/\eta|$ is the wave-length. With the above substitution, the equations in (5.5) reduce to the following pair of equations:

$$\left.\begin{array}{l} \{m\omega^2-2\phi''(\tfrac{1}{2}s)\}u+\phi''(\tfrac{1}{2}s)(1+e^{-2\pi i\eta})u' = 0 \\ \phi''(\tfrac{1}{2}s)(1+e^{2\pi i\eta})u+\{m'\omega^2-2\phi''(\tfrac{1}{2}s)\}u' = 0 \end{array}\right\}, \tag{5.7}$$

which are linear homogeneous equations, soluble only if

$$\begin{vmatrix} m\omega^2-2\phi''(\tfrac{1}{2}s) & \phi''(\tfrac{1}{2}s)(1+e^{-2\pi i\eta}) \\ \phi''(\tfrac{1}{2}s)(1+e^{2\pi i\eta}) & m'\omega^2-2\phi''(\tfrac{1}{2}s) \end{vmatrix} = 0. \tag{5.8}$$

For any given η, there are thus two admissible values for the frequency, given by the solutions of (5.8), namely,

$$\omega^2 = \begin{cases} \dfrac{\phi''(\tfrac{1}{2}s)}{mm'}\{(m+m')-[(m+m')^2-4mm'\sin^2\pi\eta]^{\frac{1}{2}}\} & \text{(acoustic)} \\ \dfrac{\phi''(\tfrac{1}{2}s)}{mm'}\{(m+m')+[(m+m')^2-4mm'\sin^2\pi\eta]^{\frac{1}{2}}\} & \text{(optical)} \end{cases}. \tag{5.9}$$

Substituting these alternative values for ω^2 in (5.7), we obtain the corresponding amplitude ratios:

$$\frac{u}{u'} = \begin{cases} \dfrac{-m'(1+e^{-2\pi i\eta})}{(m-m')-[(m+m')^2-4mm'\sin^2\pi\eta]^{\frac{1}{2}}} & \text{(acoustic)} \\ \dfrac{-m'(1+e^{-2\pi i\eta})}{(m-m')+[(m+m')^2-4mm'\sin^2\pi\eta]^{\frac{1}{2}}} & \text{(optical)} \end{cases}. \tag{5.10}$$

As usual we can take the real part of the complex solution to represent the actual displacements. This leads to a sinusoidal wave for the displacements of either kind of particle. The two waves (cf. (5.6)) describing respectively the displacements of the two types of particle have the same wave-length and frequency; moreover, they share the same arbitrary phase and amplitude factor, for the equations of motion, as we have seen, fix the ratio of the complex amplitudes u, u', and leave only a common multiplicative (complex) factor arbitrary. We note that owing to the presence of the arbitrarily disposable phase and amplitude factor, the imaginary part of the complex solution does not give further independent vibrations. Furthermore, we shall use only positive values for the frequency ω so that the solutions for η and $-\eta$ are distinct from one another. Corresponding to the two alternative complex solutions above, we have thus exactly two real modes of vibration for each given η.

The solutions for ω^2 and u/u', we notice, are both periodic functions of η with period 1. All the distinct solutions are thus obtained, if we restrict η to a unit range, which we can choose as the following:

$$-\tfrac{1}{2} \leqslant \eta < \tfrac{1}{2}. \qquad (5.11)$$

This restriction on the range of η is in fact to be expected. The trial solution (5.6) is equivalent to the specification that as we pass from one cell to the next, the phase of the motion is changed by $e^{2\pi i \eta}$. The pattern of motion specified by (5.6) is thus completely unaffected if η is altered by an integer. Hence, only by restricting η within a unit range do we obtain a *unique* correspondence between the value of η and the pattern of wave motion. As $|\eta/s|$ is the wave-number, the restriction (5.11) is equivalent to a cut-off of the wave-number at $(1/2s)$; waves with a higher wave-number do not correspond to any further distinct vibrations. This fact has already been mentioned in introducing the maximum frequency in the Debye model.

In Fig. 7 the solutions are represented by plotting the frequency against η for several different mass ratios m'/m. In general (i.e. $m' \neq m$), the frequencies fall into two distinct branches, corresponding respectively to the two alternative solutions given in (5.9) and (5.10). The lower and upper branches are usually designated as the *acoustic* and *optical branches* for the following reasons:

The acoustic branch is characterized by the fact that $\omega \to 0$ as $\eta \to 0$. We find in fact on expanding the acoustic solution in (5.9) in a Taylor's series with respect to η that

$$\omega^2 = \left(\frac{2\pi^2 \phi''(\tfrac{1}{2}s)}{m+m'} \right) \eta^2 + O(\eta^4) + \dots,$$

or
$$\omega \cong \left(\frac{\phi''(s/2)}{2(m+m')}\right)^{\frac{1}{2}} (2\pi\eta) \quad \text{(small } \eta\text{)}. \tag{5.12}$$

Thus for small η (i.e. long wave-lengths), the frequency is linearly proportional to the wave-number. Since s/η is the wave-length it follows from (5.12) that the phase velocity v of the acoustic vibrations approaches

$$v = \frac{\omega}{2\pi}\left(\frac{s}{\eta}\right) = \left(\frac{\phi''(s/2)}{2(m+m')}\right)^{\frac{1}{2}} s \tag{5.13}$$

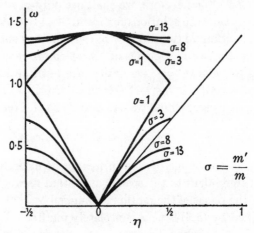

Fɪɢ. 7. Vibration frequencies (arbitrary units).

for large wave-lengths. Moreover, it follows immediately from (5.10) that as $\eta \to 0$, the corresponding amplitude ratio $u/u' \cong 1$. This means that for long acoustic waves the two particles in the same cell move practically in unison as a rigid unit.

The long waves of the acoustic branch are in fact identical with the longitudinal elastic vibrations, if the chain is regarded as an elastic string. The phase velocity of such elastic vibrations is given in terms of the Young's modulus E and the linear density ρ by the following formula:

$$v = \left(\frac{E}{\rho}\right)^{\frac{1}{2}}.$$

Since in this case ρ is obviously given by $(m+m')/s$, we find by using the value (5.3) deduced earlier for the Young's modulus that this formula becomes

$$v = \left(\frac{s^2\phi''(s/2)}{2(m+m')}\right)^{\frac{1}{2}}, \tag{5.14}$$

which is identical with (5.13).

The optical vibrations approach, on the other hand, the finite frequency (see (5.9))

$$\omega = \left\{ \frac{2(m+m')\phi''(s/2)}{mm'} \right\}^{\frac{1}{2}} \tag{5.15}$$

as $\eta \to 0$. Moreover, it follows from (5.10) that the corresponding amplitude ratio is given by

$$\frac{u}{u'} = \frac{-m'}{m} \quad \text{or} \quad mu + m'u' = 0. \tag{5.16}$$

The motions of the two particles in every cell are thus opposed to one another and the centre of mass of the cell remains stationary.

The optical vibrations of long wave-lengths are of special importance in considering the interaction of crystals with light. In general an electromagnetic wave interacts only with lattice vibrations of the same wave-length, and will be strongly affected only if its frequency is near that of the latter. The frequencies of lattice vibrations are generally in the range from 0 to 10^{13} sec.$^{-1}$, hence light waves of similar frequencies have wave-lengths larger than

$$c10^{-13} = 0 \cdot 003 \text{ cm.},$$

which, we note, are enormously large compared with the lattice constants of crystals ($\sim 10^{-8}$ cm.). Therefore the lattice vibrations which can interact appreciably with light are very long waves, with practically vanishing η ($\sim 10^{-5}$). The strength of the interaction depends on the electric oscillations associated with the lattice vibrations. As an analogue of ionic crystals, let us assume the two types of particles in the chain to be equal but oppositely charged (the chain as a whole must be electrically neutral). In the long optical vibrations, the opposed motions of the oppositely charged particles give rise to a net oscillating dipole moment of a cell, which is totally absent in the long waves of the acoustic branch, where the opposite charges move in unison. Therefore only the long waves of the optical branch will concern us in the discussion of the optical behaviour of crystals.

(The above discussion is only schematic. For if one considers an electromagnetic wave travelling along the chain, the electric field will be perpendicular to the chain, whereas the oscillatory electric moments of the optical vibrations are parallel to the chain. The formal features of the schematic discussion remain nevertheless the same in a real three-dimensional case, which we shall consider in § 8.)

We have seen that to each η there are two modes of vibration belonging respectively to the acoustic and optical branches. For an infinite chain, η can take *any value* between $-\frac{1}{2}$ and $\frac{1}{2}$; there are thus an infinite number of independent vibrations. It was seen that for the discussion of thermodynamical functions the vibrations must be normalized to a finite bulk of a crystal. The required normalization can be achieved in the present case by taking a finite chain of L cells and joining the two ends to form a ring. So long as L is large enough for the curvature of the ring to be negligible, the equations of motion (5.5) remain completely unaltered, if we identify the Lth cell with the cell 0. The solutions are the same as before, but now subject to the additional condition

$$u_L = u_0,$$

since these displacements now refer to the same particle. It follows from this condition that we must have

$$\eta L = \text{integer.} \qquad (5.17)$$

Combining this with (5.11), we find that η is restricted to the following values:
$$\frac{h}{L}, \quad h \text{ being integers: } -\tfrac{1}{2}L \leqslant h < \tfrac{1}{2}L. \qquad (5.18)$$

The total number of vibrations is thus $2L$, equal to the degrees of freedom of the chain (notice that a particle has only one degree of freedom in the present case, as its movement is restricted along the chain).

It is observed that in the one-dimensional case the finite ring arrangement is exactly equivalent to the use of the periodic boundary condition; this condition is not so easily realizable in the three-dimensional case.

When the chain is uniformly stretched, the vibrational frequencies are altered, since $\phi''(\frac{1}{2}s)$ depends on the average spacing $\frac{1}{2}s$. From the expressions (5.9) for the frequencies, we find readily that

$$\gamma = \frac{d\ln\nu}{d\ln s} = \frac{1}{2}\frac{d\ln\omega^2}{d\ln s} = \frac{s}{4}\frac{\phi'''(\frac{1}{2}s)}{\phi''(\frac{1}{2}s)}, \qquad (5.19)$$

which is independent of η and hence the same for *all* modes of vibration. In this simple case, the Mie–Grüneisen equation is thus rigorously justified.

We have seen that the elastic waves agree only with the long acoustic lattice vibrations. In Debye's model, one replaces the lattice vibrations completely by elastic waves; in this way, one effectively assumes that, in an (ω, η) diagram such as Fig. 7, all the vibration frequencies fall on the tangents to the acoustic branch at the origin. These tangents are indicated in one case in Fig. 7. Debye's model is thus only a very crude representation of the acoustic lattice vibrations. The vibrations of the optical branch are taken into account only in so far as the number of degrees of freedom is concerned; the frequency maximum in Debye's theory is chosen such that the tangents representing the elastic vibrations are cut off at points such that the total number of vibrational modes is correctly given.

We notice that if there is a large discrepancy between the masses of the two types of particles (cf. Fig. 7) the frequencies of the optical branch are confined to a relatively narrow range. These vibrations involve mainly the motion of the light particles, the heavy particles remaining largely stationary. Since in the model the light particles affect one another only through the motions of the heavy particles, the motions of different light particles in these modes are largely independent and the corresponding vibrational frequencies do not thus differ very appreciably from the vibration of a light particle between its two neighbours held fixed. In this case, an Einstein model suitably describes the vibrations of the optical branch; the Debye model can be used, in addition, to take account of the vibrations of the acoustic branch.

6. Frequency spectrum of lattice vibrations and specific heats

Despite its value as a widely usable approximation, Debye's elastic continuum theory is not sufficient for the complete elucidation of the behaviour of the specific heats, and has to be replaced by the atomistic theory of Born and Kármán,[†] published independently about the same time. This fact was first demonstrated conclusively in a series of articles by Blackman,[‡] when the success and simplicity of the continuum theory had all but obliterated the fact that the theory is only an approximation.

In the first place, one expects that any discrepancy between Debye's theory and the experimental results should be most manifest at fairly low temperatures. For specific heats at high temperatures are not sensitive to the characteristics in the frequency distributions; we have seen, for instance, that the widely different Einstein and Debye models yield specific heat values not essentially different except at low temperatures. On the other hand, Debye's theory is expected to be rigorously correct in the limit of extremely low temperatures. At such temperatures, the specific heat is contributed predominantly by the vibrations of the lowest frequencies. This is immediately clear from the expression (4.17) for the Einstein model; the formula shows that the specific heat due to

---

[†] Loc. cit., p. 55.
[‡] M. Blackman, *Proc. Roy. Soc.* A, **148**, 365, 384 (1935); A, **149**, 117, (1935); *Phil. Trans. Roy. Soc.* A, **236**, 103 (1936); *Proc. Roy. Soc.* A, **159**, 416 (1937); *Proc. Camb. Phil. Soc.* **33**, 94 (1937).

a vibration with frequency ν falls rapidly with temperature as

$$\sim T^{-2}e^{-h\nu/kT}$$

at low temperatures such that $kT \ll h\nu$. In the linear chain example, we have seen that the lowest vibration frequencies belong to the long acoustic waves, which are identical with the elastic waves used in the Debye model. This is true generally also for any three-dimensional crystal (see §§ 26, 27). It thus follows that Debye's theory must reproduce correctly the specific heat in the limit of extreme low temperatures. Blackman has shown essentially that there is a low-temperature region immediately above this extreme limit, where considerable deviations from the Debye values are expected on the basis of lattice theory and are in fact borne out by certain experimental results.

If we refer to Debye's expression (4.26) for the specific heat, we find that, for $T \ll \Theta_D$, the integral has very nearly the limiting value

$$\int_0^\infty \frac{\xi^4 e^\xi \, d\xi}{(e^\xi - 1)^2} = \frac{4\pi^4}{15},$$

and we have thus approximately

$$C_V \cong \frac{12 N n k \pi^4}{5}\left(\frac{T}{\Theta_D}\right)^3 \quad (T \ll \Theta_D). \tag{6.1}$$

This is the well-known T^3-law that the specific heat varies as the third power of the absolute temperature at very low temperatures.

It follows from the arguments given above that the T^3-law should be *rigorously* fulfilled at sufficiently low temperatures. It is true that (4.26) has been deduced under the special assumptions of elastic isotropy and a single cut-off frequency $\nu_m = k\Theta_D/h$ for the vibrational spectrum of the crystal, but it can be readily shown that the expression (6.1) for C_V at very low temperatures is independent of both these assumptions, *provided* Θ_D is chosen in accordance with (4.36) and the normalization condition (4.24). Let us consider an *anisotropic* elastic medium. The three types of elastic waves for a given wave-number $\mathbf{y}$ (no longer either transverse or longitudinal) have then in general distinct phase velocities $c_i(\theta, \phi)$ $(i = 1, 2, 3)$, which are moreover functions of the direction of $\mathbf{y}$, specified here by the polar angles θ, ϕ. We confine our attention first to the elastic waves of a particular type i, whose directions of propagation are within an infinitesimal solid angle $\sin\theta \, d\theta d\phi = d\omega$. For a piece of the crystal of volume V, the number of frequencies within the range $\nu, \nu + d\nu$

is obviously equal to V times the element of the y-space, or

$$Vy^2\,dy\,d\omega = V\frac{v^2\,dv\,d\omega}{c_i^3(\theta,\phi)}.$$

If the vibrational spectrum is assumed to be cut off at a certain maximum $|\mathbf{y}|_{\max}$, irrespective of the direction of propagation, then the contribution to specific heat from those vibrations of the type i whose propagation vectors $\mathbf{y}$ lie in the solid angle $d\omega$ is given by

$$(|\mathbf{y}|_{\max})^3 V k \times \left(\frac{T}{\Theta_i(\theta,\phi)}\right)^3 d\omega \int\limits_0^{\Theta_i(\theta,\phi)/T} \frac{\xi^4 e^\xi\,d\xi}{(e^\xi-1)^2}, \tag{6.2}$$

where, for convenience, we have written

$$\xi = h\nu/kT \quad \text{and} \quad c_i(\theta,\phi) \times |\mathbf{y}|_{\max} = (k/h)\Theta_i(\theta,\phi).$$

At very low temperatures $T \ll \Theta_i(\theta,\phi)$ we can replace, as before, the upper limit of the integral in (6.2) by ∞. Then (6.2) becomes

$$(|\mathbf{y}|_{\max})^3 V k \times \left(\frac{T}{\Theta_i(\theta,\phi)}\right)^3 d\omega \times \frac{4\pi^4}{15}.$$

Summing over the three types of waves and remembering that

$$(|\mathbf{y}|_{\max})^3 V = \frac{3}{4\pi} N n,$$

we obtain for the total specific heat per mole the expression

$$C_V = \frac{12 N n k \pi^4}{5}\, T^3 \times \frac{1}{\Theta^3},$$

where

$$\frac{1}{\Theta^3} = \left(\frac{k}{h}\right)^3 \frac{v}{n}\frac{1}{9} \int \sum_i \frac{1}{c_i^3(\theta,\phi)}\, d\omega. \tag{6.3}$$

It is immediately verified that Θ as given by (6.3) is the same as Θ_D when Θ_D is defined by (4.36) and (4.27) as has been actually done in (6.1).

Experimentally it has been found that C_V/T^3 attains a roughly constant value for a large number of substances when the temperature is lowered to the order $10°$–$50°$ K. The corresponding temperatures have thus been interpreted as the beginning of the region of low temperature within which the T^3 law is rigorously fulfilled. Blackman† showed, however, that this is not the case; the experimentally verified T^3-variations are in fact spurious and do not correspond to the low-temperature limit, which lies at a much lower temperature and extends usually not more than a few degrees above the absolute zero of temperature.

† M. Blackman, *Proc. Roy. Soc.* A, **149**, 117 (1935).

Blackman's conclusions, which are based on semi-quantitative considerations, have since been confirmed by a number of actual calculations carried out for real crystals. Following Blackman we shall use the linear chain model to discuss schematically the behaviour of the specific heat and its deviation from the continuum approximation; afterwards we shall reproduce some results of the calculations that have been made for real crystals.

Let us denote the acoustic and optical frequencies of a linear chain by ν_1, ν_2; using (5.9), we can write

$$\left.\begin{matrix}\nu_1\\\nu_2\end{matrix}\right\} = \frac{1}{2\pi}\left[\frac{(m+m')\phi''(s/2)}{mm'}\right]^{\frac{1}{2}}\left\{1\mp\left[1-\frac{4\sin^2\pi\eta}{(1+m'/m)(1+m/m')}\right]^{\frac{1}{2}}\right\}^{\frac{1}{2}}. \tag{6.4}$$

Acoustic vibrations within the frequency range ν, $\nu+\Delta\nu$ correspond to η-values in the following dual ranges: η, $\eta+\Delta\eta$ and $-\eta$, $-(\eta+\Delta\eta)$, where $\Delta\eta$ is given by

$$\Delta\eta = \left(\frac{d\eta}{d\nu_1}\right)_{\nu_1=\nu}\Delta\nu.$$

It thus follows from (5.18) that for a finite chain of L cells, there are

$$2L\Delta\eta = 2L\left(\frac{d\eta}{d\nu_1}\right)_{\nu_1=\nu}\Delta\nu$$

acoustic vibrations in the above frequency range, where the factor 2 takes account of the dual ranges for the η-values. Similarly we find that

$$2L\left(\frac{d\eta}{d\nu_2}\right)_{\nu_2=\nu}\Delta\nu$$

gives the number of optical vibrations in the same frequency interval. The heat capacity of the chain is thus given by the following expression:

$$C_V = 2Lk\int\frac{e^{h\nu/kT}}{(e^{h\nu/kT}-1)^2}(h\nu/kT)^2\left[\left(\frac{d\eta}{d\nu_1}\right)_{\nu_1=\nu}+\left(\frac{d\eta}{d\nu_2}\right)_{\nu_2=\nu}\right]d\nu. \tag{6.5}$$

Both the acoustic and optical frequencies are limited within certain ranges; in the above expression, we shall understand that the integrations of the two terms in the integral are extended over these respective ranges.

If we apply Debye's approximation to this case, the relation between the frequency and η is given by (cf. (5.12))

$$\nu_{\text{app}} = \left(\frac{\phi''}{2(m+m')}\right)^{\frac{1}{2}}\eta. \tag{6.6}$$

The number of vibrations in the frequency interval ν, $\nu+\Delta\nu$ is thus

$$2L\Delta\eta = 2L\left(\frac{d\eta}{d\nu_{\text{app}}}\right)\Delta\nu = 2L\left(\frac{2(m+m')}{\phi''}\right)^{\frac{1}{2}}\Delta\nu. \tag{6.7}$$

The frequency maximum ν_D is determined by the normalization condition

$$\int_0^{\nu_D} 2L\left(\frac{2(m+m')}{\phi''}\right)^{\frac{1}{2}} d\nu = 2L,$$

or

$$\nu_D = \left(\frac{\phi''}{2(m+m')}\right)^{\frac{1}{2}}. \tag{6.8}$$

The frequency distribution can thus be written as

$$f(\nu) = 2\left(\frac{2(m+m')}{\phi''}\right)^{\frac{1}{2}} = \frac{2}{\nu_D}, \quad 0 < \nu < \nu_D,$$
$$= 0, \quad \nu_D < \nu; \tag{6.9}$$

hence the specific heat in the continuum approximation is given by the following expression:

$$(C_V)_{\text{app}} = \frac{2Lk}{\nu_D}\int_0^{\nu_D} \frac{e^{h\nu/kT}}{(e^{h\nu/kT}-1)^2}(h\nu/kT)^2 \, d\nu = 2Lk\left(\frac{T}{\Theta_D}\right)\int_0^{\Theta_D/T} \frac{\xi^2 e^\xi}{(e^\xi-1)^2}\, d\xi, \tag{6.10}$$

where

$$\Theta_D = \frac{h\nu_D}{k} = \frac{h}{k}\left[\frac{\phi''}{2(m+m')}\right]^{\frac{1}{2}} \tag{6.11}$$

is the Debye temperature.

A convenient way of expressing the deviation of the approximate expression (6.10) from the accurate expression (6.5) is the following: If we simply equate (6.5) to (6.10), we can calculate a value for Θ_D for any particular T from this relation. We expect that the values of Θ_D calculated in this way will agree exactly with (6.11) only for very small T, for only in this limiting case is the Debye formula (6.10) exact. The deviation of Θ_D from this limiting value can be used conveniently as a measure of the inaccuracy of the approximate formula. Blackman[†] has calculated Θ_D in this way as a function of T for a number of cases and represented the results in graphical form; his graphs are reproduced in Fig. 8. We notice that the frequencies ν_1 and ν_2, and hence also the specific heat, depend only on the following two parameters: the frequency maximum (see (6.4))

$$\nu_0 = \frac{1}{2\pi}\left\{\frac{2(m+m')\phi''}{mm'}\right\}^{\frac{1}{2}},$$

and the mass ratio m'/m.

The cases considered by Blackman correspond to the same arbitrarily

† M. Blackman, *Proc. Roy. Soc.* A, **148**, 365 (1935).

fixed value of ν_0 (namely $h\nu_0/k = 200°$) and various values for the mass ratio m'/m, as indicated in the figure.

If the continuum theory were an accurate approximation for all temperatures, Θ_D should appear as a horizontal straight line equal to the

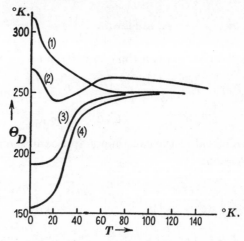

Fig. 8. Θ_D for linear diatomic chains of various atomic mass-ratios.
(1) $m'/m = 1$,   (2) $= 3$,   (3) $= 8$,   (4) $= 13$.

value given by (6.11). As regards the actual graphs in Fig. 8, we note thus in particular the following points:

(i) Θ_D is far from being independent of temperature and shows conspicuous variations at low temperatures.

(ii) Although Θ_D tends to a fairly constant value at high temperatures, it is very different from the low-temperature limiting value at $T = 0$ (equal to (6.11)). This means that although a fair representation of the specific heat at relatively high temperatures can be achieved by using the Debye formula with a suitably chosen value for the parameter Θ_D, this does not necessarily confirm the idea that the lattice vibrations can be suitably replaced by elastic waves; the formula used in this way becomes largely heuristic in significance.

(iii) It is readily seen from (6.10) that the analogue of the T^3-law in this linear case is a linear law $C_V \propto T$; the region for which this limiting law holds should appear in the graph as a region of constant Θ_D immediately adjoining $T = 0$. The curves shown indicate that this region is extremely narrow and is barely recognizable as a horizontal tangent at $T = 0$.

(iv) Curve (2) has a minimum, where Θ_D is stationary. If the corresponding temperature is sufficiently low compared with Θ_D, so that the upper limit of the integral in Debye's formula can be put equal to ∞, we have in the neighbourhood of the minimum spurious T^3-variation, quite unrelated to the real T^3-region at $T = 0$. A minimum of this type appears always to occur in the curves for real crystals, and is apparently related to the experimentally verified T^3-region.

As we shall see, all the above features are retained in the real three-dimensional cases, but in a less exaggerated way.

The general features of the lattice vibrations in three-dimensional lattices are very similar to the linear case. It will be shown in Chapter V that the equations of motion can be set up quite generally without any assumption as regards the interaction forces and can, moreover, be simplified readily. For the present purpose, we need only quote the results to be established there.

Let us consider a general lattice structure with n particles to each cell. We shall distinguish different particles in the same cell by an index $k = 1, 2,..., n$. Choosing any cell as reference, we can label different cells by a triple lattice index $l(l^1, l^2, l^3)$ exactly as in the case of the simple lattices considered in § 3. A particle in a general lattice is thus specified by the indices l and k.

If we represent the total potential energy Φ of a lattice as a function of the displacements of the lattice particles from their equilibrium positions, we can form its derivatives with respect to the components of the particle displacements $u_\alpha\binom{l}{k}$ (α-component of the displacement vector of the particle $\binom{l}{k}$, $\alpha = 1, 2, 3$). Small vibrations of a lattice are controlled by the second derivatives:

$$\left(\frac{\partial^2\Phi}{\partial u_\alpha\binom{l}{k}\partial u_\beta\binom{l'}{k'}}\right)_0 = \Phi_{\alpha\beta}\binom{l-l'}{kk'}. \tag{6.12}$$

The symbol on the right is introduced partly for simplicity, partly, what is more important, because it exhibits the fact that the second derivative depends only on the relative cell index $l-l'$ and not on the indices l and l' individually (§ 23).

If we denote the position vector of a lattice particle $\binom{l}{k}$ in the equilibrium configuration by $\mathbf{x}\binom{l}{k}$, the complex solutions of the equations of motion are of the following form:

$$\mathbf{u}\binom{l}{k} = \mathbf{u}^0(k)\exp[2\pi i\mathbf{y}.\mathbf{x}\binom{l}{k}-i\omega t], \tag{6.13}$$

where $\mathbf{u}^0(k)$ is a constant vector depending only on k. For a given $\mathbf{y}$ the circular frequency $\omega = 2\pi\nu$ is determined by the following determinantal equation:

$$|C_{\alpha\beta}\binom{\mathbf{y}}{kk'}-\omega^2\delta_{kk'}\delta_{\alpha\beta}| = 0, \tag{6.14}$$

where $C_{\alpha\beta}(^{y}_{kk'})$ is defined by

$$C_{\alpha\beta}(^{y}_{kk'}) = \frac{1}{(m_k m_{k'})^{\frac{1}{2}}} \sum_{l} \Phi_{\alpha\beta}(^{l}_{kk'}) \exp\{-2\pi i y \cdot [\mathbf{x}(^{l}_{k})-\mathbf{x}(^{0}_{k'})]\}, \quad (6.15)$$

the summation being over all integral values of l^1, l^2, l^3 and m_k, $m_{k'}$ being the particle masses; $\delta_{kk'}$, $\delta_{\alpha\beta}$ in (6.14) signify Kronecker symbols, e.g.

$$\delta_{ij} = \begin{cases} 1 & i = j \\ 0 & i \neq j. \end{cases} \quad (6.16)$$

If we order a pair of indices (α, k) in the sequence $(1, 1)$, $(2, 1)$, $(3, 1)$, $(1,2)$, ..., $(1, n)$, $(2, n)$, $(3, n)$, the elements $C_{\alpha\beta}(^{y}_{kk'})$ can be displayed in a square array (a square matrix):

$$\begin{bmatrix} C_{11}(^{y}_{11}) & C_{12}(^{y}_{11}) & C_{13}(^{y}_{11}) & \cdot & \cdot & C_{11}(^{y}_{1n}) & C_{12}(^{y}_{1n}) & C_{13}(^{y}_{1n}) \\ C_{21}(^{y}_{11}) & C_{22}(^{y}_{11}) & C_{23}(^{y}_{11}) & \cdot & \cdot & \cdot & \cdot & \cdot \\ C_{31}(^{y}_{11}) & \cdot & \cdot & \cdot & \cdot & \cdot & \cdot & \cdot \\ \cdot & \cdot & \cdot & \cdot & \cdot & \cdot & \cdot & \cdot \\ C_{31}(^{y}_{n1}) & \cdot & \cdot & \cdot & \cdot & \cdot & \cdot & C_{33}(^{y}_{nn}) \end{bmatrix}$$

$$(6.14\,\text{a})$$

The left-hand side of (6.14) is merely a short notation for the characteristic (or secular) determinant of the above matrix.

(6.14) is a $3n$-degree equation in ω^2; we can as before restrict our considerations to the positive frequencies; thus the equation gives $3n$ frequencies. To each of these frequencies we have a distinct solution of the form (6.13); it is not necessary to discuss here how the polarization vectors $\mathbf{u}^0(k)$ are to be determined. The construction of the real solutions is the same as in the linear case, and for a given $\mathbf{y}$ we obtain exactly one real progressive wave for each of the $3n$ frequencies given by equation (6.14).

Exactly as in the case of the linear chain, not all $\mathbf{y}$-values correspond to mutually distinct solutions. This is most readily seen as follows: let us write
$$\mathbf{x}(^{l}_{k}) = \mathbf{x}(l)+\mathbf{x}(k),$$

where $\mathbf{x}(l)$ is the *lattice vector*:

$$\mathbf{x}(l) = l^1\mathbf{a}_1+l^2\mathbf{a}_2+l^3\mathbf{a}_3, \quad (6.17)$$

$\mathbf{a}_1$, $\mathbf{a}_2$, $\mathbf{a}_3$ being the basic vectors forming the edges of the cells. As $\mathbf{x}(k)$ does not depend on the index l, (6.13) expresses that the phase of the motion changes progressively from cell to cell according to the factor

$$e^{2\pi i y \cdot \mathbf{x}(l)}. \quad (6.18)$$

We notice that the *reciprocal basic vectors* defined by

$$\mathbf{b}^1 = \frac{\mathbf{a}_2 \wedge \mathbf{a}_3}{|\mathbf{a}_1.\mathbf{a}_2 \wedge \mathbf{a}_3|}, \qquad \mathbf{b}^2 = \frac{\mathbf{a}_3 \wedge \mathbf{a}_1}{|\mathbf{a}_2.\mathbf{a}_3 \wedge \mathbf{a}_1|}, \qquad \mathbf{b}^3 = \frac{\mathbf{a}_1 \wedge \mathbf{a}_2}{|\mathbf{a}_3.\mathbf{a}_1 \wedge \mathbf{a}_2|}$$

$$(6.19)$$

have the property that their scalar products with the basic vectors $\mathbf{a}_1$, $\mathbf{a}_2$, $\mathbf{a}_3$ are either equal to one or zero according to the following rule:

$$\mathbf{b}^i.\mathbf{a}_j = \delta_{ij}. \qquad (6.20)$$

Let us define a *reciprocal lattice* in the $\mathbf{y}$-space with the reciprocal basic vectors; thus a lattice vector in the reciprocal lattice is of the form

$$\mathbf{y}(h) = h_1\mathbf{b}^1 + h_2\mathbf{b}^2 + h_3\mathbf{b}^3, \qquad (6.21)$$

$h(h_1, h_2, h_3)$ being integers. It follows immediately from (6.20) that the scalar product between a lattice vector and a reciprocal lattice vector is an integer:

$$\mathbf{x}(l).\mathbf{y}(h) = l^1h_1 + l^2h_2 + l^3h_3. \qquad (6.22)$$

Thus if we add to $\mathbf{y}$ any reciprocal lattice vector $\mathbf{y}(h)$, the phase factor (6.18) is not affected. Therefore *all the distinct solutions* (6.13) *are taken into account, if we restrict the values of* $\mathbf{y}$ *to one reciprocal lattice cell in the* $\mathbf{y}$-*space*, because, for any point outside the cell, there exists a point within the cell connected to it by a reciprocal lattice vector.

It is, however, not necessary always to choose a reciprocal lattice cell to limit the values of $\mathbf{y}$; in fact it often proves more convenient for practical considerations to choose an equivalent region in $\mathbf{y}$-space, which has the same volume as the reciprocal cell but a greater geometric symmetry; the only criterion to be satisfied is that no two points in the chosen region are related by a reciprocal lattice vector. We shall illustrate this point by a hexagonal network (Fig. 9 a). The simplest choice of the basic vectors $\mathbf{a}_1$, $\mathbf{a}_2$ does not exhibit the hexagonal symmetry. The reciprocal lattice (Fig. 9 b) is similarly a hexagonal network; the hexagonal symmetry is not exhibited by a reciprocal lattice cell (shaded in the figure). A more symmetric region is the hexagon (indicated in Fig. 9 b) about a lattice point bounded by the bisecting lines with its six neighbours. It is immediately clear from the figure that this region is completely equivalent to a reciprocal cell, for the respective portions of the reciprocal cell and the hexagonal region marked off by the same number with and without a prime in the figure are obviously related by reciprocal lattice displacements.

In § 4 we have seen that to obtain the frequency distribution, we can normalize the vibrations to a finite volume V by imposing the periodic boundary condition. The latter restricts y to certain discrete values, which are represented by evenly distributed points in the y-space; the density of such representative points in the y-space is exactly constant and equal to V. We notice that the volume of a reciprocal cell is given by

$$|\mathbf{b}^1.\mathbf{b}^2 \wedge \mathbf{b}^3| = \frac{|(\mathbf{a}_2 \wedge \mathbf{a}_3).[(\mathbf{a}_3 \wedge \mathbf{a}_1) \wedge (\mathbf{a}_1 \wedge \mathbf{a}_2)]|}{|\mathbf{a}_1.\mathbf{a}_2 \wedge \mathbf{a}_3|^3} = \frac{1}{|\mathbf{a}_1.\mathbf{a}_2 \wedge \mathbf{a}_3|} = \frac{1}{v_a},$$

$$(6.23)$$

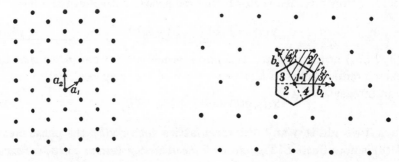

FIG. 9a. Hexagonal network.   FIG. 9b. Reciprocal lattice.

for $(\mathbf{a}_1.\mathbf{a}_2 \wedge \mathbf{a}_3)$ is equal to the volume of a lattice cell. Multiplying this by the density of representative points V, we obtain the total number of y-values:

$$V \times \frac{1}{v_a} = N,$$

where N is the number of lattice cells in V. As there are $3n$ real waves for each y, the total number of vibrational modes is

$$3nN,$$

which is exactly the total number of degrees of freedom of the lattice particles contained in V. This fact further ensures that not only do we obtain all the distinct solutions by restricting the y-values to a reciprocal cell, but the different y-values in the same cell do in fact correspond to distinct solutions. We can therefore write for the distribution function $f(\nu)$:

$$f(\nu) = \frac{V}{N} \lim_{\Delta\nu \to 0} \frac{1}{\Delta\nu} \sum_{j=1}^{3n} \iiint_{\nu \leqslant \nu_j(y_1,y_2,y_3) \leqslant \nu + \Delta\nu} dy_1 \, dy_2 \, dy_3, \qquad (6.24)$$

where the integration is over a reciprocal cell. (6.24) evidently corresponds to the normalization of $f(\nu)$ in accordance with (4.20), namely

$$\int f(\nu)\, d\nu = 3n. \tag{6.25}$$

Various methods have been suggested to calculate the frequency distribution function $f(\nu)$ for crystals. We shall now briefly describe these methods and discuss some characteristic features of $f(\nu)$ obtained by them.

The most commonly used method to calculate $f(\nu)$ is due to Blackman.[†] It consists of selecting a finite number of evenly distributed points in a reciprocal cell in the y-space, or rather an equivalent region with maximum symmetry. The $3n$ frequencies for each of these y-values are then determined from the equation (6.14). Finally the frequency scale is divided into a number of suitably chosen finite intervals and the number of frequencies falling within each interval counted. If the resulting number is plotted as ordinate against the frequencies in the corresponding interval, a step curve is obtained which may be smoothed out into a continuous curve. The vertical scale is then adjusted so as to give a curve representing the frequency distribution function $f(\nu)$ satisfying the normalization condition (6.25). For structures possessing special symmetry, the equation (6.14) is the same for groups of symmetrically placed points in the y-space so that the equation has to be solved only for a small fraction of the total number of y-values considered. For a crystal lattice with cubic symmetry such as NaCl, the equation need only be solved for points within a volume 1/48th of that of a reciprocal cell. The actual numerical work involved in such calculations remains, however, still very considerable.

Blackman has calculated the frequency distribution by this method for a two-dimensional square lattice and simple monatomic cubic lattice under the simplifying assumption that only the nearest and next nearest neighbours interact with each other. The (Θ_D, T) curve calculated from this distribution function is similar to one obtained by him for a linear chain.

The above numerical method has been somewhat refined by Houston.[‡] There are always to be found straight lines through the origin of the reciprocal space for which the frequency equation (6.14) can be solved exactly. Thus the frequency distribution $F(\nu, \theta_s, \phi_s)$ per unit solid angle can be obtained easily for several directions (θ_s, ϕ_s) in the reciprocal

† M. Blackman, *Proc. Roy. Soc.* A, **159**, 416 (1937).
‡ W. V. Houston, *Rev. Mod. Phys.* **20**, 161 (1948).

space. These can be expanded in spherical harmonics in the form

$$F(\nu, \theta_s, \phi_s) = \sum_i f_i(\nu) Y_i(\theta_s, \phi_s), \tag{6.26}$$

where by choosing the Y's properly one can allow for the symmetry of the lattice; the number of terms on the right-hand side corresponds to a certain degree of approximation. By taking this number equal to the number of directions (θ_s, ϕ_s) for which the frequency equation has been solved, one can solve for $f_i(\nu)$. The total distribution function $f(\nu)$ is then given by

$$f(\nu) = \iint F(\nu, \theta, \phi) \sin \theta \, d\theta d\phi = 4\pi f_0(\nu) Y_0.$$

Houston has applied this method to obtain the frequency distribution for a monatomic simple cubic lattice with the same force constants as those chosen by Blackman. In contrast to Blackman's finite step function Houston's curve for $f(\nu)$ has several sharp peaks where it is actually infinite (cf. p. 75). The area under the peaks, however, is finite. The curve Θ_D plotted against T from this distribution function has a much greater dip than that in Blackman's curve.

As distinct from these numerical methods for obtaining the frequency distribution, Montroll† has proposed a method for approximating the frequency distribution by analytical expressions. In order to determine all the $3nN$ circular frequencies $\omega_j^2(\mathbf{y}_l)$ one has to obtain the characteristic roots of the matrix (6.14 a) for N values of $\mathbf{y}$ in the cell of the reciprocal lattice. Let these values of $\mathbf{y}$ be denoted by $\mathbf{y}_1, \mathbf{y}_2, ..., \mathbf{y}_N$ and let $C(\mathbf{y})$ denote the matrix obtained from (6.14 a) by dividing each row of it by $(2\pi)^2$. Then, if $\omega/2\pi = \nu$, obviously $\nu_j^2(\mathbf{y}_l)$ $(j = 1, 2,..., 3n; l = 1, 2,..., N)$ are the characteristic roots of the $N \times N$ diagonal matrix

$$C = \begin{bmatrix} C(\mathbf{y}_1) & 0 & . & . & . & . & 0 \\ 0 & C(\mathbf{y}_2) & . & . & . & . & . \\ . & . & . & . & . & . & . \\ . & . & . & . & . & . & . \\ 0 & . & . & . & . & . & C(\mathbf{y}_N) \end{bmatrix}, \tag{6.27}$$

in which each diagonal element $C(\mathbf{y}_l)$ is a $3n \times 3n$ matrix as defined above. Now from a well-known theorem on matrices it follows that the sum of the kth powers of all the characteristic roots of a matrix C is given by *trace* C^k (i.e. by the sum of the diagonal elements of the kth power of C), provided we use the convention that if a characteristic root is d-fold degenerate, it is to be counted d times. Hence, denoting the average

† E. W. Montroll, *J. Chem. Phys.* **10**, 218 (1942); ibid. **11**, 481 (1943).

value of the kth power of the characteristic roots $v_j^2(\mathbf{y}_l)$ of C by μ_{2k}, we have

$$\mu_0 = \frac{1}{3nN} \, tr \, C^0 = 1,$$

and
$$\mu_{2k} = \frac{1}{3nN} \, tr \, C^k = \frac{1}{3nN} \sum_{l=1}^{N} tr\{C(\mathbf{y}_l)\}^k$$

$$= \frac{1}{3nN} \sum_{j=1}^{3n} \sum_{l=1}^{N} [v_j^2(\mathbf{y}_l)]^k \quad (k = 1, 2, 3,...), \quad (6.28)$$

so that by definition μ_{2k} is just the $2k$th moment of the distribution function. Assuming the set $\{v_j(\mathbf{y}_l)\}$ to be densely distributed, we can write

$$\mu_{2k} = \int f(v)v^{2k} \, dv \Big/ \int f(v) \, dv. \quad (6.29)$$

Now since the frequencies are the square roots of the characteristic roots of C, the distribution function $f(v)$ is an even function of v. Hence if v_m ($v_m > 0$) denotes the highest frequency, the limits of integration in (6.29) can be chosen from $-v_m$ to $+v_m$ so that the odd moments μ_{2k+1} are all zero. Following Montroll we next express the distribution function in terms of the moments μ_{2k}. Let

$$f(v) = \sum_{n=0}^{\infty} a_n P_n(v/v_m), \quad (6.30)$$

where the P's are the Legendre polynomials of the argument v/v_m and the a's are given by

$$a_n = \frac{2n+1}{2} \int_{-1}^{1} f(xv_m)P_n(x) \, dx. \quad (6.31)$$

If now
$$u_k = \int_{-1}^{1} f(xv_m)x^k \, dx, \quad (6.32)$$

then
$$\mu_k = \int_{-v_m}^{v_m} f(v)v^k \, dv \Big/ \int_{-v_m}^{v_m} f(v) \, dv = \frac{1}{6n} v_m^{k+1} u_k, \quad (6.33)$$

since
$$\int_{0}^{v_m} f(v) \, dv = 3n.$$

With the help of these relations and remembering that odd moments are all zero and that odd Legendre polynomials contain only odd powers of x, one easily obtains

$$a_1 = a_3 = a_5 = ... = a_{2k+1} = 0,$$

$$a_0 = (6n)\mu_0/2v_m, \qquad a_2 = (6n)5(3\mu_2 - \mu_0 v_m^2)/(4v_m^3)$$

and generally

$$a_{2k} = 6n \frac{4k+1}{(2k)! \, 2^{k+1} \nu_m} \left\{ \frac{d^{2k}}{dz^{2k}} (z^2-1)^{2k} \right\}_{z^k = \mu_k / \nu_m^k}. \tag{6.34}$$

Thus if the μ's are known, the distribution function $f(\nu)$ can be obtained as a series in Legendre polynomials. Since the μ's can be determined with the help of (6.27) without actually solving (6.14) for the frequencies, this method for determining $f(\nu)$ involves much less calculation than the numerical method. Another point worth mentioning in this connexion is that since a thermodynamic quantity (as a function of temperature) is generally an average over all the normal modes, it can be expressed,[†] as Montroll has shown, as a series in terms of the moments, thus avoiding the use of the distribution function altogether. Hence in cases where such series are rapidly convergent this method can be very advantageous.

Montroll[‡] and Montroll and Peasle[§] have applied this method to determine respectively the frequency distributions for monatomic simple and body-centred cubic lattices under the simplifying assumption that only the nearest and the next nearest neighbours interact with each other. The matrices $C(\mathbf{y})$ for such lattices are 3×3 matrices and explicit expressions for the moments can be easily derived. These authors determined the coefficients a_{2k} in (6.30) up to $k = 5$. Their results are in general agreement with those of Blackman[||] and Fine[††] who obtained the distribution functions for the two types of lattices by the numerical method.

A characteristic feature of the frequency distributions for monatomic cubic lattices is that they have two maxima, one near about $\nu_m/2$ and the other close to the upper frequency limit ν_m. In view of this result Brenig and Schröder[‡‡] have recently suggested that to a good approximation $f(\nu)$ can be taken of the form

$$f(\nu) = z_D(\nu) + \alpha_E \, \delta(\nu - \nu_E). \tag{6.35}$$

Here $\delta(\nu - \nu_E)$ is the Dirac δ-function and $z_D(\nu)$ is the Debye distribution function; the cut-off frequency ν_c, however, is now somewhat smaller than the Debye cut-off frequency ν_D. There are three unknowns in (6.35), namely α_E, ν_E, and ν_c. By equating the expressions for the

---

† Such an expansion was first introduced by Thirring: H. Thirring, *Phys. Zeit.* **14**, 867 (1913); ibid. **15**, 127, 180 (1914).

‡ E. W. Montroll, loc. cit. (1943).

§ E. W. Montroll and D. C. Peasle, *J. Chem. Phys.* **12**, 98 (1944).

|| M. Blackman, loc. cit. (1937).

†† P. C. Fine, *Phys. Rev.* **55**, 355 (1939).

‡‡ W. Brenig and M. Schröder, *Zeit. f. Phys.* **132**, 312 (1952).

moments μ_0, μ_2, and μ_4 of (6.35) to the corresponding expressions for these moments obtained from (6.28), one is able to determine ν_c, ν_E, and α_E uniquely. A distribution function of the type (6.35) is obviously more manageable for calculations of specific heats, etc., since it merely represents the superposition of an Einstein term on the Debye spectrum. If, however, there occur more than two maxima in the frequency distribution, one will have to superpose several Einstein terms of different frequencies ν_{E_1}, ν_{E_2},.... The determination of the various constants ν_c; ν_{E_1}, α_{E_1}; ν_{E_2}, α_{E_2};... is then more complicated.

Montroll[†] has also made a very detailed analytical investigation of the frequency distribution of a two-dimensional square lattice, whereby he is able to obtain for $f(\nu)$ closed expressions involving complete elliptic integrals for certain values of the force constants. One theoretical result of considerable interest in this investigation turns out to be that $f(\nu)$ has logarithmic singularities in this case (these singularities, unlike the singularities obtained by Houston which are due to approximate methods employed, are genuine). This result has been derived by a simpler method and extended to the three-dimensional case by van Hove.[‡] A short account of this work will be given here.

We consider the distribution function $f_j(\nu)$ for the jth branch of the normal modes,

$$f(\nu) = \sum_{j=1}^{3n} f_j(\nu).$$

Then the integral corresponding to $f_j(\nu)$ on the right-hand side of (6.24) obviously represents the volume enclosed between the two neighbouring surfaces $\nu_j(y_1, y_2, y_3) = \nu$ and $\nu_j(y_1, y_2, y_3) = \nu + \Delta\nu$. Hence if dS represents an element of area on the surface of $\nu_j(y_1, y_2, y_3) = \nu$ and Δn the element of the normal to it at dS, $f_j(\nu)$ can be written as[§]

$$f_j(\nu) = \frac{V}{N} \lim_{\Delta\nu \to 0} \frac{1}{\Delta\nu} \iint\limits_{\nu \leqslant \nu(y_1,y_2,y_3) \leqslant \nu+\Delta\nu} \Delta n \, dS = \frac{V}{N} \lim_{\Delta\nu \to 0} \iint \frac{\Delta n}{\Delta\nu} \, dS.$$

(6.36)

Now $\quad \Delta\nu = \mathrm{grad}\, \nu(y_1, y_2, y_3) \cdot \Delta \mathbf{y} = \dfrac{\partial \nu}{\partial y_1} \Delta y_1 + \dfrac{\partial \nu}{\partial y_2} \Delta y_2 + \dfrac{\partial \nu}{\partial y_3} \Delta y_3$ (6.37)

and $\qquad \dfrac{\Delta\nu}{\Delta n} = \dfrac{\partial \nu}{\partial y_1} \dfrac{\Delta y_1}{\Delta n} + \dfrac{\partial \nu}{\partial y_2} \dfrac{\Delta y_2}{\Delta n} + \dfrac{\partial \nu}{\partial y_3} \dfrac{\Delta y_3}{\Delta n}.$

[†] E. W. Montroll, *J. Chem. Phys.* **15**, 575 (1947).
[‡] L. van Hove, *Phys. Rev.* **89**, 1189 (1953).
[§] For convenience the suffix j on the function ν_j in the remaining equations will be omitted.

But $\Delta y_1/\Delta n$, $\Delta y_2/\Delta n$, $\Delta y_3/\Delta n$ in the limit $\Delta \nu$ or $\Delta n \to 0$ are the direction cosines of the normal to the surface $\nu(y_1, y_2, y_3) = \nu$. Since these are proportional to $\partial \nu/\partial y_1$, $\partial \nu/\partial y_2$, and $\partial \nu/\partial y_3$, we have

$$\frac{dn}{d\nu} = \frac{1}{\sqrt{\{(\partial \nu/\partial y_1)^2 + (\partial \nu/\partial y_2)^2 + (\partial \nu/\partial y_3)^2\}}}. \tag{6.38}$$

Hence (6.36) becomes

$$f_j(\nu) = \frac{V}{N} \int\int \frac{dS}{\sqrt{\{(\partial \nu/\partial y_1)^2 + (\partial \nu/\partial y_2)^2 + (\partial \nu/\partial y_3)^2\}}}. \tag{6.39}$$

For the two-dimensional case $f_j(\nu)$ is obviously given by

$$f_j(\nu) = \frac{V}{N} \int \frac{ds}{\sqrt{\{(\partial \nu/\partial y_1)^2 + (\partial \nu/\partial y_2)^2\}}}, \tag{6.40}$$

where ds now represents an element of arc of the curve $\nu(y_1, y_2) = \nu$.

The discussion of the analytic singularities in the frequency distribution $f_j(\nu)$ due to a particular branch j is now an easy matter. They obviously occur at 'critical' points where all derivatives $\partial \nu/\partial y_\alpha$ vanish. One can assume that the determinant

$$\left| \frac{\partial^2 \nu}{\partial y_\alpha \partial y_\beta} \right| \neq 0$$

at these points, as this could happen only for special values of the dynamical constants. This implies that the critical points are isolated and hence finite in number. In the neighbourhood of any critical point ν_c one has an expansion of ν which contains no linear terms; if it is broken off after the quadratic terms, the $\partial \nu/\partial y_\alpha$ are linear in the y_α and the integral for $f_j(\nu)$ can be evaluated by elementary methods. The result is the following: For a two-dimensional crystal there is a finite discontinuity (jump) in the curve $f_j(\nu)$ at each maximum (jump downwards) and minimum (jump upwards) and a logarithmic peak of the type $-\log|1-\nu/\nu_c|$ at a saddle point (i.e. when $\nu = \nu_c + a(y_2^2 - y_1^2)$). For a three-dimensional lattice the function $f_j(\nu)$ is continuous at a maximum or minimum, but has a discontinuous direction of the tangent, of the type

$$f_j(\nu) \sim \begin{cases} (\nu_c - \nu)^{\frac{1}{2}} & \text{for } \nu < \nu_c \\ 0 & \text{for } \nu > \nu_c \end{cases}$$

in the case of a maximum and vice versa in the case of a minimum; the behaviour at saddle points is similar.

The singularities are therefore of a rather harmless type and do not invalidate the numerical computations. It can further be shown (see van Hove, loc. cit.) that owing to the periodicity of the lattice, a number of critical points must necessarily exist for each branch, though a different number for the acoustical and optical branches.

In contrast to the work of Blackman, Montroll, Fine, and others described above, where the forces are assumed to be of short range, Lyddane and Herzfeld† (NaCl), Kellermann‡ (NaCl), Iona§ (KCl), and Helen Smith‖ (diamond) have calculated frequency distributions without assuming that the forces in the crystal are of a short range. Lyddane and Herzfeld have calculated the frequencies only for relatively few isolated values of y, from which a frequency distribution cannot be deduced. In the following we quote some of the results obtained by Kellermann, Iona, and Helen Smith, which represent the most accurate calculations available. The frequency distribution was calculated by these workers by using the numerical method of Blackman.

For his calculations on NaCl, Kellermann considers only the overlap force between the nearest neighbours; as regards the Coulomb inter-action, the ions are considered as equivalent to point-charges. The overlap force is determined with the help of the observed lattice constant and compressibility essentially as in § 3. Since in lattice vibrations the ions are displaced from their symmetric positions, the effect of the polarization of the ions, which is ignored in Kellermann's calculation, can be quite considerable; we shall obtain some idea of the inaccuracy from this source in the next section. The point-charge and overlap interactions are two-body forces; to make use of the general formulae given earlier in this section, we have only to replace Φ by the sum of such interactions between all pairs of ions in the lattice. Once the coefficients $C_{\alpha\beta}\binom{y}{kk'}$ given by (6.15) are calculated, the frequencies can be determined from (6.14). The principal difficulty in the calculation is that for the Coulomb interaction the series in (6.15) converges very slowly. In § 30 we shall see how such series can be converted into quickly convergent expressions suitable for practical calculations.

Some of Kellermann's results are represented graphically in Figs. 10–12. Fig. 10 gives the frequencies of the various branches of lattice waves propagating in the direction of any cube-diagonal. As there are two particles to each cell, there should be altogether six branches of vibrations corresponding to the $3n$ solutions of the determinantal equation (6.14). Owing, however, to the symmetry of the NaCl-structure,

† R. H. Lyddane and K. F. Herzfeld, *Phys. Rev.* **54**, 846 (1938).

‡ E. W. Kellermann, *Phil. Trans. Roy. Soc.* **238**, 513 (1940); *Proc. Roy. Soc.* A, **178**, 17 (1941).

§ M. Iona, Jr., *Phys. Rev.* **60**, 822 (1941).

‖ Helen M. J. Smith, *Phil. Trans. Roy. Soc.* **241**, 105 (1948).

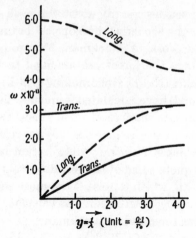

Fig. 10.

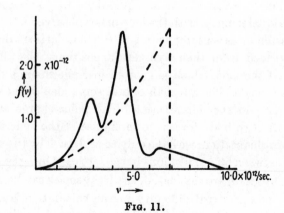

Fig. 11.

Fig. 12.

all waves along the cube diagonals are either transverse or longitudinal, all the former being double (corresponding to two independent choices for the polarization vector). The acoustic and optical vibrations are easily distinguished in the graph, as the former go through the origin. Fig. 11 gives the frequency distribution function obtained in the way described above. In the same figure the frequency distribution according to the Debye model is also given for comparison (dotted line). The calculated and experimental values of the Debye temperature Θ_D given in Fig. 12 are obtained in the same way as in the linear case, namely, by equating the Debye formula (4.26) to the calculated and experimental values of the specific heat for various temperatures.

Iona's calculations for KCl are essentially similar. He has considered the overlap forces between second neighbours as well, which are determined with the help of the infra-red dispersion frequency (see §§ 8, 9) together with the elastic data. Since the masses of the K^+ and Cl^- ions are not very different, Iona has assumed them equal in his calculations. The frequency distribution function and the calculated (Θ_D, T) curve are given respectively in Figs. 13 and 14. We notice that the Debye distribution (thin line in Fig. 13) is much closer to the calculated distribution according to the lattice theory in this case. When the masses of the ions are equal, the lattice behaves somewhat like a monatomic solid with one atom to each cell; the optical branches join smoothly with the acoustic branches so that we have apparently only three acoustic branches, the upper halves of which correspond to the optical vibrations. Such acoustic-like branches are obviously more closely approximated by the Debye model than in the more general case where the optical and acoustic vibrations form separated frequency-bands.

The nearest neighbours in the diamond lattice are joined by valence bonds, which cannot be adequately approximated by central interactions. In her considerations of the diamond lattice, Helen Smith thus makes no assumption whatever about the second derivatives of Φ correlating two nearest neighbours. She shows that all such derivatives depend only on two independent constants on account of the symmetry of the diamond structure. Besides the nearest neighbour interactions, she finds that one should, on the evidence of experiments, consider also the interaction between second neighbours, which she assumes to be central. Thus there are altogether three independent constants underlying the potential in her treatment; these constants are determined with the help of the Raman frequency and the elastic constants. Fig. 15 gives the frequencies for the waves travelling in the direction of any

cube-diagonal; as in the case of NaCl, these vibrations are either transverse or longitudinal, the former being double. One difference between
Fig. 15 for diamond and Fig. 10 for NaCl is particularly worth noticing,
namely, that all the optical branches in the diamond lattice converge to

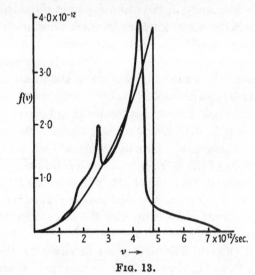

FIG. 13.

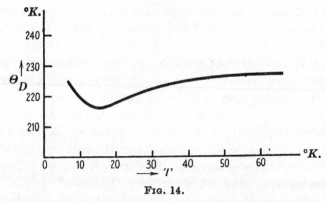

FIG. 14.

the same limit for $\lambda \to \infty$, whereas in the NaCl lattice the longitudinal
and transverse vibrations converge to different limits. As we shall see
in the next section, this difference is due to the fact that owing to the
ionic character of the NaCl lattice, there is an electric field associated
with the longitudinal vibrations, which raises the vibrational frequencies.
In Figs. 16 and 17 respectively are given the frequency distribution
function and the calculated curve and experimental values for the Debye
temperature Θ_D as a function of temperature.

We notice that in all cases there is a minimum in the (Θ_D, T) curve, as predicted by Blackman, giving rise to a spurious T^3-region. The

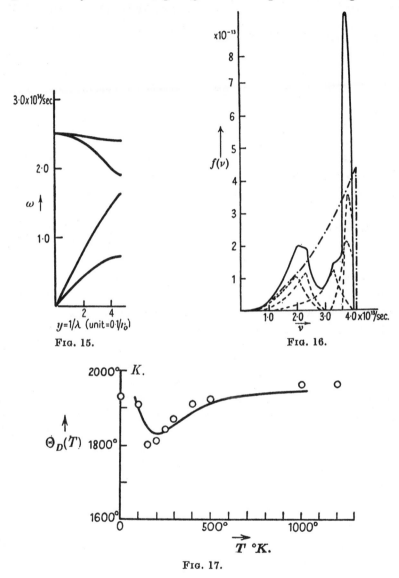

Fig. 15.

Fig. 16.

Fig. 17.

calculated values in the graphs also indicate that the true T^3-region is very narrow, appearing in the graphs merely as a horizontal approach of the Θ_D curve towards the axis $T = 0$. Clusius† has made extensive

† K. Clusius, *Zeit. f. Naturforschung*, **1**, 79 (1946).

measurements on the specific heat at low temperatures for seven alkali halides; he reports that in none of the cases has he been able to reach the proper T^3-region, although evidence of its existence is detected.

7. Long lattice vibrations of the optical branches

We have seen that long acoustic lattice vibrations are identical with elastic waves; the latter, of course, follow from the macroscopic equations of the elasticity theory. Equally the optical lattice vibrations of large wave-lengths can be considered on a macroscopic basis.† Such optical vibrations are important chiefly in the ionic crystals owing to the strong electric moments associated with the motion. The macroscopic theory is particularly simple for the diatomic crystals with optical isotropy; in this category are included most of the commonly discussed simple ionic crystals (structures: NaCl, CsCl, ZnS). In the following we shall discuss this case in detail.

In the linear chain example, we recall that, in contrast to the long acoustic vibrations where atoms in a macroscopically small section move practically in unison, in a similar section of an optical vibration atoms of one type move as a body against the atoms of the other type. Thus to describe the long optical vibrations, we require a parameter specifying the relative displacement between the positive and negative ions. For an elastic motion, the effective inertial mass for a unit volume is the density; for the optical type of motion, on the other hand, the corresponding mass is the reduced mass of the positive and negative ions $\bar{M} = \{M_+ M_-/(M_+ + M_-)\}$ divided by the volume of a lattice cell. It has been found that the most convenient parameter to choose for describing the optical type of motion is the displacement of the positive relative to the negative ions multiplied by the square root of this effective mass per unit volume; we shall denote it by $\mathbf{w}$. For the diatomic ionic crystals under discussion, the macroscopic theory is fully embodied in the following pair of equations:†

$$\dot{\mathbf{w}} = b_{11}\mathbf{w} + b_{12}\mathbf{E}, \tag{7.1}$$

$$\mathbf{P} = b_{21}\mathbf{w} + b_{22}\mathbf{E}, \tag{7.2}$$

where $\mathbf{P}$ and $\mathbf{E}$ are the dielectric polarization and electric field as defined in the usual way in Maxwell's theory. The precise interpretations of the various terms in these equations will become apparent when we deduce equations of this form from a special microscopic model in § 9; at present we can regard them as purely phenomenological equations.

† K. Huang, *E.R.A. Report L/T* 239 (1950); *Proc. Roy. Soc.* A, **208**, 352 (1951).

As macroscopic relations, the above equations are rigorous, and thus applicable whenever conditions are everywhere practically uniform over regions containing many lattice cells. The linearity of the equations is an assumption analogous to Hooke's law in the theory of elasticity and is here strictly equivalent to the usual approximation used in considering lattice vibrations, whereby terms of orders higher than the second in the potential functions are ignored. That the coefficients b_{11}, b_{12}, b_{21}, b_{22} are scalars, not tensors, follows from the condition of isotropy.

The coefficients in (7.1) and (7.2) are not all independent; it is shown in Appendix V that

$$b_{12} = b_{21}. \tag{7.3}$$

Taking account of this general relation, we can express the b-coefficients in terms of experimentally measurable quantities. It is known in the electromagnetic theory of light that the square of the refractive index is equal to the dielectric constant; the phenomenon of dispersion (the dependence of the refraction of a monochromatic wave on its frequency) follows directly from a frequency-dependent dielectric constant. For the crystals under consideration, the dielectric constant for any particular frequency ω $(= 2\pi\nu)$ can be deduced directly from (7.1) and (7.2) by considering periodic solutions:

$$\left.\begin{aligned} \mathbf{E} &= \mathbf{E}_0 \\ \mathbf{w} &= \mathbf{w}_0 \\ \mathbf{P} &= \mathbf{P}_0 \end{aligned}\right\} \times e^{-i\omega t}.$$

Thus we obtain from these equations,

$$-\omega^2 \mathbf{w} = b_{11} \mathbf{w} + b_{12} \mathbf{E},$$
$$\mathbf{P} = b_{21} \mathbf{w} + b_{22} \mathbf{E}.$$

When $\mathbf{w}$ is eliminated from these equations, it is seen that $\mathbf{P}$ and $\mathbf{E}$ are related by

$$\mathbf{P} = \left\{ b_{22} + \frac{b_{12}b_{21}}{-b_{11}-\omega^2} \right\} \mathbf{E}.$$

Comparing it with the definition of the dielectric displacement,

$$\mathbf{D} = \mathbf{E} + 4\pi\mathbf{P} = \epsilon\mathbf{E},$$

we obtain the dielectric constant

$$\epsilon = 1 + 4\pi b_{22} + \frac{4\pi b_{12}b_{21}}{-b_{11}-\omega^2}. \tag{7.4}$$

This *dispersion formula* is most conveniently written as follows:

$$\epsilon = \epsilon_\infty + \frac{\epsilon_0 - \epsilon_\infty}{1 - (\omega/\omega_0)^2}. \tag{7.5}$$

The constants in this formula are directly measurable quantities:

(i) ω_0, the *infra-red dispersion frequency* (circular), is the frequency at which the refractive index and the dielectric constant become infinitely large. As we shall see later in § 10, in practice it is measured as the absorption frequency of a thin film of the crystal.

(ii) ϵ_0, the *static dielectric constant*, is the value of the dielectric constant measured in a static field or an alternating field of a frequency very low compared with ω_0.

(iii) ϵ_∞, the *high-frequency dielectric constant*, is the value of the dielectric constant deduced from the refraction of electromagnetic waves with frequencies high compared with ω_0 so that the last term in (7.5) can be ignored.

We shall discuss the experimental aspect of the dispersion in a later section. Here we need only remark that ϵ_∞ is a constant only for frequencies small compared with the frequencies of electronic motion in the crystal; in the following we shall always understand this to be the case. This restricts the use of the above dispersion formula to the infra-red region. However, as the frequencies of electronic motion are generally several hundred times the infra-red dispersion frequency ω_0, this restriction creates no practical difficulty in determining a constant value for ϵ_∞.

Using the general relation (7.3) and comparing (7.4) with (7.5), we can express the b-coefficients in terms of the measurable constants ω_0, ϵ_0, ϵ_∞ as follows:

$$b_{11} = -\omega_0^2, \tag{7.6}$$

$$b_{12} = b_{21} = \left(\frac{\epsilon_0 - \epsilon_\infty}{4\pi}\right)^{\frac{1}{2}} \omega_0, \tag{7.7}$$

$$b_{22} = \frac{\epsilon_\infty - 1}{4\pi}. \tag{7.8}$$

The empirical values of ω_0, ϵ_0, ϵ_∞ for a number of common crystals are collected in Table 17, where for convenience of reference $\nu_0 = \omega_0/2\pi$, the ordinary frequency, and $\lambda_0 = c/\nu_0$, the corresponding vacuum wavelength, are given as well as ω_0.

The long optical vibrations can be obtained directly by considering (7.1) and (7.2) in conjunction with the equation of electrostatics:

$$\nabla . \mathbf{D} = \nabla . (\mathbf{E} + 4\pi \mathbf{P}) = 0, \tag{7.9}$$

the electric field $\mathbf{E}$ being an irrotational vector. The interpretation of the relative significance of the equations (7.1), (7.2), and (7.9) is plain: (7.9) can be interpreted as the Poisson equation giving the electric field

TABLE 17

Static and High-frequency Dielectric Constants and Dispersion Frequencies†

Substance	Structure	ω_0	ν_0	λ_0	ϵ_0	ϵ_∞
LiF	NaCl	$5\cdot78 \times 10^{13}$/sec.	$9\cdot20 \times 10^{12}$/sec.	32·6 microns	9·27	1·92
NaF	NaCl	4·64	7·40	40·6	6·0	1·74
NaCl	NaCl	3·09	4·92	61·1	5·62	2·25
NaBr	NaCl	2·52	4·01	74·7	5·99	2·62
NaI	NaCl	2·20	3·50	85·5	6·60	2·91
KCl	NaCl	2·67	4·25	70·7	4·68	2·13
KBr	NaCl	2·13	3·39	88·3	4·78	2·33
KI	NaCl	1·85	2·94	102·0	4·94	2·69
RbCl	NaCl	2·22	3·54	84·8	5	2·19
RbBr	NaCl	1·65	2·64	114·0	5	2·33
RbI	NaCl	1·45	2·31	129·5	5	2·63
CsCl	CsCl	1·85	2·95	102	7·20	2·60
CsBr	CsCl	1·41	2·25	134	6·51	2·78
TlCl	CsCl	1·61	2·58	117	31·9	5·10
TlBr	CsCl	(1·16)	(1·85)	(162)	29·8	5·10
CuCl	ZnS	3·56	5·67	53	10	3·57
CuBr	ZnS	3·31	5·27	57	8	4·08
AgCl	NaCl	(1·94)	(3·09)	(97)	12·3	4·04
AgBr	NaCl	(1·44)	(2·29)	(131)	13·1	4·62
MgO	NaCl	10·9	17·3	17·3	9·8	2·95
CaO	NaCl	6·88	10·9	27·4	11·8	3·28
SrO	NaCl	4·01	6·35	47	13·3	3·31
ZnS	ZnS	5·71	9·1	33	8·3	5·07

due to the charge density $-\nabla\cdot\mathbf{P}$ caused by dielectric polarization, the value of which is given by (7.2). (7.1) is the equation of motion: apart from a constant factor $(\overline{M}/v_a)^{\frac{1}{2}}$, the first term on the right is the local elastic restoring force, whereas the second term expresses the effect of Coulomb interaction with the charges elsewhere. The use of the method of electrostatics is equivalent to assuming a Coulomb interaction between charges in the lattice. In reality, of course, the Coulomb interaction is retarded; the effect of retardation will be considered in the next section.

Using the value of $\mathbf{P}$ given by (7.2) in (7.9), we find that

$$\nabla\cdot\mathbf{E} = \frac{-4\pi b_{21}}{1 + 4\pi b_{22}} \nabla\cdot\mathbf{w}. \tag{7.10}$$

The solution for $\mathbf{E}$ is readily obtained as follows: Let $\mathbf{w}$ be split into its solenoidal and irrotational parts,

$$\mathbf{w} = \mathbf{w}_t + \mathbf{w}_l,$$

† ϵ_0 and ϵ_∞ from K. Højendahl, *K. Danske Vidensk. Selskab*, **16**, No. 2 (1938); ω_0 from data collected by B. Szigeti, *Trans. Faraday Soc.* **45**, 155 (1949); values given in brackets are indirectly deduced from residual ray frequencies (see § 10).

where
$$\left.\begin{array}{l} \nabla . \mathbf{w}_t = 0 \quad \text{(solenoidal)} \\ \nabla \wedge \mathbf{w}_l = 0 \quad \text{(irrotational)} \end{array}\right\}. \tag{7.11}$$

Thus (7.10) can be written as

$$\nabla . \mathbf{E} = \frac{-4\pi b_{21}}{1+4\pi b_{22}} \nabla . \mathbf{w}_l. \tag{7.12}$$

An obvious solution for $\mathbf{E}$ is

$$\mathbf{E} = \frac{-4\pi b_{21}}{1+4\pi b_{22}} \mathbf{w}_l. \tag{7.13}$$

This solution is in fact unique, as $\mathbf{E}$ must be irrotational. Substituting this expression in the equation of motion (7.1) and writing $\mathbf{w}$ as the sum of $\mathbf{w}_t$ and $\mathbf{w}_l$, we find that

$$\ddot{\mathbf{w}}_t + \ddot{\mathbf{w}}_l = \left\{ b_{11} - \frac{4\pi b_{12} b_{21}}{1+4\pi b_{22}} \right\} \mathbf{w}_l + b_{11} \mathbf{w}_t.$$

Since the division of a vector function into its solenoidal and irrotational parts is unique, we can equate the solenoidal and irrotational parts on the two sides of the equation separately:

$$\ddot{\mathbf{w}}_t = b_{11} \mathbf{w}_t = -\omega_0^2 \mathbf{w}_t,$$

$$\ddot{\mathbf{w}}_l = \left\{ b_{11} - \frac{4\pi b_{12} b_{21}}{1+4\pi b_{22}} \right\} \mathbf{w}_l = -\left(\frac{\epsilon_0}{\epsilon_\infty}\right) \omega_0^2 \mathbf{w}_l,$$

where we have expressed the b-coefficients in terms of ω_0, ϵ_0, ϵ_∞, with the help of (7.6), (7.7), and (7.8). The respective general solutions of the above equations can clearly be written as

$$\mathbf{w}_t = \mathbf{w}_t(\mathbf{x}) e^{i\omega_t t} + \text{complex conj.}, \tag{7.14}$$

$$\mathbf{w}_l = \mathbf{w}_l(\mathbf{x}) e^{i\omega_l t} + \text{complex conj.}, \tag{7.15}$$

where
$$\omega_t = \omega_0, \tag{7.16}$$

$$\omega_l = \left(\frac{\epsilon_0}{\epsilon_\infty}\right)^{\frac{1}{2}} \omega_0, \tag{7.17}$$

and $\mathbf{w}_t(\mathbf{x})$, $\mathbf{w}_l(\mathbf{x})$ are arbitrary functions of the space coordinates satisfying the conditions
$$\nabla . \mathbf{w}_t(\mathbf{x}) = 0, \tag{7.18}$$

$$\nabla \wedge \mathbf{w}_l(\mathbf{x}) = 0. \tag{7.19}$$

We notice that transverse and longitudinal plane waves are special cases of $\mathbf{w}_t(\mathbf{x})$ and $\mathbf{w}_l(\mathbf{x})$, satisfying respectively (7.18) and (7.19); moreover, any arbitrary functions $\mathbf{w}_t(\mathbf{x})$ and $\mathbf{w}_l(\mathbf{x})$ satisfying (7.18), (7.19) can be represented as linear superpositions of such plane waves of different wave-numbers and directions of propagation. Thus in the macroscopic theory the independent vibrational modes can be taken as transverse and longitudinal plane waves of different wave-numbers and directions of propagation: all the transverse waves vibrate with the infra-red dispersion frequency ω_0 and all the longitudinal waves vibrate with the higher frequency ω_l given by (7.17). The reason for the difference in the frequencies of the transverse and longitudinal waves is obvious: In a transverse wave, the electric field $\mathbf{E}$ vanishes everywhere (see (7.13)) and the vibration frequency is solely determined by the local elastic restoring force represented by the first term on the right-hand side of (7.1); in a longitudinal wave, there is an electric field (given by (7.13)) which contributes an additional restoring force through the second term on the right-hand side of (7.1). In a non-ionic crystal such as diamond, the coefficient b_{12} vanishes and the motion is solely determined by the elastic restoring force, hence in the macroscopic theory all the vibrations have the same frequency. As the macroscopic theory is rigorously applicable only for wave-lengths long compared with the lattice constant, the above solutions represent the long-wave limits of the actual lattice vibrations. The difference in the long-wave limits of the optical vibrations between the non-polar diamond lattice and the ionic NaCl has already been noted in connexion with the results of Helen Smith and Kellermann. The frequency difference between the longitudinal and transverse waves in ionic crystals was first pointed out by Lyddane and Herzfeld† and Fröhlich and Mott;‡ the rigorous value of the frequency ratio, $(\epsilon_0/\epsilon_\infty)^{\frac{1}{2}}$, was derived in the first place by Lyddane, Sachs, and Teller.§ The treatment followed here by the systematic development of a macroscopic theory for the optical type of motion is that given by Huang.‖

A check on the accuracy of Kellermann's results is obtained by calculating ω_t and ω_l for NaCl with the values of ω_0, ϵ_0, ϵ_∞ given in Table 17:

$$\omega_t = 3 \cdot 09 \times 10^{13}/\text{sec.}, \qquad \omega_l = 4 \cdot 87 \times 10^{13}/\text{sec.} \quad \text{(rigorous)},$$

$$\omega_t = 2 \cdot 86 \times 10^{13}/\text{sec.}, \qquad \omega_l = 6 \cdot 02 \times 10^{13}/\text{sec.} \quad \text{(Kellermann)}.$$

The most obvious cause of the inaccuracy in Kellermann's treatment is the complete neglect of the polarization of the ions, which have been considered as point charges. The second term on the right-hand side of (7.2), for instance, is due entirely to the effect of the polarization of the ions; for clearly if the ions were not themselves polarizable, the relative

---

† R. H. Lyddane and K. F. Herzfeld, *Phys. Rev.* **54**, 846 (1938).
‡ H. Fröhlich and N. F. Mott, *Proc. Roy. Soc.* A, **171**, 496 (1939).
§ R. H. Lyddane, R. G. Sachs, and E. Teller, *Phys. Rev.* **59**, 673 (1941).
‖ K. Huang, loc. cit., p. 82.

displacement between the ions would completely fix the value of the dielectric polarization **P** and the second term on the right-hand side of (7.2) would in consequence be absent.

In numerous problems of practical interest, it is necessary to consider the motion of charges in ionic crystals. Such problems become practically intractable, if the motion of the ions is treated rigorously. For semi-quantitative purposes, it is thus useful to consider the lattice motion approximately on the basis of the macroscopic theory. It is outside the scope of the present book to consider explicitly such special problems; we shall confine ourselves to deriving certain general formulae, which can be used as the basis for treating special problems. In the following, we shall consider first the formulae suited for classical considerations; the quantum-mechanical case will be taken up in the next section.

Let us describe the charges present at any time t by the charge density function $\rho(\mathbf{x}, t)$. In the presence of the charges, the earlier equation (7.9) for the charge-free case has to be replaced by

$$\nabla . \mathbf{D} = \nabla . (\mathbf{E} + 4\pi \mathbf{P}) = 4\pi \rho. \tag{7.20}$$

We can follow the same method as used earlier in deriving the free lattice vibrations. When **P** is eliminated from (7.20) with (7.2) and $\mathbf{w}(\mathbf{x})$ written as the sum of its solenoidal and irrotational parts, we obtain

$$\nabla . \mathbf{E} = \frac{4\pi}{1 + 4\pi b_{22}} \{ -b_{21} \nabla . \mathbf{w}_l + \rho(\mathbf{x}, t) \}. \tag{7.21}$$

We can look upon it as the Poisson equation determining the electric field. If we compare it with the equation (7.12) for the charge-free case, we see that the solution of (7.21) must be equal to the sum of the solution (7.13) for the charge-free case and $1/(1 + 4\pi b_{22})$ times the Coulomb field that would be produced in vacuum by the charge density $\rho(\mathbf{x}, t)$. Hence if we introduce the vacuum field due to $\rho(\mathbf{x}, t)$,

$$\mathbf{E}_{\text{vac}}(\mathbf{x}, t) = -\nabla \int \frac{\rho(\mathbf{x}', t)}{|\mathbf{x} - \mathbf{x}'|} d\mathbf{x}', \tag{7.22}$$

the solution of (7.21) can be written as

$$\mathbf{E} = \frac{-4\pi b_{21}}{1 + 4\pi b_{22}} \mathbf{w}_l + \frac{\mathbf{E}_{\text{vac}}}{1 + 4\pi b_{22}} = -\omega_l \left[4\pi \left(\frac{1}{\epsilon_\infty} - \frac{1}{\epsilon_0} \right) \right]^{\frac{1}{2}} \mathbf{w}_l + \frac{1}{\epsilon_\infty} \mathbf{E}_{\text{vac}}. \tag{7.23}$$

Substituting this field in the equation of motion (7.1) and separating the solenoidal and irrotational parts, we obtain the following equations (note that $\mathbf{E}_{\mathrm{vac}}$ is irrotational):

$$\ddot{\mathbf{w}}_t = b_{11}\mathbf{w}_t = -\omega_0^2\mathbf{w}_t, \tag{7.24}$$

$$\ddot{\mathbf{w}}_l = \left\{b_{11} - \frac{4\pi b_{12}b_{21}}{1+4\pi b_{22}}\right\}\mathbf{w}_l + \frac{b_{12}\mathbf{E}_{\mathrm{vac}}}{1+4\pi b_{22}} = -\omega_l^2\mathbf{w}_l + \omega_l\left[\frac{1}{4\pi}\left(\frac{1}{\epsilon_\infty} - \frac{1}{\epsilon_0}\right)\right]^{\frac{1}{2}}\mathbf{E}_{\mathrm{vac}}. \tag{7.25}$$

(7.24) and (7.25) are the classical equations of motion for the lattice, when free charges are present. For any particular problem the corresponding equations of motion for the charges can be readily written down with the help of the electric field (7.23).

We notice that the solenoidal motions of the lattice are not affected by the presence of the free charges. On the one hand (7.24) is free of both $\mathbf{w}_l$ and $\rho(\mathbf{x}, t)$, and, on the other hand, as the electric field (7.23) is independent of $\mathbf{w}_l$, the latter does not enter the equations of motion of the charges. (7.23) and (7.25) constitute the basic equations for the classical treatment of the motion of charges in the lattice; (7.24) merely gives rise to free transverse vibrations which are uncoupled to either the longitudinal vibrations or the charges.

To cite a simple example, one may, for instance, readily estimate the energy loss by a charged particle to the lattice from (7.25), if one considers the particle approximately as having a constant velocity.†

8. Infra-red dispersion and the retardation effect on lattice vibrations‡

The electrostatic method used in the last section is equivalent to describing the electric interaction between the ions by the Coulomb force. Since the actual interaction does not operate instantaneously but propagates with the finite velocity of light, the method is only approximate. In a rigorous treatment, one should drop the irrotational condition on $\mathbf{E}$ and introduce besides (7.9) the remaining Maxwell equations

$$\nabla . \mathbf{H} = 0, \tag{8.1}$$

$$\nabla \wedge \mathbf{E} = -\frac{1}{c}\dot{\mathbf{H}}, \tag{8.2}$$

$$\nabla \wedge \mathbf{H} = \frac{1}{c}(\dot{\mathbf{E}} + 4\pi\dot{\mathbf{P}}). \tag{8.3}$$

In using these equations of the electromagnetic theory, we do more than take account of the retardation of the interaction forces. Consider

† Cf. H. Fröhlich and H. Pelzer, *E.R.A. Report* L/T 184 (1948), where a less accurate equation than (7.25) has been used; see, however, ibid. L/T 221.

‡ K. Huang, *Nature*, **167**, 779 (1951); *Proc. Roy. Soc.* A, **208**, 352 (1951).

the case when the lattice motion is completely suppressed by holding the ions fixed in the configuration $\mathbf{w} \equiv 0$. (7.1), the equation of motion for the ions, is then to be discarded, and (7.2) reduces to

$$\mathbf{P} = b_{22}\mathbf{E} = \frac{(\epsilon_\infty - 1)}{4\pi}\mathbf{E}; \qquad (8.4)$$

written alternatively as

$$\mathbf{D} = \mathbf{E} + 4\pi\mathbf{P} = \epsilon_\infty\mathbf{E}, \qquad (8.5)$$

the relation shows that the lattice merely functions as a normal refractive medium with the refractive index ϵ_∞. Therefore, in this case, the lattice equations used in conjunction with the electromagnetic equations give rise to transverse light waves with the constant phase velocity $c/\sqrt{\epsilon_\infty}$. In Fig. 18 a these light waves and the lattice vibrations obtained in the last section are represented on the same (ω, k) plot by dotted lines. These two groups of solutions both represent certain approximate solutions of the joint set of equations (7.1), (7.2) (lattice equations) and (7.9), (8.1), (8.2), (8.3) (electromagnetic equations). In one case, we effectively regard the ions as infinitely heavy and obtain then the radiative waves (the light waves) without lattice motion; in the other case, we ignore the retardation effect and obtain the lattice vibrations discussed in the last section, which contain no radiative energy. The two groups of diametrically opposite solutions together display the full ground covered by the six equations.

In the rigorous treatment, the above approximate solutions are inter-mixed. As we shall see, the transverse lattice vibrations (without retardation) mix with the radiative waves (the light waves) and together they become the observed optical waves (infra-red dispersion). Of these, the waves immediately below the dispersion frequency ω_0 are actually predominantly mechanical oscillations of the lattice, only a small fraction of their energy being radiative. If the small admixture of radiative energy is ignored, these waves reduce to the transverse lattice vibrations discussed in the last section. These vibrational modes thus play the dual role of being optical waves as well as lattice vibrations; for these vibra-tions, the electrostatic approximation is adequate. They are, as we shall find, essentially transverse vibrations with phase velocities small com-pared with $c/\sqrt{\epsilon_\infty}$. For the transverse lattice vibrations which were found in the last section as having phase velocities larger than $c/\sqrt{\epsilon_\infty}$ (or wave-length $\gtrsim 10^{-2}$ cm.), the electrostatic method is invalid. The admixture of radiative energy is in fact so large, that they can hardly be considered as lattice vibrations.

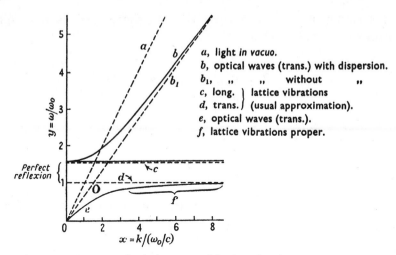

FIG. 18 a. Optical waves and lattice vibrations.

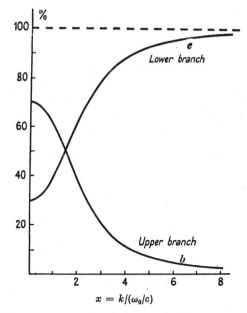

FIG. 18 b. Percentage mechanical energy in the transverse modes.

We shall also see that in the rigorous theory, we obtain the same longitudinal vibrations as in the last section. In other words, the retardation of the Coulomb force has no effect on the longitudinal vibrations of the lattice.

Let us now proceed to consider the underlying mathematics. When the obvious trial solutions

$$\left.\begin{aligned} \mathbf{w} &= \mathbf{w}_0 \\ \mathbf{P} &= \mathbf{P}_0 \\ \mathbf{E} &= \mathbf{E}_0 \end{aligned}\right\} \times \exp\{i(\mathbf{k}.\mathbf{x}-\omega t)\}$$

are substituted, the system of equations (7.1), (7.2), (7.9), (8.1), (8.2), (8.3) reduces to the following:

$$-\omega^2 \mathbf{w} = b_{11}\mathbf{w} + b_{12}\mathbf{E}, \tag{8.6}$$

$$\mathbf{P} = b_{21}\mathbf{w} + b_{22}\mathbf{E}, \tag{8.7}$$

$$\mathbf{k}.(\mathbf{E}+4\pi\mathbf{P}) = 0, \tag{8.8}$$

$$\mathbf{k}.\mathbf{H} = 0, \tag{8.9}$$

$$\mathbf{k}\wedge\mathbf{E} = \frac{\omega}{c}\mathbf{H}, \tag{8.10}$$

$$\mathbf{k}\wedge\mathbf{H} = -\frac{\omega}{c}(\mathbf{E}+4\pi\mathbf{P}). \tag{8.11}$$

To reach the physical conclusions outlined above, it is important that we should take care to obtain the complete solutions.

In the first place, we notice that, in contrast to the electrostatic treatment, the electric field can in no case vanish identically. For if $\mathbf{E}$ were to vanish, it follows from (8.10) that the magnetic field $\mathbf{H}$ vanishes. On account of (8.11), $\mathbf{P}$ also vanishes. Likewise $\mathbf{w}$ vanishes because of (8.7) and the fact that $\mathbf{E} = \mathbf{P} = 0$. Thus vanishing $\mathbf{E}$ leads only to the trivial case that $\mathbf{E} = \mathbf{H} = \mathbf{P} = \mathbf{w} = 0$.

Let us now write (8.6) as

$$\mathbf{w} = \frac{b_{12}\mathbf{E}}{-b_{11}-\omega^2}. \tag{8.12}$$

When this relation is used (8.7) reduces to

$$\mathbf{P} = \left\{\frac{b_{12}b_{21}}{-b_{11}-\omega^2}+b_{22}\right\}\mathbf{E}. \tag{8.13}$$

With the help of (8.13) we find that (8.3) can be written as

$$(\mathbf{k}.\mathbf{E})\left\{1+4\pi b_{22}+\frac{4\pi b_{12}b_{21}}{-b_{11}-\omega^2}\right\} = 0. \tag{8.14}$$

This equation admits two alternative possibilities, namely, either

Case A: $\qquad\qquad 1+4\pi b_{22}+\dfrac{4\pi b_{12}b_{21}}{-b_{11}-\omega^2} = 0,$ \hfill (8.15)

which also implies that (cf. (8.13))

$$\mathbf{E} + 4\pi\mathbf{P} = 0; \tag{8.16}$$

or

Case B: $\mathbf{k}.\mathbf{E} = 0,$

but as $\mathbf{E}$ does not vanish, we must have in this case

$$\mathbf{E} \perp \mathbf{k}. \tag{8.17}$$

The discussions for the above two alternative types of solution in relation to the remaining equations (8.9), (8.10), (8.11) have to be conducted separately.

Consider first the case A. Owing to (8.16) the equation (8.11) reduces to

$$\mathbf{k} \wedge \mathbf{H} = 0. \tag{8.18}$$

This requires that $\mathbf{H}$ either vanishes or is parallel to $\mathbf{k}$; (8.9), on the other hand, requires that $\mathbf{H}$ either vanishes or is perpendicular to $\mathbf{k}$. It follows that

$$\mathbf{H} = 0. \tag{8.19}$$

Now only (8.10) remains to be considered, which becomes in the present case

$$\mathbf{k} \wedge \mathbf{E} = 0. \tag{8.20}$$

As $\mathbf{E}$ does not vanish, $\mathbf{E}$ must be parallel to $\mathbf{k}$.

Apart from being parallel to $\mathbf{k}$, $\mathbf{E}$ is otherwise arbitrary. Once $\mathbf{E}$ is chosen, $\mathbf{w}$ and $\mathbf{P}$ are given by (8.12) and (8.13). The solutions under case A may thus be summarized as follows: the vectors are all longitudinal, namely,
$$\mathbf{w} \parallel \mathbf{P} \parallel \mathbf{E} \parallel \mathbf{k};$$

and the frequencies are given by the solution of (8.15):

$$\omega^2 = -b_{11} + \frac{4\pi b_{12} b_{21}}{1 + 4\pi b_{22}} = \left(\frac{\epsilon_0}{\epsilon_\infty}\right)\omega_0^2 \quad (= \omega_l^2!),$$

which, we note, is independent of $\mathbf{k}$. We recognize that the solutions are identical with the longitudinal lattice vibrations (the irrotational solutions) obtained in the last section by the electrostatic method. The retardation has thus no effect on the longitudinal lattice vibrations.

Consider next the case B. Since in this case $\mathbf{E}$ is perpendicular to $\mathbf{k}$, it follows from (8.10) that $\mathbf{k}$, $\mathbf{E}$, $\mathbf{H}$ form a right-handed system of orthogonal vectors in the given order, and their scalar magnitudes satisfy the following relation:

$$kE = \frac{\omega}{c}H. \tag{8.21}$$

(8.9) is now automatically fulfilled. The only equation still to be considered is (8.11), which reduces to the scalar equation

$$kH = \frac{\omega}{c}(E + 4\pi P).\tag{8.22}$$

After H and P are eliminated with (8.21) and (8.13), the equation becomes

$$\left(\frac{k^2 c^2}{\omega^2}\right)E = \left(1 + 4\pi b_{22} + \frac{4\pi b_{12} b_{21}}{-b_{11} - \omega^2}\right)E.$$

Since E does not vanish, we have

$$\frac{k^2 c^2}{\omega^2} = 1 + 4\pi b_{22} + \frac{4\pi b_{12} b_{21}}{-b_{11} - \omega^2} = \epsilon_\infty + \frac{\epsilon_0 - \epsilon_\infty}{\omega_0^2 - \omega^2}.\tag{8.23}$$

Apart from being perpendicular to $\mathbf{k}$, $\mathbf{E}$ is arbitrary. Thus to each $\mathbf{k}$ there are two independent solutions corresponding to the two possible independent choices for $\mathbf{E}$ (e.g. any two mutually perpendicular vectors, both perpendicular to $\mathbf{k}$); the corresponding values of $\mathbf{w}$ and $\mathbf{P}$ are determined by (8.12) and (8.13). To summarize, all the solutions in case B are transverse, namely,

$$\mathbf{k}, \quad (\mathbf{w} \| \mathbf{P} \| \mathbf{E}), \quad \mathbf{H},$$

are mutually perpendicular. The frequencies, given by the solutions of (8.23), are in this case functions of k. For a given wave-number k, (8.23) gives two solutions, each being, as noted above, doubly degenerate (i.e. having two independent modes).

In Fig. 18a all the solutions are represented on the (ω, k) plot by solid curves.

Remembering that the refractive index is equal to the ratio of the vacuum light velocity c to the phase velocity of an optical wave, we find that (8.23) is identical with the dispersion formula for optical waves which we have already mentioned in the last section. Since the solutions which we have obtained are complete, it follows that all the transverse modes are optical waves.

In general, when two oscillatory systems are coupled, the vibrational modes of the coupled system are mixtures of the original vibrations of the two systems. If the frequencies of the original systems are close to one another, their mutual perturbation is strong, and in the coupled modes both systems take comparable parts. If the original systems have equal frequencies, their interaction is described as being in resonance. On the other hand, if the original frequencies are very different, the mutual perturbation is small and the coupled modes are essentially like the original oscillation of one system, with a small admixture of the other, or vice versa. In Fig. 18 a the transverse solutions (solid curves) represent the coupled modes, which are mixtures of the approximate transverse solutions (dotted lines) of the same wave-number k. The interaction point marked as O corresponds to resonance between the two groups of approximate solutions. The resonance effect tails off towards the right, where the upper and lower branches of the solid curves approach the dotted lines. The underlying physical interpretation is clear: towards the right of the resonance point, the frequency of the upper branch becomes so high that the ions cannot take part in the vibrations to an appreciable extent owing to their large inertia; hence the approximate solutions corresponding to fixed ions provide a good approximation. Here the lower branch represents essentially lattice vibrations with phase velocities small compared with $c/\sqrt{\epsilon_\infty}$; by comparison, the propagation velocity of the interaction force is so large as to make the retardation effect inappreciable. These modes are therefore closely approximated by the lattice vibration calculated with unretarded forces.

Using the expression for the energy density, to be established presently, we can calculate the relative proportions of the radiative energy and mechanical lattice vibrational energy in the transverse modes. Fig. 18b gives the proportion of mechanical lattice vibrational energy for both branches of the transverse modes. We notice in particular that, towards the right of the resonance point, the lower branch of the optical waves approaches practically pure mechanical lattice vibrations and the upper branch, on the other hand, tends to purely radiative energy, both in agreement with the above interpretation. There are no modes resembling the pure mechanical transverse lattice vibrations for waves near or to the left of the resonance point $\left(k \leqslant (\omega_0\sqrt{\epsilon_\infty}/c) \sim 10^3 \text{ cm.}^{-1}\right)$ as given by the electrostatic method; for in this region (wave-lengths $\gtrsim 10^{-2}$ cm.) all modes are seen to be mixtures of radiative and mechanical energies in comparable proportions.

. The essentially mechanical nature of the transverse modes immediately below the dispersion frequency ω_0 has an important consequence on their behaviour as optical waves. The mechanical vibrations conserve energy only in the approximation where terms of third and higher order in the potential energy are ignored (linearity of the lattice equation (7.1)). In reality, each mechanical mode is coupled to the others through the higher-order terms and there is thus a small energy leakage from the mechanical vibrations. In the optical waves near ω_0 the mechanical energy density is so large in relation to the radiative energy flux that the small energy loss by the mechanical vibrations absorbs an abnormally large fraction of the energy flux, resulting in a strong damping of the waves. The experimental aspect of this phenomenon will be discussed in § 10.

We now proceed to derive the energy density. We shall do so for the more general case when free charges are present; the result can then be used to construct the Hamiltonian for charges moving in a lattice. Let the charges be represented by e_i $(i = 1, 2,..., n)$ and their position vectors by $\mathbf{x}_i$. In the presence of the charges, the electromagnetic equations (7.9), (8.1), (8.2), and (8.3) become

$$\nabla.(\mathbf{E}+4\pi\mathbf{P}) = 4\pi \sum_i e_i \delta(\mathbf{x}-\mathbf{x}_i), \tag{8.24}$$

$$\nabla.\mathbf{H} = 0, \tag{8.25}$$

$$\nabla \wedge \mathbf{E} = -\frac{1}{c}\dot{\mathbf{H}}, \tag{8.26}$$

$$\nabla \wedge \mathbf{H} = \frac{1}{c}\left(\dot{\mathbf{E}}+4\pi\dot{\mathbf{P}}+4\pi \sum_i e_i \dot{\mathbf{x}}_i \delta(\mathbf{x}-\mathbf{x}_i)\right), \tag{8.27}$$

where $\delta(\mathbf{x}-\mathbf{x}_i)$ is the three-dimensional Dirac delta-function. (8.24) is equivalent to (7.20) used in the last section; the delta-function gives the charge density due to a unit charge concentrated at one point. Similarly, $e_i\dot{\mathbf{x}}_i\delta(\mathbf{x}-\mathbf{x}_i)$ in (8.27) is equivalent to the more conventional form $\rho\mathbf{v}$.

Using (8.26) and (8.27), we can write

$$\frac{c}{4\pi}\nabla.(\mathbf{E}\wedge\mathbf{H}) = \frac{c}{4\pi}\left(\mathbf{H}.\nabla\wedge\mathbf{E}-\mathbf{E}.\nabla\wedge\mathbf{H}\right)$$

$$= -\left\{\frac{1}{4\pi}(\mathbf{H}.\dot{\mathbf{H}}+\mathbf{E}.\dot{\mathbf{E}})+\mathbf{E}.\dot{\mathbf{P}}+ \sum_i e_i \dot{\mathbf{x}}_i.\mathbf{E}\delta(\mathbf{x}-\mathbf{x}_i)\right\}. \tag{8.28}$$

Integrating both sides over an arbitrary volume, we obtain

$$\int_S \left\{\frac{c}{4\pi}(\mathbf{E}\wedge\mathbf{H})\right\}.d\mathbf{s}$$

$$= -\int_V \left\{\frac{1}{4\pi}(\mathbf{H}.\dot{\mathbf{H}}+\mathbf{E}.\dot{\mathbf{E}})+\mathbf{E}.\dot{\mathbf{P}}+\sum_i e_i \dot{\mathbf{x}}_i.\mathbf{E}\delta(\mathbf{x}-\mathbf{x}_i)\right\}d\tau, \tag{8.29}$$

where the volume integral on the left-hand side has been converted into a surface integral over the bounding surface by Green's theorem. We recognize that $(c/4\pi)(\mathbf{E}\wedge\mathbf{H})$ is the Poynting vector for electromagnetic energy flux; thus the left-hand side of (8.29) represents the rate at which electromagnetic energy flows out of the volume. (It is clear from the lattice equations (7.1), (7.2) that different volume elements are coupled in the macroscopic theory only through their electric interaction; thus the left-hand side of (8.29) also represents the rate of total energy out-flow from the volume.)

The last term on the right-hand side of (8.29) can be written as

$$- \sum_i e_i \dot{\mathbf{x}}_i . \mathbf{E}(\mathbf{x}_i), \qquad (8.30)$$

where the summation is extended over the charges contained in the volume considered. Using the equations of motion for the charges:

$$m_i \ddot{\mathbf{x}}_i = e_i \mathbf{E}(\mathbf{x}_i), \qquad (8.31)$$

we find that

$$-\frac{d}{dt}\{\tfrac{1}{2}m_i \dot{\mathbf{x}}_i^2\} = -m_i \dot{\mathbf{x}}_i . \ddot{\mathbf{x}}_i = -e_i \dot{\mathbf{x}}_i . \mathbf{E}(\mathbf{x}_i). \qquad (8.32)$$

The last term in (8.29) is thus equal to the rate of decrease of the kinetic energy of the charges within the volume. Hence it follows from the energy conservation requirement and (8.29) that

$$\frac{1}{4\pi}(\mathbf{H}.\dot{\mathbf{H}}+\mathbf{E}.\dot{\mathbf{E}})+\mathbf{E}.\dot{\mathbf{P}} \qquad (8.33)$$

must be equal to the rate of change of the energy density (including the kinetic energy of the lattice particles as well as all the electromagnetic and potential energies in the system). If we denote the energy density by U, then dU/dt must be equal to the above expression. The following expression:

$$U = \tfrac{1}{2}\dot{\mathbf{w}}^2-\tfrac{1}{2}b_{11}\mathbf{w}^2-b_{12}\mathbf{w}.\mathbf{E}-\tfrac{1}{2}b_{22}\mathbf{E}^2+\mathbf{E}.\mathbf{P}+\frac{1}{8\pi}(\mathbf{E}^2+\mathbf{H}^2)$$

$$(8.34)$$

fulfils this condition. For differentiating (8.34) with respect to time, we find that

$$\frac{dU}{dt} = \dot{\mathbf{w}}.\ddot{\mathbf{w}}-b_{11}\mathbf{w}.\dot{\mathbf{w}}-b_{12}\dot{\mathbf{w}}.\mathbf{E}-b_{12}\mathbf{w}.\dot{\mathbf{E}}-b_{22}\mathbf{E}.\dot{\mathbf{E}}+\dot{\mathbf{E}}.\mathbf{P}+\mathbf{E}.\dot{\mathbf{P}}+$$

$$+\frac{1}{4\pi}(\mathbf{E}.\dot{\mathbf{E}}+\mathbf{H}.\dot{\mathbf{H}}). \quad (8.35)$$

The last three terms are equal to (8.33). The rest of the terms can be written as follows:

$$\dot{\mathbf{w}}(\ddot{\mathbf{w}}-b_{11}\mathbf{w}-b_{12}\mathbf{E})+\dot{\mathbf{E}}(\mathbf{P}-b_{12}\mathbf{w}-b_{22}\mathbf{E}).$$

This expression vanishes because of (7.1), (7.2), and (7.3). Hence apart from an arbitrary constant (equivalent to an arbitrary choice of the zero for measuring the potential energy) (8.34) provides the requisite expression for the energy density. When $\mathbf{P}$ is eliminated by the use of (7.2) and the b-coefficients are expressed in terms of ω_0, ϵ_0, ϵ_∞, the energy density assumes the very simple form:

$$U = \tfrac{1}{2}(\dot{\mathbf{w}}^2 + \omega_0^2\,\mathbf{w}^2) + \frac{1}{8\pi}(\epsilon_\infty\,\mathbf{E}^2 + \mathbf{H}^2).\qquad(8.36)$$

For a transverse mode, the second term represents radiative energy and the first term the mechanical lattice vibration energy. The curves in Fig. 18 b are calculated by substituting the transverse solution in the respective expressions.†

The Hamiltonian of the system is obtained by integrating U over space and adding the kinetic energies of the charges, namely,

$$H = \sum_i \tfrac{1}{2}m_i\dot{\mathbf{x}}_i^2 + \int \left\{ \tfrac{1}{2}(\dot{\mathbf{w}}^2 + \omega_0^2\,\mathbf{w}) + \frac{1}{8\pi}(\epsilon_\infty\mathbf{E}^2 + \mathbf{H}^2) \right\} d\tau.\qquad(8.37)$$

The Hamiltonian includes also the energy of electromagnetic radiation. For problems where radiation effects are negligible, we can use the electrostatic approximation. $\mathbf{H}$ may then be put equal to zero, and the electric field $\mathbf{E}$ eliminated with the help of (7.23). If furthermore we write $\mathbf{w} = \mathbf{w}_t + \mathbf{w}_l$, we find that the Hamiltonian can be written as

$$H = \sum_i \tfrac{1}{2}m_i\dot{\mathbf{x}}_i^2 + \int \left\{ \tfrac{1}{2}(\dot{\mathbf{w}}_t^2 + \omega_0^2\,\mathbf{w}_t^2) + \tfrac{1}{2}(\dot{\mathbf{w}}_l^2 + \omega_0^2\,\mathbf{w}_l^2) + \right.$$

$$\left. + \frac{1}{2}\!\left(\frac{1}{\epsilon_\infty} - \frac{1}{\epsilon_0}\right)\!\epsilon_\infty\,\omega_l^2\,\mathbf{w}_l^2 - \omega_l\!\left[\frac{1}{4\pi}\!\left(\frac{1}{\epsilon_\infty} - \frac{1}{\epsilon_0}\right)\right]^{\!\frac{1}{2}}\!\mathbf{w}_l.\,\mathbf{E}_{\text{vac}} + \frac{1}{8\pi\epsilon_\infty}\mathbf{E}_{\text{vac}}^2 \right\} d\tau,\qquad(8.38)$$

where we have made use of the fact that the integral of the scalar product between a solenoidal vector (e.g. $\mathbf{w}_t$ or $\dot{\mathbf{w}}_t$) and an irrotational vector (e.g. $\mathbf{w}_l$, $\dot{\mathbf{w}}_l$, or $\mathbf{E}_{\text{vac}}$) vanishes identically.

The last term in (8.38) can be transformed as follows: $\mathbf{E}_{\text{vac}}$ is, we remember, the electric field due to the charges, if they were in vacuum; it is well known in electrostatics that

$$\int \frac{\mathbf{E}_{\text{vac}}^2}{8\pi} \, d\tau$$

is, apart from the electrostatic self-energies of the individual charges, equal to the Coulomb interaction between the charges:

$$\frac{1}{2}\sum_{i \neq j} \frac{e_i e_j}{|\mathbf{x}_i - \mathbf{x}_j|}.$$

Therefore if we omit the electrostatic self-energies of the charges from the Hamiltonian, the last term in (8.38) can be replaced by the Coulomb interaction divided by ϵ_∞. Furthermore, since ω_0^2 is equal to $\epsilon_\infty\omega_l^2/\epsilon_0$ (see

† For details see K. Huang, *Proc. Roy. Soc.* A, **208**, 352 (1951).

(7.17)) the two terms in (8.38) with $\mathbf{w}_l^2$ can be combined, giving simply $\frac{1}{2}\omega_l^2\,\mathbf{w}_l^2$. The Hamiltonian (in the electrostatic approximation) can thus be written as

$$H = \sum_i \tfrac{1}{2}m_i\dot{\mathbf{x}}_i^2 + \frac{1}{2\epsilon_\infty}\sum_{i\neq j}\frac{e_i e_j}{|\mathbf{x}_i-\mathbf{x}_j|} +$$

$$+ \int \{\tfrac{1}{2}(\dot{\mathbf{w}}_t^2+\omega_0^2\,\mathbf{w}_t^2)+\tfrac{1}{2}(\dot{\mathbf{w}}_l^2+\omega_l^2\,\mathbf{w}_l^2)\}\,d\tau -$$

$$-\omega_l\left[\frac{1}{4\pi}\left(\frac{1}{\epsilon_\infty}-\frac{1}{\epsilon_0}\right)\right]^{\frac{1}{2}}\int \mathbf{w}_l.\mathbf{E}_{\text{vac}}\,d\tau. \quad (8.39)$$

We have seen earlier that when the lattice particles are held fixed at $\mathbf{w}\equiv 0$, the lattice behaves as a normal dielectric with the dielectric constant ϵ_∞; the first two terms in (8.39) represent the Hamiltonian of the charges in such a medium. The third term represents the Hamiltonian for the lattice in the absence of free charges. The fourth term gives the interaction between the system of charges and the lattice motions. Once more we notice that only the irrotational lattice motion is coupled to the charges present; this is true, however, only in the electrostatic approximation.

For quantum-mechanical applications it is necessary to describe the lattice motion in terms of normal coordinates. Since the transverse vibrations do not interact with the charges, and are thus of no interest in the present connexion, we shall ignore $\mathbf{w}_t$ in the following considerations. Let us consider a finite volume V containing $L\times L\times L = N$ lattice cells by imposing the corresponding periodic boundary condition. Then the normal coordinates can be introduced as follows (for general methods see § 38). Express $\mathbf{w}_l$ as the Fourier series

$$\mathbf{w}_l = \frac{1}{\sqrt{V}}\sum_{\mathbf{y}}(\mathbf{y}/|\mathbf{y}|)Q(\mathbf{y})e^{2\pi i\mathbf{y}.\mathbf{x}}, \quad (8.40)$$

where, owing to the periodic condition, $\mathbf{y}$ is restricted to the following values:

$$\mathbf{y} = \frac{1}{L}(h_1\,\mathbf{b}^1+h_2\,\mathbf{b}^2+h_3\,\mathbf{b}^3), \quad (8.41)$$

h_1, h_2, h_3 being integers in the range

$$0 \leqslant |h_i| < L/2 \quad (i = 1, 2, 3). \quad (8.42)$$

The polarization vector, $\mathbf{y}/|\mathbf{y}|$, in (8.40) follows directly from the irrotational character of $\mathbf{w}_l$. For $\mathbf{w}_l$ to be real, we must have

$$Q(\mathbf{y}) = -Q^*(-\mathbf{y}). \quad (8.43)$$

Substituting (8.40) in the Hamiltonian (8.39) and carrying out the volume integral over the volume V, we get (ignoring $\mathbf{w}_l$):

$$H = \sum_i \tfrac{1}{2} m_i \dot{\mathbf{x}}_i^2 + \frac{1}{2} \sum_{i \neq j} \frac{e_i e_j}{|\mathbf{x}_i - \mathbf{x}_j|} + \frac{1}{2} \sum_{\mathbf{y}} \{\dot{Q}(\mathbf{y}) \dot{Q}^*(\mathbf{y}) + \omega_t^2\, Q(\mathbf{y}) Q^*(\mathbf{y})\} -$$

$$- \frac{\omega_l}{\sqrt{V}} \left[\frac{1}{4\pi} \left(\frac{1}{\epsilon_\infty} - \frac{1}{\epsilon_0} \right) \right]^{\frac{1}{2}} \sum_{\mathbf{y}} Q(\mathbf{y}) \int \left\{ \left(\frac{\mathbf{y}}{|\mathbf{y}|} \right) . \mathbf{E}_{\text{vac}}\, e^{2\pi i \mathbf{y} . \mathbf{x}} \right\} d\tau. \quad (8.44)$$

The real normal coordinates $q_1(\mathbf{y})$, $q_2(\mathbf{y})$ are introduced by writing

$$Q(\mathbf{y}) = \frac{1}{\sqrt{2}} \{q_1(\mathbf{y}) + i q_2(\mathbf{y})\} = -Q^*(-\mathbf{y}) \quad (8.45)$$

[in the case $\mathbf{y} = 0$, $Q(0)$ is the normal coordinate itself; for the sake of simplicity in the following we shall not represent explicitly terms in $Q(0)$], where $\mathbf{y}$ can now be restricted to the values lying on one side of an arbitrary plane through the origin of the $\mathbf{y}$-space so that only one of a pair $\mathbf{y}$, $-\mathbf{y}$ is counted. In terms of the normal coordinates, the Hamiltonian then becomes

$$H = \sum_i \tfrac{1}{2} m_i \dot{\mathbf{x}}_i^2 + \frac{1}{2} \sum_{i \neq j} \frac{e_i e_j}{|\mathbf{x}_i - \mathbf{x}_j|} + \frac{1}{2} \sum_{\mathbf{y}} \sum_{\lambda = 1,2} \{\dot{q}_\lambda^2(\mathbf{y}) + \omega_t^2\, q_\lambda(\mathbf{y})\} +$$

$$+ \frac{\omega_l}{\sqrt{V}} \left[\frac{1}{2\pi} \left(\frac{1}{\epsilon_\infty} - \frac{1}{\epsilon_0} \right) \right]^{\frac{1}{2}} \left\{ \sum_{\mathbf{y}} q_1(\mathbf{y}) \int \left[\left(\frac{\mathbf{y}}{|\mathbf{y}|} \right) . \mathbf{E}_{\text{vac}} \cos 2\pi \mathbf{y} . \mathbf{x} \right] d\tau + \right.$$

$$\left. + \sum_{\mathbf{y}} q_2(\mathbf{y}) \int \left[\left(\frac{\mathbf{y}}{|\mathbf{y}|} \right) . \mathbf{E}_{\text{vac}} \sin 2\pi \mathbf{y} . \mathbf{x} \right] d\tau \right\}. \quad (8.46)$$

This is the form in which the Hamiltonian can be used directly for quantum-mechanical considerations.

9. Atomic theory of long optical vibrations and infra-red dispersion

Having discussed the long optical vibrations and the infra-red dispersion on a phenomenological basis with (7.1) and (7.2), we can now establish the atomic theory of these phenomena simply by deducing these equations from the general atomic theory.

For the present purpose, we have to take into account a feature hitherto ignored in §§ 1 and 3, namely, that the ions are polarizable by electric fields. As explained earlier in § 3, this effect can be ignored when the ions occupy positions of symmetry, such as in many simple crystals in the equilibrium state (or subject only to elastic strains). This is no longer permissible if either a macroscopic electric field is present or if the ions are displaced from their symmetric positions, for then, as we shall see, they may experience a field even in the absence of a macroscopic field. These are precisely the circumstances contemplated in the phenomenological equations. The polarization of the ions

can be described as follows. When an ion is subject to an electric field **E**
its electron cloud is deformed and acquires an electric dipole moment **μ**.
If the field is not excessively high the induced dipole moment is proportional to the field,

$$\boldsymbol{\mu} = \alpha\mathbf{E}, \tag{9.1}$$

where α is the atomic polarizability of the ion. α is a constant as long
as the field does not vary appreciably within a period of the electronic
motion in the ion.

To derive the phenomenological equations, we have to find, for given
w and **E**, the forces acting on the ions and the dipole moment per unit
volume. The dipole moment is due partly to the displacements of the
ionic charges and partly to the induced electric moments on the ions.
When an ionic charge Ze is displaced by **u**, the net effect is as though
a charge $-Ze$ has been placed at the undisplaced position of the ion to
annihilate the original charge and a fresh charge Ze has been created at
the displaced position. The displacement is thus equivalent to the
addition of a dipole with the moment $Ze\mathbf{u}$. The induced electric moment
on an ion, on the other hand, depends on the field acting upon the ion.
In practice, this field is never altogether uniform as implied in the
formula (9.1); in the following considerations we shall always take the
value of the field at the centre of the ion. This field, which acts upon an
ion, will be designated as the *effective field*.

It is important to realize that the effective field is different from the
macroscopic electric field. The latter is not simply the total field, which
includes the field of the ion under consideration; it is the total field
averaged over the space occupied by a lattice cell. The effective field on
an ion, on the other hand, is the total field with the contribution of the
ion itself excluded and the value, as we have said, is taken at the centre
of the ion. Our first task is to calculate the value of the effective field.

The discrepancy between the effective field on an ion and the macroscopic field in the same neighbourhood is clearly due only to the contributions to these fields of the matter in the close neighbourhood of the
ion. For, in the first place, the charge sources farther away are naturally
irrelevant for the exclusion of the field due to the ion itself. In the
second place, the fields due to such distant sources do not vary appreciably in a neighbourhood of the dimensions of a lattice cell, as long as
the sources are located at distances away which are large compared
with the lattice constant a; thus the average value and the value at a
specific point need not be distinguished. Hence let us draw a sphere
around the ion under discussion with a radius R large compared with a

and proceed to calculate the difference between the effective and the macroscopic fields due to the matter contained within the sphere.

For convenience, we take an intermediate value for R such that it is large compared with a, while the macroscopic quantities $\mathbf{w}$, $\mathbf{E}$, $\mathbf{P}$ do not vary appreciably over the region of the sphere. This is possible, because the lattice constant a is to be regarded as an infinitesimal in a macroscopic treatment. Macroscopically speaking, the matter within the sphere is a uniformly polarized sphere. It is well known in electrostatics that the uniform polarization $\mathbf{P}$ is equivalent to a surface charge distribution with a density everywhere equal to the component of $\mathbf{P}$ along the outward normal to the surface. From the surface charge the macroscopic field at the centre of the sphere can be easily calculated. It is obvious from the symmetry that the resultant field must be parallel to $\mathbf{P}$; thus in considering the contributions due to different surface elements we need only consider the component in this direction. If we introduce the polar angles θ, ϕ with the polar axis in the direction of $\mathbf{P}$, the contribution due to a surface element $R^2 \sin\theta\, d\theta d\phi$ is clearly

$$\left(\frac{\mathbf{P}\cos\theta\; R^2 \sin\theta\, d\theta d\phi}{R^2}\right)(-\cos\theta) = -\mathbf{P}\cos^2\theta \sin\theta\, d\theta d\phi. \qquad (9.2)$$

Integration over the whole spherical surface gives immediately the macroscopic field due to the matter contained in the sphere as

$$-\frac{4\pi}{3}\mathbf{P}. \qquad (9.3)$$

Next we have to consider microscopically the effective field on the central ion due to all the other ions in the sphere. We are concerned, of course, only with the field arising from the displacements of the ions and their induced moments (the original field due to the undisplaced and unpolarized ions vanishes for the highly symmetric structures under consideration); we may thus imagine each ion site being occupied by a dipole equal to the sum of the displacement dipole $Ze\mathbf{u}$ and the induced dipole $\boldsymbol{\mu}$. Owing to the assumed uniformity over the spherical region, all the dipoles on either the positive or negative ion sites are identical among themselves. Unlike the macroscopic field, which depends only on the macroscopic polarization, the effective field will depend on the detailed geometrical arrangement of the dipoles. Here we are concerned only with lattices in which every ion is in a surrounding with tetrahedral symmetry, the meaning of which we shall presently explain. For such lattices, as we shall find, the effective field due to the dipoles in the sphere vanishes identically.

The tetrahedral symmetry about an ion implies that there exist Cartesian axes XYZ through the ion such that if there is an ion site with the coordinates (a, b, c), there must be identical ion sites (i.e. sites occupied by the same kind of ion) at points $(-a, -b, c)$, $(a, -b, -c)$, $(-a, b, -c)$ and the eight more points with coordinates obtained from the above by cyclic permutations of a, b, c. (These points are generated from (a, b, c) by the tetrahedral group of operations about the origin which are rotations by π about the axes X, Y, Z and rotations either way of $2\pi/3$ about the cube diagonals $\pm X = \pm Y = \pm Z$. Tetrahedral symmetry of a structure means that the geometrical arrangement is completely unaltered, if the structure is rigidly subjected to the tetrahedral operations. The simultaneous existence of the group of twelve points listed above clearly follows directly from this definition.) A dipole $\mathbf{p}$ at $\mathbf{x}_1(a, b, c)$ gives rise to a field at the origin equal to

$$-\frac{\mathbf{p}}{|\mathbf{x}_1|^3} + \frac{3\mathbf{p}.\mathbf{x}_1}{|\mathbf{x}_1|^5}\mathbf{x}_1.$$

Let $\mathbf{x}_i$, $i = 2, 3,..., 12$ denote the other eleven points which are occupied by identical dipoles; the contribution to the effective field due to the group of twelve identical dipoles can thus be written as

$$-\sum_{i=1}^{12}\frac{\mathbf{p}}{|\mathbf{x}_i|^3} + 3\sum_{i=1}^{12}\frac{\mathbf{p}.\mathbf{x}_i}{|\mathbf{x}_i|^5}\mathbf{x}_i.$$

We notice that all twelve points are at the same distance $(a^2+b^2+c^2)^{\frac{1}{2}}$ from the origin. Thus when the explicit coordinates of the points are substituted, one obtains:

$$-12\frac{\mathbf{p}}{(a^2+b^2+c^2)^{\frac{3}{2}}} + \frac{3}{(a^2+b^2+c^2)^{\frac{5}{2}}} \times$$

$$\times \begin{bmatrix} a(ap_1+bp_2+cp_3)-a(-ap_1-bp_2+cp_3)+a(ap_1-bp_2-cp_3)- \\ -a(-ap_1+bp_2-cp_3)+\text{cyclic permutations of } a, b, c \\ b(ap_1+bp_2+cp_3)-b(-ap_1-bp_2+cp_3)-b(ap_1-bp_2-cp_3)+ \\ +b(-ap_1+bp_2-cp_3)+\text{cyclic permutations of } a, b, c \\ c(ap_1+bp_2+cp_3)+c(-ap_1-bp_2+cp_3)-c(ap_1-bp_2+cp_3)- \\ -c(-ap_1+bp_2-cp_3)+\text{cyclic permutations of } a, b, c \end{bmatrix}$$

$$= -12\frac{\mathbf{p}}{(a^2+b^2+c^2)^{\frac{3}{2}}} + \frac{3}{(a^2+b^2+c^2)^{\frac{5}{2}}}\begin{bmatrix} 4a^2p_1+\text{cyclic perm.} \\ 4b^2p_2+ \quad ,, \quad ,, \\ 4c^2p_3+ \quad ,, \quad ,, \end{bmatrix}$$

$$= -\frac{12\mathbf{p}}{(a^2+b^2+c^2)^{\frac{3}{2}}} + \frac{12(a^2+b^2+c^2)}{(a^2+b^2+c^2)^{\frac{5}{2}}}\begin{bmatrix} p_1 \\ p_2 \\ p_3 \end{bmatrix} = 0,$$

where we have displayed the three components of the second term in a column. The effective field due to the ions in the sphere thus vanishes.

Since the difference between the effective and macroscopic fields is due solely to the contributions of the matter contained within the sphere, we can write

$$\mathbf{E}_{\text{eff}} - \mathbf{E} = 0 - \left(-\frac{4\pi}{3}\mathbf{P}\right),$$

or

$$\mathbf{E}_{\text{eff}} = \mathbf{E} + \frac{4\pi}{3}\mathbf{P}, \tag{9.4}$$

which, we notice, is the same for either a positive or a negative ion site.

The effective field in more complex and less symmetric structures cannot be calculated quite so simply. In Chapter V an expression for the general case will be given in the form of a quickly convergent series. In Appendix VI the above relation is alternatively derived from the general expression by imposing the special tetrahedral symmetry condition.

Once the effective field is known, the rest of the task is straightforward. Let $\mathbf{u}_{\pm}$, $\pm Ze$, and $\alpha_{\pm}$ be respectively the displacements, the ionic charges, and the atomic polarizabilities of the positive and negative ions. The effective dipole moments on the two types of ions are thus

$$Ze\mathbf{u}_{+} + \alpha_{+}\mathbf{E}_{\text{eff}}, \tag{9.5}$$

$$-Ze\mathbf{u}_{-} + \alpha_{-}\mathbf{E}_{\text{eff}}. \tag{9.6}$$

Since there are $1/v_a$ ion pairs per unit volume, the macroscopic polarization $\mathbf{P}$ is obtained by multiplying the sum of (9.5) and (9.6) by $1/v_a$:

$$\mathbf{P} = \frac{1}{v_a}[Ze(\mathbf{u}_{+} - \mathbf{u}_{-}) + (\alpha_{+} + \alpha_{-})\mathbf{E}_{\text{eff}}]. \tag{9.7}$$

After the effective field $\mathbf{E}_{\text{eff}}$ is eliminated with the relation (9.4), we find that

$$\mathbf{P} = \frac{1}{1 - (4\pi/3)(\alpha_{+} + \alpha_{-})/v_a}\left\{Ze\left(\frac{\mathbf{u}_{+} - \mathbf{u}_{-}}{v_a}\right) + \left(\frac{\alpha_{+} + \alpha_{-}}{v_a}\right)\mathbf{E}\right\}, \tag{9.8}$$

$\mathbf{u}_{+} - \mathbf{u}_{-}$, which is the displacement of the positive ions relative to the negative, is related to the parameter $\mathbf{w}$ used in the last two sections by

$$\mathbf{w} = \left(\frac{\bar{M}}{v_a}\right)^{\frac{1}{2}}(\mathbf{u}_{+} - \mathbf{u}_{-}) \qquad \left(\bar{M} = \frac{M_{+}M_{-}}{M_{+} + M_{-}}\right). \tag{9.9}$$

On expressing $(\mathbf{u}_{+} - \mathbf{u}_{-})$ in (9.8) in terms of $\mathbf{w}$, we obtain an equation which is strictly comparable with (7.2). The comparison of these equations leads directly to the following values for the coefficients b_{21}, b_{22} in the phenomenological equation:

$$b_{21} = \frac{Ze\left(\frac{1}{\bar{M}v_a}\right)^{\frac{1}{2}}}{1-\frac{4\pi}{3}\left(\frac{\alpha_++\alpha_-}{v_a}\right)}, \qquad b_{22} = \frac{\left(\frac{\alpha_++\alpha_-}{v_a}\right)}{1-\frac{4\pi}{3}\left(\frac{\alpha_++\alpha_-}{v_a}\right)}. \qquad (9.10)$$

When the positive and negative ions are relatively displaced, the overlap potential between them gives rise to a force on the ions, which for small displacements can be taken as linearly proportional to the relative displacement $(\mathbf{u}_+-\mathbf{u}_-)$ between the ions. Thus let us write

$$-k(\mathbf{u}_+-\mathbf{u}_-) \quad \text{and} \quad k(\mathbf{u}_+-\mathbf{u}_-) \qquad (9.11)$$

as the corresponding forces acting on a positive ion and a negative ion respectively. We note that the coefficient k is a simple scalar; this is a consequence of the symmetry condition that every ion is surrounded with the tetrahedral symmetry. (In general, the linear coefficient can be a tensor of the second rank, but the only tensors of this kind consistent with the tetrahedral symmetry are necessarily isotropic, in other words, equivalent to a scalar. A proof of this fact is to be found in Appendix VI.) That the forces on the two kinds of ions are equal and opposed as shown, follows directly from Newton's reaction law. Later we shall express k explicitly in terms of the overlap forces between neighbours as discussed in § 3.

Taking into account, besides the overlap forces, the forces exerted on the ions by the effective electric field $\mathbf{E}_{\text{eff}}$ we can write down the equations of motion for the two types of ions as:

$$M_+\ddot{\mathbf{u}}_+ = -k(\mathbf{u}_+-\mathbf{u}_-)+Ze\mathbf{E}_{\text{eff}}, \qquad (9.12)$$

$$M_-\ddot{\mathbf{u}}_- = k(\mathbf{u}_+-\mathbf{u}_-)-Ze\mathbf{E}_{\text{eff}}. \qquad (9.13)$$

Multiplying the equations respectively by M_- and M_+, subtracting and dividing afterwards by (M_++M_-), we find that

$$\bar{M}(\ddot{\mathbf{u}}_+-\ddot{\mathbf{u}}_-) = -k(\mathbf{u}_+-\mathbf{u}_-)+Ze\mathbf{E}_{\text{eff}}. \qquad (9.14)$$

On eliminating $\mathbf{E}_{\text{eff}}$ from this equation and (9.4), and afterwards expressing $\mathbf{P}$ in terms of $(\mathbf{u}_+-\mathbf{u}_-)$ and $\mathbf{E}$ with the help of (9.8), we obtain the following:

$$\bar{M}(\ddot{\mathbf{u}}_+-\ddot{\mathbf{u}}_-)$$

$$= \left[-k+\frac{\frac{4\pi}{3}\frac{(Ze)^2}{v_a}}{1-\frac{4\pi}{3}\left(\frac{\alpha_++\alpha_-}{v_a}\right)}\right](\mathbf{u}_+-\mathbf{u}_-)+\left[\frac{Ze}{1-\frac{4\pi}{3}\left(\frac{\alpha_++\alpha_-}{v_a}\right)}\right]\mathbf{E}. \qquad (9.15)$$

When the relative displacement $(\mathbf{u}_+ - \mathbf{u}_-)$ is expressed in terms of the parameter $\mathbf{w}$ by (9.9), (9.15) becomes

$$\ddot{\mathbf{w}} = \left[\frac{k}{\overline{M}} + \frac{\dfrac{4\pi}{3} \dfrac{(Ze)^2}{\overline{M} v_a}}{1 - \dfrac{4\pi}{3}\left(\dfrac{\alpha_+ + \alpha_-}{v_a}\right)} \right] \mathbf{w} + \left[\frac{\dfrac{Ze}{(\overline{M} v_a)^{\frac{1}{2}}}}{1 - \dfrac{4\pi}{3}\left(\dfrac{\alpha_+ + \alpha_-}{v_a}\right)} \right] \mathbf{E}, \quad (9.16)$$

which is the phenomenological equation (7.1) as deduced from the atomic theory. A comparison of these equations gives the following values for the phenomenological coefficients b_{11}, b_{12}:

$$b_{11} = -\frac{k}{\overline{M}} + \frac{\dfrac{4\pi}{3} \dfrac{(Ze)^2}{\overline{M} v_a}}{1 - \dfrac{4\pi}{3}\left(\dfrac{\alpha_+ + \alpha_-}{v_a}\right)}, \quad (9.17)$$

$$b_{12} = \frac{\dfrac{Ze}{(\overline{M} v_a)^{\frac{1}{2}}}}{1 - \dfrac{4\pi}{3}\left(\dfrac{\alpha_+ + \alpha_-}{v_a}\right)}. \quad (9.18)$$

In particular, we notice by comparing (9.18) with (9.10) that the general relation $b_{12} = b_{21}$ is fulfilled.

We shall next discuss in turn the values of b_{22}, b_{11}, b_{12} given by the microscopic theory as compared with the observed values given in the relations (7.6), (7.7), and (7.8).

b_{22} is distinguished from the other coefficients by the fact that it is not peculiar to the ionic crystals. For we remember that if the ionic motions are suppressed, the lattice behaves like a normal dielectric substance; the corresponding dielectric constant ϵ_∞ is related to b_{22} by the relation (7.8), namely,

$$\epsilon_\infty = 1 + 4\pi b_{22}.$$

The microscopic expression for b_{22} leads in fact to an expression for the dielectric constant ϵ_∞, which is typical of all substances consisting of polarizable atoms. Thus using (9.10), we find that

$$\epsilon_\infty = 1 + \frac{4\pi\left(\dfrac{\alpha_+ + \alpha_-}{v_a}\right)}{1 - \dfrac{4\pi}{3}\left(\dfrac{\alpha_+ + \alpha_-}{v_a}\right)}. \quad (9.19)$$

The above relation can be written alternatively as

$$\frac{4\pi}{3}(\alpha_+ + \alpha_-) = \left(\frac{\epsilon_\infty - 1}{\epsilon_\infty + 2}\right) v_a. \quad (9.20)$$

Formulae of this type were obtained long ago† by Clausius-Mosotti in connexion with electrostatics, by Lorenz and Lorentz in connexion with the electromagnetic theory. If we assume that the atomic polarizabilities are not altered when the substance is compressed, then $(\epsilon_\infty - 1)/(\epsilon_\infty + 2)$ becomes inversely proportional to v_a, or directly proportional to the density. Another feature of great interest is the additivity of the polarizabilities expressed in (9.20). It is the usual practice to multiply both sides of the equation by Avogadro's number N; $4\pi N\alpha_\pm/3$ are known as the *atom-refractions* of the corresponding particles, and $V(\epsilon_\infty - 1)(\epsilon_\infty + 2)$ ($V = Nv_a$ is the molar volume) as the *mole-refraction* of the compound. (9.20) is a special form of the general result‡ that the mole-refraction of a composite substance is equal to the sum of the atom-refractions of its constituents. This result is significant, of course, only if the atomic polarizabilities of the same particles in different compounds and mixtures remain essentially unaltered so that the atom-refractions can be assigned unique values; the additivity law of the refractions is in fact found closely fulfilled in many chemical compounds as well as mixtures. For the alkali halide crystals Shockley§ has determined the values of the atomic polarizabilities of the alkali and halide ions so that the mean square deviation of ϵ_∞ calculated with (9.19) from the observed values is a minimum. In Table 18 his values are compared with the values for the free ions as estimated by Pauling.

TABLE 18

Atomic Polarizabilities of the Alkali and Halide Ions

(Units 10^{-24} cm.^{3})

	Li$^+$	Na$^+$	K$^+$	Rb$^+$	Cs$^+$	F$^-$	Cl$^-$	Br$^-$	I$^-$
Crystal (Shockley)	0·045	0·28	1·13	1·79	2·85	0·86	2·92	4·12	6·41
Free (Pauling)	0·03	0·18	0·83	1·40	2·42	1·04	3·66	4·77	7·10

It appears from the above table that the positive ions become more polarizable and the negative ions less polarizable in the crystal lattices.

The values of ϵ_∞ calculated from Shockley's atomic polarizabilities are compared with the empirical values in Table 19. The agreement is seen to be remarkably close.‖

† See H. A. Lorentz, *Theory of Electrons* (Teubner, 1909), pp. 137–50.
‡ See e.g. H. A. Lorentz, loc. cit.; M. Born, *Optik* (Springer, 1933), pp. 341–5.
§ W. Shockley, *Phys. Rev.* **70**, 105 (A) (1946).
‖ Fajans and Joos (K. Fajans and G. Joos, *Zeit. f. Phys.* **23**, 1 (1924)) had concluded that the additivity law is not fulfilled in the alkali halide crystals; their argument appears, however, to be fallacious.

TABLE 19

The Calculated and Observed Values† for the High-frequency
Dielectric Constant ϵ_∞ of the Alkali Halide Crystals

		Li⁺	Na⁺	K⁺	Rb⁺	Cs⁺
F⁻	cal.	1·93	1·72	1·85	1·99	2·21
	obs.	1·92	1·74	1·85	1·93	..
Cl⁻	cal.	2·72	2·30	2·13	2·18	2·59
	obs.	2·75	2·25	2·13	2·19	2·60
Br⁻	cal.	3·20	2·62	2·34	2·34	2·76
	obs.	3·16	2·62	2·33	2·33	2·78
I⁻	cal.	4·00	3·13	2·69	2·62	3·08
	obs.	3·80	2·91	2·69	2·63	3·03

Later when we come to examine the coefficient b_{12} (or b_{21}) we shall discover reason to believe that the second term in (9.17) has to be modified. In fact, we shall find that the ionic charge Ze appearing in all the coefficients should be replaced by an effective charge, which is not *a priori* known. Hence for the moment we shall consider the value of the first term in (9.17) alone by eliminating the second term as follows: Combining (7.7) with the expression (9.18) for b_{12}, we find that

$$\omega_0^2\left(\frac{\epsilon_0-\epsilon_\infty}{4\pi}\right) = b_{12}^2 = \frac{\dfrac{Z^2e^2}{\overline{M}v_a}}{\left[1-\dfrac{4\pi}{3}\left(\dfrac{\alpha_++\alpha_-}{v_a}\right)\right]^2}.$$

With its help, the expression (9.17) for b_{11} can be written as

$$b_{11} = -\frac{k}{\overline{M}}+\frac{4\pi}{3}\omega_0^2\left(\frac{\epsilon_0-\epsilon_\infty}{4\pi}\right)\left[1-\frac{4\pi}{3}\left(\frac{\alpha_++\alpha_-}{v_a}\right)\right].$$

When the atomic polarizabilities are eliminated with (9.20), we obtain the following expression for b_{11}:

$$b_{11} = -\frac{k}{\overline{M}}+\omega_0^2\left(\frac{\epsilon_0-\epsilon_\infty}{\epsilon_\infty+2}\right).$$

After using the relation (7.6) that $b_{11} = -\omega_0^2$, we can write the above relation as one determining k:

$$\frac{k}{\overline{M}} = \omega_0^2\left(\frac{\epsilon_0+2}{\epsilon_\infty+2}\right). \tag{9.21}$$

Now let us relate k explicitly to the overlap potentials between the ions. In a displacement of all the positive ions against the negative, the

† Empirical values taken from K. Højendahl, *K. Danske Vidensk. Selskab* **16**, No. 2 (1938).

distance between second neighbours which, we note, are always ions of the same kind, is not altered. Hence if overlap potentials between third and higher neighbours are ignored, we need only consider the overlap forces between nearest neighbours for this type of displacement. Thus in this approximation we may concentrate on a single positive ion, which we may consider as situated at the origin surrounded by its nearest neighbours $\mathbf{x}(i)$ $(i = 1, 2,..., M, M$ being as in § 3 the coordination number) when the lattice is in the equilibrium state. The value of k may then be found from the fact that if the positive ion is subject to a small displacement $\mathbf{x}$ while its neighbours are held fixed, the force acting on the positive ion is in this approximation given by

$$-k\mathbf{x}. \tag{9.22}$$

If we denote the overlap potential between a positive and a negative ion by ϕ, the potential of the positive ion at any position $\mathbf{x}$ due to its neighbours is given by

$$\sum_{i=1}^{M} \phi(|\mathbf{x}(i)-\mathbf{x}|).$$

The α-component of the force acting on the ion is thus

$$-\sum_{i=1}^{M} \frac{\partial}{\partial x_\alpha} \phi(|\mathbf{x}(i)-\mathbf{x}|). \tag{9.23}$$

For small $\mathbf{x}$ the expression can be expanded with respect to $\mathbf{x}$ and terms of second and higher orders may be neglected; hence the above force component becomes

$$-\sum_{i=1}^{M} \sum_{\beta=1}^{3} \left\{\frac{\partial^2}{\partial x_\alpha \partial x_\beta} \phi(|\mathbf{x}(i)-\mathbf{x}|)\right\}_{\mathbf{x}=0} x_\beta,$$

which can also be written as

$$-\sum_{i=1}^{M} \sum_{\beta=1}^{3} \left\{\frac{\partial^2}{\partial x_\alpha(i)\partial x_\beta(i)} \phi(|\mathbf{x}_i|)\right\} x_\beta \qquad \left(\frac{\partial}{\partial x_\alpha} = -\frac{\partial}{\partial x_\alpha(i)}\right). \tag{9.24}$$

It follows from the theorem already mentioned in connexion with the tetrahedral symmetry that

$$\sum_{i} \frac{\partial^2}{\partial x_\alpha(i)\partial x_\beta(i)} \phi(|\mathbf{x}_i|)$$

must be an isotropic tensor; i.e. the expression vanishes for $\alpha \neq \beta$ and has the same value for $\alpha = \beta = 1, 2, 3$. In fact comparing (9.24) with

(9.22) we see that the diagonal elements of the tensor must be equal to k; thus we may write

$$\sum_{i=1}^{M} \frac{\partial^2}{\partial x_\alpha(i) \partial x_\beta(i)} \phi(|\mathbf{x}(i)|) = k\delta_{\alpha\beta}.$$

Putting $\alpha = \beta$ and summing over α, we find that

$$\sum_{i=1}^{M} \sum_{\alpha=1}^{3} \frac{\partial^2}{\partial x_\alpha^2(i)} \phi(|\mathbf{x}(i)|) = 3k.$$

$\sum_{\alpha=1}^{3} (\partial^2/\partial x_\alpha^2(i))$ is the Laplacian operator in the coordinates $\mathbf{x}(i)$; since $\phi(|\mathbf{x}(i)|)$ is only a function of the radial magnitude $|\mathbf{x}(i)|$ we find immediately on writing the Laplacian operators in polar coordinates that the above relation reduces simply to

$$\sum_{i=1}^{M} \left\{ \phi''(|\mathbf{x}(i)|) + \frac{2}{|\mathbf{x}(i)|} \phi'(|\mathbf{x}(i)|) \right\} = 3k.$$

As all the negative ions are equidistant from the origin, $|\mathbf{x}(i)| = r_0$, r_0 being the nearest neighbour distance, all terms in the summation are equal, and we have

$$k = \frac{M}{3} \left\{ \phi''(r_0) + \frac{2}{r_0} \phi'(r_0) \right\}. \tag{9.25}$$

In § 3 we have seen that if only overlap potentials between nearest neighbours are considered, they can be determined from the equilibrium condition and the observed compressibility. Instead, however, of calculating k from the potentials thus determined to compare with (9.21), an equivalent and more direct procedure can be followed, namely, by expressing k directly in terms of the compressibility with the help of (9.25). To show that this is possible let us rewrite some of the equations used in § 3 with the overlap potential expressed as ϕ as we have done here. Thus the energy per cell (3.10) is rewritten as

$$u(r) = -\frac{A}{r} + M\phi(r), \tag{9.26}$$

where the first term, we remember, is the Madelung energy. The equilibrium condition becomes now

$$0 = \left(\frac{du}{dr}\right)_0 = \frac{A}{r_0^2} + M\phi'(r_0). \tag{9.27}$$

Moreover, remembering that $dv/v = 3\,dr/r$, we have for the compressibility β

$$\frac{1}{\beta} = v_a \left(\frac{d^2 u}{dv^2}\right)_0 = \left\{ \frac{r_0}{3} \frac{d}{dr} \left(\frac{r}{3v} \frac{du}{dr} \right) \right\}_0 = \frac{r_0^2}{9v_a} \left(\frac{d^2 u}{dr^2}\right)_0 = \frac{r_0^2}{9v_a} \left\{ -\frac{2A}{r_0^3} + M\phi''(r_0) \right\}, \tag{9.28}$$

where we have made use of the equilibrium condition $(du/dr)_0 = 0$; this relation clearly corresponds to (3.14) in § 3.

On eliminating the Madelung term in (9.28) with the help of (9.27) we find that

$$\frac{1}{\beta} = \frac{Mr_0^2}{9v_a}\left\{\phi''(r_0) + \frac{2\phi'(r_0)}{r_0}\right\}.$$

Comparing this relation with (9.25), we obtain the following simple relation between the constant k and compressibility β:

$$\frac{1}{\beta} = \frac{Mr_0^2}{3v_a}k. \tag{9.29}$$

Our intention is to test the validity of (9.21). In view of (9.29), we can do so by verifying the relation

$$\frac{1}{\beta} = \frac{M\bar{M}r_0^2}{3v_a}\left(\frac{\epsilon_0 + 2}{\epsilon_\infty + 2}\right)\omega_0^2, \tag{9.30}$$

which is obtained by eliminating k between (9.21) and (9.29). The relation contains only measurable quantities and relates essentially the compressibility and the dispersion frequency.† To indicate to what extent it is fulfilled in actual cases, the ratios of the value of β calculated from (9.30) to the observed value are given in Table 20 for a number of crystals. The values are very close to unity for practically all the lighter alkali halides. A variety of causes is probably responsible for the larger discrepancies in the other cases. We notice that we have completely ignored the van der Waals forces in the above considerations; and in deducing (9.29) we have also left out the overlap forces between second neighbours. Some genuine deviations from the ideal ionic structure must also account for the discrepancies in some cases, such as those near the bottom of the table. It is, however, difficult to ascertain the predominant cause of error in individual cases.

† Historically the first connexion between the infra-red frequency ω_0 and the compressibility β was derived, from simple dimensional considerations, independently by E. Madelung (*Gött. Nachr.* **48** (1910); *Phys. Zeit.* **11**, 898 (1910)), W. Sutherland (*Phil. Mag.* (6) **20**, 657 (1910)), and A. Einstein (*Ann. d. Phys.* (4) **34**, 170, 590; **35**, 679 (1911)), in the form $\lambda_0 = \mathscr{C}\beta\bar{M}^{\frac{1}{2}}\rho^{\frac{1}{2}}$, where $\lambda_0 = c/\omega_0$ is the wave-length corresponding to ω_0, $\bar{M}$ a certain mean mass of the ions, ρ the density, and $\mathscr{C}$ a constant, independent of the material. W. Dehlinger (*Phys. Zeit.* **15**, 276 (1914)) obtained the dispersion formula for infra-red light for a special lattice, and M. Born (*Berl. Ber.* 604 (1918); *Phys. Zeit.* **19**, 539 (1918); 'Atomtheorie des festen Zustandes', loc. cit. 626) incorporated this in the general lattice theory and established the connexion with the compressibility; in the formula thus obtained, $1/\beta = \mathscr{C}\bar{M}r_0^2\,\omega_0^2/v_a$, the order of magnitude of the constant $\mathscr{C}$ was determined, but no attempt was made to find its numerical value. The determination of $\mathscr{C}$ was the object of modern investigations. An expression similar to that given in (9.30) was derived by K. Højendahl (*K. Dansk. Vidensk. Selskab* **16**, No. 2 (1938)); the formula (9.30) itself was found by B. Szigeti (*Proc. Roy. Soc.* A, **204**, 52 (1950)).

To discuss the microscopic expression (9.18) for b_{12}, let us form the ratio of the directly observable value as given by (7.7) to (9.18):

$$s = \omega_0 \left(\frac{\epsilon_0 - \epsilon_\infty}{4\pi}\right)^{\frac{1}{2}} \left[1 - \frac{4\pi}{3}\left(\frac{\alpha_+ + \alpha_-}{v_a}\right)\right]\frac{(\bar{M}v_a)^{\frac{1}{2}}}{Ze}.$$

We can eliminate the atomic polarizabilities with (9.20) and write the ratio s as

$$s = \omega_0 \left(\frac{\epsilon_0 - \epsilon_\infty}{4\pi}\right)^{\frac{1}{2}}\left(\frac{3}{\epsilon_\infty + 2}\right)\frac{(\bar{M}v_a)^{\frac{1}{2}}}{Ze}. \qquad (9.31)$$

TABLE 20†

Compressibility calculated from the Dispersion Frequency and Distortion Factor s

	$\beta_{\text{calc}}/\beta_{\text{obs}}$	s
LiF	1·0	0·87
NaF	0·83	0·93
NaCl	0·99	0·74
NaBr	1·13	0·69
NaI	1·05	0·71
KCl	0·96	0·80
KBr	0·95	0·76
KI	0·99	0·69
RbCl	0·89	0·84
RbBr	0·83	0·82
RbI	0·66	0·89
CsCl	0·87	0·84
CsBr	0·87	0·79
TlCl	0·51	1·08
CuCl	0·85	1·10
CuBr	0·72	1·00
MgO	0·47	0·88
CaO	..	0·88
SrO	..	0·58
ZnS	0·98	0·48

The values of s for a number of crystals calculated on the basis of (9.31) are given in Table 20. Their deviations from the value unity are quite pronounced for the lighter alkali halides, which should approximate most closely the ideal ionic model. One cause for the deviations has been particularly emphasized by Szigeti,‡ namely, the mutual distortions of neighbouring ions owing to their overlap. This explanation of the discrepancy is favoured by the fact that it will not affect (9.19) or (9.30), which, as we have seen, are reasonably well fulfilled; this we shall see in due course. Moreover, it is *a priori* clear that some such distortion must

† Values are taken from B. Szigeti, *Proc. Roy. Soc.* A, **204**, 51 (1950).
‡ B. Szigeti, op. cit., p. 52 (1950).

accompany the overlap between ions, which is responsible for the overlap forces. Thus we shall investigate how such distortions would modify the microscopic expressions for the b-coefficients and in what manner they are related to the deviation of s from unity.

Since distortions are merely rearrangements of the charge distribution, the most important effect can be described by the dipole moment associated with the rearrangement. We shall consider only the distortions occurring between nearest neighbours. Let us denote by $m(r)$ the distortion dipole moment between a positive and a negative ion; $m(r)$ is a function of their separation r and the sign is chosen such that m is counted positive if the moment is directed from the negative towards the positive ion. Exactly as we have done for the overlap forces (p. 109), we may again consider a positive ion at the origin surrounded by its neighbours at $\mathbf{x}(i)$. When the positive ion is in an arbitrary displaced position $\mathbf{x}$, the total distortion dipole moment surrounding it can clearly be written as

$$\sum_{i=1}^{M} m(|\mathbf{x}(i)-\mathbf{x}|)\left[-\left(\frac{\mathbf{x}(i)-\mathbf{x}}{|\mathbf{x}(i)-\mathbf{x}|}\right)\right], \tag{9.32}$$

where the vectors $-\{(\mathbf{x}(i)-\mathbf{x})/(|\mathbf{x}(i)-\mathbf{x}|)\}$ are unit vectors indicating the directions of the various individual dipoles, which, according to our sign convention, are directed inwards towards the positive ion at $\mathbf{x}$. We can immediately reduce our problem to the same considerations as in the case of the overlap forces, if we introduce the integral function ψ of $-m(r)$, namely,

$$\psi'(r) = -m(r). \tag{9.33}$$

It is easily verified that the α-component of the distortion moment (9.32) can be written in terms of the function ψ as

$$-\sum_{i=1}^{M} \frac{\partial}{\partial x_\alpha} \psi(|\mathbf{x}(i)-\mathbf{x}|). \tag{9.34}$$

This is identical with the expression (9.23) for the total overlap force on the ion, apart from the replacement of ϕ by ψ. Therefore we can write the distortion moment as

$$-q\mathbf{x}, \tag{9.35}$$

in analogy with (9.22), where q is given by

$$q = \frac{M}{3}\left\{\psi''(r_0)+\frac{2\psi'(r_0)}{r_0}\right\} = -\frac{M}{3}\left\{m'(r_0)+\frac{2m(r_0)}{r_0}\right\}, \tag{9.36}$$

which is a direct transcription of (9.25) with ϕ now replaced by ψ. q is thus determined jointly by the distortion moment $m(r_0)$ already present in the equilibrium state between neighbours and its rate of change with the ion separation.

In deriving the equation for the dielectric polarization **P**, we have to consider now the further contribution due to the distortion dipoles. For the relative displacement $(\mathbf{u}_+ - \mathbf{u}_-)$, the total distortion dipole moment surrounding a positive ion is, according to (9.35),

$$-q(\mathbf{u}_+ - \mathbf{u}_-). \tag{9.37}$$

Since the component dipoles are actually shared between the ion and its neighbours, (9.37) gives directly the distortion dipole moment per cell. Hence incorporating this contribution, we should replace (9.7) by

$$\mathbf{P} = \frac{1}{v_a}\{(Ze - q)(\mathbf{u}_+ - \mathbf{u}_-) + (\alpha_+ + \alpha_-)\mathbf{E}_{\text{eff}}\}. \tag{9.38}$$

We notice that the only difference from the former equation is the replacement of Ze by $Ze - q$; the rest of the former treatment remains completely unaffected. So the net effect of the distortion on the equation for **P** is that Ze is replaced by the effective charge $Ze - q$ in the expression (9.10) for b_{21}.

As regards the derivation of the equation of motion (9.16), we note that the forces exerted on the ions by the field $\mathbf{E}_{\text{eff}}$ are no longer given by the product of the respective ionic charges by the field. The actual forces can be obtained by applying the principle of virtual work. When a positive ion is displaced by **x** in the presence of a field $\mathbf{E}_{\text{eff}}$, the virtual work involving the field is

$$-\mathbf{E}_{\text{eff}}.(Ze - q)\mathbf{x}.$$

The corresponding force is thus given by

$$(Ze - q)\mathbf{E}_{\text{eff}}.$$

When a negative ion is displaced, the same considerations as used in obtaining the distortion moment for the positive ion still apply, the only difference being the sign convention for $m(r)$; the distortion moment is thus the same as in the case of the positive ion but for a reversal of sign, i.e. $q\mathbf{x}$. It follows from a similar application of the principle of virtual work that the force exerted on a negative ion by the field $\mathbf{E}_{\text{eff}}$ is given by

$$\mathbf{E}_{\text{eff}}(-Ze + q).$$

Once more we find that the effect of distortion is equivalent to the replacement of the Ze by the effective charge $Ze - q$. Apart from this change, the details of the derivation of (9.16) remain completely unaltered. The net effect on the coefficients b_{11} and b_{12} consists thus only in the replacement of Ze by $Ze - q$.

Let us now assume that this effect alone is responsible for the deviation of the ratio s from unity. Under this assumption, when Ze is replaced by $Ze-q$ in (9.18), the formula should agree exactly with (7.7), i.e. (see p. 108)

$$\frac{\dfrac{(Ze-q)}{(\bar{M}v_a)^{\frac{1}{2}}}}{1-\dfrac{4\pi}{3}\left(\dfrac{\alpha_++\alpha_-}{v_a}\right)} = \left(\frac{\epsilon_0-\epsilon_\infty}{4\pi}\right)^{\frac{1}{2}}\omega_0.$$

On eliminating the atomic polarizabilities with (9.20) and comparing the resulting formula with (9.31), we find that

$$q = Ze(1-s). \tag{9.39}$$

Using (9.36), we find the following relation between $s-1$ and the distortion moment $m(r)$:

$$(s-1)Ze = \frac{M}{3}\left\{m'(r_0)+\frac{2m(r_0)}{r_0}\right\}. \tag{9.40}$$

It is important to note that $m(r)$ refers only to the deformation due to overlap; the deformations due to the Coulomb interactions are in the above model taken into account by the atomic polarization. Thus $m(r)$ vanishes for r sufficiently large to correspond to separated ions. Hence we have

$$\int_{r_0}^{\infty} m'(r)\,dr = m(r)\,\Big|_{r_0}^{\infty} = -m(r_0). \tag{9.41}$$

There appears to be no reason to suppose that $m'(r)$ should change sign in the range r_0 to ∞, hence it follows from (9.41) that the two terms in the bracket of (9.40) are opposed in sign. However, it is reasonable to assume that the distortion should vary roughly in the same manner as the overlap forces, for instance, as $e^{-r/\rho}$, where ρ is of the order of a tenth of r_0. In that case, the $m'(r_0)$ term in (9.40) predominates and has thus the same sign as $s-1$ ($m(r_0)$ has the opposite sign). Now Table 20 shows that for all the alkali halides $s-1$ is negative, therefore $m(r_0)$ is positive and $m'(r_0)$ is negative. Thus from the sign convention of $m(r)$ follows the interpretation that in the equilibrium configuration the electron cloud of the negative ion is more strongly pushed back than that of the positive ion, resulting in a dipole directed towards the positive ion (positive $m(r_0)$). The negative sign of $m'(r_0)$ merely means that the magnitude of this dipole increases with decreasing r, or increasing overlap. This is understandable, as the electron cloud of the negative ion is very loosely held compared with that of the positive ion, so that the former will be pressed out of the overlap region more readily than the latter. If, however, there is any degree of valence binding, no such simple pictures are available; this might explain the rather irregular values for s for the crystals given near the bottom of Table 20.

In the above model, two approximations have been made in connexion with the polarization of the ions. In the first place, $\mathbf{E}_{\mathrm{eff}}$ has effectively been used as a uniform electric field over an ion; in the second place, the Coulomb field due to a polarized ion has been taken as that of a suitable point-dipole. Both approximations are unsatisfactory in describing the mutual electric effect of closely situated ions. Thus, instead of considering the distortion due to the overlap between ions, several authors† have attempted to modify the method by introducing an additional factor to the $4\pi \mathbf{P}/3$ term in the effective field. Even *a priori* this does not seem to be a reasonable procedure. For if the dipole approximation breaks down owing to the electric interaction between closely situated ions, there is no reason to suppose that the actual effect should still be proportional to the value given by the dipole approximation. In fact the factor introduced has to be assigned very different values, depending on which of the three coefficients b_{11}, b_{12}, or b_{22} is to be fitted. The reasonable fulfilment of (9.19) and (9.30) seems to indicate that the model as presented here is not very far from the truth.

10. Experimental aspect of infra-red dispersion by ionic crystals

By fitting the measured values of the refractive index in the wavelength range $\lambda = 1,800\,\text{Å}$ to $22\,\mu$ ($1\,\mu = 10^4\,\text{Å} = 10^{-4}\,\text{cm.}$), Fuchs and Wolf‡ have constructed the following empirical dispersion formula for the NaCl and KCl crystals:

$$n^2 = \epsilon = 1 + \sum_i \frac{C_i}{\nu_i^2 - \nu^2}, \qquad (10.1)$$

where the constants are ($\lambda_i = c/\nu_i$)

$$
\begin{aligned}
C_1 &= 3\!\cdot\!90 \times 10^{30}, & \lambda_1 &= 347\,\text{Å}, \\
C_2 &= 7\!\cdot\!68 \times 10^{30}, & \lambda_2 &= 1,085\,\text{Å}, \\
C_3 &= 0\!\cdot\!972 \times 10^{30}, & \lambda_3 &= 1,584\,\text{Å}, \\
C_4 &= 8\!\cdot\!37 \times 10^{25} & \lambda_4 &= 61\!\cdot\!67\,\mu
\end{aligned}
$$

for NaCl, and

$$
\begin{aligned}
C_1 &= 5\!\cdot\!066 \times 10^{30}, & \lambda_1 &= 529\!\cdot\!1\,\text{Å}, \\
C_2 &= 6\!\cdot\!044 \times 10^{30}, & \lambda_2 &= 1,082\!\cdot\!8\,\text{Å}, \\
C_3 &= 0\!\cdot\!7883 \times 10^{30}, & \lambda_3 &= 1,621\!\cdot\!4\,\text{Å}, \\
C_4 &= 4\!\cdot\!765 \times 10^{25} & \lambda_4 &= 70\!\cdot\!23\,\mu
\end{aligned}
$$

† G. Heckmann, *Zeit. f. Kristal.* **61**, 254 (1925); K. Højendahl, *K. Danske Vidensk. Selskab* **16**, No. 2 (1938); N. F. Mott and R. W. Gurney, *Electronic Processes in Ionic Crystals* (Clarendon Press, Oxford, 1940), pp. 16–25.

‡ O. Fuchs and K. L. Wolf, *Zeit. f. Phys.* **46**, 506 (1928).

for KCl. In the above formula, v_1, v_2, v_3 are absorption frequencies related to the excitation of electrons; their values are very high compared with the frequencies of infra-red radiations. The corresponding terms in the formula thus do not vary appreciably with the frequency v in the infra-red region and, for our purpose, may be replaced by their limiting values for $v = 0$. The formula then reduces to exactly the form of the infra-red dispersion formula (7.5) derived in § 7. Remembering that now $2\pi v_4 = \omega_0$, we readily find from the above constants the following values for ϵ_0, ϵ_∞, and ω_0:

$$\epsilon_0 = 5\cdot81, \qquad \epsilon_\infty = 2\cdot33, \qquad \omega_0 = 3\cdot05 \times 10^{13}/\text{sec.} \quad \text{(NaCl)}$$

$$\epsilon_0 = 4\cdot78, \qquad \epsilon_\infty = 2\cdot18, \qquad \omega_0 = 2\cdot68 \times 10^{13}/\text{sec.} \quad \text{(KCl)}$$

which are to be compared with the directly measured values quoted in Table 17, p. 85. As the constants in Fuchs and Wolf's formula are determined on the basis of data far removed to the high-frequency side of the dispersion frequency ω_0, the agreement with the directly measured values is indeed very fair.

The best-known phenomenon in connexion with the infra-red dispersion by ionic crystals is the selective reflection of radiations with frequencies in the neighbourhood of the dispersion frequency. This fact forms the basis of the important residual ray (Reststrahlen) method developed by Rubens[†] to obtain practically homogeneous infra-red radiations, whereby an initially inhomogeneous beam is repeatedly reflected from crystal plates so that in the residual beam only the narrow range of the most strongly reflected frequencies remains.

For an incident beam perpendicular to the surface of an optically isotropic medium, the fraction R reflected (*reflective power*) is given by the well-known formula:[‡]

$$R = \left| \frac{n-1}{n+1} \right|^2, \tag{10.2}$$

where n is the refractive index (whether real or complex, see later) and the vertical bars indicate the absolute magnitude. Now let us examine, on the basis of the dispersion formula (7.5), how R should vary with the frequency. According to (7.5), as ω is raised from a zero value to the dispersion frequency ω_0, the refractive index $n = \sqrt{\epsilon}$ rises from $\sqrt{\epsilon_0}$ to approach an infinite value. It thus follows from (10.2) that the reflection

[†] For references see C. Schaeffer and F. Matossi, *Das Ultrarote Spektrum* (Springer, Berlin, 1930).

[‡] See M. Born, *Optik*, p. 265 (Springer, Berlin, 1933).

should become perfect ($R = 1$) when ω_0 is reached. Immediately ω_0 is exceeded, ϵ switches from $+\infty$ to $-\infty$ and continues to have a negative value on further increase of ω from ω_0 to the value of ω satisfying the equation

$$0 = \epsilon(\omega) = \epsilon_\infty + \frac{\epsilon_0 - \epsilon_\infty}{1 - (\omega/\omega_0)^2}. \tag{10.3}$$

Within this range, the refractive index $n = \sqrt{\epsilon}$ is purely imaginary and, as follows from (10.2), there should be complete reflection. The solution of (10.3), we note, is exactly the longitudinal lattice vibration frequency $\omega_l = \omega_0(\epsilon_0/\epsilon_\infty)^{\frac{1}{2}}$. Thus, according to the dispersion formula (7.5), there is a band of perfect reflection located between the frequencies ω_0 and ω_l. Above ω_l, R decreases once more and approaches eventually the limiting value $(\sqrt{\epsilon_\infty}-1)^2/(\sqrt{\epsilon_\infty}+1)^2$.

The dispersion formula (7.5) thus offers a natural explanation of the observed selective reflection in the neighbourhood of the dispersion frequency. The quantitative features predicted by the formula are, however, by no means correct; the observed reflection is, for instance, at no frequency actually perfect. In Fig. 19, the percentage reflections by LiF, NaF, NaCl, and KCl, as calculated from the dispersion formula (in dotted lines) using the constants given in Table 17, are compared with the experimental curves (solid lines) given by Czerny[†] and Hohls.[‡] In all the cases shown, the region within which the observed reflection is large agrees well with the position of the theoretical band of perfect reflection; within this region, however, there is no agreement between theory and experiment.

This discrepancy is not confined to the reflective power alone. The corresponding effect is even more manifest in transmission experiments. For frequencies within the theoretical reflection band, what remains of the incident beam, after the strong reflection, is found to be strongly absorbed by the crystal, whereas the dispersion formula (7.5) provides for no such absorption. Whenever we have a *real* dielectric constant, as in (7.5), the dielectric polarization $\mathbf{P}$ is in phase with the field. The work done by the electromagnetic field on the medium is given by $(\mathbf{E}.\dot{\mathbf{P}})/4\pi$ per unit volume per unit time. For a sinusoidal variation with time, it is easily seen that $\dot{\mathbf{P}}$ is $\pi/2$ out of phase with $\mathbf{P}$, thus also with $\mathbf{E}$. It follows immediately on integrating $\mathbf{E}.\dot{\mathbf{P}}$ that the net energy loss by the electromagnetic field vanishes over a period.

Thus we have to conclude that although (7.5) is correct for frequencies

† M. Czerny, *Zeit. f. Phys.* **65**, 600 (1930).
‡ H. W. Hohls, *Ann. d. Phys.* **29**, 433 (1937).

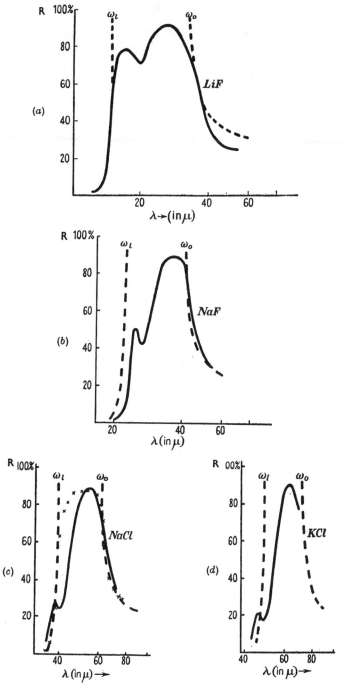

FIG. 19 a–d. Selective reflection by ionic crystals. Dotted lines = theoretical curves based on (7.5); solid lines = experimental curves; crosses in the NaCl case represent values calculated from the dispersion formula with damping (10.6).

removed from ω_0 (as shown, for instance, by Fuchs and Wolf's result), it fails in the neighbourhood of ω_0. The cause for this has already been indicated in § 8: the phenomenological equations (7.1), (7.2), from which the dispersion formula is deduced, are approximate in that we have ignored all but the linear terms. In ordinary circumstances, this is a perfectly legitimate approximation; however, the optical experiments in the neighbourhood of ω_0 are, so to speak, inordinately sensitive to the slight inaccuracy occasioned by the approximation. The linearity of the equations (7.1) and (7.2) is responsible for the result that different lattice waves are mutually independent. In reality they are coupled by the small higher-order terms, which have been ignored. Owing to the coupling the energy of a particular lattice wave slowly diffuses into the numerous other lattice vibrations and appears eventually as heat. The magnified importance of this small effect on optical waves near ω_0 is, as we have seen in § 8, due to the circumstance that the energy density in such waves is lodged predominantly in the lattice particles as mechanical energy of oscillation, rather than in the electromagnetic field. Thus the energy of the lattice particles is abnormally high relative to the electro-magnetic energy flux. In a steady state, whatever energy is lost by the lattice vibration must be supplied by the electromagnetic energy flux. Under the circumstances, a small percentage energy loss by the former through the small coupling drastically reduces the latter. This means, of course, a strong absorption of the optical wave by the medium.

A detailed discussion of the dispersion in the neighbourhood of the dispersion frequency is very complex, for we have to consider the bulk of all the lattice vibrations, which are responsible for the energy dissipa-tion. We shall return to this problem in the last chapter.

For the purpose of analysing the empirical data in the neighbourhood of ω_0, it proves convenient to use a dispersion formula which takes account of the energy dissipation in an *ad hoc* way. Namely, we introduce in the equation of motion (7.1) a simple damping term as follows:

$$\ddot{\mathbf{w}} = b_{11}\mathbf{w} - \gamma\dot{\mathbf{w}} + b_{12}\mathbf{E}, \qquad\qquad (10.4)$$

where γ is a positive constant with the dimension of frequency; the additional term represents a force always opposed to the motion. For complex periodic solutions of the type considered in §§ 7 and 8, (10.4) reduces to

$$-\omega^2\mathbf{w} = (b_{11} + i\omega\gamma)\mathbf{w} + b_{12}\mathbf{E}. \qquad\qquad (10.5)$$

The addition of the damping term is thus equivalent to the replacement of b_{11} by $b_{11}+i\omega\gamma$. Hence the dispersion formula (7.4) now becomes

$$\epsilon(\omega) = 1+4\pi b_{22}+\frac{4\pi b_{12}b_{21}}{-b_{11}-i\omega\gamma-\omega^2} = \epsilon_\infty+\frac{\epsilon_0-\epsilon_\infty}{1-(\omega/\omega_0)^2-i(\omega/\omega_0)(\gamma/\omega_0)};$$
(10.6)

and, for plane optical waves, we have now in place of (8.23) that k^2c^2/ω^2 is equal to the above expression.

For a plane wave travelling in the x-direction, the phase factor is given by

$$e^{i(kx-\omega t)} = e^{i\omega(kx/\omega-t)} = e^{i\omega[\sqrt{\epsilon(\omega)}x/c-t]}.$$
(10.7)

Exactly as for non-absorbing media (real dielectric constant), we define $\sqrt{\epsilon}$ as the refractive index, which we denote by $\bar{n}$:

$$\bar{n} = n(1+i\kappa) = \sqrt{\epsilon}.$$
(10.8)

$\bar{n}$ is now complex, n, $n\kappa$ being its real and imaginary parts. (In the literature the refractive index is sometimes written as $n(1-i\kappa)$; the alternative choices depend on the convention as regards ϵ, i.e. whether it goes with the part $\exp(-i\omega t)$ (as here) or with the part $\exp(i\omega t)$. The meanings of n and κ remain, however, the same for either of these conventions.) Squaring (10.8) and comparing with the dispersion formula (10.6) we see that

$$n^2(1-\kappa^2) = \epsilon_\infty+\frac{(\epsilon_0-\epsilon_\infty)[1-(\omega/\omega_0)^2]}{[1-(\omega/\omega_0)^2]^2+(\omega/\omega_0)^2(\gamma/\omega_0)^2},$$
(10.9)

$$2n^2\kappa = \frac{(\epsilon_0-\epsilon_\infty)(\gamma/\omega_0)(\omega/\omega_0)}{[1-(\omega/\omega_0)^2]^2+(\omega/\omega_0)^2(\gamma/\omega_0)^2}.$$
(10.10)

The definition (10.8) for $\bar{n}$ leaves its sign still ambiguous; thus a simultaneous reversal of the signs of n and $n\kappa$ does not affect (10.9) and (10.10). (10.10) shows, however, that n and $n\kappa$ must have the same sign (for $\epsilon_0 > \epsilon_\infty$); the usual convention is to choose the sign of $\bar{n}$ such that both n and κ are positive.

Writing now $\sqrt{\epsilon}$ in (10.7) as $n(1+i\kappa)$, we have

$$e^{i\omega(nx/c-t)} \times e^{-(n\kappa\omega/c)x}.$$

With the above sign convention, the first factor represents a progressive wave in the x-direction, the phase velocity being c/n. The second factor represents an exponential diminution of the amplitude in the direction of propagation. The fractional decrement of this factor in a length dx is clearly

$$2\pi\kappa n\left(\frac{dx}{2\pi c/\omega}\right).$$

As $2\pi c/\omega$ is the vacuum wave-length, $2\pi n\kappa$ gives thus the fractional decrement of amplitude per unit vacuum wave-length. n and κ are known as the *optical constants of an absorbing medium*.

Fig. 20 shows the typical behaviour of the optical constants. The curves have been computed with the numerical values of ϵ_0, ϵ_∞, γ/ω_0 for the NaCl crystal; the value 0·045 for (γ/ω_0) has been chosen so as to reproduce as closely as possible the experimental curve for the reflective power.

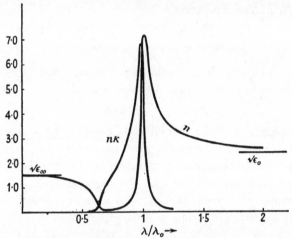

FIG. 20. Optical constants for NaCl ($\gamma/\omega_0 = 0\cdot045$).

Replacing the refractive index in (10.2) by $\bar{n} = n(1+i\kappa)$, we find that the reflective power is given in terms of the optical constants by

$$R = \frac{(n-1)^2 + n^2\kappa^2}{(n+1)^2 + n^2\kappa^2}. \tag{10.11}$$

The crosses in Fig. 19 c represent the values calculated from this formula for NaCl when the value $\gamma/\omega_0 = 0\cdot045$ is used. The general agreement with the experimental curve is seen to be quite satisfactory; but the theoretical values are too high on the immediate short-wave side of the peak and do not reproduce the secondary maximum, which is very distinct in all the experimental curves in Fig. 19.

It is evident from Fig. 19 that the frequency ω_m at which the maximum reflection occurs does not coincide with the dispersion frequency ω_0 (= long-wave end of the theoretical reflection band). Using the optical constants based on the dispersion formula (10.6), Havelock[†] has shown that if γ/ω_0 is small, the ratio ω_m/ω_0 is approximately independent of the damping constant γ and given by

$$\frac{\omega_m}{\omega_0} = \left(1 + \frac{\epsilon_0 - \epsilon_\infty}{6\epsilon_\infty - 4}\right)^{\frac{1}{2}} = \frac{\lambda_0}{\lambda_m} \qquad \left(\lambda_0 = \frac{2\pi c}{\omega_0}, \lambda_m = \frac{2\pi c}{\omega_m}\right). \tag{10.12}$$

† T. H. Havelock, *Proc. Roy. Soc.* A, **105**, 488 (1924).

Under ideal conditions the measured residual ray frequency should be equal to ω_m. The measured residual ray wave-lengths for a number of crystals are collected in Table 21 as λ_m. In the next column are given the corresponding values for λ_0 calculated with the help of Havelock's formula and the values for ϵ_0, ϵ_∞ quoted in Table 17. The agreement with the directly measured values of λ_0, which are given in the last column, is on the whole very satisfactory. The only large deviation occurs in the case of TlCl. In this case the conversion factor deviates considerably from unity; presumably Havelock's formula is no longer a good approximation.

TABLE 21

Determination of λ_0 from Residual Rays

	λ_m†	λ_0 (from (10.12))	λ_0 (measured, see Table 17)
NaCl	52 μ	60 μ	61·1 μ
KCl	63·4	72	70·7
KBr	82·6	92	88·3
KI	74·0	103	102
RbCl	74·0	85	84·8
TlCl	91·5	129	117
TlBr	117	162	..
ZnS	30·9	33	33
AgCl	81·5	97	..
AgBr	112·7	131	..

Apart from the complication involved in converting the residual ray frequency to the dispersion frequency, the residual ray frequency may at times differ appreciably from the frequency ω_m for maximum reflective power. Owing to the rather flat nature of the reflection maximum (see Fig. 19), it has been found that both the spectral distribution of the initial beam and the absorption by moisture could modify the mean frequency of the residual rays. A more desirable way of determining the dispersion frequency is by transmission experiments. Such experiments are, however, more difficult to perform, for the strong absorption near ω_0 requires the use of very thin specimens of the crystal. By the use of evaporated films with a thickness of the order 0·1 to 1·0 μ, Barnes and Czerny‡ were the first to succeed in making accurate transmission experiments throughout the absorption region for the NaCl and KCl crystals. For sufficiently thin specimens, as we shall see, the results of such experiments give directly the value of ω_0.

† From C. Schaeffer and F. Matossi, *Das Ultrarote Spektrum*, p. 306 (Springer, Berlin, 1930).

‡ R. B. Barnes and M. Czerny, *Zeit. f. Phys.* **72**, 447 (1931).

To obtain the expression for the transmission coefficient, let us consider linearly polarized light of a given frequency ω incident normally on a plate of thickness d. If x is the direction of incidence, the scalar magnitude of the electric field (the part going with $e^{-i\omega t}$) can be written as

$$
\left.
\begin{array}{ll}
\text{incident side} & (Ae^{i\omega x/c}+Be^{-i\omega x/c})e^{-i\omega t} \\
\text{within plate} & (Ce^{i\omega\bar{n}x/c}+De^{-i\omega\bar{n}x/c})e^{-i\omega t} \\
\text{far side} & Ee^{i\omega(x/c-t)}
\end{array}
\right\}. \tag{10.13}
$$

Part of the wave travels in the negative x-direction both on the incident side and within the plate owing to the reflections at the two boundaries, while on the far side only a wave in the x-direction emerges. The corresponding magnetic fields are given in (10.14) (cf. (8.21), where k should now be replaced by $(\omega/c)\times$ refractive index, i.e. (ω/c) outside the plate and $(\omega\bar{n}/c)$ within):

$$
\left.
\begin{array}{ll}
\text{incident side} & (Ae^{i\omega x/c}+Be^{-i\omega x/c})e^{-i\omega t} \\
\text{within plate} & \bar{n}(Ce^{i\omega\bar{n}x/c}+De^{-i\omega\bar{n}x/c})e^{-i\omega t} \\
\text{far side} & Ee^{i\omega[(x/c)-t]}
\end{array}
\right\}. \tag{10.14}
$$

The requirement that the electric and magnetic fields should all be continuous across the two boundaries gives four linear homogeneous equations in the coefficients A, B, C, D, E, from which one obtains quite readily that

$$
\frac{E}{A} = \frac{4\bar{n}}{(1+\bar{n})^2 e^{-i\omega\bar{n}d/c}-(1-\bar{n})^2 e^{i\omega\bar{n}d/c}}. \tag{10.15}
$$

For very thin plates such that $2\pi d \ll \lambda = 2\pi c/\omega$, we can expand the exponential functions in the denominator:

$$
\frac{E}{A} = \frac{1}{1-(i\omega d/2c)(1+\bar{n}^2)+\ldots}. \tag{10.16}
$$

The radiation intensity is proportional to the square of the electric (or magnetic) field; thus the transmission coefficient, which is defined as the intensity ratio of the emerging to the incident radiation, is given by

$$
D = \left|\frac{E}{A}\right|^2 = \frac{1}{1+(i\omega d/2c)(\bar{n}^{*2}-\bar{n}^2)+\ldots}. \tag{10.17}
$$

Hence for very thin plates the minimum transmission occurs at the frequency for which

$$
i\omega(\bar{n}^{*2}-\bar{n}^2) = i\omega(\epsilon^*-\epsilon)
$$

is a maximum. When the value for ϵ given by (10.6) is substituted, the condition for maximum becomes

$$0 = \frac{d}{d\omega} \frac{\omega^2}{[\omega_0^2 - \omega^2]^2 + \gamma^2 \omega^2} = \frac{2\omega(\omega_0^2 + \omega^2)(\omega_0^2 - \omega^2)}{\{[\omega_0^2 - \omega^2]^2 + \gamma^2 \omega^2\}^2}. \qquad (10.18)$$

The solution is clearly $\omega = \omega_0$, showing that as $d \to 0$, the minimum transmission occurs exactly at the dispersion frequency. For films with a thickness of the order $0 \cdot 1$ to $1 \cdot 0$ μ, it is found experimentally that the

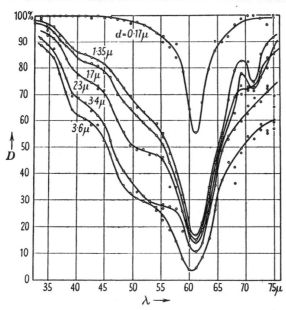

Fig. 21. Transmission curves for NaCl (Barnes and Czerny) for films of various thickness.

transmission minimum does not any longer shift appreciably with the thickness, giving thus directly the value of ω_0. In Fig. 21 are reproduced the transmission curves for NaCl obtained by Barnes and Czerny[†] for films of various thicknesses. The values for ω_0 quoted in Table 17 were obtained by transmission experiments by Barnes[‡] and Parodi.[§]

As shown by Figs. 19 and 21, both the transmission and the reflection reveal secondary variations in the neighbourhood of ω_0, which are not to be found in the dispersion formula (10.6). Part of the variations in the transmission curves is, however, only apparent; namely, they have been shown to be due to interference effects in accordance with (10.17),

† R. B. Barnes and M. Czerny, loc. cit.
‡ R. B. Barnes, Zeit. f. Phys. 75, 723 (1932).
§ M. Parodi, C.R. Acad. Sci. (Paris), 204, 1111, 1636; 205, 906, 1224 (1937); 206, 1717 (1938).

when the finite thickness d is taken into account. This is, for instance, the case with the violent variations on the long wave-length side of the curves; the irregularities on the short-wave side, however, correspond to genuine secondary variations in the optical constants. As shown in Fig. 22, these secondary variations are revealed as genuine minima in transmission, when the effect of the finite slit width used in the experiment is corrected for. Similar secondary structures in the dispersion

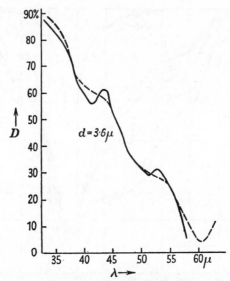

Fig. 22. Dotted curve as in Fig. 21; solid curve after correction for interference.

have been observed in all the alkali halides measured by Barnes[†] and in many oxides by Parodi[‡] and Tolksdorf.[§] Of the many oxides, MgO in particular has been subject to careful experiments by Strong, by Fock, and by Barnes, Brattain, Seitz, and Wilmot.[||] Their measurements show the existence of a strong secondary transmission minimum on the long-wave side ($\sim 24\,\mu$) of the principal minimum ($\sim 17\cdot3\,\mu$). Less distinct but still unmistakable maxima in the optical constant $n\kappa$ on the long-wave side of λ_0 have also been observed by Cartwright and Czerny[††] in NaCl and KCl. Barnes and collaborators[‡‡] have further reported

† R. B. Barnes, loc. cit.   ‡ M. Parodi, loc. cit.
§ S. Tolksdorf, *Zeit. f. Phys. Chem.* **132**, 161 (1928).
|| J. Strong, *Phys. Rev.* **37**, 72; **38**, 1565 (1931); J. Fock, *Zeit. f. Phys.* **90**, 44 (1934);
R. B. Barnes, R. Brattain, and F. Seitz, *Phys. Rev.* **48**, 582 (1935); J. C. Wilmot, *Proc. Phys. Soc.* **63**, 389 (1950).
†† C. H. Cartwright and M. Czerny, *Zeit. f. Phys.* **90**, 7, 457 (1934).
‡‡ R. B. Barnes, R. Brattain, and F. Seitz, loc. cit.

that on higher resolution many fine peaks are revealed in the transmission curves of MgO (nearly 40 in number); their results do not appear, however, to be confirmed by other investigators.

Apart from the above characteristic deviations from the dispersion formula (10.6), Czerny and Mentzel[†] have also observed that the value of $n\kappa$ begins, somewhere removed to the short-wave side of λ_0, to fall away much more rapidly than is to be expected from the dispersion formula (10.6).

As (10.6) is only an *ad hoc* formula designed to reproduce energy absorption in the simplest manner, these deviations in details are hardly surprising. When we develop the dispersion theory more properly by taking into account the coupling between the lattice vibration modes, we shall see that the presence of secondary structures in the optical constants near the dispersion frequency is only to be expected. Owing, however, to the excessive computational work that will be required, no satisfactory theoretical results are as yet available for a quantitative comparison with the experiments.

TABLE 22

Inner vibrations of CO_3^{--}[‡]

$NaCl.MgCO_3.Na_2CO_3$	$6.82\,\mu$	$11.46\,\mu$	$14.14\,\mu$
$MgCO_3$	6.69	11.25	13.78
$CaCO_3$	6.56	11.38	14.16
$MnCO_3$	6.76	11.38	14.04
$FeCO_3$ uniaxial	6.77	11.53	13.54
$ZnCO_3$	6.78	11.44	13.92
$(Ca, Mg)CO_3$	6.90	11.45	14.70
$CaCO_3$	6.64	11.54	14.40
$SrCO_3$	6.78	11.62	14.28
$BaCO_3$ biaxial	6.85	11.61	14.48
$PbCO_3$	7.06	12.00	14.92 and 15.2
Na_2CO_3	7.00	12.48	14.65

In more complex crystals the principal difference is the presence of a number of dispersion frequencies, and a certain dependence on the polarization due to anisotropy of the crystal. In chemical compounds, some particles in the cell may form chemical radicals such as CO_3^{--}, SO_3^{--}, etc.; such particles are more tightly bound to one another than to the lattice. Therefore some branches of the lattice vibrations reduce essentially to

† M. Czerny, *Zeit. f. Phys.* **65**, 600 (1930); A. Mentzel, ibid. **88**, 178 (1934).

‡ From C. Schaeffer and F. Matossi, *Das Ultrarote Spektrum*, Table 54 (Springer, Berlin, 1930).

the internal vibrations within the radicals; the absorption lines corresponding to such vibrations often persist even in solution. On the other hand, the lattice type of vibration is due to relative motion between the radicals and the metal ions; the corresponding absorption lines disappear on the break-up of the lattice structure. Owing to the stronger binding forces within the chemical radicals, the corresponding absorption lines lie on the long-wave side of those of the lattice type, and are readily distinguished by their characteristic frequencies, which are not much affected by the binding to the lattice. In Table 22 are given the three reflection maxima describing mainly the internal vibrations of the radical CO_3^{--} in various crystals; their differences in different crystals are seen to be relatively slight.

ELASTICITY AND STABILITY

11. Homogeneous deformation and the elastic constants

IF a lattice is deformed such that the resulting structure remains a perfect lattice, the deformation is *homogeneous*.

A homogeneous deformation can be built up as follows. First we subject the coordinates of all particles in the lattice to a linear homogeneous transformation: namely, a particle initially at $\mathbf{x}$ is moved to $\mathbf{x}'$, where

$$x'_\alpha = x_\alpha + \sum_\beta u_{\alpha\beta} x_\beta \quad (\alpha, \beta = 1, 2, 3), \tag{11.1}$$

$u_{\alpha\beta}$ being constants, which we shall call the deformation parameters. We have written the linear transformation in the above form so that the second term represents the displacement of the particle from $\mathbf{x}$ to $\mathbf{x}'$. If the lattice is composite, with n particles per cell labelled by

$$k = 1, 2,..., n,$$

we may further displace all particles of each type by the same vector $\mathbf{u}(k)$. A particle $\binom{l}{k}$ (kth particle in l-cell) is thus subject to the total displacement

$$u_\alpha\binom{l}{k} = u_\alpha(k) + \sum_\beta u_{\alpha\beta} x_\beta\binom{l}{k}, \tag{11.2}$$

where $\mathbf{x}\binom{l}{k}$ denotes the initial position vector of $\binom{l}{k}$. Evidently $\mathbf{x}\binom{l}{k}$ may be decomposed as follows:

$$\mathbf{x}\binom{l}{k} = \mathbf{x}(l) + \mathbf{x}(k), \tag{11.3}$$

where $\mathbf{x}(l)$ is the simple Bravais lattice vector:

$$\mathbf{x}(l) = l^1\mathbf{a}_1 + l^2\mathbf{a}_2 + l^3\mathbf{a}_3 \tag{11.4}$$

and $\mathbf{x}(k)$ is the position vector of the kth particle in the zero-cell ($l = 0$). The decomposition (11.3) will be useful later on.

A general homogeneous deformation of a composite lattice is thus characterized by the n vectors $\mathbf{u}(k)$ and the tensor $\mathbf{u}_{\alpha\beta}$. We are interested in the case when the vectors $\mathbf{u}(k)$ are small compared with the interatomic distances and the (dimensionless) tensor components $u_{\alpha\beta}$ small compared with unity. In this case, as we shall presently see, the energy density can be expressed as a Taylor expansion and the terms of the second order determine completely the elastic properties.

We shall assume that the lattice particles interact with central forces (the general case will be considered in Part II). Mathematically it is simpler to regard the potential energy between two particles as a function of the square of their separation; a potential function regarded in this way will be denoted by $\psi(r^2)$, while $\phi(r)$ will be retained to denote as before the potential energy as a function of the distance r. Consider the displacement of two particles such that the position vector of one relative to the other is changed from $\mathbf{x}$ to $\mathbf{x}+\Delta\mathbf{x}$; the change in the square of their separation is given by

$$|\mathbf{x}+\Delta\mathbf{x}|^2-|\mathbf{x}|^2 = 2\sum_\alpha x_\alpha \Delta x_\alpha + \sum_\alpha (\Delta x_\alpha)^2.$$

For $\Delta\mathbf{x} \ll \mathbf{x}$, the corresponding change in the potential energy can be expressed as a Taylor series in ascending powers of the components of $\Delta\mathbf{x}$:

$$\psi'(|\mathbf{x}|^2)\Big[2\sum_\alpha x_\alpha \Delta x_\alpha + \sum_\alpha (\Delta x_\alpha)^2\Big]+2\psi''(|\mathbf{x}|^2)\Big[\sum_\alpha x_\alpha \Delta x_\alpha\Big]^2. \quad (11.5)$$

For the present discussion, terms of the third and higher orders can be ignored.

Since the lattice structure remains perfect in a homogeneous deformation, all cells in the lattice are mutually equivalent. Accordingly the energy change per cell may be calculated directly; namely, we consider the change in the potential energy between $\binom{0}{k}$ and all other lattice particles, and then sum $\binom{0}{k}$ over the zero-cell. Thus as the position vector of $\binom{l'}{k'}$ relative to $\binom{0}{k}$ is changed from

$$\mathbf{x}\binom{l'}{k'}-\mathbf{x}\binom{0}{k}$$

to $$\mathbf{x}\binom{l'}{k'}-\mathbf{x}\binom{0}{k}+\mathbf{u}\binom{l'}{k'}-\mathbf{u}\binom{0}{k},$$

we find readily with the help of (11.5) that the corresponding energy change per cell is

$$\tfrac{1}{2}\sum_k \sum_{l'k'} \Big\{ \psi'_{kk'}(|\mathbf{x}\binom{l'}{k'}-\mathbf{x}\binom{0}{k}|^2)\Big[2\sum_\alpha (x_\alpha\binom{l'}{k'}-x_\alpha\binom{0}{k})(u_\alpha\binom{l'}{k'}-u_\alpha\binom{0}{k})+$$

$$+ \sum_\alpha (u_\alpha\binom{l'}{k'}-u_\alpha\binom{0}{k})^2\Big]+2\psi''_{kk'}(|\mathbf{x}\binom{l'}{k'}-\mathbf{x}\binom{0}{k}|^2)\times$$

$$\times\Big[\sum_\alpha (x_\alpha\binom{l'}{k'}-x_\alpha\binom{0}{k})(u_\alpha\binom{l'}{k'}-u_\alpha\binom{0}{k})\Big]^2\Big\}, \quad (11.6)$$

where $\psi_{kk'}$ refers to the potential function between particles of the types k and k'. In the summation over $\binom{l'}{k'}$ we need not explicitly exclude the case $\binom{l'}{k'}=\binom{0}{k}$ (which is clearly meaningless) if we understand that the ψ-functions vanish for zero value of the argument. The factor $\tfrac{1}{2}$ takes account of the fact that the interaction energy between $\binom{l'}{k'}$ and $\binom{0}{k}$ must be considered as being shared between them.

For a homogeneous deformation, we have from (11.2) that

$$u_\alpha\binom{l'}{k'}-u_\alpha\binom{0}{k} = u_\alpha(k')-u_\alpha(k)+\sum_\beta u_{\alpha\beta}\, x_\beta\binom{l'}{k'k}, \quad (11.7)$$

where we have written for simplicity

$$\mathbf{x}\begin{pmatrix} l'l \\ k'k \end{pmatrix} = \mathbf{x}\begin{pmatrix} l' \\ k' \end{pmatrix} - \mathbf{x}\begin{pmatrix} 0 \\ k \end{pmatrix}. \tag{11.8}$$

When (11.7) is substituted in (11.6) and the products in the various terms are worked out, we find that the energy density u due to the deformation can be written as

$$
\begin{aligned}
u = \frac{1}{v_a}\Bigl\{ &-2\sum_{k\alpha}\Bigl(u_\alpha(k)+\sum_\beta u_\beta(k)u_{\beta\alpha}\Bigr)\sum_{l'k'}[\psi'x_\alpha]_{\mathbf{x}(^{l'}_{k'k})}+\\
&+\sum_{\alpha\beta}\Bigl(u_{\alpha\beta}+\tfrac{1}{2}\sum_\gamma u_{\gamma\alpha}u_{\gamma\beta}\Bigr)\sum_{l'k'k}[\psi'x_\alpha x_\beta]_{\mathbf{x}(^{l'}_{k'k})}+\\
&+\sum_{k\alpha}\sum_{k'\beta} u_\alpha(k)u_\beta(k')\Bigl(\delta_{kk'}\delta_{\alpha\beta}\sum_{l'k'}[\psi']_{\mathbf{x}(^{l'}_{k'k})}-\delta_{\alpha\beta}\sum_{l'}[\psi']_{\mathbf{x}(^{l'}_{k'k})}+\\
&+2\delta_{kk'}\sum_{l'k'}[\psi''x_\alpha x_\beta]_{\mathbf{x}(^{l'}_{k'k})}-2\sum_{l'}[\psi''x_\alpha x_\beta]_{\mathbf{x}(^{l'}_{k'k})}\Bigr)-\\
&-4\sum_{k\alpha}\sum_{\beta\gamma} u_\alpha(k)u_{\beta\gamma}\sum_{l'k'}[\psi''x_\alpha x_\beta x_\gamma]_{\mathbf{x}(^{l'}_{k'k})}+\\
&+\sum_{\alpha\beta\gamma\lambda} u_{\alpha\gamma}u_{\beta\lambda}\sum_{l'k'k}[\psi''x_\alpha x_\gamma x_\beta x_\lambda]_{\mathbf{x}(^{l'}_{k'k})}\Bigr\}, \quad (11.9)
\end{aligned}
$$

where the energy per cell has been multiplied by $1/v_a$ to give the energy density. The arguments for the functions enclosed in the square brackets are indicated in the lower right corner in each case; moreover ψ is understood to refer to $\psi_{kk'}$, if the argument indicated refers to the relative position vector between particles of the types k and k' such as $\mathbf{x}(^{l'}_{k'k})$.

It is not altogether obvious by inspection that (11.9) follows from the substitution of (11.7) in (11.6); our experience shows that it is a useful exercise to verify this result in detail. In this connexion one should take note of certain symmetry characters of the following sums:

(a)	$\sum_{l'}[\psi']_{\mathbf{x}(^{l'}_{k'k})}$	symmetric
(b)	$\sum_{l'}[\psi'x_\alpha]_{\mathbf{x}(^{l'}_{k'k})}$	antisymmetric
(c)	$\sum_{l'}[\psi'x_\alpha x_\beta]_{\mathbf{x}(^{l'}_{k'k})}$	symmetric
(d)	$\sum_{l'}[\psi''x_\alpha x_\beta]_{\mathbf{x}(^{l'}_{k'k})}$	symmetric
(e)	$\sum_{l'}[\psi''x_\alpha x_\beta x_\gamma]_{\mathbf{x}(^{l'}_{k'k})}$	antisymmetric
(f)	$\sum_{l'}[\psi''x_\alpha x_\gamma x_\beta x_\lambda]_{\mathbf{x}(^{l'}_{k'k})}$	symmetric

$$(11.10)$$

Take (b) as an example, which can be written explicitly as

$$\sum_{l'}[\psi'x_\alpha]_{\mathbf{x}(^{l'}_{k'k})} = \sum_{l'}\psi'(|\mathbf{x}(l')+\mathbf{x}(k')-\mathbf{x}(k)|^2)\{x_\alpha(l')+x_\alpha(k')-x_\alpha(k)\}, \tag{11.11}$$

where we have split the vector $\mathbf{x}(^{l'}_{k'k}) = \mathbf{x}(^{l'}_{k'}) - \mathbf{x}(^{0}_{k})$ in accordance with (11.3). We may clearly replace $\mathbf{x}(l')$ in the summand on the right-hand side by $-\mathbf{x}(l')$, for as l' goes over all integral values, $-\mathbf{x}(l')$, exactly as $\mathbf{x}(l')$, goes through all points on the simple Bravais lattice. Thus we have

$$\sum_{l'} [\psi' x_\alpha]_{\mathbf{x}(^{l'}_{k'k})} = \sum_{l'} \psi'(|-\mathbf{x}(l')+\mathbf{x}(k')-\mathbf{x}(k)|^2)\{-x_\alpha(l')+x_\alpha(k')-x_\alpha(k)\}$$

$$= -\sum_{l'} \psi'(|\mathbf{x}(l')+\mathbf{x}(k)-\mathbf{x}(k')|^2)\{x_\alpha(l')+x_\alpha(k)-x_\alpha(k')\}. \quad (11.12)$$

Apart from the negative sign, the right-hand side is the same as (11.11) with k and k' interchanged; it follows thus that the sum (b) is antisymmetric in k and k'. The same argument can be applied to all the sums in (11.10); the result is obvious: a sum is symmetric or antisymmetric in k and k' according as whether there is an even or an odd number of x-factors appearing in the sum. In (11.10) the symmetry characters of the sums are indicated; they have been taken into account in writing down (11.9).

If the lattice is initially in equilibrium, the corresponding energy density is a minimum; the terms in (11.9) which are linear in $\mathbf{u}(k)$ and $u_{\alpha\beta}$ must thus vanish. Hence we must have

$$\sum_{l'k} [\psi' x_\alpha]_{\mathbf{x}(^{l'}_{k'k})} = 0, \quad (11.13)$$

$$\sum_{l'k'k} [\psi' x_\alpha x_\beta]_{\mathbf{x}(^{l'}_{k'k})} = 0. \quad (11.14)$$

These constitute the equilibrium conditions for a lattice free from external forces. It follows from the antisymmetry of (11.10) (b) (in k and k') that if we sum the right-hand side of (11.13) over all values of k, the result vanishes identically. This shows that of the n equations (11.13) for each given α, only $(n-1)$ are mutually independent, so that (11.13) gives altogether $3(n-1)$ independent equations. For a given shape and size of the lattice cell, these equations determine the *relative* positions of the n particles in a cell. In view of the symmetry in α and β, (11.14) gives six independent equations, which exactly suffice to determine the shape and size of a cell; for a cell is specified by six parameters, which may for instance be taken as the lengths of the three edges and the three angles between them.

Taking into account the equilibrium conditions, we can write the energy density as

$$u = \tfrac{1}{2} \sum_{k\alpha} \sum_{k'\beta} \{^{kk'}_{\alpha\beta}\} u_\alpha(k) u_\beta(k') + \sum_{k\alpha} \sum_{\beta\gamma} \{^{k}_{\alpha}, \beta\gamma\} u_\alpha(k) u_{\beta\gamma} + \tfrac{1}{2} \sum_{\alpha\beta\gamma\lambda} \{\alpha\gamma\beta\lambda\} u_{\alpha\gamma} u_{\beta\lambda},$$

$$(11.15)$$

where for brevity the coefficients are expressed in the form of the brackets which are defined as follows:

$$\{^{kk'}_{\alpha\beta}\} = \frac{2}{v_a}\Big\{\delta_{\alpha\beta}\delta_{kk'}\sum_{l'k'}[\psi']_{\mathbf{x}(^{l'}_{k'k})} - \delta_{\alpha\beta}\sum_{l'}[\psi']_{\mathbf{x}(^{l'}_{k'k})} +$$
$$+ 2\delta_{kk'}\sum_{l'k'}[\psi''x_\alpha x_\beta]_{\mathbf{x}(^{l'}_{k'k})} - 2\sum_{l'}[\psi''x_\alpha x_\beta]_{\mathbf{x}(^{l'}_{k'k})}\Big\}, \quad (11.16)$$

$$\{^{k}_{\alpha}, \beta\gamma\} = -\frac{4}{v_a}\sum_{l'k'}[\psi''x_\alpha x_\beta x_\gamma]_{\mathbf{x}(^{l'}_{k'k})}, \quad (11.17)$$

$$\{\alpha\gamma\beta\lambda\} = \frac{2}{v_a}\sum_{l'k'k}[\psi''x_\alpha x_\beta x_\gamma x_\lambda]_{\mathbf{x}(^{l'}_{k'k})}. \quad (11.18)$$

The brackets are subject to certain obvious symmetry relations and identities. The first and third terms on the right of (11.16) are evidently symmetric in k and k'. The same symmetry holds also for the second and fourth terms; this follows directly from the symmetry characters of the corresponding sums in (11.10). Since every term in (11.16) is obviously also symmetric in the Cartesian indices α and β, we have the double symmetry in k, k' and α, β:

$$\{^{kk'}_{\alpha\beta}\} = \{^{k'k}_{\alpha\beta}\} = \{^{kk'}_{\beta\alpha}\}. \quad (11.19)$$

When we sum (11.16) over all values of k', we find that the first term on the right is cancelled by the second and the third term by the fourth. Thus we have the identities

$$\sum_{k'}\{^{kk'}_{\alpha\beta}\} \equiv 0. \quad (11.20)$$

The symmetry relations (11.19) reduce the maximum number of independent coefficients of this type to $3n(n+1)$. (11.20) provides $6n$ independent identities which further reduce the number to $3n(n-1)$.

It is obvious from the definition (11.17) that

$\{^{k}_{\alpha}, \beta\gamma\}$ is completely symmetric in all the Cartesian indices. (11.21)

Moreover it follows directly from the antisymmetry of the sum (11.10) (e) that

$$\sum_{k}\{^{k}_{\alpha}, \beta\gamma\} = \frac{-4}{v_a}\sum_{kk'}\sum_{l'}[\psi''x_\alpha x_\beta x_\gamma]_{\mathbf{x}(^{l'}_{k'k})} \equiv 0. \quad (11.22)$$

The number of independent coefficients consistent with (11.21) is $10n$; this number is reduced to $10(n-1)$ by the ten independent identical relations (11.22).

From the definition (11.18) follows directly that

$\{\alpha\gamma\beta\lambda\}$ is completely symmetric in all four Cartesian indices. (11.23)

This requirement leaves only fifteen of the coefficients mutually independent.

It is evident from the symmetry of all the coefficients in (11.15) with respect to the Cartesian indices α, β,... that the energy density depends only on the symmetrized parameters

$$u_{\alpha\beta} + u_{\beta\alpha}.$$

Following Voigt† we may thus introduce instead of $u_{\alpha\beta}$:

$$\left. \begin{aligned} s_\rho &= u_{\alpha\beta} + u_{\beta\alpha}, \quad \beta \neq \alpha \\ &= u_{\alpha\beta}, \qquad\qquad \beta = \alpha \end{aligned} \right\}, \tag{11.24}$$

where the indices $\rho = 1, 2,..., 6$ are related to the tensor indices α, β as follows:

ρ	1	2	3	4	5	6	
(α, β)	11	22	33	23 (32)	31 (13)	12 (21)	(11.25)

We can express the energy density (11.15) in terms of s_ρ as

$$u = \tfrac{1}{2} \sum_{k\alpha} \sum_{k'\beta} \{{}^{kk'}_{\alpha\beta}\} u_\alpha(k) u_\beta(k') + \sum_{k\alpha} \sum_\rho \{{}^{k}_\alpha, \rho\} u_\alpha(k) s_\rho + \tfrac{1}{2} \sum_{\rho\sigma} \{\rho\sigma\} s_\rho s_\sigma, \tag{11.26}$$

where the brackets are the same as before apart from a direct transcription of the tensor indices into the Voigt indices in accordance with (11.25).

The components s_ρ, we remember, describe the type of deformation expressed in (11.1). They are identical with the elastic strains in the classical elasticity theory. They specify to the first order (see later) the size and shape of a macroscopic specimen as well as a lattice cell.

It follows from the identities (11.20) and (11.22) that the energy density (11.26) is unaffected if an arbitrary vector is added to all the vectors $\mathbf{u}(k)$. This shows that the energy density depends only on the differences between the vectors $\mathbf{u}(k)$; these differences represent the relative shifts between the different component Bravais lattices. Such shifts are microscopic in magnitude and do not affect the macroscopic dimensions of a specimen. We may thus describe the deformations due respectively to $\mathbf{u}(k)$ and s_ρ as the *internal* and *external* strains; the latter, as we have seen, are identical with the elastic strains.

In the energy density, the internal and external strains are coupled together through the terms containing their products. Thus as a body is elastically deformed a certain internal strain is in general induced; the internal strain is such as to make the energy density a minimum for the

† W. Voigt, *Lehrbuch der Kristallphysik* (Teubner, Berlin, 1910).

given elastic strain components s_ρ. Thus the $\mathbf{u}(k)$ are determined by the condition for a stationary value of the energy density:

$$0 = \frac{\partial u}{\partial u_\alpha(k)} = \sum_{k'\beta} \{_{\alpha\beta}^{kk'}\} u_\beta(k') + \sum_\rho \{_\alpha^k, \rho\} s_\rho. \qquad (11.27)$$

Owing to the identities (11.20) and (11.22), the sum of the right-hand side of this equation over all values of k vanishes identically; and only $3(n-1)$ of the equations are mutually independent. The solutions are accordingly arbitrary in that an arbitrary vector can be added to all the vectors $\mathbf{u}(k)$ without affecting the equations; in other words, the equations determine only the differences between the different vectors $\mathbf{u}(k)$.

The internal strains can be eliminated from the energy density (11.26) with the equations (11.27). We remember that the energy density depends only on the differences between the different $\mathbf{u}(k)$, and (11.27) determine only these differences. Thus we may, without loss of generality, put $\mathbf{u}(1) = 0$ in both (11.26) and (11.27) and eliminate $\mathbf{u}(k)$ $(k = 2, 3, ..., n)$ from (11.26) with the $3(n-1)$ independent equations in (11.27) corresponding to $k = 2, 3, ..., n$. When the internal strains are eliminated, the energy density becomes a quadratic expression in s_ρ, which we may write as

$$u = \tfrac{1}{2} \sum_{\rho\sigma} c_{\rho\sigma} s_\rho s_\sigma \quad (c_{\rho\sigma} = c_{\sigma\rho}). \qquad (11.28)$$

This result is now strictly comparable with the strain energy function in the classical elasticity theory; the elastic stresses S_ρ are given by the derivatives of the energy density with respect to the elastic strains:

$$S_\rho = \frac{\partial u}{\partial s_\rho} = \sum_\sigma c_{\rho\sigma} s_\sigma. \qquad (11.29)$$

This expresses the general *Hooke's law* that the elastic stresses are linear functions of the elastic strains, $c_{\rho\sigma}$ being the elastic constants.

If it so happens that no internal strains are induced by the elastic strains, the energy density (11.26) reduces directly to

$$u = \tfrac{1}{2} \sum_{\rho\sigma} \{\rho\sigma\} s_\rho s_\sigma \quad (\{\rho\sigma\} \to \{\alpha\beta\gamma\lambda\}, \ \rho \to \alpha, \beta; \ \sigma \to \gamma, \lambda).$$

In this special case the elastic constants are thus given directly by the brackets defined in (11.18):

$$c_{\rho\sigma} = c_{\sigma\rho} = \{\rho\sigma\} \quad (\{\rho\sigma\} \to \{\alpha\beta\gamma\lambda\}, \ \rho \to \alpha, \beta; \ \sigma \to \gamma, \lambda). \qquad (11.30)$$

We have seen that the maximum number of independent brackets of this kind is fifteen. It is easily verified that in this case the complete symmetry of $\{\alpha\gamma\beta\lambda\}$ in the four tensor indices leads to the following relations between the elastic constants (11.30):

$$c_{23} = c_{44}, \qquad c_{31} = c_{55}, \qquad c_{12} = c_{66} \left.\right\}$$
$$c_{14} = c_{56}, \qquad c_{25} = c_{64}, \qquad c_{36} = c_{45} \left.\right\} \qquad (11.31)$$

These are known as the *Cauchy relations*. As $c_{\rho\sigma} = c_{\sigma\rho}$, there are in the general case twenty-one independent elastic constants; when the Cauchy relations hold, the maximum number of independent elastic constants is reduced to fifteen.

A lattice point P is said to be a centre of symmetry, if all lattice particles can be grouped into identical pairs such that the partners in a pair are reflection images of one another through the point P. In other words, an inversion operation applied to the lattice with respect to the point P leaves the lattice invariant. If a lattice is such that every lattice particle occupies a centre of symmetry, then no internal strain will be induced by elastic deformations. Let us examine the coefficients $\{^k_\alpha, \beta\gamma\}$ defined in (11.17). Since $\binom{0}{k}$ is a centre of symmetry, the lattice points $\binom{l}{k'}$ can be grouped into pairs occupied by identical particles, their position vectors $\mathbf{x}\binom{l}{k'k}$ from $\binom{0}{k}$ being equal and opposite to one another. Owing to the odd number of times the components of $\mathbf{x}\binom{l}{k'k}$ occur in the summand of (11.17), the contributions due to a pair exactly cancel, so the coefficients $\{^k_\alpha, \beta\gamma\}$ vanish identically. Thus from (11.27) it follows that the internal strain can be put equal to zero. This result is readily understandable in physical terms. It is fairly evident that a particle at a centre of symmetry can experience no net force from the totality of the other particles. Moreover it is easily verified that a centre of symmetry remains one if the lattice is subject to an external strain (cf. (11.1)). This means that in the above type of lattice a pure external strain automatically leaves all the lattice particles in equilibrium; hence no internal strain occurs.

Therefore the Cauchy relations will be fulfilled if the lattice structure is such that every lattice particle occupies a centre of symmetry and if the particles interact with central forces.

The above treatment of the elastic properties cannot be applied to ionic crystals, for the coefficients $\{^{kk'}_{\alpha\beta}\}$ turn out to be essentially indeterminate. Thus corresponding to the Coulomb interaction between the ions, we have

$$\psi'_{kk'}(r^2) = \frac{d}{dr^2}\left(\frac{e_k e_{k'}}{r}\right) = -\frac{1}{2}\frac{e_k e_{k'}}{r^3}.$$

The value of the second term in (11.16) is

$$-\frac{2}{v_a}\delta_{\alpha\beta}\sum_{l'}[\psi']_{\mathbf{x}\binom{l'}{k'k}} = \delta_{\alpha\beta}\frac{1}{v_a}\sum_{l'}\frac{e_k e_{k'}}{|\mathbf{x}(l')+\mathbf{x}(k)-\mathbf{x}(k')|^3}.$$

For large l', we can ignore $\mathbf{x}(k')-\mathbf{x}(k)$ in comparison with $\mathbf{x}(l')$ and replace the summation by an integration:

$$\int dl' = \frac{1}{v_a}\int d\mathbf{x}(l').$$

Hence for the part of the above sum due to $|\mathbf{x}(l')| > R \gg$ lattice constant, we have approximately

$$\delta_{\alpha\beta}\frac{e_k e_{k'}}{v_a^2}\int_R^\infty\frac{d\mathbf{x}}{|\mathbf{x}|^3} = \delta_{\alpha\beta}\frac{4\pi e_k e_{k'}}{v_a^2}\ln|\mathbf{x}|\Big|_R^\infty,$$

which is divergent. Similarly we find that the fourth term in (11.16) is also divergent in this case for $\alpha = \beta$ and has a sign opposed to the above term. The values of the coefficients $\{^{kk'}_{\alpha\beta}\}$ are thus completely indeterminate.

This anomaly can be understood in physical terms as follows: $\{^{kk'}_{\alpha\beta}\}$ represents the force (α-component) on the ions k (per unit volume) caused by a unit displacement of all the k' ions in the β-direction. In terms of Maxwell's theory, the displacement of the k' ions produces a uniform dielectric polarization. If we are considering a finite specimen, the polarization creates a surface charge; the latter gives rise to a macroscopic electric field which exerts forces on the k ions. The electric field, and thus also the forces exerted, depend essentially on the shape of the specimen. In the above treatment, we have tacitly assumed that the lattice is infinitely extended; in other words, we are considering formally the limiting case of an infinite specimen. As the forces depend on the shape of a specimen, they have no unique limiting value for an infinitely large and arbitrary specimen.

However, in the special case when no internal strain is induced, the above-mentioned difficulty no longer arises and the above theory may then be applied. As we have seen, this is the case if the lattice structure is such that every lattice particle is at a centre of symmetry.

It is obvious that in all alkali halide lattices, both the alkali and halide ions are at centres of symmetry. Thus the above treatment applies; and the Cauchy relations will be fulfilled, if the ions interact with central forces. Owing to the cubic symmetry of the alkali halide crystals, the non-vanishing elastic constants are related as follows:

$$\left.\begin{array}{l} c_{11} = c_{22} = c_{33} \\ c_{12} = c_{23} = c_{31} \\ c_{44} = c_{55} = c_{66} \end{array}\right\}. \tag{11.32}$$

Thus the Cauchy relations reduce to the single relation

$$c_{12} = c_{44}.$$

Experimental values for c_{12} and c_{44} are available for LiF, NaCl, KCl, KBr; they are given in Table 23 together with the ratio c_{44}/c_{12}. The deviations of c_{44}/c_{12} from unity are seen to be fairly small except in the case of LiF. As mentioned in § 1, Löwden finds by using a first-order wave-mechanical perturbation method that a part of the lattice cohesive energy cannot be ascribed to two-body interactions; in elastic deformations this part causes deviations from the Cauchy relations. However, his theoretically calculated values for c_{44}/c_{12}, though showing the correct signs of the deviations from unity, differ in fact more strongly from the experimental values than the latter from the value unity.

TABLE 23†

Deviations from Cauchy's Relation

	c_{12}	c_{44}	c_{44}/c_{12}
LiF	$4 \cdot 04 \times 10^{11}$ dynes/cm.²	$5 \cdot 54 \times 10^{11}$ dynes/cm.²	1·37
NaCl	1·23 (1·17)	1·26 (1·34)	1·03 (1·14)
KCl	0·60 (0·60)	0·03 (0·67)	1·05 (1·12)
KBr	0·54	0·508	0·94

Let us return once more to the general expression for the energy density (11.9). In general, it refers to the lattice being initially under strain. The corresponding stresses in the initial configuration are obtained directly by differentiating (11.9); namely,

$$S_\alpha(k) = -\left(\frac{\partial u}{\partial u_\alpha(k)}\right)_0 = \frac{2}{v_a} \sum_{l'k'} [\psi' x_\alpha]_{\mathbf{x}\binom{l'}{k'k}}, \tag{11.33}$$

$$S_\rho = \left(\frac{\partial u}{\partial s_\rho}\right)_0 = \frac{1}{v_a} \sum_{l'kk'} [\psi' x_\alpha x_\beta]_{\mathbf{x}\binom{l'}{k'k}} \quad (\rho \to \alpha, \beta). \tag{11.34}$$

$S_\alpha(k)$ represents the external forces to be applied on the k ions per unit volume to keep the lattice in this strained configuration; whereas S_ρ

† The values given refer to room temperatures; for LiF, NaCl, KBr see H. B. Huntington, *Phys. Rev.* **72**, 321 (1947); for KCl see M. Durand, ibid. **50**, 449 (1936). The values given in brackets are values extrapolated to the absolute zero of temperature by Durand (loc. cit.).

represents the elastic stress components in the lattice. The equilibrium conditions (11.13) and (11.14) given earlier for a free crystal are seen to be equivalent to the requirement that no external forces act on the ions and the elastic stresses vanish.

A general homogeneous deformation, as we have seen, is described by $3n+9$ parameters, namely, the three components of each of the n vectors $\mathbf{u}(k)$ ($k = 1, 2,..., n$) and the nine deformation parameters $u_{\alpha\beta}$. The homogeneous deformations evidently include as special cases the translations and rotations of the lattice as a whole; these motions, which are characteristic of rigid bodies and correspond to six degrees of freedom, clearly do not affect the energy density. Thus it should be possible to choose the $3n+9$ parameters suitably so that only $3n+3$ of these express genuine strains of the lattice; and the energy density will then be a function of the latter only. In fact, we have seen above that the energy density (11.26) depends only on the differences $\mathbf{u}(k)-\mathbf{u}(1)$ ($k = 2, 3,..., n$) and the six elastic strains s_ρ; making in all exactly $3(n+1)$ parameters as required.

s_ρ and $\mathbf{u}(k)-\mathbf{u}(1)$ ($k = 2, 3,..., n$) are adequate strain parameters, however, only to the first order of approximation. Owing to the equilibrium conditions, the strains only appear in (11.26) in the second-order terms. Since the third- and higher-order terms have been ignored, we have thus only been concerned with these quantities to the first order of approximation. It becomes evident that these parameters do not completely describe the state of strain in the general case, if we return to the more general expression for the energy density (11.9). For (11.9) no longer depends only on the symmetrized parameters $u_{\alpha\beta}+u_{\beta\alpha}$ ($= s_\rho$).

The form of the first two terms in (11.9) suggests that we should introduce in the general case as strain parameters the vectors $\bar{\mathbf{u}}(k)$ and the symmetric tensor $\bar{u}_{\alpha\beta} = \bar{u}_{\beta\alpha}$, defined as follows:

$$\bar{u}_\alpha(k) = u_\alpha(k) + \sum_\beta u_\beta(k)u_{\beta\alpha}, \tag{11.35}$$

$$\bar{u}_{\alpha\beta} = \bar{u}_{\beta\alpha} = \tfrac{1}{2}\{u_{\alpha\beta}+u_{\beta\alpha}+ \sum_\gamma u_{\gamma\alpha}u_{\gamma\beta}\}. \tag{11.36}$$

It is easily verified that by adding terms of the third order only we can write (11.9) as

$$u = \sum_{k\alpha} \bar{u}_\alpha(k)\left\{-\frac{2}{v_a}\sum_{l'k'}[\psi'x_\alpha]_{\mathsf{x}\binom{l'}{k'k}}\right\} + \sum_{\alpha\beta} \bar{u}_{\alpha\beta}\left\{\frac{1}{v_a}\sum_{l'k'k}[\psi'x_\alpha x_\beta]_{\mathsf{x}\binom{l'}{k'k}}\right\} +$$

$$+\tfrac{1}{2}\sum_{k\alpha}\sum_{k'\beta}\{{}^{kk'}_{\alpha\beta}\}\bar{u}_\alpha(k)\bar{u}_\beta(k') + \sum_{k\alpha}\sum_{\beta\gamma}\{{}^{k}_{\alpha},\beta\gamma\}\bar{u}_\alpha(k)\bar{u}_{\beta\gamma} + \tfrac{1}{2}\sum_{\alpha\gamma}\sum_{\beta\lambda}\{\alpha\gamma\beta\lambda\}\bar{u}_{\alpha\gamma}\bar{u}_{\beta\lambda}. \tag{11.37}$$

Taking account of the antisymmetry of (11.10) (*b*) and the identities (11.20), (11.22), we can show as before that the energy density is unaffected by the addition of the same vector to all the vectors $\bar{\mathbf{u}}(k)$ and may thus be considered as a function of the $n-1$ vectors $\bar{\mathbf{u}}(k)-\bar{\mathbf{u}}(1)$ ($k = 2, 3,..., n$). Therefore in this general case $\bar{\mathbf{u}}(k)-\bar{\mathbf{u}}(1)$ together with $\bar{u}_{\alpha\beta} = \bar{u}_{\beta\alpha}$ constitute the $3(n+1)$ strain parameters.

The above considerations are, of course, still restricted by the fact that higher-order terms in the energy density have been ignored; the above strain parameters might, for instance, be correct only up to the second-order terms. Later in § 36 we shall, however, find that these strain parameters are in fact the adequate parameters in the general case. We notice that the parameters $\mathbf{u}(k)$ and s_ρ are merely the first-order approximations to $\bar{\mathbf{u}}(k)$ and $\bar{u}_{\alpha\beta} = \bar{u}_{\beta\alpha}$.

12. Mechanical stability of simple lattices

The equilibrium conditions (11.13) and (11.14) express only the requirement that the energy density has a stationary value at equilibrium. For a lattice to be stable, the energy density (11.15) must be a positive definite quadratic form so that the energy is raised by any small strains. Using the results of the last section, we shall present some stability considerations for a few simple lattice types due to Born and his collaborators.[†]

Consider first the three cubic Bravais lattices, namely, the simple cubic, body-centred cubic, and the face-centred cubic lattices. All three lattices can be described in terms of a common framework as follows. Let $\mathbf{a}_1, \mathbf{a}_2, \mathbf{a}_3$ be three mutually perpendicular vectors of the same length a. The lattice points in all three lattices can be expressed as

$$l_1\mathbf{a}_1+l_2\mathbf{a}_2+l_3\mathbf{a}_3, \qquad (12.1)$$

where l_1, l_2, l_3 are integers subject to the following conditions for the three cases:

  (i) s.c.   no restriction;
  (ii) b.c.c. l_1, l_2, l_3 either all even or all odd;
  (iii) f.c.c. $l_1+l_2+l_3$ even.

We note that the vectors $\mathbf{a}_1$, $\mathbf{a}_2$, $\mathbf{a}_3$ are not basic lattice vectors for the two latter lattices.

For simple lattices (11.1) represents the only type of homogeneous

† M. Born, *Proc. Camb. Phil. Soc.* **36**, 160 (1940); R. D. Misra, ibid. **36**, 173 (1940); S. C. Power, ibid. **38**, 62 (1942); M. Born, ibid. **38**, 82 (1942); ibid. **40**, 262 (1944); and also other papers in the series.

deformation; thus only the last term remains in the energy density and
is equal directly to the strain-energy function

$$\tfrac{1}{2} \sum_{\rho\sigma} c_{\rho\sigma} s_\rho s_\sigma;$$

the corresponding elastic constants are related to the brackets defined
in (11.18) by (11.30). If we arrange the coefficients in the above quadratic
form in the form of a matrix

$$\begin{bmatrix} c_{11} & c_{12} & c_{13} & c_{14} & c_{15} & c_{16} \\ c_{21} & c_{22} & c_{23} & c_{24} & c_{25} & c_{26} \\ c_{31} & c_{32} & c_{33} & c_{34} & c_{35} & c_{36} \\ c_{41} & c_{42} & c_{43} & c_{44} & c_{45} & c_{46} \\ c_{51} & c_{52} & c_{53} & c_{54} & c_{55} & c_{56} \\ c_{61} & c_{62} & c_{63} & c_{64} & c_{65} & c_{66} \end{bmatrix}$$

then, according to a well-known theorem in algebra, the above quadratic
form is positive definite if the determinants of the matrices of successive
orders as marked out above (the *principal minors*) are all positive.

   In the present case, the particle indices k, k', etc., can be omitted and
the summation in (11.18) is over all lattice points as given by (12.1). Thus,
if the Cartesian axes are chosen parallel to the vectors $\mathbf{a}_1$, $\mathbf{a}_2$, $\mathbf{a}_3$, we find
that

$$\{\alpha\beta\gamma\lambda\} = \frac{2a^4}{v_a} \sum_l l_\alpha l_\beta l_\gamma l_\lambda \psi''(a^2(l_1^2 + l_2^2 + l_3^2)), \qquad (12.2)$$

where the summation over $l(l_1, l_2, l_3)$ is subject to the respective restrictions (i), (ii), (iii) for the three types of lattices. (12.2) vanishes identically if one of the indices differs from all the rest. Take the case when α
is different from β, γ, λ. The terms with $l_\alpha = 0$ in (12.2) obviously vanish.
The rest of the terms can always be grouped into pairs with indices l_α, l_β,
l_γ, l_λ and $-l_\alpha, l_\beta, l_\gamma, l_\lambda$, for such a pair will be either admitted or excluded
together in all the three cases (i), (ii), (iii). The contributions due to a
pair evidently cancel; and so (12.2) vanishes. Hence, for non-vanishing
coefficients, either all four indices are equal or they comprise two pairs
of equal indices; i.e. the coefficients must be of either of the two forms:

$$(a) \qquad \{\alpha\alpha\beta\beta\} = \frac{2a^4}{v_a} \sum_l l_\alpha^2 l_\beta^2 \psi''(a^2(l_1^2 + l_2^2 + l_3^2)) \quad \alpha \neq \beta$$

$$(b) \qquad \{\alpha\alpha\alpha\alpha\} = \frac{2a^4}{v_a} \sum_l l_\alpha^4 \psi''(a^2(l_1^2 + l_2^2 + l_3^2)) \qquad\qquad\qquad (12.3)$$

Owing to the complete symmetry of the structures with respect to the
three Cartesian directions, the values of both (a) and (b) are independent
of the value of α or β.

The number of independent elastic constants for cubic Bravais lattices is the same as in the case of the alkali halides. As already mentioned in connexion with the latter, the cubic symmetry leads to only three independent elastic constants (cf. (11.32)) which may be taken as c_{11}, c_{12}, c_{44}; the Cauchy relations require further that $c_{12} = c_{44}$. These facts can be directly verified for the cubic Bravais lattices with the help of the above result, namely, that the sums (a), (b) are independent of the Cartesian indices and all other coefficients vanish. For the independent elastic constants c_{11} and c_{12}, we have

$$c_{11} = \{1111\} = \frac{2a^4}{v_a} \sum_l l_1^4 \psi''(a^2(l_1^2+l_2^2+l_3^2)), \qquad (12.4)$$

$$c_{12} = \{1122\} = \frac{2a^4}{v_a} \sum_l l_1^2 l_2^2 \psi''(a^2(l_1^2+l_2^2+l_3^2)). \qquad (12.5)$$

The matrix of the quadratic coefficients has the form

$$\begin{bmatrix} c_{11} & c_{12} & c_{12} & 0 & 0 & 0 \\ c_{12} & c_{11} & c_{12} & 0 & 0 & 0 \\ c_{12} & c_{12} & c_{11} & 0 & 0 & 0 \\ 0 & 0 & 0 & c_{12} & 0 & 0 \\ 0 & 0 & 0 & 0 & c_{12} & 0 \\ 0 & 0 & 0 & 0 & 0 & c_{12} \end{bmatrix}.$$

The principal minors are

$$c_{12}, \ c_{12}^2, \ c_{12}^3, \ c_{12}^3 c_{11}, \ c_{12}^3(c_{11}^2-c_{12}^2), \ c_{12}^3(c_{11}-c_{12})^2(c_{11}+2c_{12}).$$

For these to be positive, only the two following conditions need be fulfilled:

$$c_{12} > 0, \qquad c_{11}-c_{12} > 0. \qquad (12.6)$$

In considering the summations in (12.4) and (12.5) let us group the lattice points on the s.c. lattice according to their distances from the origin; points in different groups lie on successive shells with radii r_1, r_2, r_3, etc. In the b.c.c. and f.c.c. lattices some of these shells are completely excluded. It is easily verified that in the b.c.c. lattice the first, second, fifth, sixth shells, etc., are missing and in the f.c.c. lattice the first, third, fifth shells, etc., are missing. Given below are the values of $c_{11}-c_{12}$ and c_{12} calculated from (12.4) and (12.5):

	$\dfrac{2v_a}{a^4}(c_{11}-c_{12})$	$\dfrac{2v_a}{a^4}c_{12}$
s.c.	$8\psi''(r_1^2)+16\psi''(r_2^2)+\dots$	$0\psi''(r_1^2)+16\psi''(r_2^2)+\dots$
b.c.c.	$0\psi''(r_3^2)+128\psi''(r_4^2)+\dots$	$32\psi''(r_3^2)+256\psi''(r_8^2)+\dots$
f.c.c.	$16\psi''(r_2^2)+128\psi''(r_4^2)+\dots$	$16\psi''(r_2^2)+288\psi''(r_6^2)+\dots$

$$(12.7)$$

Bearing in mind that the innermost (non-vanishing) shells in the s.c., b.c.c., f.c.c. are respectively the first, the third, and the second shells, we observe immediately a qualitative difference between the s.c. and the b.c.c. lattices on the one hand and the f.c.c. on the other; namely, in the two former lattices the innermost shell contributes nothing to one of the two quantities $c_{11}-c_{12}$ and c_{12}. Hence these lattices will be unstable if ψ'' turns out to be negative for all but the innermost shell. This is in fact most likely to be the case.

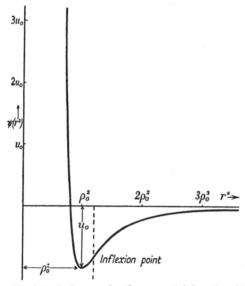

FIG. 23. A typical curve for the potential function $\psi(r^2)$.

A typical curve for the potential function ψ is depicted in Fig. 23. The potential function has a minimum $-u_0$ at $r = \rho_0$, at which distance the attractive and repulsive forces balance one another. The nearest neighbour distance (equal to the radius of the innermost non-vanishing shell) of a lattice in equilibrium is in general very close to ρ_0, since the lattice particles interact predominantly with their nearest neighbours. From the figure we see that, from the point $r^2 = \rho_0^2$ outward, the ψ function soon passes through an inflexion point, where the second derivative ψ'' changes from positive to negative values. Thus if the shell next to the innermost one falls beyond the inflexion point, c_{12} in the s.c. case and $c_{11}-c_{12}$ in the b.c.c. case will be negative, and the lattices in consequence unstable. Since the region between ρ_0^2 and the inflexion is very narrow, such lattices with particles interacting with central forces are most likely to be unstable. In a case where the region between ρ_0^2

and the inflexion point is relatively wide, the b.c.c. structure has a much better chance of being stable than the s.c. lattice; for the ratio of the r^2 of the second non-vanishing shell to that of the innermost shell is

$$r_4^2/r_3^2 = 4/3 \qquad (12.8)$$

in the b.c.c. lattice and

$$r_2^2/r_1^2 = 2 \qquad (12.9)$$

in the s.c. lattice. The separation between the two shells is thus much narrower in the b.c.c. lattice.

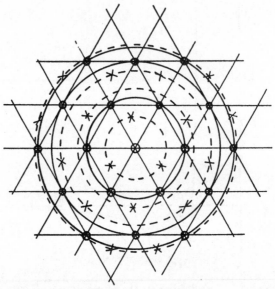

Fig. 24. Projections of the atomic positions in the two hexagonal lattices.

A similar analysis has been carried out by Born† for two hexagonal lattices. Consider first the hexagonal Bravais lattice in which the lattice particles form planes of hexagonal network and atoms in different planes are directly aligned vertically. The atomic positions on a hexagonal plane are indicated by small circles in Fig. 24. The basic vectors of the lattice can be chosen as

$$\mathbf{a}_1 = (a, 0, 0), \qquad \mathbf{a}_2 = \left(\tfrac{1}{2}a, \frac{\sqrt{3}a}{2}, 0\right), \qquad \mathbf{a}_3 = (0, 0, \gamma a), \quad (12.10)$$

where the Cartesian components are indicated in the brackets. The vectors $\mathbf{a}_1$, $\mathbf{a}_2$ generate the plane of hexagonal network and $\mathbf{a}_3$ displaces the atoms in a perpendicular direction with the separation γa between neighbouring planes.

† M. Born, *Proc. Camb. Phil. Soc.* **38**, 82 (1942); ibid. **40**, 262 (1944).

For a simple lattice only the equilibrium conditions (11.14) remain. The summation in (11.14) is in this case over the following lattice points:

$$\mathbf{x}(l) = l_1\,\mathbf{a}_1 + l_2\,\mathbf{a}_2 + l_3\,\mathbf{a}_3 = \left(l_1 + \tfrac{1}{2}l_2,\ \frac{\sqrt{3}}{2}l_2,\ \gamma l_3\right)a.$$

The consideration of symmetry shows quite readily that for $\alpha \neq \beta$ (11.14) is automatically fulfilled and for $\alpha = \beta$ we obtain only two independent conditions:

$$\left.\begin{aligned}
(a) \quad &\sum_l [\psi' x_1^2]_{\mathbf{x}(l)} = \sum_l [\psi' x_2^2]_{\mathbf{x}(l)} = 0\\
(b) \quad &\sum_l [\psi' x_3^2]_{\mathbf{x}(l)} = 0
\end{aligned}\right\}. \qquad (12.11)$$

As before, we group the lattice points into successive shells. If γ is not too far removed from unity, the first few shells have the following values for the squares of their radii:

$$a^2,\ \gamma^2 a^2,\ (1+\gamma^2)a^2,\ 3a^2,\ 4a^2,\ (3+\gamma^2)a^2,\dots.$$

When the contributions due to the different shells are worked out, the equilibrium conditions are found to be

$$\left.\begin{aligned}
(a) \quad &\psi'(a^2) + 2\psi'((1+\gamma^2)a^2) + 3\psi'(3a^2) + \dots = 0\\
(b) \quad &\psi'(\gamma^2 a^2) + 12\psi'((1+\gamma^2)a^2) + \dots = 0
\end{aligned}\right\}. \qquad (12.12)$$

For any value of γ the first term given has the smallest argument in both (a) and (b). For forces of short range, we need retain only this term and thus obtain approximately

$$\psi'(a^2) = 0, \qquad \psi'(\gamma^2 a^2) = 0.$$

These conditions require that $a = \rho_0$ and $\gamma = 1$, where ρ_0, we remember, denotes the distance at which ψ is a minimum. The condition that $\gamma = 1$ means that the atoms vertically above and below an atom are at the same distance away as its neighbours in its hexagonal plane. We notice that this is precisely the configuration of the atoms, if they behaved like rigid spheres.

With the help of the Cauchy relations, which evidently hold in this case, and the considerations of symmetry, it can be shown that the energy density has the following form:

$$u = \tfrac{1}{2}\{c_{11}(s_1^2 + s_2^2 + \tfrac{2}{3}s_1 s_2 + \tfrac{1}{3}s_6^2) + c_{33} s_3^2 + c_{44}(s_4^2 + s_5^2 + 2s_1 s_3 + 2s_2 s_3)\}. \qquad (12.13)$$

Clearly the lattice will be unstable if c_{44} is negative; for the energy density can then be lowered by a strain s_4 (or s_5). When the contributions to $\{2233\}$ (cf. (11.18)) due to various shells are worked out, we find that

$$c_{44} = \{2233\} = \frac{a^4}{v_a}\{0\psi''(\rho_0^2) + 6\psi''(2\rho_0^2) + \dots\}, \qquad (12.14)$$

where we have put in the values $a = \rho_0$, $\gamma = 1$, $v_a = (\sqrt{3}/2)\rho_0^3$. The innermost shell which has the radius ρ_0 is seen to contribute nothing to c_{44}. It follows from the same argument as before that most likely c_{44} is negative and the lattice is thus unstable. The ratio of the square of the radius of the second non-vanishing shell to that of the innermost shell in this case is 2. Comparing with (12.9), we may conclude that the hexagonal Bravais lattice is as unlikely to be stable as the s.c. lattice.

Adding to each atom in the above lattice a similar atom at a vector distance

$$\tfrac{1}{3}\mathbf{a}_1 + \tfrac{1}{3}\mathbf{a}_2 + \tfrac{1}{2}\mathbf{a}_3 \tag{12.15}$$

removed from the former, we obtain a composite lattice with two similar atoms per cell. Using the notation of the last section, we can label the original atoms in the Bravais lattice and the additional atoms respectively by $k = 1$ and 2. The atoms designated as 2 form similar hexagonal networks halfway between the hexagonal planes of the atoms designated as 1. Their projections are indicated by crosses superimposed on Fig. 24.

About any lattice point $k = 1$ are shells of atoms 1 with the squares of their radii given as before by

$$a^2,\ \gamma^2 a^2,\ (1+\gamma^2)a^2,\ 3a^2,\ \text{etc.}; \tag{12.16}$$

the squares of the radii of the additional shells of atoms 2 have the values

$$\left(\frac{1}{3}+\frac{\gamma^2}{4}\right)a^2,\ \left(\frac{4}{3}+\frac{\gamma^2}{4}\right)a^2,\ \left(\frac{7}{3}+\frac{\gamma^2}{4}\right)a^2,\ \text{etc.} \tag{12.17}$$

(12.16) also gives the squares of the radii of the shells of atoms 2 about an atom 2, and (12.17) gives the corresponding values for the shells of atoms 1 about an atom 2.

It follows readily from the symmetry of the structure that the equilibrium conditions (11.13) and the conditions (11.14) for $\alpha \neq \beta$ are identically fulfilled in this case. For $\alpha = \beta$ (11.14) gives two independent conditions:

$$0 = \sum_{l'kk'} [\psi' x_3^2]_{\times\binom{l'}{kk}} = \frac{2a^2\gamma^2}{v_a}\left\{\tfrac{3}{2}\psi'\left(\frac{a^2}{3}+\frac{\gamma^2 a^2}{4}\right)+\tfrac{3}{2}\psi'\left(\frac{4a^2}{3}+\frac{\gamma^2 a^2}{4}\right)+...\right\}, \tag{12.18}$$

$$0 = \sum_{l'kk'} [\psi' x_1^2]_{\times\binom{l'}{kk}} = \sum_{l'kk'} [\psi' x_2^2]_{\times\binom{l'}{kk}} = \frac{2a^2}{v_a}\left\{3\psi'(a^2)+\psi'\left(\frac{a^2}{3}+\frac{\gamma^2 a^2}{4}\right)+ \right.$$
$$\left. +4\psi'\left(\frac{4a^2}{3}+\frac{\gamma^2 a^2}{4}\right)+...\right\}. \tag{12.19}$$

For relevant values of γ (see below) the first term given has the smallest argument in both of the above equations. Thus for forces of short range we retain only this term and obtain approximately

$$\psi'\left(\frac{a^2}{3}+\frac{\gamma^2 a^2}{4}\right)=0, \qquad \psi'(a^2)=0.$$

These conditions require that $a=\rho_0$ and

$$\frac{1}{3}+\frac{\gamma^2}{4}=1 \quad \text{or} \quad \gamma=(8/3)^{\frac{1}{2}}=1\cdot633.$$

It is easily verified that this value of γ corresponds to the configuration in which each atom has three atoms each on the hexagonal planes above and below it, all at the distance ρ_0 away. Thus every atom has twelve nearest neighbours at the distance ρ_0, six in its own hexagonal plane and three each on the neighbouring planes. The structure is known as the hexagonal close-packed structure, for it is an arrangement of hard spheres in the closest packing.

Owing to the symmetry of the structure, a great number of the brackets (11.16), (11.17), (11.18) vanish identically and many of the rest are mutually related. It is found that the energy density (11.15) can be expressed as

$$u=\tfrac{1}{2}\{P(s_1^2+s_2^2+\tfrac{2}{3}s_1 s_2+\tfrac{1}{3}s_6^2)+Qs_3^2+R(s_4^2+s_5^2+2s_1 s_3+2s_2 s_3)+$$
$$+2U(s_1 u_2-s_2 u_2+s_6 u_1)+V(u_1^2+u_2^2)+Wu_3^2\}, \quad (12.20)$$

where $\mathbf{u}$ is the difference $\mathbf{u}(2)-\mathbf{u}(1)$ describing the internal strain and

$$P=\sqrt{2}\,\rho_0\{10\psi''(\rho_0^2)+16\psi''(2\rho_0^2)+81\psi''(3\rho_0^2)+...\}$$

$$Q=\frac{\sqrt{2}}{3}\rho_0\{32\psi''(\rho_0^2)+32\psi''(2\rho_0^2)+\tfrac{512}{3}\psi''(3\rho_0^2)+...\}$$

$$R=\frac{\sqrt{2}}{3}\rho_0\{8\psi''(\rho_0^2)+32\psi''(2\rho_0^2)+112\psi''(3\rho_0^2)+...\}$$

$$U=(\tfrac{2}{3})^{\frac{1}{2}}\{-2\psi''(\rho_0^2)+16\psi''(2\rho_0^2)-40\psi''(3\rho_0^2)+...\}$$

$$V=\frac{\sqrt{2}}{\rho_0}\left\{4\psi''(\rho_0^2)+16\psi''(2\rho_0^2)+\frac{12}{\rho_0}\psi'(2\rho_0^2)+...\right\}$$

$$W=\frac{\sqrt{2}}{\rho_0}\left\{16\psi''(\rho_0^2)+16\psi''(2\rho_0^2)+\frac{12}{\rho_0}\psi'(2\rho_0^2)+...\right\}.$$

(12.21)

In the above expressions we have put in the values $\gamma^2=8/3$, $a=\rho_0$ and $v_a=\gamma\sqrt{3}a^3/2=\sqrt{2}\,\rho_0^3$ for the close-packed structure.

We notice that the strains s_4, s_5, u_3 appear only in square terms. For stability, the corresponding coefficients must be positive, i.e.

$$R>0, \qquad W>0.$$

The rest of the terms are in two groups, namely

$$\tfrac{1}{2}\{\tfrac{1}{3}Ps_6^2 + 2Us_6u_1 + Vu_1^2\}$$

and

$$\tfrac{1}{2}\{P(s_1^2 + s_2^2 + \tfrac{2}{3}s_1s_2) + Qs_3^2 + R(2s_1s_3 + 2s_2s_3) + 2U(s_1u_2 - s_2u_2) + Vu_2^2\}.$$

Apart from a constant factor, the corresponding quadratic coefficients are

$$\begin{pmatrix} \tfrac{1}{3}P & U \\ U & V \end{pmatrix}, \tag{12.22}$$

and

$$\begin{pmatrix} P & \tfrac{1}{3}P & R & U \\ \tfrac{1}{3}P & P & R & -U \\ R & R & Q & 0 \\ U & -U & 0 & V \end{pmatrix}. \tag{12.23}$$

The principal minors of (12.22) and (12.23) are respectively

$$\tfrac{1}{3}P, \qquad \tfrac{1}{3}PV - U^2,$$

and

$$P, \ \tfrac{8}{9}P^2, \ \tfrac{4}{3}P(\tfrac{2}{3}PQ - R^2), \ 4(\tfrac{2}{3}PQ - R^2)(\tfrac{1}{3}PV - U^2).$$

Considering only the interactions between the nearest neighbours, we have, from (12.21),

$$P = 10\sqrt{2}\,\rho_0\psi''(\rho_0^2), \qquad Q = \frac{32\sqrt{2}}{3}\rho_0\psi''(\rho_0^2),$$

$$R = \frac{8\sqrt{2}}{3}\rho_0\psi''(\rho_0^2), \qquad U = -\frac{2\sqrt{2}}{3}\psi''(\rho_0^2), \tag{12.24}$$

$$V = \frac{4\sqrt{2}}{\rho_0}\psi''(\rho_0^2), \qquad W = \frac{16\sqrt{2}}{\rho_0}\psi''(\rho_0^2),$$

and thus

$$\tfrac{1}{3}PV - U^2 = 24(\psi''(\rho_0^2))^2, \qquad \tfrac{2}{3}PQ - R^2 = 128(\psi''(\rho_0^2))^2.$$

All the stability conditions are evidently fulfilled for this case as $\psi''(\rho_0^2) > 0$ (see Fig. 23).

The structure provides a simple example where internal strains are induced by elastic deformations. For the determination of the internal strain **u** we have in the present case:

$$0 = \frac{\partial u}{\partial u_1} = Us_6 + Vu_1,$$

$$0 = \frac{\partial u}{\partial u_2} = U(s_1 - s_2) + Vu_2, \tag{12.25}$$

$$0 = \frac{\partial u}{\partial u_3} = Wu_3.$$

When the internal strains are eliminated from the energy density (12.20) with the help of (12.25), we obtain the strain energy function

$$u = \frac{1}{2}\left\{\left(P - \frac{U^2}{V}\right)(s_1^2 + s_2^2) + 2\left(\tfrac{1}{3}P + \frac{U^2}{V}\right)s_1 s_2 + Q s_3^2 + 2R s_1 s_3 + \right.$$
$$\left. + R(s_4^2 + s_5^2) + \left(\frac{P}{3} - \frac{U^2}{V}\right)s_6^2 + 2R s_2 s_3\right\}.$$

$$(12.26)$$

The non-vanishing elastic constants are thus given by

$$c_{11} = c_{22} = P - \frac{U^2}{V}, \qquad c_{12} = \tfrac{1}{3}P + \frac{U^2}{V}, \qquad c_{33} = Q,$$

$$c_{13} = c_{23} = c_{44} = c_{55} = R, \qquad c_{66} = \tfrac{1}{3}P - \frac{U^2}{V} = \tfrac{1}{2}(c_{11} - c_{12}).$$

$$(12.27)$$

We observe that all the Cauchy relations (11.31) except $c_{12} = c_{66}$ are fulfilled in this case. Evidently if the latter relation is also to be fulfilled, the constant U must vanish; in view of (12.25) this occurs only if no internal strains are induced by elastic deformations.

If only the interactions between the nearest neighbours are considered, we see from (12.24) that all the constants P, Q, R, U, V, W, and hence also all the elastic constants, are proportional to $\psi''(\rho_0^2)$. In fact, on substituting (12.24) in the expressions (12.27) for the elastic constants, we find that the elastic constants are in the following ratios to one another:

$$c_{33} : c_{11} : c_{12} : c_{13} = 32 : 29 : 11 : 8.$$

The only non-metallic crystal of the correct symmetry for which the complete set of elastic constants is known seems to be beryl, $Be_3Al_2(SiO_3)_6$. It is interesting to observe that although it has a very complicated structure, the ratios between the elastic constants

$$c_{33} : c_{11} : c_{12} : c_{13} = 28 \cdot 6 : 32 \cdot 6 : 11 \cdot 6 : 8$$

are fairly close to the above calculated values and the Cauchy relation $c_{13} = c_{44}$ is very closely fulfilled, as in the above theoretical model.

The basis of the above considerations of stability is essentially qualitative.† In view of the short range of the atomic forces it has been assumed that the nearest neighbour interactions predominate, so that in the equilibrium configuration the nearest neighbour distance is practically equal to ρ_0. Moreover, it has been supposed that the second neighbours most probably lie beyond the inflexion point of the potential

† A more elaborate calculation which takes account of the Fermi energy for metals has been published by F. R. N. Nabarro and J. H. O. Varley, *Proc. Camb. Phil. Soc.* **48**, 316 (1952).

function $\psi(r^2)$. Such considerations have been verified quantitatively by Misra[†] for the cubic Bravais lattices, using a potential function of the following form:

$$\psi(r^2) = \phi(r) = \frac{A}{r^m} - \frac{B}{r^n} \quad (m > n), \tag{12.28}$$

where the first and second terms correspond respectively to the repulsive and attractive forces. The potential function has the general form as already depicted in Fig. 23. It is convenient to express the constants A and B in terms of the energy minimum $-u_0$ and the equilibrium distance ρ_0; the latter are related to A and B by the following relations:

$$-u_0 = \psi(\rho_0^2) = \frac{A}{\rho_0^m} - \frac{B}{\rho_0^n}, \tag{12.29}$$

$$0 = \psi'(\rho_0^2) = -\frac{m}{2} \frac{A}{\rho_0^{m+2}} + \frac{n}{2} \frac{B}{\rho_0^{n+2}}. \tag{12.30}$$

When A and B are eliminated from (12.28) with (12.29), (12.30), we find the following:

$$\psi(r^2) = u_0 \frac{mn}{m-n} \left\{ \frac{1}{m} \left(\frac{\rho_0}{r} \right)^m - \frac{1}{n} \left(\frac{\rho_0}{r} \right)^n \right\}. \tag{12.31}$$

$\psi'(r^2)$, $\psi''(r^2)$ are thus given by

$$\psi'(r^2) = \frac{mn}{2(m-n)} \frac{u_0}{\rho_0^2} \left\{ -\left(\frac{\rho_0}{r} \right)^{m+2} + \left(\frac{\rho_0}{r} \right)^{n+2} \right\}, \tag{12.32}$$

$$\psi''(r^2) = \frac{mn}{4(m-n)} \frac{u_0}{\rho_0^4} \left\{ (m+2)\left(\frac{\rho_0}{r} \right)^{m+4} - (n+2)\left(\frac{\rho_0}{r} \right)^{n+4} \right\}. \tag{12.33}$$

When (12.33) is substituted in the expressions (12.4), (12.5) for the elastic constants c_{11} and c_{12} of the cubic Bravais lattices, we obtain

$$c_{11} = \left(\frac{u_0}{2v_a} \right) \frac{mn}{m-n} \left\{ (m+2)\left(\frac{\rho_0}{a} \right)^m \sum_l \frac{l_1^4}{(l_1^2+l_2^2+l_3^2)^{(m+4)/2}} - \right.$$
$$\left. -(n+2)\left(\frac{\rho_0}{a} \right)^n \sum_l \frac{l_1^4}{(l_1^2+l_2^2+l_3^2)^{(n+4)/2}} \right\}, \tag{12.34}$$

$$c_{12} = \left(\frac{u_0}{2v_a} \right) \frac{mn}{m-n} \left\{ (m+2)\left(\frac{\rho_0}{a} \right)^m \sum_l \frac{l_1^2 l_2^2}{(l_1^2+l_2^2+l_3^2)^{(m+4)/2}} - \right.$$
$$\left. -(n+2)\left(\frac{\rho_0}{a} \right)^n \sum_l \frac{l_1^2 l_2^2}{(l_1^2+l_2^2+l_3^2)^{(n+4)/2}} \right\}. \tag{12.35}$$

[†] R. D. Misra, *Proc. Camb. Phil. Soc.* **36**, 173 (1940).

The stability conditions (12.6) thus become

$$(m+2)\left(\frac{\rho_0}{a}\right)^m \sum_l \frac{l_1^2 l_2^2}{(l_1^2+l_2^2+l_3^2)^{(m+4)/2}} > (n+2)\left(\frac{\rho_0}{a}\right)^n \sum_l \frac{l_1^2 l_2^2}{(l_1^2+l_2^2+l_3^2)^{(n+4)/2}},$$

$$\tag{12.36}$$

$$(m+2)\left(\frac{\rho_0}{a}\right)^m \left\{ \sum_l \frac{l_1^4}{(l_1^2+l_2^2+l_3^2)^{(m+4)/2}} - \sum_l \frac{l_1^2 l_2^2}{(l_1^2+l_2^2+l_3^2)^{(m+4)/2}} \right\}$$

$$> (n+2)\left(\frac{\rho_0}{a}\right)^n \left\{ \sum_l \frac{l_1^4}{(l_1^2+l_2^2+l_3^2)^{(n+4)/2}} - \sum_l \frac{l_1^2 l_2^2}{(l_1^2+l_2^2+l_3^2)^{(n+4)/2}} \right\}. \tag{12.37}$$

The parameter a specifying the dimension of the lattice cell is determined by the equilibrium conditions. Owing to the cubic symmetry of the lattices, the equilibrium conditions (11.14) are identically fulfilled for $\alpha \neq \beta$ and, for $\alpha = \beta = 1, 2, 3$, the sum in (11.14) has the same value. Thus, putting $\alpha = \beta = 1$ and using the expression (12.32) for $\psi'(r^2)$, we obtain the single equilibrium condition

$$\left(\frac{u_0}{2v_a}\right)\frac{mn}{(m-n)}\left\{ -\left(\frac{\rho_0}{a}\right)^m \sum_l \frac{l_1^2}{(l_1^2+l_2^2+l_3^2)^{(m+2)/2}} + \right.$$

$$\left. +\left(\frac{\rho_0}{a}\right)^n \sum_l \frac{l_1^2}{(l_1^2+l_2^2+l_3^2)^{(n+2)/2}} \right\} = 0$$

or 　　$$\left(\frac{\rho_0}{a}\right)^m \sum_l \frac{l_1^2}{(l_1^2+l_2^2+l_3^2)^{(m+2)/2}} = \left(\frac{\rho_0}{a}\right)^n \sum_l \frac{l_1^2}{(l_1^2+l_2^2+l_3^2)^{(n+2)/2}}. \tag{12.38}$$

Dividing the stability conditions (12.36) and (12.37) by (12.38), we find

$$\frac{(m+2)\left\{ \sum_l \dfrac{l_1^2 l_2^2}{(l_1^2+l_2^2+l_3^2)^{(m+4)/2}} \right\}}{\left\{ \sum_l \dfrac{l_1^2}{(l_1^2+l_2^2+l_3^2)^{(m+2)/2}} \right\}} > \frac{(n+2)\left\{ \sum_l \dfrac{l_1^2 l_2^2}{(l_1^2+l_2^2+l_3^2)^{(n+4)/2}} \right\}}{\left\{ \sum_l \dfrac{l_1^2}{(l_1^2+l_2^2+l_3^2)^{(n+2)/2}} \right\}},$$

$$\tag{12.39}$$

$$\frac{(m+2)\left\{ \sum_l \dfrac{l_1^4}{(l_1^2+l_2^2+l_3^2)^{(m+4)/2}} - \sum_l \dfrac{l_1^2 l_2^2}{(l_1^2+l_2^2+l_3^2)^{(m+4)/2}} \right\}}{\left\{ \sum_l \dfrac{l_1^2}{(l_1^2+l_2^2+l_3^2)^{(m+2)/2}} \right\}}$$

$$> \frac{(n+2)\left\{ \sum_l \dfrac{l_1^4}{(l_1^2+l_2^2+l_3^2)^{(n+4)/2}} - \sum_l \dfrac{l_1^2 l_2^2}{(l_1^2+l_2^2+l_3^2)^{(n+4)/2}} \right\}}{\left\{ \sum_l \dfrac{l_1^2}{(l_1^2+l_2^2+l_3^2)^{(n+2)/2}} \right\}},$$

$$\tag{12.40}$$

which, we observe, depend now only on the exponents m, n in the force law and are independent of the values of u_0 and ρ_0.

In view of the cubic symmetry of the lattices we clearly have the relations ($p = $ any number)

$$\sum_l \frac{l_1^2}{(l_1^2+l_2^2+l_3^2)^{p/2}} = \frac{1}{3}\sum_l \frac{(l_1^2+l_2^2+l_3^2)}{(l_1^2+l_2^2+l_3^2)^{p/2}} = \frac{1}{3}\sum_l \frac{1}{(l_1^2+l_2^2+l_3^2)^{(p-2)/2}},$$

and

$$\sum_l \frac{l_1^2 l_2^2}{(l_1^2+l_2^2+l_3^2)^{p/2}} = \frac{1}{6}\sum_l \frac{(l_1^2+l_2^2+l_3^2)^2-l_1^4-l_2^4-l_3^4}{(l_1^2+l_2^2+l_3^2)^{p/2}}$$

$$= \frac{1}{6}\sum_l \frac{1}{(l_1^2+l_2^2+l_3^2)^{(p-4)/2}} - \frac{1}{2}\sum_l \frac{l_1^4}{(l_1^2+l_2^2+l_3^2)^{p/2}}.$$

Hence we can express all the lattice sums in (12.39) and (12.40) in terms of lattice sums of the form

$$S_p^{(0)} = \sum_l \frac{1}{(l_1^2+l_2^2+l_3^2)^{p/2}}, \qquad S_p^{(4)} = \sum_l \frac{l_1^4}{(l_1^2+l_2^2+l_3^2)^{p/2}}; \quad (12.41)$$

namely, we can write (12.39) and (12.40) as

$$\left(\frac{m+2}{2}\right)\left(1 - 3\frac{S_{m+4}^{(4)}}{S_m^{(0)}}\right) > \left(\frac{n+2}{2}\right)\left(1 - 3\frac{S_{n+4}^{(4)}}{S_n^{(0)}}\right), \quad (12.42)$$

$$\left(\frac{m+2}{2}\right)\left(9\frac{S_{m+4}^{(4)}}{S_m^{(0)}} - 1\right) > \left(\frac{n+2}{2}\right)\left(9\frac{S_{n+4}^{(4)}}{S_n^{(0)}} - 1\right). \quad (12.43)$$

The values of the two expressions

$$A(p) = \left(\frac{p+2}{2}\right)\left\{9\frac{S_{p+4}^{(4)}}{S_p^{(0)}} - 1\right\}, \quad B(p) = \left(\frac{p+2}{2}\right)\left\{1 - 3\frac{S_{p+4}^{(4)}}{S_p^{(0)}}\right\}$$

$$(12.44)$$

have been calculated by Misra for various values of p and for all three cubic Bravais lattices. They are plotted against p in Fig. 25. Since $m > n$ the stability conditions (12.42) and (12.43) are equivalent to the requirement that the curves should increase monotonically with increasing p. The curves given in Fig. 25 show that the f.c.c. lattice is always stable and the s.c. lattice is always unstable. One of the curves for the b.c.c. lattice is seen to decrease monotonically except for very small p. Since the rapidly varying repulsive force in reality corresponds to an exponent $m \sim 10$, the structure is thus very unlikely to be stable for atoms interacting with central forces.

Of course, a structure may be stable with respect to homogeneous deformations, yet unstable for other types of small deformations. It is most convenient to consider the totality of all small deformations in terms of the normal coordinates. If it is found by solving the equations of motion that the frequencies of all the normal modes are real, then the lattice is stable for all small deformations; otherwise the lattice is unstable. For an imaginary frequency means that the system, subject to a small displacement, will disrupt exponentially with time instead of executing an oscillatory motion about the equilibrium configuration.

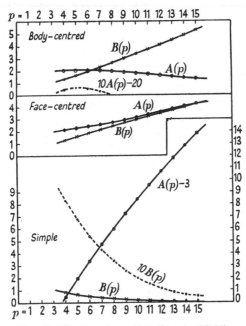

Fig. 25. The functions $A(p)$, $B(p)$ in (12.44).

We have already seen that the long acoustic waves are essentially determined by the elastic properties; the stability against homogeneous deformations in fact only ensures that the frequencies of the long lattice waves will be real.

In § 6 we have quoted the result that the frequencies of the normal vibrations are determined by the equation (6.14),

$$|C_{\alpha\beta}(\genfrac{}{}{0pt}{}{y}{kk'}) - \omega^2 \delta_{\alpha\beta}\delta_{kk'}| = 0.$$

In order that the solutions for ω^2 should be positive, the condition is that all the principal minors of the $3n \times 3n$ matrix

$$[C_{\alpha\beta}(\genfrac{}{}{0pt}{}{y}{kk'})]$$

should be positive. If the lattice is to be stable for all small deformations, the above condition must be fulfilled for all values of the wave-number vector **y**. This problem has been considered by Born[†] for the model of a linear chain of identical particles. It is found that the stability for all waves follows from the stability against homogeneous deformations, if the second derivative of the interaction potential is positive between nearest neighbours and negative between all other neighbours. As we have seen, this latter provision is probably always fulfilled in reality. A three-dimensional analogue has been developed by Power,[‡] who considers specifically the f.c.c. lattice. For this lattice she shows that the stability against homogeneous deformation ensures also stability against all deformations, if only the interactions between the nearest neighbours are taken into account.

13. Relative stability and polymorphism

The consideration of mechanical stability alone cannot decide the actual structure of a crystal. The actual structure is determined by the Gibbs free energy (often called today free enthalpy)

$$U + pV - TS; \qquad (13.1)$$

the structure, which is thermodynamically the most stable, has the lowest free energy.

At zero pressure and the absolute zero of temperature, a solid thus crystallizes in the structure with the lowest energy. As alternative structures frequently have energies which are only slightly higher, theoretical discussion of the relative stability between different structures is often subject to considerable uncertainty, and has not been altogether successful in individual cases. Without attempting to give a general survey[§] of the subject, we shall discuss a few simple instances to illustrate some factors which are important in determining the structure of a crystal.

For structures with sufficient symmetry to suppress polarization effects, the Madelung energy and the overlap energy are the most important energy terms. Hund[||] has thus considered the relative stability between alternative coordinate lattices, taking into account only these energy terms. As an example, let us consider the relative stability of the three structures CsCl, NaCl, and ZnS for ionic compounds of the type X^+Y^- (the zinc blende and *zincite* structures of ZnS

[†] M. Born, *Proc. Camb. Phil. Soc.* **36**, 160 (1940), Appendix.
[‡] S. C. Power, *Proc. Camb. Phil. Soc.* **38**, 62 (1942).
[§] For a more comprehensive survey see L. Pauling, *The Nature of the Chemical Bond*, chap. x (2nd ed., Cornell Univ. Press (1948)).
[||] F. Hund, *Zeit. f. Phys.* **34**, 833 (1925).

are essentially so little different that their difference may be ignored in the following discussion). We shall present Hund's considerations in a somewhat simplified manner. Thus we take into account only the overlap energy between nearest neighbours; for the overlap energy we use the inverse power expression (3.1 a). The energy per cell can be written as

$$u = -\frac{\alpha(ze)^2}{r} + \frac{M\lambda_{+-}}{r^n}, \tag{13.2}$$

where the Madelung constant, denoted here by α, and the coordination number M have the following values for the three structures:

	CsCl	NaCl	ZnS†
α	1·7627	1·7476	1·6381
M	8	6	4

$$\tag{13.3}$$

Remembering the equilibrium condition

$$\left(\frac{du}{dr}\right)_{r_0} = 0, \tag{13.4}$$

we find that

$$u(r_0) = -\left(\frac{\alpha^n}{M}\right)^{1/(n-1)}\left\{\left(\frac{n-1}{n}\right)\left[\frac{(ze)^{2n}}{n\lambda_{+-}}\right]^{1/(n-1)}\right\}. \tag{13.5}$$

Since the second factor is the same for all structures, the relative stability depends only on the value of α^n/M; the structure having the highest value for α^n/M is the most stable. For very small values of n the value of M predominates in determining the relative magnitude of α^n/M, so in this extreme case we have in the order of stability (from the most to the least stable):

ZnS, NaCl, CsCl   (small n).

For sufficiently large values of n, on the other hand, the effect of α preponderates and the above ordering is exactly reversed:

CsCl, NaCl, ZnS   (large n).

Let α, M and α', M' denote the constants for two alternative structures. Consider the number p defined by the following equation:

$$\frac{\alpha^p}{M} = \frac{\alpha'^p}{M'}. \tag{13.6}$$

Clearly, for $n > p$ one structure is the more stable and for $n < p$ the

† The values refer to the zinc blende structure; for the zincite structure
$$\alpha = 1\cdot639, \qquad M = 4.$$

other is the more stable. Writing (13.6) as

$$p = \frac{\log(M/M')}{\log(\alpha/\alpha')},$$

we can immediately calculate the value of p for the three pairs of structures ZnS—NaCl, ZnS—CsCl, and NaCl—CsCl; the results are

	ZnS—NaCl	ZnS—CsCl	NaCl—CsCl
p	6·3	9·5	33

From these values and our knowledge of the relative stability for the extreme values of n we can form the following table:

$$n < 6·3 \qquad \text{ZnS, NaCl, CsCl;}$$
$$6·3 < n < 9·5 \qquad \text{NaCl, ZnS, CsCl;}$$
$$9·5 < n < 33 \qquad \text{NaCl, CsCl, ZnS;} \qquad (13.7)$$
$$33 < n \qquad \text{CsCl, NaCl, ZnS;}$$

here the three structures are given in the order of decreasing stability. We notice that, according to this simple theory, for n between 6·3 and 33 the NaCl structure is the most stable.

In Table 24 the actual structures of a number of diatomic salts are indicated by the corresponding coordination numbers (8 for CsCl, 6 for

TABLE 24
Crystal Structures of Diatomic Ionic Crystals

	F 1·36	Cl 1·81	Br 1·95	I 2·11		O 1·40	S 1·84	Se 1·98	Te 2·21
Li 0·60	6	6	6	6	Be 0·31	4	4	4	4
						0·22	0·17	0·16	0·14
Na 0·95	6	6	6	6	Mg 0·65	6	6	6	4
						0·47	0·36	0·33	0·30
K 1·33	6	6	6	6	Ca 0·99	6	6	6	6
						0·71	0·54	0·50	0·45
Rb 1·49	6	6	6	6	Sr 1·13	6	6	6	6
						0·81	0·62	0·57	0·51
Cs 1·69	6	8	8	8	Ba 1·35	6	6	6	6
						0·97	0·73	0·68	0·61
Cu 0·96	..	4	4	4	Zn 0·74	4	4	4	4
Ag 1·26	6	6	6	4	Cd 0·97	6	4	4	4
					Hg 1·10	..	4	4	4
Tl ..	..	8	8	8	Mn ..	6	6 (4)	6 (4)	..
					Fe ..	6	..	..	..
					Co ..	6	..	..	..
					Ni ..	6	..	..	..
					Pb ..	..	6	6	6

Pauling's crystal radii are given beside the ions and, for the alkali-earth compounds, the radius ratio (r_+/r_-) is given under the coordination number.

NaCl, 4 for ZnS). The preponderance of the NaCl structure is evidently related to the fact that the exponent n is generally in the neighbourhood of 10 (cf. § 3).

Although in a general way the above theory is confirmed by the observed structures quoted in Table 24, it is by no means adequate to account for the structures in all individual cases. It is clearly out of the question to consider the salts with the CsCl structure on the above basis; for the required value of n would have to be greater than 30. Nor indeed, as we shall presently see, are the ZnS structures to be generally interpreted as due to small values of n.

The overlap energy between second neighbours can no longer be ignored if the negative ions are large compared with the positive ions. The most important effect of the interaction between second neighbours can be crudely reproduced by considering the ions as rigid spheres. In this crude model the energy is simply equal to the Madelung term

$$-\frac{\alpha(ze)^2}{r};$$

the radii of the ions effectively take over the function of the overlap forces in that the radii determine the value of the nearest neighbour distance r and hence also the energy. In order to see how the values of the radii affect the energy, let us imagine that, starting with $r_+ = r_-$, r_+ is progressively reduced while r_- is kept fixed (r_+, r_- being respectively the radii of the positive and negative ions). For each particular structure there is a critical ratio $\rho = r_+/r_-$ of the radii, at which the negative ions that are second neighbours come into contact. Evidently before this critical ratio is reached the nearest neighbour distance is $r = r_+ + r_-$; the corresponding energy

$$-\frac{\alpha(ze)^2}{(r_+ + r_-)}$$

thus decreases with decreasing r_+. Once, however, r_+ is reduced below the value ρr_- the lattice dimension becomes a constant which is determined by the distance $2r_-$ between the second neighbours. Accordingly, the energy remains at the constant value

$$-\frac{\alpha(ze)^2}{r_-(1+\rho)}.$$

Let us now compare the energies of the above three structures. The values for the critical ratio of the three structures are readily found to be

	CsCl	NaCl	ZnS
ρ	$\sqrt{3}-1 = 0{\cdot}732$	$\sqrt{2}-1 = 0{\cdot}414$	$\sqrt{\tfrac{3}{2}}-1 = 0{\cdot}225$

For $r_+ > (\sqrt{3}-1)r_-$ the second neighbours are not in contact and the nearest neighbour distance is equal to $r_+ + r_-$ in all three structures. The corresponding values of the lattice energy are thus proportional to the Madelung constants; hence we have, in the order of decreasing stability,

$$\text{CsCl, NaCl, ZnS.}$$

For $r_+ < (\sqrt{3}-1)r_-$ the energy of the CsCl structure remains at the constant value

$$-\frac{1 \cdot 7627(ze)^2}{1 \cdot 7321 r_-}, \tag{13.8}$$

whereas the energies of the NaCl and ZnS structures, given by

$$-\frac{1 \cdot 7476(ze)^2}{r_+ + r_-}, \tag{13.9}$$

$$-\frac{1 \cdot 6381(ze)^2}{r_+ + r_-}, \tag{13.10}$$

continue to decrease with decreasing r_+. Equating (13.8) and (13.9), we find that

$$\frac{r_+}{r_-} = \frac{1 \cdot 7321 \times 1 \cdot 7476}{1 \cdot 7627} - 1 = 0 \cdot 717. \tag{13.11}$$

For a ratio of the radii smaller than this value, the NaCl structure becomes more stable than the CsCl structure. Similarly by equating (13.8) and (13.10) we find that for a radius ratio smaller than

$$\frac{r_+}{r_-} = \frac{1 \cdot 6381 \times 1 \cdot 7321}{1 \cdot 7627} - 1 = 0 \cdot 609, \tag{13.12}$$

the ZnS structure becomes also more stable than the CsCl structure. For $r_+ < (\sqrt{2}-1)r_-$ the energy of the NaCl structure ceases to decrease so that the ZnS structure becomes eventually the most stable. Using the same argument as before, we find that the ZnS structure becomes more stable than the NaCl structure for a radius ratio smaller than

$$\frac{r_+}{r_-} = \frac{1 \cdot 6381 \times 1 \cdot 4142}{1 \cdot 7476} - 1 = 0 \cdot 325. \tag{13.13}$$

Using (13.11), (13.12), and (13.13), we find that for various values of the radius ratio we can arrange the structures in the order of decreasing stability as follows:

$$0{\cdot}717 < \frac{r_+}{r_-} \qquad\qquad \text{CsCl, NaCl, ZnS};$$

$$0{\cdot}609 < \frac{r_+}{r_-} < 0{\cdot}717 \quad \text{NaCl, CsCl, ZnS};$$

$$0{\cdot}325 < \frac{r_+}{r_-} < 0{\cdot}609 \quad \text{NaCl, ZnS, CsCl};$$

$$\frac{r_+}{r_-} < 0{\cdot}325 \quad \text{ZnS, NaCl, CsCl}.$$

(13.14)

In Table 24 Pauling's crystal radii for the ions are quoted. The corresponding values for the radius ratio are given for the compounds of the alkali-earth ions. The strict correlation between the appearance of the ZnS structure and the rule (13.14) shows clearly that for these compounds the stability of the ZnS structure must be largely due to the strong repulsion between second neighbours.

Clearly (13.7) and (13.14) each cover only one special factor which is important for the consideration of relative stability. Nor indeed, by considering both of these factors together, can we hope to account for the observed crystal structure in all individual cases. In fact, all theoretical calculations carried out so far have failed to account for the stability of the CsCl structure of the three caesium salts. In particular May[†] has carried out such calculations for the CsCl crystal, using the energy expression due to Born, Mayer, and Huggins; the energy expression, we remember (see § 3), includes the dipole and quadrupole van der Waals potentials as well as the overlap energy between second neighbours. Yet he found an energy for the NaCl structure 1 per cent. lower as compared with the CsCl structure. At 718° K, the CsCl crystal undergoes a phase transition from the CsCl structure to the NaCl structure. Assuming that the energy difference between the two structures is approximately equal to the latent heat at the transition, May found that the van der Waals potentials as estimated by Mayer have to be raised by at least a factor 3·5 in order to reproduce correctly the energy difference. Since Mayer's estimates are not expected to be so seriously in error, May's results show that further physical factors will have to be taken into consideration before we can hope to account for the stability of the CsCl structure.

The caesium salts are by no means the only instances for which elaborate theoretical calculations have failed to account for the observed

† A. May, *Phys. Rev.* **52**, 339 (1937); *Phys. Rev.* **54**, 629 (1938).

crystal structure. Mayer† has calculated the lattice energies for the halides of Ag and Tl both for the observed structure and for a hypothetical structure. It was found that for AgI, TlCl, and TlBr the calculated energy is lower for the hypothetical structure. In the case of AgI the calculated values for the lattice constant and the cohesive energy are in such marked disagreement with the empirical values that Mayer considered the results as a clear indication of the partial formation of valence bonds. Such possibilities make it very difficult to predict theoretically the stable structure with certainty.

The structure most stable at the absolute zero of temperature and zero pressure need not, of course, be stable under varied conditions of temperature and pressure. Bridgman's extensive investigations‡ on high pressure phenomena have in fact shown that polymorphic transition (phase transition involving change of structure) is of common occurrence; of 150-odd substances examined up to 1931 such transitions were detected in approximately 40 within the temperature range 20–200° C. and under pressures up to 12,000 kg./cm.² (1·033 kg./cm.² = 1 atmosphere). Let us consider the possibility of polymorphism under high pressure, at the absolute zero of temperature. The Gibbs free energy (13.1) reduces at $T = 0°$ K. to the quantity (enthalpy)

$$U+pV. \tag{13.15}$$

If there is a polymorphic transition to an alternative structure at p the two structures are in thermodynamical equilibrium at this pressure and the corresponding free energies must thus be equal:

$$U+pV = U'+pV', \tag{13.16}$$

where U', V' denote the energy and volume of the second structure. Let us write (13.16) as

$$p = \frac{U'-U}{V-V'}. \tag{13.17}$$

The right-hand side of (13.17) is, of course, a function of pressure. However, as a first approximation, we can simply ignore the pressure dependence of the expression on the right and use its value corresponding to zero pressure:

$$p \cong \frac{U'_0-U_0}{V_0-V'_0}, \tag{13.18}$$

where the suffix 0 signifies the value at zero pressure. Later we shall see that (13.18) is a crude but usable approximation.

† J. E. Mayer, *Journ. Chem. Phys.* 1, 327 (1933).
‡ P. W. Bridgman, *The Physics of High Pressure*, chap. viii (Bell & Sons, London (1949)).

Let us once more take the diatomic compounds X^+Y^- as an example and use the simple energy expression (13.2). It follows immediately from the equilibrium condition and (13.2) that the nearest neighbour distance at zero pressure is given by

$$r_0 = \left(\frac{nM\lambda_{+-}}{\alpha(ze)^2}\right)^{1/(n-1)} \tag{13.19}$$

and the energy per cell can be written as

$$u(r_0) = -\frac{\alpha(ze)^2}{r_0}\left(\frac{n-1}{n}\right). \tag{13.20}$$

Since the volume per cell is proportional to the third power of the nearest neighbour distance, let us denote the volume per cell by sr^3 and $s'r'^3$ respectively for the two structures. Thus using (13.19) and (13.20) in (13.18), we obtain the following approximate expression for the transition pressure:

$$p \cong \frac{\alpha(ze)^2}{sr_0^4} F_n, \tag{13.21}$$

with F_n defined by

$$F_n = \left(\frac{n-1}{n}\right)\left\{\frac{1-\left(\frac{\alpha'}{\alpha}\right)^{n/(n-1)}\left(\frac{M}{M'}\right)^{1/(n-1)}}{1-\frac{s'}{s}\left(\frac{M'\alpha}{M\alpha'}\right)^{3/(n-1)}}\right\},$$

where the quantities with and without prime refer respectively to the high and low pressure modifications of the crystal.

We have seen that it follows from the simple energy expression (13.2) that practically all ionic compounds X^+Y^- should have the NaCl structure at $p = 0$. Thus (13.21), which is based on (13.2), is at best only applicable to cases where the actual structure at $p = T = 0$ is of the NaCl type. Hence let us consider possible polymorphic transitions to the ZnS and CsCl structures. The values of F_n for such transitions are given in Table 25 for a few values of the exponent n. It follows from the negative values of F_n for the transition NaCl → ZnS that the corresponding transition pressure is negative. In other words, a crystal with the NaCl structure at zero pressure will undergo a polymorphic transition into the ZnS structure only under tension. On the other hand the positive values of F_n for the NaCl → CsCl transition show that such transitions will occur under pressure.

Using the values of n and r_0 given in Table 9 (remember that $n = r_0/\rho - 1$, p. 27) and the corresponding values of F_n obtained by interpolation from Table 25, we can calculate readily the pressure for the transition $NaCl \to CsCl$ with (13.21) for the alkali halides. The transition pressures calculated in this way are given in Table 26.

TABLE 25

Values of F_n

(See (13.21))

Transitions	s	s'	$n = 7$	8	9	10	11
$NaCl \to ZnS$	2	$16/3\sqrt{3}$	$-0\cdot022$	$-0\cdot043$	$-0\cdot055$	$-0\cdot064$	$-0\cdot070$
$NaCl \to CsCl$	2	$8/3\sqrt{3}$	$0\cdot28$	$0\cdot20$	$0\cdot16$	$0\cdot13$	$0\cdot11$

TABLE 26

Pressure for Polymorphic Transition in the Alkali Halides

(Pressure in kg./cm.²)

		Li^+	Na^+	K^+	Rb^+	Cs^+
F^-	(13.21)	310,000	200,000	88,000	68,000	35,000
Cl^-	(13.21)	140,000	74,000	36,000	31,000	..
	exp.	..	20,000	20,000	5,500	..
	Jacobs	..	..	74,000	39,000	..
Br^-	(13.21)	105,000	53,000	29,000	25,000	..
	exp.	..	..	19,000	5,000	..
	Jacobs	..	..	59,000	30,000	..
I^-	(13.21)	68,000	39,000	21,000	15,000	..
	exp.	..	..	18,000	4,000	..
	Jacobs	..	..	49,000	22,000	..

In the pressure range so far achieved (up to 100,000 kg./cm.²) polymorphic transition has been found only in six of the alkali halide crystals. The experiments have been carried out in the temperature range 20–200° C. The transition pressures were found to be practically independent of temperature (this is usually not the case with other crystals). It is thus fairly safe to assume that the transition pressures at the absolute zero of temperature are not appreciably different from the room-temperature values. The experimentally measured values for the transition pressure are also given in Table 26. We see that there is no quantitative agreement between the theoretical and the experimental values. From the fact that CsCl, CsBr, and CsI actually crystallize in the CsCl structure at low temperatures (whereas theory predicts the NaCl structure), we may infer that the NaCl–CsCl transition actually occurs in these salts under negative pressures. In other words, for these salts the theoretical transition pressures are too high (algebraically). Table 26 shows that this is still the case with the rubidium salts. For the potassium salts the theoretical and experimental values are in rough agreement. Experimentally no transition has been found in the case of NaI up to a pressure of 100,000 kg./cm.², so it is almost certain that none of the salts of Na and Li has a transition in this range. The theoretical transition pressures are thus too low for these salts. These facts bring into evidence a systematic tendency for the theoretical transition pressures to be too high for the salts of the heavier metallic elements and too low for the salts of the lighter metallic elements.

We should expect the consideration of the van der Waals potentials to correct the theoretical results in the right direction. For the van der Waals potentials favour the CsCl structure as compared with the NaCl structure, as is evident from May's results. As the van der Waals potentials increase rapidly from the lithium to the caesium salts, they should thus lower the transition pressures of the salts of the heavier metallic elements as compared with the salts of the lighter metallic elements. This anticipation was borne out by the calculations made by Jacobs,[†] who used the energy expression of Born, Mayer, and Huggins for his calculations. His results are also given in Table 26. His values are in no better agreement with the observed values, but they do in fact display a more marked increase of the transition pressure from the caesium to the lithium salts. In particular, he found that no transition should occur in NaI. Transition pressures for the alkali halide crystals have also been calculated by Löwden,[‡] who used direct wave-mechanical methods in a few instances and the exponential overlap potential for the rest. His results are practically the same as the values calculated above with (13.21). The close agreement is clearly due to the fact that the overlap energy between second neighbours and the van der Waals potentials have been neglected in both treatments.

We can easily verify that the approximation involved in (13.21) is not responsible for the discrepancy between the theoretical and experimental results. A second approximation for the transition pressure can be obtained by taking into account, to the first order, the pressure dependence of the right-hand side of (13.17). Thus expanding

$$p = -\frac{du}{dv}$$

† R. B. Jacobs, *Phys. Rev.* **54**, 468 (1938).
‡ Per-Olav Löwden, Univ. Uppsala Diss. (Almqvist und Wiksells Boktryckeri, A.B., Uppsala, 1948), *Theoretical Investigation into Some Properties of Ionic Crystals.*

with respect to the volume difference $v-v_0$, we obtain to the first order

$$p = -\left(\frac{d^2u}{dv^2}\right)_0 (v-v_0),$$

or
$$v = v_0 - \frac{p}{(d^2u/dv^2)_0}. \tag{13.22}$$

Similarly, by expanding the energy u, we find that

$$u = u_0 + \frac{1}{2}\left(\frac{d^2u}{dv^2}\right)_0 (v-v_0)^2 = u_0 + \frac{1}{2}\frac{p^2}{(d^2u/dv^2)_0}, \tag{13.23}$$

where $v-v_0$ has been eliminated with the help of (13.22). Differentiating (13.2), we obtain

$$\left(\frac{d^2u}{dv^2}\right)_0 = \frac{r_0^2}{9v_0^2}\left\{-\frac{2\alpha(ze)^2}{r_0^3} + \frac{M\lambda_{+-}n(n+1)}{r_0^{n+2}}\right\}.$$

After λ_{+-} is eliminated by means of (13.19) a comparison with (13.20) shows that

$$\left(\frac{d^2u}{dv^2}\right)_0 = -\frac{nu_0}{9v_0^2}. \tag{13.24}$$

Using (13.22), (13.23) in (13.17), we obtain

$$p = \frac{u_0'-u_0-\dfrac{4\cdot5p^2}{n}\left[\dfrac{v_0'^2}{u_0'}-\dfrac{v_0^2}{u_0}\right]}{v_0-v_0'-\dfrac{9p}{n}\left[\dfrac{v_0'^2}{u_0'}-\dfrac{v_0^2}{u_0}\right]}. \tag{13.25}$$

If we write $\quad \Delta u_0 = u_0'-u_0, \quad \Delta v_0 = v_0'-v_0,$

we have approximately

$$\frac{v_0'^2}{u_0'}-\frac{v_0^2}{u_0} = \frac{2v_0\Delta v_0}{u_0} - \frac{v_0^2\Delta u_0}{u_0^2}.$$

Hence (13.25) can be rewritten as

$$p = \left(\frac{\Delta u_0}{-\Delta v_0}\right)\left\{\frac{1-\dfrac{4\cdot5p^2}{n}\left[\dfrac{2v_0}{u_0}\left(\dfrac{\Delta v_0}{\Delta u_0}\right)-\dfrac{v_0^2}{u_0^2}\right]}{1+\dfrac{9p}{n}\left[\dfrac{2v_0}{u_0}-\dfrac{v_0^2}{u_0^2}\left(\dfrac{\Delta u_0}{\Delta v_0}\right)\right]}\right\}. \tag{13.26}$$

Since in the present notation the first approximation p^0 is given by

$$p^0 = \left(\frac{\Delta u_0}{-\Delta v_0}\right), \tag{13.27}$$

we see that the second approximation differs from the first by the second factor in (13.26). In this correction factor, let us replace p by the first approximation $p^0 = -(\Delta u_0/\Delta v_0)$; thus we find that the correction factor is given approximately by

$$\left\{ \frac{1 + \dfrac{4 \cdot 5}{n}\left[\dfrac{2 p_0 v_0}{u_0} + \left(\dfrac{p_0 v_0}{u_0}\right)^2\right]}{1 + \dfrac{9}{n}\left[\dfrac{2 p_0 v_0}{u_0} + \left(\dfrac{p_0 v_0}{u_0}\right)^2\right]} \right\} \cong 1 - \frac{4 \cdot 5}{n}\left[\frac{2 p_0 v_0}{u_0} + \left(\frac{p_0 v_0}{u_0}\right)^2\right].$$

Comparing (13.20) and (13.21), we find that

$$\frac{p_0 v_0}{u_0} = -\frac{n F_n}{n-1}.$$

Hence the correction factor can be written as

$$1 + \frac{9}{(n-1)} F_n - \frac{4 \cdot 5 n}{(n-1)^2} F_n^2.$$

From the values of F_n given in Table 25 we see that the first approximation may involve an error of 10–30 per cent.; the inaccuracy is certainly not sufficiently large to account for the discrepancy between the theoretical and experimental values given in Table 26.

PART II

GENERAL THEORIES

IV

QUANTUM MECHANICAL FOUNDATION

14. Quantum mechanics of molecular systems†

LET us consider a system of nuclei and electrons and denote the properties of the former by capital letters (mass M, coordinates X, and momenta P), and of the latter by small letters (m, x, p). The kinetic energy of the nuclei is the operator

$$T_N = \sum \frac{1}{2M} P^2 = - \sum \frac{\hbar^2}{2M} \left(\frac{\partial^2}{\partial X^2} \right), \tag{14.1}$$

and that of the electrons

$$T_E = \sum \frac{1}{2m} p^2 = - \sum \frac{\hbar^2}{2m} \left(\frac{\partial^2}{\partial x^2} \right). \tag{14.2}$$

The total Coulomb energy of nuclei and electrons will be represented by $U(x, X)$. We further introduce the abbreviation

$$T_E + U = H_0 \left(x, \frac{\partial}{\partial x}, X \right). \tag{14.3}$$

H_0, we note, does not contain the momenta P of the nuclei and can be considered as the Hamiltonian of the electrons for fixed nuclei. Since the kinetic energy of the nuclei is usually small owing to their large masses, H_0 can be taken as the zeroth approximation for the actual Hamiltonian of the system:

$$H = T_E + U + T_N = H_0 + T_N, \tag{14.4}$$

from which the solutions may be found by the method of perturbation theory, regarding T_N as small.

The expansion parameter must clearly be some power of the mass ratio m/M_0, where M_0 can be taken as any one of the nuclear masses or their mean. It has been found that the correct choice is

$$\kappa = \left(\frac{m}{M_0} \right)^{\frac{1}{4}}. \tag{14.5}$$

† M. Born and R. Oppenheimer, *Ann. d. Phys.* **84**, 457 (1927).

Thus we write

$$T_N = \kappa^4 H_1\left(\frac{\partial}{\partial X}\right), \qquad H_1\left(\frac{\partial}{\partial X}\right) = -\sum\left(\frac{M_0}{M}\right)\frac{\hbar^2}{2m}\left(\frac{\partial^2}{\partial X^2}\right). \qquad (14.6)$$

H_1, however, is not of the same order of magnitude as H_0, as we shall presently see.

Now the total Hamiltonian (14.4) has the form

$$H = H_0 + \kappa^4 H_1 \qquad (14.7)$$

and the Schrödinger equation is

$$(H - E)\psi(x, X) = 0. \qquad (14.8)$$

Let us assume that the equation for the electronic motion, with the nuclei in arbitrary fixed positions,

$$(H_0 - E^0)\phi(x, X) = 0, \qquad (14.9)$$

is solved. Both the eigenvalues and eigenfunctions depend on the nuclear coordinates as parameters. Thus we denote them as

$$E^0 = \Phi_n(X); \qquad \phi = \phi_n(x, X), \qquad (14.10)$$

where n is the *electronic quantum number*. Considering the functions (14.10) to be known for a certain nuclear configuration X^0 and all neighbouring configurations, we shall try to solve the exact equation (14.8) under the assumption that the nuclear motion is confined to a small vicinity of X^0, so that $X - X^0$ can be considered as small. We express this supposition by writing

$$X - X^0 = \kappa u \qquad (14.11)$$

and using u as the nuclear coordinate. We shall see that the perturbation method can be consistently carried out only if the primary configuration X^0 is properly chosen. Expanding the functions (14.10), we obtain

$$\left. \begin{array}{l} \Phi_n(X) = \Phi_n(X^0 + \kappa u) = \Phi_n^{(0)} + \kappa\Phi_n^{(1)} + \kappa^2\Phi_n^{(2)} + \cdots \\ \phi_n(x, X) = \phi_n(x, X^0 + \kappa u) = \phi_n^{(0)} + \kappa\phi_n^{(1)} + \kappa^2\phi_n^{(2)} + \cdots \end{array} \right\} \qquad (14.12)$$

We notice that $\Phi_n^{(0)}$, $\phi_n^{(0)}$ are independent of u, $\Phi_n^{(1)}$, $\phi_n^{(1)}$ linear in u, $\Phi_n^{(2)}$, $\phi_n^{(2)}$ quadratic in u, etc. Similarly we can write

$$H_0(x, \partial/\partial x, X) = H_0(x, \partial/\partial x, X^0 + \kappa u) = H_0^{(0)} + \kappa H_0^{(1)} + \kappa^2 H_0^{(2)} + \cdots, \qquad (14.13)$$

where $H_0^{(r)}$ are operators with respect to x and homogeneous functions (degree indicated by the superscript) of u. Substituting (14.12) and

(14.13) in (14.9), we obtain the following series of equations by equating the coefficients of different powers of κ to zero:

(a) $\quad (H_0^{(0)} - \Phi_n^{(0)})\phi_n^{(0)} = 0,$

(b) $\quad (H_0^{(0)} - \Phi_n^{(0)})\phi_n^{(1)} = -(H_0^{(1)} - \Phi_n^{(1)})\phi_n^{(0)},$

(c) $\quad (H_0^{(0)} - \Phi_n^{(0)})\phi_n^{(2)} = -(H_0^{(1)} - \Phi_n^{(1)})\phi_n^{(1)} - (H_0^{(2)} - \Phi_n^{(2)})\phi_n^{(0)},$

$$\cdots \cdots \cdots \cdots \cdots \cdots \cdots \cdots \cdots \quad (14.14)$$

As
$$\frac{\partial}{\partial X} = \frac{1}{\kappa}\frac{\partial}{\partial u}, \quad (14.15)$$

it follows from (14.1) that the kinetic energy of the nuclei consists of only one term of the order κ^2, namely

$$T_N = \kappa^4 H_1 = \kappa^2 H_1^{(2)}; \qquad H_1^{(2)} = -\sum \left(\frac{M_0}{M}\right)\frac{\hbar^2}{2m}\left(\frac{\partial^2}{\partial u^2}\right). \quad (14.16)$$

Combining (14.16) with (14.13), we obtain for the exact Hamiltonian of the system

$$H = H_0^{(0)} + \kappa H_0^{(1)} + \kappa^2(H_0^{(2)} + H_1^{(2)}) + \kappa^3 H_0^{(3)} + \dots . \quad (14.17)$$

Different coefficients in an expansion with respect to κ can be considered to be of the same order of magnitude if the wave function $\psi(x, u)$ is essentially different from zero only in a domain which extends roughly as far in u as in x. We shall suppose this to be the case and proceed to solve the exact wave equation (14.8) by the usual perturbation method. Thus we write

$$\left.\begin{aligned} E &= \Phi_n^{(0)} + \kappa E_n^{(1)} + \kappa^2 E_n^{(2)} + \dots \\ \psi &= \psi_n^{(0)} + \kappa\psi_n^{(1)} + \kappa^2\psi_n^{(2)} + \dots \end{aligned}\right\} \quad (14.18)$$

Using (14.17) and (14.18) in the wave equation, we obtain the successive equations:

(a) $\quad (H_0^{(0)} - \Phi_n^{(0)})\psi_n^{(0)} = 0,$

(b) $\quad (H_0^{(0)} - \Phi_n^{(0)})\psi_n^{(1)} = -(H_0^{(1)} - E_n^{(1)})\psi_n^{(0)},$

(c) $\quad (H_0^{(0)} - \Phi_n^{(0)})\psi_n^{(2)} = -(H_0^{(1)} - E_n^{(1)})\psi_n^{(1)} - (H_0^{(2)} + H_1^{(2)} - E_n^{(2)})\psi_n^{(0)},$

$$\cdots \cdots \cdots \cdots \cdots \cdots \cdots \cdots \cdots \quad (14.19)$$

It follows from (14.14)(a) that $\phi_n^{(0)}(x) = \phi_n(x, X^0)$ is a solution of the zero-order equation (14.19)(a). But evidently we can multiply $\phi_n^{(0)}$ by an arbitrary function in u. Hence we have

$$\psi_n^{(0)}(x, u) = \chi^{(0)}(u)\phi_n^{(0)}(x), \quad (14.20)$$

where $\chi^{(0)}(u)$ is as yet completely arbitrary; but, as we shall see, it will be determined by the higher-order equations.

Consider next the first-order equation (14.19)(b). The equation is a linear inhomogeneous differential equation with respect to x; the solubility condition is that the inhomogeneous part (the right-hand side of the equation) must be orthogonal to the solution $\phi_n^{(0)}$ of the homogeneous equation. Thus we must have

$$\int \phi_n^{(0)}(x)(H_0^{(1)} - E_n^{(1)})\psi_n^{(0)}(x, u)\, dx$$
$$= \chi^{(0)}(u) \int \phi_n^{(0)}(x)(H_0^{(1)} - E_n^{(1)})\phi_n^{(0)}(x)\, dx = 0. \quad (14.21)$$

On the other hand, multiplying (14.14)(b) by $\phi_n^{(0)}$ and integrating over x, we find that

$$-\int \phi_n^{(0)}(H_0^{(1)} - \Phi_n^{(1)})\phi_n^{(0)}\, dx = \int \phi_n^{(0)}(H_0^{(0)} - \Phi_n^{(0)})\phi_n^{(1)}\, dx = 0. \quad (14.22)$$

Comparing (14.21) with (14.22), we obtain

$$\Phi_n^{(1)} = E_n^{(1)}. \quad (14.23)$$

The eigenvalue E, hence also $E_n^{(0)}$, $E_n^{(1)}$,..., must be constants independent of u, whereas $\Phi_n^{(1)}$ is a linear homogeneous function of u. Thus (14.23) can be fulfilled only if $\Phi_n^{(1)}$ vanishes identically:

$$\Phi_n^{(1)} \equiv \sum_i \left(\frac{\partial \Phi_n(X)}{\partial X_i}\right)_{X^0} u_i \equiv 0.$$

In other words, X^0 must be an equilibrium configuration where

$$\left(\frac{\partial \Phi_n(X)}{\partial X_i}\right)_{X^0} = 0. \quad (14.24)$$

With this required choice of X^0,

$$E_n^{(1)} = 0. \quad (14.25)$$

After putting $E_n^{(1)} = \Phi_n^{(1)} = 0$ in (14.19)(b) and (14.14)(b), we find by comparing these equations that $\chi^{(0)}(u)\phi_n^{(1)}(x, u)$ is a solution of the inhomogeneous equation (14.19)(b). To this solution we may add any solution of the corresponding homogeneous equation. Hence we have

$$\psi_n^{(1)} = \chi^{(0)}(u)\phi_n^{(1)}(x, u) + \chi^{(1)}(u)\phi_n^{(0)}(x), \quad (14.26)$$

where $\chi^{(1)}$ is an arbitrary function of u.

When (14.25) and (14.26) are substituted, the second-order equation (14.19)(c) becomes

$$(H_0^{(0)} - \Phi_n^{(0)})\psi_n^{(2)} = -H_0^{(1)}\chi^{(0)}\phi_n^{(1)} - (H_0^{(2)} + H_1^{(2)} - E_n^{(2)})\chi^{(0)}\phi_n^{(0)} - H_0^{(1)}\chi^{(1)}\phi_n^{(0)}. \quad (14.27)$$

From this equation we subtract $\chi^{(1)}$ times (14.14)(b) and $\chi^{(0)}$ times

(14.14) (c). Remembering that $H_0^{(r)}$ does not operate on u, we can write the resulting equation as

$$(H_0^{(0)} - \Phi_n^{(0)})\{\psi_n^{(2)} - \chi^{(0)}\phi_n^{(2)} - \chi^{(1)}\phi_n^{(1)}\} = -(H_1^{(2)} + \Phi_n^{(2)} - E_n^{(2)})\chi^{(0)}\phi_n^{(0)}.$$

(14.28)

The solubility condition for this equation is

$$\int \phi_n^{(0)}(H_1^{(2)} + \Phi_n^{(2)} - E_n^{(2)})\chi^{(0)}\phi_n^{(0)} \, dx = 0,$$

or, since $(H_1^{(2)} + \Phi_n^{(2)} - E_n^{(2)})\chi^{(0)}$ is independent of x,

$$(H_1^{(2)} + \Phi_n^{(2)} - E_n^{(2)})\chi^{(0)}(u) = 0.$$

(14.29)

If the approximation is broken off at this point (14.29) is the equation determining the nuclear motion. When the equation is multiplied by κ^2, $\kappa^2 H_1^{(2)}$ represents the kinetic energy of the nuclei and $\kappa^2 \Phi_n^{(2)}(u)$ acts as a potential function for the nuclear motion. $\kappa^2 E_n^{(2)}$ is the corresponding eigenvalue. Since $\Phi_n^{(2)}(u)$ is a homogeneous quadratic function of the nuclear coordinates, the equation leads to harmonic vibrations of the nuclei, which we shall discuss explicitly in the next section. We shall call this approximation the *harmonic approximation*. In the harmonic approximation the wave function of the system is determined only to the zero order; this zero-order wave function is given by the product of the nuclear wave function $\chi^{(0)}(u)$ and the electronic wave function $\phi_n^0(x, X^0)$. The eigenvalue is the sum of the eigenvalue $\Phi_n(X^0)$ for the electronic motion (with nuclei at X^0) and the energy of nuclear vibration in the effective potential $\Phi_n^{(2)}(u)$.

Many important crystal properties are, however, directly at variance with the supposition that the nuclei move in a harmonic potential. To discuss such properties one has to consider the higher approximations. The harmonic approximation presents us with a very simple description of the motion of the system; namely, the nuclei move in accordance with an effective potential function, and the electrons move as though the nuclei remain fixed at X^0 and they affect the nuclei only in so far as the effective potential function (for the nuclei) is dependent on the electronic quantum number n. In developing the higher approximations, we have particularly in mind the question as to how much of the simplicity of the harmonic approximation will still be preserved.

The mathematical consideration of the higher approximations is given in Appendix VII. It is shown there that the second-order term of the wave function has the form

$$\psi_n^{(2)}(x, u) = \chi^{(0)}(u)\phi_n^{(2)}(x, u) + \chi^{(1)}(u)\phi_n^{(1)}(x, u) + \chi^{(2)}(u)\phi_n^{(0)}(x).$$

(14.30)

The functions $\chi^{(1)}$ and $\chi^{(2)}$ satisfy the differential equations

$$(H_1^{(2)}+\Phi_n^{(2)}-E_n^{(2)})\chi^{(1)}(u) = -(\Phi_n^{(3)}-E_n^{(3)})\chi^{(0)}(u), \qquad (14.31)$$

$$(H_1^{(2)}+\Phi_n^{(2)}-E_n^{(2)})\chi^{(2)}(u) = -(\Phi_n^{(3)}-E_n^{(3)})\chi^{(1)}(u)-(\Phi_n^{(4)}+C-E_n^{(4)})\chi^{(0)}(u),$$
$$(14.32)$$

where C is a constant. Thus if the approximation is broken off after the second-order term, the wave function is given by

$$\psi_n(x,u) = \psi_n^{(0)}+\kappa\psi_n^{(1)}+\kappa^2\psi_n^{(2)}$$
$$= \chi^{(0)}(u)\{\phi_n^{(0)}(x)+\kappa\phi_n^{(1)}(x,u)+\kappa^2\phi_n^{(2)}(x,u)\}+$$
$$+\kappa\chi^{(1)}(u)\{\phi_n^{(0)}(x)+\kappa\phi_n^{(1)}(x,u)\}+\kappa^2\chi^{(2)}(u)\{\phi_n^{(0)}(x)\}. \quad (14.33)$$

Adding terms of higher orders, we can write the above wave function to the same order of accuracy in the modified form

$$\psi_n(x,u) = \big(\chi^{(0)}(u)+\kappa\chi^{(1)}(u)+\kappa^2\chi^{(2)}(u)\big)\phi_n(x,X). \qquad (14.34)$$

This wave function has a simple interpretation. The first factor describes the nuclear motion and the second factor shows that during the nuclear motion the electrons move as though the nuclei were fixed in their instantaneous positions. We say that the electrons follow the nuclear motion *adiabatically*. In an adiabatic motion, an electron does not make transitions from one state to others; instead, an electronic state itself is deformed progressively by the nuclear displacements. If we proceed to an even higher approximation than the above, the electronic motion will no longer be adiabatic; thus we shall call the above approximation the *adiabatic approximation*.

In the adiabatic approximation, as in the harmonic approximation, an effective potential function exists for the nuclear motion. An inspection of (14.29), together with (14.31) and (14.32), shows that these equations are identical with the equations which would be obtained if the perturbation method is applied to a system with the Hamiltonian

$$H_1^{(2)}(\partial/\partial u)+\Phi_n^{(2)}(u)+\kappa\Phi_n^{(3)}(u)+\kappa^2[\Phi_n^{(4)}(u)+C]. \qquad (14.35)$$

If we multiply (14.35) by κ^2 the first term $\kappa^2 H_1^{(2)}$ represents the kinetic energy of the nuclei and the rest of the terms, namely,

$$\kappa^2\Phi_n^{(2)}(u)+\kappa^3\Phi_n^{(3)}(u)+\kappa^4[\Phi_n^{(4)}(u)+C]$$

can be interpreted as an *effective potential function*. We notice that the effective potential function contains terms including up to the fourth power of the nuclear displacements. We shall find that all important crystal properties can at least be understood formally on the supposition that the nuclei move in accordance with such a potential function.

In Appendix VII it is shown that the third-order term in the wave function is of the form

$$\psi_n^{(3)}(x, u) = \chi^{(0)}(u)\phi_n^{(3)}(x, u) + \chi^{(1)}(u)\phi_n^{(2)}(x, u) + \chi^{(2)}(u)\phi_n^{(1)}(x, u) +$$
$$+ \chi^{(3)}(u)\phi_n^{(0)}(x) + F(x, u), \quad (14.36)$$

where $F(x, u)$ is an involved function of x and u and does not contain x solely through a factor of the type $\phi_n^{(r)}(x, u)$. Thus once we proceed beyond the second-order term in the wave function (or the fourth-order terms in the Hamiltonian), the simple features of the harmonic approximation and the adiabatic approximation are lost. Thus, for instance, it is not permissible to treat formally the dynamics of nuclei on the basis of a potential function which contains fifth or higher powers of the parameter κ (but see Appendix VIII).

The preceding considerations have to be amended for molecules of normal size; for these can rotate freely in space with a frequency of the same or even higher order than the vibrational frequencies. There are among the nuclear coordinates always six which describe translations and rotations and cannot be assumed *a priori* to be confined to the vicinity of certain fixed values. The three translational coordinates are, of course, trivial. The three rotational coordinates lead to terms in T_N of the order κ^3 and κ^4, so that we have, instead of (14.16),

$$T_N = \kappa^2 H_1^{(2)} + \kappa^3 H_1^{(3)} + \kappa^4 H_1^{(4)}.$$

The systematic solution of the perturbation equations then leads to the wave equation for the rotational motion of the molecule and to the interaction between rotation, oscillation, and electronic motion.

It is, however, well known that the energy terms of the rotational spectrum are proportional to the reciprocals of the moments of inertia. They become extremely small for large molecules and can be completely neglected for macroscopic solids. In fact the rotational parameters lose here the quality of quantum coordinates and become quantities which can be chosen at will. Therefore the theory as developed above is adequate for the treatment of solids. It may, however, happen that molecules in a crystal are so loosely bound to their neighbours that they can rotate within the structure. This is, of course, quite distinct from the rotation of the structure as a whole; if parts of the structure are in the first approximation free to rotate, one is confronted with degeneracy of a particular kind. Although this phenomenon is of considerable practical importance and has been treated by approximate methods, it is inconvenient to include it in a systematic theory. In the following discussions this particular case is not considered.

15. Normal coordinates

Let us consider the nuclear motion on the basis of the harmonic approximation. We shall denote the effective potential function for the nuclei simply by Φ. Φ refers, of course, to a particular electronic state; it is, however, not necessary to specify the latter explicitly. We shall distinguish different nuclei of the system by an index $k = 1, 2,..., n$, n being the total number of nuclei in the system. Capital letters for the coordinates, momenta, and masses of the nuclei will be used only in the rare cases where it is necessary to distinguish them from quantities referring to electrons. Thus we denote the mass of the nucleus k by m_k, its rectangular coordinates by $x_\alpha(k)$ ($\alpha = 1, 2, 3$) and its displacement from $x_\alpha^0(k)$ (written symbolically as X^0 in the last section) by $u_\alpha(k)$.

Let us introduce the notation

$$\Phi_\alpha(k) = \left(\frac{\partial \Phi}{\partial x_\alpha(k)}\right)_0, \qquad \Phi_{\alpha\beta}(k, k') = \left(\frac{\partial^2 \Phi}{\partial x_\alpha(k)\partial x_\beta(k')}\right)_0, \qquad (15.1)$$

where the index 0 signifies the configuration $x_\alpha^0(k)$. The condition (14.24) becomes

$$\Phi_\alpha(k) = 0, \qquad (15.2)$$

which determines the equilibrium configuration $x_\alpha^0(k)$. The effective potential function Φ ($\sim \kappa^2\Phi_n^{(2)}$ of the last section) is given by

$$\Phi = \tfrac{1}{2}\sum_{kk'}\sum_{\alpha\beta}\Phi_{\alpha\beta}(k, k')u_\alpha(k)u_\beta(k'). \qquad (15.3)$$

It is often convenient to use

$$D_{\alpha\beta}(k, k') = \frac{1}{(m_k m_{k'})^{\frac{1}{2}}}\Phi_{\alpha\beta}(k, k') \qquad (15.4)$$

instead of $\Phi_{\alpha\beta}(k, k')$, and the *reduced displacements*

$$w_\alpha(k) = (m_k)^{\frac{1}{2}}u_\alpha(k) \qquad (15.5)$$

instead of $u_\alpha(k)$. If the pair (k, α) is considered as a single index, (15.4) defines a $3n \times 3n$ matrix D, which we shall call the *dynamical matrix*. The kinetic and potential energies for nuclear motion can clearly be written as

$$T_N = \tfrac{1}{2}\sum_k\sum_\alpha P_\alpha^2(k), \qquad (15.6)$$

$$\Phi = \tfrac{1}{2}\sum_{kk'}\sum_{\alpha\beta}D_{\alpha\beta}(k, k')w_\alpha(k)w_\beta(k'), \qquad (15.7)$$

where
$$P_\alpha(k) = -i\hbar \frac{\partial}{\partial w_\alpha(k)} \qquad (15.8)$$

represents the canonical momentum corresponding to $w_\alpha(k)$. We notice that the nuclear masses do not appear explicitly in the above expressions; it is in order to achieve this simplicity that the dynamical matrix and the reduced displacements are introduced.

The coordinates $w_\alpha(k)$ are dynamically coupled by the cross-product terms in Φ. Hence we introduce a set of new coordinates q_j ($j = 1, 2,..., 3n$) by the transformation

$$q_j = \sum_k \sum_\alpha e_\alpha(k|j) w_\alpha(k). \qquad (15.9)$$

The transformation coefficients $e_\alpha(k|j)$ are determined as follows: we consider the $3n$ simultaneous equations

$$\omega^2 e_\alpha(k) = \sum_{k'} \sum_\beta D_{\alpha\beta}(k, k') e_\beta(k'), \qquad (15.10)$$

where the $3n$ quantities $e_\alpha(k)$ are the unknowns and ω^2 is to be determined so that the equations are soluble. The equations are linear and homogeneous; it follows from a well-known theorem in algebra that, for the equations to be soluble, the determinant formed of the coefficients must vanish (the *secular equation* of the matrix; compare (6.14)):

$$|D_{\alpha\beta}(k, k') - \omega^2 \delta_{kk'} \delta_{\alpha\beta}| = 0. \qquad (15.11)$$

(15.11) is a $3n$-degree equation for ω^2. Let us denote the solutions by ω_j^2, where $j = 1, 2,..., 3n$. For each of these values for ω^2 (15.10) yields a set of $e_\alpha(k)$, which we denote by $e_\alpha(k|j)$; thus ω_j^2 and the corresponding $e_\alpha(k|j)$ satisfy the equations

$$\omega_j^2 e_\alpha(k|j) = \sum_{k'} \sum_\beta D_{\alpha\beta}(k, k') e_\beta(k'|j). \qquad (15.12)$$

The quantities $e_\alpha(k|j)$ are not uniquely determined in this way. Multiplication of $e_\alpha(k|j)$ by a common factor, for instance, does not affect (15.12). One encounters additional arbitrariness in the determination of $e_\alpha(k|j)$, if some of the ω_j^2 ($j = 1, 2,..., 3n$) are equal (*degeneracy*). It can be shown, however, that $e_\alpha(k|j)$ consistent with (15.12) can always be chosen such that

$$\sum_k \sum_\alpha e_\alpha(k|j) e_\alpha(k|j') = \delta_{jj'}, \qquad \sum_j e_\alpha(k|j) e_\beta(k'|j) = \delta_{\alpha\beta} \delta_{kk'}. \quad (15.13)$$

In fact, if there is no degeneracy, (15.12) together with (15.13) determines the $(3n)^2$ quantities $e_\alpha(k|j)$ uniquely. If there is degeneracy, certain arbitrariness in the choice of $e_\alpha(k|j)$ still remains. But for our purpose it suffices to know that there always exist quantities $e_\alpha(k|j)$ which satisfy both (15.12) and (15.13). Such a set (no matter whether it is the only possible set, or one of many possible sets) may then be used to define the transformation (15.9). The quantities $e_\alpha(k|j)$ define a $3n \times 3n$ matrix with the indices (k, α) and j, both going over $3n$ values. (15.13) merely signifies that the matrix is non-singular and orthogonal. Multiplying (15.9) by $e_\beta(k'|j)$ and summing over j we find, on using (15.13), the inverse transformation

$$w_\beta(k') = \sum_j e_\beta(k'|j)q_j. \qquad (15.14)$$

The new coordinates q_j are known as *normal coordinates*.

The normal coordinates are dynamically independent of one another. For with the help of (15.9), (15.14), (15.13), and (15.12) we find that the kinetic energy (15.6) and the potential energy (15.7), when expressed in terms of the normal coordinates, become

$$T_N = \tfrac{1}{2} \sum_j P_j^2, \qquad (15.15)$$

$$\Phi = \tfrac{1}{2} \sum_j \omega_j^2 q_j^2, \qquad (15.16)$$

where

$$P_j = -i\hbar \frac{\partial}{\partial q_j}. \qquad (15.17)$$

In the above expressions there are no cross terms between different normal coordinates.

The wave equation for nuclear motion has now the form

$$(T_N + \Phi - \epsilon)\chi = \left\{ \sum_j \tfrac{1}{2}(P_j^2 + \omega_j^2 q_j^2) - \epsilon \right\}\chi = 0. \qquad (15.18)$$

Since the Hamiltonian $T_N + \Phi$ is a sum of terms, each depending on only one coordinate, the wave equation is separable in the usual way. Namely, putting

$$\chi = \chi_1(q_1)\chi_2(q_2) \dots \chi_j(q_j) \dots \chi_{3n}(q_{3n}), \qquad (15.19)$$

we find that (15.18) resolves into the equations

$$\{\tfrac{1}{2}(P_j^2 + \omega_j^2 q_j^2) - \epsilon_j\}\chi_j(q_j) = 0, \qquad (15.20)$$

where ϵ is related to ϵ_j by

$$\epsilon = \sum_{j=1}^{3n} \epsilon_j. \qquad (15.21)$$

(15.20) is the familiar wave equation for a simple harmonic oscillator with circular frequency ω_j. An admissible solution is characterized by an integral quantum number $\geqslant 0$. If we denote the quantum number for q_j by v_j, the eigenvalue ϵ_j can have the values

$$\epsilon_j(v_j) = (v_j + \tfrac{1}{2})\hbar\omega_j \qquad v_j = 0, 1, 2, \dots .$$

The corresponding wave functions we shall denote by

$$\chi(v_j, q_j), \qquad v_j = 0, 1, 2, \dots .$$

A state of the whole system is to be specified by the set of $3n$ quantum numbers $(v_1, v_2,..., v_{3n})$. Sometimes, for the sake of simplicity, such a set of $3n$ quantities will be denoted symbolically by a single letter. Thus we write

$$\epsilon(v) \equiv \epsilon(v_1, v_2,..., v_{3n}) = \epsilon_1(v_1) + \epsilon_2(v_2) + \epsilon_3(v_3)... + \epsilon_{3n}(v_{3n}) \Big\}$$
$$\chi(v, q) \equiv \chi(v_1, v_2,...; q_1, q_2,...) = \chi(v_1; q_1)\chi(v_2; q_2),..., \chi(v_{3n}; q_{3n}) \Big\}$$

$$(15.22)$$

The transition frequency (circular) connected with a transition from a state v to another state v' can be defined as

$$\omega(v, v') = \frac{1}{\hbar}\big(\epsilon(v) - \epsilon(v')\big) = \sum_{j=1}^{3n} (v_j - v'_j)\omega_j. \qquad (15.23)$$

In order to describe a transition simply, we say that the oscillator j jumps by $0, \pm 1, \pm 2,...,$ if $v'_j = v_j, v_j \pm 1, v_j \pm 2,....$ The matrix elements of q and p for a single oscillator making a transition from v to v' are given by the formulae

$$\langle v'|q|v \rangle = \left(\frac{\hbar}{2\omega}\right)^{\frac{1}{2}} \{\delta_{v', v+1}(v')^{\frac{1}{2}} + \delta_{v'+1, v}(v)^{\frac{1}{2}}\} \Big\}$$
$$\langle v'|p|v \rangle = i\left(\frac{\hbar\omega}{2}\right)^{\frac{1}{2}} \{\delta_{v', v+1}(v')^{\frac{1}{2}} - \delta_{v'+1, v}(v)^{\frac{1}{2}}\} \Big\}$$

$$(15.24)$$

(real wave functions for the oscillator are used). The matrix elements of q or p are thus non-vanishing only for transitions in which the oscillator jumps by one, either up or down. With the help of (15.24) the matrix elements of any arbitrary product between the normal coordinates (or also the momenta) can be worked out by the rule of matrix multiplication. A moment's consideration shows, for instance, that the following rule holds generally: the matrix element for a product of the form $q_j^\alpha q_{j'}^\beta q_{j''}^\gamma....$ is non-vanishing only for a transition in which the oscillator j jumps by $0, \pm 2,..., \pm \alpha$ for even α, or by $\pm 1, \pm 3,..., \pm \alpha$ for odd α, and the oscillator j' jumps by $0, \pm 2,..., \pm \beta$ for even β, or by $\pm 1, \pm 3,..., \pm \beta$ for odd β, etc.

In Table 27 are collected the matrix elements for a few simple products of the normal coordinates. A transition can clearly be specified by the quantum number v of the initial state and the transition frequency $\omega(v, v')$. In the table the matrix elements are given as functions of v, for given values of $\omega(v, v')$. The abbreviation

$$c_j = \left(\frac{\hbar}{2\omega_j}\right)^{\frac{1}{2}} \qquad (15.25)$$

is used in the table to save space.

TABLE 27

Matrix Elements of Products of Normal Coordinates

$\omega_{vv'}$	0	$-\omega_j$	ω_j	$-2\omega_j$	$2\omega_j$	$-\omega_j-\omega_{j'}$	$\omega_j+\omega_{j'}$	$\omega_j-\omega_{j'}$
$\langle v'\|q_j\|v\rangle$	0	$c_j(v_j+1)^{\frac{1}{2}}$	$c_j v_j^{\frac{1}{2}}$	0	0	0	0	0
$\langle v'\|q_j q_{j'}\|v\rangle$	0	0	0	0	0	$c_j c_{j'}(v_j+1)^{\frac{1}{2}} \times (v_{j'}+1)^{\frac{1}{2}}$	$c_j c_{j'} v_j^{\frac{1}{2}} v_{j'}^{\frac{1}{2}}$	$c_j c_{j'} v_j^{\frac{1}{2}}(v_{j'}+1)^{\frac{1}{2}}$
$\langle v'\|q_j^2\|v\rangle$	$c_j^2(2v_j+1)$	0	0	$c_j^2(v_j+1)^{\frac{1}{2}}(v_j+2)^{\frac{1}{2}}$	$c_j^2 v_j^{\frac{1}{2}}(v_j-1)^{\frac{1}{2}}$	0	0	0
$\langle v'\|q_j q_{j'}^2\|v\rangle$	0	$c_j c_{j'}^2(2v_{j'}+1) \times (v_j+1)^{\frac{1}{2}}$	$c_j c_{j'}^2(2v_{j'}+1)v_j^{\frac{1}{2}}$	0	0	0	0	0
$\langle v'\|q_j^3\|v\rangle$	0	$3c_j^3(v_j+1)^{\frac{3}{2}}$	$3c_j^3 v_j^{\frac{3}{2}}$	0	0	0	0	0
$\langle v'\|q_j^4\|v\rangle$	$3c_j^4(2v_j^2+ +2v_j+1)$	0	0	$2c_j^4(2v_j+3) \times (v_j+1)^{\frac{1}{2}}(v_j+2)^{\frac{1}{2}}$	$2c_j^4(2v_j-1)v_j^{\frac{1}{2}} \times (v_j-1)^{\frac{1}{2}}$	0	0	0

16. Statistical mechanics of systems of oscillators

Physically observable properties of molecular systems are often average values over the thermal motion. For the discussion of such properties the *partition function*

$$Z = \sum_l e^{-\epsilon_l/kT} \quad (k = \text{Boltzmann constant}) \tag{16.1}$$

provides a convenient starting-point; l in the formula is a quantum number distinguishing all stationary states of the system and ϵ_l represents the eigenvalue of the state l. The free energy is related to Z by

$$F = -kT \ln Z, \tag{16.2}$$

from which the entropy S and the energy E follow according to the well-known thermodynamical formulae

$$\left. \begin{aligned} S &= -\frac{\partial F}{\partial T} = k \ln Z + \frac{kT}{Z}\frac{\partial Z}{\partial T} \\ E &= F + TS = \frac{kT^2}{Z}\frac{\partial Z}{\partial T} = \frac{\sum_l \epsilon_l e^{-\epsilon_l/kT}}{\sum_l e^{-\epsilon_l/kT}} \end{aligned} \right\}. \tag{16.3}$$

The last formula shows that the probability of finding the system in the state l is given by the so-called Boltzmann factor (normalized)

$$W_l = \frac{e^{-\epsilon_l/kT}}{\sum_l e^{-\epsilon_l/kT}}. \tag{16.4}$$

Let us consider the case when ϵ_l depends on certain macroscopic parameters f_α (e.g. external field components). If the system is in the state l the work required to achieve a small change δf_α is given by $(\partial \epsilon_l/\partial f_\alpha)\,\delta f_\alpha$. $\partial \epsilon_l/\partial f_\alpha$ may be described as a generalized force conjugate to the parameter f_α. It may not, of course, be a force in the usual sense; thus if f_α is an angle the conjugate force represents a moment of force, if f_α is an electric field component the conjugate force represents the corresponding component of an electric moment (see § 18). The observed value F_α of a force is the thermal average calculated in accordance with (16.4):

$$F_\alpha = \sum_l \frac{\partial \epsilon_l}{\partial f_\alpha} W_l = \left(\frac{\partial F}{\partial f_\alpha}\right)_T. \tag{16.5}$$

In forming the partition function for a molecular system we can ignore all but the lowest electronic state, for a temperature of at least the order 10^4 degrees will be required to excite appreciably the higher electronic states (i.e. for the corresponding thermal probability (16.4) to be appreciable as compared with unity). Thus adding the energy Φ^0 of the lowest electronic state (with nuclei at the equilibrium configuration X^0) to the eigenvalues for nuclear motion (15.22), we find for ϵ_l the expression

$$\Phi^0 + \sum_j \hbar\omega_j(v_j + \tfrac{1}{2}).\tag{16.6}$$

In the formula, the set of nuclear vibrational quantum numbers v $(v_1, v_2,..., v_{3n})$ replaces the quantum number l in (16.1); thus the summation in (16.1) is over all $3n$ quantum numbers $v_1, v_2,..., v_{3n}$. If we introduce the abbreviation

$$\beta_j = \hbar\omega_j/kT,\tag{16.7}$$

we readily find that the partition function factorizes in this case into the following product of factors:

$$Z = e^{-\Phi^0/kT} \prod_j e^{-\frac{1}{2}\beta_j} z_j,\tag{16.8}$$

where z_j is the partition function of a single oscillator without the zero-point energy $\tfrac{1}{2}\hbar\omega_j$, obtainable from

$$z = \sum_{v=0}^{\infty} e^{-\beta v} = \frac{1}{1-e^{-\beta}}\tag{16.9}$$

by substituting β_j for β. Introducing (16.9) in (16.8) we have

$$Z = e^{-\Phi^0/kT} \prod_j \frac{1}{2\sinh\frac{1}{2}\beta_j}.\tag{16.10}$$

Using (16.10) in (16.2) and (16.3), we obtain the following expressions for the thermodynamical functions:

$$F = \Phi^0 + kT \sum_j \ln(2\sinh\tfrac{1}{2}\beta_j) = \Phi^0 + kT \sum_j \left(\tfrac{1}{2}\beta_j + \ln(1-e^{-\beta_j})\right);$$

$$S = k \sum_j \left(\tfrac{1}{2}\beta_j \coth\tfrac{1}{2}\beta_j - \ln(2\sinh\tfrac{1}{2}\beta_j)\right) = k \sum_j \left(\frac{\beta_j}{e^{\beta_j}-1} - \ln(1-e^{-\beta_j})\right);$$

$$E = \Phi^0 + kT \sum_j \tfrac{1}{2}\beta_j \coth\tfrac{1}{2}\beta_j = \Phi^0 + kT \sum_j \left(\tfrac{1}{2}\beta_j + \frac{\beta_j}{e^{\beta_j}-1}\right).$$

$$\tag{16.11}$$

Like the partition function, the thermal probability also factorizes into a product of factors:

$$W(v) = \prod_j w_j(v_j),\tag{16.12}$$

where w_j is clearly the thermal probability for a single oscillator and is obtainable from

$$w(v) = \frac{e^{-\beta v}}{\sum_{v=0}^{\infty} e^{-\beta v}} = e^{-\beta v}(1-e^{-\beta}) = \frac{e^{-\beta v}}{z}\tag{16.13}$$

by substituting v_j, β_j for v, β. All physical properties depend on thermal averages of simple combinations of the matrix elements of functions of the oscillator amplitudes. If the functions are expanded into power series, their matrix elements can be reduced to sums of those given in Table 27. As we have seen, the matrix elements can be specified by giving the initial state and the transition frequency; what we shall require is the thermal average over the initial state for a fixed transition frequency. (16.12) shows that the thermal averages over different oscillators are independent of one another, hence, in view of the values given in Table 27, we need calculate only the thermal averages of various powers of the quantum number of a single oscillator, such as

$$\langle v \rangle = \frac{\sum_{v=0}^{\infty} v e^{-\beta v}}{\sum_{v=0}^{\infty} e^{-\beta v}} = -\frac{1}{z}\frac{dz}{d\beta} = \frac{1}{e^{\beta}-1}, \tag{16.14}$$

$$\langle v^2 \rangle = \frac{\sum_{v=0}^{\infty} v^2 e^{-\beta v}}{\sum_{v=0}^{\infty} e^{-\beta v}} = \frac{1}{z}\frac{d^2z}{d\beta^2} = \frac{e^{\beta}+1}{(e^{\beta}-1)^2}. \tag{16.15}$$

TABLE 28

Thermal Average Values

Diagonal elements

$\{\langle v\|q_j^2\|v\rangle\}_{Av}$	$\overline{q_j^2}$
$\{\langle v\|q_j^2 q_{j'}^2\|v\rangle\}_{Av}$	$\overline{q_j^2}\,\overline{q_{j'}^2}$
$\{\langle v\|q_j^4\|v\rangle\}_{Av}$	$3(\overline{q_j^2})^2$
$\{[\langle v\|q_j^2\|v\rangle]^2\}_{Av}$	$2(\overline{q_j^2})^2 - c_j^4$

Non-diagonal elements $(\omega_{vv'} \neq 0)$

$\omega_{vv'} =$	ω_j	$-\omega_j$
$\{\langle v\|q_j\|v'\rangle\langle v'\|q_j\|v\rangle\}_{Av}$	$C_j e^{-\beta_j}$	C_j
$\{\langle v\|q_j\|v'\rangle\langle v'\|q_j q_{j'}^2\|v\rangle\}_{Av}$	$C_j e^{-\beta_j}\overline{q_{j'}^2}$	$C_j\overline{q_{j'}^2}$
$\{\langle v\|q_j\|v'\rangle\langle v'\|q_j^3\|v\rangle\}_{Av}$	$3C_j e^{-\beta_j}\overline{q_j^2}$	$3C_j\overline{q_j^2}$

$\omega_{vv'} =$	$2\omega_j$	$-2\omega_j$
$\{\langle v\|q_j^2\|v'\rangle\langle v'\|q_j^2\|v\rangle\}_{Av}$	$2C_j^2 e^{-2\beta_j}$	$2C_j^2$

$\omega_{vv'} =$	$\omega_j + \omega_{j'}$	$-(\omega_j+\omega_{j'})$	$\omega_j - \omega_{j'}$	$-(\omega_j-\omega_{j'})$
$\{\langle v\|q_j q_{j'}\|v'\rangle\langle v'\|q_j q_{j'}\|v\rangle\}_{Av}$	$C_j C_{j'} e^{-(\beta_j+\beta_{j'})}$	$C_j C_{j'}$	$C_j C_{j'} e^{-\beta_j}$	$C_j C_{j'} e^{-\beta_{j'}}$

With the help of (16.14), (16.15) and the values given in Table 27, one readily obtains the values given in Table 28, which contains the non-vanishing averages of simple combinations of matrix elements

ordered according to the transition frequency. In the table, the following
abbreviations are used:

$$C_j = \frac{c_j^2}{1-e^{-\beta_j}} = \frac{\hbar/2\omega_j}{1-e^{-\hbar\omega_j/kT}}, \qquad (16.16)$$

$$\overline{q_j^2} = c_j^2\left(\frac{e^{\beta_j}+1}{e^{\beta_j}-1}\right) = c_j^2 \coth \tfrac{1}{2}\beta_j = \frac{1}{\omega_j^2}\left(\tfrac{1}{2}\hbar\omega_j + \frac{\hbar\omega_j}{e^{\hbar\omega_j/kT}-1}\right) = \frac{\bar{\epsilon}_j}{\omega_j^2}. \qquad (16.17)$$

The latter quantity is the mean square amplitude of the oscillator with
the average energy $\bar{\epsilon}_j$.

17. Statistical mechanics of a molecular system under external forces

Let us now consider the case when the energy levels are functions of
certain macroscopic parameters. For small values of the parameters f_α,
it is generally possible to represent the Hamiltonian as a power series
in f_α in the form

$$H = H_0 + H_1 + H_2 + ..., \qquad (17.1)$$

where

$$H_0 = \Phi_0 + \tfrac{1}{2}\sum_j (p_j^2 + \omega_j^2 q_j^2),$$

$$H_1 = \sum_\alpha g^\alpha(q) f_\alpha, \qquad (17.2)$$

$$H_2 = \tfrac{1}{2}\sum_{\alpha\beta} g^{\alpha\beta}(q) f_\alpha f_\beta, \qquad g^{\alpha\beta}(q) = g^{\beta\alpha}(q).$$

The quantities q_j are coordinates which for given values of the parameters completely specify the configuration of the system; p_j are the
conjugate momenta $-i\hbar\,\partial/\partial q_j$. For $f_\alpha = 0$, H and q_j reduce respectively
to H_0 and the corresponding normal coordinates.

The coefficients in H_1, H_2, etc., are functions of q_j. Let us represent
these coefficients by power series in q_j:

$$g^\alpha(q) = g_0^\alpha + \sum_j g_j^\alpha q_j + \tfrac{1}{2}\sum_{jj'} g_{jj'}^\alpha q_j q_{j'} + ... \quad (g_{jj'}^\alpha = g_{j'j}^\alpha);$$

$$g^{\alpha\beta}(q) = g_0^{\alpha\beta} + \sum_j g_j^{\alpha\beta} q_j + \tfrac{1}{2}\sum_{jj'} g_{jj'}^{\alpha\beta} q_j q_{j'} + ... \quad (g_{jj'}^{\alpha\beta} = g_{j'j}^{\alpha\beta}). \qquad (17.3)$$

In the following discussion we shall take into account only terms up to
second order in q_j.

In order to obtain the partition function and the free energy as
functions of the macroscopic parameters, we have to find the eigenvalues of the Hamiltonian (17.1). If we wish to discuss how the forces
F_α depend on the values of f_α or vice versa, and how the parameters f_α
vary with temperature, we have at least to know the terms in the free

energy up to the second order in the parameters f_α. In obtaining the eigenvalues of (17.1), we shall retain only terms up to second order in the parameters f_α and ignore the higher terms. Let us divide the Hamiltonian into two parts. The first part includes H_0 and the terms in H_1, H_2, which are lower than the second order in q_j, namely,

$$H^{\mathrm{I}} = \Phi_0 + \tfrac{1}{2} \sum_j (p_j^2 + \omega_j^2 q_j^2) + \sum_\alpha g_0^\alpha f_\alpha + \sum_\alpha \sum_j g_j^\alpha f_\alpha q_j +$$
$$+ \tfrac{1}{2} \sum_{\alpha\beta} g_0^{\alpha\beta} f_\alpha f_\beta + \tfrac{1}{2} \sum_{\alpha\beta} \sum_j g_j^{\alpha\beta} f_\alpha f_\beta q_j. \quad (17.4)$$

The second part includes the terms in H_1, H_2, which are of the second order in q_j:

$$H^{\mathrm{II}} = \tfrac{1}{2} \sum_\alpha \sum_{jj'} g_{jj'}^\alpha f_\alpha q_j q_{j'} + \tfrac{1}{4} \sum_{\alpha\beta} \sum_{jj'} g_{jj'}^{\alpha\beta} f_\alpha f_\beta q_j q_{j'}. \quad (17.5)$$

All the rest of the terms in H are of third or higher orders in either the parameters f_α or the coordinates q_j.

To the desired degree of accuracy we have thus

$$H = H^{\mathrm{I}} + H^{\mathrm{II}}, \quad (17.6)$$

where we may express H^{I} and H^{II} in powers of q_j as follows:

$$H^{\mathrm{I}} = a_0 + \sum_j a_j q_j + \tfrac{1}{2} \sum_j (p_j^2 + \omega_j^2 q_j^2),$$
$$H^{\mathrm{II}} = \tfrac{1}{2} \sum_{jj'} a_{jj'} q_j q_{j'}, \quad (17.7)$$

the a coefficients being defined by

$$a_0 = \Phi_0 + \sum_\alpha g_0^\alpha f_\alpha + \tfrac{1}{2} \sum_{\alpha\beta} g_0^{\alpha\beta} f_\alpha f_\beta,$$
$$a_j = \sum_\alpha g_j^\alpha f_\alpha + \tfrac{1}{2} \sum_{\alpha\beta} g_j^{\alpha\beta} f_\alpha f_\beta, \quad (17.8)$$
$$a_{jj'} = \sum_\alpha g_{jj'}^\alpha f_\alpha + \tfrac{1}{2} \sum_{\alpha\beta} g_{jj'}^{\alpha\beta} f_\alpha f_\beta.$$

Now introduce instead of q_j the variables

$$q_j' = q_j + \frac{a_j}{\omega_j^2}. \quad (17.9)$$

The conjugate momenta are clearly unaltered:

$$p_j' = -i\hbar \frac{\partial}{\partial q_j'} = -i\hbar \frac{\partial}{\partial q_j} = p_j. \quad (17.10)$$

H^{I} goes over into

$$H^{\mathrm{I}} = a_0 - \frac{1}{2} \sum_j \frac{a_j^2}{\omega_j^2} + \frac{1}{2} \sum_j (p_j'^2 + \omega_j^2 q_j'^2). \quad (17.11)$$

H^{II} becomes

$$H^{\mathrm{II}} = \tfrac{1}{2} \sum_{jj'} a_{jj'} \left(q_j' q_{j'}' - \frac{2 a_j q_{j'}'}{\omega_j^2} \right). \quad (17.12)$$

We note that both a_j and $a_{jj'}$ begin with first-order terms in f_α; thus in writing down H^{II} we have ignored terms including the product $a_{jj'} a_j a_{j'}$.

H^I, taken by itself, describes an assembly of oscillators with the frequencies ω_j, exactly as discussed in § 15; the additional energy Φ^0 in that section being now replaced by the expression

$$a_0 - \frac{1}{2} \sum_j \frac{a_j^2}{\omega_j^2}. \tag{17.13}$$

The eigenvalues of H^I are thus

$$\epsilon^{(0)}(v) = a_0 - \frac{1}{2} \sum_j \frac{a_j^2}{\omega_j^2} + \sum_j \hbar\omega_j(v_j + \tfrac{1}{2}) \tag{17.14}$$

and the wave functions are as given in (15.22) with the variables q_j' as arguments. The eigenvalues of the complete Hamiltonian can be obtained by the perturbation method, wherein H^I and H^{II} are regarded respectively as the unperturbed Hamiltonian and the perturbation term. The eigenvalues may be written accordingly as

$$\epsilon(v) = \epsilon^{(0)}(v) + \epsilon^{(1)}(v) + \epsilon^{(2)}(v) + \ldots \tag{17.15}$$

where the superscript indicates the order in the perturbation method. Since the perturbation term H^{II} begins with linear terms in the parameters f_α, $\epsilon^{(1)}(v)$, $\epsilon^{(2)}(v)$ will respectively begin with linear and second-order terms in f_α. $\epsilon^{(3)}(v),\ldots$, beginning with third- and higher-order terms in f_α, may be ignored. Substituting $\epsilon(v) = \epsilon^{(0)} + \epsilon^{(1)} + \epsilon^{(2)}$ in the partition function (16.1) and expanding with respect to $(\epsilon^{(1)} + \epsilon^{(2)})/kT$, we find that, to the desired degree of accuracy,

$$Z = \sum_v e^{-\epsilon^{(0)}(v)/kT} \left\{ 1 - \frac{1}{kT}\left(\epsilon^{(1)}(v) + \epsilon^{(2)}(v)\right) + \frac{1}{2}\frac{[\epsilon^{(1)}(v)]^2}{k^2 T^2} \right\}$$

$$= Z_0 \left\{ 1 - \frac{1}{kT}\left[\langle\epsilon^{(1)}(v)\rangle_{Av} + \langle\epsilon^{(2)}(v)\rangle_{Av}\right] + \frac{1}{2(kT)^2}\langle[\epsilon^{(1)}(v)]^2\rangle_{Av} \right\}, \tag{17.16}$$

where

$$Z_0 = \sum_v e^{-\epsilon^{(0)}(v)/kT} = \exp\left\{ -\frac{1}{kT}\left[a_0 - \frac{1}{2}\sum_j \frac{a_j^2}{\omega_j^2} \right] \right\} \left\{ \prod_j \frac{1}{2 \sinh \tfrac{1}{2}\beta_j} \right\} \tag{17.17}$$

is the same partition function as in the last section apart from the replacement of Φ^0 by the expression (17.13). The averages in (17.16) are over the thermal distribution (16.4). After substituting (17.16) in

(16.2), we obtain the free energy by an expansion with respect to $\langle\epsilon^{(1)}(v)\rangle_{Av}$ and $\langle\epsilon^{(2)}(v)\rangle_{Av}$:

$$F = F_0+\langle\epsilon^{(1)}(v)\rangle_{Av}+\langle\epsilon^{(2)}(v)\rangle_{Av}+\frac{1}{2kT}\{\langle\epsilon^{(1)}(v)\rangle^2_{Av}-\langle[\epsilon^{(1)}(v)]^2\rangle_{Av}\},$$

(17.18)

where

$$F_0 = -kT\ln Z_0 = a_0-\frac{1}{2}\sum_j\frac{a_j^2}{\omega_j^2}+kT\sum_j\ln(2\sinh\tfrac{1}{2}\beta_j). \quad (17.19)$$

The explicit expressions for $\epsilon^{(1)}(v)$ and $\epsilon^{(2)}(v)$ follow from a straight-forward application of the perturbation method. We note in particular that the second term in (17.12) can be directly ignored. This term, being second order in f_α(cf. (17.8)), need only be considered for the calculation of the first-order perturbation energy $\epsilon^{(1)}(v)$. Its contribution to $\epsilon^{(1)}(v)$ vanishes, however, as the diagonal matrix elements of q_j are zero. We readily find that

$$\epsilon^{(1)}(v) = \tfrac{1}{2}\sum_j a_{jj}q_j^2(0), \quad (17.20)$$

$$\epsilon^{(2)}(v) = \frac{1}{8\hbar}\sum_j\frac{(a_{jj})^2}{\omega_j}\{[q_j^2(2\omega_j)]^2-[q_j^2(-2\omega_j)]^2\}+$$

$$+\frac{1}{2\hbar}\sum_{j,j'}'(a_{jj'})^2\left\{\frac{[q_j(\omega_j)]^2[q_{j'}(\omega_{j'})]^2}{\omega_j+\omega_{j'}}+\frac{[q_j(\omega_j)]^2[q_{j'}(-\omega_{j'})]^2}{\omega_j-\omega_{j'}}+\right.$$

$$\left.+\frac{[q_j(-\omega_j)]^2[q_{j'}(\omega_{j'})]^2}{-\omega_j+\omega_{j'}}+\frac{[q_j(-\omega_j)]^2[q_{j'}(-\omega_{j'})]^2}{-\omega_j-\omega_{j'}}\right\}, \quad (17.21)$$

where the matrix elements of q_j and q_j^2 are specified by the transition frequencies (e.g. $q_j^2(0)$ means diagonal element of q_j^2), the initial quantum numbers $v_1, v_2,..., v_{3n}$ being omitted for the sake of simplicity. The prime over the second summation in (17.21) excludes the case $j = j'$. The thermal averages of $\epsilon^{(1)}(v)$, $\epsilon^{(2)}(v)$, and $[\epsilon^{(1)}(v)]^2$ over the quantum number v can now be written down with the help of Table 28; after some simplification, we find that

$$\langle\epsilon^{(1)}(v)\rangle_{Av} = \tfrac{1}{2}\sum_j a_{jj}\overline{q_j^2}, \quad (17.22)$$

$$\langle\epsilon^{(2)}(v)\rangle_{Av} = -\tfrac{1}{8}\sum_j a_{jj}^2\left(\frac{\overline{q_j^2}}{\omega_j^2}\right)+\tfrac{1}{2}\sum_{jj'}'a_{jj'}^2\left\{\frac{\overline{q_j^2}}{\omega_j^2-\omega_{j'}^2}\right\}, \quad (17.23)$$

$$\langle[\epsilon^{(1)}(v)]^2\rangle_{Av} = \tfrac{1}{4}\sum_{jj'}'a_{jj}a_{j'j'}\overline{q_j^2}\,\overline{q_{j'}^2}+\tfrac{1}{4}\sum_j a_{jj}^2\left\{2(\overline{q_j^2})^2-\frac{\hbar^2}{4\omega_j^2}\right\}$$

$$= \{\langle\epsilon^{(1)}(v)\rangle_{Av}\}^2+\tfrac{1}{4}\sum_j a_{jj}^2\left((\overline{q_j^2})^2-\frac{\hbar^2}{4\omega_j^2}\right). \quad (17.24)$$

When these expressions and (17.8) are substituted in (17.18), we find that the free energy, arranged in powers of the macroscopic parameters, is given by

$$F = A^0 + \sum_\alpha A^\alpha f_\alpha + \tfrac{1}{2} \sum_{\alpha\beta} A^{\alpha\beta} f_\alpha f_\beta, \qquad (17.25)$$

where the coefficients are the following functions of temperature:

$$A^0 = \Phi_0 + kT \sum_j \ln(2\sinh \tfrac{1}{2}\beta_j),$$

$$A^\alpha = g_0^\alpha + \tfrac{1}{2} \sum_j g_{jj}^\alpha \overline{q_j^2},$$

$$A^{\alpha\beta} = g_0^{\alpha\beta} - \sum_j \frac{g_j^\alpha g_j^\beta}{\omega_j^2} + \sum_j \left\{ \tfrac{1}{2}g_{jj}^{\alpha\beta} - \frac{1}{4\omega_j^2} g_{jj}^\alpha g_{jj}^\beta + \sum_{j'}{}' \frac{g_{jj'}^\alpha g_{jj'}^\beta}{\omega_j^2 - \omega_{j'}^2} \right\} \overline{q_j^2} -$$

$$- \frac{1}{4kT} \sum_j g_{jj}^\alpha g_{jj}^\beta \left[(\overline{q_j^2})^2 - \frac{\hbar^2}{4\omega_j^2} \right]. \qquad (17.26)$$

We recall that the mean square amplitudes $\overline{q_j^2}$ are given by the following temperature function:

$$\overline{q_j^2} = \frac{\hbar}{2\omega_j} \coth \frac{\hbar\omega_j}{2kT}.$$

Owing to the fact that we retain only terms up to the second order in q_j in the Hamiltonian, we can use an *alternative method* to obtain the above result. As we shall see, by an orthogonal transformation followed by a translational transformation, we can obtain new coordinates Q_j, in terms of which the Hamiltonian (17.6) has the form

$$H = H^{\mathrm{I}} + H^{\mathrm{II}} = C + \tfrac{1}{2} \sum_j (P_j^2 + \Omega_j^2 Q_j^2), \qquad (17.27)$$

where P_j are the conjugate momenta $-i\hbar\partial/\partial Q_j$. The system may thus always be regarded as a system of harmonic oscillators with frequencies Ω_j depending on the macroscopic parameters f_α. Once the constant term C and the frequencies Ω_j are found, the free energy can be directly written down with the help of (16.11), namely

$$F = C + kT \sum_j \ln(2\sinh \tfrac{1}{2}B_j) \qquad \left(B_j = \frac{\hbar\Omega_j}{kT} \right). \qquad (17.28)$$

Adding (17.11) and (17.12), we have for the complete Hamiltonian:

$$H = H^{\mathrm{I}} + H^{\mathrm{II}} = a_0 - \frac{1}{2} \sum_j \frac{a_j^2}{\omega_j^2} - \sum_{jj'} \frac{a_{jj'} a_j q_{j'}'}{\omega_j^2} +$$

$$+ \tfrac{1}{2} \sum_j (p_j'^2 + \omega_j^2 q_j'^2) + \tfrac{1}{2} \sum_{jj'} a_{jj'} q_j' q_{j'}'. \qquad (17.29)$$

Let us seek to reduce the second-order terms

$$\tfrac{1}{2} \sum_j \omega_j^2 q_j'^2 + \tfrac{1}{2} \sum_{jj'} a_{jj'} q_j' q_{j'}' \qquad (17.30)$$

to a sum of squares $\qquad \frac{1}{2} \sum_j \Omega_j^2 Q_j'^2$ (17.31)

by an orthogonal transformation

$$Q_j' = \sum_{j'} b_{j'j} q_{j'}'.$$ (17.32)

The problem is exactly the same as that considered in § 15. Thus Ω_j^2 and $b_{jj'}$ must satisfy the secular equations

$$\Omega_j^2 b_{j'j} = \sum_{j'} H_{j'j''} b_{j''j},$$ (17.33)

where $H_{j'j'}$ are the quadratic coefficients in (17.30):

$$H_{j'j'} = \omega_j^2 \delta_{j'j'} + a_{j'j'}.$$ (17.34)

We can solve (17.33) by the perturbation method. The results are well known, for instance, in quantum mechanics, where Ω_j^2 appears usually as the energy eigenvalue. We note that the non-diagonal elements of (17.34) begin with first-order terms in f_α, hence we require the eigenvalue Ω^2 calculated to the second order in the perturbation method, namely,

$$\Omega_j^2 = \omega_j^2 + a_{jj} + \sum_{j'} \frac{(a_{jj'})^2}{\omega_j^2 - \omega_{j'}^2}.$$ (17.35)

Taking the square root we obtain by expansion, to the same order of accuracy,

$$\Omega_j = \omega_j + \frac{a_{jj}}{2\omega_j} + \frac{1}{2\omega_j} \sum_{j'} \frac{(a_{jj'})^2}{\omega_j^2 - \omega_{j'}^2} - \frac{(a_{jj})^2}{8\omega_j^3}.$$ (17.36)

When the coordinates Q_j' are introduced in (17.29), we find up to second-order terms in f_α (note that $q_{j'}'$ is equal to $Q_{j'}'$ in the zeroth order of f_α):

$$H^{\mathrm{I}} + H^{\mathrm{II}} = a_0 - \frac{1}{2} \sum_j \frac{a_j^2}{\omega_j^2} - \sum_{jj'} \frac{a_{jj'} a_j}{\omega_j^2} Q_j' + \frac{1}{2} \sum_j (P_j'^2 + \Omega_j^2 Q_j'^2).$$ (17.37)

The terms linear in Q_j' can be removed by a translational transformation from Q_j' to $Q_j = Q_j' - \sum_{j'} a_{j'j} a_{j'}/(\omega_{j'} \Omega_j)^2$, which leaves the momenta unaltered. Since $a_{j'j} a_{j'}$ begins with second-order terms in f_α, this transformation, apart from cancelling the linear terms, introduces only terms of the fourth order in f_α, which can be ignored. Hence in terms of Q_j the Hamiltonian becomes

$$H = H^{\mathrm{I}} + H^{\mathrm{II}} = a_0 - \frac{1}{2} \sum_j \frac{a_j^2}{\omega_j^2} + \frac{1}{2} \sum_j (P_j^2 + \Omega_j^2 Q_j^2).$$ (17.38)

This is of the form (17.27).

(17.36) shows that $\Omega_j - \omega_j$ begins with linear terms in f_α; thus let us write

$$B_j = (B_j - \beta_j) + \beta_j$$

in (17.28) and expand the free energy with respect to $(B_j - \beta_j)$ up to second-order terms:

$$F = C + kT \sum_j \ln(2 \sinh \tfrac{1}{2}\beta_j) + \tfrac{1}{2}kT \sum_j (B_j - \beta_j)\coth \tfrac{1}{2}\beta_j -$$
$$- \tfrac{1}{8}kT \sum_j (B_j - \beta_j)^2 \mathrm{cosech}^2 \tfrac{1}{2}\beta_j$$

$$= C + kT \sum_j \ln(2 \sinh \tfrac{1}{2}\beta_j) + \sum_j \omega_j(\Omega_j - \omega_j)\overline{q_j^2} -$$
$$- \frac{1}{2kT} \sum_j \left[(\overline{q_j^2})^2 - \frac{\hbar^2}{4\omega_j^2} \right] \omega_j^2 (\Omega_j - \omega_j)^2, \quad (17.39)$$

where use has been made of (16.17). When the values of Ω_j and C given by (17.36), (17.38) are substituted, we have

$$F = a_0 - \frac{1}{2} \sum_j \frac{a_j^2}{\omega_j^2} + kT \sum_j \ln(2 \sinh \tfrac{1}{2}\beta_j) + \tfrac{1}{2} \sum_j a_{jj} \overline{q_j^2} -$$
$$- \frac{1}{8} \sum_j \frac{a_{jj}^2 \overline{q_j^2}}{\omega_j^2} + \frac{1}{2} {\sum_{jj'}}' \frac{a_{jj'}^2 \overline{q_j^2}}{\omega_j^2 - \omega_{j'}^2} - \frac{1}{8kT} \sum_j a_{jj}^2 \left[(\overline{q_j^2})^2 - \frac{\hbar^2}{4\omega_j^2} \right]. \quad (17.40)$$

It is readily seen that when the a coefficients are eliminated with the help of (17.8), (17.40) becomes identical with the expression (17.25) obtained by the first method.

We recall that Grüneisen's method as explained in § 4 is based upon the idea underlying the second of the above methods. There we have taken the small change of volume of a crystal as the only macroscopic parameter; and we have seen that Grüneisen's approximation can be obtained by assuming the expression

$$\frac{d \ln \nu_i}{dV}$$

to have the same value for all oscillators. The analogue of the above expression in the general case is

$$\left(\frac{\partial \ln \Omega_j}{\partial f_\alpha} \right)_0 = \frac{g_{jj}^\alpha}{2\omega_j^2}, \quad (17.41)$$

where the right-hand side is obtained by differentiating (17.36) and using (17.8).

In view of (16.5), we can obtain the forces F_α by differentiating (17.25):

$$F_\alpha = A^\alpha + \sum_\beta A^{\alpha\beta} f_\beta. \quad (17.42)$$

Let us suppose that the parameters have been chosen such that zero values of f_α correspond to the static equilibrium configuration of a lattice. Thus if we are considering volume changes as in § 4, the parameter will be $V-V_0$, V_0 being the volume of the crystal in static equilibrium. (17.42) describes the static case, if we put the mean square vibrational amplitude $\overline{q_j^2}$ equal to zero in the expressions for A^α and $A^{\alpha\beta}$; thus in this case A^α reduces to g_0^α (see (17.26)). Since the system is in equilibrium (i.e. $F_\alpha = 0$) in this case for zero values of f_α, we find on putting both F_α and f_α equal to zero in (17.42) that $g_0^\alpha = 0$. Thus with this choice of the parameters the values of A^α reduce to (cf. (17.26))

$$A^\alpha = \tfrac{1}{2} \sum_j g_{jj}^\alpha \overline{q_j^2}. \tag{17.43}$$

The quantities A^α, we note, arise solely in virtue of the zero-point and thermal vibrations of the nuclei. (17.42) shows that $-A^\alpha$ has the same effect as an imposed force. Thus if there is no force acting on the system the parameters have the values given by

$$-A^\alpha = \sum_\beta A^{\alpha\beta} f_\beta, \tag{17.44}$$

exactly as though forces $-A_\alpha$ had been applied to the system. We may call $-A^\alpha$ the *vibrational forces*.

Let us consider the cases where it is permissible to generalize Grüneisen's approximation by assuming (17.41) to have the same value for all oscillators in the system, i.e.

$$g_{jj}^\alpha = \lambda^\alpha \omega_j^2. \tag{17.45}$$

On using (17.45) in (17.43) we find that the vibrational forces become

$$-A^\alpha = -\lambda^\alpha \sum_j \omega_j^2 \overline{q_j^2} = -\lambda^\alpha \sum_j \bar{\epsilon}_j \quad \text{(see (16.17))}. \tag{17.46}$$

In this case the vibrational forces are simply proportional to the total vibrational energy. The temperature-dependent part of $A^{\alpha\beta}$ (cf. (17.26)) is usually in the nature of a small correction. If the temperature-dependence of $A^{\alpha\beta}$ is ignored, it follows from (17.44) and (17.46) that, if no external forces are acting, the values of f_α are proportional to the total vibrational energy.

If we denote A^α as a function of the absolute temperature by $A^\alpha(T)$, $-A^\alpha(T)+A^\alpha(0)$ may be called the *thermal force*. Similarly the values of f_α, in the absence of external forces, are definite functions $f_\alpha(T)$ of temperature. When the approximation (17.45) is applicable, the thermal forces are proportional to the thermal energy

$$\sum_j (\bar{\epsilon}_j - \tfrac{1}{2}\hbar\omega_j). \tag{17.47}$$

Provided that the temperature-dependence of $A^{\alpha\beta}$ can be ignored, we have

$$-A^\alpha(T) + A^\alpha(0) = \sum_\beta A^{\alpha\beta}(f_\beta(T) - f_\beta(0)). \qquad (17.48)$$

The thermal increments of the parameters, $f_\beta(T) - f_\beta(0)$, are thus proportional to the thermal energy.

The expressions for the entropy and energy follow directly from the free energy (17.25). They can be written as the following series in the macroscopic parameters:

$$S = -\frac{\partial F}{\partial T} = S^0 + \sum_\alpha S^\alpha f_\alpha + \tfrac{1}{2}\sum_{\alpha\beta} S^{\alpha\beta} f_\alpha f_\beta + \cdots,$$

$$E = F + TS = E^0 + \sum_\alpha E^\alpha f_\alpha + \tfrac{1}{2}\sum_{\alpha\beta} E^{\alpha\beta} f_\alpha f_\beta + \cdots, \qquad (17.49)$$

where S^0 and E^0 are the expressions given in (16.11) and the other coefficients are obtained readily from (17.26) with the help of (16.17):

$$\left.\begin{aligned}
S^\alpha &= -\frac{\partial A^\alpha}{\partial T} = -\frac{1}{2kT^2}\sum_j \frac{g_{jj}^\alpha}{\omega_j^2}[\bar\epsilon_j^2 - (\tfrac{1}{2}\hbar\omega_j)^2] \\[2mm]
E^\alpha &= A^\alpha + TS^\alpha = g_0^\alpha + \frac{1}{2}\sum_j \frac{g_{jj}^\alpha}{\omega_j^2}\left(\bar\epsilon_j - \frac{[\bar\epsilon_j^2 - (\tfrac{1}{2}\hbar\omega_j)^2]}{kT}\right)
\end{aligned}\right\}. \qquad (17.50)$$

$$\left.\begin{aligned}
S^{\alpha\beta} &= -\frac{\partial A^{\alpha\beta}}{\partial T} = -\frac{1}{kT^2}\sum_j \frac{1}{\omega_j^2}\left(\tfrac{1}{2}g_{jj}^{\alpha\beta} + \sum_{j'}' \frac{g_{jj'}^\alpha g_{jj'}^\beta}{\omega_j^2 - \omega_{j'}^2}\right) \times \\[2mm]
&\quad \times[\bar\epsilon_j^2 - (\tfrac{1}{2}\hbar\omega_j)^2] + \frac{1}{2k^2T^3}\sum_j \frac{g_{jj}^\alpha g_{jj}^\beta}{\omega_j^4}\,\bar\epsilon_j[\bar\epsilon_j^2 - (\tfrac{1}{2}\hbar\omega_j)^2] \\[3mm]
E^{\alpha\beta} &= A^{\alpha\beta} + TS^{\alpha\beta} = g_0^{\alpha\beta} - \frac{1}{2}\sum_j \frac{g_j^\alpha g_j^\beta}{\omega_j^2} + \\[2mm]
&\quad + \sum_j \frac{1}{\omega_j^2}\left(\tfrac{1}{2}g_{jj}^{\alpha\beta} - \frac{1}{4}\frac{g_{jj}^\alpha g_{jj}^\beta}{\omega_j^2} + \sum_{j'}' \frac{g_{jj'}^\alpha g_{jj'}^\beta}{\omega_j^2 - \omega_{j'}^2}\right) \times \\[2mm]
&\quad \times\left\{\bar\epsilon_j - \frac{[\bar\epsilon_j^2 - (\tfrac{1}{2}\hbar\omega_j)^2]}{kT}\right\} - \frac{1}{2kT}\sum_j \frac{g_{jj}^\alpha g_{jj}^\beta}{\omega_j^4}[\bar\epsilon_j^2 - (\tfrac{1}{2}\hbar\omega_j)^2] + \\[2mm]
&\quad + \frac{1}{2(kT)^2}\sum_j \frac{g_{jj}^\alpha g_{jj}^\beta}{\omega_j^4}\,\bar\epsilon_j[\bar\epsilon_j^2 - (\tfrac{1}{2}\hbar\omega_j)^2]
\end{aligned}\right\}. \qquad (17.51)$$

18. Static polarizability and polarizability in variable fields

Let us consider first a simple theorem. Suppose that we have a Hamiltonian $H(\xi)$, which depends on a parameter ξ. The theorem states that

$$\frac{\partial}{\partial\xi}\,\epsilon_l(\xi) = \int \psi_l^*(\xi)\left(\frac{\partial H(\xi)}{\partial\xi}\right)\psi_l(\xi)\,d\tau, \qquad (18.1)$$

where ϵ_l is an eigenvalue of the system in question and ψ_l the corresponding wave function, both, of course, being functions of ξ. The integration is over all coordinates of the system.

Remembering that

$$\epsilon_l(\xi) = \int \psi_l^*(\xi)H(\xi)\psi_l(\xi)\,d\tau \qquad (18.2)$$

we have

$$\frac{\partial}{\partial\xi}\,\epsilon_l(\xi) = \lim_{\delta\xi\to 0}\frac{1}{\delta\xi}\left\{ \int \left(\psi_l(\xi)+\frac{\partial\psi_l(\xi)}{\partial\xi}\,\delta\xi+...\right)^*\left(H(\xi)+\frac{\partial H(\xi)}{\partial\xi}\,\delta\xi+...\right)\times\right.$$
$$\left.\times\left(\psi_l(\xi)+\frac{\partial\psi_l(\xi)}{\partial\xi}\,\delta\xi+...\right)d\tau - \int \psi_l^*(\xi)H(\xi)\psi_l(\xi)\,d\tau\right\}. \qquad (18.3)$$

We can neglect terms of the second and higher orders in the first integral, and write

$$\frac{\partial}{\partial\xi}\,\epsilon_l(\xi) = \lim_{\delta\xi\to 0}\frac{1}{\delta\xi}\left\{\delta\xi \int \psi_l^*(\xi)\left(\frac{\partial H(\xi)}{\partial\xi}\right)\psi_l(\xi)\,d\tau\,+\right.$$
$$+\left[\int \left(\psi_l(\xi)+\frac{\partial\psi_l(\xi)}{\partial\xi}\,\delta\xi\right)^* H(\xi)\left(\psi_l(\xi)+\frac{\partial\psi_l(\xi)}{\partial\xi}\,\delta\xi\right)d\tau\,-\right.$$
$$\left.\left.-\int \psi_l^*(\xi)H(\xi)\psi_l(\xi)\,d\tau\right]\right\}. \qquad (18.4)$$

The expression given in the square bracket is the change of the expectation value of $H(\xi)$ due to a small variation $\delta\xi(\partial\psi_l/\partial\xi)$ in the wave function, and thus vanishes to the first order according to the variational principle. Hence when the limit is taken, (18.4) reduces directly to (18.1).

Consider now any molecular system and subject it to a uniform static electric field $\mathbf{E}$. The corresponding Hamiltonian, eigenvalues, and wave functions contain $\mathbf{E}$ as a parameter; accordingly we denote them respectively by $H(\mathbf{E})$, $\epsilon_l(\mathbf{E})$, and $\psi_l(\mathbf{E})$. Thus $H(0)$, $\epsilon_l(0)$, and $\psi_l(0)$ refer to the molecular system in the absence of field.

The Hamiltonian $H(\mathbf{E})$ has the form

$$H(\mathbf{E}) = H(0)-\mathbf{M}.\mathbf{E}, \qquad (18.5)$$

where $\mathbf{M}$ is the operator representing the electric dipole moment of the system, given by the sum of (charge $\times$ position vector) over all particles in the system. In view of (18.1), we find that in this case

$$\frac{\partial}{\partial E_\alpha}\,\epsilon_l(\mathbf{E}) = -\int \psi_l^*(\mathbf{E})\mathbf{M}_\alpha\psi_l(\mathbf{E})\,d\tau. \qquad (18.6)$$

The integral on the right-hand side is the value of the moment of the system in state l. (18.6) thus shows that the electric dipole moment is the 'force' conjugate to the parameter $-\mathbf{E}$ (see § 16). In particular, it follows from (16.5) that once the free energy is known as a function of the applied field $\mathbf{E}$, the observed electric moment can be obtained by differentiating the free energy with respect to $-\mathbf{E}$.

Taking the interaction term $-\mathbf{M}.\mathbf{E}$ as a perturbation, we can immediately write down $\epsilon_l(\mathbf{E})$ as a Taylor series in the components of $\mathbf{E}$ with the help of well-known results of perturbation theory. Thus, up to the second order of accuracy, we have

$$\epsilon_l(\mathbf{E}) = \epsilon_l(0) - \sum_\alpha \langle l|M_\alpha|l\rangle E_\alpha + \sum_{r\neq l}\sum_{\alpha\beta} \frac{\langle l|M_\alpha|r\rangle\langle r|M_\beta|l\rangle}{\epsilon_l(0)-\epsilon_r(0)} E_\alpha E_\beta,$$
(18.7)

where both the eigenvalues $\epsilon_l(0)$, $\epsilon_r(0)$ and the matrix elements on the right-hand side refer to the unperturbed molecular system.

Using (18.7) in (18.6) we find that the electric moment is given by the expression

$$\langle l|M_\alpha|l\rangle + \sum_\beta \left\{ \frac{1}{\hbar}\sum_{r\neq l} \frac{\langle l|M_\alpha|r\rangle\langle r|M_\beta|l\rangle+\langle l|M_\beta|r\rangle\langle r|M_\alpha|l\rangle}{\omega_{rl}} \right\} E_\beta, \quad (18.8)$$

where we have introduced the transition frequency of the unperturbed system

$$\omega_{rl} = \frac{1}{\hbar}\left[\epsilon_r(0)-\epsilon_l(0)\right] \qquad (18.9)$$

in place of the energy difference $\epsilon_r(0)-\epsilon_l(0)$. The first term in (18.8) is independent of the field and represents the permanent dipole moment of the system; the second term represents the moment induced by the applied field. The coefficient in the second term (enclosed in curly brackets) is the (α,β) component of a tensor, which we shall call the *static polarizability*. Let us introduce the following notation for it:

$$P^{ll}_{\alpha\beta}(0) = \frac{1}{\hbar}\sum_{r\neq l} \frac{\langle l|M_\alpha|r\rangle\langle r|M_\beta|l\rangle+\langle l|M_\beta|r\rangle\langle r|M_\alpha|l\rangle}{\omega_{rl}}, \qquad (18.10)$$

where l indicates the state to which the polarizability refers. The repetition of the index and the argument 0 are introduced to bring the formula in line with the results which we shall obtain in connexion with variable fields. Using (18.10) we can rewrite (18.7) and (18.8) as follows:

$$\epsilon_l(\mathbf{E}) = \epsilon_l(0) - \sum_\alpha \langle l|M_\alpha|l\rangle E_\alpha - \tfrac{1}{2}\sum_{\alpha\beta} P^{ll}_{\alpha\beta}(0)E_\alpha E_\beta. \qquad (18.11)$$

$$\int \psi_l^*(\mathbf{E})M_\alpha\psi_l(\mathbf{E})\,d\tau = \langle l|M_\alpha|l\rangle + \sum_\beta P^{ll}_{\alpha\beta}(0)E_\beta. \qquad (18.12)$$

Next consider the system in a periodic electric field:

$$\mathbf{E}(t) = \mathbf{E}^- e^{-i\omega t} + \mathbf{E}^+ e^{i\omega t} \qquad (\mathbf{E}^- = (\mathbf{E}^+)^*), \qquad (18.13)$$

where $\mathbf{E}^- = (\mathbf{E}^+)^*$ is an arbitrary constant vector, the components of which may be complex. (18.13) represents an elliptical vibration of the electric vector; thus if the vector $\mathbf{E}(t)$ is graphically represented, its end point describes an elliptical orbit with the circular frequency $|\omega|$ (note that in (18.13) we do not specify whether ω is positive or negative). The Hamiltonian of the system is still given by (18.5),† but the electric field is now the time-dependent function (18.13). Thus we have to consider the time-dependent Schrödinger equation

$$\{H(0) - \mathbf{M} \cdot \mathbf{E}^- e^{-i\omega t} - \mathbf{M} \cdot \mathbf{E}^+ e^{i\omega t}\}\Psi = i\hbar \frac{\partial}{\partial t} \Psi. \qquad (18.14)$$

Let us regard the electric terms as a perturbation and consider trial solutions of the following form:

$$\Psi_l = \psi_l(0)e^{-i\epsilon_l(0)t/\hbar} + [\psi_l^- \, e^{-i\omega t} + \psi_l^+ \, e^{i\omega t}]e^{-i\epsilon_l(0)t/\hbar}. \qquad (18.15)$$

The first term evidently satisfies the unperturbed equation

$$H(0)\psi_l(0)e^{-i\epsilon_l(0)t/\hbar} = i\hbar \frac{\partial}{\partial t}(\psi_l(0)e^{-i\epsilon_l(0)t/\hbar}). \qquad (18.16)$$

Let us suppose, subject to later verification, that the other terms in (18.15) are of the first order with respect to the perturbation. After substituting (18.15) in (18.14) we find, upon ignoring terms of second order and using (18.16), that

$$H(0)[\psi_l^- \, e^{-i\omega t} + \psi_l^+ \, e^{i\omega t}] - [\mathbf{M} \cdot \mathbf{E}^- e^{-i\omega t} + \mathbf{M} \cdot \mathbf{E}^+ e^{i\omega t}]\psi_l(0)$$
$$= (\epsilon_l(0) + \hbar\omega)\psi_l^- \, e^{-i\omega t} + (\epsilon_l(0) - \hbar\omega)\psi_l^+ \, e^{i\omega t}. \qquad (18.17)$$

(18.17) is equivalent to the following pair of equations which are independent of t:

$$[H(0) - \epsilon_l(0) \pm \hbar\omega]\psi_l^\pm = \mathbf{M} \cdot \mathbf{E}^\pm \psi_l(0), \qquad (18.18)$$

where either all the upper or all the lower signs are to be taken.

Since the unperturbed wave functions $\psi_l(0)$ form a complete set, we can express $\psi_l^\pm$ as follows:

$$\psi_l^\pm = \sum_r a_r^\pm \psi_r(0). \qquad (18.19)$$

When these series are substituted in (18.18), the following equations are obtained:

$$\hbar \sum_r a_r^\pm [\omega_{rl} \pm \omega]\psi_r(0) = \mathbf{M} \cdot \mathbf{E}^\pm \psi_l(0), \qquad (18.20)$$

---

† A variable electric field is always associated with a magnetic field. The interaction of the latter with the molecular system is, however, small ($\sim v/c$) and may be ignored without causing appreciable error.

where we have used the fact that $H(0)\psi_r(0) = \epsilon_r(0)\psi_r(0)$ and expressed $\epsilon_r(0)-\epsilon_l(0)$ in terms of the transition frequency ω_{rl} defined in (18.9). After multiplying (18.20) by $\psi_s^*(0)$ and integrating over the particle coordinates, we find on using the orthogonal relations between different wave functions that

$$a_s^\pm = \frac{1}{\hbar}\sum_\alpha \frac{\langle s|M_\alpha|l\rangle E_\alpha^\pm}{\omega_{sl}\pm\omega}. \tag{18.21}$$

$\psi_l^\pm$ are thus linear in the field amplitudes; this justifies the supposition that they represent first-order terms.

The wave functions given by (18.15) together with (18.19) and (18.21) are, of course, correct only to the first order of accuracy. However, as long as the frequency $\pm\omega$ of the applied field is not too close to any of the transition frequencies ω_{sl}, the wave functions are adequate for considering the electric moment induced by the field. The electric dipole moment $\mathbf{m}(t)$ of the system is obtained by forming the expectation value:

$$\mathbf{m}(t) = \int \Psi_l^* \mathbf{M}\Psi_l\, d\tau = \langle l|\mathbf{M}|l\rangle + e^{-i\omega t}\sum_r [\langle l|\mathbf{M}|r\rangle a_r^- +\langle r|\mathbf{M}|l\rangle(a_r^+)^*]+$$
$$+e^{i\omega t}\sum_r [\langle r|\mathbf{M}|l\rangle(a_r^-)^*+\langle l|\mathbf{M}|r\rangle a_r^+], \tag{18.22}$$

where we have used (18.15) and (18.19) for the wave function and neglected all terms of the second order. When the values of the a coefficients given by (18.21) are substituted, (18.22) can be expressed as follows:

$$m_\alpha(t) = \int \Psi_l^* M_\alpha \Psi_l\, d\tau$$
$$= \langle l|M_\alpha|l\rangle + \sum_\beta \{P_{\alpha\beta}^{ll}(-\omega)E_\beta^-\, e^{-i\omega t}+P_{\alpha\beta}^{ll}(\omega)E_\beta^+\, e^{i\omega t}\}, \tag{18.23}$$

where

$$P_{\alpha\beta}^{ll}(\omega) = \frac{1}{\hbar}\sum_r \left\{\frac{\langle l|M_\alpha|r\rangle\langle r|M_\beta|l\rangle}{\omega_{rl}+\omega}+\frac{\langle l|M_\beta|r\rangle\langle r|M_\alpha|l\rangle}{\omega_{rl}-\omega}\right\} \tag{18.24}$$

(note that the term $r = l$ vanishes; it is sometimes convenient to exclude it explicitly in the summation). We observe that $P_{\alpha\beta}^{ll}(\omega)$ satisfies the following relations:

$$\left.\begin{array}{l}P_{\alpha\beta}^{ll}(\omega) = [P_{\alpha\beta}^{ll}(-\omega)]^*\\ P_{\alpha\beta}^{ll}(\omega) = [P_{\beta\alpha}^{ll}(\omega)]^*\end{array}\right\}. \tag{18.25}$$

The first relation merely guarantees that the electric moment (18.23) is real (note that $\mathbf{E}^- = (\mathbf{E}^+)^*$). The second relation shows that $P_{\alpha\beta}^{ll}(\omega)$ is the (α,β) component of a *Hermitian* tensor, which we shall call the *polarizability*. If we put $\omega = 0$, (18.24) reduces to (18.10); the static polarizability thus represents a special case of the polarizability. We note, however, that the relation (18.6) has no counterpart in the general case, for in a time-dependent field eigenvalues can no longer be defined.

The first-order wave function (18.15), and hence also the polarizability (18.24), become progressively less accurate as $\pm\omega$ approaches one of the transition frequencies. In particular we find that (18.24) provides for no energy absorption. Thus let us consider the work done by a field $\mathbf{E}(t)$ in a time interval δt on a dipole moment $\mathbf{m}(t)$:

$$\mathbf{E}(t)\left(\frac{d\mathbf{m}(t)}{dt}\right)\delta t. \tag{18.26}$$

If we use (18.13) and (18.23) respectively for $\mathbf{E}(t)$ and $\mathbf{m}(t)$ in the above formula we obtain (the permanent moment in (18.23) can clearly be ignored)

$$i\omega\left\{-\sum_{\alpha\beta}P^{ll}_{\alpha\beta}(-\omega)E^-_\alpha\,E^-_{\bar\beta}\,e^{-2i\omega t}+\sum_{\alpha\beta}P^{ll}_{\alpha\beta}(\omega)E^-_\alpha\,E^+_\beta-\right.$$
$$\left.-\sum_{\alpha\beta}P^{ll}_{\alpha\beta}(-\omega)E^+_\alpha\,E^-_{\bar\beta}+\sum_{\alpha\beta}P^{ll}_{\alpha\beta}(\omega)E^+_\alpha\,E^+_\beta\,e^{2i\omega t}\right\}. \tag{18.27}$$

When the expression is integrated over a period, the time-dependent terms drop out leaving

$$2\pi i\left\{\sum_{\alpha\beta}P^{ll}_{\alpha\beta}(\omega)E^-_\alpha\,E^+_\beta-\sum_{\alpha\beta}P^{ll}_{\alpha\beta}(-\omega)E^+_\alpha\,E^-_{\bar\beta}\right\}. \tag{18.28}$$

It follows immediately from the Hermitian character (see (18.25)) of the polarizability that the work done over a period vanishes.

It is, however, well known that a molecular system can absorb radiative energy in the neighbourhood of its transition frequencies. In order to take account of this important effect, let us add to the polarizability another part which we shall denote by $R^{ll}_{\alpha\beta}(\omega)$. Since the induced dipole must always be a real quantity, we must have, similarly to (18.25),

$$R^{ll}_{\alpha\beta}(\omega)=[R^{ll}_{\alpha\beta}(-\omega)]^*. \tag{18.29}$$

Moreover, since we have seen that a Hermitian polarizability tensor gives no energy absorption, and as $R^{ll}_{\alpha\beta}(\omega)$ is designed just to take account of the absorption, we may assume that

$$R^{ll}_{\alpha\beta}(\omega)=-[R^{ll}_{\beta\alpha}(\omega)]^*=-R^{ll}_{\beta\alpha}(-\omega) \tag{18.30}$$

(i.e. $R^{ll}_{\alpha\beta}(\omega)$ is anti-Hermitian). For any tensor can always be decomposed into a Hermitian and an anti-Hermitian part.

We expect that $R^{ll}_{\alpha\beta}(\omega)$ has an appreciable value only if either of $\pm\omega$

is close to a transition frequency ω_{rl}. Thus let us subject the molecular system (in state l) to the following electric field:

$$\mathbf{E}(t) = \int_{\omega_{rl}-\Delta}^{\omega_{rl}+\Delta} \{\mathbf{E}(-\omega)e^{-i\omega t} + \mathbf{E}(\omega)e^{i\omega t}\}\, d\omega, \qquad (18.31)$$

where

$$\mathbf{E}(-\omega) = \mathbf{E}^*(\omega) \qquad (18.32)$$

is an arbitrary function of ω. (18.31) describes an electric pulse, which has approximately the frequency ω_{rl} and lasts a time of the order $1/\Delta$. We shall calculate on the one hand the work done by the pulse on the system with the help of $R^{ll}_{\alpha\beta}(\omega)$ and on the other hand the energy absorbed (or released) by the system by means of a quantum transition from state l to state r. The expression for $R^{ll}_{\alpha\beta}(\omega)$ is then obtained by equating the two results.

The induced dipole moment corresponding to the modified polarizability is

$$m_\alpha(t) = \sum_\beta \int_{\omega_{rl}-\Delta}^{\omega_{rl}+\Delta} \{(P^{ll}_{\alpha\beta}(-\omega) + R^{ll}_{\alpha\beta}(-\omega))E_\beta(-\omega)e^{-i\omega t} +$$
$$+ (P^{ll}_{\alpha\beta}(\omega) + R^{ll}_{\alpha\beta}(\omega))E_\beta(\omega)e^{i\omega t}\}\, d\omega. \quad (18.33)$$

To calculate the work done by the pulse, we have to substitute this expression and the electric field (18.31) in (18.26) and integrate the resulting expression over time. We may, however, omit directly the terms in (18.33) due to $P^{ll}_{\alpha\beta}(\omega)$, since we have seen that these terms contribute nothing. Thus the total work done by the pulse can be written as

$$i \sum_{\alpha\beta} \int_{-\infty}^{\infty} dt \left\{ \int_{\omega_{rl}-\Delta}^{\omega_{rl}+\Delta} (E_\alpha(-\omega')e^{-i\omega't} + E_\alpha(\omega')e^{i\omega't})\, d\omega' \right\} \times$$

$$\times \left\{ \int_{\omega_{rl}-\Delta}^{\omega_{rl}+\Delta} (-R^{ll}_{\alpha\beta}(-\omega)E_\beta(-\omega)e^{-i\omega t} + R^{ll}_{\alpha\beta}(\omega)E_\beta(\omega)e^{i\omega t})\omega\, d\omega \right\}$$

$$= 2\pi i \sum_{\alpha\beta} \int_{\omega_{rl}-\Delta}^{\omega_{rl}+\Delta} \omega\, d\omega \int_{\omega_{rl}-\Delta}^{\omega_{rl}+\Delta} d\omega' \{-R^{ll}_{\alpha\beta}(-\omega)E_\alpha(-\omega')E_\beta(-\omega)\delta(\omega'+\omega) +$$
$$+ R^{ll}_{\alpha\beta}(\omega)E_\alpha(-\omega')E_\beta(\omega)\delta(\omega-\omega') - R^{ll}_{\alpha\beta}(-\omega)E_\alpha(\omega')E_\beta(-\omega)\delta(\omega'-\omega) +$$
$$+ R^{ll}_{\alpha\beta}(\omega)E_\alpha(\omega')E_\beta(\omega)\delta(\omega'+\omega)\}, \quad (18.34)$$

where we have carried out the integration with respect to t and made use of the following relation:

$$\int_{-\infty}^{\infty} e^{\pm ixt}\, dt = 2\pi\, \delta(x) \quad (\delta(x) = \text{Dirac delta-function}). \quad (18.35)$$

The terms in (18.34) with the factor $\delta(\omega+\omega')$ contribute nothing, for the region of integration does not cover points with $\omega = -\omega'$. Thus after carrying out the integration with respect to ω', we obtain

$$2\pi i \sum_{\alpha\beta} \int_{\omega_{rl}-\Delta}^{\omega_{rl}+\Delta} \{R_{\alpha\beta}^{u}(\omega)E_\alpha(-\omega)E_\beta(\omega) - R_{\alpha\beta}^{u}(-\omega)E_\alpha(\omega)E_\beta(-\omega)\}\omega\, d\omega.$$

(18.36)

Remembering (18.30), we see that the first term in (18.36) is equal to the second. Hence we can write the work done by the pulse as

$$4\pi i \sum_{\alpha\beta} \int_{\omega_{rl}-\Delta}^{\omega_{rl}+\Delta} R_{\alpha\beta}^{u}(\omega)E_\alpha(-\omega)E_\beta(\omega)\omega\, d\omega.$$

(18.37)

Now let us calculate the probability for the system to make a transition to the state r under the influence of the electric pulse. If we regard the interaction $-\mathbf{M}.\mathbf{E}$ as a perturbation, it follows from a well-known formula in perturbation theory that the transition probability is given to the first order of accuracy by

$$\frac{1}{\hbar^2} \left| \int_{-\infty}^{\infty} \langle r|\mathbf{M}.\mathbf{E}|l\rangle e^{i\omega_{rl}t}\, dt \right|^2$$

$$= \frac{1}{\hbar^2} \left| \sum_\alpha \langle r|M_\alpha|l\rangle \int_{-\infty}^{\infty} dt \int_{\omega_{rl}-\Delta}^{\omega_{rl}+\Delta} [E_\alpha(-\omega)e^{-i(\omega-\omega_{rl})t} + E_\alpha(\omega)e^{i(\omega+\omega_{rl})t}]\, d\omega \right|^2$$

$$= \frac{4\pi^2}{\hbar^2} \left| \sum_\alpha \langle r|M_\alpha|l\rangle \int_{\omega_{rl}-\Delta}^{\omega_{rl}+\Delta} [E_\alpha(-\omega)\delta(\omega-\omega_{rl}) + E_\alpha(\omega)\delta(\omega+\omega_{rl})]\, d\omega \right|^2,$$

(18.38)

where the integration over t has been carried out with the help of (18.35). Evidently the second term contributes nothing; thus on carrying out the integration with respect to ω we obtain

$$\frac{4\pi^2}{\hbar^2} \sum_{\alpha\beta} \langle r|M_\beta|l\rangle^* E_\beta^*(-\omega_{rl})\langle r|M_\alpha|l\rangle E_\alpha(-\omega_{rl})$$

$$= \frac{4\pi^2}{\hbar^2} \sum_{\alpha\beta} \langle l|M_\beta|r\rangle\langle r|M_\alpha|l\rangle E_\alpha(-\omega_{rl})E_\beta(\omega_{rl}). \quad (18.39)$$

The energy absorbed by the system during a transition is $\hbar\omega_{rl}$. Multiplying (18.39) by $\hbar\omega_{rl}$ we obtain for the average rate of energy absorption

$$\frac{4\pi^2\omega_{rl}}{\hbar} \sum_{\alpha\beta} \langle l|M_\beta|r\rangle\langle r|M_\alpha|l\rangle E_\alpha(-\omega_{rl})E_\beta(\omega_{rl}). \quad (18.40)$$

(18.37) and (18.40) must clearly be equal. Thus we find that we have to set $R_{\alpha\beta}^{u}(\omega)$ equal to the expression

$$-\frac{i\pi}{\hbar} \langle l|M_\beta|r\rangle\langle r|M_\alpha|l\rangle \delta(\omega-\omega_{rl}) \quad (18.41)$$

in the frequency range in question, namely,

$$\omega_{rl} - \Delta < \omega < \omega_{rl} + \Delta. \qquad (18.42)$$

We may also express this in an alternative way, namely, that for ω in the range

$$-\omega_{rl} - \Delta < \omega < -\omega_{rl} + \Delta \qquad (18.43)$$

$R^{ll}_{\alpha\beta}(-\omega)$ is equal to

$$-\frac{i\pi}{\hbar} \langle l|M_\beta|r\rangle\langle r|M_\alpha|l\rangle \delta(-\omega - \omega_{rl}) = -\frac{i\pi}{\hbar} \langle l|M_\beta|r\rangle\langle r|M_\alpha|l\rangle \delta(\omega + \omega_{rl}).$$
$$(18.44)$$

According to the condition (18.29) the complex conjugate of $R^{ll}_{\alpha\beta}(-\omega)$ must be equal to $R^{ll}_{\alpha\beta}(\omega)$. Thus it follows that in the range (18.43) $R^{ll}_{\alpha\beta}(\omega)$ is given by the complex conjugate of (18.44):

$$\frac{i\pi}{\hbar} \langle l|M_\alpha|r\rangle\langle r|M_\beta|l\rangle \delta(\omega + \omega_{rl}). \qquad (18.45)$$

Adding (18.45) to (18.41) and summing over all possible transition frequencies, we obtain finally

$$R^{ll}_{\alpha\beta}(\omega) = \frac{i\pi}{\hbar} \sum_r \{\langle l|M_\alpha|r\rangle\langle r|M_\beta|l\rangle \delta(\omega + \omega_{rl}) - \langle l|M_\beta|r\rangle\langle r|M_\alpha|l\rangle \delta(\omega - \omega_{rl})\}.$$
$$(18.46)$$

We notice that the expression is anti-Hermitian, in agreement with the original assumption (18.30).

By adding (18.46) to (18.24) we obtain a polarizability tensor which also takes account of absorption.† The polarizability thus obtained is based on certain first approximations. (18.24) is based on the first-order wave function (18.15), and (18.46) is based on the first-order approximation (18.38) for the transition probability. Thus the Hermitian part (18.24) remains inaccurate in the immediate neighbourhood of the transition frequencies. And the anti-Hermitian part (18.46) does not describe the frequency-dependence very realistically. The delta-functions in (18.46) lead to absorptions at the sharply defined frequencies $|\omega_{rl}|$, whereas in reality appreciable absorption takes place over a small but finite neighbourhood of a transition frequency. In other words, every delta-function in (18.46) should have been replaced

---

† One sees that the sum of (18.24) and (18.46) can be simply obtained from (18.24) by replacing $(\omega_{rl} + \omega)^{-1}$ and $(\omega_{rl} - \omega)^{-1}$ by

$$(\omega_{rl} + \omega)^{-1} + i\pi\delta(\omega_{rl} + \omega) \quad \text{and} \quad (\omega_{rl} - \omega)^{-1} - i\pi\delta(\omega_{rl} - \omega)$$

respectively. A deeper understanding of this can be obtained by regarding the optical effects as collisions of atoms with photons; cf. P. A. M. Dirac, *Quantum Mechanics* (Oxford, Clarendon Press, 3rd ed., 1947), chap. III, sect. 15, p. 61, and chap. VIII, sect. 50, p. 193).

by a function of ω which has an appreciable value in a small neighbour-hood of $\omega = 0$. We shall, however, make serious use of the polarizability only in cases where the transition frequencies are so close together as to form practically a continuum. In such cases the above inaccuracies are not serious owing to two circumstances. Firstly, since the Hermitian part (18.24) is respectively positive and negative immediately below and above a transition frequency ω_{rl} (and immediately above and below $-\omega_{rl}$) the inaccuracies due to closely situated transition frequencies mutually cancel. Secondly, though the shapes of the functions which should have replaced the delta-functions in (18.46) are different from that of the singular delta-function, their integrals are equal to unity as for the delta-function. This fact makes it possible to use (18.46) for a continuous distribution of transition frequencies without much error.

The Maxwell equations give a complete account of the refractive properties of a material medium, once the relation between the Maxwell displacement vector and the macroscopic electric field is known. Since the difference between the displacement vector and the macroscopic field is by definition equal to 4π times the electric dipole moment per unit volume (dielectric polarization), we can discuss the refractive properties with the help of the polarizability. In other words, the expression for the polarizability such as we have obtained provides the basis for a theory of dispersion. Certain points should, however, be borne in mind in this connexion. If one is considering a medium which is composed of distinctly separated molecules such as in a gas, then in order to obtain the dielectric polarization one has only to divide the molecular moment as given by the polarizability by the average volume occupied by a molecule; the field acting on the system differs in general, however, from the macroscopic field (see § 9). The difference can be ignored only if the density is very small; otherwise one has first to obtain a relation between the macroscopic field and the field $\mathbf{E}$ that acts on a molecule. In the case of crystalline solids, the whole crystal forms a single molecular system. Evidently we cannot apply the results obtained above directly to the system; thus if we are considering the passage of an optical wave, there will be many wave-lengths in the crystal and our assumption of a constant field amplitude $\mathbf{E}^- = (\mathbf{E}^+)^*$ no longer applies. Actually what we wish to find is not the electric moment of the whole system, but a relation between the electric moment of a portion of the system in a small volume element (small compared with the wave-length of the optical wave) and the macroscopic field in the same neighbourhood. It is one of the principal aims of Chapter V to show how such a volume element can be virtually isolated so that a polarizability of the form deduced above can be used.

19. The Rayleigh and Raman scattering of light

Let us give a brief description of the radiation emitted by an oscillating electric dipole moment

$$\mathbf{m}(t) = \mathbf{m}^- e^{-i\omega t} + \mathbf{m}^+ e^{i\omega t}, \qquad (19.1)$$

where

$$\mathbf{m}^- = (\mathbf{m}^+)^* \qquad (19.2)$$

is an arbitrary complex vector. At a large distance $\mathbf{R}$ from the dipole the electric and magnetic fields are given respectively (at time $t + (R/c)$) by

$$\mathbf{E}\left(t + \frac{R}{c}\right) = \frac{1}{R^3 c^2}[\mathbf{R} \wedge \{\mathbf{R} \wedge \ddot{\mathbf{m}}(t)\}] = -\frac{\omega^2}{R^3 c^2}[\mathbf{R} \wedge \{\mathbf{R} \wedge \mathbf{m}(t)\}]. \quad (19.3)$$

$$\mathbf{H}\left(t + \frac{R}{c}\right) = -\frac{1}{R^2 c^2}[\mathbf{R} \wedge \ddot{\mathbf{m}}(t)] = \frac{\omega^2}{R^2 c^2}[\mathbf{R} \wedge \mathbf{m}(t)]. \qquad (19.4)$$

The fields are evidently equal in magnitude and perpendicular to one another. The scalar magnitude of the Poynting vector

$$\mathbf{S} = \frac{c}{4\pi}[\mathbf{E} \wedge \mathbf{H}]$$

can thus be written as

$$S = \frac{c}{4\pi}\mathbf{E}^2 = \frac{\omega^4}{4\pi R^2 c^3}\left|\frac{\mathbf{R}}{R} \wedge \left[\frac{\mathbf{R}}{R} \wedge \mathbf{m}(t)\right]\right|^2. \qquad (19.5)$$

We note that the vector product in the above expression represents the projection of $\mathbf{m}(t)$ on a plane perpendicular to the vector $\mathbf{R}$. Thus if $\mathbf{n}^1$, $\mathbf{n}^2$ are two mutually perpendicular unit vectors, both perpendicular to $\mathbf{R}$, we can evidently write (19.5) as follows:

$$S = \frac{\omega^4}{4\pi R^2 c^3}\sum_{i=1,2}\sum_{\alpha\beta} n_\alpha^i n_\beta^i m_\alpha(t)m_\beta(t). \qquad (19.6)$$

If we are interested in one linearly polarized component of the radiation only, for instance, the component with the electric vector in the direction of $\mathbf{n}^1$, we have simply to omit the first summation sign in (19.6) and put $i = 1$.

On substituting (19.1) in (19.6) and taking the average value over a period of vibration, we obtain the mean value

$$\bar{S} = \frac{\omega^4}{2\pi R^2 c^3}\sum_i\sum_{\alpha\beta} n_\alpha^i n_\beta^i m_\alpha^+ m_\beta^-. \qquad (19.7)$$

(19.3) and (19.4) taken together show that $\mathbf{E}$, $\mathbf{H}$, and $\mathbf{R}$ taken in this order form a right-handed system of orthogonal vectors; hence the energy flow $\mathbf{S}$ is radially outward. The average rate of energy flow through a solid angle $d\Omega$ is obtained by multiplying (19.7) by the area $R^2 d\Omega$; i.e.

$$R^2\bar{S}\,d\Omega = \frac{\omega^4}{2\pi c^3}\left\{\sum_i \sum_{\alpha\beta} n^i_\alpha n^i_\beta m^+_\alpha m^-_\beta\right\}d\Omega. \tag{19.8}$$

The total rate of radiation is obtained by integrating the above expression over all angles. If we use polar coordinates, the vectors $\mathbf{n}^1$ and $\mathbf{n}^2$ can be chosen as follows:

$$n^1_1 = \cos\theta\cos\phi, \qquad n^1_2 = \cos\theta\sin\phi, \qquad n^1_3 = -\sin\theta;$$

$$n^2_1 = -\sin\phi, \qquad n^2_2 = \cos\phi, \qquad n^2_3 = 0. \tag{19.9}$$

Using these values in (19.8), we readily find that

$$\int R^2\bar{S}\,d\Omega = \frac{4\omega^4}{3c^3}\sum_\alpha m^+_\alpha m^-_\alpha. \tag{19.10}$$

(18.13) describes the electric field over a molecular system exposed to a beam of elliptically polarized light, provided that the dimension of the molecular system is small compared with the wave-length of the light wave. The radiation emitted by the induced moment represents the light scattered by the molecular system. Thus we arrive at a complete description of the scattered light by making the following substitutions:

$$m^+_\alpha = \sum_\beta P^{ll}_{\alpha\beta}(\omega)E^+_\beta, \qquad m^-_\alpha = \sum_\beta P^{ll}_{\alpha\beta}(-\omega)E^-_\beta \tag{19.11}$$

(provided that ω is not too close to the transition frequencies) in the above formulae for the radiation by a dipole. This type of light scattering is known as the *Rayleigh scattering*. We note that the scattered light has the same frequency as the incident light; moreover, as the polarizability is unambiguously defined (in particular, independent of the arbitrary phases of the wave functions used in forming the matrix elements in the polarizability (18.24)), the induced moment and hence the scattered light bears a definite phase relation with the incident light. In other words, the Rayleigh scattering is coherent.

The phenomenon of optical refraction and the Rayleigh scattering are closely related. The former simply describes the result of the interference between the incident light and the scattered waves from all sub-units of a medium, if we assume the sub-units to be uniformly distributed. The course of a refracted beam is thus defined by means of destructive interference of waves in other directions. Only in a medium where appreciable randomness occurs in the disposition of the scattering units can one observe the Rayleigh scattering directly.

In the Rayleigh scattering a molecular system persists in a fixed quantum state and the scattering is due to the periodic deformation of the state by the electric field in the incident light. What is known as the *Raman scattering*, on the other hand, is associated with a quantum transition in the system. It is most easily understood in terms of quantum electrodynamics, wherein we regard light as composed of photons. Let us return to the electric field (18.13). We can, without loss of generality, specify that $\omega > 0$; it is more convenient to do so in discussing the Raman scattering. The field (18.13) may now be interpreted as representing the effect of incident photons with energy $\hbar\omega$. If, as a result of scattering a photon, the molecular system goes over from an initial state l to a final state l', the energy difference is compensated by a frequency change in the scattered photon. Thus to conserve energy the scattered photon must have the modified frequency $\omega + \Delta\omega$ satisfying the relation

$$\epsilon_l(0) + \hbar\omega = \epsilon_{l'}(0) + \hbar(\omega + \Delta\omega),$$

or
$$\omega + \Delta\omega = \omega + \omega_{ll'}. \tag{19.12}$$

With the help of this relation we can then discuss the Raman scattering by treating the electromagnetic field classically as above.

In this semi-classical theory the Raman scattering, exactly like the Rayleigh scattering, can be described in terms of an induced electric moment; the latter is to be obtained by considering the transition electric moment between two states defined as follows:

$$\int \{\Psi_m^* \mathbf{M}\Psi_l + \Psi_l^* \mathbf{M}\Psi_m\} \, d\tau. \tag{19.13}$$

Let us write down explicitly the wave functions Ψ_l and Ψ_m^* in the presence of the electric field (18.13) with the help of (18.15), (18.19), and (18.21):

$$\Psi_l = e^{-i\epsilon_l(0)t/\hbar}\left\{\psi_l(0) + \frac{1}{\hbar}\sum_\beta \sum_r \left[\frac{\langle r|M_\beta|l\rangle}{\omega_{rl}-\omega}\psi_r(0)E_\beta^- \, e^{-i\omega t} + \right.\right.$$
$$\left.\left. + \frac{\langle r|M_\beta|l\rangle}{\omega_{rl}+\omega}\psi_r(0)E_\beta^+ \, e^{i\omega t}\right]\right\},$$

$$\Psi_m^* = e^{i\epsilon_m(0)t/\hbar}\left\{\psi_m^*(0) + \frac{1}{\hbar}\sum_\beta \sum_r \left[\frac{\langle m|M_\beta|r\rangle}{\omega_{rm}-\omega}\psi_r^*(0)E_\beta^+ \, e^{i\omega t} + \right.\right.$$
$$\left.\left. + \frac{\langle m|M_\beta|r\rangle}{\omega_{rm}+\omega}\psi_r^*(0)E_\beta^- \, e^{-i\omega t}\right]\right\}. \tag{19.14}$$

Upon substituting these wave functions in (19.13) and ignoring second-order terms, we obtain

$$\int \{\Psi_m^* M_\alpha \Psi_l + \Psi_l^* M_\alpha \Psi_m\}\, d\tau = \langle m|M_\alpha|l\rangle e^{-i\omega_{lm}t} + \langle l|M_\alpha|m\rangle e^{i\omega_{lm}t} +$$

$$+\frac{1}{\hbar} \sum_\beta \sum_r \left\{ \frac{\langle m|M_\beta|r\rangle\langle r|M_\alpha|l\rangle}{\omega_{rm}-\omega} E_\beta^+ e^{i(\omega-\omega_{lm})t} + \right.$$

$$+\frac{\langle m|M_\beta|r\rangle\langle r|M_\alpha|l\rangle}{\omega_{rm}+\omega} E_\beta^- e^{-i(\omega+\omega_{lm})t} +$$

$$+\frac{\langle m|M_\alpha|r\rangle\langle r|M_\beta|l\rangle}{\omega_{rl}-\omega} E_\beta^- e^{-i(\omega+\omega_{lm})t} + \frac{\langle m|M_\alpha|r\rangle\langle r|M_\beta|l\rangle}{\omega_{rl}+\omega} E_\beta^+ e^{i(\omega-\omega_{lm})t} +$$

$$+\frac{\langle l|M_\beta|r\rangle\langle r|M_\alpha|m\rangle}{\omega_{rl}-\omega} E_\beta^+ e^{i(\omega+\omega_{lm})t} + \frac{\langle l|M_\beta|r\rangle\langle r|M_\alpha|m\rangle}{\omega_{rl}+\omega} E_\beta^- e^{-i(\omega-\omega_{lm})t} +$$

$$\left. +\frac{\langle l|M_\alpha|r\rangle\langle r|M_\beta|m\rangle}{\omega_{rm}-\omega} E_\beta^- e^{-i(\omega-\omega_{lm})t} + \frac{\langle l|M_\alpha|r\rangle\langle r|M_\beta|m\rangle}{\omega_{rm}+\omega} E_\beta^+ e^{i(\omega+\omega_{lm})t} \right\},$$

$$(19.15)$$

where the first four terms in the curly brackets come from the first term on the left-hand side and the remaining four terms are obtained simply by interchanging the indices l and m (note that $\omega_{ml} = -\omega_{lm}$).

The first two terms on the right-hand side of (19.15) are independent of the electric field; together they give the dipole moment describing the spontaneous emission for the transition from the higher to the lower of the two states l and m. The other terms describe the Raman scattering associated with either of the two transitions $l \to m$ or $m \to l$. With the help of (19.12) we can easily separate the two respective types of terms. Thus the terms describing scattering associated with the transition $l \to m$ must have the frequency $\omega + \omega_{lm}$ and must hence appear with the time factor $\exp\{i(\omega+\omega_{lm})t\}$ or $\exp\{-i(\omega+\omega_{lm})t\}$. Similarly the time factors for the scattering $m \to l$ must be the same as above with the indices interchanged, i.e. $\exp\{\pm i(\omega-\omega_{lm})t\}$. Using this criterion, we see readily that the second, third, fifth, and eighth terms in the curly brackets in (19.15) describe the scattering $l \to m$ and the remaining terms describe the scattering $m \to l$. Furthermore, we notice that the two groups of terms are transformed into one another if we interchange the indices l and m. Hence we can write consistently the electric moment describing the Raman scattering associated with a transition

$l \to m$ as follows:

$$m_\alpha(t) = \sum_\beta [P_{\alpha\beta}^{lm}(\omega)]^* E_\beta^- e^{-i(\omega+\omega_{lm})t} + \sum_\beta P_{\alpha\beta}^{lm}(\omega) E_\beta^+ e^{i(\omega+\omega_{lm})t}, \quad (19.16)$$

where

$$P_{\alpha\beta}^{lm}(\omega) = \frac{1}{\hbar} \sum_r \left\{ \frac{\langle l|M_\alpha|r\rangle\langle r|M_\beta|m\rangle}{\omega_{rm}+\omega} + \frac{\langle l|M_\beta|r\rangle\langle r|M_\alpha|m\rangle}{\omega_{rl}-\omega} \right\} \quad (\omega > 0).$$
$$(19.17)$$

(19.17) defines the ($\alpha\beta$) component of a tensor, which we may call the *transition polarizability* from state l to state m. We notice that if we put m equal to l (19.17) reduces to the polarizability (18.24); and the latter, as we have seen, reduces further to the static polarizability (18.10) if we put ω equal to zero.

Unlike the Rayleigh scattering, the Raman scattering is incoherent. Apart from the fact that the scattered radiation has a modified frequency, the transition polarizability depends on the arbitrary phases of the wave functions used in forming the matrix elements; in actual calculations we have to average over such arbitrary phases. Thus the Raman scattering from two independent molecular systems is not subject to the effect of interference.

Using (19.16) in (19.8), we find that the energy scattered (Raman scattering) per unit time into a solid angle $d\Omega$ is given by the following expression:

$$R^2 \bar{S} \, d\Omega = \frac{(\omega+\omega_{lm})^4}{2\pi c^3} \sum_{i=1,2} \sum_{\alpha\beta\gamma\lambda} n_\alpha^i n_\beta^i [P_{\alpha\gamma}^{lm}(\omega)]^* P_{\beta\lambda}^{lm}(\omega) E_\gamma^- E_\lambda^+ \, d\Omega. \quad (19.18)$$

Similarly from (19.10) we obtain the total Raman scattering per unit time:

$$\int R^2 \bar{S} \, d\Omega = \frac{4(\omega+\omega_{lm})^4}{3c^3} \sum_{\alpha\gamma\lambda} [P_{\alpha\gamma}^{lm}(\omega)]^* P_{\alpha\lambda}^{lm}(\omega) E_\gamma^- E_\lambda^+. \quad (19.19)$$

If we are interested only in one linearly polarized component of the radiation scattered in a given direction we can omit the summation over i in (19.18), as we have explained before.

If the spectrum of the scattered light from a molecular system (in a definite state l) exposed to a monochromatic beam is analysed we obtain a series of lines corresponding to transitions to various final states. The lines corresponding to transitions to final states higher than the initial state have frequencies lower than that of the incident beam; these are known as the *Stokes lines*. The lines corresponding to transitions to final states lower than the initial state have frequencies higher than that of the incident beam, and are known as the *anti-Stokes lines*.

20. Placzek's approximation

The results obtained in the last two sections can be simplified with the help of the adiabatic approximation. Thus we may write a wave function of a molecular system in two factors as follows:

$$\psi_{nv}(x, X) = \chi_{nv}(X)\phi_n(x, X). \tag{20.1}$$

$\phi_n(x, X)$, we remember, is a wave function for the electrons moving in the field of the nuclei, which are held fixed in an arbitrary configuration X, n being the corresponding quantum number. The eigenvalue for the electronic motion is a function of X, which we have denoted by $\Phi_n(X)$ in § 14. $\chi_{nv}(X)$, on the other hand, represents a wave function for the nuclei moving in the effective potential $\Phi_n(X) - \Phi_n(X^0)$ (apart from a fourth-order constant, see § 14), v being the quantum number for the nuclear motion. The eigenvalue for the state represented by (20.1) is the sum

$$\Phi_n(X^0) + \epsilon_{nv}, \tag{20.2}$$

where the eigenvalue ϵ_{nv} for the nuclear motion is small compared with the differences $\Phi_{n'}(X^0) - \Phi_n(X^0)$ between different electronic levels.

Let us consider the permanent electric moment and the polarizability of the system in the lowest electronic level, the nuclei being held fixed in a configuration X so that only the electrons are free to move. Both quantities are evidently functions of X and will be denoted respectively by $\mathbf{M}(X)$ and $P_{\alpha\beta}(\omega, X)$. Remembering that, with the nuclei so constrained, the electronic wave functions and eigenvalues are given respectively by $\phi_n(x, X)$ and $\Phi_n(X)$, we obtain immediately from (18.23) and (18.24)

$$\mathbf{M}(X) = \int \phi_0^*(x, X)\mathbf{M}(x, X)\phi_0(x, X)\, dx, \tag{20.3}$$

$$P_{\alpha\beta}(\omega, X) = \frac{1}{\hbar}\sum_{n \neq 0}\left\{(\omega_{n0} + \omega)^{-1}\int \phi_0^*(x, X)M_\alpha(x, X)\phi_n(x, X)\, dx \times\right.$$
$$\times \int \phi_n^*(x', X)M_\beta(x', X)\phi_0(x', X)\, dx' +$$
$$+ (\omega_{n0} - \omega)^{-1}\int \phi_0^*(x, X)M_\beta(x, X)\phi_n(x, X)\, dx \times$$
$$\left.\times \int \phi_n^*(x', X)M_\alpha(x', X)\phi_0(x', X)\, dx'\right\}, \tag{20.4}$$

where we may put approximately

$$\omega_{n0} = \frac{1}{\hbar}(\Phi_n(X^0) - \Phi_0(X^0)), \tag{20.5}$$

so long as $\pm\omega$ are not too close to any of the frequencies ω_{n0}. We shall see that if the functions $\mathbf{M}(X)$ and $P_{\alpha\beta}(\omega, X)$ are considered formally as known, the electronic wave functions $\phi_n(x, X)$ will no longer be required for the discussion of optical effects.

As pointed out before, at ordinary temperatures a molecular system is practically always in a state belonging to the lowest electronic level. This means that we shall require the polarizability only for states belonging to the lowest electronic level and the transition polarizability from similar initial states. Furthermore we shall discuss only the Raman scattering due to a change in the nuclear motion alone. Hence for our purposes we need only consider the expression (19.17) for the transition polarizability $P_{\alpha\beta}^{lm}(\omega)$ where both l and m belong to the lowest electronic level; we may thus leave out the electronic quantum number and write the transition polarizability as $P_{\alpha\beta}^{vv'}(\omega)$, where v, v' signify nuclear states in the lowest electronic level. From $P_{\alpha\beta}^{vv'}(\omega)$ the polarizability and the static polarizability are obtained by putting successively $v = v'$ and $\omega = 0$.

According to (19.17) we have

$$P_{\alpha\beta}^{vv'}(\omega) = \frac{1}{\hbar} \sum_{n''} \sum_{v''} \left\{ \frac{\langle 0v|M_\alpha|n''v''\rangle\langle n''v''|M_\beta|0v'\rangle}{\omega_{n''v'',0v'}+\omega} + \right.$$

$$\left. + \frac{\langle 0v|M_\beta|n''v''\rangle\langle n''v''|M_\alpha|0v'\rangle}{\omega_{n''v'',0v}-\omega} \right\}, \quad (20.6)$$

where the quantum numbers in (19.17) appear now as pairs of quantum numbers. Let us split the summation over n'' into two parts corresponding to the two cases: $n'' = 0$ and $n'' \neq 0$. For the latter part, we may approximate $\omega_{n''v'',0v'}$ in the denominator by $\omega_{n''0}$ defined in (20.5); this approximation is permissible so long as $\pm\omega$ is not close to any of the frequencies $\omega_{n''0}$. Thus (20.6) becomes

$$P_{\alpha\beta}^{vv'}(\omega) = \frac{1}{\hbar} \sum_{v''} \left\{ \frac{\langle 0v|M_\alpha|0v''\rangle\langle 0v''|M_\beta|0v'\rangle}{\omega_{v''v'}+\omega} + \frac{\langle 0v|M_\beta|0v''\rangle\langle 0v''|M_\alpha|0v'\rangle}{\omega_{v''v}-\omega} \right\} +$$

$$+ \frac{1}{\hbar} \sum_{n''\neq 0} \left\{ \frac{\sum_{v''}\langle 0v|M_\alpha|n''v''\rangle\langle n''v''|M_\beta|0v'\rangle}{\omega_{n''0}+\omega} + \frac{\sum_{v''}\langle 0v|M_\beta|n''v''\rangle\langle n''v''|M_\alpha|0v'\rangle}{\omega_{n''0}-\omega} \right\},$$

$$(20.7)$$

where $\omega_{v''v'}$, $\omega_{v''v}$ in the first sum denote the transition frequencies between nuclear vibrational states belonging to the lowest electronic level. In the above formula the moment operator is a function $\mathbf{M}(x, X)$ of both electronic and nuclear coordinates; and the matrix elements are to be formed with the complete wave functions of the form (20.1).

However, with the help of (20.3) we see immediately that the first sum in (20.7) can be written as follows:

$$\frac{1}{\hbar}\sum_{v''}\left\{\frac{\langle v|M_\alpha(X)|v''\rangle\langle v''|M_\beta(X)|v'\rangle}{\omega_{v''v'}+\omega}+\frac{\langle v|M_\beta(X)|v''\rangle\langle v''|M_\alpha(X)|v'\rangle}{\omega_{v''v}-\omega}\right\},$$

(20.8)

where the matrix elements are now to be formed with the vibrational wave functions $\chi_{0v}(X)$ of the nuclei. Comparing this formula with (19.17) we see that, as regards this part, the system behaves as a purely nuclear system, where the moment operator is given by $M(X)$, and the wave functions and eigenvalues for the nuclear motion are those for the lowest electronic level. (20.8) is usually known as the *ionic part of the polarizability*.

As regards the remaining part in (20.7), let us examine a typical term

$$\frac{\sum_{v''}\langle 0v|M_\alpha|n''v''\rangle\langle n''v''|M_\beta|0v'\rangle}{\omega_{n''0}+\omega}.$$

(20.9)

Expressing the matrix elements explicitly, we may write (20.9) in the following form:

$$\frac{1}{\omega_{n''0}+\omega}\sum_{v''}\int dx\int dX\int dx'\int dX'\{\phi_0^*(x,X)\chi_{0v}^*(X)M_\alpha(x,X)\phi_{n''}(x,X)\times$$

$$\times\chi_{n''v''}(X)\phi_{n''}^*(x',X')\chi_{n''v''}^*(X')M_\beta(x',X')\phi_0(x',X')\chi_{0v'}(X')\}. \quad (20.10)$$

Since the $\chi_{n''v''}(X)$ for various values of v'' (fixed value of n'') form a complete set of functions in X, we have the completeness relation

$$\sum_{v''}\chi_{n''v''}(X)\chi_{n''v''}^*(X')=\delta(X-X').$$

(20.11)

When this relation is used in (20.10) and the integration over X' is carried out, (20.10) becomes

$$\frac{1}{\omega_{n''0}+\omega}\int dx\int dx'\int dX\{\phi_0^*(x,X)\chi_{0v}^*(X)M_\alpha(x,X)\phi_{n''}(x,X)\phi_{n''}^*(x',X)\times$$

$$\times M_\beta(x',X)\phi_0(x',X)\chi_{0v'}(X)\}$$

$$=\int\chi_{0v}^*(X)\Big\{(\omega_{n''0}+\omega)^{-1}\int\phi_0^*(x,X)M_\alpha(x,X)\phi_{n''}(x,X)\,dx\times$$

$$\times\int\phi_{n''}^*(x',X)M_\beta(x',X)\phi_0(x',X)\,dx'\Big\}\chi_{0v'}(X)\,dX. \quad (20.12)$$

After expressing all terms of the second sum in (20.7) in this way, we find upon comparing with (20.4) that this part of the polarizability is simply the matrix element of the function $P_{\alpha\beta}(\omega, X)$ with respect to the vibrational wave functions $\chi_{0v}(X)$, $\chi_{0v'}(X)$, i.e.

$$\langle v|P_{\alpha\beta}(\omega, X)|v'\rangle. \tag{20.13}$$

This is known as the *electronic part of the polarizability*.

The contributions to the static polarizability ($v' = v$, $\omega = 0$) by the ionic part (20.8) and the electronic part (20.13) are roughly of the same order of magnitude in ionic crystals. Hence for $|\omega| \gg \omega_{vv'}$ (20.8) becomes very small compared with (20.13) and may be ignored. Experimentally, the frequency of the incident light used in measuring the Raman scattering is always high compared with the transition frequencies $\omega_{vv'}$ between the vibrational states. Hence for practical purposes we can omit the ionic part when we are considering the transition polarizability, so that we have

$$P_{\alpha\beta}^{vv'}(\omega) = \langle v|P_{\alpha\beta}(\omega, X)|v'\rangle \quad (v \neq v'). \tag{20.14}$$

Using this expression in (19.18) and (19.19) we obtain for the angular and total Raman scattering, respectively, the following expressions:

$$\frac{(\omega+\omega_{vv'})^4}{2\pi c^3} \sum_i \sum_{\alpha\beta} \sum_{\gamma\lambda} n_\alpha^i n_\beta^i \langle v'|P_{\alpha\gamma}^*(\omega, X)|v\rangle\langle v|P_{\beta\lambda}(\omega, X)|v'\rangle E_\gamma^- E_\lambda^+ \, d\Omega, \tag{20.15}$$

$$\frac{4(\omega+\omega_{vv'})^4}{3c^3} \sum_{\alpha\gamma\lambda} \langle v'|P_{\alpha\gamma}^*(\omega, X)|v\rangle\langle v|P_{\alpha\lambda}(\omega, X)|v'\rangle E_\gamma^- E_\lambda^+. \tag{20.16}$$

In order to obtain the experimentally observed intensities, we have to average the above expressions over the thermal distribution for the quantum number v of the initial state.

Adding (20.8) and (20.13) and putting $v' = v$, we have for the polarizability:

$$P_{\alpha\beta}^{vv}(\omega) = \langle v|P_{\alpha\beta}(\omega, X)|v\rangle +$$
$$+\frac{1}{\hbar} \sum_{v''}{}' \left\{ \frac{\langle v|M_\alpha(X)|v''\rangle\langle v''|M_\beta(X)|v\rangle}{\omega_{v''v}+\omega} + \frac{\langle v|M_\beta(X)|v''\rangle\langle v''|M_\alpha(X)|v\rangle}{\omega_{v''v}-\omega} \right\}. \tag{20.17}$$

For the consideration of the refractive properties in the infra-red region the electronic and ionic parts are comparable in magnitude, so neither can be ignored. However, as $|\omega|$ in the infra-red region is small compared with the electronic transition frequencies ω_{n0}, we can put $\omega = 0$ in the electronic part without much error (see the expression for $P_{\alpha\beta}(\omega, X)$). Moreover, if we expand $P_{\alpha\beta}(\omega, X)$ as a Taylor's series in $X-X^0$, the constant term is by far the most important (note that the same argument does not apply to $M(X)$, for in the ionic part only the non-diagonal matrix elements of $M(X)$ occur; to these a constant term contributes nothing). Therefore in the infra-red region we have approximately

$$P_{\alpha\beta}^{vv}(\omega) = P_{\alpha\beta}(0, X^0) +$$
$$+ \frac{1}{\hbar} \sum_{v'' \neq v} \left\{ \frac{\langle v|M_\alpha(X)|v''\rangle\langle v''|M_\beta(X)|v\rangle}{\omega_{v''v}+\omega} + \frac{\langle v|M_\beta(X)|v''\rangle\langle v''|M_\alpha(X)|v\rangle}{\omega_{v''v}-\omega} \right\}.$$

$$(20.18)$$

The electronic part is practically a constant, independent of both the frequency and the temperature (for it does not depend on v).

In order to include the effect of absorption we have only to add the expression (18.46) for $R_{\alpha\beta}^{ll}(\omega)$ to the polarizability (20.18). For the discussion of infra-red optics we may restrict the summation in (18.46) to the states belonging to the lowest electronic level, for the other terms only lead to absorption at the much higher electronic frequencies ω_{n0}. The matrix elements in the terms retained are all expressible in terms of the function $M(X)$, leading to the formula

$$R_{\alpha\beta}^{vv}(\omega) = \frac{i\pi}{\hbar} \sum_{v''} \{\langle v|M_\alpha(X)|v''\rangle\langle v''|M_\beta(X)|v\rangle\delta(\omega+\omega_{v''v}) -$$
$$- \langle v|M_\beta(X)|v''\rangle\langle v''|M_\alpha(X)|v\rangle\delta(\omega-\omega_{v''v})\}. \quad (20.19)$$

In actual applications we have to average (20.18) and (20.19) over the thermal distribution of the quantum number v.

21. Expansion of the optical parameters and the classification of optical effects

In Placzek's approximation the various optical effects depend only on $M(X)$, $P_{\alpha\beta}(\omega, X)$, and the wave functions for nuclear motion in the lowest electronic level. The motion of the nuclei is usually discussed in this connexion in the harmonic approximation with the help of the normal coordinates. A survey of the optical effects and their classification is obtained in the first place by expanding the functions $M(X)$ and $P_{\alpha\beta}(\omega, X)$ in Taylor's series with respect to the normal coordinates as follows:

$$M(X) = M(X^0) + \sum_j M(j)q_j + \tfrac{1}{2}\sum_{jj'} M(j,j')q_j q_{j'} + \cdots, \quad (21.1)$$

$$P_{\alpha\beta}(\omega, X) = P_{\alpha\beta}(\omega, X^0) + \sum_j P_{\alpha\beta}(j)q_j + \tfrac{1}{2}\sum_{jj'} P_{\alpha\beta}(j,j')q_j q_{j'} + \cdots, \quad (21.2)$$

where the coefficients are understood to be symmetric in the indices:

$$M(j,j') = M(j',j), \qquad P_{\alpha\beta}(j,j') = P_{\alpha\beta}(j',j), \quad \text{etc.} \quad (21.3)$$

We observe that the formulae for both the intensity of the Raman scattering and the polarizability tensors (see (20.15), (20.16), (20.18), and (20.19)) depend on products of matrix elements of the type

$$\langle v|A(q)|v'\rangle\langle v'|B(q)|v\rangle, \tag{21.4}$$

where $A(q)$ and $B(q)$ are components of $\mathbf{M}(X)$ or $P_{\alpha\beta}(\omega, X)$ and thus, according to (21.1) and (21.2), may be expressed as Taylor's series:

$$\left.\begin{aligned} A(q) &= A_0 + \sum_j A_j q_j + \tfrac{1}{2} \sum_{jj'} A_{jj'} q_j q_{j'} + \dots \quad (A_{jj'} = A_{j'j}) \\ B(q) &= B_0 + \sum_j B_j q_j + \tfrac{1}{2} \sum_{jj'} B_{jj'} q_j q_{j'} + \dots \quad (B_{jj'} = B_{j'j}) \end{aligned}\right\}. \tag{21.5}$$

We notice in particular that in order to carry out the requisite thermal average in the optical formulae, we have to average products of the form (21.4) over v for a fixed transition frequency $\omega_{vv'}$. For, in the case of the Raman scattering, we are interested in the thermal average of the intensity of a given line (i.e. of a fixed frequency shift). In the case of the polarizabilities the above products are multiplied by coefficients (viz. $1/(\omega_{vv'}\pm\omega)$ in (20.18) and $\delta(\omega\pm\omega_{v'v})$ in (20.19)) which depend only on the transition frequency, and hence one has only to take the thermal average of the products for fixed transition frequencies.

In the harmonic approximation the quantum number v stands for the set of oscillator quantum numbers v_j ($j = 1, 2,\dots, 3n$), one for each normal coordinate. Transitions can be classified into different orders according to the total number of jumps made by all the oscillators; thus a transition from $v(v_1, v_2,\dots, v_{3n})$ to $v'(v_1', v_2',\dots, v_{3n}')$ is said to be of the order r, where

$$r = \sum_j |v_j - v_j'|. \tag{21.6}$$

In this transition the matrix element of a product $q_j^\alpha q_j^\beta q_{j'}^\gamma \dots$ vanishes, unless the exponents α, β, γ, etc. are respectively equal to or greater than $|v_j - v_j'|$, $|v_{j'} - v_{j'}'|$, etc. It follows immediately that, for an rth-order transition, terms in $A(q)$ and $B(q)$ of lower than the rth order contribute nothing to their matrix elements; the major contributions to the matrix elements come from the terms of the rth order:

$$q_j^{|v_j - v_j'|} q_{j'}^{|v_{j'} - v_{j'}'|}\dots,$$

as terms of still higher order are much smaller.

First-order transitions are due to a single jump (either upwards or downwards) by one of the oscillators, the transition frequencies being $\pm\omega_j$ ($j = 1, 2,\dots, 3n$). When higher-order contributions are neglected, the value of (21.4) for first-order transitions is clearly

$$\begin{aligned} \langle v|A(q)|v'\rangle\langle v'|B(q)|v\rangle &= \langle v_j|A_j q_j|v_j\pm1\rangle\langle v_j\pm1|B_j q_j|v_j\rangle \\ &= A_j B_j\langle v_j|q_j|v_j\pm1\rangle\langle v_j\pm1|q_j|v_j\rangle. \tag{21.7} \end{aligned}$$

The upper and lower signs correspond respectively to the transition frequencies $-\omega_j$ and ω_j. With the help of Table 28, p. 180, we readily find the thermal average values:

First-order Transitions

$$
\begin{array}{c|c}
\{\langle v|A(q)|v'\rangle\langle v'|B(q)|v\rangle\}_{Av} & \omega_{vv'} \\
\hline
A_j\,B_j\left\{\begin{matrix} e^{-\beta_j C_j} \\ C_j \end{matrix}\right. & \begin{matrix} \omega_j \\ -\omega_j \end{matrix}
\end{array}
\tag{21.8}
$$

Second-order transitions are due to either a double jump by a single oscillator or single jumps by two different oscillators; the transition frequencies being respectively the *overtone frequencies* $\pm 2\omega_j$ or the combination frequencies $\pm(\omega_j+\omega_{j'})$ (*summation frequencies*) and $\pm(\omega_j-\omega_{j'})$ (*difference frequencies*). For the second-order transitions we need retain only the second-order terms in $A(q)$, $B(q)$ in (21.4); in this way we find

$$
\begin{array}{c|c}
\langle v|A(q)|v'\rangle\langle v'|B(q)|v\rangle & \omega_{vv'} \\
\hline
\tfrac{1}{4}A_{jj}\,B_{jj}\langle v_j|q_j^2|v_j\pm2\rangle\langle v_j\pm2|q_j^2|v_j\rangle & \mp 2\omega_j \\
\{A_{jj'}\langle v_j|q_j|v_j\pm1\rangle\langle v_{j'}|q_{j'}|v_{j'}\pm1\rangle\}\{B_{jj'}\langle v_j\pm1|q_j|v_j\rangle\langle v_{j'}\pm1|q_{j'}|v_{j'}\rangle\} & \mp\omega_j\mp\omega_{j'} \\
\{A_{jj'}\langle v_j|q_j|v_j\pm1\rangle\langle v_{j'}|q_{j'}|v_{j'}\mp1\rangle\}\{B_{jj'}\langle v_j\pm1|q_j|v_j\rangle\langle v_{j'}\mp1|q_{j'}|v_{j'}\rangle\} & \mp\omega_j\pm\omega_{j'}
\end{array}
\tag{21.9}
$$

where in each case either all upper or all lower signs are to be taken. The thermal average values, obtained with the help of Table 28, are given below:

Second-order Transitions

$$
\begin{array}{c|cl}
\{\langle v|A(q)|v'\rangle\langle v'|B(q)|v\rangle\}_{Av} & \omega_{vv'} & \\
\hline
\tfrac{1}{2}A_{jj}\,B_{jj}\left\{\begin{matrix}e^{-2\beta_j}C_j^2 \\ C_j^2\end{matrix}\right. & \left.\begin{matrix}2\omega_j \\ -2\omega_j\end{matrix}\right\} & \text{overtones} \\
A_{jj'}\,B_{jj'}\left\{\begin{matrix}e^{-(\beta_j+\beta_{j'})}C_j\,C_{j'} \\ C_j\,C_{j'}\end{matrix}\right. & \left.\begin{matrix}(\omega_j+\omega_{j'}) \\ (-\omega_j-\omega_{j'})\end{matrix}\right\} & \text{summation frequencies} \\
A_{jj'}\,B_{jj'}\left\{\begin{matrix}e^{-\beta_j}C_j\,C_{j'} \\ e^{-\beta_{j'}}C_j\,C_{j'}\end{matrix}\right. & \left.\begin{matrix}(\omega_j-\omega_{j'}) \\ (-\omega_j+\omega_{j'})\end{matrix}\right\} & \text{difference frequencies}
\end{array}
\tag{21.10}
$$

If for a particular transition the value given in (21.8) or (21.10) vanishes, we describe the transition as being *forbidden* for the particular optical effect concerned. Since higher-order contributions have been consistently neglected in the above formulae, the contribution of a forbidden transition is not zero, but only abnormally small. Thus if $\mathbf{M}(j)$ (see (21.1)) vanishes, the first-order transitions $\pm\omega_j$ are forbidden in connexion with the polarizabilities; we have then to proceed to the third-order terms in $\mathbf{M}(X)$ to obtain a finite contribution.

Optical effects can be classified according to the orders of the underlying transitions. Thus the first-order Raman scattering consists of lines with the frequencies $|\omega\pm\omega_j|$ and the second-order Raman scattering consists of lines with the frequencies $|\omega\pm2\omega_j|$, $|\omega\pm(\omega_j+\omega_{j'})|$, and $|\omega\pm(\omega_j-\omega_{j'})|$. Similarly, if we use the polarizability (20.18) to discuss the refractive properties, we obtain dispersion of the first order corresponding to the dispersion frequency $|\omega_j|$, and dispersion of the second order corresponding to the dispersion frequencies $|2\omega_j|$ and $|\omega_j\pm\omega_{j'}|$. One can write down the intensity of the Raman scattering and the dispersion formula of the first and second orders by a straightforward calculation, substituting (21.8) and (21.10) in the formulae (20.15), (20.16), (20.18), and (20.19). In crystals, however, there are very restrictive selection rules so that a great majority of the transitions are forbidden. Hence we shall not discuss the optical formulae explicitly until Chapter VII, where they will be developed in detail specifically for crystalline lattices.

One particular result of interest may, however, be considered here, namely, that the first-order polarizability is independent of temperature. Thus if we discard in (20.18) all except the terms corresponding to first-order transitions and carry out the thermal average with the help of (21.8), we find that the first-order polarizability is given by

$$\{P^{vv}_{\alpha\beta}(\omega)\}_{Av} = P_{\alpha\beta}(0, X^0) +$$

$$+ \frac{1}{\hbar}\sum_j M_\alpha(j)M_\beta(j)\left\{\frac{e^{-\beta_j}}{-\omega_j+\omega} + \frac{e^{-\beta_j}}{-\omega_j-\omega} + \frac{1}{\omega_j+\omega} + \frac{1}{\omega_j-\omega}\right\}C_j$$

$$= P_{\alpha\beta}(0, X^0) + \frac{1}{\hbar}\sum_j M_\alpha(j)M_\beta(j)(1-e^{-\beta_j})C_j\left(\frac{2\omega_j}{\omega_j^2-\omega^2}\right). \quad (21.11)$$

By using the explicit expression (16.16) for C_j we find that $\{P^{vv}_{\alpha\beta}(\omega)\}_{Av}$ reduces to the temperature-independent expression

$$\{P^{vv}_{\alpha\beta}(\omega)\}_{Av} = P_{\alpha\beta}(0, X^0) + \sum_j \frac{M_\alpha(j)M_\beta(j)}{\omega_j^2-\omega^2}. \quad (21.12)$$

In a similar way we find for the anti-Hermitian part (20.19)

$$\{R_{\alpha\beta}^{vv}(\omega)\}_{Av} = \frac{i\pi}{\hbar} \sum_j M_\alpha(j)M_\beta(j) \times$$

$$\times \{e^{-\beta_j}\delta(\omega-\omega_j)-e^{-\beta_j}\delta(\omega+\omega_j)+\delta(\omega+\omega_j)-\delta(\omega-\omega_j)\}C_j$$

$$= \frac{i\pi}{2} \sum_j M_\alpha(j)M_\beta(j)\left(\frac{1}{\omega_j}\right)\{\delta(\omega+\omega_j)-\delta(\omega-\omega_j)\}, \qquad (21.13)$$

which is likewise independent of temperature.

One may note that in ionic crystals the first-order terms in the expansion of $M(X)$ are expected to be particularly large. Thus consider the idealized model where the ions are rigid structures, undeformed by the displacements of the ions. We may express this model mathematically as follows: Let Z_k be the atomic number of the nucleus k, and $X_\alpha(k)$ its coordinates. The assumption of rigid ions means that to every nucleus k belong permanently $Z_k - z_k$ (z_k = the ionic charge) electrons with co-ordinates $x_\alpha(k, s)$ ($s = 1, 2,..., Z_k-z_k$) such that the electronic wave function $\phi_0(x, X)$ is a function $\phi(x')$ of only the relative coordinates

$$x'_\alpha(k, s) = x_\alpha(k, s) - X_\alpha(k). \qquad (21.14)$$

The electric moment operator $M(x, X)$ is clearly

$$M(x, X) = e \sum_k \left\{ Z_k X(k) - \sum_{s=1}^{Z_k-z_k} x(k, s) \right\} = e \sum_k z_k X(k) - e \sum_{k,s} x'(k, s).$$
$$(21.15)$$

Constructing the function $M(X)$ according to (20.3) with the electronic wave function $\phi(x')$, we have

$$M(X) = \sum_k (z_k e)u(k) + M(X^0), \qquad (21.16)$$

where $M(X^0)$ is the constant term

$$e \sum_k z_k X^0 - e \sum_{k,s} \int \phi^*(x')x'(k, s)\phi(x') \, dx'. \qquad (21.17)$$

(21.16) shows that apart from the irrelevant constant term the expansion of $M(X)$ consists only of linear terms (this is, of course, still the case if the expansion is with respect to the normal coordinates, as they are related to $u(k)$ by linear transformations); the linear terms give simply the electric moment due to the displacements of the ionic charges. In other words, the ionic charges contribute directly only to the first-order effects. First-order effects in non-ionic crystals, and higher-order effects in ionic as well as non-ionic crystals, on the other hand, are due to deformation of the electron clouds.

V

THE METHOD OF LONG WAVES

22. The geometry of perfect lattices

FOR the general discussion in this and the following chapters a systematic scheme of notation is essential. Thus let us put together in a systematic way the general notation, which has already been introduced somewhat haphazardly in Part I (particularly §§ 6, 11).

We begin with the simplest periodic structure in space, namely, the Bravais lattice, which is built up from three *basic vectors* $\mathbf{a}_1$, $\mathbf{a}_2$, $\mathbf{a}_3$ (non-coplanar). The lattice points defined by

$$\mathbf{x}(l) = l^1\mathbf{a}_1 + l^2\mathbf{a}_2 + l^3\mathbf{a}_3 \quad (l^1, l^2, l^3 \text{ being integers}) \tag{22.1}$$

occupy the corners of the *lattice cells*, which are parallelepipeds bounded by the edges $\mathbf{a}_1$, $\mathbf{a}_2$, $\mathbf{a}_3$. The integers (l^1, l^2, l^3), known as the *cell indices*, will sometimes be denoted for simplicity by a single letter l. If the lattice points of a Bravais lattice are occupied by identical atoms, we have a simple crystal lattice.

Crystal lattices in general are of a composite structure, consisting of a number of interpenetrating Bravais lattices of identical structure (i.e. with identical basic vectors); the lattice points of different constituent Bravais lattices may or may not be occupied by the same type of atom (e.g. diamond and NaCl respectively). Within a lattice cell of a composite lattice, we find thus a number of atoms, one from each constituent Bravais lattice; they are described as forming the *basis* of the lattice. If we visualize a simple lattice as the periodic repetition of a single atom, we can look upon a composite lattice as the repetition of the atoms in the basis as a group.

In general, the position vectors of the nuclei in a perfect crystal lattice may thus be represented as follows:

$$\mathbf{x}\binom{l}{k} = \mathbf{x}\binom{l}{-} + \mathbf{x}\binom{-}{k} = \mathbf{x}(l) + \mathbf{x}(k), \tag{22.2}$$

where the dashes can be omitted, as indicated, if there is no doubt of the meaning. k is the *base index*, which distinguishes different nuclei in a cell and has the values $0, 1, ..., (n-1)$, n being the number of nuclei in the basis. For definiteness, it is sometimes convenient to choose the origin such that

$$\mathbf{x}\binom{-}{0} = 0. \tag{22.3}$$

Let us express $\mathbf{x}(k)$ in terms of the basic vectors as follows:

$$\mathbf{x}(k) = \lambda^1(k)\mathbf{a}_1 + \lambda^2(k)\mathbf{a}_2 + \lambda^3(k)\mathbf{a}_3 \qquad (22.4)$$

and require that

$$0 \leqslant \lambda^i(k) < 1 \quad (i = 1, 2, 3). \qquad (22.5)$$

If we call the cell, bounded by $\mathbf{a}_1$, $\mathbf{a}_2$, $\mathbf{a}_3$ drawn from the origin, the *zero-cell*, the above requirement merely means that the cell indices are zero for all the nuclei found in the zero-cell.

Any position vector (dimension length) may conveniently be expressed in terms of the basic vectors as follows:

$$\mathbf{x}(\xi) = \xi^1\mathbf{a}_1 + \xi^2\mathbf{a}_2 + \xi^3\mathbf{a}_3, \qquad (22.6)$$

the components (ξ^1, ξ^2, ξ^3) being dimensionless. The reciprocal basic vectors

$$\mathbf{b}^1 = \frac{\mathbf{a}_2 \wedge \mathbf{a}_3}{\mathbf{a}_1.\mathbf{a}_2 \wedge \mathbf{a}_3}, \qquad \mathbf{b}^2 = \frac{\mathbf{a}_3 \wedge \mathbf{a}_1}{\mathbf{a}_2.\mathbf{a}_3 \wedge \mathbf{a}_1}, \qquad \mathbf{b}^3 = \frac{\mathbf{a}_1 \wedge \mathbf{a}_2}{\mathbf{a}_3.\mathbf{a}_1 \wedge \mathbf{a}_2} \qquad (22.7)$$

have on the other hand the dimension of the inverse of length. The two complementary sets of basic vectors satisfy the following relations:

$$\mathbf{b}^\alpha.\mathbf{a}_\beta = \delta_{\alpha\beta} \quad \left(\text{the Kronecker symbol } \delta_{\alpha\beta} = \begin{cases} 0, & \alpha \neq \beta \\ 1, & \alpha = \beta \end{cases} \right). \qquad (22.8)$$

A vector with the dimension of the inverse of length is sometimes described as belonging to the reciprocal space; such a vector may be expressed in terms of the reciprocal basic vectors as follows:

$$\mathbf{y}(\eta) = \eta_1\mathbf{b}^1 + \eta_2\mathbf{b}^2 + \eta_3\mathbf{b}^3, \qquad (22.9)$$

the components (η_1, η_2, η_3) being dimensionless numbers. If the components are integral, it is convenient to use a different set of letters $h(h_1, h_2, h_3)$. The vectors $\mathbf{y}(h)$ for all possible values of h give rise to a Bravais lattice in the reciprocal space, known as the *reciprocal lattice*.

The components of $\mathbf{x}(\xi)$ and $\mathbf{y}(\eta)$ can be obtained by scalar multiplication with the respective complementary set of basic vectors as follows:

$$\mathbf{b}^\alpha.\mathbf{x}(\xi) = \mathbf{b}^\alpha.(\xi^1\mathbf{a}_1 + \xi^2\mathbf{a}_2 + \xi^3\mathbf{a}_3) = \xi^\alpha, \qquad (22.10)$$

$$\mathbf{a}_\alpha.\mathbf{y}(\eta) = \mathbf{a}_\alpha.(\eta_1\mathbf{b}^1 + \eta_2\mathbf{b}^2 + \eta_3\mathbf{b}^3) = \eta_\alpha. \qquad (22.11)$$

The above relations follow directly from (22.8). Similarly, we find with the help of (22.8) that

$$\mathbf{x}(\xi).\mathbf{y}(\eta) = \xi^1\eta_1 + \xi^2\eta_2 + \xi^3\eta_3 = (\xi\eta). \qquad (22.12)$$

Readers familiar with tensor calculus will recognize that $\mathbf{a}_1$, $\mathbf{a}_2$, $\mathbf{a}_3$ can be regarded as a set of covariant basic vectors and the vectors $\mathbf{b}^1$, $\mathbf{b}^2$, $\mathbf{b}^3$ as the corresponding contravariant basic vectors. ξ^i, η_i are respectively the contra- and covariant components. The upper and lower indices are in each case introduced in accordance with the usual conventions in tensor calculus. With the basic vectors we can form the metric coefficients

$$g_{\alpha\beta} = \mathbf{a}_\alpha . \mathbf{a}_\beta, \qquad (22.13)$$

$$g^{\alpha\beta} = \mathbf{b}^\alpha . \mathbf{b}^\beta, \qquad (22.14)$$

$$g^\beta_\alpha = \mathbf{a}_\alpha . \mathbf{b}^\beta \quad (= \delta_{\alpha\beta}). \qquad (22.15)$$

Using these coefficients, we can write

$$\mathbf{x}(\xi) . \mathbf{x}(\xi') = \sum_{\alpha\beta} g_{\alpha\beta} \xi^\alpha \xi'^\beta; \qquad (22.16)$$

$$\mathbf{y}(\eta) . \mathbf{y}(\eta') = \sum_{\alpha\beta} g^{\alpha\beta} \eta_\alpha \eta'_\beta. \qquad (22.17)$$

It follows from (22.12) that the scalar product between a lattice vector $\mathbf{x}(l)$ and a reciprocal lattice vector $\mathbf{y}(h)$ is an integer. Owing to this fact, as we have seen in § 6, the reciprocal lattice vectors are useful in specifying lattice waves satisfying the periodic boundary condition. In the following, we give some further examples where the reciprocal lattice vectors are useful.

We notice that if $\mathbf{y}(h)$ is a reciprocal lattice vector, the function

$$\exp\{2\pi i \mathbf{y}(h) . \mathbf{x}\} \qquad (22.18)$$

is periodic in $\mathbf{x}$ (with respect to $\mathbf{a}_1$, $\mathbf{a}_2$, $\mathbf{a}_3$). For evidently, if we add to $\mathbf{x}$ a lattice vector $\mathbf{x}(l)$, the function is unchanged. In fact the functions (22.18) corresponding to all possible reciprocal lattice vectors form a complete set of functions, in terms of which all such periodic functions can be expanded. Consider thus a periodic function $f(\mathbf{x})$. If we use the components ξ^1, ξ^2, ξ^3 of $\mathbf{x}$ as arguments, the function $f(\xi^1, \xi^2, \xi^3)$ is by definition periodic in all three variables ξ^1, ξ^2, ξ^3 with the period unity. Hence we may expand it in a Fourier series as follows:

$$f(\xi^1, \xi^2, \xi^3) = \sum_h g(h_1, h_2, h_3) e^{2\pi i (h_1 \xi^1 + h_2 \xi^2 + h_3 \xi^3)}, \qquad (22.19)$$

where the coefficients are given by

$$g(h_1, h_2, h_3) = \int_0^1 d\xi^1 \int_0^1 d\xi^2 \int_0^1 d\xi^3 \{f(\xi^1, \xi^2, \xi^3) e^{-2\pi i (h_1 \xi^1 + h_2 \xi^2 + h_3 \xi^3)}\}. \qquad (22.20)$$

We may regard the indices (h_1, h_2, h_3) as the components of a reciprocal lattice vector $\mathbf{y}(h)$. It then follows immediately from (22.12) that (22.19) can also be written directly in terms of the vector $\mathbf{x}$ as follows:

$$f(\mathbf{x}) = \sum_h g(\mathbf{y}) e^{2\pi i \mathbf{y}(h) . \mathbf{x}}. \qquad (22.21)$$

Similarly we can put (22.20) in the form

$$g(\mathbf{y}) = \frac{1}{v_a} \int\limits_{\text{cell}} f(\mathbf{x}) e^{-2\pi i \mathbf{y} \cdot \mathbf{x}} \, d\mathbf{x}, \qquad (22.22)$$

where v_a is the volume of a lattice cell and $d\mathbf{x}$ denotes as usual the volume element $dx_1 \, dx_2 \, dx_3$. The factor $1/v_a$ arises from the Jacobian introduced by the change of the integration variables from ξ^1, ξ^2, ξ^3 to x_1, x_2, x_3. For if we denote the Cartesian components of the basic vectors by superscripts, we have

$$\begin{aligned}
x_1 &= \xi^1 a_1^1 + \xi^2 a_2^1 + \xi^3 a_3^1, \\
x_2 &= \xi^1 a_1^2 + \xi^2 a_2^2 + \xi^3 a_3^2, \\
x_3 &= \xi^1 a_1^3 + \xi^2 a_2^3 + \xi^3 a_3^3.
\end{aligned} \qquad (22.23)$$

To change from $d\xi^1 d\xi^2 d\xi^3$ to $d\mathbf{x} = dx_1 \, dx_2 \, dx_3$ we have to introduce the Jacobian $\partial(\xi^1, \xi^2, \xi^3)/\partial(x_1, x_2, x_3)$, which is the inverse of

$$\frac{\partial(x_1, x_2, x_3)}{\partial(\xi^1, \xi^2, \xi^3)} = \begin{vmatrix} a_1^1 & a_1^2 & a_1^3 \\ a_2^1 & a_2^2 & a_2^3 \\ a_3^1 & a_3^2 & a_3^3 \end{vmatrix} = \mathbf{a}_1 . \mathbf{a}_2 \wedge \mathbf{a}_3 = v_a. \qquad (22.24)$$

(22.21) provides the most convenient expression for the Fourier expansion of a function which has the periodicity of a crystal lattice.

Any three arbitrarily chosen lattice points on a Bravais lattice define a *crystal plane*. Since all lattice points on a Bravais lattice are structurally equivalent, the three points must repeat themselves indefinitely over the plane; a crystal plane in an ideal infinite lattice must thus always contain an infinite number of lattice points. If through every lattice point of the lattice we draw a plane parallel to a given crystal plane, the lattice points are divided into a system of parallel crystal planes. It follows from the mutual equivalence of the lattice points that the planes thus obtained are equidistant. Such a system of crystal planes is specified by the Miller indices defined in the following way. Consider two lattice points separated by $\mathbf{a}_1$. Since both points themselves lie on crystal planes, the connecting vector $\mathbf{a}_1$ is cut into a number of equal sections by the intervening planes. The number of sections h_1 is then the first Miller index. Similarly by considering the division of $\mathbf{a}_2, \mathbf{a}_3$ we obtain two more indices h_2, h_3.

With a set of crystal planes we can associate a reciprocal lattice vector, which has the corresponding Miller indices as components. The equations describing the planes can then be written down directly in the form

$$\mathbf{x} . \mathbf{y}(h) = \xi^1 h_1 + \xi^2 h_2 + \xi^3 h_3 = \text{integer}. \qquad (22.25)$$

To show that these are the correct equations, we notice in the first place that the planes (22.25) do in fact contain all the lattice points in the lattice, for obviously every lattice vector $\mathbf{x}(l)$ satisfies an equation of the form (22.25), the corresponding integer on the right-hand side being

$$l^1 h_1 + l^2 h_2 + l^3 h_3.$$

If we pass from a lattice point $\mathbf{x}(l)$ to $\mathbf{x}(l)+\mathbf{a}_1$, the above integer is increased by h_1. This means that the second point is on the h_1th plane from the first point, and thus the connecting vector $\mathbf{a}_1$ is cut into h_1 sections by the intervening planes. It follows that the integers h_1, h_2, h_3 are indeed the Miller indices.

We have introduced the Bravais lattice by regarding the basic vectors as being given. The reverse problem of defining the basic vectors for a given Bravais lattice has no unique solution. In fact, the number of possible choices for the basic vectors is unlimited. Thus, take any crystal plane in a Bravais lattice; the lattice points in the plane form a two-dimensional lattice. If two basic vectors for this two-dimensional lattice are chosen as $\mathbf{a}_1$, $\mathbf{a}_2$, then *any* vector joining a lattice point on this plane to a lattice point on one of its two neighbouring planes can be chosen as $\mathbf{a}_3$. The number of such choices is clearly unlimited. In a general treatment of crystal theories the choice of the basic vectors is usually immaterial. In special applications the choice is dictated by the consideration of convenience.

23. The infinite lattice model and general invariance relations

In reality we are, of course, always concerned with crystals of finite dimensions. The non-uniform conditions near the surface of a finite crystal, however, give rise only to specific surface effects, and are irrelevant for the discussion of bulk properties. We can avoid making superfluous assumptions about the surface conditions by imagining the crystal lattice to be infinitely extended in all directions.

Careful considerations are, however, required for the legitimate use of the infinite lattice model. Take, for instance, the condition of equilibrium. We have seen in the last chapter that in a molecular system the motion of the nuclei has to be described by their displacements from the equilibrium configuration, where every nucleus is free of forces. It is peculiar to an infinite lattice that, in order to specify the equilibrium configuration, it is not sufficient simply to require that every nucleus is in equilibrium. The equilibrium condition in an infinite lattice is in fact two-fold:

(i) every nucleus is in equilibrium;
(ii) the configuration corresponds to vanishing stresses.   (23.1)

The necessity of requiring (ii) as well as (i) can be seen as follows: The infinite lattice model is essentially an idealization of the conditions inside a finite crystal, where the direct influence of the surface is negligible. In a finite crystal, where every nucleus is in equilibrium (including those near the surface), all the stresses vanish automatically throughout the crystal. Thus if this condition is to be reproduced in the infinite lattice model the condition (ii) must be fulfilled. In order to see that this condition is not automatically fulfilled in the infinite lattice we have only to return to the linear chain considered in § 5. We have seen that so long as the particles in the chain are equally spaced they are in equilibrium; nevertheless there is in general a tension in the chain. In this particular example the condition (ii) is equivalent to the requirement that the tension vanishes. We shall discuss the equilibrium condition further in later sections.

In different types of solids the forces holding the lattice together differ widely in nature. Apart, however, from the metals, where the electronic levels are infinitely close to one another, the adiabatic approximation is generally applicable. Hence we shall formulate the general theory on the basis of the adiabatic approximation, without making other special assumptions concerning the forces. In this way many general results can be obtained which are valid for all crystals except the metals.

We have seen that, in the adiabatic approximation, the motion of the nuclei and its effect on the electrical and optical properties are characterized by certain functions of the nuclear coordinates, namely, the effective potential function, the functions $M(X)$ and $P_{\alpha\beta}(\omega, X)$. Quite apart from specific physical and geometrical characteristics true for special lattices, the above functions are subject to a number of general restrictions which follow from certain invariance conditions. The invariance conditions fall into two classes:

(a) The basic property of lattice periodicity stipulates that if the whole lattice is displaced by a lattice vector $\mathbf{x}(l)$ the lattice coincides once more with itself.

(b) Physical quantities transform covariantly (e.g. the potential energy, moment, and polarizability as scalar, vector, and tensor, respectively) for *rigid* displacements of the lattice. In other words, the physical quantities are invariant whatever the displacement, if we always refer them to Cartesian axes which are oriented similarly to the crystal.   (23.2)

In the rest of the section we shall develop these conditions explicitly.

In the last chapter we have represented the potential function, the electric moment $\mathbf{M}(X)$, and the polarizability $\mathbf{P}_{\alpha\beta}(\omega, X)$ as Taylor's series. In the next chapter we shall see that the use of these expansions in connexion with the infinite lattice model requires some care, for the actual values of the above functions for the whole crystal are infinite; they must thus be normalized to a finite volume eventually. (We cannot, however, begin by considering these functions as being already normalized to unit volume, for this is possible only for a completely homogeneous configuration, and we shall have to discuss configurations where the nuclei are arbitrarily displaced. The final normalization will refer to a lattice which is thermodynamically but not microscopically homogeneous.) For the moment we have only to observe that the constant terms in the Taylor expansions are meaningless divergent quantities; in their place we shall require the potential energy, moment, and polarizability per unit volume for the equilibrium configuration. If the expansions are made with respect to the nuclear displacements $\mathbf{u}(_k^l)$, we denote the coefficients of expansion as follows:

$$\Phi_\alpha(_k^l) = \left(\frac{\partial\Phi}{\partial u_\alpha(_k^l)}\right)_0, \qquad \Phi_{\alpha\beta}(_k^l {}_{k'}^{l'}) = \left(\frac{\partial^2\Phi}{\partial u_\alpha(_k^l)\partial u_\beta(_{k'}^{l'})}\right)_0,$$

$$\Phi_{\alpha\beta\gamma}(_k^l {}_{k'}^{l'} {}_{k''}^{l''}) = \left(\frac{\partial^3\Phi}{\partial u_\alpha(_k^l)\partial u_\beta(_{k'}^{l'})\partial u_\gamma(_{k''}^{l''})}\right)_0,$$

$$M_{\alpha,\beta}(_k^l) = \left(\frac{\partial M_\alpha(X)}{\partial u_\beta(_k^l)}\right)_0, \qquad M_{\alpha,\beta\gamma}(_k^l {}_{k'}^{l'}) = \left(\frac{\partial^2 M_\alpha(X)}{\partial u_\beta(_k^l)\partial u_\gamma(_{k'}^{l'})}\right)_0, \quad \text{etc.}$$

$$P_{\alpha\beta,\gamma}(_k^l) = \left(\frac{\partial P_{\alpha\beta}(\omega, X)}{\partial u_\gamma(_k^l)}\right)_0, \qquad P_{\alpha\beta,\gamma\lambda}(_k^l {}_{k'}^{l'}) = \left(\frac{\partial^2 P_{\alpha\beta}(\omega, X)}{\partial u_\gamma(_k^l)\partial u_\lambda(_{k'}^{l'})}\right)_0, \quad \text{etc.}$$

$$(23.3)$$

In contrast to the constant terms in the expansions, these coefficients are well-defined finite quantities. Thus $-\Phi_\alpha(_k^l)$ is the force on $(_k^l)$ in the configuration 0; $-\Phi_{\alpha\beta}(_k^l {}_{k'}^{l'})$ is, to the first order of accuracy, the α-component of the force on $(_k^l)$ due to a unit displacement of $(_{k'}^{l'})$ in the β-direction, etc. In the general theory the physical properties of a crystal are thus described in terms of the above coefficients together with the normalized values of the potential function, the moment $\mathbf{M}(X^0)$, and the polarizability $\mathbf{P}_{\alpha\beta}(\omega, X^0)$ for the equilibrium configuration.

The invariance conditions (23.2) impose certain identical relations on the above coefficients, which we shall proceed to derive.

From the periodicity requirement it immediately follows that *if the same set of integers is added to all the cell indices of a coefficient, the value of the latter is unchanged.* For the addition of $l(l^1, l^2, l^3)$ is equivalent to the displacement of the whole lattice by the lattice vector $\mathbf{x}(l)$. Thus the first-order coefficients must be independent of the cell index l altogether, and the higher-order coefficients can depend only on the relative cell indices $l-l'$, $l-l''$, etc. We can exhibit this fact explicitly by using the following alternative notation for the expansion coefficients:

$$\Phi_\alpha(k) = \Phi_\alpha(\tbinom{l}{k}), \qquad \Phi_{\alpha\beta}(\tbinom{l-l'}{kk'}) = \Phi_{\alpha\beta}(\tbinom{l\ l'}{k\ k'}), \qquad \Phi_{\alpha\beta\gamma}(\tbinom{l-l'l-l'}{kk'k''}) = \Phi_{\alpha\beta\gamma}(\tbinom{l\ l'\ l'}{k\ k'\ k'}),$$
$$\text{etc.,} \qquad (23.4)$$

$$M_{\alpha,\beta}(k) = M_{\alpha,\beta}(\tbinom{l}{k}), \qquad M_{\alpha,\beta\gamma}(\tbinom{l-l'}{kk'}) = M_{\alpha,\beta\gamma}(\tbinom{l\ l'}{k\ k'}), \text{ etc.} \qquad (23.5)$$

$$P_{\alpha\beta,\gamma}(k) = P_{\alpha\beta,\gamma}(\tbinom{l}{k}), \qquad P_{\alpha\beta,\gamma\lambda}(\tbinom{l-l'}{kk'}) = P_{\alpha\beta,\gamma\lambda}(\tbinom{l\ l'}{k\ k'}), \text{ etc.} \qquad (23.6)$$

This notation is particularly useful for the first- and second-order coefficients; for the higher-order coefficients we have often to use the original notation of (23.3), while bearing in mind the invariance with respect to the addition of an arbitrary l to all cell indices.

If the lattice is displaced as a whole by an arbitrary vector $\boldsymbol{\epsilon}$, the potential function must obviously remain unchanged. On the other hand, to the first order in $\boldsymbol{\epsilon}$, we find from the displacements of the nuclei (i.e. all $\mathbf{u}(\tbinom{l}{k})$ equal to $\boldsymbol{\epsilon}$) that each cell contributes to the change in the potential function by the identical amount

$$\sum_{k\alpha} \Phi_\alpha(k)\epsilon_\alpha. \qquad (23.7)$$

Since this must vanish for all values of $\boldsymbol{\epsilon}$, we have

$$\sum_k \Phi_\alpha(k) = 0. \qquad (23.8)$$

Next consider a homogeneous deformation of the lattice about the lattice point $(\tbinom{l}{k})$, where the nuclear displacements are given by

$$u_\alpha(\tbinom{l'}{k'}) = \sum_\beta u_{\alpha\beta}\{x_\beta(\tbinom{l'}{k'}) - x_\beta(\tbinom{l}{k})\}. \qquad (23.9)$$

After the homogeneous deformation (see § 11) the structure is still a perfect lattice. Hence a relation exactly analogous to (23.8) also holds for the lattice after the deformation; the relation may be written as

$$\sum_k \frac{\partial\Phi}{\partial u_\alpha(\tbinom{l}{k})} = 0, \qquad (23.10)$$

where the derivatives refer to the homogeneously deformed configuration. Expressing the derivatives as expansions in the displacements (23.9), we have

$$\sum_k \left\{ \Phi_\alpha(k) - \sum_{l'k'}\sum_{\beta\gamma} \Phi_{\alpha\beta}(\tbinom{l-l'}{kk'})u_{\beta\gamma}\,x_\gamma(\tbinom{l-l'}{kk'}) + \ldots \right\} = 0, \qquad (23.11)$$

where we have written

$$\mathbf{x}\binom{l-l'}{kk'} = \mathbf{x}\binom{l}{k} - \mathbf{x}\binom{l'}{k'}. \tag{23.12}$$

Since (23.11) holds for arbitrary values of $u_{\alpha\beta}$, the various coefficients of the power series in $u_{\alpha\beta}$ must vanish identically. The constant term merely leads to (23.8). The linear terms lead to the new identities

$$\sum_{lkk'} \Phi_{\alpha\beta}\binom{l}{kk'} x_\gamma\binom{l}{kk'} = 0. \tag{23.13}$$

Let us consider the value of the derivative $\partial\Phi/\partial u_\alpha\binom{l}{k}$ for some special configurations, which are derived from the equilibrium configuration by certain displacements of the nuclei.

Suppose that all the nuclei are displaced from the equilibrium configuration by the same vector

$$\mathbf{u}\binom{l}{k} = \boldsymbol{\epsilon}. \tag{23.14}$$

The corresponding value of the derivative $\partial\Phi/\partial u_\alpha\binom{l}{k}$ is given by the expansion

$$\Phi_\alpha(k) + \sum_{l'k'\beta} \Phi_{\alpha\beta}\binom{l\ \ l'}{k\ \ k'}\epsilon_\beta + \tfrac{1}{2}\sum_{l'k'\beta}\sum_{l''k''\gamma} \Phi_{\alpha\beta\gamma}\binom{l\ \ l'\ \ l''}{k\ \ k'\ \ k''}\epsilon_\beta\epsilon_\gamma + \dots . \tag{23.15}$$

Now (23.14) evidently represents a translation of the lattice as a whole. Since the value of $\partial\Phi/\partial u_\alpha\binom{l}{k}$ cannot be affected by a translation, (23.15) must be independent of the values of ϵ_α ($\alpha = 1, 2, 3$). Therefore the coefficients of the linear and all higher-order terms in (23.15) must vanish; hence we have the series of relations

$$\sum_{l'k'} \Phi_{\alpha\beta}\binom{l-l'}{kk'} = \sum_{lk} \Phi_{\alpha\beta}\binom{l}{kk'} = 0,$$

$$\sum_{l'k'}\sum_{l''k''} \Phi_{\alpha\beta\gamma}\binom{l\ \ l'\ \ l''}{k\ \ k'\ \ k''} = 0, \text{ etc.} \tag{23.16}$$

Next suppose that the nuclei are displaced as follows:

$$u_\beta\binom{l'}{k'} = \sum_\gamma \omega_{\beta\gamma}\left(x_\gamma\binom{l'}{k'} - x_\gamma\binom{l}{k}\right) = -\sum_\gamma \omega_{\beta\gamma} x_\gamma\binom{l-l'}{kk'}, \tag{23.17}$$

where the parameters $\omega_{\beta\gamma}$ are elements of an infinitesimal antisymmetric matrix:

$$\omega_{\beta\gamma} = -\omega_{\gamma\beta}. \tag{23.18}$$

Expressing $\partial\Phi/\partial u_\alpha\binom{l}{k}$ as an expansion in the nuclear displacements, we have

$$\frac{\partial\Phi}{\partial u_\alpha\binom{l}{k}} = \Phi_\alpha(k) - \sum_{\gamma l'k'} \Phi_{\alpha\beta}\binom{l-l'}{kk'}\omega_{\beta\gamma} x_\gamma\binom{l-l'}{kk'} + \dots . \tag{23.19}$$

(23.17), we note, represents to the first order of accuracy the displacements of the nuclei in an infinitesimal rotation of the lattice about $\binom{l}{k}$; the transformation matrix for the rotation being given to the same order of accuracy by

$$\delta_{\beta\gamma} + \omega_{\beta\gamma}. \tag{23.20}$$

Since for a rigid rotation of the lattice $\partial\Phi/\partial u_\alpha(^l_k)$ must behave as the α-component of a vector,

$$\frac{\partial\Phi}{\partial u_\alpha(^l_k)} = \sum_\beta (\delta_{\alpha\beta}+\omega_{\alpha\beta})\Phi_\beta(^l_k) = \Phi_\alpha(k)+\sum_\beta \omega_{\alpha\beta}\Phi_\beta(k). \quad (23.21)$$

Equating the two alternative expressions (23.19) and (23.21) for the derivative of Φ, we obtain the relation

$$\sum_\beta \omega_{\alpha\beta}\Phi_\beta(k) = -\sum_{l'k'\gamma} \Phi_{\alpha\beta}(^{\;l-l'}_{kk'})\omega_{\beta\gamma}x_\gamma(^{\;l-l'}_{kk'})+\cdots$$

$$= -\sum_{lk'\gamma} \Phi_{\alpha\beta}(^l_{kk'})\omega_{\beta\gamma}x_\gamma(^l_{kk'})+\cdots. \quad (23.22)$$

As this relation must be fulfilled identically to the first order of accuracy, we can equate the coefficients of the linear terms in the parameters $\omega_{\alpha\beta} = -\omega_{\beta\alpha}$. We can do so most conveniently by differentiating (23.22) with respect to $\omega_{\mu\nu} = -\omega_{\nu\mu}$ and putting the parameters $\omega_{\alpha\beta}$ equal to zero afterwards. In this way we obtain

$$\delta_{\alpha\mu}\Phi_\nu(k)-\delta_{\alpha\nu}\Phi_\mu(k) = -\sum_{l'k'}\{\Phi_{\alpha\mu}(^l_{kk'})x_\nu(^l_{kk'})-\Phi_{\alpha\nu}(^l_{kk'})x_\mu(^l_{kk'})\}. \quad (23.23)$$

All the identical relations which we have derived for the potential function Φ are evidently valid for any scalar function relating to a lattice.

Moreover, during translations, the components of vectors and tensors remain invariant exactly as scalars. Hence all the identities deduced above for scalars from the consideration of translations are valid also for the components of vectors and tensors. Thus applying (23.8), (23.13), and (23.16) to the α-component of $\mathbf{M}(X)$ and the (α,β)-component $P_{\alpha\beta}(\omega, X)$ of the polarizability, we obtain the relations

$$\sum_k M_{\alpha,\beta}(k) = 0, \quad (23.24)$$

$$\sum_{lk'} M_{\alpha,\beta\gamma}(^l_{kk'}) = 0, \quad (23.25)$$

$$\sum_{lkk'} M_{\alpha,\beta\gamma}(^l_{kk'})x_\lambda(^l_{kk'}) = 0, \quad (23.26)$$

$$\sum_k P_{\alpha\beta,\gamma}(k) = 0, \quad (23.27)$$

$$\sum_{lk} P_{\alpha\beta,\gamma\lambda}(^l_{kk'}) = 0, \quad (23.28)$$

$$\sum_{lkk'} P_{\alpha\beta,\gamma\lambda}(^l_{kk'})x_\mu(^l_{kk'}) = 0. \quad (23.29)$$

By considering the change in the derivative $\partial M_\alpha(X)/\partial u_\beta(^l_k)$ caused by an infinitesimal rotation, we can obtain an identity analogous to (23.23). The only difference in this case is that during the rotation $\partial M_\alpha(X)/\partial u_\beta(^l_k)$ transforms as the (α,β)-component of a Cartesian tensor.

Thus corresponding to (23.22) we find that, to the first order,

$$\sum_\gamma \omega_{\alpha\gamma} M_{\gamma,\beta}(k) + \sum_\gamma \omega_{\beta\gamma} M_{\alpha,\gamma}(k) = -\sum_{lk'\gamma\lambda} M_{\alpha,\beta\gamma}(\tfrac{l}{kk'})\omega_{\gamma\lambda} x_\lambda(\tfrac{l}{kk'}), \quad (23.30)$$

a relation to be fulfilled identically. Differentiating (23.30) with respect to $\omega_{\mu\nu} = -\omega_{\nu\mu}$ and putting all the parameters $\omega_{\mu\nu}$ equal to zero, we obtain the identities

$$\delta_{\alpha\mu} M_{\nu,\beta}(k) - \delta_{\alpha\nu} M_{\mu,\beta}(k) + \delta_{\beta\mu} M_{\alpha,\nu}(k) - \delta_{\beta\nu} M_{\alpha,\mu}(k)$$
$$= -\sum_{lk'} \{ M_{\alpha,\beta\mu}(\tfrac{l}{kk'}) x_\nu(\tfrac{l}{kk'}) - M_{\alpha,\beta\nu}(\tfrac{l}{kk'}) x_\mu(\tfrac{l}{kk'}) \}. \quad (23.31)$$

The identities obtained for $\mathbf{M}(X)$ and $\mathbf{P}_{\alpha\beta}(\omega, X)$ are clearly valid respectively for any vector and tensor relating to a lattice.

24. Lattice waves

With the nuclei in arbitrary displaced positions $\mathbf{x}(\tfrac{l}{k}) + \mathbf{u}(\tfrac{l}{k})$, we have the equations

$$m_k \ddot{u}_\alpha(\tfrac{l}{k}) = -\frac{\partial \Phi}{\partial u_\alpha(\tfrac{l}{k})} \quad (24.1)$$

(note that $\partial\Phi/\partial u_\alpha(\tfrac{l}{k})$ refers to the configuration of the nuclei in their displaced positions). We shall consider the motion of the nuclei in the harmonic approximation. Thus if Φ is written as a Taylor series in the nuclear displacements, the third- and higher-order terms are to be ignored. Accordingly, we find that

$$\frac{\partial \Phi}{\partial u_\alpha(\tfrac{l}{k})} = \Phi_\alpha(k) + \sum_{l'k'\beta} \Phi_{\alpha\beta}(\tfrac{l\ l'}{k\ k'}) u_\beta(\tfrac{l'}{k'}). \quad (24.2)$$

The constant term $\Phi_\alpha(k)$ represents the α-component of the force on a particle k in the equilibrium configuration; thus according to the condition (23.1) (i),

$$\Phi_\alpha(k) = 0. \quad (24.3)$$

Hence upon substituting (24.2) in (24.1) we have

$$m_k \ddot{u}_\alpha(\tfrac{l}{k}) = -\sum_{l'k'\beta} \Phi_{\alpha\beta}(\tfrac{l\ l'}{k\ k'}) u_\beta(\tfrac{l'}{k'}). \quad (24.4)$$

(24.4) represents a system of simultaneous linear differential equations, infinite in number. Owing to the basic property of lattice periodicity, we shall find that an immediate reduction can be effected by using the wave solutions:†

$$u_\alpha(\tfrac{l}{k}) = \frac{1}{\sqrt{m_k}} w_\alpha(k) \exp\{2\pi i \mathbf{y} \cdot \mathbf{x}(\tfrac{l}{k}) - i\omega t\}, \quad (24.5)$$

where $\mathbf{y}$ is an arbitrary vector in the reciprocal space and can be interpreted as a wave-number vector ($|\mathbf{y}|$ = wave-number, parallel to the

† Here the symbol $w_\alpha(k)$ for the amplitude does not necessarily imply that $w_\alpha(k)$ is independent of $\mathbf{y}$.

wave normal). Upon substituting (24.5) in (24.4) and dividing the resulting equation by $\exp\{2\pi i y . x(^l_k) - i\omega t\}$, we find that the infinite number of equations reduce to the $3n$ linear homogeneous equations in the $3n$ unknowns $w_\alpha(k)$ ($k = 0, 1, 2,..., n-1; \alpha = 1, 2, 3$):

$$\omega^2 w_\alpha(k) = \sum_{k'\beta} C_{\alpha\beta}(^{\,y}_{kk'}) w_\beta(k'), \qquad (24.6)$$

where the coefficients are defined as

$$C_{\alpha\beta}(^{\,y}_{kk'}) = \frac{1}{(m_k m_{k'})^{\frac12}} \sum_{l'} \Phi_{\alpha\beta}(^{l-l'}_{kk'}) \exp\{-2\pi i y . (x(^l_k) - x(^{l'}_{k}))\}$$

$$= \frac{\exp\{-2\pi i y . [x(k) - x(k')]\}}{(m_k m_{k'})^{\frac12}} \sum_{l} \Phi_{\alpha\beta}(^l_{kk'}) e^{-2\pi i(\eta l)}, \quad (24.7)$$

$\eta(\eta_1, \eta_2, \eta_3)$ being the components of y.

The possibility of the above reduction is a direct result of the periodic property of a lattice. For to the latter is due the fact that

$$\Phi_{\alpha\beta}(^l_k \,^{l'}_{k'})$$

depends only on $(l-l')$. But for this fact the coefficients (24.7) would not be independent of the lattice-cell index l and (24.6) would be inadmissible.

The equations (24.6) lead to the familiar condition for solubility:

$$|\omega^2\delta_{\alpha\beta}\delta_{kk'} - C_{\alpha\beta}(^{\,y}_{kk'})| = 0 \qquad (24.8)$$

which has been quoted and discussed in § 6. (24.8) is an equation of $3n$th degree in ω^2; we shall denote the $3n$ solutions by $\omega^2(^y_j)$ ($j = 1, 2,..., 3n$). For each of these values of ω^2 (24.6) yields a set of values for $w_\alpha(k)$, which we shall denote by $w_\alpha(k|^y_j)$. For $\omega^2(^y_j)$ and $w_\alpha(k|^y_j)$ we have thus the following relations:

$$|\omega^2(^y_j)\delta_{\alpha\beta}\delta_{kk'} - C_{\alpha\beta}(^{\,y}_{kk'})| = 0, \qquad (24.9)$$

$$\omega^2(^y_j)w_\alpha(k|^y_j) = \sum_{k'\beta} C_{\alpha\beta}(^{\,y}_{kk'}) w_\beta(k'|^y_j). \qquad (24.10)$$

It is obvious from their definition that

$$\Phi_{\alpha\beta}(^l_k \,^{l'}_{k'}) = \Phi_{\beta\alpha}(^{l'}_{k'} \,^l_k). \qquad (24.11)$$

In terms of the alternative notation (23.4), (24.11) can be written as

$$\Phi_{\alpha\beta}(^l_{kk'}) = \Phi_{\beta\alpha}(^{-l}_{k'k}). \qquad (24.12)$$

Taking the complex conjugate of (24.7), we have

$$C^*_{\alpha\beta}(^{\,y}_{kk'}) = \frac{\exp\{-2\pi i y . [x(k') - x(k)]\}}{(m_k m_{k'})^{\frac12}} \sum_{l} \Phi_{\alpha\beta}(^l_{kk'}) e^{2\pi i(\eta l)}. \quad (24.13)$$

Using (24.12) in (24.13) and introducing $l' = -l$ as the summation index, we find immediately that

$$C^*_{\alpha\beta}(^{y}_{kk'}) = \frac{\exp\{-2\pi i y . [x(k')-x(k)]\}}{(m_k m_{k'})^{\frac{1}{2}}} \sum_{l'} \Phi_{\beta\alpha}(^{l'}_{k'k}) e^{-2\pi i(\eta l')} = C_{\beta\alpha}(^{y}_{k'k}).$$

(24.14)

$C_{\alpha\beta}(^{y}_{kk'})$ defines a $3n \times 3n$ matrix with the indices (α, k), (β, k'); (24.14) shows that the matrix is Hermitian. Hence, according to a well-known theorem in algebra, the solutions $\omega^2(^{y}_{j})$ of (24.8) (*secular equation* of the matrix $C_{\alpha\beta}(^{y}_{kk'})$) are all real. This shows that the vibrational frequencies must be either real or purely imaginary. As we have seen in § 12, in order that ω^2 should be positive so that the frequencies are real, the principal minors of the matrix $C_{\alpha\beta}(^{y}_{kk'})$ must be all positive; and such must be the case if the lattice is to be stable.

We have already discussed the solutions in some detail in § 6. It has been seen that we can use the real part of the complex solutions (24.5) to represent the real lattice waves, and all the distinct lattice waves are obtained if we consider only positive frequencies $\omega(^{y}_{j})$, and y-values within a suitably chosen region of volume $1/v_a$ in the reciprocal space. A possible choice, for instance, is the region bounded by the corners $(\eta_1, \eta_2, \eta_3) = (\pm\frac{1}{2}, \pm\frac{1}{2}, \pm\frac{1}{2})$, where each combination of the signs is to be taken in turn.

Since

$$C^*_{\alpha\beta}(^{y}_{kk'}) = C_{\alpha\beta}(^{-y}_{kk'})$$

(24.15)

and $\omega^2(^{y}_{j})$ is always real, we find upon taking the complex conjugates of (24.9) and (24.10) that

$$|\omega^2(^{y}_{j})\delta_{\alpha\beta}\,\delta_{kk'} - C_{\alpha\beta}(^{-y}_{kk'})| = 0$$

(24.16)

and

$$\omega^2(^{y}_{j})w^*_\alpha(k\,|^{y}_{j}) = \sum_{k'\beta} C_{\alpha\beta}(^{-y}_{kk'})w^*_\beta(k'\,|^{y}_{j}).$$

(24.17)

These equations show that we can choose

$$\left. \begin{aligned} \omega(^{-y}_{j}) &= \omega(^{y}_{j}) \\ w_\alpha(k\,|^{-y}_{j}) &= w^*_\alpha(k\,|^{y}_{j}) \end{aligned} \right\}.$$

(24.18)

The real lattice waves $(^{y}_{j})$ and $(^{-y}_{j})$ are thus two identical sinusoidal waves travelling respectively in the directions of y and $-$y.

25. Failure of the method of homogeneous deformation, and the method of long waves

By considering homogeneous deformations we have obtained in § 11 the elastic constants for lattice models in which the particles interact with central forces. The same procedure can no longer be followed in the general theory, as in the general theory it is not possible to write down the energy density for a homogeneous deformation. Thus if we express Φ as a Taylor series in the displacements (11.2) we shall find that there is no way to normalize the expression to a finite volume. In other words, we obtain in this way only a divergent expression, from which the finite energy density due to the deformation cannot be deduced.

To obtain a better understanding of this difficulty, let us consider the formalism of the general theory, using the linear chain discussed in § 5 as an example. The potential function is in this case given by the formal sum

$$\Phi = \tfrac{1}{2} \sum_{l} \sum_{l' \neq l} \phi(|x_l - x_{l'}|), \tag{25.1}$$

where x_l, $x_{l'}$ are the coordinates of the particles l and l' measured along the length of the chain. From a standard configuration, where the particles are evenly spaced with the interval $s/2$, let us calculate the derivatives of Φ with respect to the particle coordinates. By straightforward differentiation of (25.1) one finds that the first derivatives of Φ are equal to zero; this expresses the evident fact that in such a uniform configuration the particles are all in equilibrium. By further differentiation it is readily found that the second derivatives of Φ depend only on $\phi''(s/2)$, the third derivatives depend only on $\phi'''(s/2)$, etc. The point to note is that all the derivatives of Φ are independent of $\phi'(s/2)$. The quantity $\phi'(s/2)$ thus completely eludes the formalism of the general theory, for the latter describes a lattice solely through the derivatives of Φ.

If the chain is uniformly stretched so that s becomes $s + \delta s$, the energy per cell contains a term

$$\delta s\, \phi'(s/2). \tag{25.2}$$

As it is inherently impossible to express such a quantity in the general theory, it is hardly surprising that in the general case we cannot obtain an unambiguous expression for the energy density.

Moreover, this example shows that the equilibrium condition (23.1) (ii) cannot be written down explicitly in the general theory. For, as we have seen in § 5, the tension in the chain is equal to $\phi'(s/2)$; the equilibrium condition will thus have to be

$$\phi'(s/2) = 0.$$

In the two following sections we shall show that these difficulties can be avoided by considering long acoustic lattice vibrations. The basic idea is simple: since in setting up the equations of motion (24.4) we have imposed only the equilibrium condition (23.1) (i), the corresponding long acoustic lattice waves $(\omega(\begin{smallmatrix} y \\ j \end{smallmatrix}) \to 0$ as $y \to 0)$ must represent elastic waves in a medium which may be under certain homogeneous stresses. We shall show that by comparing the lattice waves with the elastic waves obtained from elasticity theory, we obtain not only the elastic constants but also the expressions for the anisotropic stresses in the configuration with respect to which the derivatives $\Phi_{\alpha\beta}(\begin{smallmatrix} l & l' \\ k & k' \end{smallmatrix})$ are defined. Thus in the general case the equilibrium condition (23.1) (ii) can be explicitly introduced in part by requiring these stresses to vanish. Like the tension in the linear case, the isotropic pressure finds no expression in the general theory, and has to be implicitly understood to vanish in any application of the results of the theory.

In § 11 it was shown that the results obtained by the method of homogeneous deformation are divergent for ionic lattices except in certain special cases of high symmetry. In fact, apart from these latter cases, for which we shall not make special provision in the following discussion, the elastic properties of ionic lattices have to be considered apart from those of other lattices whatever method we may choose to employ. The reason is that ionic lattices are *in general* piezoelectric, and the elastic properties cannot be considered in isolation from the electrical effects. In other words, for ionic lattices a pure elasticity theory does not exist.

A broader understanding of this point can be obtained by approaching the problem in a different way. We observe that the basis of all macroscopic theories in connexion with a material medium is the assumption that the 'response' by the medium is of a local character. Thus, in fluid mechanics we regard the specific volume as determined by the temperature and pressure *at the same point*; in elasticity theory we regard the strain as determined by the stress *at the same point*; etc. The task of an atomic theory is usually to provide the relations between the local parameters (e.g. specific volume, temperature, and pressure; elastic strain and stress components; etc.). From the viewpoint of the atomic theory, the above basic assumption is justified by the idea that the sphere of influence of atomic particles is microscopic in dimension, as is the mean free path, range of force, etc. Thus, as a prerequisite for an elasticity theory, the range of interatomic forces concerned must be microscopic. For ionic crystals this is not the case. As we have seen, the divergent results in § 11 are related to the fact that the forces on the ions depend on the shape of the specimen, however large it may be; this implies that the ions at the surface exert a finite influence on the ions in the interior of the specimen. In particular this means that the forces on the ions cannot be completely determined by the local strain. Thus the divergent results are not to be seen as the consequence of the method of homogeneous deformation; they in fact express a genuine *impasse* from the point of view of elasticity theory.

Thanks, however, to the Maxwell theory, this difficulty is resolved without giving up the basic assumption. One has to introduce besides the strain an additional local parameter, the macroscopic electric field; together they do completely determine the forces acting on the lattice particles. In this way the local basis of the macroscopic treatment is restored. Of course the macroscopic field itself is ultimately determined in part by conditions elsewhere, but this difficulty is overcome by the use of the Maxwell equations, which are local in character.

A situation of this kind has already been met before in the discussion of the long optical vibrations given in § 7. Since in any small neighbourhood the motion is the same in both the longitudinal and transverse modes (viz. a relative oscillation between the positive and negative ions), the difference in their frequencies signifies that the atomic forces cannot be completely local in character. In the phenomenological treatment this difference is brought about by the macroscopic field; the higher frequency of the longitudinal vibrations is seen to be due to the reinforcement of the restoring force by the macroscopic field, which is absent in the transverse vibrations. We may further note that the macroscopic field does not carry the entire electric interaction between the ions. It has been seen in § 9 that the electric field on an ion is split into the macroscopic field and the Lorentz field $4\pi\mathbf{P}/3$. The latter describes that part of the interaction which is uniquely determined by local conditions; in character, this part is thus the same as the atomic forces responsible for elastic properties in non-ionic crystals. From this viewpoint the macroscopic field is not so much characterized by its being an electric field as by the fact that it carries the part of the forces on the particles which is not determined by local conditions alone.

The situation is closely analogous when we consider the elastic properties of ionic lattices. In order to use the Maxwell equations we have to introduce the dielectric polarization as well as the macroscopic field. Thus, instead of Hooke's law, which relates the elastic strain and stress components, we require for the discussion of the elastic properties of ionic lattices certain relations between the four types of parameter describing respectively the elastic strain and stress, the macroscopic electric field, and the dielectric polarization. The requisite relations, as usually assumed in the macroscopic theory, are the following:

$$S_\rho = \sum_\sigma c_{\rho\sigma} s_\sigma - \sum_\beta e_{\beta\rho} E_\beta, \tag{25.3}$$

$$P_\alpha = \sum_\rho e_{\alpha\rho} s_\rho + \sum_\beta a_{\alpha\beta} E_\beta, \tag{25.4}$$

where the notation follows Part I (S_ρ, s_ρ = the stress and strain components with Voigt's indices, $\mathbf{P}$ = the dielectric polarization, $\mathbf{E}$ = the macroscopic field). The second term on the right-hand side of (25.3) represents the *piezoelectric stress* due to the macroscopic field $\mathbf{E}$; the first term on the right-hand side of (25.4) represents the *piezoelectric polarization* due to the elastic strain s_ρ. The parallelism between these equations and the phenomenological equations in § 7 is obvious; the only difference is that we are here concerned with the elastic strain and stress instead of the internal strain and the corresponding restoring force considered there.

The piezoelectric terms in the above equations couple the mechanical parameters s_ρ and S_ρ to the electrical parameters $\mathbf{P}$ and $\mathbf{E}$. Hence the elastic properties of ionic lattices cannot be discussed in isolation from the electrical effects, and a theory of the elastic properties is necessarily linked with a theory of the piezoelectric and dielectric effects.

If we apply the method of homogeneous deformation to a finite specimen of an ionic crystal, the induced piezoelectric polarization produces a macroscopic field which depends on the shape of the specimen. It follows that we cannot apply the method to an infinite lattice model. In principle one could develop correctly an atomic theory by applying the method of homogeneous deformation to a finite specimen of a specified shape (with certain special assumptions about the forces) and interpret the results with the help of the equations (25.3) and (25.4). A much more acceptable alternative is, however, to develop the long-wave method suitably for ionic lattices. To this end, we have, on the one hand, to obtain the elastic waves from the macroscopic theory, taking proper account of the piezoelectric coupling, i.e. using (25.3) and (25.4) instead of the usual Hooke's law. On the other hand, we have to express the lattice waves in such a way as to be suitable for comparison with the results of the macroscopic theory. We shall develop these considerations in §§ 30–32, and it will be seen that in this way we obtain the coefficients $a_{\alpha\beta}$, $e_{\alpha\rho}$ as well as the elastic constants.

26. Long acoustic vibrations

For small values of $\mathbf{y}$, (24.6) can be solved by a perturbation method due originally to Born.† The corresponding solutions, which represent

† See M. Born, *Atomtheorie des festen Zustandes*, 2nd ed., pp. 578–87 (1923).

lattice waves of large wave-length, provide exactly what will be needed for the discussion of macroscopic properties in the general theory (the elastic, piezoelectric, and dielectric properties).

Before we develop the perturbation method one point is worthy of note: if $\mathbf{y}$ is varied continuously one would expect that each of the $3n$ solutions of (24.6), which we have denoted by $w_\alpha(k|^{\mathbf{y}}_j)$, $j = 1, 2,..., 3n$, will also vary in a continuous way. This is generally true, so that the $3n$ solutions lead to $3n$ branches of solutions in this way, except, however, in the immediate neighbourhood of the point $\mathbf{y} = 0$. That the dependence of the solutions on the parameter $\mathbf{y}$ is not regular at the point $\mathbf{y} = 0$ becomes immediately obvious if we examine the elastic waves. Consider the ideally simple case of an isotropic medium. From elasticity theory we know that to a given wave-number vector $\mathbf{y}$ there are two transverse waves and one longitudinal wave. This means that as we approach the point $\mathbf{y} = 0$ through $\mathbf{y}$-values pointing in a fixed direction, the polarization vectors of the three solutions remain throughout respectively parallel and perpendicular to the given direction. The limits of the polarization vectors are thus different if the point $\mathbf{y} = 0$ is approached from different directions; in other words, no unique limit exists for the solutions at $\mathbf{y} = 0$. It follows that the solutions $w_\alpha(k|^{\mathbf{y}}_j)$ cannot be represented in the form of Taylor series in the components y_1, y_2, y_3 as independent parameters.

The difficulty is avoided by considering only a one-dimensional continuum of solutions, which belong to the same branch (i.e. with a fixed j) and have wave-number vectors in the same direction. If we use the magnitude $|\mathbf{y}|$ as a parameter to specify the solutions in such a restricted group, the solutions do vary continuously with $|\mathbf{y}|$ down to the point $\mathbf{y} = 0$. Accordingly, we shall develop the perturbation method by writing the wave-number vector as

$$\epsilon\mathbf{y} \qquad (26.1)$$

and solving the corresponding equation (24.6) by expanding all quantities depending on the wave-number vector with respect to ϵ. This form of expansion is evidently equivalent to an expansion with respect to the magnitude of the wave-number vector for a fixed direction of the latter. ϵ can be taken as a formal expansion parameter to be put equal to unity in the results.

Writing the wave-number in (24.7) as $\epsilon\mathbf{y}$ and expanding with respect to ϵ, we have

$$C_{\alpha\beta}\left(^{\epsilon\mathbf{y}}_{kk'}\right) = C^{(0)}_{\alpha\beta}(kk') + i\epsilon \sum_\gamma C^{(1)}_{\alpha\beta,\gamma}(kk')y_\gamma + \tfrac{1}{2}\epsilon^2 \sum_{\gamma\lambda} C^{(2)}_{\alpha\beta,\gamma\lambda}(kk')y_\gamma y_\lambda + ..., \quad (26.2)$$

where

$$C^{(0)}_{\alpha\beta}(kk') = \frac{1}{(m_k m_{k'})^{\frac{1}{2}}} \sum_l \Phi_{\alpha\beta}(^l_{kk'}) = C^{(0)}_{\beta\alpha}(k'k), \qquad (26.3)$$

$$C^{(1)}_{\alpha\beta,\gamma}(kk') = \frac{-2\pi}{(m_k m_{k'})^{\frac{1}{2}}} \sum_l \Phi_{\alpha\beta}(^l_{kk'}) x_\gamma(^l_{kk'}) = -C^{(1)}_{\beta\alpha,\gamma}(k'k), \qquad (26.4)$$

$$C^{(2)}_{\alpha\beta,\gamma\lambda}(kk') = \frac{-4\pi^2}{(m_k m_{k'})^{\frac{1}{2}}} \sum_l \Phi_{\alpha\beta}(^l_{kk'}) x_\gamma(^l_{kk'}) x_\lambda(^l_{kk'})$$

$$= C^{(2)}_{\alpha\beta,\lambda\gamma}(kk') = C^{(2)}_{\beta\alpha,\gamma\lambda}(k'k). \qquad (26.5)$$

Remembering the equilibrium condition that $\Phi_\alpha(k) = 0$, we readily find with the help of (23.16), (23.23), and (23.13) the following relations:

$$\sum_{k'} \sqrt{m_{k'}}\, C^{(0)}_{\alpha\beta}(kk') = \sum_{k'} \sqrt{m_{k'}}\, C^{(0)}_{\beta\alpha}(k'k) = 0, \qquad (26.6)$$

$$\sum_{k'} \sqrt{m_{k'}}\, C^{(1)}_{\alpha\beta,\gamma}(kk') = \sum_{k'} \sqrt{m_{k'}}\, C^{(1)}_{\alpha\gamma,\beta}(kk'), \qquad (26.7)$$

$$\sum_{kk'} (m_k m_{k'})^{\frac{1}{2}} C^{(1)}_{\alpha\beta,\gamma}(kk') = 0. \qquad (26.8)$$

Similarly we represent the solutions on an acoustic branch j by the following expansions:

$$\omega(^{y}_{j}) = \epsilon\omega^{(1)}(^{y}_{j}) + \tfrac{1}{2}\epsilon^2\omega^{(2)}(^{y}_{j}) + ..., \qquad (26.9)$$

$$w_\alpha(k|^{y}_{j}) = w^{(0)}_\alpha(k|^{y}_{j}) + i\epsilon w^{(1)}_\alpha(k|^{y}_{j}) + \tfrac{1}{2}\epsilon^2 w^{(2)}_\alpha(k|^{y}_{j}) + ... \qquad (26.10)$$

For the acoustic waves the frequencies approach zero as y approaches zero; hence (26.9) begins with a linear term in ϵ.

On substituting the series (26.2), (26.9), and (26.10) in (24.10), and equating the coefficients of various powers of ϵ to zero, we obtain the perturbation equations

$$0 = \sum_{k'\beta} C^{(0)}_{\alpha\beta}(kk') w^{(0)}_\beta(k'|^{y}_{j}), \qquad (26.11)$$

$$0 = \sum_{k'\beta\gamma} C^{(1)}_{\alpha\beta,\gamma}(kk') y_\gamma\, w^{(0)}_\beta(k'|^{y}_{j}) + \sum_{k'\beta} C^{(0)}_{\alpha\beta}(kk') w^{(1)}_\beta(k'|^{y}_{j}), \qquad (26.12)$$

$$[\omega^{(1)}(^{y}_{j})]^2 w^{(0)}_\alpha(k|^{y}_{j}) = \tfrac{1}{2} \sum_{k'\beta\gamma\lambda} C^{(2)}_{\alpha\beta,\gamma\lambda}(kk') y_\gamma y_\lambda w^{(0)}_\beta(k'|^{y}_{j}) -$$

$$- \sum_{k'\beta\gamma} C^{(1)}_{\alpha\beta,\gamma}(kk') y_\gamma w^{(1)}_\beta(k'|^{y}_{j}) + \tfrac{1}{2} \sum_{k'\beta} C^{(0)}_{\alpha\beta}(kk') w^{(2)}_\beta(k'|^{y}_{j}). \qquad (26.13)$$

The left-hand side vanishes in both the zero- and first-order equations, because $\omega^2(^{y}_{j})$ is of the second order in ϵ. We note that the coefficients in the above equations are all real, so we shall be concerned with only real solutions.

The zero-order equation (26.11) has non-trivial solutions of the form

$$w^{(0)}_\alpha(k|^{y}_{j}) = \sqrt{m_k}\, u_\alpha(j), \qquad (26.14)$$

where $\mathbf{u}(j)$ can be any arbitrary vector in space. The fact that
(26.14) satisfies (26.11) follows directly from the relation (26.6). There
are evidently three independent solutions of the above type, corresponding to any three mutually perpendicular vectors chosen respectively as $\mathbf{u}(j)$. We have thus three branches of acoustic vibrations,
which we shall denote by $j = 1, 2, 3$. Apart from the fact that they
must be independent, the vectors $\mathbf{u}(j)$ for the three branches remain
completely arbitrary in this approximation. They will first be determined when we come to consider the second-order equations.

To obtain the solutions of the first-order equations we shall have to
use a well-known theorem in algebra, which may be put as follows.
Consider the s equations in s unknowns, $x_1, x_2, x_3,..., x_s$:

$$\sum_{n=1}^{s} A_{mn} x_n = C_m \quad (m = 1, 2,..., s), \tag{26.15}$$

where the left- and right-hand sides are known respectively as the
homogeneous and inhomogeneous parts of the equations. The equations
can be solved in the usual way, for instance by the use of the matrix
inverse to A_{mn}, only if A_{mn} is non-singular, i.e. if

$$|A_{mn}| \neq 0.$$

If one or more solutions $x_m(j)$ exist for the associated homogeneous
equations obtained by equating the homogeneous part of (26.15) to
zero:

$$\sum_{n=1}^{s} A_{mn} x_n(j) = 0 \tag{26.16}$$

(where j distinguishes different solutions when there is more than one
solution), the matrix A_{mn} is singular. The inhomogeneous equations
(26.15) are then not in general soluble. The necessary and sufficient
condition for the inhomogeneous equations to be soluble is that

$$\sum_{m=1}^{s} x_m(j) C_m = 0 \tag{26.17}$$

for all j. These conditions can be interpreted as orthogonality relations
between C_m and $x_m(j)$, both being regarded as general Cartesian vectors
with s components.

When the zero-order solution (26.14) is substituted in (26.12), we
obtain the first-order equations in the form

$$\sum_{k'\beta} C^{(0)}_{\alpha\beta}(kk') w^{(1)}_{\beta}(k'|_j^{\mathbf{y}}) = -\sum_{k'\beta\gamma} \sqrt{m_{k'}}\, C^{(1)}_{\alpha\beta,\gamma}(kk') y_\gamma u_\beta(j), \tag{26.18}$$

in which $w^{(1)}_{\beta}(k'|_j^{\mathbf{y}})$ are the unknowns. The left- and right-hand members
correspond respectively to the homogeneous and inhomogeneous parts

in (26.15). The associated homogeneous equations are seen to be identical
with the zero-order equations (26.11). Since the zero-order equations
have solutions of the form (26.14), where all three components of $\mathbf{u}(j)$ are
arbitrary, it is easily seen that the condition for solubility (26.17) reduces
in this case to the requirement that the inhomogeneous part of
(26.18) multiplied by $\sqrt{m_k}$ and summed over k must vanish, i.e.

$$\sum_{\beta\gamma}\left\{\sum_{kk'}(m_k m_{k'})^{\frac{1}{2}}C^{(1)}_{\alpha\beta,\gamma}(kk')\right\}y_\gamma u_\beta(j) = 0. \qquad (26.19)$$

Owing to (26.8) this condition is identically fulfilled.

Before discussing the solution of (26.18) it is instructive to observe
that the equation has an interpretation closely related to the homogeneous deformation of the lattice. Within a region small compared
with the wave-length of a long wave, the lattice is essentially in a state
of homogeneous strain. Thus, within such a region, the particle displacements due to the zero-order wave

$$u^{(0)}_\alpha(\mathbf{x}) = \frac{1}{\sqrt{m_k}}w^{(0)}_\alpha(k|^{\mathbf{y}}_j)e^{2\pi i\epsilon\mathbf{y}\cdot\mathbf{x}} = u_\alpha(j)e^{2\pi i\epsilon\mathbf{y}\cdot\mathbf{x}}, \qquad (26.20)$$

can be described to the first order by a homogeneous deformation of the
type (11.1) (i.e. a pure external strain), the deformation parameters
being

$$u_{\alpha\beta} = \frac{\partial u_\alpha}{\partial x_\beta} = 2\pi i\epsilon y_\beta u_\alpha(j)e^{2\pi i\epsilon\mathbf{y}\cdot\mathbf{x}}, \qquad (26.21)$$

where the exponential factor can be regarded as a constant within the
region considered. Using (26.21) we can write (26.18) as

$$\sum_{lk'\beta}\Phi_{\alpha\beta}(^{\,l}_{kk'})u^{(1)}_\beta(k') = -\sum_l\Phi_{\alpha\beta}(^{\,l}_{kk'})\left\{-\sum_\gamma u_{\beta\gamma}x_\gamma(^{\,l}_{kk'})\right\}, \qquad (26.22)$$

where
$$u^{(1)}_\beta(k') = \frac{i\epsilon}{\sqrt{m_k}}w^{(1)}_\beta(k'|^{\mathbf{y}}_j)e^{2\pi i\epsilon\mathbf{y}\cdot\mathbf{x}}. \qquad (26.23)$$

(26.23) evidently represents the particle displacements due to the first-
order wave (the part of the wave associated with the first-order term
in (26.10)). The right-hand member of (26.22) is readily seen to be the
force acting on a particle k due to the external strain (26.21). (26.23) expresses certain rigid displacements (the exponential factor to be regarded
as a constant in the region under consideration) of the constituent
Bravais lattices, describing thus a state of internal strain (see § 11). The
left-hand member of (26.22) is the negative of the force on a particle k
due to this strain. (26.22) states thus that the forces arising from the
two types of strain must be in balance.

The interpretation of (26.18) is now clear: the zero-order wave
subjects each local region of the lattice to a practically homogeneous

external strain. The first-order wave, given by the solution $w_\alpha^{(1)}(k|_j^y)$ of (26.18), describes the internal strain thereby induced to maintain the inner elastic equilibrium. (26.18) is thus the exact analogue of (11.27), which determines the internal strain of a lattice subject to a given homogeneous elastic strain (= external strain) in the special case of central interaction.

Like the inhomogeneous part, the homogeneous part of (26.18) multiplied by $\sqrt{m_k}$ and summed over k vanishes. It follows that if we multiply (26.18) by $\sqrt{m_k}$ and sum over k, the resulting equation is identically satisfied. This shows that, of the n equations ($k = 0, 1,..., n-1$) for a given value of α, only $n-1$ can be independent; in the following we may thus consider only the $3(n-1)$ equations with the indices $\alpha = 1, 2, 3$; $k = 1, 2,..., n-1$. This is closely related to the fact that we can always add to the solution of an inhomogeneous equation any solution of the corresponding homogeneous equation. In view of this latter fact, we can assume without loss of generality that

$$w_\alpha^{(1)}(0|_j^y) = 0, \quad \alpha = 1, 2, 3. \tag{26.24}$$

For this can always be achieved by the addition or subtraction of a suitable solution of the form (26.14); any difference thus caused can be incorporated in the zero-order wave.

(26.18) thus reduces to a system of $3(n-1)$ equations in $3(n-1)$ unknowns. In general, these equations are independent; in other words, the $(3n-3) \times (3n-3)$ matrix $C_{\alpha\beta}^{(0)}(kk')$ ($k, k' = 1,..., n-1$ and $\alpha = 1, 2, 3$) is in general non-singular. Let us denote its inverse matrix by $\Gamma^{(3n-3)}$, the elements of which satisfy by definition the following relations:

$$\sum_{k'\beta} \Gamma_{\alpha\beta}^{(3n-3)}(kk') C_{\beta\gamma}^{(0)}(k'k'') = \delta_{kk''}\cdot\delta_{\alpha\beta} = \sum_{k'\beta} C_{\alpha\beta}^{(0)}(kk') \Gamma_{\beta\gamma}^{(3n-3)}(k'k''). \tag{26.25}$$

Upon multiplying (26.18) by $\Gamma_{\mu\alpha}^{(3n-3)}(k''k)$ and summing over $\alpha = 1, 2, 3$, $k = 1, 2,..., n-1$, we obtain

$$w_\mu^{(1)}(k''|_j^y) = -\sum_{k=1}^{n-1} \sum_\alpha \Gamma_{\mu\alpha}^{(3n-3)}(k''k) \sum_{k'=0}^{n-1} \sum_{\beta\gamma} \sqrt{m_{k'}}\, C_{\alpha\beta,\gamma}^{(1)}(kk') y_\gamma u_\beta(j). \tag{26.26}$$

Formally it is convenient to introduce a $3n \times 3n$ matrix Γ by bordering $\Gamma^{(3n-3)}$ with zeros as follows:

$$\left. \begin{aligned} \Gamma_{\alpha\beta}(kk') &= \Gamma_{\alpha\beta}^{(3n-3)}(kk'), \quad k, k' \neq 0 \\ &= 0 \qquad\qquad\quad \text{otherwise} \end{aligned} \right\}. \tag{26.27}$$

Using Γ, we can write (26.24) and (26.26) together as follows:

$$w_\alpha^{(1)}(k|{}^y_j) = - \sum_{k'\mu} \Gamma_{\alpha\mu}(kk') \sum_{k''} \sum_{\beta\gamma} \sqrt{m_{k'}} \cdot C_{\mu\beta,\gamma}^{(1)}(k'k'')y_\gamma u_\beta(j), \quad (26.28)$$

where the base indices go over all the n values $0, 1,\ldots, n-1$.

Since $C_{\alpha\beta}^{(0)}(kk')$ is symmetric in $\binom{k}{\alpha}$ and $\binom{k'}{\beta}$, $\Gamma^{(3n-3)}$ is a symmetric matrix; and hence also Γ:

$$\Gamma_{\alpha\beta}(kk') = \Gamma_{\beta\alpha}(k'k). \quad (26.29)$$

When the zero-order solution (26.14) and the first-order solution (26.28) are substituted, the second-order equation (26.13) can be written as

$$\tfrac{1}{2} \sum_{k'\beta} C_{\alpha\beta}^{(0)}(kk')w_\beta^{(2)}(k'|{}^y_j) = [\omega^{(1)}({}^y_j)]^2\sqrt{m_k}\, u_\alpha(j) -$$

$$- \tfrac{1}{2} \sum_{k'\beta\gamma\lambda} \sqrt{m_{k'}}\, C_{\alpha\beta,\gamma\lambda}^{(2)}(kk')y_\gamma y_\lambda u_\beta(j) -$$

$$- \sum_{k'} \sum_{\mu\gamma} C_{\alpha\mu,\gamma}^{(1)}(kk')y_\gamma \sum_{k''\nu} \Gamma_{\mu\nu}(k'k'') \sum_{k'''} \sum_{\beta\lambda} C_{\nu\beta,\lambda}^{(1)}(k''k''')\sqrt{m_{k'''}}\, y_\lambda u_\beta(j), \quad (26.30)$$

where $w_\beta^{(2)}(k'|{}^y_j)$ are the unknowns and the right-hand member represents the inhomogeneous part. The solubility condition is obtained as before by multiplying the inhomogeneous part by $\sqrt{m_k}$, summing over k, and equating the resulting expression to zero. The condition thus obtained can be written

$$\left(\frac{\sum_k m_k}{v_a}\right)[\omega^{(1)}({}^y_j)]^2 u_\alpha(j) = 4\pi^2 \sum_\beta \left\{ \sum_{\gamma\lambda} [\alpha\beta, \gamma\lambda]y_\gamma y_\lambda + \sum_{\gamma\lambda} (\alpha\gamma, \beta\lambda)y_\gamma y_\lambda \right\} u_\beta(j), \quad (26.31)$$

where the coefficients denoted by the brackets are defined by

$$[\alpha\beta, \gamma\lambda] = \frac{1}{8\pi^2 v_a} \sum_{kk'} (m_k m_{k'})^{\frac{1}{2}} C_{\alpha\beta,\gamma\lambda}^{(2)}(kk'), \quad (26.32)$$

$$(\alpha\gamma, \beta\lambda) = \frac{-1}{4\pi^2 v_a} \sum_{kk'} \sum_{\mu\nu} \Gamma_{\mu\nu}(kk')\left(\sum_{k''} C_{\mu\alpha,\gamma}^{(1)}(kk'')\sqrt{m_{k''}}\right)\left(\sum_{k'''} C_{\nu\beta,\lambda}^{(1)}(k'k''')\sqrt{m_{k'''}}\right). \quad (26.33)$$

Use has been made of the symmetry relation (26.4) in writing (26.33) in this form. The brackets satisfy the symmetry relations

$$[\alpha\beta, \gamma\lambda] = [\beta\alpha, \gamma\lambda] = [\alpha\beta, \lambda\gamma], \quad (26.34)$$

$$(\alpha\beta, \gamma\lambda) = (\beta\alpha, \gamma\lambda) = (\gamma\lambda, \alpha\beta). \quad (26.35)$$

(26.34) follows directly from (26.5). (26.35) follows from (26.7) and the fact that Γ is a symmetric matrix. We note that the round brackets possess all the symmetry displayed by the square brackets and are in addition symmetric with respect to an interchange between the first and second pairs of indices.

(26.31) determines the polarization vector of the zero-order wave, which has thus far remained arbitrary. The equation is directly comparable with the macroscopic equation describing elastic waves, as we shall see in the next section.

27. The elastic constants for non-ionic crystals†

In dynamical problems of elasticity theory the elastic deformation concerned, though not homogeneous as regards the medium as a whole, can be considered as homogeneous in a small neighbourhood about any point. Thus, if $\mathbf{u}(\mathbf{x})$ is the elastic displacement at a point $\mathbf{x}$, we have in a small neighbourhood of a point $\mathbf{x}^0$

$$u_\alpha(\mathbf{x}^0 + \delta\mathbf{x}) = u_\alpha(\mathbf{x}^0) + \sum_\gamma \frac{\partial u_\alpha}{\partial x_\gamma} \delta x_\gamma. \qquad (27.1)$$

The first term on the right represents merely a translation of the small region as a whole and only the second term represents an elastic deformation. A comparison of (27.1) with (11.1) shows that about the point $\mathbf{x}^0$ the medium is subject to a homogeneous elastic deformation with the parameters $u_{\alpha\gamma}$, given by

$$u_{\alpha\gamma} = \frac{\partial u_\alpha}{\partial x_\gamma}. \qquad (27.2)$$

In § 11 we have already seen that the elastic strain of a homogeneously deformed specimen is to the first order described by six strain components s_ρ, which are related to the parameters $u_{\alpha\gamma}$ by (11.24). In dynamical problems the strain components are defined locally by substituting (27.2) in (11.24), namely,

$$s_\rho = \frac{\partial u_\alpha}{\partial x_\gamma} + \frac{\partial u_\gamma}{\partial x_\alpha} \quad (\alpha \neq \gamma),$$

$$= \frac{\partial u_\alpha}{\partial x_\gamma} \quad (\alpha = \gamma). \qquad (27.3)$$

The stress components S_ρ are similarly functions of position and are related to the local strain components by Hooke's law (11.29).

The equations of motion are most readily obtained by using the stress and strain components in tensor notation; these are defined (see (11.25)) by

$$S_{\alpha\gamma} = S_{\gamma\alpha} = S_\rho,$$

$$s_{\alpha\gamma} = s_{\gamma\alpha} = \frac{1}{2}\left(\frac{\partial u_\alpha}{\partial x_\gamma} + \frac{\partial u_\gamma}{\partial x_\alpha}\right) = \begin{cases} \frac{1}{2}s_\rho & (\alpha \neq \gamma) \\ s_\rho & (\alpha = \gamma). \end{cases} \qquad (27.4)$$

$S_{\alpha\gamma}$ thus defined represents the α-component of the force exerted on the medium which is on the negative side of a unit surface, normal to the

† K. Huang, *Proc. Roy. Soc.* A, **203**, 178 (1950).

γ-direction, by the medium on the positive side. $s_{\alpha\gamma}$ is defined in the above way so that Hooke's law (11.29) can be simply expressed in tensor notation, as follows:

$$S_{\alpha\gamma} = \sum_{\beta\lambda} c_{\alpha\gamma,\beta\lambda} s_{\beta\lambda}, \qquad (27.5)$$

where the elastic constants in tensor notation are obtained from the elastic constants in Voigt's notation by a simple transcription of the indices in accordance with (11.25). Thus by definition $c_{\alpha\gamma,\beta\lambda}$ is symmetric in both α, γ and β, λ; taking account of the fact that $c_{\rho\sigma} = c_{\sigma\rho}$, we have thus the symmetry relations

$$c_{\alpha\gamma,\beta\lambda} = c_{\gamma\alpha,\beta\lambda} = c_{\beta\lambda,\alpha\gamma}. \qquad (27.6)$$

Since the divergence of the stress tensor $S_{\alpha\beta}$ is equal to the force per unit volume, the equations of motion can be written as

$$\rho\ddot{u}_{\alpha} = \sum_{\gamma} \frac{\partial S_{\alpha\gamma}}{\partial x_{\gamma}}, \qquad (27.7)$$

where ρ is the mass density. Using Hooke's law (27.5) and afterwards expressing $s_{\alpha\beta}$ in terms of the displacement vector $\mathbf{u}(\mathbf{x})$ by (27.4), we have

$$\rho\ddot{u}_{\alpha} = \sum_{\beta\gamma\lambda} c_{\alpha\gamma,\beta\lambda} \frac{\partial^2 u_{\beta}}{\partial x_{\gamma}\,\partial x_{\lambda}}. \qquad (27.8)$$

Consider now a plane elastic wave

$$\mathbf{u}(\mathbf{x}) = \bar{\mathbf{u}} \exp\{2\pi i \mathbf{y}.\,\mathbf{x} - i\omega t\}. \qquad (27.9)$$

Substituting (27.9) in (27.8), we obtain

$$\rho\omega^2 \bar{u}_{\alpha} = 4\pi^2 \sum_{\beta} \left\{ \sum_{\gamma\lambda} c_{\alpha\gamma,\beta\lambda} y_{\gamma} y_{\lambda} \right\} \bar{u}_{\beta}. \qquad (27.10)$$

It is seen that this equation is identical in form with the equation (26.31) for the long acoustic lattice waves. It is, however, important to note that (27.10) specifically refers to elastic waves in a medium initially free of stresses; a medium, even if only subject to homogeneous stresses, cannot be described with twenty-one elastic constants as above (see next section). Thus (27.10) is comparable with (26.31), only if the latter describes lattice waves in a lattice initially free of stresses: in other words, only if the coefficients $\Phi_{\alpha\beta}\binom{l}{kk'}$, which underlie the brackets in (26.31), are actually defined with respect to a reference configuration satisfying the equilibrium condition (23.1) (ii). We shall presently find that for (27.10) and (26.31) to be compatible, certain relations must be fulfilled by the square brackets $[\alpha\beta, \gamma\lambda]$; from this we may infer that these relations express restrictions on the derivatives, which are in some way related to the condition of equilibrium (23.1) (ii). The exact significance of these relations will be considered in the next section.

For (27.10) and (26.31) to be identical we must have

$$\sum_{\gamma\lambda} c_{\alpha\gamma,\beta\lambda} y_\gamma y_\lambda = \sum_{\gamma\lambda} \{[\alpha\beta,\gamma\lambda] + (\alpha\gamma,\beta\lambda)\} y_\gamma y_\lambda \qquad (27.11)$$

for any value of y; it follows that

$$c_{\alpha\gamma,\beta\lambda} + c_{\alpha\lambda,\beta\gamma} = 2[\alpha\beta,\gamma\lambda] + (\alpha\gamma,\beta\lambda) + (\alpha\lambda,\beta\gamma). \qquad (27.12)$$

These relations must be satisfied subject to the symmetry relations (26.34), (26.35), and (27.6) for the brackets and the elastic constants.

Let us regard the brackets as known, and consider (27.12) as equations determining the elastic constants $c_{\alpha\gamma,\beta\lambda}$. We observe that the round brackets satisfy exactly the same symmetry relations with respect to permutation of the indices as the elastic constants. Thus we can eliminate the round brackets from our consideration by introducing instead of $c_{\alpha\gamma,\beta\lambda}$ the quantities $d_{\alpha\gamma,\beta\lambda}$ defined by

$$c_{\alpha\gamma,\beta\lambda} = d_{\alpha\gamma,\beta\lambda} + (\alpha\gamma,\beta\lambda), \qquad (27.13)$$

so that (27.12) becomes

$$d_{\alpha\gamma,\beta\lambda} + d_{\alpha\lambda,\beta\gamma} = 2[\alpha\beta,\gamma\lambda]. \qquad (27.14)$$

The quantities $d_{\alpha\gamma,\beta\lambda}$ must evidently satisfy the same symmetry relations as $c_{\alpha\gamma,\beta\lambda}$ and $(\alpha\gamma,\beta\lambda)$. We express these symmetry relations in the twofold form:

$$\left.\begin{array}{ll} (a) & d_{\alpha\gamma,\beta\lambda} = d_{\gamma\alpha,\beta\lambda} \\ (b) & d_{\alpha\gamma,\beta\lambda} = d_{\beta\lambda,\alpha\gamma} \end{array}\right\}. \qquad (27.15)$$

Now we can regard $d_{\alpha\gamma,\beta\lambda}$ as the unknowns and seek the solutions of (27.14) consistent with the symmetry relations (27.15).

We shall show that in fact there is only one solution of (27.14) consistent with (27.15) (a). Let us suppose that there are two distinct solutions $d'_{\alpha\gamma,\beta\lambda}$ and $d''_{\alpha\gamma,\beta\lambda}$ of (27.14), both consistent with (27.15) (a). Their difference $\Delta d_{\alpha\gamma,\beta\lambda} = d'_{\alpha\gamma,\beta\lambda} - d''_{\alpha\gamma,\beta\lambda}$ must be symmetric in the first pair of indices:

$$\Delta d_{\alpha\gamma,\beta\lambda} = \Delta d_{\gamma\alpha,\beta\lambda}, \qquad (27.16)$$

and satisfy the equation

$$\Delta d_{\alpha\gamma,\beta\lambda} + \Delta d_{\alpha\lambda,\beta\gamma} = 0, \qquad (27.17)$$

which is obtained by substituting $d'_{\alpha\gamma,\beta\lambda}$ and $d''_{\alpha\gamma,\beta\lambda}$ respectively in (27.14) and taking the difference. Relabelling α and γ respectively as γ and α, we can write (27.17) as

$$\Delta d_{\gamma\alpha,\beta\lambda} + \Delta d_{\gamma\lambda,\beta\alpha} = 0. \qquad (27.18)$$

It follows from (27.16) that the first term in (27.17) is equal to the first term in (27.18). Thus, subtracting (27.18) from (27.17), we find that

$$\Delta d_{\alpha\lambda,\beta\gamma} - \Delta d_{\gamma\lambda,\beta\alpha} = 0. \qquad (27.19)$$

In view of (27.16) we can permute the first pair of indices in both terms; relabelling afterwards λ, α, γ respectively as α, γ, λ, we obtain

$$\Delta d_{\alpha\gamma,\beta\lambda} - \Delta d_{\alpha\lambda,\beta\gamma} = 0. \qquad (27.20)$$

Adding (27.20) to (27.17), we find that

$$\Delta d_{\alpha\gamma,\beta\lambda} = 0. \qquad (27.21)$$

$d'_{\alpha\gamma,\beta\lambda}$ and $d''_{\alpha\gamma,\beta\lambda}$ are thus necessarily equal. In other words, with (27.15) (a) alone as auxiliary conditions, the solution of (27.14) is already unique.

It is easily verified, with the help of the symmetry relations (26.34) for the square brackets, that

$$d_{\alpha\gamma,\beta\lambda} = [\alpha\beta, \gamma\lambda] + [\beta\gamma, \alpha\lambda] - [\beta\lambda, \alpha\gamma] \qquad (27.22)$$

satisfies (27.14) and is symmetric in the first pair of indices. (27.22) thus represents the only solution of (27.14) which is also consistent with the symmetry requirement (27.15) (a).

The solution is not in general compatible with the remaining symmetry requirement (27.15) (b). In fact, upon substituting (27.22) in (27.15) (b), one obtains the conditions

$$[\beta\gamma, \alpha\lambda] - [\beta\lambda, \alpha\gamma] = [\alpha\lambda, \beta\gamma] - [\alpha\gamma, \beta\lambda]. \qquad (27.23)$$

Relabelling β, γ, α, λ respectively as γ, β, λ, α, one can write (27.23) as

$$[\gamma\beta, \lambda\alpha] - [\gamma\alpha, \lambda\beta] = [\lambda\alpha, \gamma\beta] - [\lambda\beta, \gamma\alpha]. \qquad (27.24)$$

Adding (27.24) to (27.23) and remembering the symmetry properties (26.34) of the square brackets, we find that

$$[\beta\gamma, \alpha\lambda] = [\alpha\lambda, \beta\gamma]. \qquad (27.25)$$

These compatibility conditions, as we have mentioned earlier, must be related in some way to the equilibrium condition (23.1) (ii) for the lattice. We know, for instance, that if the coefficients $\Phi_{\alpha\beta}\binom{l\ l'}{k\ k'}$ do actually refer to the lattice configuration satisfying (23.1) (ii), (27.10) and (26.31) must on physical grounds be identical, and (27.25) must then be automatically fulfilled.

Under the supposition that the lattice is actually free of stress so that (27.25) is fulfilled, the expressions for the elastic constants are obtained by substituting (27.22) in (27.13):

$$c_{\alpha\gamma,\beta\lambda} = [\alpha\beta, \gamma\lambda] + [\beta\gamma, \alpha\lambda] - [\beta\lambda, \alpha\gamma] + (\alpha\gamma, \beta\lambda). \qquad (27.26)$$

It is easily verified with the help of the symmetry properties (27.25) that this expression for $c_{\alpha\gamma,\beta\lambda}$ satisfies the necessary symmetry requirements (27.6) as well as the equation (27.12).

It is easily seen that the above treatment does not apply to ionic crystals. Thus let e_k be the ionic charge of an ion k; the Coulomb interaction between the ions contributes to Φ the terms

$$\frac{1}{2} \sum_{\binom{l}{k}} \sum_{\binom{l'}{k'} \neq \binom{l}{k}} \frac{e_k e_{k'}}{|\mathbf{x}\binom{l}{k} + \mathbf{u}\binom{l}{k} - \mathbf{x}\binom{l'}{k'} - \mathbf{u}\binom{l'}{k'}|}. \tag{27.27}$$

If we denote the corresponding contributions to the second derivatives of Φ by $\Phi^C_{\alpha\beta}\binom{l\ l'}{k\ k'}$, we find by straightforward differentiation of (27.27) that, excepting $l = 0$, $k' = k$,

$$\Phi^C_{\alpha\beta}\binom{l}{kk'} = \Phi^C_{\alpha\beta}\binom{l\ 0}{k\ k'} = -e_k e_{k'}\left\{\frac{\partial^2}{\partial x_\alpha \partial x_\beta} \frac{1}{|\mathbf{x}|}\right\}_{\mathbf{x} = \mathbf{x}\binom{l}{kk'}}, \tag{27.28}$$

whereas for $l = 0$, $k = k'$,

$$\Phi^C_{\alpha\beta}\binom{0}{kk} = \Phi^C_{\alpha\beta}\binom{l\ l}{k\ k} = e_k \sum_{lk'}' e_{k'}\left\{\frac{\partial^2}{\partial x_\alpha \partial x_\beta} \frac{1}{|\mathbf{x}|}\right\}_{\mathbf{x} = \mathbf{x}\binom{l}{kk'}}, \tag{27.29}$$

where the prime over the summation excludes the term $\binom{l}{k'} = \binom{0}{k}$. We notice in particular that $\Phi^C_{\alpha\beta}\binom{l}{kk'}$ given by (27.28) falls off with the inverse third power of the distance $|\mathbf{x}\binom{l}{kk'}|$; thus we immediately see, for instance, that $C^{(0)}_{\alpha\beta}(kk')$ involves a divergent sum.

The failure of the above method for ionic crystals is indeed expected for the reason given in § 25, namely that the long lattice waves are no longer pure elastic waves such as are described by (27.7)–(27.10).

28. Equilibrium conditions (vanishing stresses) and further invariance relations†

Certain points in the above discussion still remain to be clarified. In particular, taking account of the symmetry relations (26.34) satisfied by the square brackets, we find that (27.25) imposes fifteen independent conditions on these brackets; whereas the requirement of vanishing stresses should not give rise to more than six conditions.

Since in the formulation of the equations in § 24 the equilibrium conditions (23.1) (ii) have not been introduced, the equation (26.31) may in fact describe lattice waves in a lattice which is under stresses (homogeneous, for otherwise (24.3) could not have been satisfied). Therefore in the following we shall compare (26.31) directly with the equations for elastic waves in a medium subject to a system of arbitrary homogeneous stresses.

† K. Huang, *Proc. Roy. Soc.* A, **203**, 178 (1950).

Consider thus an elastic medium which is initially under stress, and denote the position of points in the medium by $\mathbf{x}$. Let the medium be subject to the following homogeneous strain from this initial configuration:

$$u_\alpha(\mathbf{x}) = \sum_\beta u_{\alpha\beta} x_\beta. \tag{28.1}$$

The energy density u due to the deformation can be expressed as a Taylor series thus:

$$u = \sum_{\alpha\beta} S_{\alpha\beta} u_{\alpha\beta} + \tfrac{1}{2} \sum_{\alpha\gamma} \sum_{\beta\lambda} S_{\alpha\gamma,\beta\lambda} u_{\alpha\gamma} u_{\beta\lambda} + \dots, \tag{28.2}$$

where

$$S_{\alpha\gamma,\beta\lambda} = S_{\beta\lambda,\alpha\gamma}. \tag{28.3}$$

It is obvious that if the deformed medium is rotated rigidly the energy density must remain invariant. As can be shown,† it follows from this invariance requirement that

$$S_{\alpha\beta} = S_{\beta\alpha}, \tag{28.4}$$

$$S_{\alpha\lambda} \delta_{\beta\gamma} - S_{\gamma\lambda} \delta_{\beta\alpha} + S_{\alpha\gamma,\beta\lambda} - S_{\gamma\alpha,\beta\lambda} = 0. \tag{28.5}$$

We notice that $S_{\alpha\beta} = (\partial u / \partial u_{\alpha\beta})$ are simply the initial stress components in the medium (cf. (11.29) where the derivation of the stress components from the energy density is given in Voigt's notation). For a medium free of stresses we have to put $S_{\alpha\beta} = 0$. Then the energy density becomes quadratic in the strain components, and $S_{\alpha\gamma,\beta\lambda}$ becomes identical with the elastic constant $c_{\alpha\gamma,\beta\lambda}$. It is seen that in this case (28.3) and (28.5) reduce to the symmetry relations (27.6) for the elastic constants.

The equations of motion can be deduced with the help of the Lagrangian function constructed from the energy density (28.2). It can be shown† that, if only terms up to the second degree in the wave number are considered, the wave equation is the same as (27.10) with the elastic constants $c_{\alpha\gamma,\beta\lambda}$ replaced by $S_{\alpha\gamma,\beta\lambda}$, namely,

$$\rho\omega^2 \bar{u}_\alpha = 4\pi^2 \sum_\beta \left\{ \sum_{\gamma\lambda} S_{\alpha\gamma,\beta\lambda} y_\gamma y_\lambda \right\} \bar{u}_\beta. \tag{28.6}$$

The difference of the present case from the stress-free case is not so much in the wave equation as in the different symmetry relations which $S_{\alpha\gamma,\beta\lambda}$ has to fulfil.

Comparison of (28.6) with the lattice-wave equation (26.31) gives

$$S_{\alpha\gamma,\beta\lambda} + S_{\alpha\lambda,\beta\gamma} = 2[\alpha\beta, \gamma\lambda] + (\alpha\gamma, \beta\lambda) + (\alpha\lambda, \beta\gamma). \tag{28.7}$$

(28.7), together with the symmetry requirements (28.3), (28.4), and (28.5), may be regarded as a system of equations determining $S_{\alpha\beta}$ and $S_{\alpha\gamma,\beta\lambda}$.

† See K. Huang, *Proc. Roy. Soc.* A, **203**, 178, Appendix (1950).

Instead of $S_{\alpha\gamma,\beta\lambda}$ it is convenient to introduce new unknowns $x_{\alpha\beta,\gamma\lambda}$ which are defined as follows:

$$S_{\alpha\gamma,\beta\lambda} = \{[\alpha\beta,\gamma\lambda]+[\beta\gamma,\alpha\lambda]-[\beta\lambda,\alpha\gamma]+(\alpha\gamma,\beta\lambda)\}+x_{\alpha\beta,\gamma\lambda}. \qquad (28.8)$$

We have seen in the last section that the expression in the curly brackets satisfies equations of the form (28.7) and is, moreover, symmetric in the indices α and γ. Hence on substituting (28.8) in (28.7), we find that

$$x_{\alpha\beta,\gamma\lambda} = -x_{\alpha\beta,\lambda\gamma}. \qquad (28.9)$$

$x_{\alpha\beta,\gamma\lambda}$ must thus be antisymmetric in the last pair of indices. The symmetry relations (28.5) require that

$$S_{\alpha\lambda}\delta_{\beta\gamma}-S_{\gamma\lambda}\delta_{\beta\alpha}+x_{\alpha\beta,\gamma\lambda}-x_{\gamma\beta,\alpha\lambda} = 0, \qquad (28.10)$$

and the relations (28.3) become

$$x_{\alpha\beta,\gamma\lambda}+[\beta\gamma,\alpha\lambda]-[\beta\lambda,\alpha\gamma] = x_{\beta\alpha,\lambda\gamma}+[\alpha\lambda,\beta\gamma]-[\alpha\gamma,\beta\lambda]. \qquad (28.11)$$

For $\beta = \gamma \neq \alpha$, (28.10) reduces to

$$S_{\alpha\lambda} = x_{\beta\beta,\alpha\lambda}-x_{\alpha\beta,\beta\lambda} \quad (\beta \neq \alpha). \qquad (28.12)$$

The expression on the right must thus be independent of β so long as $\beta \neq \alpha$. For $\alpha \neq \lambda$ we may put $\beta = \lambda$ in the above formula and obtain

$$S_{\alpha\lambda} = x_{\lambda\lambda,\alpha\lambda}-x_{\alpha\lambda,\lambda\lambda} = x_{\lambda\lambda,\alpha\lambda} \quad (\alpha \neq \lambda), \qquad (28.13)$$

where $x_{\alpha\lambda,\lambda\lambda}$ vanishes owing to the antisymmetry requirement (28.9). For $\alpha = \lambda$ the first term in (28.12) vanishes for the same reason, leaving

$$S_{\alpha\alpha} = -x_{\alpha\beta,\beta\alpha} \quad (\beta \neq \alpha). \qquad (28.14)$$

This implies, of course, that $x_{\alpha\beta,\beta\alpha}$ is independent of β if $\beta \neq \alpha$.

Now that all the stress components $S_{\alpha\beta}$ are expressed in terms of the quantities $x_{\alpha\gamma,\beta\lambda}$ let us proceed to determine the latter. Clearly out of the four indices of $x_{\alpha\gamma,\beta\lambda}$ at least two must be identical; we shall consider the solutions for $x_{\alpha\gamma,\beta\lambda}$ according to the positions occupied by the pair of identical indices:

(a) $x_{\alpha\beta,\gamma\gamma}$: It follows directly from the antisymmetry relation (28.9) for the last pair of indices that

$$x_{\alpha\beta,\gamma\gamma} = 0. \qquad (28.15)$$

(b) $x_{\alpha\alpha,\gamma\lambda}$: On putting $\beta = \alpha$ in (28.11), we find, with the help of (28.9), that

$$x_{\alpha\alpha,\gamma\lambda} = [\alpha\lambda,\alpha\gamma]-[\alpha\gamma,\alpha\lambda]. \qquad (28.16)$$

(c) $x_{\alpha\beta,\beta\lambda}$: With the restriction that $\alpha \neq \lambda$, $\alpha \neq \beta$, we can combine (28.12) with (28.13) and obtain

$$x_{\alpha\beta,\beta\lambda} = x_{\beta\beta,\alpha\lambda}-S_{\alpha\lambda} = x_{\beta\beta,\alpha\lambda}-x_{\lambda\lambda,\alpha\lambda} \quad (\alpha \neq \lambda, \alpha \neq \beta).$$

Both terms on the right are of the description given in (b); thus by suitable relabelling of the indices in (28.16) we can express the right-hand member of the formula in terms of square brackets:

$$x_{\alpha\beta,\beta\lambda} = [\beta\lambda,\beta\alpha]-[\beta\alpha,\beta\lambda]-[\lambda\lambda,\lambda\alpha]+[\lambda\alpha,\lambda\lambda] \quad (\alpha \neq \lambda, \alpha \neq \beta). \quad (28.17)$$

(d) $x_{\alpha\beta,\gamma\beta}$: In view of the antisymmetry in the last pair of indices we can reduce this case to the one above. Thus, on relabelling λ as γ in (28.17), we find, for $\alpha \neq \gamma, \alpha \neq \beta$,

$$x_{\alpha\beta,\gamma\beta} = -x_{\alpha\beta,\beta\gamma} = -[\beta\gamma,\beta\alpha]+[\beta\alpha,\beta\gamma]+[\gamma\gamma,\gamma\alpha]-[\gamma\alpha,\gamma\gamma]$$
$$(\alpha \neq \lambda, \alpha \neq \beta). \quad (28.18)$$

(e) $x_{\alpha\beta,\gamma\alpha}$: On putting $\lambda = \alpha$ in (28.11), we have

$$x_{\alpha\beta,\gamma\alpha}+[\beta\gamma,\alpha\alpha]-[\beta\alpha,\alpha\gamma] = x_{\beta\alpha,\alpha\gamma}+[\alpha\alpha,\beta\gamma]-[\alpha\gamma,\beta\alpha]. \quad (28.19)$$

For $\beta \neq \alpha, \beta \neq \gamma$, $x_{\beta\alpha,\alpha\gamma}$ on the right-hand side is of the type described in (c) and may thus be eliminated with the help of (28.17) (relabel α, β, λ as β, α, γ). In this way we obtain

$$x_{\alpha\beta,\gamma\alpha} = [\alpha\alpha,\beta\gamma]-[\beta\gamma,\alpha\alpha]-[\gamma\gamma,\gamma\beta]+[\gamma\beta,\gamma\gamma] \quad (\beta \neq \alpha, \beta \neq \gamma). \quad (28.20)$$

(f) $x_{\alpha\beta,\alpha\lambda}$: For $\beta \neq \alpha, \beta \neq \lambda$, we can reduce this case to the one above with the help of the antisymmetry relation (28.9):

$$x_{\alpha\beta,\alpha\lambda} = -x_{\alpha\beta,\lambda\alpha} = -[\alpha\alpha,\beta\lambda]+[\beta\lambda,\alpha\alpha]+[\lambda\lambda,\lambda\beta]-[\lambda\beta,\lambda\lambda]$$
$$(\beta \neq \alpha, \beta \neq \lambda), \quad (28.21)$$

where (28.20) has been used with γ relabelled as λ.

Taking account of the restrictions indicated in the brackets given after the above formulae we find that the above list covers all possible forms of $x_{\alpha\gamma,\beta\lambda}$ except the following:

$$x_{\alpha\beta,\beta\alpha} = -x_{\alpha\beta,\alpha\beta}.$$

These excepted cases are related to the stress components as follows (see (28.14)):

$$S_{11} = -x_{12,21} = -x_{13,31} = x_{12,12} = x_{13,13},$$
$$S_{22} = -x_{23,32} = -x_{21,12} = x_{23,23} = x_{21,21}, \quad (28.22)$$
$$S_{33} = -x_{31,13} = -x_{32,23} = x_{31,31} = x_{32,32}.$$

The differences between these stress components are determined by (28.11), for we can write

$$S_{11}-S_{22} = x_{21,12}-x_{12,21}, \quad (28.23)$$
$$S_{22}-S_{33} = x_{32,23}-x_{23,32}, \quad (28.24)$$
$$S_{33}-S_{11} = x_{13,31}-x_{31,13}. \quad (28.25)$$

The right-hand members of the above equations may then be expressed in terms of the square brackets by putting $\alpha = \lambda = 2$, $\beta = \gamma = 1$; $\alpha = \lambda = 3$, $\beta = \gamma = 2$; and $\alpha = \lambda = 1$, $\beta = \lambda = 3$ respectively in (28.11):

$$S_{11} - S_{22} = [22, 11] - [11, 22], \tag{28.26}$$

$$S_{22} - S_{33} = [33, 22] - [22, 33], \tag{28.27}$$

$$S_{33} - S_{11} = [11, 33] - [33, 11]. \tag{28.28}$$

On introducing the average pressure

$$p = -\tfrac{1}{3}(S_{11} + S_{22} + S_{33}), \tag{28.29}$$

we find, on combining (28.29) with (28.26), (28.27), that

$$S_{11} = -p + \tfrac{1}{3}\{[22, 11] + [33, 11] - [11, 22] - [11, 33]\}. \tag{28.30}$$

It is easily seen that the expressions for S_{22} and S_{33} can be written down by cyclic permutation of the indices in (28.30). From (28.22) it follows that all the quantities $x_{\alpha\beta,\beta\alpha} = -x_{\alpha\beta,\alpha\beta}$ can be similarly expressed in terms of the square brackets and the average pressure p.

Our description of the solution is completed by giving the stress components $S_{\alpha\lambda}$ ($\alpha \neq \lambda$) in terms of the square brackets. Let us put $\alpha = \lambda$ in (28.16) and relabel γ as α. According to (28.13) the result is equal to $S_{\alpha\lambda}$, hence

$$S_{\alpha\lambda} = [\lambda\lambda, \lambda\alpha] - [\alpha\lambda, \lambda\lambda] \quad (\alpha \neq \lambda). \tag{28.31}$$

Since $S_{\alpha\gamma,\beta\lambda}$ can be expressed in terms of the quantities $x_{\alpha\beta,\gamma\lambda}$ with the help of (28.8), we have thus determined $S_{\alpha\gamma,\beta\lambda}$ as well as $S_{\alpha\beta}$, all in terms of the square brackets and the average pressure. These will be a solution of the equations (28.3), (28.4), (28.5), (28.7) only under the assumption that a solution does exist. This has to be verified by substituting in these equations the expressions obtained above for $S_{\alpha\beta}$ and $S_{\alpha\gamma,\beta\lambda}$. In fact one finds that these equations are not identically satisfied; instead one obtains ten independent relations by this substitution. These are thus the conditions for the existence of a solution. Since on physical grounds the solution must exist (i.e. the results of the lattice and elastic theory must agree at the long-wave limit), the ten relations must thus be identities necessarily fulfilled by the square brackets. The identities are

$$[21, 13] = [31, 12] \text{ and cyclic permutations}, \tag{28.32}$$

$$[22, 23] - [23, 22] = [33, 32] - [32, 33] \text{ and cyclic permutations}, \tag{28.33}$$

$$[33, 23] - [23, 33] = [11, 23] - [23, 11] \text{ and cyclic permutations}, \tag{28.34}$$

$$[22, 11] + [33, 22] + [11, 33] = [11, 22] + [22, 33] + [33, 11]. \tag{28.35}$$

Since these identities have been obtained without making any special assumptions apart from the rotational invariance of the elastic energy density expressed by (28.4) and (28.5), they must clearly express the corresponding invariance conditions in the lattice theory. These relations will consequently be automatically fulfilled in any special example.

Apart from the ten invariance relations the solution gives also the explicit expressions for the five anisotropic stress components: $S_{11}-S_{22}$, $S_{22}-S_{33}$, S_{12}, S_{23}, S_{31} in terms of the square brackets ((28.26), (28.27), and (28.31)).

The significance of the fifteen conditions in (27.25) is now easily understood. If these conditions are fulfilled we find immediately, with the help of (28.26), (28.27), and (28.31), that all the five anisotropic stress components vanish; and, moreover, the above ten invariance relations are satisfied. Conversely, if we set the expressions for the five anisotropic stress components to zero, we can easily show, with the help of the ten invariance relations, that all the conditions (27.25) are fulfilled. The conditions embodied in (27.25) are thus equivalent to the five conditions for vanishing anisotropic stress components and the ten invariance relations.

The above treatment does not give an expression for the pressure. This is very much as we should expect in view of the linear example discussed in § 25; clearly, like the tension in the linear example, the pressure is not determined by the derivatives of the total potential function Φ.

29. Central forces

It is instructive to apply the results obtained in the last two sections to lattice models where particles interact with central forces. As in § 11, we regard the potential energy between two particles of the types k and k' respectively as a function of the square of their distance apart and denote the function by $\psi_{kk'}$. The formal total potential function for the lattice is thus given by

$$\Phi = \tfrac{1}{2} \sum_{lk} \sum_{l'k'} \psi_{kk'}(|\mathbf{x}(^{l-l'}_{kk'})+\mathbf{u}(^{l}_{k})-\mathbf{u}(^{l'}_{k'})|^2), \qquad (29.1)$$

where, as in § 11, instead of excluding $(^{l'}_{k'}) = (^{l}_{k})$ we make the formal provision that ψ_{kk} and all its derivatives are equal to zero for zero value of the argument.

The derivatives of Φ can be obtained by straightforward differentiation of the above expression. In carrying out the differentiation we need only note that the differentiation of Φ with respect to $u_{\alpha}(^{l'}_{k'})$ selects out of the double summation in (29.1) only the terms with either $\binom{l}{k}$ or $\binom{l'}{k'}$ equal to $\binom{l''}{k''}$. The same is true in obtaining the second derivatives with respect to displacements of the same particle; the second derivatives with respect to the displacements $u_\alpha\binom{l''}{k''}$, $u_\beta\binom{l'''}{k'''}$ of two different particles $\binom{l''}{k''}$ and $\binom{l'''}{k'''}$ are, however, contributed only by the two identical terms $\binom{l}{k} = \binom{l''}{k''}$, $\binom{l'}{k'} = \binom{l'''}{k'''}$ and $\binom{l}{k} = \binom{l'''}{k'''}$, $\binom{l'}{k'} = \binom{l''}{k''}$ in the above sum. The explicit expressions for the first and second derivatives can be written

$$\Phi_\alpha(k) = 2\sum_{l'k'}[x_\alpha\psi']_{x\binom{l'}{kk'}}, \tag{29.2}$$

$$\Phi_{\alpha\beta}\binom{l}{kk'} = \{-2\delta_{\alpha\beta}\psi' - 4x_\alpha x_\beta\psi''\}_{x\binom{l}{kk'}} \tag{29.3}$$

(the case $l = 0$, $k' = k$ excluded),

$$\Phi_{\alpha\beta}\binom{0}{kk} = \sum_{l'k'}\{2\delta_{\alpha\beta}\psi' + 4x_\alpha x_\beta\psi''\}_{x\binom{l'}{kk'}}, \tag{29.4}$$

where, as in § 11, the arguments are indicated at the lower right corner, and ψ refers to $\psi_{kk'}$ if the argument indicated concerns two particles of the types k and k' respectively. The formal understanding that $\psi_{kk}(0)$, $\psi'_{kk}(0)$, etc., are equal to zero is evidently equivalent to excluding the term $\binom{l'}{k'} = \binom{0}{k}$ in (29.2) and (29.4).

Substitution of (29.3) and (29.4) in (26.3), (26.4), and (26.5) gives

$$C^{(0)}_{\alpha\beta}(kk') = \frac{1}{(m_k m_{k'})^{\frac{1}{2}}}\left\{-2\delta_{\alpha\beta}\sum_l[\psi']_{x\binom{l}{kk'}} - 4\sum_l[x_\alpha x_\beta\psi'']_{x\binom{l}{kk'}}\right\}\quad (k \neq k'), \tag{29.5}$$

$$C^{(0)}_{\alpha\beta}(kk) = \frac{1}{m_k}\left\{2\delta_{\alpha\beta}\sum_{k'(\neq k)}\sum_l[\psi']_{x\binom{l}{kk'}} + 4\sum_{k'(\neq k)}\sum_l[x_\alpha x_\beta\psi'']_{x\binom{l}{kk'}}\right\},$$

$$C^{(1)}_{\alpha\beta,\gamma}(kk') = \frac{4\pi}{(m_k m_{k'})^{\frac{1}{2}}}\left\{\delta_{\alpha\beta}\sum_l[x_\gamma\psi']_{x\binom{l}{kk'}} + 2\sum_l[x_\alpha x_\beta x_\gamma\psi'']_{x\binom{l}{kk'}}\right\}, \tag{29.6}$$

$$C^{(2)}_{\alpha\beta,\gamma\lambda}(kk') = \frac{8\pi^2}{(m_k m_{k'})^{\frac{1}{2}}}\left\{\delta_{\alpha\beta}\sum_l[x_\gamma x_\lambda\psi']_{x\binom{l}{kk'}} + 2\sum_l[x_\alpha x_\beta x_\gamma x_\lambda\psi'']_{x\binom{l}{kk'}}\right\}. \tag{29.7}$$

In writing down (29.6), (29.7), we note that (29.4) contributes nothing. The equilibrium condition (23.1) (i) now reads

$$\Phi_\alpha(k) = 2\sum_{l'k'}[x_\alpha\psi']_{x\binom{l'}{kk'}} = 0, \tag{29.8}$$

which is seen to be completely equivalent to (11.13), in view of the symmetry relation (11.10) (b). Moreover, with the help of the relations (11.10), it is easily verified that the above expressions for the coefficients $C^{(0)}_{\alpha\beta}(kk')$, $C^{(1)}_{\alpha\beta,\gamma}(kk')$, and $C^{(2)}_{\alpha\beta,\gamma\lambda}(kk')$ fulfil the general relations given in (26.3)–(26.8).

On substituting (29.7) in (26.32) we obtain for the square brackets:

$$[\alpha\beta,\gamma\lambda] = \frac{1}{v_a}\Big\{\delta_{\alpha\beta}\sum_{lkk'}[x_\gamma x_\lambda \psi']_{\mathbf{x}(\substack{l\\kk'})} + 2\sum_{lkk'}[x_\alpha x_\beta x_\gamma x_\lambda \psi'']_{\mathbf{x}(\substack{l\\kk'})}\Big\}, \quad (29.9)$$

which evidently satisfies the symmetry relations (26.34). Furthermore, it is readily verified that the ten additional invariance relations (28.32)–(28.35) deduced in the last section are also satisfied by the above expression.

(29.9) used in (28.26), (28.27), and (28.31) gives for the anisotropic stress components:

$$S_{11}-S_{22} = \frac{1}{v_a}\sum_{lkk'}(x_1^2(\substack{l\\kk'})-x_2^2(\substack{l\\kk'}))\psi'(|\mathbf{x}(\substack{l\\kk'})|^2), \quad (29.10)$$

$$S_{22}-S_{33} = \frac{1}{v_a}\sum_{lkk'}(x_2^2(\substack{l\\kk'})-x_3^2(\substack{l\\kk'}))\psi'(|\mathbf{x}(\substack{l\\kk'})|^2), \quad (29.11)$$

$$S_{\alpha\lambda} = \frac{1}{v_a}\sum_{lkk'}x_\alpha(\substack{l\\kk'})x_\lambda(\substack{l\\kk'})\psi'(|\mathbf{x}(\substack{l\\kk'})|^2) \quad (\alpha \neq \lambda). \quad (29.12)$$

The average pressure can, of course, be obtained now that we have a specific model. Let us consider a uniform compression, whereby all distances in the lattice are reduced by the same factor, which we denote by $(1-\epsilon)$. By considering the interaction energy of each particle in the zero-cell with all other particles in the lattice and summing over the cell, we obtain for the energy change per cell

$$\tfrac{1}{2}\sum_k\sum_{l'k'}\{\psi_{kk'}((1-\epsilon)^2|\mathbf{x}(\substack{l'\\k'k})|^2)-\psi_{kk'}(|\mathbf{x}(\substack{l'\\k'k})|^2)\}$$

$$= -\epsilon\sum_k\sum_{l'k'}|\mathbf{x}(\substack{l'\\k'k})|^2\psi'_{kk'}(|\mathbf{x}(\substack{l'\\k'k})|^2)+O(\epsilon^2)+\dots. \quad (29.13)$$

This must be equal to the work done on a cell by the pressure, which, to the first order in ϵ, is given by

$$(\text{volume decrement per cell})\times p = 3\epsilon v_a p. \quad (29.14)$$

Equating (29.13) and (29.14) in the limit $\epsilon \to 0$, one finds that

$$\tfrac{1}{3}(S_{11}+S_{22}+S_{33}) = -p = \frac{1}{3v_a}\sum_{l'kk'}|\mathbf{x}(\substack{l'\\k'k})|^2\psi'(|\mathbf{x}(\substack{l'\\k'k})|^2). \quad (29.15)$$

Combining this expression with (29.10) and (29.11), we find that the same formula as (29.12),

$$S_{\alpha\beta} = \frac{1}{v_a}\sum_{lkk'}x_\alpha(\substack{l\\kk'})x_\beta(\substack{l\\kk'})\psi'(|\mathbf{x}(\substack{l\\kk'})|^2), \quad (29.16)$$

holds for $\alpha = \beta$ as well.

The equilibrium conditions (23.1) (ii) require that all the stress components vanish, and hence read in the present case

$$\sum_{lkk'} x_\alpha(_{kk'}^{\ l})x_\beta(_{kk'}^{\ l})\psi'(|\mathbf{x}(_{kk'}^{\ l})|^2) = 0. \tag{29.17}$$

These conditions, we note, are identical with the conditions given previously as (11.14).

When the equilibrium conditions (29.17) are fulfilled, the expressions (29.9) for the square brackets reduce to

$$[\alpha\beta, \gamma\lambda] = \frac{2}{v_a} \sum_{lkk'} [x_\alpha x_\beta x_\gamma x_\lambda \psi'']_{\mathbf{x}(_{kk'}^{\ l})}. \tag{29.18}$$

Let us consider the elastic constants for lattices where every lattice particle occupies a centre of symmetry (see § 11). For this special case the expression (29.6) for $C^{(1)}_{\alpha\beta,\gamma}(kk')$ vanishes identically; it follows from (26.33) that all the round brackets vanish in this case. In view of this and the fact that the square brackets given now by (29.18) are symmetric in all four indices, we find that the elastic constants (27.26) become in this case identical with the square brackets:

$$c_{\alpha\gamma,\beta\lambda} = [\alpha\beta, \gamma\lambda] = \frac{2}{v_a} \sum_{lkk'} [x_\alpha x_\beta x_\gamma x_\lambda \psi'']_{\mathbf{x}(_{kk'}^{\ l})}, \tag{29.19}$$

which has been obtained in § 11 as (11.30).

30. Coulomb field in a dipole lattice—Ewald's method and separation of the macroscopic field

The divergent results given by the long-wave method in the case of ionic crystals indicate not only that the lattice waves are no longer comparable with pure elastic waves, but also that the perturbation method developed in § 26 has to be modified. In this section we shall establish a result which is the basis of the modified treatment; the essential mathematical technique required for this purpose was due originally to Ewald.[†]

Consider the dipoles

$$\mathbf{p}(l) = \mathbf{p}e^{2\pi i \mathbf{y} \cdot \mathbf{x}(l)} \tag{30.1}$$

distributed over a simple Bravais lattice $\mathbf{x}(l) = l^1\mathbf{a}_1 + l^2\mathbf{a}_2 + l^3\mathbf{a}_3$. If the wave number $\mathbf{y}$ is small compared with the inverse of the dimensions of a lattice cell so that $\mathbf{p}(l)$ changes very little from one cell to the next, the lattice can be imagined as a polarized continuum with the following

---

† P. P. Ewald, *Dissertation* (München, 1912); *Ann. d. Phys.* **54**, 519, 557 (1917); **64**, 253 (1921); *Nach. Ges. Wiss. Göttingen*, 55 (1938).

macroscopic dielectric polarization:

$$P(x) = \frac{p}{v_a} e^{2\pi i y \cdot x}. \tag{30.2}$$

The corresponding macroscopic electric field can be determined in the same way as in § 7 with the help of the equation of electrostatics:

$$\nabla . (E(x) + 4\pi P(x)) = 0, \tag{30.3}$$

where $E(x)$ is an irrotational field. We have to use (30.2) for $P(x)$ and obtain the value for $E(x)$ from (30.3). Let us split $P(x)$ into two parts $P_{\parallel}(x)$ and $P_{\perp}(x)$ respectively parallel and perpendicular to y; it is easily verified that $P_{\parallel}(x)$ is irrotational and $P_{\perp}(x)$ is solenoidal. Hence $P_{\perp}(x)$ contributes nothing to (30.3), which may thus be written as

$$\nabla . (E(x) + 4\pi P_{\parallel}(x)) = 0. \tag{30.4}$$

Since both $E(x)$ and $P_{\parallel}(x)$ are irrotational, we have, moreover,

$$\nabla \wedge (E(x) + 4\pi P_{\parallel}(x)) = 0. \tag{30.5}$$

It is a well-known theorem in vector analysis that if both the divergence and curl of a vector vanish everywhere, the vector field itself must vanish identically. Thus it follows from (30.4) and (30.5) that

$$E(x) = -4\pi P_{\parallel}(x). \tag{30.6}$$

Expressing $P_{\parallel}(x)$ explicitly with the help of (30.2), we have

$$E(x) = E e^{2\pi i y \cdot x}, \tag{30.7}$$

where the amplitude is given by

$$E = -\frac{4\pi}{v_a} \left(\frac{y}{|y|}\right)\left(\frac{y \cdot p}{|y|}\right). \tag{30.8}$$

The actual Coulomb field in the lattice is, however, different from the macroscopic field. The field at a point x due to the single dipole $p(l)$ at $x(l)$ is given by (α-component of the field)

$$\sum_{\beta} p_{\beta}(l) \frac{\partial^2}{\partial x_{\alpha} \partial x_{\beta}} \left\{\frac{1}{|x(l) - x|}\right\}. \tag{30.9}$$

Summing over all dipoles in the lattice, we obtain for the Coulomb field at x

$$\sum_{\beta} p_{\beta} \frac{\partial^2}{\partial x_{\alpha} \partial x_{\beta}} \sum_{l} \frac{e^{2\pi i y \cdot x(l)}}{|x(l) - x|}. \tag{30.10}$$

Our purpose is to transform (30.10) so that it becomes the sum of the macroscopic field (30.7) and an *inner field*, and then to show that the inner field, as a function of $\mathbf{y}$, is regular at $\mathbf{y} = 0$, whereas the macroscopic field is not.

Following Ewald, we introduce the identity

$$\frac{2}{\sqrt{\pi}} \int_0^\infty e^{-|\mathbf{x}(l)-\mathbf{x}|^2 \rho^2} \, d\rho = \frac{1}{|\mathbf{x}(l)-\mathbf{x}|} \tag{30.11}$$

to give an integral representation of $1/|\mathbf{x}(l)-\mathbf{x}|$. Thus, using (30.11), we can write (30.10) in the form

$$\sum_\beta p_\beta \frac{\partial^2}{\partial x_\alpha \, \partial x_\beta} \int_0^\infty \left\{ \frac{2}{\sqrt{\pi}} \sum_l e^{-|\mathbf{x}(l)-\mathbf{x}|^2 \rho^2 + 2\pi i \mathbf{y} \cdot (\mathbf{x}(l)-\mathbf{x})} \right\} e^{2\pi i \mathbf{y} \cdot \mathbf{x}} \, d\rho. \tag{30.12}$$

The expression given in the curly brackets is a periodic function of $\mathbf{x}$ with the periodicity of the lattice; for if we replace $\mathbf{x}$ by $\mathbf{x} + \mathbf{x}(l)$, we can restore the function to its original form simply by introducing $l - \bar{l}$ (i.e. $l^1 - \bar{l}^1$, $l^2 - \bar{l}^2$, $l^3 - \bar{l}^3$) instead of l as the summation index. The expression may thus be represented by a Fourier series. According to (22.22), the Fourier coefficients are given by

$$g(h_1, h_2, h_3) = \frac{1}{v_a} \int_{\text{zero-cell}} \sum_l \left\{ \frac{2}{\sqrt{\pi}} e^{-|\mathbf{x}(l)-\mathbf{x}|^2 \rho^2 + 2\pi i \mathbf{y} \cdot (\mathbf{x}(l)-\mathbf{x})} \right\} e^{-2\pi i \mathbf{y}(h) \cdot \mathbf{x}} \, d\mathbf{x}, \tag{30.13}$$

where the integration could be over any arbitrary lattice cell, but we have chosen for definiteness the zero-cell. After interchanging the order of integration and summation we may introduce, for a particular term l, $\mathbf{x}' = \mathbf{x} - \mathbf{x}(l)$ as the integration variable, the range of integration for $\mathbf{x}'$ being evidently over the cell $-l$. Hence we can write (30.13) as

$$g(h_1, h_2, h_3) = \frac{1}{v_a} \sum_l \int_{-l\text{-cell}} \left\{ \frac{2}{\sqrt{\pi}} e^{-|\mathbf{x}'|^2 \rho^2 - 2\pi i (\mathbf{y}(h)+\mathbf{y}) \cdot \mathbf{x}'} \right\} d\mathbf{x}', \tag{30.14}$$

where we have used the fact that $\exp\{-2\pi i \mathbf{y}(h) \cdot \mathbf{x}(l)\}$ is equal to unity (§ 22). The sum in (30.14) is evidently equivalent to an integration over all space, and the resulting integral is easily evaluated, giving

$$g(h_1, h_2, h_3) = \frac{1}{v_a} \int_{\text{all space}} \left\{ \frac{2}{\sqrt{\pi}} e^{-|\mathbf{x}|^2 \rho^2 - 2\pi i (\mathbf{y}(h)+\mathbf{y}) \cdot \mathbf{x}} \right\} d\mathbf{x}$$

$$= \frac{2\pi}{v_a} \frac{1}{\rho^3} \exp\left\{ -\frac{\pi^2}{\rho^2} |\mathbf{y}(h)+\mathbf{y}|^2 \right\}. \tag{30.15}$$

The Fourier expansion of the periodic function given in the curly brackets of (30.12) can thus be written down explicitly as

$$\frac{2}{\sqrt{\pi}} \sum_{l} e^{-|\mathbf{x}(l)-\mathbf{x}|^2\rho^2+2\pi i\mathbf{y}\cdot(\mathbf{x}(l)-\mathbf{x})} = \sum_{h} g(h_1,h_2,h_3)e^{2\pi i\mathbf{y}(h)\cdot\mathbf{x}}$$

$$= \frac{2\pi}{v_a}\sum_{h}\frac{1}{\rho^3}e^{-(\pi^2/\rho^2)|\mathbf{y}(h)+\mathbf{y}|^2+2\pi i\mathbf{y}(h)\cdot\mathbf{x}}. \tag{30.16}$$

(30.16) is known as the theta-function transformation. The appearance of this important transformation in crystal physics we owe entirely to the work of Ewald. It is clearly evident that the series on the two sides of (30.16) are rapidly convergent for large and small values of ρ respectively. If we divide the integral in (30.12) into two parts and use the two alternative expressions offered by (30.16) for the respective integrands as follows:

$$\sum_{\beta}p_\beta\frac{\partial^2}{\partial x_\alpha \partial x_\beta}\left\{\frac{2}{\sqrt{\pi}}\sum_{l}\int_{R}^{\infty} e^{-|\mathbf{x}(l)-\mathbf{x}|^2\rho^2+2\pi i\mathbf{y}\cdot\mathbf{x}(l)}\,d\rho+\right.$$

$$\left.+\frac{2\pi}{v_a}\sum_{h}\int_{0}^{R} e^{-(\pi^2/\rho^2)|\mathbf{y}(h)+\mathbf{y}|^2+2\pi i(\mathbf{y}(h)+\mathbf{y})\cdot\mathbf{x}}\frac{1}{\rho^3}d\rho\right\}, \tag{30.17}$$

we can, by proper choice of the dividing-point R, secure fast convergence for both series in (30.17). This procedure provides a powerful method for calculating lattice vibrations in ionic crystals[†] and certain lattice sums in general.

However, Ewald's method is of even greater theoretical significance in providing a way of separating the macroscopic field from the actual Coulomb field. Introducing for simplicity the functions

$$G(x) = \frac{e^{-x}}{x}, \qquad H(x) = \frac{2}{\sqrt{\pi}}\frac{1}{x}\int_{x}^{\infty} e^{-x^2}\,dx, \tag{30.18}$$

we may rewrite the Coulomb field (30.17) as

$$\sum_{\beta}p_\beta\frac{\partial^2}{\partial x_\alpha \partial x_\beta}\left\{\frac{1}{\pi v_a|\mathbf{y}|^2}\exp\left[-\frac{\pi^2|\mathbf{y}|^2}{R^2}+2\pi i\mathbf{y}\cdot\mathbf{x}\right]+\right.$$

$$+R\sum_{l} H(R|\mathbf{x}(l)-\mathbf{x}|)\exp[2\pi i\mathbf{y}\cdot\mathbf{x}(l)]+$$

$$\left.+\frac{\pi}{v_a}\frac{1}{R^2}\sum_{h}' G(\pi^2|\mathbf{y}(h)+\mathbf{y}|^2/R^2)\exp[2\pi i(\mathbf{y}(h)+\mathbf{y})\cdot\mathbf{x}]\right\}, \tag{30.19}$$

where the term $h = 0$ in (30.17) is written separately, and the prime over the summation of h indicates that the term $h = 0$ is now to be omitted.

† See, for instance, E. W. Kellermann, *Phil. Trans. Roy. Soc. A*, **238**, 513 (No. 798) (1940).

After the differentiation is carried out, the field becomes

$$\sum_{\beta} p_{\beta}\left\{-\frac{4\pi}{v_a}\frac{y_{\alpha}y_{\beta}}{|\mathbf{y}|^2}\exp\left[-\frac{\pi^2|\mathbf{y}|^2}{R^2}+2\pi i\mathbf{y}.\mathbf{x}\right]+\right.$$

$$+R^3\sum_{l}H_{\alpha\beta}(R(\mathbf{x}(l)-\mathbf{x}))\exp[2\pi i\mathbf{y}.\mathbf{x}(l)]-$$

$$-\frac{4\pi^3}{R^2v_a}\sideset{}{'}\sum_{h}(y_{\alpha}(h)+y_{\alpha})(y_{\beta}(h)+y_{\beta})\times$$

$$\left.\times G(\pi^2|\mathbf{y}(h)+\mathbf{y}|^2/R^2)\exp[2\pi i(\mathbf{y}(h)+\mathbf{y}).\mathbf{x}]\right\}, \quad (30.20)$$

where
$$H_{\alpha\beta}(\mathbf{x}) = \frac{\partial^2}{\partial x_{\alpha}\,\partial x_{\beta}}H(|\mathbf{x}|). \quad (30.21)$$

Let us write the first term in (30.20) in two parts as follows:

$$-\frac{4\pi}{v_a}\frac{y_{\alpha}}{|\mathbf{y}|}\left(\frac{\mathbf{y}.\mathbf{p}}{|\mathbf{y}|}\right)e^{2\pi i\mathbf{y}.\mathbf{x}}+\frac{4\pi}{v_a}\sum_{\beta}\frac{y_{\alpha}y_{\beta}p_{\beta}}{|\mathbf{y}|^2}\{1-e^{-\pi^2|\mathbf{y}|^2/R^2}\}e^{2\pi i\mathbf{y}.\mathbf{x}}, \quad (30.22)$$

where the first part is seen to be identical with the macroscopic electric field given by (30.7). Hence we can write the Coulomb field (30.20) in the form

$$E_{\alpha}e^{2\pi i\mathbf{y}.\mathbf{x}}+\sum_{\beta}p_{\beta}\left\{\frac{4\pi}{v_a}\frac{y_{\alpha}y_{\beta}}{|\mathbf{y}|^2}[1-e^{-\pi^2|\mathbf{y}|^2/R^2}]e^{2\pi i\mathbf{y}.\mathbf{x}}+\right.$$

$$+R^3\sum_{l}H_{\alpha\beta}(R(\mathbf{x}(l)-\mathbf{x}))\exp[2\pi i\mathbf{y}.\mathbf{x}(l)]-$$

$$-\frac{4\pi^3}{R^2v_a}\sideset{}{'}\sum_{h}(y_{\alpha}(h)+y_{\alpha})(y_{\beta}(h)+y_{\beta})\times$$

$$\left.\times G(\pi^2|\mathbf{y}(h)+\mathbf{y}|^2/R^2)\exp[2\pi i(\mathbf{y}(h)+\mathbf{y}).\mathbf{x}]\right\}, \quad (30.23)$$

where $\mathbf{E}$ is given by (30.8). The first term of (30.23) represents the macroscopic field and the rest the inner field.

From a purely mathematical point of view, the significance of separating the macroscopic field is the following. Both sums (over l and h) in (30.20), we note, are regular functions of $\mathbf{y}$ as $\mathbf{y}$ approaches zero; as series, they converge rapidly, and every term in these sums has a perfectly defined limit for $\mathbf{y}=0$. The first term of (30.20), on the other hand, is not a regular function of $\mathbf{y}$ at $\mathbf{y}=0$ owing to the appearance of $|\mathbf{y}|^2$ in the denominator. The factor $y_{\alpha}y_{\beta}/|\mathbf{y}|^2$, for instance, has no unique limit; its limiting value depends entirely on the direction by which the point $\mathbf{y}=0$ is approached. In subdividing this term in (30.22), we have separated off a part which is easily seen to be a regular function of $\mathbf{y}$, for the factor $\{1-\exp[-\pi^2|\mathbf{y}|^2/R^2]\}$ varies as $|\mathbf{y}|^2$ for small values of $\mathbf{y}$, and thus cancels $|\mathbf{y}|^2$ in the denominator of the term. From this point of view, in the final expression (30.23) for the field, the first term, which represents the macroscopic field, absorbs all that is not regular at $\mathbf{y} = 0$; the inner field represented by the rest of the terms is, on the other hand, a regular function of $\mathbf{y}$ and possesses an unambiguous limit at the point $\mathbf{y} = 0$.

The above result is readily generalized to the case of a composite lattice, where a lattice point $\binom{l}{k}$ is occupied by a dipole

$$\mathbf{p}\binom{l}{k'} = \mathbf{p}(k')e^{2\pi i\mathbf{y}\cdot\mathbf{x}\binom{l}{k'}}; \qquad (30.24)$$

for, as far as the field due to a particular component Bravais lattice k' is concerned, the problem is the same as before except that, for this field, the amplitude $\mathbf{p}$ (cf. (30.1)) is now given by $\mathbf{p}(k')\exp\{2\pi i\mathbf{y}.\mathbf{x}(k')\}$ and, moreover, as the origin of the component lattice is at $\mathbf{x}(k')$, to use the above results we have to replace the argument $\mathbf{x}$ by $\mathbf{x}-\mathbf{x}(k')$. In this connexion, we should remember that $\mathbf{E}$ in (30.23) is determined by the amplitude of the dipole as in (30.8). After making the above replacements in (30.23) and summing over different component lattices, we obtain for the Coulomb field in the general case the expression

$$E_\alpha e^{2\pi i\mathbf{y}\cdot\mathbf{x}} + \sum_{k'}\sum_\beta p_\beta(k')\bigg\{\frac{4\pi}{v_a}\frac{y_\alpha y_\beta}{|\mathbf{y}|^2}[1-e^{-\pi^2|\mathbf{y}|^2/R^2}]e^{2\pi i\mathbf{y}\cdot\mathbf{x}} +$$

$$+ R^3\sum_{l'}H_{\alpha\beta}\Big(R\big(\mathbf{x}\binom{l'}{k'}-\mathbf{x}\big)\Big)\exp[2\pi i\mathbf{y}\cdot\mathbf{x}\binom{l'}{k'}] -$$

$$- \frac{4\pi^3}{v_a}\frac{e^{2\pi i\mathbf{y}\cdot\mathbf{x}}}{R^2}\sum_h{}'(y_\alpha(h)+y_\alpha)(y_\beta(h)+y_\beta)\times$$

$$\times G(\pi^2|\mathbf{y}(h)+\mathbf{y}|^2/R^2)\exp[2\pi i\mathbf{y}(h)\cdot(\mathbf{x}-\mathbf{x}(k'))]\bigg\}, \quad (30.25)$$

where the amplitude of the macroscopic field is now given by

$$E_\alpha = -\frac{4\pi}{v_a}\bigg(\frac{y_\alpha}{|\mathbf{y}|}\bigg)\bigg\{\bigg(\frac{\mathbf{y}}{|\mathbf{y}|}\bigg)\cdot\sum_{k'}\mathbf{p}(k')\bigg\}. \qquad (30.26)$$

The function $H_{\alpha\beta}(\mathbf{x})$ is singular at $\mathbf{x} = 0$. Hence the Coulomb field (30.25) diverges at the lattice points. Take, for instance, the field at $\binom{0}{k}$; the term $l = 0$, $k' = k$ in (30.25) is divergent. This is clearly due to the contribution from the dipole situated at $\binom{0}{k}$. What we shall actually require for later purposes is the field at a lattice point such as $\binom{0}{k}$ due to all other dipoles; this field has been called the *exciting field* (erregendes Feld) by Ewald. To obtain the exciting field at $\binom{0}{k}$, we subtract the field due to the dipole at $\binom{0}{k}$:

$$\sum_\beta p_\beta(k)e^{2\pi i\mathbf{y}\cdot\mathbf{x}(k)}\frac{\partial^2}{\partial x_\alpha\,\partial x_\beta}\frac{1}{|\mathbf{x}(k)-\mathbf{x}|} \qquad (30.27)$$

from the term $k' = k$, $l = 0$ in (30.25), obtaining

$$\sum_\beta p_\beta(k)e^{2\pi i y \cdot x(k)}\frac{\partial^2}{\partial x_\alpha \partial x_\beta}\left\{RH\big(R|x(k)-x|\big)-\frac{1}{|x(k)-x|}\right\}$$

$$=\sum_\beta p_\beta(k)e^{2\pi i y \cdot x(k)}\frac{\partial^2}{\partial x_\alpha \partial x_\beta}\left\{\frac{-1}{|x(k)-x|}\frac{2}{\sqrt{\pi}}\int_0^{R|x(k)-x|}e^{-x^2}\,dx\right\}. \quad (30.28)$$

It is easily verified by expanding the integrand e^{-x^2} that the function in the curly brackets is regular at $|x(k)-x| = 0$. If we introduce the function

$$H^0(x) = \frac{-2}{x\sqrt{\pi}}\int_0^x e^{-x^2}\,dx, \quad (30.29)$$

the net effect of subtracting the contribution of the dipole at $\binom{0}{k}$ is equivalent to replacing the function $H_{\alpha\beta}(x)$ by $H^0_{\alpha\beta}(x)$ in the term $l = 0$, $k' = k$ in (30.25). After making this replacement and putting in the coordinates of $\binom{0}{k}$, namely $x(k)$, for x, we obtain for the exciting field at $\binom{0}{k}$ the expression

$$E_\alpha e^{2\pi i y \cdot x(k)}+e^{2\pi i y \cdot x(k)}\sum_{k'\beta}Q_{\alpha\beta}(\tfrac{y}{kk'})p_\beta(k'), \quad (30.30)$$

where, for later convenience, we have introduced the symbol

$$Q_{\alpha\beta}(\tfrac{y}{kk'}) = \frac{4\pi}{v_a}\left(\frac{y_\alpha y_\beta}{|y|^2}\right)\{1-e^{-\pi^2|y|^2/R^2}\}+R^3\sum_{l'}H_{\alpha\beta}(Rx(\tfrac{l'}{k'k}))\exp[2\pi i y . x(\tfrac{l'}{k'k})]-$$

$$-\frac{4\pi^3}{R^2v_a}\sum_h{}'(y_\alpha(h)+y_\alpha)(y_\beta(h)+y_\beta)\times$$

$$\times G(\pi^2|y(h)+y|^2/R^2)\exp[2\pi i y(h).\big(x(k)-x(k')\big)]. \quad (30.31)$$

It is to be understood that, for the case $k' = k$, $H_{\alpha\beta}(x)$ has to be replaced by $H^0_{\alpha\beta}(x)$ in the term $l = 0$.

It may be remarked that the above expression for the field is perfectly rigorous whatever the value of y, although only for small values of y has the first term in (30.30) a simple macroscopic significance.

The coefficients $Q_{\alpha\beta}(\tfrac{y}{kk'})$ satisfy the relations

$$Q_{\alpha\beta}(\tfrac{y}{kk'}) = Q_{\beta\alpha}(\tfrac{y}{kk'}), \quad (30.32)$$

$$Q_{\alpha\beta}(\tfrac{\bar y}{kk'}) = Q^*_{\alpha\beta}(\tfrac{y}{kk'}), \quad (30.33)$$

$$Q_{\alpha\beta}(\tfrac{y}{k'k}) = Q^*_{\alpha\beta}(\tfrac{y}{kk'}). \quad (30.34)$$

The relation (30.32) is immediately evident from (30.31). To prove (30.33) we replace y by $-y$ in (30.31); the result is readily seen to be equal to $Q^*_{\alpha\beta}(\tfrac{y}{kk'})$, if we introduce $h' = -h$ as the summation index in

the last term. The relation (30.34) can be proved with the help of the fact that
$$H_{\alpha\beta}(-\mathbf{x}) = H_{\alpha\beta}(\mathbf{x}). \tag{30.35}$$

For the proof, we need evidently consider only the term
$$R^3 \sum_{l'} H_{\alpha\beta}(R\mathbf{x}(_{k'k}^{l'}))\exp\{2\pi i \mathbf{y}\cdot\mathbf{x}(_{k'k}^{l'})\} \tag{30.36}$$

in $Q_{\alpha\beta}(_{kk'}^{\mathbf{y}})$. Without affecting the value of (30.36), we can write l' as $-l'$. Hence on interchanging k and k', we can write the result as
$$R^3 \sum_{l'} H_{\alpha\beta}(R\mathbf{x}(_{kk'}^{-l'}))\exp\{2\pi i \mathbf{y}\cdot\mathbf{x}(_{kk'}^{-l'})\}. \tag{30.37}$$

Since $\mathbf{x}(_{kk'}^{-l'}) = -\mathbf{x}(_{k'k}^{l'})$, it follows directly from (30.35) that (30.37) is equal to the complex conjugate of (30.36).

31. Acoustic vibrations in ionic lattices (rigid ion model)†

In this section we shall assume the ions to be spherical and rigid (not polarizable or deformable), so that as regards their Coulomb interaction the ions are equivalent to point charges. Let us write
$$\Phi_{\alpha\beta}(_{kk'}^{l}) = \Phi_{\alpha\beta}^{C}(_{kk'}^{l}) + \Phi_{\alpha\beta}^{N}(_{kk'}^{l}), \tag{31.1}$$

where the two terms represent respectively the Coulomb and non-Coulomb contributions. Similarly we can divide the coefficients $C_{\alpha\beta}(_{kk'}^{\mathbf{y}})$ into their Coulomb and non-Coulomb parts; the latter we shall denote by $C_{\alpha\beta}^{N}(_{kk'}^{\mathbf{y}})$. In the Coulomb part of $C_{\alpha\beta}(_{kk'}^{\mathbf{y}})$, namely,
$$\frac{1}{(m_k m_{k'})^{\frac{1}{2}}} \sum_{l'} \Phi_{\alpha\beta}^{C}(_{kk'}^{l\,-l'})\exp\{-2\pi i \mathbf{y}\cdot(\mathbf{x}(_{k}^{l}) - \mathbf{x}(_{k'}^{l'}))\}, \tag{31.2}$$

we can evidently put $l = 0$, obtaining thus the equivalent expression
$$\frac{e^{-2\pi i \mathbf{y}\cdot\mathbf{x}(k)}}{(m_k m_{k'})^{\frac{1}{2}}} \sum_{l'} \Phi_{\alpha\beta}^{C}(_{kk'}^{-l'})\exp\{2\pi i \mathbf{y}\cdot\mathbf{x}(_{k'}^{l'})\}. \tag{31.3}$$

Using this expression, we write the lattice wave equation (24.10) in the form
$$\omega^2(_j^{\mathbf{y}})w_\alpha(k|_j^{\mathbf{y}}) = \sum_{k'\beta} C_{\alpha\beta}^{N}(_{kk'}^{\mathbf{y}})w_\beta(k'|_j^{\mathbf{y}}) + \frac{1}{\sqrt{m_k}}\sum_{\beta}\Phi_{\alpha\beta}^{C}(_{kk}^{0})\frac{1}{\sqrt{m_k}}w_\beta(k|_j^{\mathbf{y}}) +$$
$$+ \frac{e^{-2\pi i \mathbf{y}\cdot\mathbf{x}(k)}}{\sqrt{m_k}}\sum_{\beta}\sideset{}{'}\sum_{l'k'}\Phi_{\alpha\beta}^{C}(_{kk'}^{-l'})\frac{1}{\sqrt{m_{k'}}}w_\beta(k'|_j^{\mathbf{y}})e^{2\pi i \mathbf{y}\cdot\mathbf{x}(_{k'}^{l'})}. \tag{31.4}$$

In view of the basic difference between $\Phi_{\alpha\beta}^{C}(_{kk}^{0})$ and the other coefficients $\Phi_{\alpha\beta}^{C}(_{kk'}^{-l'})$ (see the explicit expressions given previously in (27.28) and (27.29)), terms involving the former are written separately in (31.4),

† K. Huang, *Phil. Mag.* 40, 733 (1949).

where the prime over the summation in the last term indicates that the terms with $l' = 0$, $k' = k$ are to be omitted.

Both of the Coulomb terms in (31.4) can be transformed with the help of the results obtained in the last section. Consider first the second term on the right-hand side of the equation. Using for $\Phi^C_{\alpha\beta}(\substack{0\\kk})$ the explicit expression (27.29), in which we now introduce $l' = -l$ as the summation index, we find that the term can be written as follows:

$$\frac{e_k}{\sqrt{m_k}} \sum_\beta \sum_{l'k'}{}' \frac{e_{k'}}{\sqrt{m_k}} w_\beta(k|\substack{\mathbf{y}\\j}) \left\{ \frac{\partial^2}{\partial x_\alpha\, \partial x_\beta} \frac{1}{|\mathbf{x}|} \right\}_{\mathbf{x}(k) - \mathbf{x}(\substack{l'\\k'})}$$

$$= \frac{e_k}{\sqrt{m_k}} \sum_\beta \sum_{l'k'}{}' \frac{e_{k'}}{\sqrt{m_k}} w_\beta(k|\substack{\mathbf{y}\\j}) \left\{ \frac{\partial^2}{\partial x_\alpha\, \partial x_\beta} \frac{1}{|\mathbf{x}(\substack{l'\\k'}) - \mathbf{x}|} \right\}_{\mathbf{x}(k)}. \qquad (31.5)$$

We observe that the sum in this expression is equal to the exciting field at $(\substack{0\\k})$ (position vector $= \mathbf{x}(k)$) in a dipole lattice where a lattice point $(\substack{l'\\k'})$ is occupied by a dipole

$$p_\beta(\substack{l'\\k'}) = \frac{e_{k'}}{\sqrt{m_k}} w_\beta(k|\substack{\mathbf{y}\\j}). \qquad (31.6)$$

This array of dipoles is a special case of (30.24), corresponding to

$$\left. \begin{array}{l} \mathbf{y} = 0 \quad \text{(not to be confused with } \mathbf{y} \text{ in (31.6), which in this} \\ \qquad\qquad \text{connexion functions as a fixed index)} \\[2mm] p_\beta(k') = \dfrac{e_{k'}}{\sqrt{m_k}} w_\beta(k|\substack{\mathbf{y}\\j}) \end{array} \right\}. \qquad (31.7)$$

Substituting (31.7) in (30.30), we obtain for the corresponding exciting field at $(\substack{0\\k})$:

$$\sum_{k'\beta}{}' Q_{\alpha\beta}(\substack{0\\kk'}) \frac{e_{k'}}{\sqrt{m_k}} w_\beta(k|\substack{\mathbf{y}\\j}), \qquad (31.8)$$

where, we note, the term describing the macroscopic field in (30.30) vanishes, for (30.26) gives in this case

$$E_\alpha \sim \sum_{k'} e_{k'} = 0$$

(net charge per cell must vanish). When the sum in (31.5) is replaced by (31.8), the term becomes

$$\sum_{k'\beta}{}' \frac{e_k e_{k'}}{m_k} Q_{\alpha\beta}(\substack{0\\kk'}) w_\beta(k|\substack{\mathbf{y}\\j}). \qquad (31.9)$$

Putting $l = -l'$ in (27.28) we can rewrite the formula thus:

$$\Phi^C_{\alpha\beta}(\substack{-l'\\kk'}) = -e_k e_{k'} \left\{ \frac{\partial^2}{\partial x_\alpha\, \partial x_\beta} \frac{1}{|\mathbf{x}|} \right\}_{\mathbf{x}(k) - \mathbf{x}(\substack{l'\\k'})} = -e_k e_{k'} \left\{ \frac{\partial^2}{\partial x_\alpha\, \partial x_\beta} \frac{1}{|\mathbf{x}(\substack{l'\\k'}) - \mathbf{x}|} \right\}_{\mathbf{x}(k)}. $$
$$(31.10)$$

When (31.10) is substituted, the last term in the equation (31.4) becomes

$$-\frac{e_k e^{-2\pi i \mathbf{y} \cdot \mathbf{x}(k)}}{\sqrt{m_k}} \sum_\beta \sideset{}{'}\sum_{l'k'} \frac{e_{k'}}{\sqrt{m_{k'}}} w_\beta(k'|\mathbf{y}_j) e^{2\pi i \mathbf{y} \cdot \mathbf{x}(\mathbf{k}'')} \left\{ \frac{\partial^2}{\partial x_\alpha \, \partial x_\beta} \frac{1}{|\mathbf{x}(\mathbf{k}'')-\mathbf{x}|} \right\}_{\mathbf{x}(k)}.$$

$$(31.11)$$

The sum in this expression is evidently equal to the exciting field at $\binom{0}{k}$ in a dipole lattice described by (30.24) with

$$p_\beta(k') = \frac{e_{k'}}{\sqrt{m_{k'}}} w_\beta(k'|\mathbf{y}_j). \tag{31.12}$$

Substituting (31.12) in (30.30), we find that the exciting field at $\binom{0}{k}$ is given by

$$E_\alpha e^{2\pi i \mathbf{y} \cdot \mathbf{x}(k)} + e^{2\pi i \mathbf{y} \cdot \mathbf{x}(k)} \sum_{k'\beta} Q_{\alpha\beta}(\mathbf{y}_{kk'}) \frac{e_{k'}}{\sqrt{m_{k'}}} w_\beta(k'|\mathbf{y}_j), \tag{31.13}$$

where (see (30.26))

$$E_\alpha = -\frac{4\pi}{v_a}\left(\frac{y_\alpha}{|\mathbf{y}|}\right) \sum_\beta \left(\frac{y_\beta}{|\mathbf{y}|}\right) \sum_{k'} \frac{e_{k'}}{\sqrt{m_{k'}}} w_\beta(k'|\mathbf{y}_j). \tag{31.14}$$

The term given in (31.11) can thus be written as

$$\frac{-e_k}{\sqrt{m_k}} E_\alpha - \sum_{k'\beta} \frac{e_k e_{k'}}{(m_k m_{k'})^{\frac{1}{2}}} Q_{\alpha\beta}(\mathbf{y}_{kk'}) w_\beta(k'|\mathbf{y}_j). \tag{31.15}$$

When the Coulomb terms are replaced by their respective equivalents (31.9) and (31.15), the equation (31.4) becomes

$$\omega^2(\mathbf{y}_j) w_\alpha(k|\mathbf{y}_j) = \sum_{k'\beta} C^N_{\alpha\beta}(\mathbf{y}_{kk'}) w_\beta(k'|\mathbf{y}_j) + \sum_{k'\beta} \frac{e_k e_{k'}}{m_k} Q_{\alpha\beta}(\mathbf{0}_{kk'}) w_\beta(k|\mathbf{y}_j) -$$

$$-\frac{e_k}{\sqrt{m_k}} E_\alpha - \sum_{k'\beta} \frac{e_k e_{k'}}{(m_k m_{k'})^{\frac{1}{2}}} Q_{\alpha\beta}(\mathbf{y}_{kk'}) w_\beta(k'|\mathbf{y}_j). \tag{31.16}$$

The reason behind the close connexion between the lattice wave equation and the Coulomb field in dipole lattices is fairly evident. In the harmonic approximation the Coulomb force on a particle $\binom{0}{k}$ is the sum of two contributions: one due to the field change experienced by $\binom{0}{k}$ owing to its own shifted position $u_\beta(\binom{0}{k}) = (m_k)^{-\frac{1}{2}} w_\beta(k|\mathbf{y}_j)\exp\{2\pi i \mathbf{y} \cdot \mathbf{x}(k)\}$, the other due to the field created by the displacements

$$u_\beta(\mathbf{k}'') = (m_{k'})^{-\frac{1}{2}} w_\beta(k'|\mathbf{y}_j)\exp\{2\pi i \mathbf{y} \cdot \mathbf{x}(\mathbf{k}'')\}$$

of the other ions. These contributions are described respectively by the second and third terms on the right-hand side of (31.4). The field produced by the ion displacements $u_\beta(\mathbf{k}'')$ is equivalent to that of dipoles $e_{k'} u_\beta(\mathbf{k}'')$, so we find that the last term in (31.4) can be expressed in terms of the exciting field at $\mathbf{x}(k)$ in the dipole lattice (31.12). The field change experienced by $\binom{0}{k}$ due to its own displacement $\mathbf{u}(\binom{0}{k})$ is clearly equal to the field created at $\mathbf{x}(k)$ by the other ions, if the latter were all displaced by $-\mathbf{u}(^{0}_{k})$. Thus we find that the second term on the right-hand side of (31.4) can be expressed in terms of the exciting field at $\mathbf{x}(k)$ in the dipole lattice (31.7).

On substituting (31.14) in (31.16) and comparing the resulting equation with (24.10), we find that in this case

$$C_{\alpha\beta}(^{\mathbf{y}}_{kk'}) = C^{N}_{\alpha\beta}(^{\mathbf{y}}_{kk'}) + \frac{4\pi}{v_a}\left(\frac{y_\alpha y_\beta}{|\mathbf{y}|^2}\right)\frac{e_k e_{k'}}{(m_k m_{k'})^{\frac{1}{2}}} + \delta_{kk'}\frac{e_k}{m_k}\sum_{k''} e_{k''} Q_{\alpha\beta}(^{0}_{kk''}) - $$
$$- \frac{e_k e_{k'}}{(m_k m_{k'})^{\frac{1}{2}}} Q_{\alpha\beta}(^{\mathbf{y}}_{kk'}). \quad (31.17)$$

The perturbation method developed in § 26 breaks down in this case; owing to the second term on the right-hand side of (31.17), it is no longer permissible to assume an expansion of $C_{\alpha\beta}(^{\mathbf{y}}_{kk'})$ of the form (26.2), where the zero-order term is given as independent of $\mathbf{y}$. To avoid this difficulty, we leave the above singular term explicitly in the lattice-wave equation and introduce instead of $C_{\alpha\beta}(^{\mathbf{y}}_{kk'})$ the modified coefficients

$$\bar{C}_{\alpha\beta}(^{\mathbf{y}}_{kk'}) = C^{N}_{\alpha\beta}(^{\mathbf{y}}_{kk'}) + \delta_{kk'}\frac{e_k}{m_k}\sum_{k''} e_{k''} Q_{\alpha\beta}(^{0}_{kk''}) - \frac{e_k e_{k'}}{(m_k m_{k'})^{\frac{1}{2}}} Q_{\alpha\beta}(^{\mathbf{y}}_{kk'}). \quad (31.18)$$

Thus we write the lattice-wave equation (31.16) in the form

$$\omega^2(^{\mathbf{y}}_{j})w_\alpha(k|^{\mathbf{y}}_{j}) = \sum_{k'\beta} \bar{C}_{\alpha\beta}(^{\mathbf{y}}_{kk'})w_\beta(k'|^{\mathbf{y}}_{j}) - \frac{e_k}{\sqrt{m_k}} E_\alpha. \quad (31.19)$$

Replacing $\mathbf{y}$ by $\epsilon\mathbf{y}$ in (31.18) and expanding with respect to ϵ. we have

$$\bar{C}_{\alpha\beta}(^{\epsilon\mathbf{y}}_{kk'}) = \bar{C}^{(0)}_{\alpha\beta}(kk') + i\epsilon \sum_{\gamma} \bar{C}^{(1)}_{\alpha\beta,\gamma}(kk')y_\gamma + \tfrac{1}{2}\epsilon^2 \sum_{\gamma\lambda} \bar{C}^{(2)}_{\alpha\beta,\gamma\lambda}(kk')y_\gamma y_\lambda + \dots. \quad (31.20)$$

The expansion coefficients can be obtained with the help of the explicit expression (30.31) for $Q_{\alpha\beta}(^{\mathbf{y}}_{kk'})$; their exact expressions are not required for the following treatment, but are given below for convenience of reference:

$$\bar{C}^{(0)}_{\alpha\beta}(kk') = \frac{1}{(m_k m_{k'})^{\frac{1}{2}}}\sum_{l} \Phi^{N}_{\alpha\beta}(^{l}_{kk'}) + \delta_{kk'}\frac{e_k}{m_k}\Bigg\{ R^3 \sum_{l''k''} e_{k''} H_{\alpha\beta}\big(R\mathbf{x}(^{l''}_{k''k})\big) - $$
$$- \frac{4\pi^3}{R^2 v_a}\sum_{k''} e_{k''} {\sum_{h}}' y_\alpha(h)y_\beta(h)G\big(\pi^2|\mathbf{y}(h)|^2/R^2\big) \times $$
$$\times \exp\{2\pi i\mathbf{y}(h).(\mathbf{x}(k)-\mathbf{x}(k''))\}\Bigg\} - $$
$$- \frac{e_k e_{k'}}{(m_k m_{k'})^{\frac{1}{2}}}\Bigg\{ R^3 \sum_{l} H_{\alpha\beta}\big(R\mathbf{x}(^{l}_{kk'})\big) - $$
$$- \frac{4\pi^3}{R^2 v_a}{\sum_{h}}' y_\alpha(h)y_\beta(h)G\big(\pi^2|\mathbf{y}(h)|^2/R^2\big)\exp\{2\pi i\mathbf{y}(h).(\mathbf{x}(k)-\mathbf{x}(k'))\}\Bigg\}, $$
$$(31.21)$$

$$\bar{C}^{(1)}_{\alpha\beta,\gamma}(kk') = \frac{-2\pi}{(m_k m_{k'})^{\frac{1}{2}}} \sum_l \Phi^N_{\alpha\beta}(\substack{l \\ kk'})x_\gamma(\substack{l \\ kk'}) -$$

$$-\frac{2\pi e_k e_{k'} R^3}{(m_k m_{k'})^{\frac{1}{2}}} \sum_{l'} H_{\alpha\beta}(R\mathbf{x}(\substack{l' \\ k'k}))x_\gamma(\substack{l' \\ k'k}) -$$

$$-\frac{4\pi^3 i e_k e_{k'}}{R^2 v_a(m_k m_{k'})^{\frac{1}{2}}} \sideset{}{'}\sum_h \Big\{(y_\alpha(h)\delta_{\beta\gamma}+y_\beta(h)\delta_{\alpha\gamma})G(\pi^2|\mathbf{y}(h)|^2/R^2)+$$

$$+\frac{2\pi^2}{R^2}y_\alpha(h)y_\beta(h)y_\gamma(h)G'(\pi^2|\mathbf{y}(h)|^2/R^2)\Big\}\exp\{2\pi i\mathbf{y}(h).[\mathbf{x}(k)-\mathbf{x}(k')]\};$$

$$\text{(31.22)}$$

$$\bar{C}^{(2)}_{\alpha\beta,\gamma\lambda}(kk') = \frac{-4\pi^2}{(m_k m_{k'})^{\frac{1}{2}}} \sum_l \Phi^N_{\alpha\beta}(\substack{l \\ kk'})x_\gamma(\substack{l \\ kk'})x_\lambda(\substack{l \\ kk'}) -$$

$$-\frac{4\pi^3 e_k e_{k'}}{R^2 v_a(m_k m_{k'})^{\frac{1}{2}}} (\delta_{\alpha\gamma}\delta_{\beta\lambda}+\delta_{\alpha\lambda}\delta_{\beta\gamma}) +$$

$$+\frac{4\pi^2 R^3 e_k e_{k'}}{(m_k m_{k'})^{\frac{1}{2}}} \sum_{l'} H_{\alpha\beta}(R\mathbf{x}(\substack{l' \\ k'k}))x_\gamma(\substack{l' \\ k'k})x_\lambda(\substack{l' \\ k'k}) +$$

$$+\frac{4\pi^3 e_k e_{k'}}{R^2 v_a(m_k m_{k'})^{\frac{1}{2}}} \sideset{}{'}\sum_h \Big\{(\delta_{\alpha\gamma}\delta_{\beta\lambda}+\delta_{\alpha\lambda}\delta_{\beta\gamma})G(\pi^2|\mathbf{y}(h)|^2/R^2)+$$

$$+\frac{4\pi^4}{R^4}y_\alpha(h)y_\beta(h)y_\gamma(h)y_\lambda(h)G''(\pi^2|\mathbf{y}(h)|^2/R^2)+$$

$$+\frac{2\pi^2}{R^2}\{y_\alpha(h)y_\beta(h)\delta_{\gamma\lambda}+y_\alpha(h)y_\gamma(h)\delta_{\beta\lambda}+y_\alpha(h)y_\lambda(h)\delta_{\beta\gamma}+y_\beta(h)y_\gamma(h)\delta_{\alpha\lambda}+$$

$$+y_\gamma(h)y_\lambda(h)\delta_{\alpha\beta}+y_\beta(h)y_\lambda(h)\delta_{\alpha\gamma}\}G'(\pi^2|\mathbf{y}(h)|^2/R^2)\Big\} \times$$

$$\times \exp\{2\pi i\mathbf{y}(h).(\mathbf{x}(k)-\mathbf{x}(k'))\}. \quad \text{(31.23)}$$

The above expansion coefficients are very similar to the corresponding coefficients considered in § 26; thus we notice the following:

With the help of (30.32), (30.33), (30.34), we find that

$$\bar{C}_{\alpha\beta}(\substack{-y \\ kk'}) = \bar{C}^*_{\alpha\beta}(\substack{y \\ kk'}), \quad \text{(31.24)}$$

and
$$\bar{C}_{\alpha\beta}(\substack{y \\ kk'}) = \bar{C}^*_{\beta\alpha}(\substack{y \\ k'k}). \quad \text{(31.25)}$$

(31.24) with (31.20) shows immediately that the expansion coefficients $\bar{C}^{(0)}_{\alpha\beta}(kk')$, $\bar{C}^{(1)}_{\alpha\beta,\gamma}(kk')$, and $\bar{C}^{(2)}_{\alpha\beta,\gamma\lambda}(kk')$ are real quantities.

In view of the above, it follows from (31.25) and (31.20) that the symmetry relations given in (26.3), (26.4), and (26.5) are valid also for $\bar{C}^{(0)}_{\alpha\beta}(kk')$, $\bar{C}^{(1)}_{\alpha\beta,\gamma}(kk')$, and $\bar{C}^{(2)}_{\alpha\beta,\gamma\lambda}(kk')$ respectively.

Multiplying (31.17) by $\sqrt{m_{k'}}$ and summing over k', we have

$$\sum_{k'} \sqrt{m_{k'}}\, C_{\alpha\beta}(\substack{y\\kk'}) = \sum_{k'} \sqrt{m_{k'}}\, \bar{C}_{\alpha\beta}(\substack{y\\kk'}) + \frac{4\pi}{v_a}\left(\frac{y_\alpha y_\beta}{|\mathbf{y}|^2}\right)\frac{e_k}{\sqrt{m_k}}\sum_{k'} e_{k'}$$

$$= \sum_{k'} \sqrt{m_{k'}}\, \bar{C}_{\alpha\beta}(\substack{y\\kk'}), \qquad (31.26)$$

where we have used the fact that the net charge per cell vanishes. It follows from this relation that the identities (26.6), (26.7), and (26.8) are also valid for $\bar{C}^{(0)}_{\alpha\beta}(kk')$ and $\bar{C}^{(1)}_{\alpha\beta,\gamma}(kk')$.

To solve the equation (31.19) we replace $\mathbf{y}$ by $\epsilon\mathbf{y}$ in the equation; for $\omega(\substack{y\\j})$ and $w_\alpha(k|\substack{y\\j})$ we use the same expansions as in § 26, namely (26.9) and (26.10). The amplitude E_α of the macroscopic field, we remember, is given by (31.14). On substituting (26.10) in (31.14) we obtain the following expansion for the macroscopic field:

$$E_\alpha = E^{(0)}_\alpha + i\epsilon E^{(1)}_\alpha + \tfrac{1}{2}\epsilon^2 E^{(2)}_\alpha + ..., \qquad (31.27)$$

where

$$E^{(i)}_\alpha = -\frac{4\pi}{v_a}\left(\frac{y_\alpha}{|\mathbf{y}|}\right)\sum_\beta \left(\frac{y_\beta}{|\mathbf{y}|}\right)\sum_{k'}\frac{e_{k'}}{\sqrt{m_{k'}}}w^{(i)}_\beta(k'|\substack{y\\j}) \quad (i = 0, 1, 2,...). \qquad (31.28)$$

Using the expansions (31.20), (26.9), (26.10), and (31.27) in the lattice-wave equation (31.19), we obtain the following perturbation equations of various orders:

$$\sum_{k'\beta} \bar{C}^{(0)}_{\alpha\beta}(kk')w^{(0)}_\beta(k'|\substack{y\\j}) = \frac{e_k}{\sqrt{m_k}}E^{(0)}_\alpha, \qquad (31.29)$$

$$\sum_{k'\beta} \bar{C}^{(0)}_{\alpha\beta}(kk')w^{(1)}_\beta(k'|\substack{y\\j}) = -\sum_{k'\beta\gamma} \bar{C}^{(1)}_{\alpha\beta,\gamma}(kk')y_\gamma w^{(0)}_\beta(k'|\substack{y\\j}) + \frac{e_k}{\sqrt{m_k}}E^{(1)}_\alpha, \qquad (31.30)$$

$$\sum_{k'\beta} \bar{C}^{(0)}_{\alpha\beta}(kk')w^{(2)}_\beta(k'|\substack{y\\j}) = -\sum_{k'\beta\gamma\lambda} \bar{C}^{(2)}_{\alpha\beta,\gamma\lambda}(kk')y_\gamma y_\lambda w^{(0)}_\beta(k'|\substack{y\\j}) +$$

$$+2\sum_{k'\beta\gamma} \bar{C}^{(1)}_{\alpha\beta,\gamma}(kk')y_\gamma w^{(1)}_\beta(k'|\substack{y\\j}) + \frac{e_k}{\sqrt{m_k}}E^{(2)}_\alpha + 2[\omega^{(1)}(\substack{y\\j})]^2 w^{(0)}_\alpha(k|\substack{y\\j}). \qquad (31.31)$$

The non-trivial solutions of the zero-order equations (31.29) are of the same form as in § 26, namely

$$w^{(0)}_\beta(k'|\substack{y\\j}) = \sqrt{m_{k'}}\, u_\beta(j). \qquad (31.32)$$

(31.32) is a solution of (31.29) because firstly the relation (26.6) is also valid for $\bar{C}^{(0)}_{\alpha\beta}(kk')$, and secondly

$$E^{(0)}_\alpha = -\frac{4\pi}{v_a}\left(\frac{y_\alpha}{|\mathbf{y}|}\right)\left[\sum_\beta \left(\frac{y_\beta}{|\mathbf{y}|}\right)u_\beta(j)\right]\sum_{k'} e_{k'} = 0. \qquad (31.33)$$

When (31.32) is substituted in (31.30) the first-order equations become

$$\sum_{k'\beta} \bar{C}^{(0)}_{\alpha\beta}(kk')w^{(1)}_{\beta}(k'|{}^{\mathbf{y}}_{\mathbf{j}}) = -\sum_{k'\beta\gamma} \sqrt{m_{k'}}\, \bar{C}^{(1)}_{\alpha\beta,\gamma}(kk')y_{\gamma}\,u_{\beta}(j) + \frac{e_k}{\sqrt{m_k}}\, E^{(1)}_{\alpha}.$$

$$(31.34)$$

Despite the relation (31.28) it is convenient to regard $E^{(1)}_{\alpha}$ as an independent quantity so that only the explicitly appearing $w^{(1)}_{\beta}(k'|{}^{\mathbf{y}}_{\mathbf{j}})$ in the equations are considered as unknowns. The expression on the right-hand side of (31.34) is thus the inhomogeneous part. Since (26.8) is valid also for $\bar{C}^{(1)}_{\alpha\beta,\gamma}(kk')$, we find that, exactly as in § 26, the solubility condition for the first-order equations is identically satisfied, namely, that the inhomogeneous part multiplied by $\sqrt{m_k}$ and summed over k vanishes identically (the additional term depending on $E^{(1)}_{\alpha}$ does not affect the situation, for owing to the electrical neutrality of a cell, it vanishes when multiplied by $\sqrt{m_k}$ and summed over k).

Let us compare (31.34) with the corresponding equation (26.18) in the non-ionic case. Apart from the term containing $E^{(1)}_{\alpha}$ in the former, the terms in (31.34) and (26.18) exactly correspond with one another and are to be interpreted in the same way. Thus, apart from a factor $1/\sqrt{m_k}$, the first term on the right-hand side of (31.34) is the force on a particle k due to the external strain caused by the zero-order wave; the term on the left-hand side is the counter-force due to the induced internal strain, which is described in the perturbation method by the first-order wave. Taking into account the additional term depending on $E^{(1)}_{\alpha}$, we see that (31.34) describes the balance of forces in a volume element in a state of homogeneous strain (both external and internal) and subject at the same time to an electric field. The point to note is that not the whole of the Coulomb interaction contributes to the forces due to the strains, but only the part corresponding to the inner field discussed in the last section. Although $E^{(1)}_{\alpha}$ can be expressed in terms of $w^{(1)}_{\beta}(k'|{}^{\mathbf{y}}_{\mathbf{j}})$, the term containing $E^{(1)}_{\alpha}$ cannot be interpreted as a force due to the internal strain like the term on the left-hand side of the equation. For the coefficients of $w^{(1)}_{\beta}(k'|{}^{\mathbf{y}}_{\mathbf{j}})$ would then depend on $\mathbf{y}$ and render the interpretation meaningless.

We express the solution of (31.34) formally in the same way as in § 26, namely by constructing a matrix $\Gamma_{\alpha\beta}(kk')$ satisfying the relations

$$\Gamma_{\alpha\beta}(kk') = 0 \quad \text{if either or both of } k,\, k' = 0;$$

$$\sum_{k'\beta} \Gamma_{\alpha\beta}(kk')\bar{C}^{(0)}_{\beta\gamma}(k'k'') = \delta_{\alpha\gamma}\delta_{kk''} = \sum_{k'\beta} \bar{C}^{(0)}_{\alpha\beta}(kk')\Gamma_{\beta\gamma}(k'k'')$$

$$(k,\, k',\, k'' = 1,\, 2,\, 3,..., n-1). \quad (31.35)$$

Using these relations, we can write

$$w_\alpha^{(1)}(k|_j^y) = -\sum_{k'\mu} \Gamma_{\alpha\mu}(kk') \sum_{k''\beta\gamma} \sqrt{m_{k''}}\; \bar{C}_{\mu\beta,\gamma}^{(1)}(k'k'')y_\gamma\, u_\beta(j) +$$

$$+ \sum_{k'\beta} \Gamma_{\alpha\beta}(kk') \frac{e_{k'}}{\sqrt{m_{k'}}}\, E_\beta^{(1)}. \quad (31.36)$$

When the inhomogeneous part of the second-order equation (31.31) is multiplied by $\sqrt{m_k}$ and summed over k, the term containing $E_\alpha^{(2)}$ drops out owing to the fact that the net charge of a cell vanishes. When the expressions (31.32) and (31.36) for the zero- and first-order solutions are used, we find that the condition for solubility can be written as follows:

$$\left(\frac{\sum_k m_k}{v_a}\right)[\omega^{(1)}(_j^y)]^2 u_\alpha(j) = 4\pi^2 \sum_\beta \sum_{\gamma\lambda} \{[\alpha\beta,\gamma\lambda]+(\alpha\gamma,\beta\lambda)\}y_\gamma\, y_\lambda u_\beta(j) -$$

$$- 2\pi \sum_\beta \left\{\sum_\gamma [\beta,\alpha\gamma]y_\gamma\right\} E_\beta^{(1)}, \quad (31.37)$$

where the brackets are defined by

$$[\alpha\beta,\gamma\lambda] = \frac{1}{8\pi^2 v_a} \sum_{kk'} (m_k m_{k'})^{\frac{1}{2}} \bar{C}_{\alpha\beta,\gamma\lambda}^{(2)}(kk'), \quad (31.38)$$

$$(\alpha\gamma,\beta\lambda) = \frac{-1}{4\pi^2 v_a} \sum_{kk'} \sum_{\mu\nu} \Gamma_{\mu\nu}(kk')\left(\sum_{k''} \bar{C}_{\mu\alpha,\gamma}^{(1)}(kk'')\sqrt{m_{k''}}\right)\left(\sum_{k'''} \bar{C}_{\nu\beta,\lambda}^{(1)}(kk''')\sqrt{m_{k'''}}\right), \quad (31.39)$$

$$[\beta,\alpha\gamma] = \frac{1}{2\pi v_a} \sum_{kk'\mu} \sqrt{m_k}\, \bar{C}_{\alpha\mu,\gamma}^{(1)}(kk')\left(\sum_{k''} \Gamma_{\mu\beta}(k'k'')\frac{e_{k''}}{\sqrt{m_{k''}}}\right). \quad (31.40)$$

By exactly the same arguments as used in § 26, we find that

$$[\alpha\beta,\gamma\lambda] = [\beta\alpha,\gamma\lambda] = [\alpha\beta,\lambda\gamma], \quad (31.41)$$

$$(\alpha\gamma,\beta\lambda) = (\gamma\alpha,\beta\lambda) = (\beta\lambda,\alpha\gamma). \quad (31.42)$$

Moreover, it follows from (26.7) that

$$[\beta,\alpha\gamma] = [\beta,\gamma\alpha]. \quad (31.43)$$

32. The elastic and piezoelectric constants and the dielectric tensor

It follows from the discussion given in § 25 that if we do not take account of special symmetry properties explicitly, we must consider ionic crystals in general as piezoelectric. To obtain the appropriate equations of motion for ionic crystals in the macroscopic theory, we have thus to use (25.3) and (25.4) in place of Hooke's law. Let us write these equations in tensor notation as follows:

$$S_{\alpha\gamma} = \sum_{\beta\lambda} c_{\alpha\gamma,\beta\lambda} s_{\beta\lambda} - \sum_{\beta} e_{\beta,\alpha\gamma} E_{\beta} \quad (e_{\beta,\alpha\gamma} = e_{\beta,\gamma\alpha}), \tag{32.1}$$

$$P_{\alpha} = \sum_{\beta\gamma} e_{\alpha,\beta\gamma} s_{\beta\gamma} + \sum_{\beta} a_{\alpha\beta} E_{\beta}, \tag{32.2}$$

where the coefficients are identical with the coefficients in Voigt's notation apart from a transcription of the indices. Using (32.1) we have

$$\rho \ddot{u}_{\alpha} = \sum_{\gamma} \frac{\partial S_{\alpha\gamma}}{\partial x_{\gamma}} = \sum_{\beta\gamma\lambda} c_{\alpha\gamma,\beta\lambda} \frac{\partial^2 u_{\beta}}{\partial x_{\gamma} \partial x_{\lambda}} - \sum_{\beta\gamma} e_{\beta,\alpha\gamma} \frac{\partial E_{\beta}}{\partial x_{\gamma}}. \tag{32.3}$$

In a piezoelectric crystal, an elastic wave

$$u_{\alpha}(\mathbf{x},t) = \bar{u}_{\alpha} \exp\{2\pi i \mathbf{y}.\mathbf{x} - i\omega t\} \tag{32.4}$$

is associated with an electric field

$$E_{\alpha} = \bar{E}_{\alpha} \exp\{2\pi i \mathbf{y}.\mathbf{x} - i\omega t\}. \tag{32.5}$$

Substituting (32.4) and (32.5) in the equations of motion given by (32.3), we obtain the equation

$$\rho \omega^2 \bar{u}_{\alpha} = 4\pi^2 \sum_{\beta} \left(\sum_{\gamma\lambda} c_{\alpha\gamma,\beta\lambda} y_{\gamma} y_{\lambda} \right) \bar{u}_{\beta} + 2\pi i \sum_{\beta} \left(\sum_{\gamma} e_{\beta,\alpha\gamma} y_{\gamma} \right) \bar{E}_{\beta}. \tag{32.6}$$

If we put the formal parameter ϵ in the perturbation method equal to unity, the particle displacements and the macroscopic field in an acoustic lattice vibration are represented respectively by the expressions

$$u_{\alpha}\binom{l}{k} = \frac{1}{\sqrt{m_k}} w_{\alpha}(k|\mathbf{y}_j)e^{2\pi i \mathbf{y}.\mathbf{x}\binom{l}{k} - i\omega\binom{\mathbf{y}}{j}t}$$

$$= \frac{1}{\sqrt{m_k}} (w_{\alpha}^{(0)}(k|\mathbf{y}_j) + i w_{\alpha}^{(1)}(k|\mathbf{y}_j) + \ldots)e^{2\pi i \mathbf{y}.\mathbf{x}\binom{l}{k} - i\omega\binom{\mathbf{y}}{j}t}, \tag{32.7}$$

$$E_{\alpha}(\mathbf{x},t) = (E_{\alpha}^{(0)} + i E_{\alpha}^{(1)} + \ldots)e^{2\pi i \mathbf{y}.\mathbf{x} - i\omega\binom{\mathbf{y}}{j}t}. \tag{32.8}$$

The quantities in the macroscopic theory are in every case to be compared with the lowest-order non-vanishing term in the lattice theory. Thus the elastic displacement (32.4) is to be compared with the zero-order term of (32.7); whereas the field (32.5) has to be compared with the first-order term in (32.8), since the zero-order term vanishes (see (31.33)). In other words, $\bar{u}_{\alpha}$ and $\bar{E}_{\alpha}$ in (32.6) are to be identified with $u_{\alpha}(j)$ and $i E_{\alpha}^{(1)}$ in (31.37). We see that these equations become identical if

$$\sum_{\gamma\lambda} c_{\alpha\gamma,\beta\lambda} y_{\gamma} y_{\lambda} \equiv \sum_{\gamma\lambda} [\alpha\beta, \gamma\lambda] y_{\gamma} y_{\lambda} + \sum_{\gamma\lambda} (\alpha\gamma, \beta\lambda) y_{\gamma} y_{\lambda}, \tag{32.9}$$

$$\sum_{\gamma} e_{\beta,\alpha\gamma} y_{\gamma} \equiv \sum_{\gamma} [\beta, \alpha\gamma] y_{\gamma}. \tag{32.10}$$

Since the brackets in (32.9) satisfy the same symmetry relations (31.41), (31.42) as the corresponding brackets in the non-ionic case, by applying the same arguments as used in §§ 27, 28 to (32.9), we obtain

$$c_{\alpha\gamma,\beta\lambda} = [\alpha\beta, \gamma\lambda] + [\beta\gamma, \alpha\lambda] - [\beta\lambda, \alpha\gamma] + (\alpha\gamma, \beta\lambda), \qquad (32.11)$$

where, we remember, the brackets differ from the non-ionic case in the replacement of the coefficients $C_{\alpha\beta}^{(0)}(kk')$, $C_{\alpha\beta,\gamma}^{(1)}(kk')$, $C_{\alpha\beta,\gamma\lambda}^{(2)}(kk')$ by the modified coefficients with the overhead bars.

Since both $e_{\beta,\alpha\gamma}$ and $[\beta, \alpha\gamma]$ are symmetric in α and γ, we obtain from (32.10) the consistent result

$$e_{\beta,\alpha\gamma} = [\beta, \alpha\gamma]. \qquad (32.12)$$

Also associated with the elastic wave (32.4) is a polarization wave

$$P_\alpha(\mathbf{x}, t) = \bar{P}_\alpha \exp\{2\pi i \mathbf{y} \cdot \mathbf{x} - i\omega t\}. \qquad (32.13)$$

The use of (32.4), (32.5), and (32.13) in (32.2) leads to the following relation:

$$\bar{P}_\alpha = 2\pi i \sum_\beta \left(\sum_\gamma e_{\alpha,\beta\gamma} y_\gamma \right) \bar{u}_\beta + \sum_\beta a_{\alpha\beta} \bar{E}_\beta. \qquad (32.14)$$

The dielectric polarization in the lattice wave is given by

$$\frac{1}{v_a} \left\{ \sum_k \frac{e_k}{\sqrt{m_k}} w_\alpha(k|\genfrac{}{}{0pt}{}{\mathbf{y}}{j}) \right\} e^{2\pi i \mathbf{y} \cdot \mathbf{x} - i\omega(\genfrac{}{}{0pt}{}{\mathbf{y}}{j})t}$$

$$= \frac{1}{v_a} \left\{ u_\alpha(j) \sum_k e_k + i \sum_k \frac{e_k}{\sqrt{m_k}} w_\alpha^{(1)}(k|\genfrac{}{}{0pt}{}{\mathbf{y}}{j}) + \dots \right\} e^{2\pi i \mathbf{y} \cdot \mathbf{x} - i\omega(\genfrac{}{}{0pt}{}{\mathbf{y}}{j})t}. \qquad (32.15)$$

The zero-order term in (32.15) evidently vanishes; (32.14) is thus to be compared with the first-order term

$$\frac{i}{v_a} \sum_k \frac{e_k}{\sqrt{m_k}} w_\alpha^{(1)}(k|\genfrac{}{}{0pt}{}{\mathbf{y}}{j})$$

$$= -\frac{i}{v_a} \sum_\beta \left\{ \sum_\gamma y_\gamma \left[\sum_{kk'} \sum_\mu \frac{e_k}{\sqrt{m_k}} \Gamma_{\alpha\mu}(kk') \sum_{k''} \sqrt{m_{k''}} \bar{C}_{\mu\beta,\gamma}^{(1)}(k'k'') \right] \right\} u_\beta(j) +$$

$$+ \frac{i}{v_a} \sum_\beta \left\{ \sum_{kk'} \frac{e_k e_{k'}}{(m_k m_{k'})^{\frac{1}{2}}} \Gamma_{\alpha\beta}(kk') \right\} E_\beta^{(1)}, \qquad (32.16)$$

where we have used the expression (31.36) for the first-order displacement. Equating (32.14) and (32.16), and remembering that $\bar{u}_\beta = u_\beta(j)$ and $\bar{E}_\beta = iE_\beta^{(1)}$, we find that

$$e_{\alpha,\beta\gamma} = \frac{-1}{2\pi v_a} \sum_{kk'} \sum_\mu \frac{e_k}{\sqrt{m_k}} \Gamma_{\alpha\mu}(kk') \sum_{k''} \sqrt{m_{k''}} \bar{C}_{\mu\beta,\gamma}^{(1)}(k'k''), \qquad (32.17)$$

$$a_{\alpha\beta} = \frac{1}{v_a} \sum_{kk'} \frac{e_k e_{k'}}{(m_k m_{k'})^{\frac{1}{2}}} \Gamma_{\alpha\beta}(kk'). \qquad (32.18)$$

With the help of the antisymmetry relation (26.4) (valid also for $\bar{C}^{(1)}_{\alpha\beta,\gamma}(kk')$) and the fact that Γ is a symmetric matrix, we readily see that (32.17) is in agreement with the previous expression (32.12) for the piezoelectric constants (see (31.40)).

The coefficients $a_{\alpha\beta}$ are components of the dielectric susceptibility tensor. Using (32.18), we find for the components of the dielectric tensor

$$\epsilon_{\alpha\beta} = \delta_{\alpha\beta} + 4\pi a_{\alpha\beta} = \delta_{\alpha\beta} + \frac{4\pi}{v_a} \sum_{kk'} \frac{e_k e_{k'}}{(m_k m_{k'})^{\frac{1}{2}}} \Gamma_{\alpha\beta}(kk'). \tag{32.19}$$

33. Phenomenological discussion of the dispersion formula for complex lattices

In contrast to the theory of elasticity, let us consider the macroscopic theory for phenomena involving essentially only internal strain. For such phenomena, we shall suppose, an energy density u can be defined at every point in the medium such that it is a quadratic function of the macroscopic electric field and the displacements $\mathbf{u}(k)$ (for particles of type k) at the same point. (Evidently, for a macroscopic treatment to be possible, conditions must be practically uniform in regions of microscopic dimensions; in speaking of a 'point', we refer to a macroscopically small region which yet contains a large number of lattice cells.) Thus we write

$$u = \tfrac{1}{2} \sum_{kk'} \sum_{\alpha\beta} f\left(\begin{smallmatrix} kk' \\ \alpha\beta \end{smallmatrix}\right) u_\alpha(k) u_\beta(k') + \sum_{k\alpha} \sum_{\beta} f\left(\begin{smallmatrix} k \\ \alpha\beta \end{smallmatrix}\right) u_\alpha(k) E_\beta - \tfrac{1}{2} \sum_{\alpha\beta} f(\alpha\beta) E_\alpha E_\beta, \tag{33.1}$$

where

$$f\left(\begin{smallmatrix} kk' \\ \alpha\beta \end{smallmatrix}\right) = f\left(\begin{smallmatrix} k'k \\ \beta\alpha \end{smallmatrix}\right), \qquad f(\alpha\beta) = f(\beta\alpha). \tag{33.2}$$

Upon introducing the mass density of particles k,

$$\rho_k = \frac{m_k}{v_a}, \tag{33.3}$$

we can write down the equations of motion from (33.1) as follows:

$$\rho_k \ddot{u}_\alpha(k) = -\frac{\partial u}{\partial u_\alpha(k)} = -\sum_{k'} \sum_{\beta} f\left(\begin{smallmatrix} kk' \\ \alpha\beta \end{smallmatrix}\right) u_\beta(k') - \sum_{\beta} f\left(\begin{smallmatrix} k \\ \alpha\beta \end{smallmatrix}\right) E_\beta. \tag{33.4}$$

Moreover, by differentiating the energy density with respect to $-\mathbf{E}$, we obtain the electric moment per unit volume (see § 18), or the dielectric polarization,

$$P_\alpha = -\frac{\partial u}{\partial E_\alpha} = -\sum_{k} \sum_{\beta} f\left(\begin{smallmatrix} k \\ \beta\alpha \end{smallmatrix}\right) u_\beta(k) + \sum_{\beta} f(\alpha\beta) E_\beta. \tag{33.5}$$

The state of internal strain is evidently not affected if all the displacements $\mathbf{u}(k)$ are increased by the same vector $\mathbf{s}$. It follows that

$$u = \tfrac{1}{2}\sum_{kk'}\sum_{\alpha\beta} f\binom{kk'}{\alpha\beta}(u_\alpha(k)+s_\alpha)(u_\beta(k')+s_\beta) +$$

$$+ \sum_k\sum_{\alpha\beta} f\binom{k}{\alpha}\beta)(u_\alpha(k)+s_\alpha)E_\beta - \tfrac{1}{2}\sum_{\alpha\beta} f(\alpha\beta)E_\alpha E_\beta \quad (33.6)$$

must be independent of $\mathbf{s}$. From the linear terms in $\mathbf{s}$ we obtain the condition

$$\sum_{kk'}\sum_\beta f\binom{kk'}{\alpha\beta}u_\beta(k') + \sum_k\sum_\beta f\binom{k}{\alpha}\beta)E_\beta = 0. \quad (33.7)$$

Since this must be true for any values of $\mathbf{u}(k)$ and $\mathbf{E}$, we find the following relations between the coefficients:

$$\sum_k f\binom{kk'}{\alpha\beta} = \sum_k f\binom{k'k}{\beta\alpha} = 0, \quad (33.8)$$

$$\sum_k f\binom{k}{\alpha}\beta) = 0. \quad (33.9)$$

The terms in (33.6) which are quadratic in $\mathbf{s}$ lead to no further new relations.

Consider first the solutions of (33.4) in the absence of any electric field. It is convenient to introduce instead of the displacements $\mathbf{u}(k)$ the new variables

$$\mathbf{w}(k) = \sqrt{\rho_k}\,\mathbf{u}(k). \quad (33.10)$$

The equation (33.4) can thus be written in this case as

$$\ddot{w}_\alpha(k) = -\sum_{k'}\sum_\beta g\binom{kk'}{\alpha\beta}w_\beta(k'), \quad (33.11)$$

where the g-coefficients are defined by

$$g\binom{kk'}{\alpha\beta} = \frac{1}{(\rho_k\,\rho_{k'})^{\frac{1}{2}}} f\binom{kk'}{\alpha\beta}. \quad (33.12)$$

Putting in (33.11) the periodic solutions

$$w_\alpha(k) = e_\alpha(k)e^{-i\omega t} \quad (33.13)$$

we find that the quantities $e_\alpha(k)$ and the frequency ω must satisfy the equations

$$\omega^2 e_\alpha(k) = \sum_{k'}\sum_\beta g\binom{kk'}{\alpha\beta}e_\beta(k'). \quad (33.14)$$

These equations are of the familiar type which we have encountered on several occasions; in particular, according to the discussion given in § 15, the equations admit $3n$ independent solutions $\omega_j, e_\alpha(k|j)$ $(j = 1, 2,..., 3n)$ which fulfil the orthogonality relations

$$\sum_k\sum_\alpha e_\alpha(k|j)e_\alpha(k|j') = \delta_{jj'}; \quad \sum_j e_\alpha(k|j)e_\beta(k'|j) = \delta_{kk'}\delta_{\alpha\beta}. \quad (33.15)$$

Three of the solutions, which we shall designate by $j = 1, 2, 3$, are distinguished from the rest by the fact that their frequencies ω_j are equal to zero. Thus if $\mathbf{u}(1)$, $\mathbf{u}(2)$, and $\mathbf{u}(3)$ are any three mutually perpendicular unit vectors, we can put

$$e_\alpha(k|j) = \sqrt{\rho_k}\, u_\alpha(j), \qquad \omega_j = 0, \qquad j = 1, 2, 3. \tag{33.16}$$

With the help of the relations (33.8) it is readily verified that these are solutions of (33.14).

Remembering that the displacements $\mathbf{u}(k)$ for a solution j are given by

$$\frac{1}{\sqrt{\rho_k}}\, e_\alpha(k|j) e^{-i\omega_j t},$$

we find from (33.5) that the solution is associated with the following dielectric polarization:

$$\mathbf{M}(j)e^{-i\omega_j t}, \tag{33.17}$$

where the amplitude is given by

$$M_\alpha(j) = -\sum_k \sum_\beta f\binom{k}{\beta}\alpha) \frac{1}{\sqrt{\rho_k}} e_\beta(k|j). \tag{33.18}$$

It follows from (33.9) and (33.16) that

$$\mathbf{M}(j) = 0 \quad \text{for } j = 1, 2, 3. \tag{33.19}$$

The dispersion formula is to be obtained by considering the effect of a periodic electric field

$$\mathbf{E} = \overline{\mathbf{E}} e^{-i\omega t}. \tag{33.20}$$

Let the forced oscillation of the lattice be

$$\mathbf{u}(k) = \frac{1}{\sqrt{\rho_k}}\, \overline{\mathbf{w}}(k) e^{-i\omega t}. \tag{33.21}$$

Upon substituting (33.20) and (33.21) in the equations of motion (33.4), we find that

$$\omega^2 \overline{w}_\alpha(k) = \sum_{k'} \sum_\beta g\binom{kk'}{\alpha\beta} \overline{w}_\beta(k') + \frac{1}{\sqrt{\rho_k}} \sum_\beta f\binom{k}{\alpha}\beta) \overline{E}_\beta. \tag{33.22}$$

Since the $3n$ sets of values $e_\alpha(k|1)$, $e_\alpha(k|2)$,..., $e_\alpha(k|3n)$ are linearly independent, we can express the quantities $\overline{w}_\alpha(k)$ as follows:

$$\overline{w}_\alpha(k) = \sum_j a_j\, e_\alpha(k|j). \tag{33.23}$$

Owing to the fact that $e_\alpha(k|j)$ is a solution of (33.14) corresponding to $\omega = \omega_j$, we find that when (33.23) is substituted, the equation (33.22) becomes

$$\sum_j (\omega^2 - \omega_j^2) a_j\, e_\alpha(k|j) = \frac{1}{\sqrt{\rho_k}} \sum_\beta f\binom{k}{\alpha}\beta) \overline{E}_\beta. \tag{33.24}$$

Multiplying (33.24) by $e_\alpha(k|j')$ and summing over k, α, we obtain with the help of the orthogonality relations (33.15) the following values for the coefficients a_j:

$$a_j = \frac{1}{\omega^2 - \omega_j^2} \sum_\beta \left\{ \sum_k \sum_\alpha e_\alpha(k|j) \frac{1}{\sqrt{\rho_k}} f(\genfrac{}{}{0pt}{}{k}{\alpha}\beta) \right\} \bar{E}_\beta. \tag{33.25}$$

In view of (33.18) the above expression can also be written

$$a_j = \frac{\mathbf{M}(j) \cdot \mathbf{E}}{\omega_j^2 - \omega^2}. \tag{33.26}$$

The dielectric polarization induced by the periodic field is obtained by substituting the above solution (given by (33.21), (33.23), and (33.26)) and the field (33.20) in (33.5). When use is made of (33.18) we find that the induced dielectric polarization is given by the expression

$$P_\alpha = \sum_\beta \left\{ \frac{M_\alpha(j) M_\beta(j)}{\omega_j^2 - \omega^2} + f(\alpha\beta) \right\} E_\beta. \tag{33.27}$$

The expression given in the curly brackets is the $(\alpha\beta)$ component of the dielectric susceptibility tensor, from which we obtain the dielectric tensor

$$\epsilon_{\alpha\beta} = \delta_{\alpha\beta} + 4\pi \left\{ f(\alpha\beta) + \sum_j \frac{M_\alpha(j) M_\beta(j)}{\omega_j^2 - \omega^2} \right\}. \tag{33.28}$$

It follows from (33.19) that the terms $j = 1, 2, 3$ contribute nothing to the formula.

The dispersion formula (33.28) depends on the coefficients $f(\genfrac{}{}{0pt}{}{kk'}{\alpha\beta})$, $f(\genfrac{}{}{0pt}{}{k}{\alpha}\beta)$, $f(\alpha\beta)$. The values of these coefficients can be determined by considering the long optical vibrations of the lattice and comparing the resulting equations with the corresponding equations in the lattice theory, which we shall derive in the next section. For a long optical vibration, we write for the displacement vectors:

$$\mathbf{u}(k) = \frac{1}{\sqrt{\rho_k}} \bar{\mathbf{w}}(k) e^{2\pi i y \cdot x - i\omega t}. \tag{33.29}$$

The electric field and dielectric polarization associated with the vibration are given by similar expressions:

$$\mathbf{E} = \bar{\mathbf{E}} e^{2\pi i y \cdot x - i\omega t}, \tag{33.30}$$

$$\mathbf{P} = \bar{\mathbf{P}} e^{2\pi i y \cdot x - i\omega t}. \tag{33.31}$$

Substituting (33.29), (33.30), (33.31) in (33.4) and (33.5), we obtain the following:

$$\omega^2 \bar{w}_\alpha(k) = \sum_{k'\beta} g\left(\begin{smallmatrix}k\ k'\\ \alpha\ \beta\end{smallmatrix}\right) \bar{w}_\beta(k') + \frac{1}{\sqrt{\rho_k}} \sum_\beta f\left(\begin{smallmatrix}k\\ \alpha\end{smallmatrix}\beta\right) \bar{E}_\beta, \tag{33.32}$$

$$\bar{P}_\alpha = -\sum_{k\beta} f\left(\begin{smallmatrix}k\\ \beta\end{smallmatrix}\alpha\right) \frac{1}{\sqrt{\rho_k}} \bar{w}_\beta(k) + \sum_\beta f(\alpha\beta) \bar{E}_\beta. \tag{33.33}$$

The electric field and dielectric polarization are related by (30.8):

$$\bar{E}_\alpha = -4\pi \left(\frac{y_\alpha}{|\mathbf{y}|}\right) \sum_\beta \left(\frac{y_\beta}{|\mathbf{y}|}\right) \bar{P}_\beta. \tag{33.34}$$

The equations (33.32), (33.33), (33.34) completely determine the optical vibrations of a lattice in the limit of very large wave-length; but unlike the particular case (optically isotropic diatomic crystals) discussed in § 7, the solutions cannot be given explicitly in the general case. A few general features are, however, worth noticing. After substituting (33.34) in (33.33), we can express $\bar{P}_\alpha$ in terms of the quantities $\bar{w}_\alpha(k)$, as follows:

$$\bar{P}_\alpha = -\sum_\gamma S(\alpha\gamma) \sum_{k\beta} f\left(\begin{smallmatrix}k\\ \beta\end{smallmatrix}\gamma\right) \frac{1}{\sqrt{\rho_k}} \bar{w}_\beta(k), \tag{33.35}$$

where $S(\alpha\gamma)$ represents the $(\alpha\gamma)$-component of the matrix inverse to the 3×3 matrix defined by

$$\delta_{\alpha\beta} + 4\pi \sum_\gamma f(\alpha\gamma)\left(\frac{y_\gamma y_\beta}{|\mathbf{y}|^2}\right). \tag{33.36}$$

With the help of (33.34) and (33.35), we can eliminate the electric field from (33.32), obtaining

$$\omega^2 \bar{w}_\alpha(k) = \sum_{k'\beta} g\left(\begin{smallmatrix}k\ k'\\ \alpha\ \beta\end{smallmatrix}\right) \bar{w}_\beta(k') +$$

$$+ \frac{4\pi}{\sqrt{\rho_k}} \left\{ \sum_\mu f\left(\begin{smallmatrix}k\\ \alpha\end{smallmatrix}\mu\right)\left(\frac{y_\mu}{|\mathbf{y}|}\right) \right\} \left\{ \sum_{k'} \sum_{\lambda\gamma\beta} \left(\frac{y_\lambda}{|\mathbf{y}|}\right) S(\lambda\gamma) f\left(\begin{smallmatrix}k'\\ \beta\end{smallmatrix}\beta\right) \frac{1}{\sqrt{\rho_{k'}}} \bar{w}_\beta(k') \right\}. \tag{33.37}$$

It follows from (33.8) and (33.9) that the equations (33.37) admit three independent solutions of the same form as (33.16). Owing to the vanishing frequency, these do not represent genuine dynamical solutions. Therefore, as expected, we obtain from (33.37) $3(n-1)$ vibrational solutions, which represent the long-wave limits of the $3(n-1)$ optical branches of lattice vibrations.

We notice that the equations (33.37) and (33.14) are by no means the same; thus the optical vibrations of a lattice do not in general approach the dispersion frequencies ω_j in the limit of long wave-length. Moreover, as the second term on the right-hand side of (33.37) depends on $\mathbf{y}/|\mathbf{y}|$, the limiting frequencies for optical vibrations travelling in different directions are not in general the same. In a long optical vibration, each volume element is subject to a uniform electric field, which is produced by the dielectric polarization associated with the wave; the additional term in (33.37) takes account of the effect of this field. However, in certain special cases such as the diatomic crystals considered in § 7, we find that the long optical vibrations are either longitudinal or transverse; the macroscopic field vanishes in the latter so that the corresponding frequency is equal to the dispersion frequency.

34. Long optical vibrations in ionic lattices (rigid ion model)

In the perturbation method the optical vibrations are distinguished from the acoustic vibrations by the fact that the zero-order term in the frequency expansion is no longer equal to zero; thus we have to replace the expansion (26.9) used previously for $\omega(\genfrac{}{}{0pt}{}{\mathbf{y}}{j})$ by

$$\omega(\genfrac{}{}{0pt}{}{\mathbf{y}}{j}) = \omega^{(0)}(\genfrac{}{}{0pt}{}{\mathbf{y}}{j}) + \epsilon\omega^{(1)}(\genfrac{}{}{0pt}{}{\mathbf{y}}{j}) + \dots. \tag{34.1}$$

Perturbation equations of various orders can be obtained as before by replacing $\mathbf{y}$ by $\epsilon\mathbf{y}$ in the basic equation (31.19) and using for the various quantities appearing in the equation the appropriate expansions. In the present case, we are primarily interested in obtaining the equations comparable with the phenomenological theory as a means for determining the coefficients underlying the dispersion formula (33.28). For this purpose we shall require only the zero-order equations

$$[\omega^{(0)}(\genfrac{}{}{0pt}{}{\mathbf{y}}{j})]^2 w_\alpha^{(0)}(k|\genfrac{}{}{0pt}{}{\mathbf{y}}{j}) = \sum_{k'\beta} \bar{C}_{\alpha\beta}^{(0)}(kk') w_\beta^{(0)}(k'|\genfrac{}{}{0pt}{}{\mathbf{y}}{j}) - \frac{e_k}{\sqrt{m_k}} E_\alpha^{(0)}. \tag{34.2}$$

The dielectric polarization associated with a lattice wave is given generally by the expression

$$\frac{1}{v_a}\left\{\sum_k \frac{e_k}{\sqrt{m_k}} w_\alpha(k|\genfrac{}{}{0pt}{}{\mathbf{y}}{j})\right\} e^{2\pi i \mathbf{y}.\mathbf{x} - i\omega t}. \tag{34.3}$$

In the perturbation method the amplitude factor is thus represented by the series

$$\mathbf{P}^{(0)} + i\epsilon\mathbf{P}^{(1)} + \dots, \tag{34.4}$$

where the expansion corresponds term by term with the expansion for $\mathbf{w}(k|\genfrac{}{}{0pt}{}{\mathbf{y}}{j})$. In particular,

$$P_\alpha^{(0)} = \frac{1}{v_a}\sum_k \frac{e_k}{\sqrt{m_k}} w_\alpha^{(0)}(k|\genfrac{}{}{0pt}{}{\mathbf{y}}{j}). \tag{34.5}$$

The expression for $E_\alpha^{(0)}$ given by (31.28) may thus also be written as follows:

$$E_\alpha^{(0)} = -4\pi \left(\frac{y_\alpha}{|\mathbf{y}|}\right) \sum_\beta \left(\frac{y_\beta}{|\mathbf{y}|}\right) P_\beta^{(0)}. \tag{34.6}$$

The equations (34.2), (34.5), and (34.6) are directly comparable with the equations (33.32), (33.33), and (33.34) of the phenomenological theory. The frequency ω and the amplitude factors $\mathbf{E}$, $\overline{\mathbf{P}}$ in the macroscopic discussion are to be identified with the corresponding lowest-order terms $\omega^{(0)}(\begin{smallmatrix} y \\ j \end{smallmatrix})$, $\mathbf{E}^{(0)}$, $\mathbf{P}^{(0)}$ in the perturbation method. The relation between $\bar{w}_\alpha(k)$ and $w_\alpha^{(0)}(k|\begin{smallmatrix} y \\ j \end{smallmatrix})$ follows immediately from a comparison of the corresponding particle displacements, namely $\bar{w}_\alpha(k)/\sqrt{\rho_k}$ and $w_\alpha^{(0)}(k|\begin{smallmatrix} y \\ j \end{smallmatrix})/\sqrt{m_k}$. Remembering that $\rho_k = m_k/v_a$, we have thus to identify $w_\alpha^{(0)}(k|\begin{smallmatrix} y \\ j \end{smallmatrix})$ with $v_a^{\frac{1}{2}} \bar{w}_\alpha(k)$. Bearing in mind the correspondence between ω, $\mathbf{E}$, $\overline{\mathbf{P}}$, $v_a^{\frac{1}{2}} \bar{w}_\alpha(k)$ and $\omega^{(0)}(\begin{smallmatrix} y \\ j \end{smallmatrix})$, $\mathbf{E}^{(0)}$, $\mathbf{P}^{(0)}$, $w_\alpha^{(0)}(k|\begin{smallmatrix} y \\ j \end{smallmatrix})$, we see that the equations (33.32), (33.33), (33.34) in the phenomenological theory become identical with (34.2) (34.5), and (34.6) for the following values of the coefficients in the phenomenological theory:

$$g(\begin{smallmatrix} k & k' \\ \alpha & \beta \end{smallmatrix}) = \frac{v_a}{(m_k m_{k'})^{\frac{1}{2}}} f(\begin{smallmatrix} k & k' \\ \alpha & \beta \end{smallmatrix}) = \bar{C}_{\alpha\beta}^{(0)}(kk'), \tag{34.7}$$

$$f(\begin{smallmatrix} k \\ \alpha \end{smallmatrix} \beta) = -\frac{e_k}{v_a}\delta_{\alpha\beta}, \tag{34.8}$$

$$f(\alpha\beta) = 0. \tag{34.9}$$

So long as we assume that the ions are not polarizable, the dielectric polarization is completely determined by the positions of the ions and hence the second term on the right-hand side of (33.5) will vanish identically. (34.9) represents thus only a special result, due to our assumption of rigid ions.

Let us consider the static dielectric tensor obtained by putting $\omega = 0$ in the dispersion formula (33.28):

$$\epsilon_{\alpha\beta}(0) = \delta_{\alpha\beta} + 4\pi\left\{f(\alpha\beta) + \sum_j \frac{M_\alpha(j)M_\beta(j)}{\omega_j^2}\right\}. \tag{34.10}$$

Substituting (33.18) in the formula and using (34.9), we get

$$\epsilon_{\alpha\beta}(0) = \delta_{\alpha\beta} + 4\pi \sum_{kk'} \sum_{\gamma\lambda} f(\begin{smallmatrix} k \\ \gamma \end{smallmatrix}\alpha)f(\begin{smallmatrix} k' \\ \lambda \end{smallmatrix}\beta)\frac{1}{(\rho_k \rho_{k'})^{\frac{1}{2}}}\left\{\sum_j \frac{e_\gamma(k|j)e_\lambda(k'|j)}{\omega_j^2}\right\}. \tag{34.11}$$

We remember that $e_\alpha(k''|j)$ satisfies the equations

$$\omega_j^2 e_\alpha(k''|j) = \sum_{k'\lambda} g(\begin{smallmatrix} k'' & k' \\ \alpha & \lambda \end{smallmatrix})e_\lambda(k'|j). \tag{34.12}$$

After dividing (34.12) by ω_j^2, we multiply the equation by $e_\gamma(k|j)$ and sum over j; upon using the orthogonality relations (33.15), we find that

$$\delta_{\gamma\alpha}\delta_{kk^-} = \sum_{k'\lambda} g\!\left(\begin{smallmatrix} k^- & k' \\ \alpha & \lambda \end{smallmatrix}\right)\!\left\{ \sum_j \frac{e_\gamma(k|j)e_\lambda(k'|j)}{\omega_j^2} \right\}. \qquad (34.13)$$

In view of (34.7) we can, with the help of the matrix Γ introduced in (31.35), write (34.13) as follows:

$$\sum_j \frac{e_\gamma(k|j)e_\lambda(k'|j)}{\omega_j^2} = \sum_{k''\alpha} \Gamma_{\lambda\alpha}(k'k'')\delta_{\gamma\alpha}\delta_{kk^-} = \Gamma_{\lambda\gamma}(k'k). \qquad (34.14)$$

Using this expression and the value for $f(\begin{smallmatrix} k \\ \alpha \end{smallmatrix}\beta)$ given by (34.8), we find that (34.11) reduces to

$$\epsilon_{\alpha\beta}(0) = \delta_{\alpha\beta} + \frac{4\pi}{v_a} \sum_{kk'} \frac{e_k e_{k'}}{(m_k m_{k'})^{\frac12}} \Gamma_{\alpha\beta}(kk'). \qquad (34.15)$$

This is identical with the dielectric tensor (32.19), which we have obtained in connexion with the acoustic vibrations. Since the frequency approaches zero for long acoustic vibrations, it is thus the static dielectric tensor that appears in the treatment of the acoustic vibrations.

35. Polarizable ions

As in § 9, we shall assume that the electronic polarization of an ion is equivalent to a point-dipole, which is proportional to the electric field at the centre of the ion.

When electronic polarization is taken into account, it is no longer convenient to consider the equations (24.10) for lattice waves by working out the Coulomb part of the coefficients $C_{\alpha\beta}(\begin{smallmatrix} y \\ kk' \end{smallmatrix})$ directly according to (24.7). For the displacement of one ion polarizes all ions in its neighbourhood and alters thereby their mutual interaction energy (in effect, a many-body interaction); it becomes thus difficult to obtain the explicit expressions for the coefficients $\Phi_{\alpha\beta}^C(\begin{smallmatrix} l & l' \\ k & k' \end{smallmatrix})$.

Let us examine the physical interpretation of the right-hand side of (24.10). We can evidently express (24.7) as follows:

$$C_{\alpha\beta}(\begin{smallmatrix} y \\ kk' \end{smallmatrix}) = \frac{e^{-2\pi i y \cdot x(k)}}{(m_k m_{k'})^{\frac12}} \sum_{l'} \Phi_{\alpha\beta}(\begin{smallmatrix} 0 & l' \\ k & k' \end{smallmatrix})e^{2\pi i y \cdot x(\begin{smallmatrix} l' \\ k' \end{smallmatrix})}. \qquad (35.1)$$

When this expression is substituted, the right-hand side of (24.10) becomes

$$\frac{e^{-2\pi i y \cdot x(k)}}{\sqrt{m_k}}\left\{ -\sum_{l'k'\beta} \Phi_{\alpha\beta}(\begin{smallmatrix} 0 & l' \\ k & k' \end{smallmatrix})\frac{1}{\sqrt{m_{k'}}}\, w_\beta(k'|\begin{smallmatrix} y \\ j \end{smallmatrix})e^{2\pi i y \cdot x(\begin{smallmatrix} l' \\ k' \end{smallmatrix})} \right\}. \qquad (35.2)$$

The expression given in the curly brackets is clearly the force acting on the particle $(\substack{0\\k})$, if the particles in the lattice are subject to the following displacements:

$$\mathbf{u}(\substack{l'\\k'}) = \frac{1}{\sqrt{m_{k'}}}\,\mathbf{w}(k'|\substack{\mathbf{y}\\j})e^{2\pi i\mathbf{y}\,\cdot\,\mathbf{x}(\substack{l'\\k'})}. \qquad (35.3)$$

In the following, we shall calculate explicitly that part of this force which is due to the Coulomb interaction between the ions.

The displacements (35.3) induce in the ions electronic dipole moments which show the same phase relations as the displacements. If we represent the induced moments by

$$\boldsymbol{\mu}(\substack{l\\k}) = \boldsymbol{\mu}(k|\substack{\mathbf{y}\\j})e^{2\pi i\mathbf{y}\,\cdot\,\mathbf{x}(\substack{l\\k})}, \qquad (35.4)$$

the Coulomb field in the lattice is the same as that in a dipole lattice of the type (30.24) with

$$\mathbf{p}(k') = \boldsymbol{\mu}(k'|\substack{\mathbf{y}\\j}) + \frac{e_{k'}}{\sqrt{m_{k'}}}\,\mathbf{w}(k'|\substack{\mathbf{y}\\j}), \qquad (35.5)$$

where the second term on the right-hand side describes the dipole moments due to the shift of the ionic charges. The macroscopic electric field in the lattice is thus given by

$$\mathbf{E}e^{2\pi i\mathbf{y}\,\cdot\,\mathbf{x}}, \qquad (35.6)$$

where the amplitude is obtained by substituting (35.5) in (30.26):

$$E_\alpha = -\frac{4\pi}{v_a}\left(\frac{y_\alpha}{|\mathbf{y}|}\right)\sum_{k'\beta}\left(\frac{y_\beta}{|\mathbf{y}|}\right)\left\{\mu_\beta(k'|\substack{\mathbf{y}\\j}) + \frac{e_{k'}}{\sqrt{m_{k'}}}\,w_\beta(k'|\substack{\mathbf{y}\\j})\right\}. \qquad (35.7)$$

The Coulomb field at the centre of the ion $(\substack{0\\k})$ is the sum of the two following terms:

(i) the exciting field at $\mathbf{x}(k)$ in the above dipole lattice,

$$\left\{E_\alpha + \sum_{k'\beta}Q_{\alpha\beta}(\substack{\mathbf{y}\\kk'})\left[\mu_\beta(k'|\substack{\mathbf{y}\\j}) + \frac{e_{k'}}{\sqrt{m_{k'}}}\,w_\beta(k'|\substack{\mathbf{y}\\j})\right]\right\}e^{2\pi i\mathbf{y}\,\cdot\,\mathbf{x}(k)}, \qquad (35.8)$$

which is obtained by substituting (35.5) in (30.30);

(ii) the field change at the centre of $(\substack{0\\k})$ owing to its displacement,

$$\mathbf{u}(\substack{0\\k}) = \frac{1}{\sqrt{m_k}}\,\mathbf{w}(k|\substack{\mathbf{y}\\j})e^{2\pi i\mathbf{y}\,\cdot\,\mathbf{x}(k)} \quad \text{(see (35.3))}. \qquad (35.9)$$

The field is evidently equal to the field created at $\mathbf{x}(k)$ by displacing all other ions by $-\mathbf{u}(\substack{0\\k})$. Hence it is also equal to the exciting field at $\mathbf{x}(k)$ in a dipole lattice of the type (30.24) with

$$\left.\begin{aligned}&\mathbf{y} = 0 &&\text{(not to be confused with } \mathbf{y} \text{ in (35.9))}\\ &\mathbf{p}(k') = -e_{k'}\,\mathbf{u}(\substack{0\\k}) \;\;\text{for all } k'\end{aligned}\right\}. \qquad (35.10)$$

Putting (35.10) in (30.30) and afterwards using (35.9), we get

$$-\frac{1}{\sqrt{m_k}}e^{2\pi i y \cdot \mathbf{x}(k)}\sum_{\beta}w_{\beta}(k|\tbinom{y}{j})\sum_{k'}e_{k'}Q_{\alpha\beta}(\tbinom{0}{kk'}) \tag{35.11}$$

(the use of (30.26) immediately shows that the term representing the macroscopic field in (30.30) vanishes in this case).

The total electric field at the centre of $\tbinom{0}{k}$ is the sum of (35.8) and (35.11) and can thus be written

$$\left\{E_{\alpha}+\sum_{k'\beta}\left[Q_{\alpha\beta}(\tbinom{y}{kk'})e_{k'}-\delta_{kk'}\sum_{k''}Q_{\alpha\beta}(\tbinom{0}{kk''})e_{k''}\right]\frac{1}{\sqrt{m_{k'}}}w_{\beta}(k'|\tbinom{y}{j})+\right.$$
$$\left.+\sum_{k'\beta}Q_{\alpha\beta}(\tbinom{y}{kk'})\mu_{\beta}(k'|\tbinom{y}{j})\right\}e^{2\pi i y \cdot \mathbf{x}(k)}. \tag{35.12}$$

By assumption, the electronic dipole moment of $\tbinom{0}{k}$ is proportional to the field (35.12). Hence we can write

$$\mu_{\alpha}(k|\tbinom{y}{j})=\alpha_k\left\{E_{\alpha}+\sum_{k'\beta}\left[Q_{\alpha\beta}(\tbinom{y}{kk'})e_{k'}-\delta_{kk'}\sum_{k''}Q_{\alpha\beta}(\tbinom{0}{kk''})e_{k''}\right]\frac{1}{\sqrt{m_{k'}}}w_{\beta}(k'|\tbinom{y}{j})+\right.$$
$$\left.+\sum_{k'\beta}Q_{\alpha\beta}(\tbinom{y}{kk'})\mu_{\beta}(k'|\tbinom{y}{j})\right\}, \tag{35.13}$$

where α_k is the electronic polarizability of ions of type k.

The force acting on the particle $\tbinom{0}{k}$ is twofold:

(i) the force exerted on the ionic charge e_k by the field (35.12),

$$e_k\left\{E_{\alpha}+\sum_{k'\beta}\left[Q_{\alpha\beta}(\tbinom{y}{kk'})e_{k'}-\delta_{kk'}\sum_{k''}Q_{\alpha\beta}(\tbinom{0}{kk''})e_{k''}\right]\frac{1}{\sqrt{m_{k'}}}w_{\beta}(k'|\tbinom{y}{j})+\right.$$
$$\left.+\sum_{k'\beta}Q_{\alpha\beta}(\tbinom{y}{kk'})\mu_{\beta}(k'|\tbinom{y}{j})\right\}e^{2\pi i y \cdot \mathbf{x}(k)}; \tag{35.14}$$

(ii) the force exerted on the dipole $\mu(\tbinom{0}{k})$ by the field of all other ions.

Since the dipole moment is linear in the particle displacements, to calculate this force in the harmonic approximation, we can consider all ions as though they are in their undisplaced positions $\mathbf{x}(\tbinom{l}{k})$. Let us subject the dipole to a virtual displacement $\mathbf{u}$ and consider the virtual energy. The latter can be calculated as the interaction energy between the dipole and the field at $\mathbf{x}(k)$ created by displacing all other ions by $-\mathbf{u}$. Using a dipole lattice of the type (30.24) with $\mathbf{y}=0$ and $\mathbf{p}(k')=-e_{k'}\mathbf{u}$, we readily find the following expression for the virtual energy:

$$\sum_{\alpha}\mu_{\alpha}(\tbinom{0}{k})\sum_{k'\beta}Q_{\alpha\beta}(\tbinom{0}{kk'})e_{k'}u_{\beta}.$$

Hence the force (α-component) on the dipole is given by

$$-\sum_{\beta}\mu_{\beta}(\tbinom{0}{k})\sum_{k'}Q_{\beta\alpha}(\tbinom{0}{kk'})e_{k'}=-\left\{\sum_{k'\beta}e_{k'}Q_{\alpha\beta}(\tbinom{0}{kk'})\mu_{\beta}(k|\tbinom{y}{j})\right\}e^{2\pi i y \cdot \mathbf{x}(k)}, \tag{35.15}$$

where we have substituted (35.4) for $\mu(^0_k)$ and made use of the fact that $Q_{\alpha\beta}(^0_{kk'})$ is symmetric in α, β (see (30.32)).

Adding (35.14) and (35.15) and multiplying the result by

$$-m_k^{-\frac{1}{2}}\exp\{-2\pi i\mathbf{y}.\mathbf{x}(k)\},$$

we get the total Coulomb contribution to the right-hand side of (24.10) (see (35.2)). Hence the equations for lattice waves can be written in the present case

$$\omega^2(^{\mathbf{y}}_j)w_\alpha(k|^{\mathbf{y}}_j)$$

$$= \sum_{k'\beta}\left\{C^N_{\alpha\beta}(^{\mathbf{y}}_{kk'}) - \frac{1}{(m_k m_{k'})^{\frac{1}{2}}}\left[e_k e_{k'} Q_{\alpha\beta}(^{\mathbf{y}}_{kk'}) - \delta_{kk'}\sum_{k''} e_k e_{k''} Q_{\alpha\beta}(^0_{kk''})\right]\right\}w_\beta(k'|^{\mathbf{y}}_j) -$$

$$- \frac{1}{\sqrt{m_k}}\sum_{k'\beta}\left\{e_k Q_{\alpha\beta}(^{\mathbf{y}}_{kk'}) - \delta_{kk'}\sum_{k''} e_{k''} Q_{\alpha\beta}(^0_{kk''})\right\}\mu_\beta(k'|^{\mathbf{y}}_j) - \frac{e_k}{\sqrt{m_k}}E_\alpha. \quad (35.16)$$

The equations (35.13) and (35.16), together with the expression (35.7) for the macroscopic field, provide the complete set of relations for determining the lattice vibrations.

The perturbation method can be developed in very much the same way as before. In place of the expansion for $\bar{C}_{\alpha\beta}(^{\mathbf{y}}_{kk'})$ we have now the expansions

$$C^N_{\alpha\beta}(^{\epsilon\mathbf{y}}_{kk'}) = C^N_{\alpha\beta}(^0_{kk'}) + i\epsilon\sum_\gamma C^{(1)N}_{\alpha\beta,\gamma}(kk')y_\gamma + ..., \quad (35.17)$$

$$Q_{\alpha\beta}(^{\epsilon\mathbf{y}}_{kk'}) = Q_{\alpha\beta}(^0_{kk'}) + i\epsilon\sum_\gamma Q^{(1)}_{\alpha\beta,\gamma}(kk')y_\gamma + ... \quad (35.18)$$

Moreover we have to introduce the expansion for the electronic polarization:

$$\mu(k|^{\epsilon\mathbf{y}}_j) = \mu^{(0)}(k|^{\mathbf{y}}_j) + i\epsilon\mu^{(1)}(k|^{\mathbf{y}}_j) + ... \quad (35.19)$$

For the optical vibrations we find that in the zero-order approximation (35.16), (35.13), and (35.7) lead to the equations

$$[\omega^{(0)}(^{\mathbf{y}}_j)]^2 w^{(0)}_\alpha(k|^{\mathbf{y}}_j)$$

$$= \sum_{k'\beta}\left\{C^N_{\alpha\beta}(^0_{kk'}) - \frac{1}{(m_k m_{k'})^{\frac{1}{2}}}\left[e_k e_{k'} Q_{\alpha\beta}(^0_{kk'}) - \delta_{kk'}\sum_{k''} e_k e_{k''} Q_{\alpha\beta}(^0_{kk''})\right]\right\}w^{(0)}_\beta(k'|^{\mathbf{y}}_j) -$$

$$- \frac{1}{\sqrt{m_k}}\sum_{k'\beta}\left\{e_k Q_{\alpha\beta}(^0_{kk'}) - \delta_{kk'}\sum_{k''} e_{k''} Q_{\alpha\beta}(^0_{kk''})\right\}\mu^{(0)}_\beta(k'|^{\mathbf{y}}_j) - \frac{e_k}{\sqrt{m_k}}E^{(0)}_\alpha, \quad (35.20)$$

$$\mu^{(0)}_\alpha(k|^{\mathbf{y}}_j) = \alpha_k\sum_{k'\beta} Q_{\alpha\beta}(^0_{kk'})\mu^{(0)}_\beta(k'|^{\mathbf{y}}_j) +$$

$$+ \alpha_k\sum_{k'\beta}\left[Q_{\alpha\beta}(^0_{kk'})e_{k'} - \delta_{kk'}\sum_{k''} Q_{\alpha\beta}(^0_{kk''})e_{k''}\right]\frac{1}{\sqrt{m_{k'}}}w^{(0)}_\beta(k'|^{\mathbf{y}}_j) + \alpha_k E^{(0)}_\alpha,$$

$$(35.21)$$

$$E^{(0)}_\alpha = -\frac{4\pi}{v_a}\left(\frac{y_\alpha}{|\mathbf{y}|}\right)\sum_{k'\beta}\left(\frac{y_\beta}{|\mathbf{y}|}\right)\left\{\mu^{(0)}_\beta(k'|^{\mathbf{y}}_j) + \frac{e_{k'}}{\sqrt{m_{k'}}}w^{(0)}_\beta(k'|^{\mathbf{y}}_j)\right\}. \quad (35.22)$$

Upon dividing (35.21) by α_k and introducing the matrix R inverse to the $3n \times 3n$ matrix defined by

$$\frac{1}{\alpha_k}\delta_{\alpha\beta}\delta_{kk'} - Q_{\alpha\beta}(\substack{0\\kk'}) \tag{35.23}$$

we can rewrite (35.21) in the form

$$\mu_\alpha^{(0)}(k|\substack{y\\j}) = \sum_{k'''k'}\sum_{\beta\gamma} R_{\alpha\gamma}(kk''')\Big\{Q_{\gamma\beta}(\substack{0\\k'''k'})e_{k'} - \delta_{k'''k'}\sum_{k''}Q_{\gamma\beta}(\substack{0\\k'''k''})e_{k''}\Big\}\frac{1}{\sqrt{m_{k'}}}w_\beta^{(0)}(k'|\substack{y\\j}) +$$
$$+ \sum_\beta\Big\{\sum_{k'}R_{\alpha\beta}(kk')\Big\}E_\beta^{(0)}. \tag{35.24}$$

With the help of (35.24) we eliminate the electronic polarization from (35.20), obtaining

$$[\omega^{(0)}(\substack{y\\j})]^2 w_\alpha^{(0)}(k|\substack{y\\j})$$

$$= \sum_{k'\beta}\Big\{C_{\alpha\beta}^N(\substack{0\\kk'}) - \frac{1}{(m_k m_{k'})^{\frac{1}{2}}}\Big[e_k e_{k'}Q_{\alpha\beta}(\substack{0\\kk'}) - \delta_{kk'}\sum_{k''}e_k e_{k''}Q_{\alpha\beta}(\substack{0\\kk''})\Big] -$$

$$- \frac{1}{(m_k m_{k'})^{\frac{1}{2}}}\sum_{k_1 k_2}\sum_{\gamma\lambda}\Big[e_k Q_{\alpha\gamma}(\substack{0\\kk_1}) - \delta_{kk_1}\sum_{k''}e_{k''}Q_{\alpha\gamma}(\substack{0\\kk''})\Big] \times$$

$$\times R_{\gamma\lambda}(k_1 k_2)\Big[Q_{\lambda\beta}(\substack{0\\k_2k'})e_{k'} - \delta_{k_2k'}\sum_{k''}e_{k''}Q_{\lambda\beta}(\substack{0\\k_2k''})\Big]\Big\}w_\beta^{(0)}(k'|\substack{y\\j}) +$$

$$+ \sum_\beta\Big\{-\frac{e_k}{\sqrt{m_k}}\delta_{\alpha\beta} - \frac{1}{\sqrt{m_k}}\sum_{k_1\gamma}\Big[e_k Q_{\alpha\gamma}(\substack{0\\kk_1}) - \delta_{kk_1}\sum_{k''}e_{k''}Q_{\alpha\gamma}(\substack{0\\kk''})\Big] \times$$

$$\times \sum_{k'}R_{\gamma\beta}(k_1 k')\Big\}E_\beta^{(0)}. \tag{35.25}$$

This is directly comparable with the equation (33.32) of the phenomenological theory.

The zero-order amplitude for the dielectric polarization is given by

$$P_\alpha^{(0)} = \frac{1}{v_a}\sum_k\Big\{\mu_\alpha^{(0)}(k|\substack{y\\j}) + \frac{e_k}{\sqrt{m_k}}w_\alpha^{(0)}(k|\substack{y\\j})\Big\}$$

$$= \frac{1}{v_a}\sum_{k'\beta}\Big\{\sum_{k'''}\sum_\gamma\Big[\sum_k R_{\alpha\gamma}(kk''')\Big]\Big[Q_{\gamma\beta}(\substack{0\\k'''k'})e_{k'} - \delta_{k'''k'}\sum_{k''}Q_{\gamma\beta}(\substack{0\\k'''k''})e_{k''}\Big]\frac{1}{\sqrt{m_{k'}}} +$$

$$+ \frac{e_{k'}}{\sqrt{m_{k'}}}\delta_{\alpha\beta}\Big\}w_\beta^{(0)}(k'|\substack{y\\j}) + \frac{1}{v_a}\sum_\beta\Big\{\sum_{kk'}R_{\alpha\beta}(kk')\Big\}E_\beta^{(0)}. \tag{35.26}$$

The relation (35.22) can be rewritten as

$$E_\alpha^{(0)} = -4\pi\Big(\frac{y_\alpha}{|\mathbf{y}|}\Big)\sum_\beta\Big(\frac{y_\beta}{|\mathbf{y}|}\Big)P_\beta^{(0)}. \tag{35.27}$$

(35.26) and (35.27) are respectively comparable with (33.33) and (33.34) of the phenomenological theory.

We remember that $\omega^{(0)}(\substack{y\\j})$, $\mathbf{E}^{(0)}$, $\mathbf{P}^{(0)}$, $w_\alpha^{(0)}(k|\substack{y\\j})$ are to be identified with ω, $\mathbf{E}$, $\bar{\mathbf{P}}$, $v_a^{\frac{1}{2}}\bar{w}_\alpha(k)$ respectively in the phenomenological theory. Thus (35.25) and (33.32) become identical if

$$g(\substack{k\ k'\\ \alpha\ \beta}) = C_{\alpha\beta}^N(\substack{0\\kk'}) - \frac{1}{(m_k m_{k'})}\left[e_k e_{k'} Q_{\alpha\beta}(\substack{0\\kk'}) - \delta_{kk'}\sum_{k''} e_k e_{k''} Q_{\alpha\beta}(\substack{0\\kk''})\right] -$$

$$- \frac{1}{(m_k m_{k'})^{\frac{1}{2}}}\sum_{k_1 k_2}\sum_{\gamma\lambda}\left[e_k Q_{\alpha\gamma}(\substack{0\\kk_1}) - \delta_{kk_1}\sum_{k''} e_{k''} Q_{\alpha\gamma}(\substack{0\\kk''})\right] \times$$

$$\times R_{\gamma\lambda}(k_1 k_2)\left[Q_{\lambda\beta}(\substack{0\\k_2k'})e_{k'} - \delta_{k_2k'}\sum_{k''} e_{k''} Q_{\lambda\beta}(\substack{0\\k_2k''})\right], \quad (35.28)$$

$$f(\substack{k\\ \alpha\ \beta}) = -\frac{1}{v_a}\left\{e_k\delta_{\alpha\beta} + \sum_{k_1\gamma}\left[e_k Q_{\alpha\gamma}(\substack{0\\kk_1}) - \delta_{kk_1}\sum_{k''} e_{k''} Q_{\alpha\gamma}(\substack{0\\kk''})\right]\left[\sum_{k'} R_{\gamma\beta}(k_1 k')\right]\right\}.$$
$$(35.29)$$

On the other hand, in order that (35.26) should be identical with the corresponding equation (33.33), we must have

$$f(\substack{k\\ \beta}\alpha) = -\frac{1}{v_a}\left\{\sum_{k_1}\sum_{\gamma}\left[\sum_{k'} R_{\alpha\gamma}(k'k_1)\right]\left[Q_{\gamma\beta}(\substack{0\\k_1k})e_k - \delta_{k_1k}\sum_{k''} e_{k''} Q_{\gamma\beta}(\substack{0\\k_1k''})\right] + e_k\delta_{\alpha\beta}\right\},$$
$$(35.30)$$

$$f(\alpha\beta) = \frac{1}{v_a}\sum_{kk'} R_{\alpha\beta}(kk'). \quad\quad (35.31)$$

(30.33) shows that $Q_{\alpha\beta}(\substack{0\\kk'})$ is real; thus it follows from (30.32) and (30.34) that $Q_{\alpha\beta}(\substack{0\\kk'})$ is symmetric in both α, β and k, k'. This also shows that R, which is the inverse of (35.23), is a symmetric matrix. Bearing these facts in mind, we see readily that (35.30) and (35.29) are completely equivalent. It is, moreover, evident that the above expressions satisfy the general relations (33.8) and (33.9) given in the phenomenological theory.

The perturbation method can be developed without difficulty for long acoustic vibrations by treating the basic equations (35.13) and (35.16) in parallel at each stage of the perturbation procedure. The formulae involved are, however, very lengthy, so we shall not consider the details here.

We have assumed that the polarizability of an ion can be represented by a scalar α_k (isotropy). We note that the above formulae still apply if the polarizability of an ion is a tensor (anisotropy); we have only to replace the factor $\delta_{\alpha\beta}/\alpha_k$ in the first term of (35.23) by the inverse of the polarizability tensor of the ion k and redefine the matrix R accordingly.

THE FREE ENERGY

36. The specification of finite strains

CONSIDER any two lattice points $\binom{l}{k}$ and $\binom{l'}{k'}$ in a perfect lattice. Upon taking the scalar product of

$$\mathbf{x}\binom{l}{k}-\mathbf{x}\binom{l'}{k'} = \sum_{\mu} \mathbf{a}_{\mu}\{(l^{\mu}+\lambda^{\mu}(k))-(l'^{\mu}+\lambda^{\mu}(k'))\}$$

with itself, we find that the square of the distance between the two points is given by (see (22.13))

$$\sum_{\mu\nu} g_{\mu\nu}\{(l^{\mu}-l'^{\mu})+[\lambda^{\mu}(k)-\lambda^{\mu}(k')]\}\{(l^{\nu}-l'^{\nu})+[\lambda^{\nu}(k)-\lambda^{\nu}(k')]\}. \quad (36.1)$$

Thus the distance between any two lattice points is completely determined by the following parameters:

$$g_{\mu\nu} = g_{\nu\mu} \qquad \text{(metrical coefficients)}, \qquad (36.2)$$

$$\lambda^1(k),\ \lambda^2(k),\ \lambda^3(k) \quad \text{(contravariant components of } \mathbf{x}(k)\text{)}. \qquad (36.3)$$

In other words the geometrical structure of a perfect lattice can be completely specified by the above parameters.

Let us consider an arbitrary lattice and specify the Cartesian components of the basic vectors $\mathbf{a}_{\mu}$ by superscripts; thus $\mathbf{a}_{\mu} = (a_{\mu}^1, a_{\mu}^2, a_{\mu}^3)$. Now subject the lattice to an external strain, whereby the lattice points are displaced as follows:

$$u_{\alpha}\binom{l}{k} = \sum_{\beta} u_{\alpha\beta} x_{\beta}\binom{l}{k}. \qquad (36.4)$$

The deformation is equivalent to a change of the basic vectors from $\mathbf{a}_{\mu}$ to $\breve{\mathbf{a}}_{\mu}$, the latter being defined by

$$\breve{a}_{\mu}^{\alpha} = a_{\mu}^{\alpha} + \sum_{\beta} u_{\alpha\beta} a_{\mu}^{\beta}. \qquad (36.5)$$

For, corresponding to the new basic vectors $\breve{\mathbf{a}}_{\mu}$, a lattice point $\binom{l}{k}$ has the position vector

$$\breve{\mathbf{x}}\binom{l}{k} = \sum_{\mu} (l^{\mu}+\lambda^{\mu}(k))\breve{\mathbf{a}}_{\mu}, \qquad (36.6)$$

which implies that $\binom{l}{k}$ has been displaced by

$$\breve{x}_{\alpha}\binom{l}{k}-x_{\alpha}\binom{l}{k} = \sum_{\mu} (l^{\mu}+\lambda^{\mu}(k))(\breve{a}_{\mu}^{\alpha}-a_{\mu}^{\alpha}) = \sum_{\beta\mu} u_{\alpha\beta}(l^{\mu}+\lambda^{\mu}(k))a_{\mu}^{\beta} = \sum_{\beta} u_{\alpha\beta} x_{\beta}\binom{l}{k}$$
$$(36.7)$$

in agreement with (36.4). In other words, in an external strain the contravariant components $\lambda^{\mu}(k)$ of $\mathbf{x}(k)$ are undisturbed; moreover, as the lattice structure depends in general only on these parameters and the metrical coefficients, an external strain can be completely described by giving the changes in the metrical coefficients. Using (36.5), we find that the metrical coefficients of the deformed lattice are given by

$$\breve{g}_{\mu\nu} = \breve{\mathbf{a}}_\mu \cdot \breve{\mathbf{a}}_\nu = \sum_\alpha \left(a_\mu^\alpha + \sum_\gamma u_{\alpha\gamma} a_\mu^\gamma\right)\left(a_\nu^\alpha + \sum_\lambda u_{\alpha\lambda} a_\nu^\lambda\right). \qquad (36.8)$$

Subtracting from (36.8) the original metrical coefficients

$$g_{\mu\nu} = \sum_\alpha a_\mu^\alpha a_\nu^\alpha,$$

we obtain after appropriate relabelling of the summation indices the following:

$$\Delta g_{\mu\nu} = \breve{g}_{\mu\nu} - g_{\mu\nu} = \sum_{\alpha\beta} a_\mu^\alpha \left(u_{\alpha\beta} + u_{\beta\alpha} + \sum_\gamma u_{\gamma\alpha} u_{\gamma\beta}\right) a_\nu^\beta. \qquad (36.9)$$

Thus the strain depends on the parameters $u_{\alpha\beta}$ only through the expressions

$$\bar{u}_{\alpha\beta} = \tfrac{1}{2}\left\{u_{\alpha\beta} + u_{\beta\alpha} + \sum_\gamma u_{\gamma\alpha} u_{\gamma\beta}\right\} = \bar{u}_{\beta\alpha}, \qquad (36.10)$$

which may hence be used as parameters to specify an external strain. These parameters are by definition symmetric in the two indices; there are thus altogether six independent parameters of this kind.

$u_{\alpha\beta}$, $\bar{u}_{\alpha\beta}$ and the components a_μ^α, $\breve{a}_\mu^\alpha$ of the basic vectors $\mathbf{a}_\mu$, $\breve{\mathbf{a}}_\mu$ can be regarded as elements of 3×3 matrices U, $\bar{U}$, A, and $\breve{A}$ respectively, where the upper and lower indices of a_μ^α, $\breve{a}_\mu^\alpha$ are taken as the row and column indices respectively. In matrix notation (36.5) can be written as

$$\breve{A} = (1 + U)A \qquad (36.11)$$

(the number 1 stands for the unit matrix). Forming the determinants of both sides of the above relation, we get

$$|\breve{A}| = |(1+U)A| = |1+U||A|, \qquad (36.12)$$

where the vertical bars signify the determinant of the matrix enclosed. $|A|$ and $|\breve{A}|$, we note, are identically equal to $\mathbf{a}_1 \cdot \mathbf{a}_2 \wedge \mathbf{a}_3$ and $\breve{\mathbf{a}}_1 \cdot \breve{\mathbf{a}}_2 \wedge \breve{\mathbf{a}}_3$; thus they represent respectively the volume per cell before and after the deformation. Hence it follows from (36.12) that $|1+U|$ is the volume expansion ratio. Clearly for physically realizable deformations $|1+U|$ cannot vanish. Moreover, as every finite deformation U must be built up continuously through physically realizable steps, during which $|1+U|$ cannot change its sign, we must have

$$|1+U| > 0. \qquad (36.13)$$

We can write (36.10) in matrix notation as follows:

$$\bar{U} = \tfrac{1}{2}\{U + \tilde{U} + \tilde{U}U\}, \qquad (36.14)$$

where the transpose of a matrix is indicated by a wavy line. Let us consider two deformations U and U', which are such that the matrices $\overline{U}$, $\overline{U}'$ constructed according to (36.14) are identical. U and U' thus fulfil the following relation:

$$(1+\tilde{U})(1+U) = 1+2\overline{U} = 1+2\overline{U}' = (1+\tilde{U}')(1+U'). \quad (36.15)$$

Since two such deformations lead to the same strain of the lattice, the corresponding deformed lattices can differ only in their orientation. This can be shown explicitly by considering the matrix

$$\Theta = (1+U')(1+U)^{-1} \quad (36.16)$$

(owing to (36.13), the inverse $(1+U)^{-1}$ exists), which relates the two deformations as follows:

$$(1+U') = \Theta(1+U). \quad (36.17)$$

Multiplying (36.16) by its transpose and using (36.15), one finds readily that

$$\tilde{\Theta}\Theta = 1. \quad (36.18)$$

Moreover, it follows from (36.13) and (36.17) that the determinant of Θ is positive. Hence Θ is an orthogonal matrix which, operating on a vector (regarded as a column matrix) represents a proper rotation. The relation (36.17) thus shows that the deformation U' is equivalent to the deformation U followed by a rotation.

In a general homogeneous deformation, the lattice points are displaced as follows:

$$u_\alpha\binom{l}{k} = \sum_\beta u_{\alpha\beta}\, x_\beta\binom{l}{k}+u_\alpha(k). \quad (36.19)$$

The deformation can be built up in two steps corresponding respectively to the first and second terms on the right-hand side of (36.19). The deformation carried out in the first step is of the type already discussed above; we have seen that it leads to the new basic vectors $\breve{a}_\mu$ given by (36.5). Since the displacements $\mathbf{u}(k)$ in the second step do not affect the basic vectors, $\breve{a}_\mu$ are still the basic vectors after the complete deformation (36.19); accordingly the changes of the metrical coefficients are also given by (36.9) in the general case. It follows from our discussion above that the contravariant components $\lambda^1(k)$, $\lambda^2(k)$, $\lambda^3(k)$ of $\mathbf{x}(k)$ are not altered during the first step of the deformation. As the displacements $\mathbf{u}(k)$ give directly the changes in $\mathbf{x}(k)$ in the second step of the deformation, during which $\breve{a}_\mu$ are the basic vectors, we have

$$u_\alpha(k) = \Delta x_\alpha(k) = \sum_\mu \Delta\lambda^\mu(k)\breve{a}_\mu^\alpha. \quad (36.20)$$

Regarding $\Delta\lambda^\mu(k)$ ($\mu = 1, 2, 3$) and $u_\alpha(k)$ ($\alpha = 1, 2, 3$) as elements of column matrices Λ_k and U_k respectively, we can write (36.20) in matrix notation thus:

$$U_k = \breve{A}\Lambda_k. \quad (36.21)$$

Multiplying (36.21) by the inverse of $\check{A}$ and using (36.11), we find that

$$\Lambda_k = \check{A}^{-1}U_k = [(1+U)A]^{-1}U_k = A^{-1}(1+U)^{-1}U_k. \quad (36.22)$$

This relation can be rewritten as follows:

$$\left.\begin{aligned}\Lambda_k &= A^{-1}(1+U)^{-1}(1+\check{U})^{-1}(1+\check{U})U_k \\ &= A^{-1}[(1+\check{U})(1+U)]^{-1}(1+\check{U})U_k \\ &= A^{-1}[1+2\bar{U}]^{-1}(1+\check{U})U_k\end{aligned}\right\}. \quad (36.23)$$

Since A^{-1} is completely defined in terms of the known basic vectors of the undeformed lattice, and $\bar{U}$ is determined by the external strain parameters $\bar{u}_{\alpha\beta}$, we have thus only to know the column matrix $(1+\check{U})U_k$, the elements of which are

$$\bar{u}_\alpha(k) = u_\alpha(k) + \sum_\beta u_{\beta\alpha}u_\beta(k). \quad (36.24)$$

Once $\bar{u}_\alpha(k)$ as well as $\bar{u}_{\alpha\beta}$ is given, the changes in $g_{\mu\nu}$ and $\lambda^\mu(k)$ are determined by (36.9) and (36.23); the structure of the deformed lattice is thereby completely specified. In the discussion of central forces in § 11, we have verified that $\bar{u}_\alpha(k)$ and $\bar{u}_{\alpha\beta}$ are adequate strain parameters to the second order of accuracy, if they are small compared with unity. The present discussion shows that they are appropriate strain parameters irrespective of the magnitude of the strain.

It is quite clear why we cannot simply use the vectors $\mathbf{u}(k)$ in conjunction with $\bar{u}_{\alpha\beta}$ to specify a general homogeneous strain. If we carry out the deformations in two successive steps as above, two deformations with the same parameters $\bar{u}_{\alpha\beta}$ in general lead to differently oriented (otherwise identical) structures. The same subsequent displacements $\mathbf{u}(k)$ will evidently then lead to different structures in the two cases.

Clearly the volume expansion ratio $|1+U|$ must be completely determined by the external strain. The explicit expression of $|1+U|$ in terms of the strain parameters $\bar{u}_{\alpha\beta}$ can be obtained as follows. Forming the determinants of both sides of the identity

$$(1+\check{U})(1+U) = 2\bar{U}+1, \quad (36.25)$$

we get

$$|1+\check{U}||1+U| = |2\bar{U}+1|. \quad (36.26)$$

Since the determinant of a matrix is equal to that of its transpose, (36.26) gives the following expression for the expansion ratio:

$$|2\bar{U}+1|^{\frac{1}{2}}, \quad (36.27)$$

which depends only on the parameters $\bar{u}_{\alpha\beta}$.

37. Phenomenological discussion of the free energy of a lattice

In a thermodynamical treatment, we consider a homogeneous strain as completely specified by the external strain parameters $\bar{u}_{\alpha\beta}$ alone, the internal strain being determined automatically by the internal coordinates used to describe the thermal motion. We may observe the contrast of the situation to the static treatment in § 11, which determines the elastic constants about the equilibrium configuration. In the static treatment, it is necessary to take account explicitly of the internal strain induced by a given external strain, whereas an adequate thermodynamical treatment makes allowance for this effect implicitly. Thus the expression for the free energy which we shall discuss corresponds to the strain-energy function (11.28) of the static treatment. Hence we may refer to the external strain as the elastic strain and the parameters $\bar{u}_{\alpha\beta}$ as the elastic strain parameters. This brings our terminology completely into line with the usual usage in elasticity theory, where a given elastic strain implies also its accompanying internal strain although the latter does not appear explicitly in the theory.

Speaking from a phenomenological point of view, we can freely alter the elastic strain by the application of mechanical stresses, but we have no direct access to the internal strain. A partial control over the internal strain may, however, be exercised by applying electric fields. In the following, we shall consider the free energy of a lattice which is in the presence of a field **E** as well as subject to an arbitrary elastic strain.

It is to be remarked that the free energy of such a lattice cannot be represented simply as a function of the field **E** and the elastic strain parameters $\bar{u}_{\alpha\beta}$. For **E** and $\bar{u}_{\alpha\beta}$ together do not describe unambiguously a thermodynamical state of the system, for the simple reason that the strain parameters give no indication as to how the deformed crystal is oriented relative to the field. Thus, instead of **E**, we shall consider its contravariant components in the deformed lattice; it is clear that the free energy must be the same for two similarly deformed specimens (i.e. with same parameters $\bar{u}_{\alpha\beta}$), each being subject to an electric field such that the contravariant field components are the same in both cases.

We remember that the quantities

$$\Delta\lambda^1(k), \quad \Delta\lambda^2(k), \quad \Delta\lambda^3(k) \tag{37.1}$$

discussed in the last section are the contravariant components of the vectors $\mathbf{u}(k)$. Thus upon replacing the Cartesian components of $\mathbf{u}(k)$ (given as a column matrix) on the right-hand side of (36.23) by the Cartesian components of **E**, we find immediately that the contravariant

components of **E** in a deformed lattice are given by the elements of the
column matrix

$$A^{-1}[1+2\overline{U}]^{-1}(1+\tilde{U})\begin{pmatrix}E_1\\E_2\\E_3\end{pmatrix}. \tag{37.2}$$

Since $[1+2\overline{U}]^{-1}$ is completely determined by the strain parameters, we
can clearly introduce the elements of the column matrix

$$(1+\tilde{U})\begin{pmatrix}E_1\\E_2\\E_3\end{pmatrix}, \tag{37.3}$$

namely, $$\overline{E}_\alpha = E_\alpha + \sum_\beta u_{\beta\alpha} E_\beta, \tag{37.4}$$

as thermodynamical parameters in addition to $\bar{u}_{\alpha\beta}$ and the temperature
T, if an electric field is present. We notice that the parameters $\overline{\mathbf{E}}$ and
$\bar{u}_{\alpha\beta}$ make no reference to the absolute orientation of the lattice and the
field.

We shall write the free energy per unit volume of a deformed lattice
as follows:

$$\frac{1}{|1+2\overline{U}|^{\frac{1}{2}}} F(\bar{u}_{\alpha\beta}, \overline{\mathbf{E}}, T), \tag{37.5}$$

where the factor $1/|1+2\overline{U}|^{\frac{1}{2}}$ is introduced essentially as a normalization
factor so that F represents the free energy for a portion of the medium
which *originally* occupies unit volume in the undeformed lattice. In the
following discussion it is convenient to consider F formally as a function
of both $\bar{u}_{\alpha\beta}$ and $\bar{u}_{\beta\alpha}$, although by definition $\bar{u}_{\alpha\beta} \equiv \bar{u}_{\beta\alpha}$. This leaves, of
course, the form of F to a certain extent arbitrary; but this arbitrariness
does not affect the discussion in any way. That is, whatever form F
may take within the permitted latitude, the results which we shall obtain
remain the same.

We shall proceed to derive from the free energy the expressions for
the stress components and the dielectric polarization in a specimen
which is in an arbitrary state of deformation $u_{\alpha\beta}$, and subject at the same
time to an electric field **E**. Let us subject the specimen to an infinitesimal
elastic deformation whereby the elastic displacement at a point $\breve{\mathbf{x}}$ in the
specimen is given by $$d\breve{x}_\alpha = \sum_\gamma v_{\alpha\gamma} \breve{x}_\gamma, \tag{37.6}$$

the deformation parameters $v_{\alpha\gamma}$ being infinitesimal. If we split up $v_{\alpha\gamma}$ as
follows: $$v_{\alpha\gamma} = \sigma_{\alpha\gamma} + \omega_{\alpha\gamma}, \tag{37.7}$$

where
$$\sigma_{\alpha\gamma} = \tfrac{1}{2}(v_{\alpha\gamma} + v_{\gamma\alpha}) = \sigma_{\gamma\alpha}, \tag{37.8}$$

$$\omega_{\alpha\gamma} = \tfrac{1}{2}(v_{\alpha\gamma} - v_{\gamma\alpha}) = -\omega_{\gamma\alpha}, \tag{37.9}$$

the symmetric part $\sigma_{\alpha\gamma}$ describes an infinitesimal elastic strain, whereas the antisymmetric part $\omega_{\alpha\gamma}$ represents an infinitesimal rotation.

We shall suppose that the infinitesimal deformation is carried out with the field **E** kept constant and consider the work done during the process on a unit volume of the medium. With the stress components denoted as before by $S_{\alpha\gamma} \equiv S_{\gamma\alpha}$, the work done in building up the infinitesimal strain is given by

$$\sum_{\alpha\gamma} S_{\alpha\gamma}\,\sigma_{\alpha\gamma}. \tag{37.10}$$

If the specimen is electrically polarized, work is expended in carrying out the infinitesimal rotation against the torque exerted by the field. The amount of work required is equal to the change of potential energy due to the rotation of the dielectric polarization **P** in the field **E**. The rotation of **P** brings about a vectorial change in the moment given by

$$\sum_{\gamma} \omega_{\alpha\gamma} P_{\gamma}, \tag{37.11}$$

the scalar product of which with $-$**E** gives directly the change in potential energy
$$-\sum_{\alpha\gamma} E_{\alpha} P_{\gamma}\,\omega_{\alpha\gamma}. \tag{37.12}$$

Adding (37.10) and (37.12), we obtain the total work done on unit volume of the medium:

$$\Delta W = \sum_{\alpha\gamma} S_{\alpha\gamma}\,\sigma_{\alpha\gamma} - \sum_{\alpha\gamma} E_{\alpha} P_{\gamma}\,\omega_{\alpha\gamma}. \tag{37.13}$$

If the infinitesimal deformation is carried out isothermally the work ΔW is equal to the change in free energy:

$$\frac{\Delta F}{|1+2\overline{\overline{U}}|^{\frac{1}{2}}} = \frac{1}{|1+2\overline{\overline{U}}|^{\frac{1}{2}}} \left\{ \sum_{\mu\nu} \left(\frac{\partial F}{\partial \bar{u}_{\mu\nu}}\right) \Delta \bar{u}_{\mu\nu} + \sum_{\mu} \left(\frac{\partial F}{\partial \bar{\bar{E}}_{\mu}}\right) \Delta \bar{E}_{\mu} \right\}. \tag{37.14}$$

The normalization factor is to be treated as a constant in this connexion, for we are considering the free energy of a definite portion of the medium, namely, a portion that occupies unit volume prior to the infinitesimal deformation.

In view of the definition (36.10) of the external strain parameters, we have

$$\Delta \bar{u}_{\mu\nu} = \tfrac{1}{2}\{\Delta u_{\mu\nu} + \Delta u_{\nu\mu} + \sum_{\alpha} \Delta u_{\alpha\mu} u_{\alpha\nu} + \sum_{\alpha} u_{\alpha\mu} \Delta u_{\alpha\nu}\}. \tag{37.15}$$

Since the field is kept constant, it follows from (37.4) that

$$\Delta \bar{E}_{\mu} = \sum_{\alpha} \Delta u_{\alpha\mu} E_{\alpha}. \tag{37.16}$$

The changes $\Delta u_{\mu\nu}$ in the deformation parameters $u_{\mu\nu}$ can be calculated as follows: a point $\breve{x}$ in the specimen prior to the infinitesimal deformation is related to its position x in the completely undeformed lattice by

$$\breve{x}_\mu = x_\mu + \sum_\nu u_{\mu\nu} x_\nu. \qquad (37.17)$$

According to (37.6) the infinitesimal deformation carries the point into the following position:

$$\begin{aligned}
\breve{x}_\mu + d\breve{x}_\mu &= \breve{x}_\mu + \sum_\gamma v_{\mu\gamma} \breve{x}_\gamma \\
&= x_\mu + \sum_\nu u_{\mu\nu} x_\nu + \sum_\gamma v_{\mu\gamma} x_\gamma + \sum_{\gamma\nu} v_{\mu\gamma} u_{\gamma\nu} x_\nu \\
&= x_\mu + \sum_\nu \left[u_{\mu\nu} + v_{\mu\nu} + \sum_\gamma v_{\mu\gamma} u_{\gamma\nu} \right] x_\nu. \qquad (37.18)
\end{aligned}$$

Comparing (37.18) with (37.17), we observe that the infinitesimal deformation changes $u_{\mu\nu}$ into

$$u_{\mu\nu} + v_{\mu\nu} + \sum_\gamma v_{\mu\gamma} u_{\gamma\nu}.$$

In other words,
$$\Delta u_{\mu\nu} = v_{\mu\nu} + \sum_\gamma v_{\mu\gamma} u_{\gamma\nu}. \qquad (37.19)$$

By suitably relabelling the indices μ, ν this relation may be used to eliminate $\Delta u_{\mu\nu}$, etc., from (37.15). Thus we find that (37.15) can be written as follows:

$$\begin{aligned}
\Delta \bar{u}_{\mu\nu} = \tfrac{1}{2} \Big\{ & v_{\mu\nu} + \sum_\gamma v_{\mu\gamma} u_{\gamma\nu} + v_{\nu\mu} + \sum_\gamma v_{\nu\gamma} u_{\gamma\mu} + \sum_\alpha v_{\alpha\mu} u_{\alpha\nu} + \\
& + \sum_{\alpha\gamma} v_{\alpha\gamma} u_{\gamma\mu} u_{\alpha\nu} + \sum_\alpha u_{\alpha\mu} v_{\alpha\nu} + \sum_{\alpha\gamma} u_{\alpha\mu} v_{\alpha\gamma} u_{\gamma\nu} \Big\}. \qquad (37.20)
\end{aligned}$$

On the right-hand side we rewrite the summation index γ as α in the fourth term, α as γ in the fifth term, and α, γ as γ, α respectively in the sixth term. It is then observed that (37.20) can be written in the alternative form

$$\begin{aligned}
\Delta \bar{u}_{\mu\nu} = \tfrac{1}{2} \Big\{ & v_{\mu\nu} + v_{\nu\mu} + \sum_\gamma (v_{\mu\gamma} + v_{\gamma\mu}) u_{\gamma\nu} + \sum_\alpha (v_{\alpha\nu} + v_{\nu\alpha}) u_{\alpha\mu} + \\
& + \sum_{\alpha\gamma} (v_{\alpha\gamma} + v_{\gamma\alpha}) u_{\alpha\mu} u_{\gamma\nu} \Big\}, \qquad (37.21)
\end{aligned}$$

which is clearly also equivalent to

$$\begin{aligned}
\Delta \bar{u}_{\mu\nu} &= \tfrac{1}{2} \sum_{\alpha\gamma} (\delta_{\alpha\mu} + u_{\alpha\mu})(\delta_{\gamma\nu} + u_{\gamma\nu})(v_{\alpha\gamma} + v_{\gamma\alpha}) \\
&= \sum_{\alpha\gamma} (\delta_{\alpha\mu} + u_{\alpha\mu})(\delta_{\gamma\nu} + u_{\gamma\nu}) \sigma_{\alpha\gamma}. \qquad (37.22)
\end{aligned}$$

Relabelling μ, ν in (37.19) as α, μ respectively and using the relation in (37.16), we get

$$\Delta \bar{E}_\mu = \sum_\alpha \left\{ v_{\alpha\mu} + \sum_\gamma v_{\alpha\gamma} u_{\gamma\mu} \right\} E_\alpha$$
$$= \sum_{\alpha\gamma} E_\alpha (\delta_{\gamma\mu} + u_{\gamma\mu}) v_{\alpha\gamma}$$
$$= \sum_{\alpha\gamma} E_\alpha (\delta_{\gamma\mu} + u_{\gamma\mu})(\sigma_{\alpha\gamma} + \omega_{\alpha\gamma}). \qquad (37.23)$$

After substituting (37.22) and (37.23) in the change of free energy (37.14) and equating the latter to the work done ΔW given by (37.13), we get

$$\sum_{\alpha\gamma} S_{\alpha\gamma} \sigma_{\alpha\gamma} - \sum_{\alpha\gamma} E_\alpha P_\gamma \omega_{\alpha\gamma}$$
$$= \frac{1}{|1+2\bar{U}|^{\frac{1}{2}}} \left\{ \sum_{\alpha\gamma} \sigma_{\alpha\gamma} \left[\sum_{\mu\nu} (\delta_{\alpha\mu} + u_{\alpha\mu})(\delta_{\gamma\nu} + u_{\gamma\nu}) \left(\frac{\partial F}{\partial \bar{u}_{\mu\nu}} \right) + \right. \right.$$
$$\left. + E_\alpha \sum_\mu (\delta_{\gamma\mu} + u_{\gamma\mu}) \left(\frac{\partial F}{\partial \bar{\bar{E}}_\mu} \right) \right] + \sum_{\alpha\gamma} \omega_{\alpha\gamma} \left[E_\alpha \sum_\mu (\delta_{\gamma\mu} + u_{\gamma\mu}) \left(\frac{\partial F}{\partial \bar{\bar{E}}_\mu} \right) \right] \right\}.$$
$$(37.24)$$

This relation must be fulfilled identically by all values of $\sigma_{\alpha\gamma}$ and $\omega_{\alpha\gamma}$ consistent with the symmetry requirements (37.8) and (37.9). It follows that

$$S_{\alpha\gamma} \equiv \tfrac{1}{2}(S_{\alpha\gamma} + S_{\gamma\alpha}) = \frac{1}{2|1+2\bar{U}|^{\frac{1}{2}}} \left\{ \sum_{\mu\nu} (\delta_{\alpha\mu} + u_{\alpha\mu})(\delta_{\gamma\nu} + u_{\gamma\nu}) \left(\frac{\partial F}{\partial \bar{u}_{\mu\nu}} \right) + \right.$$
$$+ \sum_{\mu\nu} (\delta_{\gamma\mu} + u_{\gamma\mu})(\delta_{\alpha\nu} + u_{\alpha\nu}) \left(\frac{\partial F}{\partial \bar{u}_{\mu\nu}} \right) + E_\alpha \sum_\mu (\delta_{\gamma\mu} + u_{\gamma\mu}) \left(\frac{\partial F}{\partial \bar{\bar{E}}_\mu} \right) +$$
$$\left. + E_\gamma \sum_\mu (\delta_{\alpha\mu} + u_{\alpha\mu}) \left(\frac{\partial F}{\partial \bar{\bar{E}}_\mu} \right) \right\}, \qquad (37.25)$$

$$E_\gamma P_\alpha - E_\alpha P_\gamma = \frac{1}{|1+2\bar{U}|^{\frac{1}{2}}} \left\{ E_\alpha \sum_\mu (\delta_{\gamma\mu} + u_{\gamma\mu}) \left(\frac{\partial F}{\partial \bar{\bar{E}}_\mu} \right) - \right.$$
$$\left. - E_\gamma \sum_\mu (\delta_{\alpha\mu} + u_{\alpha\mu}) \left(\frac{\partial F}{\partial \bar{\bar{E}}_\mu} \right) \right\}. \qquad (37.26)$$

Let us rewrite the summation indices μ, ν respectively as ν, μ in the second term in the curly brackets of (37.25) and combine the term with the first term in the brackets; thus (37.25) can be rewritten as

$$S_{\alpha\gamma} = \frac{1}{2|1+2\bar{U}|^{\frac{1}{2}}} \left\{ \sum_{\mu\nu} (\delta_{\alpha\mu} + u_{\alpha\mu})(\delta_{\gamma\nu} + u_{\gamma\nu}) \left[\frac{\partial F}{\partial \bar{u}_{\mu\nu}} + \frac{\partial F}{\partial \bar{u}_{\nu\mu}} \right] + \right.$$
$$\left. + E_\alpha \sum_\mu (\delta_{\gamma\mu} + u_{\gamma\mu}) \left(\frac{\partial F}{\partial \bar{\bar{E}}_\mu} \right) + E_\gamma \sum_\mu (\delta_{\alpha\mu} + u_{\alpha\mu}) \left(\frac{\partial F}{\partial \bar{\bar{E}}_\mu} \right) \right\}. \qquad (37.27)$$

It follows from our earlier discussion in § 18 that the dielectric polarization should be equal to the derivative of the free energy of unit volume of the medium with respect to $-\mathbf{E}$. Remembering that the normalization factor is to be kept constant, we find on differentiating (37.5) that the dielectric polarization is given by

$$P_\alpha = -\frac{1}{|1+2\bar{U}|^{\frac{1}{2}}}\left(\frac{\partial F}{\partial E_\alpha}\right) = -\frac{1}{|1+2\bar{U}|^{\frac{1}{2}}}\sum_\mu \left(\frac{\partial F}{\partial \bar{E}_\mu}\right)\left(\frac{\partial \bar{E}_\mu}{\partial E_\alpha}\right). \quad (37.28)$$

Using (37.4), we get thus

$$P_\alpha = -\frac{1}{|1+2\bar{U}|^{\frac{1}{2}}}\sum_\mu (\delta_{\alpha\mu}+u_{\alpha\mu})\left(\frac{\partial F}{\partial \bar{E}_\mu}\right), \quad (37.29)$$

which, we note, is in complete agreement with (37.26).

(37.27) and (37.29) are the general expressions for the stress components and dielectric polarization in an arbitrarily deformed specimen in the presence of an electric field.

At a given temperature T a free crystal (i.e. in the absence of either stress or field) assumes the structure for which the free energy is a minimum. The corresponding strain parameters, which we shall denote by $\bar{u}^T_{\alpha\beta}$, are thus determined by

$$\left(\frac{\partial F}{\partial \bar{u}_{\alpha\beta}}\right)_{\bar{u}_{\alpha\beta}}=\bar{u}^T_{\alpha\beta}, \mathbf{E}=0 = 0. \quad (37.30)$$

The parameters $\bar{u}^T_{\alpha\beta}$ completely determine the thermal expansion; the volume expansion, for instance, is given by

$$|1+2\bar{U}^T|^{\frac{1}{2}}. \quad (37.31)$$

Clearly, consistent with the same parameters $\bar{u}^T_{\alpha\beta}$, the possible sets of deformation parameters $u_{\alpha\beta}$ $((1+\tilde{U})(1+U)=1+2\bar{U}^T)$ are unlimited in number. The condition of minimum free energy leaves the orientation of the specimen completely arbitrary, and the different sets of $u_{\alpha\beta}$ merely describe the specimen in various orientations. We shall denote an arbitrarily chosen set of such parameters by $u^T_{\alpha\beta}$. When we discuss the properties of a crystal at a given temperature T, the use of a particular set of parameters $u^T_{\alpha\beta}$ is equivalent to a choice of the orientation of the specimen.

In the following discussion the derivatives of F taken at the state of the free crystal at temperature T (i.e. $\bar{u}_{\alpha\beta}=\bar{u}^T_{\alpha\beta}$, $\mathbf{E}=0$) will be indicated simply by a suffix T. Thus, the dielectric polarization (37.29) of a free crystal at a finite temperature T, can be written as

$$P_\alpha^{\mathrm{pyro}}(T) = -\frac{1}{|1+2\overline{U}^T|^{\frac{1}{2}}} \sum_\mu (\delta_{\alpha\mu}+u_{\alpha\mu}^T)\left(\frac{\partial F}{\partial \overline{\overline{E}}_\mu}\right)_T, \qquad (37.32)$$

which is known as the *pyroelectric moment* (more precisely, it is the sum of the pyroelectric moment and permanent electric moment).

In considering the normal mechanical and electrical properties of a crystal we are concerned with elastic strains and electric fields so small that the stress and the dielectric polarization depend on them practically linearly. This means that for such considerations the elastic strains and electric fields can be considered effectively as infinitesimal and the calculations accordingly confined to the first order of accuracy. Let us subject a specimen at temperature T to an elastic deformation described by the parameters $s_{\alpha y}$ and an electric field $\mathbf{E}$, wherein we may consider both $s_{\alpha y}$ and $\mathbf{E}$ as infinitesimal. Expanding the right-hand side of (37.27) about $\bar{u}_{\alpha\beta} = \bar{u}_{\alpha\beta}^T$, $u_{\alpha\beta} = u_{\alpha\beta}^T$ and $\mathbf{E} = \overline{\mathbf{E}} = 0$, and remembering (37.30), we obtain the following:

$$S_{\alpha y} = \tfrac{1}{2}|1+2\overline{U}^T|^{-\frac{1}{2}}\times$$

$$\times \Bigg\{ \sum_{\mu\nu\sigma\omega} (\delta_{\alpha\mu}+u_{\alpha\mu}^T)(\delta_{\gamma\nu}+u_{\gamma\nu}^T)\left[\left(\frac{\partial^2 F}{\partial\bar{u}_{\sigma\omega}\,\partial\bar{u}_{\mu\nu}}\right)_T+\left(\frac{\partial^2 F}{\partial\bar{u}_{\sigma\omega}\,\partial\bar{u}_{\nu\mu}}\right)_T\right]\Delta\bar{u}_{\sigma\omega}+$$

$$+ \sum_{\mu\nu\sigma} (\delta_{\alpha\mu}+u_{\alpha\mu}^T)(\delta_{\gamma\nu}+u_{\gamma\nu}^T)\left[\left(\frac{\partial^2 F}{\partial\overline{E}_\sigma\,\partial\bar{u}_{\mu\nu}}\right)_T+\left(\frac{\partial^2 F}{\partial\overline{E}_\sigma\,\partial\bar{u}_{\nu\mu}}\right)_T\right]\overline{E}_\sigma+$$

$$+ \sum_{\beta\mu} \left[\delta_{\alpha\beta}(\delta_{\gamma\mu}+u_{\gamma\mu}^T)\left(\frac{\partial F}{\partial\overline{\overline{E}}_\mu}\right)_T+\delta_{\gamma\beta}(\delta_{\alpha\mu}+u_{\alpha\mu}^T)\left(\frac{\partial F}{\partial\overline{\overline{E}}_\mu}\right)_T\right]E_\beta\Bigg\}, \qquad (37.33)$$

where terms higher than the first order are ignored. Since $\mathbf{E}$ is of the first order, the last two terms in (37.27) should remain the same in (37.33) apart from the additional suffix T to the derivatives of F; we have, however, modified their appearance in (37.33) by introducing a Kronecker factor for later convenience. The relation between $\Delta\bar{u}_{\sigma\omega}$ and the elastic deformation parameters $s_{\beta\lambda}$ can be written down directly with the help of the earlier relation (37.22); we have

$$\Delta\bar{u}_{\sigma\omega} = \tfrac{1}{2}\sum_{\beta\lambda} (\delta_{\beta\sigma}+u_{\beta\sigma}^T)(\delta_{\lambda\omega}+u_{\lambda\omega}^T)(s_{\beta\lambda}+s_{\lambda\beta}). \qquad (37.34)$$

On the other hand, since the field is of the first order, it follows from (37.4) that to the first order of accuracy

$$\bar{E}_\sigma = E_\sigma + \sum_\beta u^T_{\beta\sigma} E_\beta$$

$$= \sum_\beta (\delta_{\beta\sigma} + u^T_{\beta\sigma}) E_\beta. \tag{37.35}$$

Upon using (37.34) and (37.35) in (37.33), we get

$$S_{\alpha\gamma} = \frac{1}{2|1+2\bar{\bar{U}}^T|^{\frac{1}{2}}} \left\{ \frac{1}{2} \sum_{\beta\lambda} s_{\beta\lambda} \sum_{\mu\nu\sigma\omega} (\delta_{\alpha\mu}+u^T_{\alpha\mu})(\delta_{\gamma\nu}+u^T_{\gamma\nu}) \left[\left(\frac{\partial^2 F}{\partial \bar{u}_{\sigma\omega} \partial \bar{u}_{\mu\nu}} \right)_T + \right. \right.$$

$$+ \left. \left(\frac{\partial^2 F}{\partial \bar{u}_{\sigma\omega} \partial \bar{u}_{\nu\mu}} \right)_T \right] (\delta_{\beta\sigma}+u^T_{\beta\sigma})(\delta_{\lambda\omega}+u^T_{\lambda\omega}) +$$

$$+ \frac{1}{2} \sum_{\beta\lambda} s_{\lambda\beta} \sum_{\mu\nu\sigma\omega} (\delta_{\alpha\mu}+u^T_{\alpha\mu})(\delta_{\gamma\nu}+u^T_{\gamma\nu}) \left[\left(\frac{\partial^2 F}{\partial \bar{u}_{\sigma\omega} \partial \bar{u}_{\mu\nu}} \right)_T + \right.$$

$$+ \left. \left(\frac{\partial^2 F}{\partial \bar{u}_{\sigma\omega} \partial \bar{u}_{\nu\mu}} \right)_T \right] (\delta_{\beta\sigma}+u^T_{\beta\sigma})(\delta_{\lambda\omega}+u^T_{\lambda\omega}) +$$

$$+ \sum_\beta E_\beta \sum_{\mu\nu\sigma} (\delta_{\alpha\mu}+u^T_{\alpha\mu})(\delta_{\gamma\nu}+u^T_{\gamma\nu}) \left[\left(\frac{\partial^2 F}{\partial \bar{E}_\sigma \partial \bar{u}_{\mu\nu}} \right)_T + \right.$$

$$+ \left. \left(\frac{\partial^2 F}{\partial \bar{E}_\sigma \partial \bar{u}_{\nu\mu}} \right)_T \right] (\delta_{\beta\sigma}+u^T_{\beta\sigma}) +$$

$$+ \sum_\beta E_\beta \sum_\mu \left[\delta_{\alpha\beta}(\delta_{\gamma\mu}+u^T_{\gamma\mu}) \left(\frac{\partial F}{\partial \bar{E}_\mu} \right)_T + \right.$$

$$+ \left. \left. \delta_{\gamma\beta}(\delta_{\alpha\mu}+u^T_{\alpha\mu}) \left(\frac{\partial F}{\partial \bar{E}_\mu} \right)_T \right] \right\}. \tag{37.36}$$

After rewriting the summation indices λ, β, σ, ω respectively as β, λ, ω, σ in the second sum on the right-hand side, we can combine the latter with the first sum; it is then found that (37.36) can be written as follows:

$$S_{\alpha\gamma} = \sum_{\beta\lambda} c_{\alpha\gamma,\beta\lambda} s_{\beta\lambda} - \sum_\beta e_{\beta,\alpha\gamma} E_\beta, \tag{37.37}$$

where the coefficients are defined by

$$c_{\alpha\gamma,\beta\lambda} = \frac{1}{4|1+2\bar{\bar{U}}^T|^{\frac{1}{2}}} \sum_{\mu\nu\sigma\omega} (\delta_{\alpha\mu}+u^T_{\alpha\mu})(\delta_{\gamma\nu}+u^T_{\gamma\nu}) \left[\left(\frac{\partial^2 F}{\partial \bar{u}_{\sigma\omega} \partial \bar{u}_{\mu\nu}} \right)_T + \right.$$

$$+ \left. \left(\frac{\partial^2 F}{\partial \bar{u}_{\sigma\omega} \partial \bar{u}_{\nu\mu}} \right)_T + \left(\frac{\partial^2 F}{\partial \bar{u}_{\omega\sigma} \partial \bar{u}_{\mu\nu}} \right)_T + \left(\frac{\partial^2 F}{\partial \bar{u}_{\omega\sigma} \partial \bar{u}_{\nu\mu}} \right)_T \right] \times$$

$$\times (\delta_{\beta\sigma}+u^T_{\beta\sigma})(\delta_{\lambda\omega}+u^T_{\lambda\omega}), \tag{37.38}$$

and

$$e_{\beta,\alpha\gamma} = \frac{-1}{2|1+2\bar{U}^T|^{\frac{1}{2}}}\Bigg\{ \delta_{\alpha\beta}\sum_{\mu}(\delta_{\gamma\mu}+u_{\gamma\mu}^T)\left(\frac{\partial F}{\partial \bar{\bar{E}}_\mu}\right)_T +$$

$$+\delta_{\gamma\beta}\sum_{\mu}(\delta_{\alpha\mu}+u_{\alpha\mu}^T)\left(\frac{\partial F}{\partial \bar{\bar{E}}_\mu}\right)_T + \sum_{\mu\nu\sigma}(\delta_{\alpha\mu}+u_{\alpha\mu}^T)(\delta_{\gamma\nu}+u_{\gamma\nu}^T)\times$$

$$\times\left[\left(\frac{\partial^2 F}{\partial\bar{\bar{E}}_\sigma\,\partial\bar{u}_{\mu\nu}}\right)_T + \left(\frac{\partial^2 F}{\partial\bar{\bar{E}}_\sigma\,\partial\bar{u}_{\nu\mu}}\right)_T\right](\delta_{\beta\sigma}+u_{\beta\sigma}^T)\Bigg\}. \quad (37.39)$$

It is easily verified that the above coefficients satisfy the symmetry relations

$$c_{\alpha\gamma,\beta\lambda} = c_{\gamma\alpha,\beta\lambda} = c_{\beta\lambda,\alpha\gamma}, \qquad (37.40)$$

$$e_{\beta,\alpha\gamma} = e_{\beta,\gamma\alpha}. \qquad (37.41)$$

These are the elastic and piezoelectric constants, which we have discussed in Chapter V in connexion with a lattice in static equilibrium.

We consider next the dielectric polarization. Expanding the right-hand side of (37.29) about the state of the free crystal at temperature T, we obtain to the first order of accuracy

$$P_\alpha = P_\alpha^{\text{pyro}}(T)[1-\tfrac{1}{2}\Delta(\ln|1+2\bar{U}|)] -$$

$$-\frac{1}{|1+2\bar{U}^T|^{\frac{1}{2}}}\Bigg\{ \sum_{\mu}\Delta u_{\alpha\mu}\left(\frac{\partial F}{\partial\bar{\bar{E}}_\mu}\right)_T + \sum_{\mu\sigma\omega}(\delta_{\alpha\mu}+u_{\alpha\mu}^T)\left(\frac{\partial^2 F}{\partial\bar{u}_{\mu\omega}\,\partial\bar{\bar{E}}_\mu}\right)_T\Delta\bar{u}_{\sigma\omega} +$$

$$+ \sum_{\mu\sigma}(\delta_{\alpha\mu}+u_{\alpha\mu}^T)\left(\frac{\partial^2 F}{\partial\bar{\bar{E}}_\sigma\,\partial\bar{\bar{E}}_\mu}\right)_T\bar{E}_\sigma\Bigg\}, \quad (37.42)$$

where some terms have been expressed in terms of the pyroelectric moment with the help of (37.32). In strict analogy with (37.19), we find that in this case

$$\Delta u_{\alpha\mu} = s_{\alpha\mu} + \sum_{\gamma}s_{\alpha\gamma}u_{\gamma\mu}^T$$

$$= \sum_{\gamma}s_{\alpha\gamma}(\delta_{\gamma\mu}+u_{\gamma\mu}^T),$$

which can be further rewritten as follows:

$$\Delta u_{\alpha\mu} = \tfrac{1}{2}\sum_{\gamma}(s_{\alpha\gamma}+s_{\gamma\alpha})(\delta_{\gamma\mu}+u_{\gamma\mu}^T) + \tfrac{1}{2}\sum_{\gamma}(s_{\alpha\gamma}-s_{\gamma\alpha})(\delta_{\gamma\mu}+u_{\gamma\mu}^T)$$

$$= \tfrac{1}{2}\sum_{\beta\gamma}s_{\beta\gamma}\delta_{\alpha\beta}(\delta_{\gamma\mu}+u_{\gamma\mu}^T) + \tfrac{1}{2}\sum_{\beta\gamma}s_{\gamma\beta}\delta_{\alpha\beta}(\delta_{\gamma\mu}+u_{\gamma\mu}^T) +$$

$$+ \tfrac{1}{2}\sum_{\gamma}(s_{\alpha\gamma}-s_{\gamma\alpha})(\delta_{\gamma\mu}+u_{\gamma\mu}^T). \quad (37.43)$$

In (37.42) we substitute the expressions (37.43), (37.35), and (37.34), in the last of which we rewrite the summation index λ as γ; thus we obtain

$$P_\alpha = P_\alpha^{\mathrm{pyro}}(T)[1-\tfrac{1}{2}\Delta(\ln|1+2\overline{U}|)]-\tfrac{1}{2}|1+2\overline{U}^T|^{-\frac{1}{2}}\times$$

$$\times\Bigg\{ \sum_{\gamma\mu}(s_{\alpha\gamma}-s_{\gamma\alpha})(\delta_{\gamma\mu}+u_{\gamma\mu}^T)\Big(\frac{\partial F}{\partial\overline{E}_\mu}\Big)_T+$$

$$+\sum_{\beta\gamma}s_{\beta\gamma}\bigg[\delta_{\alpha\beta}\sum_\mu(\delta_{\gamma\mu}+u_{\gamma\mu}^T)\Big(\frac{\partial F}{\partial\overline{E}_\mu}\Big)_T\bigg]+$$

$$+\sum_{\beta\gamma}s_{\gamma\beta}\bigg[\delta_{\alpha\beta}\sum_\mu(\delta_{\gamma\mu}+u_{\gamma\mu}^T)\Big(\frac{\partial F}{\partial\overline{E}_\mu}\Big)_T\bigg]+$$

$$+\sum_{\beta\gamma}s_{\beta\gamma}\sum_{\mu\sigma\omega}(\delta_{\alpha\mu}+u_{\alpha\mu}^T)\Big(\frac{\partial^2 F}{\partial\bar{u}_{\sigma\omega}\partial\overline{E}_\mu}\Big)_T(\delta_{\beta\sigma}+u_{\beta\sigma}^T)(\delta_{\gamma\omega}+u_{\gamma\omega}^T)+$$

$$+\sum_{\beta\gamma}s_{\gamma\beta}\sum_{\mu\sigma\omega}(\delta_{\alpha\mu}+u_{\alpha\mu}^T)\Big(\frac{\partial^2 F}{\partial\bar{u}_{\sigma\omega}\partial\overline{E}_\mu}\Big)_T(\delta_{\beta\sigma}+u_{\beta\sigma}^T)(\delta_{\gamma\omega}+u_{\gamma\omega}^T)+$$

$$+2\sum_\beta E_\beta\sum_{\mu\sigma}(\delta_{\alpha\mu}+u_{\alpha\mu}^T)\Big(\frac{\partial^2 F}{\partial\overline{E}_\sigma\partial\overline{E}_\mu}\Big)_T(\delta_{\beta\sigma}+u_{\beta\sigma}^T)\Bigg\}. \qquad (37.44)$$

We notice that the second term on the right-hand side can be expressed in terms of the pyroelectric moment (37.32). Moreover, we can rewrite the summation indices γ, β respectively as β, γ in the third sum in the curly brackets and combine it with the second sum; similarly we can rewrite the indices γ, β, σ, ω respectively as β, γ, ω, σ in the fifth sum in the curly brackets and combine it with the fourth. It is then readily observed that (37.44) can be written as follows:

$$P_\alpha = P_\alpha^{\mathrm{pyro}}(T)[1-\tfrac{1}{2}\Delta(\ln|1+2\overline{U}|)]+\tfrac{1}{2}\sum_\gamma(s_{\alpha\gamma}-s_{\gamma\alpha})P_\gamma^{\mathrm{pyro}}(T)+$$

$$+\sum_{\beta\gamma}e_{\alpha,\beta\gamma}s_{\beta\gamma}+\sum_\beta a_{\alpha\beta}E_\beta, \qquad (37.45)$$

where

$$a_{\alpha\beta} = -\frac{1}{|1+2\overline{U}^T|^{\frac{1}{2}}}\sum_{\mu\sigma}(\delta_{\alpha\mu}+u_{\alpha\mu}^T)\Big(\frac{\partial^2 F}{\partial\overline{E}_\sigma\partial\overline{E}_\mu}\Big)_T(\delta_{\beta\sigma}+u_{\beta\sigma}^T)=a_{\beta\alpha}$$

$$(37.46)$$

are the components of the dielectric susceptibility tensor.

The term containing $\tfrac{1}{2}(s_{\alpha\gamma}-s_{\gamma\alpha})$ in (37.45) is of little interest; it describes the small rotation of the pyroelectric moment due to the antisymmetric part of the deformation parameters $s_{\alpha\gamma}$. The term with $\Delta(\ln|1+2\overline{U}|)$ is often overlooked; it describes a pure volume effect. Thus if a specimen preserves its total moment when expanded, its dielectric polarization decreases in inverse proportion to its volume. The above term represents this effect in connexion with the pyroelectric moment.

In (37.37) and (37.45) the elastic strain and electric field components are used as the independent variables. For some purposes it is convenient to consider the stress and field components as the independent variables determining the elastic strain and the dielectric polarization. Let us write (37.37) and (37.45) in Voigt's notation:

$$S_\rho = \sum_\sigma c_{\rho\sigma} s_\sigma - \sum_\beta e_{\beta\rho} E_\beta, \tag{37.47}$$

$$P_\alpha = P_\alpha^{\text{pyro}}[1 - \tfrac{1}{2}\Delta(\ln|1 + 2U|)] + \tfrac{1}{2}\sum_\gamma (s_{\alpha\gamma} - s_{\gamma\alpha})P_\alpha^{\text{pyro}} +$$
$$+ \sum_\sigma e_{\alpha\sigma} s_\sigma + \sum_\beta a_{\alpha\beta} E_\beta, \tag{37.48}$$

where the constants $c_{\rho\sigma}$ and $e_{\alpha\sigma}$ are the same as $c_{\alpha\gamma,\beta\lambda}$ and $e_{\alpha,\beta\gamma}$ written in Voigt's index notation (see p. 134). Denoting by $b_{\rho\sigma}$ the *elastic moduli* defined by

$$\sum_\sigma b_{\rho\sigma} c_{\sigma\rho'} = \delta_{\rho\rho'} = \sum_\sigma c_{\rho\sigma} b_{\sigma\rho'}, \tag{37.49}$$

we can put (37.47) in the alternative form

$$s_\rho = \sum_\sigma b_{\rho\sigma} S_\sigma + \sum_\beta \left(\sum_\sigma b_{\rho\sigma} e_{\beta\sigma} \right) E_\beta, \tag{37.50}$$

which, substituted in (37.48), gives

$$P_\alpha = P_\alpha^{\text{pyro}}[1 - \tfrac{1}{2}\Delta(\ln|1 + 2U|)] + \tfrac{1}{2}\sum_\gamma (s_{\alpha\gamma} - s_{\gamma\alpha})P_\gamma^{\text{pyro}} +$$
$$+ \sum_\sigma \left(\sum_\rho e_{\alpha\rho} b_{\rho\sigma} \right) S_\sigma + \sum_\beta \left(a_{\alpha\beta} + \sum_{\rho\sigma} e_{\alpha\rho} b_{\rho\sigma} e_{\beta\sigma} \right) E_\beta. \tag{37.51}$$

A piezoelectric crystal shows the phenomenon of *electro-striction*; i.e. the application of an electric field induces an elastic strain. The strain is given directly by (37.50) if we put $S_\sigma = 0$; the coefficients

$$\sum_\sigma b_{\rho\sigma} e_{\beta\sigma} \tag{37.52}$$

are the *electro-strictional moduli*. Electro-striction can be suppressed by the application of suitable mechanical stress, known as the *electro-strictional stress*; the latter is evidently given by (37.47) when we put the strain components equal to zero.

The third term on the right-hand side of (37.51) gives the piezoelectric polarization in terms of the stress components; the corresponding coefficients

$$\sum_\rho e_{\alpha\rho} b_{\rho\sigma} \tag{37.53}$$

are known as the *piezoelectric moduli*. The last term in (37.51) represents the dielectric polarization induced by a field in the absence of mechanical stress. The corresponding coefficients

$$a_{\alpha\beta} + \sum_{\rho\sigma} e_{\alpha\rho} b_{\rho\sigma} e_{\beta\sigma} \tag{37.54}$$

define, similarly to (37.46), a dielectric susceptibility tensor; (37.46) refers to the case when the crystal is clamped at zero strain whereas (37.54) refers to a free crystal. The corresponding dielectric constants are known respectively as the *clamped constants* and the *free constants*. The distinction between the two cases clearly exists only for piezoelectric crystals.

The equations (37.50) and (37.51) can equally well be written in tensor notation. In replacing the Voigt indices by tensor indices we have only to define the elastic moduli in tensor notation as follows:

$$b_{\alpha\gamma,\beta\lambda} = \tfrac{1}{4}(1+\delta_{\alpha\gamma})b_{\rho\sigma}(1+\delta_{\beta\lambda}) \quad (\rho \sim (\alpha\gamma),\ \sigma \sim (\beta\lambda)). \quad (37.55)$$

That (37.55) provides the correct relations follows directly from the fact that

$$\tfrac{1}{2}(1+\delta_{\alpha\gamma}) = \begin{cases} 1 & \alpha = \gamma, \\ \tfrac{1}{2} & \alpha \neq \gamma. \end{cases}$$

38. Normal coordinates of a lattice

The effective potential function Φ of a lattice, represented as a Taylor series in the nuclear displacements $\mathbf{u}\binom{l}{k}$, contains no first-order terms; and the second-order terms are given by

$$\tfrac{1}{2}\sum_{lk\alpha}\sum_{l'k'\beta}\Phi_{\alpha\beta}\binom{l\ l'}{k\ k'}u_{\alpha}\binom{l}{k}u_{\beta}\binom{l'}{k'}. \quad (38.1)$$

We have seen in § 15 that in discussing the normal coordinates we are concerned only with these second-order terms. For an infinite lattice (38.1) is in general a divergent expression. We shall normalize it to a finite volume by imposing the periodic condition as follows (for the interpretation of the periodic boundary condition see § 4):

Imagine the lattice partitioned into blocks of $L \times L \times L = N$ cells, with L lattice cells along each edge; the partitions form, so to speak, a macro-lattice with the basic vectors $L\mathbf{a}_1, L\mathbf{a}_2, L\mathbf{a}_3$, the blocks of N cells being the corresponding macro-cells. We impose the periodic condition by requiring that the pattern of motion should be the same in all macro-cells. In other words, we require that

$$\mathbf{u}\binom{l}{k} = \mathbf{u}\binom{\bar{l}}{k}, \quad (38.2)$$

if $\binom{l}{k}$ and $\binom{\bar{l}}{k}$ occupy equivalent positions in their respective macro-cells, i.e. if (l^1, l^2, l^3) differ respectively from $(\bar{l}^1, \bar{l}^2, \bar{l}^3)$ by integral multiples of L. Corresponding to the macro-cells, we subdivide the summation over l in (38.1) into partial summations; clearly all the partial sums are equal to one another. We may then attribute to a volume of N cells one such partial sum, which we may write as follows:

$$\tfrac{1}{2}\sum_{lk\alpha}^{N}\sum_{l'k'\beta}\Phi_{\alpha\beta}\binom{l\ l'}{k\ k'}u_{\alpha}\binom{l}{k}u_{\beta}\binom{l'}{k'}, \quad (38.3)$$

where the letter N indicates that l is summed over the N cells

$$l^1, l^2, l^3 = 0, 1, 2,..., L-1 \quad \text{(the generating volume).} \qquad (38.4)$$

For lack of a better name, we shall call the region of N cells defined by (38.4) the *generating volume*; it may be considered as forming the zero-cell in the macro-lattice.

For sufficiently large N (38.3) approximates closely to the potential energy of an actual finite crystal of N cells (error $\sim 1/N^{\frac{1}{4}}$), because the coefficients

$$\Phi_{\alpha\beta}(^{l\ l'}_{k\ k'})$$

become inappreciable when $|\mathbf{x}(^l_k) - \mathbf{x}(^{l'}_{k'})|$ exceeds what may be roughly called the range of force. Thus, if we imagine the finite crystal to be the generating volume removed from the infinite lattice, we note that the terms in (38.3) with l' beyond the generating volume form only an inappreciable fraction ($\sim 1/N^{\frac{1}{4}}$) of the total, so long as the dimension of the volume is large compared with the range of force. Similarly, the modification of the coefficients $\Phi_{\alpha\beta}(^{l\ l'}_{k\ k'})$ near the free surface affects an equally insignificant number of terms. Moreover, when (38.3) is used to describe a finite crystal, the periodic condition (38.2) is clearly no longer a restriction on the movement of the nuclei in the crystal.

As in § 15, we introduce the *dynamical matrix*

$$D_{\alpha\beta}(^{l-l'}_{k\ k'}) = D_{\alpha\beta}(^{l\ l'}_{k\ k'}) = \frac{1}{(m_k m_{k'})^{\frac{1}{2}}} \Phi_{\alpha\beta}(^{l\ l'}_{k\ k'}), \qquad (38.5)$$

and the reduced displacements

$$w_\alpha(^l_k) = \sqrt{m_k}\, u_\alpha(^l_k). \qquad (38.6)$$

The second-order terms in the potential energy given by (38.3) thus become

$$\tfrac{1}{2} \sum_{lk\alpha}^{N} \sum_{l'k'\beta} D_{\alpha\beta}(^{l-l'}_{k\ k'}) w_\alpha(^l_k) w_\beta(^{l'}_{k'}). \qquad (38.7)$$

The kinetic energy normalized to the same volume is readily seen to be

$$\tfrac{1}{2} \sum_{lk\alpha}^{N} m_k \dot{u}_\alpha^2(^l_k) = \tfrac{1}{2} \sum_{lk\alpha}^{N} \dot{w}_\alpha^2(^l_k). \qquad (38.8)$$

The normal coordinates are to be chosen so that the sum of (38.7) and (38.8), in terms of the coordinates, has the form of the Hamiltonian of $3nN$ independent simple oscillators. Before introducing the normal coordinates, let us consider a few mathematical propositions.

For every given reciprocal lattice vector $\mathbf{y}(h)$ we can construct a function of $\mathbf{x}(l)$ as follows:

$$\frac{1}{\sqrt{N}} \exp\{2\pi i \mathbf{y}(h) . \mathbf{x}(l)/L\} = \frac{1}{\sqrt{N}} \exp\{2\pi i (h_1 l^1 + h_2 l^2 + h_3 l^3)/L\}. \qquad (38.9)$$

It is obvious that (38.9) is unaltered if we add integral multiples of L to either the components (l^1, l^2, l^3) of $\mathbf{x}(l)$ or the components (h_1, h_2, h_3) of $\mathbf{y}(h)$. This means, on the one hand, that as a function of $\mathbf{x}(l)$ (38.9) satisfies the periodic condition, and, on the other hand, that all such distinct functions are obtained if we consider only the $\mathbf{y}(h)$ in the range (suppose L even):

$$-\frac{L}{2} \leqslant h_1, h_2, h_3 < \frac{L}{2}.$$

Thus if we write $\mathbf{y}(h)/L$ as $\mathbf{y}(h/L)$ the latter may be interpreted as a wave-number vector, whose components $h_1/L, h_2/L, h_3/L$ can be restricted to the range

$$-\frac{1}{2} \leqslant \frac{h_1}{L}, \frac{h_2}{L}, \frac{h_3}{L} < \frac{1}{2}. \tag{38.10}$$

There are exactly N wave-number vectors $\mathbf{y}(h/L)$ consistent with (38.10); these we shall refer to as the *permitted wave-numbers*. Their representative points in the reciprocal space are evenly distributed over a volume equal to the volume of a reciprocal cell, namely, $1/v_a$. The density (dimension = volume) of such points is thus $Nv_a = V$; to each of these N points we have a distinct function of $\mathbf{x}(l)$ of the type (38.9).

Consider the following function in the reciprocal space:

$$\Delta(\mathbf{y}) = \frac{1}{N} \sum_{l}^{N} \exp\{2\pi i \mathbf{y} . \mathbf{x}(l)\}, \tag{38.11}$$

where the summation is over the N cells in the generating volume and $\mathbf{y}$ is any point in the reciprocal space. Writing $\mathbf{y}$ and $\mathbf{x}(l)$ in terms of their respective components (η_1, η_2, η_3) and (l^1, l^2, l^3), we find upon carrying out the summation that

$$\Delta(\mathbf{y}) = \frac{1}{N} \sum_{l}^{N} \exp\{2\pi i(\eta_1 l^1 + \eta_2 l^2 + \eta_3 l^3)\}$$
$$= \left(\frac{1-e^{2\pi i L\eta_1}}{1-e^{2\pi i\eta_1}}\right)\left(\frac{1-e^{2\pi i L\eta_2}}{1-e^{2\pi i\eta_2}}\right)\left(\frac{1-e^{2\pi i L\eta_3}}{1-e^{2\pi i\eta_3}}\right). \tag{38.12}$$

It follows that the function $\Delta(y)$ vanishes at all points $\mathbf{y}$ with components of the form

$$(\eta_1, \eta_2, \eta_3) = \left(\frac{h_1}{L}, \frac{h_2}{L}, \frac{h_3}{L}\right),$$

with the exception of the reciprocal lattice points, at which the denominators as well as the numerators in (38.12) vanish. On the other hand, at the reciprocal lattice points every term in the sum of (38.11) is equal to unity. Hence we have the relations

$$\Delta(\mathbf{y}) = \begin{cases} 1 & \text{for } \mathbf{y} = \mathbf{y}(h), \\ 0 & \text{for } \mathbf{y} = \mathbf{y}(h/L), \text{ the components} \\ & \qquad (h_1/L, h_2/L, h_3/L) \text{ not all integers.} \end{cases}$$

(38.13)

Multiplying (38.9) by the complex conjugate of

$$\frac{1}{\sqrt{N}} \exp\{2\pi i \mathbf{y}(h') . \mathbf{x}(l)/L\}$$

and summing $\mathbf{x}(l)$ over the generating volume, we get

$$\sum_l^N \left\{ \frac{1}{\sqrt{N}} \exp[2\pi i \mathbf{y}(h') . \mathbf{x}(l)/L] \right\}^* \left\{ \frac{1}{\sqrt{N}} \exp[2\pi i \mathbf{y}(h) . \mathbf{x}(l)/L] \right\}$$
$$= \Delta\left[-\mathbf{y}\left(\frac{h'}{L}\right) + \mathbf{y}\left(\frac{h}{L}\right) \right]. \quad (38.14)$$

Under the restriction (38.10),

$$-\mathbf{y}\left(\frac{h'}{L}\right) + \mathbf{y}\left(\frac{h}{L}\right) = \mathbf{y}\left(\frac{h-h'}{L}\right)$$

is a reciprocal lattice vector only for the case $\mathbf{y}(h) = \mathbf{y}(h')$. Thus it follows from (38.13) that

$$\sum_l^N \left\{ \frac{1}{\sqrt{N}} \exp[2\pi i \mathbf{y}' . \mathbf{x}(l)] \right\}^* \left\{ \frac{1}{\sqrt{N}} \exp[2\pi i \mathbf{y} . \mathbf{x}(l)] \right\} = \delta_{\mathbf{y}'\mathbf{y}}$$

(38.15)

if $\mathbf{y}$ and $\mathbf{y}'$ are two permitted wave-numbers. (38.15) expresses the orthonormal relations between the N distinct functions of the type (38.9).

It can be shown by similar arguments that if l and l' are two cells both in the generating volume, we have the inverse orthonormal relations

$$\sum_{\mathbf{y}}^N \left\{ \frac{1}{\sqrt{N}} \exp[2\pi i \mathbf{y} . \mathbf{x}(l')/L] \right\}^* \left\{ \frac{1}{\sqrt{N}} \exp[2\pi i \mathbf{y} . \mathbf{x}(l)/L] \right\} = \delta_{ll'}$$

(38.16)

where the letter N signifies that $\mathbf{y}$ is summed over the N permitted wave-numbers.

Using the functions (38.9), we express the reduced displacements in terms of a set of complex variables ($\mathbf{y}$, $\mathbf{y}'$ are now permitted wave numbers only)

$$w_\alpha^*(k|\mathbf{y}) = w_\alpha(k|-\mathbf{y}) \qquad (38.17)$$

as follows:

$$w_\alpha\binom{l}{k} = \frac{1}{\sqrt{N}} \sum_\mathbf{y}^N w_\alpha(k|\mathbf{y})e^{2\pi i \mathbf{y} \cdot \mathbf{x}(l)}. \qquad (38.18)$$

The condition (38.17) ensures that the displacements are real. It follows from the relations (38.15) and (38.16) that the transformation between the variables $w_\alpha\binom{l}{k}$ and $w_\alpha(k|\mathbf{y})$ is unitary and we have the inverse transformation formulae

$$w_\alpha(k|\mathbf{y}) = \frac{1}{\sqrt{N}} \sum_l^N w_\alpha\binom{l}{k}e^{-2\pi i \mathbf{y} \cdot \mathbf{x}(l)}. \qquad (38.19)$$

Expressed in terms of the complex variables, the potential function (38.7) becomes

$$\tfrac{1}{2} \sum_{lk\alpha}^N \sum_{l'k'\beta} \sum_\mathbf{y}^N \sum_{\mathbf{y}'}^N D_{\alpha\beta}\binom{l \, l'}{k \, k'}w_\alpha(k|\mathbf{y})w_\beta(k'|\mathbf{y}')e^{2\pi i[\mathbf{y} \cdot \mathbf{x}(l) + \mathbf{y}' \cdot \mathbf{x}(l')]}$$

$$= \tfrac{1}{2} \sum_{lk\alpha}^N \sum_{k'\beta} \sum_\mathbf{y}^N \sum_{\mathbf{y}'}^N w_\alpha(k|\mathbf{y})w_\beta(k'|\mathbf{y}')e^{2\pi i(\mathbf{y}+\mathbf{y}') \cdot \mathbf{x}(l)} \times$$

$$\times \left\{ \sum_{l'} D_{\alpha\beta}\binom{l-l'}{k \, \, k'}e^{-2\pi i \mathbf{y}'[\mathbf{x}(l) - \mathbf{x}(l')]} \right\}. \qquad (38.20)$$

The expression given in the curly brackets is independent of the value of l; this is readily seen when $\bar{l} = l - l'$ is introduced as the summation index. We shall denote the expression as follows:

$$D_{\alpha\beta}\binom{\mathbf{y}'}{kk'} = \sum_{\bar{l}} D_{\alpha\beta}\binom{\bar{l}}{kk'}e^{-2\pi i \mathbf{y}' \cdot \mathbf{x}(\bar{l})}. \qquad (38.21)$$

When the summation over l is carried out, (38.20) becomes

$$\tfrac{1}{2} \sum_{k\alpha} \sum_{k'\beta} \sum_\mathbf{y}^N \sum_{\mathbf{y}'}^N w_\alpha(k|\mathbf{y})w_\beta(k'|\mathbf{y}')\delta_{-\mathbf{y}\mathbf{y}'} D_{\alpha\beta}\binom{\mathbf{y}'}{kk'}, \qquad (38.22)$$

where we have made use of (38.15). Carrying out the summation over $\mathbf{y}'$ and using (38.17), we obtain the potential energy in the form

$$\tfrac{1}{2} \sum_\mathbf{y}^N \sum_{k\alpha} \sum_{k'\beta} w_\alpha^*(k|\mathbf{y}) D_{\alpha\beta}\binom{\mathbf{y}}{kk'})w_\beta(k'|\mathbf{y}). \qquad (38.23)$$

By means of the unitary transformation (38.18) we have thus reduced the quadratic form (38.7) into N complex forms, each characterized by a $3n \times 3n$ matrix $D_{\alpha\beta}\binom{\mathbf{y}}{kk'}$, which is Hermitian (see below). For a Hermitian matrix $D_{\alpha\beta}\binom{\mathbf{y}}{kk'}$ there exist $3n$ sets of quantities $e_\alpha(k|\overset{\mathbf{y}}{j})$ (the *eigenvectors*) and $\omega^2(\overset{\mathbf{y}}{j})$ (the *eigenvalues*), $j = 1, 2,..., 3n$, which satisfy the equations

$$\sum_{k'\beta} D_{\alpha\beta}\binom{\mathbf{y}}{kk'})e_\beta(k'|\overset{\mathbf{y}}{j}) = \omega^2(\overset{\mathbf{y}}{j})e_\alpha(k|\overset{\mathbf{y}}{j}) \qquad (38.24)$$

and fulfil the direct and inverse orthonormal relations

$$\sum_{k\alpha} e_\alpha^*(k|_j^y)e_\alpha(k|_j^y) = \delta_{jj'}, \qquad \sum_j e_\beta^*(k'|_j^y)e_\alpha(k|_j^y) = \delta_{kk'}\delta_{\alpha\beta}. \tag{38.25}$$

The *complex normal coordinates* $Q(_j^y)$ are introduced by the unitary transformation

$$w_\alpha(k|\mathbf{y}) = \sum_j e_\alpha(k|_j^y)Q(_j^y), \tag{38.26}$$

the inverse formula being

$$Q(_j^y) = \sum_{k\alpha} e_\alpha^*(k|_j^y)w_\alpha(k|\mathbf{y}). \tag{38.27}$$

Expressed in terms of $Q(_j^y)$, the potential function (38.23) can be reduced as follows:

$$\tfrac{1}{2}\sum_\mathbf{y}^N\sum_j\sum_{j'} Q^*(_j^y)Q(_{j'}^y) \sum_{k\alpha}\sum_{k'\beta} e_\alpha^*(k|_j^y)D_{\alpha\beta}(_{kk'}^y)e_\beta(k'|_{j'}^y)$$

$$= \tfrac{1}{2}\sum_\mathbf{y}^N\sum_j\sum_{j'} Q^*(_j^y)Q(_{j'}^y)\omega^2(_{j'}^y) \sum_{k\alpha} e_\alpha^*(k|_j^y)e_\alpha(k|_{j'}^y)$$

$$= \tfrac{1}{2}\sum_\mathbf{y}^N\sum_j \omega^2(_j^y)Q^*(_j^y)Q(_j^y), \tag{38.28}$$

where we have made use of (38.24) and (38.25). Let us write the kinetic energy (38.8) as

$$\tfrac{1}{2}\sum_{lk\alpha}^N \dot{w}_\alpha^*(_k^l)\dot{w}_\alpha(_k^l).$$

Upon expressing it in terms of the complex normal coordinates with the help of (38.18) and (38.26) and using the orthonormal relations (38.15) (38.25), we obtain the kinetic energy in the form

$$\tfrac{1}{2}\sum_\mathbf{y}^N\sum_j \dot{Q}^*(_j^y)\dot{Q}(_j^y). \tag{38.29}$$

The close connexion between the above discussion and the derivation of the lattice waves in § 24 is evident. Thus (24.10), (38.24) are completely equivalent. The coefficients $C_{\alpha\beta}(_{kk'}^y)$ in § 24 differ from $D_{\alpha\beta}(_{kk'}^y)$ only by a phase factor $\exp\{-2\pi i\mathbf{y}.[\mathbf{x}(k)-\mathbf{x}(k')]\}$; accordingly the corresponding eigenvectors $w_\alpha(k|_j^y)$ differ from $e_\alpha(k|_j^y)$ by a compensating factor $\exp\{-2\pi i\mathbf{y}.\mathbf{x}(k)\}$ (i.e. apart from an arbitrary normalization factor) and the eigenvalues $\omega^2(_j^y)$ are the same in the two cases. The phase factors have been chosen differently in the two cases purely for reasons of convenience: the choice made in § 24 is more convenient for discussing the long acoustic waves, whereas the present choice leads to simpler formulae involving the normal coordinates. It follows immediately from (24.14) and (24.18) that $D_{\alpha\beta}(_{kk'}^y)$ is Hermitian and we may assume

$$e_\alpha(k|_j^y) = e_\alpha^*(k|_{\bar{j}}^{-\mathbf{y}}). \tag{38.30}$$

Thus it follows from (38.27) and (38.17) that

$$Q^*(\tfrac{y}{j}) = \sum_{k\alpha} e_\alpha(k|\tfrac{y}{j}) w_\alpha^*(k|\mathbf{y}) = \sum_{k\alpha} e_\alpha^*(k|\tfrac{-y}{j}) w_\alpha(k|-\mathbf{y}) = Q(\tfrac{-y}{j}). \quad (38.31)$$

The complex normal coordinates $Q(\tfrac{y}{j})$ describe the amplitudes of plane lattice waves represented in the complex form. The complex waves described respectively by $Q(\tfrac{y}{j})$ and its complex conjugate $Q^*(\tfrac{y}{j}) = Q(\tfrac{-y}{j})$ can be combined into two independent real waves: the most usual ways to do this consist in forming either two standing waves relatively shifted by a quarter wave, or two progressive waves travelling in opposite directions. In the following we shall derive the two types of normal coordinates corresponding respectively to the two types of real waves.

We notice in the first place that the $3n$ coordinates $Q(\tfrac{y}{j})$ with $\mathbf{y} = 0$ are real, for it follows from (38.31) that

$$Q^*(\tfrac{0}{j}) = Q(\tfrac{0}{j}). \quad (38.32)$$

These coordinates may thus be used directly as real normal coordinates.

Let us divide the permitted wave-numbers into two groups by an arbitrary plane through the origin of the reciprocal space; a pair of wave-numbers $\mathbf{y}$ and $-\mathbf{y}$ thus lie on opposite sides of the plane. Using the complex normal coordinates $Q(\tfrac{y}{j})$ with $\mathbf{y}$ lying on one side of the plane, we introduce the real normal coordinates $q_1(\tfrac{y}{j})$ and $q_2(\tfrac{y}{j})$ as follows:

$$Q(\tfrac{y}{j}) = \frac{1}{\sqrt{2}} (q_1(\tfrac{y}{j}) + i q_2(\tfrac{y}{j})). \quad (38.33)$$

The relations between $q_1(\tfrac{y}{j})$, $q_2(\tfrac{y}{j})$ and the complex coordinates with wave-numbers lying on the opposite side of the plane are completely determined by (38.31):

$$Q(\tfrac{-y}{j}) = \frac{1}{\sqrt{2}} (q_1(\tfrac{y}{j}) - i q_2(\tfrac{y}{j})). \quad (38.34)$$

In both the potential energy (38.28) and the kinetic energy (38.29), the terms corresponding to a pair of wave-numbers $\mathbf{y}$ and $-\mathbf{y}$ are equal owing to (38.31) and (24.18). Thus upon using (38.33) and (38.34) we can write the sum of the potential and kinetic energies in terms of the real normal coordinates as

$$\tfrac{1}{2} \sum_{\mathbf{y}}^{N/2} \sum_{j} \sum_{\lambda=1,2} \{\dot{q}_\lambda^2(\tfrac{y}{j}) + \omega^2(\tfrac{y}{j}) q_\lambda^2(\tfrac{y}{j})\}, \quad (38.35)$$

where $N/2$ signifies that $\mathbf{y}$ is summed over the permitted wave-numbers lying on one side of the plane through the origin of the reciprocal space. The form of (38.35) confirms that $q_1(\tfrac{y}{j})$, $q_2(\tfrac{y}{j})$ are normal coordinates of the system.

Substituting (38.26) in (38.18), we get

$$w_\alpha\binom{l}{k} = \frac{1}{\sqrt{N}} \sum_{\mathbf{y}}^{N} \sum_{j} Q\binom{\mathbf{y}}{j} e_\alpha(k|\genfrac{}{}{0pt}{}{\mathbf{y}}{j}) e^{2\pi i \mathbf{y} \cdot \mathbf{x}(l)}. \qquad (38.36)$$

When $Q\binom{\mathbf{y}}{j}$ are expressed in terms of the real normal coordinates, (38.36) becomes

$$w_\alpha\binom{l}{k} = \frac{1}{\sqrt{N}} \sum_{\mathbf{y}}^{N/2} \sum_{j} \Big\{ \frac{1}{\sqrt{2}} [e_\alpha(k|\genfrac{}{}{0pt}{}{\mathbf{y}}{j}) e^{2\pi i \mathbf{y} \cdot \mathbf{x}(l)} + e_\alpha^*(k|\genfrac{}{}{0pt}{}{\mathbf{y}}{j}) e^{-2\pi i \mathbf{y} \cdot \mathbf{x}(l)}] q_1\binom{\mathbf{y}}{j} +$$

$$+ \frac{i}{\sqrt{2}} [e_\alpha(k|\genfrac{}{}{0pt}{}{\mathbf{y}}{j}) e^{2\pi i \mathbf{y} \cdot \mathbf{x}(l)} - e_\alpha^*(k|\genfrac{}{}{0pt}{}{\mathbf{y}}{j}) e^{-2\pi i \mathbf{y} \cdot \mathbf{x}(l)}] q_2\binom{\mathbf{y}}{j} \Big\}. \qquad (38.37)$$

If we denote the argument of $e_\alpha(k|\genfrac{}{}{0pt}{}{\mathbf{y}}{j})$ by $\delta_\alpha(k|\genfrac{}{}{0pt}{}{\mathbf{y}}{j})$, we see that the coefficients of $q_1\binom{\mathbf{y}}{j}$ and $q_2\binom{\mathbf{y}}{j}$ are respectively proportional to

$$\cos[2\pi \mathbf{y} \cdot \mathbf{x}(l) + \delta_\alpha(k|\genfrac{}{}{0pt}{}{\mathbf{y}}{j})]$$

and
$$\sin[2\pi \mathbf{y} \cdot \mathbf{x}(l) + \delta_\alpha(k|\genfrac{}{}{0pt}{}{\mathbf{y}}{j})].$$

These coordinates thus describe the amplitudes of two standing waves relatively shifted by a quarter wave. We shall refer to this type of normal coordinates as the *normal coordinates of the first kind*.

Unlike the coordinates above, the real normal coordinates which describe progressive waves are not geometrical coordinates specifying the configuration of the system and thus cannot be derived from the nuclear displacements alone. They are coordinates which can be obtained by a canonical transformation. We shall, however, introduce these coordinates in an elementary way as follows: We write

$$Q\binom{\mathbf{y}}{j} = a_+(\genfrac{}{}{0pt}{}{-\mathbf{y}}{j}) + a_-(\genfrac{}{}{0pt}{}{\mathbf{y}}{j}), \qquad (38.38)$$

where a_+ and a_- are auxiliary variables related to the new real normal coordinates $q\binom{\mathbf{y}}{j}$ by

$$a_+(\genfrac{}{}{0pt}{}{-\mathbf{y}}{j}) = \frac{1}{2}\Big\{ \frac{\dot{q}(\genfrac{}{}{0pt}{}{-\mathbf{y}}{j})}{\omega\binom{\mathbf{y}}{j}} + iq(\genfrac{}{}{0pt}{}{-\mathbf{y}}{j}) \Big\}, \qquad (38.39)$$

$$a_-(\genfrac{}{}{0pt}{}{\mathbf{y}}{j}) = \frac{1}{2}\Big\{ \frac{\dot{q}(\genfrac{}{}{0pt}{}{\mathbf{y}}{j})}{\omega\binom{\mathbf{y}}{j}} - iq(\genfrac{}{}{0pt}{}{\mathbf{y}}{j}) \Big\}. \qquad (38.40)$$

We notice that a_+ and a_- are complex conjugates of one another; hence (38.38) is consistent with the reality condition (38.31) for the complex coordinates whatever the value of $q\binom{\mathbf{y}}{j}$.

Using the fact that

$$\ddot{q}(\tbinom{y}{j}) = -\omega^2(\tbinom{y}{j})q(\tbinom{y}{j}), \tag{38.41}$$

which will be verified later, we find on differentiating (38.39) and (38.40) with respect to time that

$$\dot{a}_+(\tbinom{-y}{j}) = \tfrac{1}{2}\{-\omega(\tbinom{y}{j})q(\tbinom{-y}{j})+i\dot{q}(\tbinom{-y}{j})\} = i\omega(\tbinom{y}{j})a_+(\tbinom{-y}{j}), \tag{38.42}$$

$$\dot{a}_-(\tbinom{y}{j}) = \tfrac{1}{2}\{-\omega(\tbinom{y}{j})q(\tbinom{y}{j})-i\dot{q}(\tbinom{y}{j})\} = -i\omega(\tbinom{y}{j})a_-(\tbinom{y}{j}). \tag{38.43}$$

With the help of the relations (38.42) and (38.43), we can write the sum of the potential energy (38.28) and the kinetic energy (38.29) in terms of the coordinates $q(\tbinom{y}{j})$ as follows:

$$\tfrac{1}{2}\sum_y^N \sum_j \{\dot{Q}^*(\tbinom{y}{j})\dot{Q}(\tbinom{y}{j})+\omega^2(\tbinom{y}{j})Q^*(\tbinom{y}{j})Q(\tbinom{y}{j})\}$$

$$= \tfrac{1}{2}\sum_y^N \sum_j \{\omega^2(\tbinom{y}{j})[a_+(\tbinom{-y}{j})-a_-(\tbinom{y}{j})]^*[a_+(\tbinom{-y}{j})-a_-(\tbinom{y}{j})]+$$

$$+\omega^2(\tbinom{y}{j})[a_+(\tbinom{-y}{j})+a_-(\tbinom{y}{j})]^*[a_+(\tbinom{-y}{j})+a_-(\tbinom{y}{j})]\}$$

$$= \sum_y^N \sum_j \omega^2(\tbinom{y}{j})\{a_+^*(\tbinom{-y}{j})a_+(\tbinom{-y}{j})+a_-^*(\tbinom{y}{j})a_-(\tbinom{y}{j})\}$$

$$= \tfrac{1}{4}\sum_y^N \sum_j \{\dot{q}^2(\tbinom{-y}{j})+\omega^2(\tbinom{y}{j})q^2(\tbinom{-y}{j})+\dot{q}^2(\tbinom{y}{j})+\omega^2(\tbinom{y}{j})q^2(\tbinom{y}{j})\}$$

$$= \tfrac{1}{2}\sum_y^N \sum_j \{\dot{q}^2(\tbinom{y}{j})+\omega^2(\tbinom{y}{j})q^2(\tbinom{y}{j})\}. \tag{38.44}$$

The form of this expression shows that $q(\tbinom{y}{j})$ are normal coordinates of the system, thereby also justifying the relation (38.41).

Let us consider the case when only one of these coordinates $(\tbinom{y}{j})$ is excited. Upon substituting (38.38) in (38.36) and putting a_+, a_- for all other modes to zero, we obtain for the reduced displacements

$$w_\alpha(\tbinom{l}{k}) = \frac{1}{\sqrt{N}}\{a_-(\tbinom{y}{j})e_\alpha(k|\tbinom{y}{j})e^{2\pi i y\cdot x(l)}+a_+(\tbinom{y}{j})e_\alpha(k|\tbinom{-y}{j})e^{-2\pi i y\cdot x(l)}\}. \tag{38.45}$$

$q(\tbinom{y}{j})$, as a function of time, can in general be written in the form

$$q(\tbinom{y}{j}) = c\{e^{i[\omega(\tbinom{y}{j})t+\delta]}+e^{-i[\omega(\tbinom{y}{j})t+\delta]}\}, \tag{38.46}$$

the corresponding values for a_+ and a_- being

$$a_+(\tbinom{y}{j}) = ice^{i[\omega(\tbinom{y}{j})t+\delta]}, \qquad a_-(\tbinom{y}{j}) = -ice^{-i[\omega(\tbinom{y}{j})t+\delta]}. \tag{38.47}$$

When (38.47) is substituted in (38.45) it is immediately observed that the reduced displacements are proportional to

$$\sin[2\pi y\cdot x(l)-\omega(\tbinom{y}{j})t+\delta_\alpha(k|\tbinom{y}{j})-\delta].$$

The coordinate $q(\tbinom{y}{j})$ describes thus a progressive wave travelling in the direction of y. We shall refer to the coordinates $q(\tbinom{y}{j})$ as the *normal coordinates of the second kind*. They were first introduced by Peierls.

The coordinates $Q(^{0}_{j})$ are to be included among both types of normal coordinates. Dropping the summation over y from (38.36) and putting y equal to zero, we get

$$w_\alpha(^l_k) = \frac{1}{\sqrt{N}} \sum_j e_\alpha(k|^0_j)Q(^0_j), \qquad (38.48)$$

which shows that the nuclear displacements described by these co-ordinates are independent of the cell index l. In other words, $Q(^0_j)$ describe relative oscillations between the constituent Bravais lattices. For charged lattice particles they are associated with electric polarization; it is clear from the discussion given in the last chapter that in such cases the coordinates cannot be uniquely defined, i.e. in such cases the coefficients $e_\alpha(k|^0_j)$ and the $\omega^2(^0_j)$ must be ambiguous. Later we shall see how these coordinates are to be chosen in such circumstances.

39. Normalization of physical parameters, selection rules and expansion methods

The crystal properties which we shall discuss are mainly determined by the changes

$$\Phi(X)-\Phi(X^0), \quad \mathbf{M}(X)-\mathbf{M}(X^0), \quad P_{\alpha\beta}(\omega, X)-P_{\alpha\beta}(\omega, X^0)$$

in the potential function, electric moment, and polarizability due to the displacements of the nuclei. As in the case of the second-order terms in Φ, their values can be normalized to be finite by imposing the periodic condition on the nuclear displacements. Thus we denote the corresponding normalized values for N cells by $\delta\Phi$, $\delta\mathbf{M}$, $\delta P_{\alpha\beta}$; then we have for $\delta\Phi$:

$$\delta\Phi = \sum_{lk\alpha}^{N} \Phi_\alpha(^l_k)u_\alpha(^l_k) + \tfrac{1}{2}\sum_{lk\alpha}^{N}\sum_{l'k'\beta} \Phi_{\alpha\beta}(^{l\ l'}_{k\ k'})u_\alpha(^l_k)u_\beta(^{l'}_{k'}) +$$

$$+ \tfrac{1}{6}\sum_{lk\alpha}^{N}\sum_{l'k'\beta}\sum_{l'k''\gamma} \Phi_{\alpha\beta\gamma}(^{l\ l'\ l''}_{k\ k'\ k''})u_\alpha(^l_k)u_\beta(^{l'}_{k'})u_\gamma(^{l''}_{k''}) + ..., \qquad (39.1)$$

where the summations with the index N are restricted to the N cells in the generating volume and the nuclear displacements satisfy the periodic condition (38.2). The linear terms ($=0$, owing to the equilibrium conditions) are formally retained so that the results which we shall obtain for $\delta\Phi$ can be used to write down the similar results for $\delta\mathbf{M}$ and $\delta P_{\alpha\beta}$.

When $\delta\Phi$ is expressed as an expansion in the complex normal co-ordinates $Q(^y_j)$ many terms in the expansion vanish identically. We shall presently see that the *selection rules* are such that we can write the expansion in the following form:

$$\delta\Phi = \sqrt{N}\sum_{\mathbf{y}j}^{N}\Delta(\mathbf{y})\Phi(\tfrac{\mathbf{y}}{j})Q(\tfrac{\mathbf{y}}{j}) + \tfrac{1}{2}\sum_{\mathbf{y}j}^{N}\sum_{\mathbf{y}'j'}^{N}\Delta(\mathbf{y}+\mathbf{y}')\Phi(\tfrac{\mathbf{y}}{j}\tfrac{\mathbf{y}'}{j'})Q(\tfrac{\mathbf{y}}{j})Q(\tfrac{\mathbf{y}'}{j'}) +$$

$$+\frac{1}{6N^{\frac{1}{2}}}\sum_{\mathbf{y}j}^{N}\sum_{\mathbf{y}'j'}^{N}\sum_{\mathbf{y}''j''}^{N}\Delta(\mathbf{y}+\mathbf{y}'+\mathbf{y}'')\Phi(\tfrac{\mathbf{y}}{j}\tfrac{\mathbf{y}'}{j'}\tfrac{\mathbf{y}''}{j''})Q(\tfrac{\mathbf{y}}{j})Q(\tfrac{\mathbf{y}'}{j'})Q(\tfrac{\mathbf{y}''}{j''}) + ..., \quad (39.2)$$

where Δ is the function defined in (38.11). In view of (38.13) a term $Q(\tfrac{\mathbf{y}}{j})Q(\tfrac{\mathbf{y}'}{j'})...Q(\tfrac{\mathbf{y}^{(l)}}{j^{(l)}})$ is present only if the sum of the corresponding wavenumbers is a reciprocal lattice vector.

In order to see how the Δ-factors in (39.2) arise, let us consider the third-order terms as an example. When the nuclear displacements are expressed in terms of the complex normal coordinates by (see (38.36))

$$u_{\alpha}(\tfrac{l}{k}) = \frac{1}{\sqrt{m_k}}w_{\alpha}(\tfrac{l}{k}) = \frac{1}{(Nm_k)^{\frac{1}{2}}}\sum_{\mathbf{y}j}^{N}Q(\tfrac{\mathbf{y}}{j})e_{\alpha}(k|\tfrac{\mathbf{y}}{j})e^{2\pi i\mathbf{y}\cdot\mathbf{x}(l)}, \quad (39.3)$$

we can write the third-order terms in (39.1) as follows:

$$\frac{1}{6N^{\frac{1}{2}}}\sum_{\mathbf{y}j}^{N}\sum_{\mathbf{y}'j'}^{N}\sum_{\mathbf{y}''j''}^{N}Q(\tfrac{\mathbf{y}}{j})Q(\tfrac{\mathbf{y}'}{j'})Q(\tfrac{\mathbf{y}''}{j''})\sum_{lk\alpha}^{N}\sum_{l'k'\beta}\sum_{l''k''\gamma}\Phi_{\alpha\beta\gamma}(\tfrac{l}{k}\tfrac{l'}{k'}\tfrac{l''}{k''})\times$$

$$\times\frac{1}{(m_k m_{k'} m_{k''})^{\frac{1}{2}}}e_{\alpha}(k|\tfrac{\mathbf{y}}{j})e_{\beta}(k'|\tfrac{\mathbf{y}'}{j'})e_{\gamma}(k''|\tfrac{\mathbf{y}''}{j''})\exp\{2\pi i[\mathbf{y}\cdot\mathbf{x}(l)+\mathbf{y}'\cdot\mathbf{x}(l')+\mathbf{y}''\cdot\mathbf{x}(l'')]\},$$

$$(39.4)$$

where, as explained in § 23, we can subtract the same index l from all the cell indices of $\Phi_{\alpha\beta\gamma}(\tfrac{l}{k}\tfrac{l'}{k'}\tfrac{l''}{k''})$ without altering the value of the latter. On the right-hand side of (39.4) let us write

$$\mathbf{x}(l') = \mathbf{x}(l'-l)+\mathbf{x}(l), \qquad \mathbf{x}(l'') = \mathbf{x}(l''-l)+\mathbf{x}(l)$$

and introduce $l'-l$, $l''-l$ as the summation indices. Writing these summation indices again as l' and l'', we obtain the following:

$$\frac{1}{6N^{\frac{1}{2}}}\sum_{\mathbf{y}j}^{N}\sum_{\mathbf{y}'j'}^{N}\sum_{\mathbf{y}''j''}^{N}Q(\tfrac{\mathbf{y}}{j})Q(\tfrac{\mathbf{y}'}{j'})Q(\tfrac{\mathbf{y}''}{j''})\sum_{l}^{N}\exp[2\pi i(\mathbf{y}+\mathbf{y}'+\mathbf{y}'')\cdot\mathbf{x}(l)]\times$$

$$\times\left\{\sum_{k\alpha}\sum_{l'k'\beta}\sum_{l''k''\gamma}\Phi_{\alpha\beta\gamma}(\tfrac{0}{k}\tfrac{l'}{k'}\tfrac{l''}{k''})\frac{1}{(m_k m_{k'} m_{k''})^{\frac{1}{2}}}e_{\alpha}(k|\tfrac{\mathbf{y}}{j})e_{\beta}(k'|\tfrac{\mathbf{y}'}{j'})e_{\gamma}(k''|\tfrac{\mathbf{y}''}{j''})\times\right.$$

$$\left.\times\exp\{2\pi i[\mathbf{y}'\cdot\mathbf{x}(l')+\mathbf{y}''\cdot\mathbf{x}(l'')]\}\right\}. \quad (39.5)$$

The expression in the curly brackets is independent of l; thus expressing the result of the summation over l in terms of the Δ-function defined in

(38.11), we obtain the third-order terms exactly in the form given in (39.2), where

$$\Phi\!\begin{pmatrix} \mathbf{y}\ \mathbf{y}'\ \mathbf{y}'' \\ j\ j'\ j'' \end{pmatrix} = \sum_{k\alpha}\sum_{l'k'\beta}\sum_{l''k''\gamma} \frac{1}{(m_k m_{k'} m_{k''})^{\frac12}}\, \Phi_{\alpha\beta\gamma}\!\begin{pmatrix} 0\ l'\ l'' \\ k\ k'\ k'' \end{pmatrix} e_\alpha(k|{}^{\mathbf{y}}_{j}) e_\beta(k'|{}^{\mathbf{y}'}_{j'}) e_\gamma(k''|{}^{\mathbf{y}''}_{j''}) \times$$

$$\times \exp\{2\pi i[\mathbf{y}'\cdot\mathbf{x}(l') + \mathbf{y}''\cdot\mathbf{x}(l'')]\}. \quad (39.6)$$

Despite the asymmetric appearance of (39.6) the coefficients

$$\Delta(\mathbf{y}+\mathbf{y}'+\mathbf{y}'')\Phi\!\begin{pmatrix} \mathbf{y}\ \mathbf{y}'\ \mathbf{y}'' \\ j\ j'\ j'' \end{pmatrix}$$

are in fact completely symmetric in the indices $({}^{\mathbf{y}}_{j})$, $({}^{\mathbf{y}'}_{j'})$, $({}^{\mathbf{y}''}_{j''})$. To prove this, we have only to show that (39.6) is symmetric in the case

$$\mathbf{y}+\mathbf{y}'+\mathbf{y}'' = \mathbf{y}(h) \quad \text{(a reciprocal lattice vector)}, \quad (39.7)$$

for which $\Delta(\mathbf{y}+\mathbf{y}'+\mathbf{y}'') \neq 0$. From the fact that

$$\Phi_{\alpha\beta\gamma}\!\begin{pmatrix} 0\ l'\ l'' \\ k\ k'\ k'' \end{pmatrix} = \Phi_{\alpha\gamma\beta}\!\begin{pmatrix} 0\ l''\ l' \\ k\ k''\ k' \end{pmatrix}$$

the symmetry of (39.6) between $({}^{\mathbf{y}'}_{j'})$ and $({}^{\mathbf{y}''}_{j''})$ is immediately obvious. In order to show that (39.6) is also symmetric between $({}^{\mathbf{y}}_{j})$ and $({}^{\mathbf{y}'}_{j'})$ we interchange $(0, k, \alpha)$ with (l', k', β) in the derivative of Φ and afterwards relabel k, α, k', β respectively as k', β, k, α, thus converting (39.6) into

$$\Phi\!\begin{pmatrix} \mathbf{y}\ \mathbf{y}'\ \mathbf{y}'' \\ j\ j'\ j'' \end{pmatrix} = \sum_{k'\beta}\sum_{l'}\sum_{k\alpha}\sum_{l''k''\gamma} \Phi_{\alpha\beta\gamma}\!\begin{pmatrix} l'\ 0\ l'' \\ k\ k'\ k'' \end{pmatrix} \frac{1}{(m_k m_{k'} m_{k''})^{\frac12}}\, e_\beta(k'|{}^{\mathbf{y}}_{j}) e_\alpha(k|{}^{\mathbf{y}'}_{j'}) e_\gamma(k''|{}^{\mathbf{y}''}_{j''}) \times$$

$$\times \exp\{2\pi i[\mathbf{y}'\cdot\mathbf{x}(l') + \mathbf{y}''\cdot\mathbf{x}(l'')]\}.$$

Subtracting l' from the cell indices in the derivative of Φ and using (39.7) to write $\mathbf{y}'$ in the exponential factor as $\mathbf{y}(h)-\mathbf{y}-\mathbf{y}''$, we get

$$\Phi\!\begin{pmatrix} \mathbf{y}\ \mathbf{y}'\ \mathbf{y}'' \\ j\ j'\ j'' \end{pmatrix} = \sum_{k'\beta}\sum_{l'}\sum_{k\alpha}\sum_{l''k''\gamma} \Phi_{\alpha\beta\gamma}\!\begin{pmatrix} 0\ -l'\ l''-l' \\ k\ k'\ k'' \end{pmatrix} \frac{1}{(m_k m_{k'} m_{k''})^{\frac12}} \times$$

$$\times e_\beta(k'|{}^{\mathbf{y}}_{j}) e_\alpha(k|{}^{\mathbf{y}'}_{j'}) e_\gamma(k''|{}^{\mathbf{y}''}_{j''}) \exp\{2\pi i[\mathbf{y}\cdot\mathbf{x}(-l') + \mathbf{y}''\cdot\mathbf{x}(l''-l')]\},$$

where, we notice, $\mathbf{y}(h)$ contributes nothing to the exponential factor as $\mathbf{y}(h)\cdot\mathbf{x}(l')$ is an integer. Now introducing $-l', l''-l'$ as summation indices and afterwards writing them simply as l', l'', we have

$$\Phi\!\begin{pmatrix} \mathbf{y}\ \mathbf{y}'\ \mathbf{y}'' \\ j\ j'\ j'' \end{pmatrix} = \sum_{k\alpha}\sum_{l'k'\beta}\sum_{l''k''\gamma} \Phi_{\alpha\beta\gamma}\!\begin{pmatrix} 0\ l'\ l'' \\ k\ k'\ k'' \end{pmatrix} \frac{1}{(m_k m_{k'} m_{k''})^{\frac12}}\, e_\alpha(k|{}^{\mathbf{y}'}_{j'}) e_\beta(k'|{}^{\mathbf{y}}_{j}) e_\gamma(k''|{}^{\mathbf{y}''}_{j''}) \times$$

$$\times \exp\{2\pi i[\mathbf{y}\cdot\mathbf{x}(l') + \mathbf{y}''\cdot\mathbf{x}(l'')]\}.$$

Comparing with (39.6) we see that the right-hand side is equal to $\Phi(^{y\ y}_{j\ j}\,^{y'}_{j'})$. In other words, the above relation can be written as

$$\Phi(^{y\ y\ y'}_{j\ j\ j'}) = \Phi(^{y'\ y\ y}_{j'\ j\ j}),$$

which is exactly the symmetry relation required.

The above arguments given for the third-order terms are clearly general. The coefficients of the sth-order terms in the expansion (39.2) are thus of the following form:

$$\frac{1}{s!}\,\frac{1}{N^{(s-2)/2}}\,\Delta(\mathbf{y}+\mathbf{y}'+\mathbf{y}''+...+\mathbf{y}^{(s)})\Phi(^{y\ y'}_{j\ j'}...^{y^{(s)}}_{j^{(s)}}),\qquad(39.8)$$

where

$$\Phi(^{y\ y'}_{j\ j'}...^{y^{(s)}}_{j^{(s)}}) = \sum_{k\alpha}\sum_{l'k'\beta}...\sum_{l^{(s)}k^{(s)}\gamma}\Phi_{\alpha\beta...\gamma}(^{0\ l'}_{k\ k'}...^{l^{(s)}}_{k^{(s)}})\,\frac{1}{(m_k\,m_{k'}...m_{k^{(s)}})^{\frac{1}{2}}}\times$$

$$e_\alpha(k|^y_j)e_\beta(k'|^{y'}_{j'})...e_\gamma(k^{(s)}|^{y^{(s)}}_{j^{(s)}})\exp\{2\pi i[\mathbf{y}'.\mathbf{x}(l')+...+\mathbf{y}^{(s)}.\mathbf{x}(l^{(s)})]\}\quad(39.9)$$

is symmetric in the s sets of indices if $\mathbf{y}+\mathbf{y}'+\mathbf{y}''...+\mathbf{y}^{(s)} = \mathbf{y}(h)$, and the coefficients (39.8) are symmetric in the indices in all cases. Moreover, in view of (38.30), we have evidently the general relation

$$\Phi(^{-y}_{j}\ ^{-y'}_{j'}...\ ^{-y^{(s)}}_{j^{(s)}}) = \Phi^*(^{y\ y'}_{j\ j'}...^{y^{(s)}}_{j^{(s)}}).\qquad(39.10)$$

Owing to the restricted range of the permitted wave-numbers, the only surviving first- and second-order terms in the expansion (39.2) correspond respectively to $\mathbf{y} = 0$ and $\mathbf{y}' = -\mathbf{y}$. In the case of $\delta\Phi$ these terms are trivial: the first-order terms are equal to zero, owing to the equilibrium condition $(\Phi_\alpha(^l_k) = \Phi_\alpha(k) = 0)$; the second-order terms reduce to (38.28). On the other hand, for the electric moment and polarizability we have explicitly up to second-order terms:

$$\delta M_\alpha = \sqrt{N}\sum_j M_\alpha(^0_j)Q(^0_j)+\tfrac{1}{2}\sum_y\sum_{jj'}^N M_\alpha(^y_j\ ^{-y}_{j'})Q(^y_j)Q(^{-y}_{j'})+...,\quad(39.11)$$

$$\delta P_{\alpha\beta} = \sqrt{N}\sum_j P_{\alpha\beta}(^0_j)Q(^0_j)+\tfrac{1}{2}\sum_y\sum_{jj'}^N P_{\alpha\beta}(^y_j\ ^{-y}_{j'})Q(^y_j)Q(^{-y}_{j'})+...,\quad(39.12)$$

where the coefficients can be written down in strict analogy with (39.9). Thus the zero-order coefficients written in terms of the notation introduced in (23.5), (23.6) are

$$M_\alpha(^0_j) = \sum_{k\beta} M_{\alpha,\beta}(k)\frac{1}{\sqrt{m_k}}e_\beta(k|^0_j),\qquad(39.13)$$

$$P_{\alpha\beta}(^0_j) = \sum_{k\gamma} P_{\alpha\beta,\gamma}(k)\frac{1}{\sqrt{m_k}}e_\gamma(k|^0_j),\qquad(39.14)$$

and the second-order coefficients are given by

$$M_\alpha(\substack{y \\ j} \substack{-y \\ j'}) = \sum_{k\beta} \sum_{l'k'\gamma} M_{\alpha,\beta\gamma}(\substack{0 \ l' \\ k \ k'}) \frac{1}{(m_k m_{k'})^{\frac{1}{2}}} e_\beta(k|\substack{y \\ j}) e_\gamma(k'|\substack{-y \\ j'}) e^{-2\pi i y \cdot \mathbf{x}(l')}$$

$$= \sum_l \sum_{k\beta} \sum_{k'\gamma} M_{\alpha,\beta\gamma}(\substack{l \\ kk'}) \frac{1}{(m_k m_{k'})^{\frac{1}{2}}} e_\beta(k|\substack{y \\ j}) e_\gamma(k'|\substack{-y \\ j'}) e^{2\pi i y \cdot \mathbf{x}(l)},$$

$$(39.15)$$

$$P_{\alpha\beta}(\substack{y \\ j} \substack{-y \\ j'}) = \sum_{k\gamma} \sum_{l'k'\lambda} P_{\alpha\beta,\gamma\lambda}(\substack{0 \ l' \\ k \ k'}) \frac{1}{(m_k m_{k'})^{\frac{1}{2}}} e_\gamma(k|\substack{y \\ j}) e_\lambda(k'|\substack{-y \\ j'}) e^{-2\pi i y \cdot \mathbf{x}(l')}$$

$$= \sum_l \sum_{k\gamma} \sum_{k'\lambda} P_{\alpha\beta,\gamma\lambda}(\substack{l \\ kk'}) \frac{1}{(m_k m_{k'})^{\frac{1}{2}}} e_\gamma(k|\substack{y \\ j}) e_\lambda(k'|\substack{-y \\ j'}) e^{2\pi i y \cdot \mathbf{x}(l)}.$$

$$(39.16)$$

For the second-order coefficients we have the following symmetry relations:

$$M_\alpha(\substack{y \\ j} \substack{-y \\ j'}) = M_\alpha(\substack{-y \\ j'} \substack{y \\ j}) = M_\alpha^*(\substack{-y \\ j} \substack{y \\ j'}), \qquad (39.17)$$

$$P_{\alpha\beta}(\substack{y \\ j} \substack{-y \\ j'}) = P_{\alpha\beta}(\substack{-y \\ j'} \substack{y \\ j}) = P_{\alpha\beta}^*(\substack{-y \\ j} \substack{y \\ j'}). \qquad (39.18)$$

In writing down the expansion (39.2) we have made use only of the fact that the lattice is a periodic structure. Therefore the same expansion can also be used in connexion with a homogeneously deformed lattice. That is, if we represent the nuclear displacements from a homogeneously deformed configuration by (39.3), the corresponding change in Φ can be written as a similar expansion:

$$\delta\Phi^{\text{def}} = \sqrt{N} \sum_{yj}^{N} \Delta(\mathbf{y})\Phi^{\text{def}}(\substack{y \\ j})Q(\substack{y \\ j}) + \frac{1}{2} \sum_{yj}^{N} \sum_{y'j'}^{N} \Delta(\mathbf{y}+\mathbf{y}')\Phi^{\text{def}}(\substack{y \ y' \\ j \ j'})Q(\substack{y \\ j})Q(\substack{y' \\ j'}) +$$

$$+ \frac{1}{6N^{\frac{1}{2}}} \sum_{yj}^{N} \sum_{y'j'}^{N} \sum_{y''j''}^{N} \Delta(\mathbf{y}+\mathbf{y}'+\mathbf{y}'')\Phi^{\text{def}}(\substack{y \ y' \ y'' \\ j \ j' \ j''})Q(\substack{y \\ j})Q(\substack{y' \\ j'})Q(\substack{y'' \\ j''}) + ..., \quad (39.19)$$

where the coefficients possess the same symmetry properties as discussed above and are related to the derivatives of Φ taken at the deformed configuration by

$$\Phi^{\text{def}}(\substack{y \ y' \\ j \ j'} ... \substack{y^{(s)} \\ j^{(s)}}) = \sum_{k\alpha} \sum_{l'k'\beta} ... \sum_{l^{(s)}k^{(s)}\gamma} \Phi^{\text{def}}_{\alpha\beta...\gamma}(\substack{0 \ l' \\ k \ k'} ... \substack{l^{(s)} \\ k^{(s)}}) \frac{1}{(m_k m_{k'}...m_{k^{(s)}})^{\frac{1}{2}}} \times$$

$$\times e_\alpha(k|\substack{y \\ j}) e_\beta(k'|\substack{y' \\ j'}) ... e_\gamma(k^{(s)}|\substack{y^{(s)} \\ j^{(s)}}) \exp\{2\pi i[\mathbf{y}' \cdot \mathbf{x}(l') + ... + \mathbf{y}^{(s)} \cdot \mathbf{x}(l^{(s)})]\}. \quad (39.20)$$

40. The normalized Hamiltonian

As we have seen (§ 17), in order to obtain the free energy for a system which depends on certain macroscopic parameters, we require the Hamiltonian expressed as a series in the macroscopic parameters and certain internal coordinates q_j. The internal coordinates describe the thermal motion and are to be so chosen that they reduce to the normal coordinates when the macroscopic parameters vanish ($\sim$ the equilibrium configuration in the absence of external forces). When considering an elastically strained lattice, we cannot directly introduce the strain parameters $\bar{u}_{\alpha\beta} \equiv \bar{u}_{\beta\alpha}$ as the macroscopic parameters; for then it would be impossible to write down the Hamiltonian explicitly. Instead, we have to begin by using the deformation parameters $u_{\alpha\beta}$. With $u_{\alpha\beta}$ as the macroscopic parameters, q_j may be considered as certain coordinates which specify further nuclear displacements $\mathbf{u}\binom{l}{k}$ from the strained configuration described by $u_{\alpha\beta}$. We can evidently choose the internal coordinates by using the relation (39.3), namely,

$$\mathbf{u}\binom{l}{k} = \frac{1}{(Nm_k)^{\frac{1}{2}}} \sum_{\mathbf{y}j}^{N} Q\binom{\mathbf{y}}{j}\mathbf{e}(k|\tfrac{\mathbf{y}}{j})e^{2\pi i \mathbf{y}\cdot\mathbf{x}(l)}, \qquad (40.1)$$

whereby the corresponding real coordinates $Q\binom{0}{j}$ and $q_\lambda\binom{\mathbf{y}}{j}$ ($\mathbf{y} \neq 0$, related to $Q\binom{\mathbf{y}}{j}$ by (38.33)) are then taken as the internal coordinates q_j.

For the moment we shall consider the complex coordinates $Q\binom{\mathbf{y}}{j}$ as the internal coordinates. Our task is thus to express the Hamiltonian as a series in $u_{\alpha\beta}$ and $Q\binom{\mathbf{y}}{j}$, on the supposition that the lattice is first subject to an external strain described by $u_{\alpha\beta}$ and then the nuclei are further displaced according to (40.1). As in § 17, we shall consider only terms up to the second order both in the macroscopic parameters and in the coordinates $Q\binom{\mathbf{y}}{j}$.

Consider first the part of Φ depending on the coordinates $Q\binom{\mathbf{y}}{j}$. This part gives the change in Φ due to the displacements (40.1) from the homogeneously deformed configuration $u_{\alpha\beta}$; its value, normalized to N lattice cells, is thus equal to the expression $\delta\Phi^{\text{def}}$ given by (39.19), if the deformed configuration in (39.19) and (39.20) is interpreted as the configuration described by $u_{\alpha\beta}$. Using the nuclear displacements (36.4), we have

$$\Phi^{\text{def}}_{\alpha\beta\ldots}\binom{0\ \ l'}{k\ \ k'}\ldots) = \Phi_{\alpha\beta\ldots}\binom{0\ \ l'}{k\ \ k'}\ldots) + \sum_{l\mu} \Phi_{\mu\alpha\beta\ldots}\binom{l\ \ 0\ \ l'}{k\ \ k\ \ k'}\ldots)\sum_{\gamma} u_{\mu\gamma}x_\gamma\binom{l}{k} +$$

$$+ \tfrac{1}{2}\sum_{l\mu}\sum_{l'\nu} \Phi_{\mu\nu\alpha\beta\ldots}\binom{l\ \ l'\ \ 0\ \ l'}{k\ \ k'\ \ k\ \ k'}\ldots)\Big\{\sum_{\gamma} u_{\mu\gamma}x_\gamma\binom{l}{k}\Big\}\Big\{\sum_{\lambda} u_{\nu\lambda}x_\lambda\binom{l'}{k'}\Big\} + \ldots . \quad (40.2)$$

With the right-hand side regarded as a series expansion in the parameters $u_{\alpha\beta}$, the corresponding coefficients may be considered as certain transforms of the derivatives of Φ, obtained by the multiplication with a lattice vector such as $x_\gamma\binom{l}{k}$ followed by a summation over the indices $\binom{l}{k}$. We shall denote such transforms by replacing the summed indices in the derivative of Φ by horizontal bars and enclosing the

Cartesian index introduced by the lattice vector with the Cartesian index originally belonging to the summed indices. Using this convention, we can write (40.2) as an expansion in $u_{\alpha\beta}$ as follows

$$\Phi^{\text{def}}_{\alpha\beta\ldots}(^{0\ l'}_{k\ k'}\ldots) = \Phi_{\alpha\beta}(^{0\ l'}_{k\ k'}\ldots) + \sum_{\mu\gamma}\Phi_{(\mu\gamma)\alpha\beta\ldots}(^{-\ 0\ l'}_{\ \ k\ k'}\ldots)u_{\mu\gamma} +$$

$$+\tfrac{1}{2}\sum_{\mu\gamma}\sum_{\nu\lambda}\Phi_{(\mu\gamma)(\nu\lambda)\alpha\beta\ldots}(^{-\ -\ 0\ l'}_{\ \ \ \ k\ k'}\ldots)u_{\mu\gamma}u_{\nu\lambda}+\ldots. \quad (40.3)$$

Substituting (40.3) in (39.20) and using the latter in (39.19), we obtain $\delta\Phi^{\text{def}}$ in the form of a series in $u_{\alpha\beta}$ as well as in $Q(^{y}_{j})$. A suitable notation for the coefficients is obtained by regarding the relations (39.20) and (39.9) as defining a transform leading from the indices $\alpha\beta\ldots$ ($^{0\ l'}_{k\ k'}\ldots$) *en bloc* to the indices $(^{y\ y'}_{j\ j'}\ldots)$. With this understanding, we can write the expansion as follows:

$$\delta\Phi^{\text{def}} = \sqrt{N}\sum_{\alpha\beta}\sum_{j}\Phi_{(\alpha\beta)}(^{-\ 0}_{\ \ j})u_{\alpha\beta}\,Q(^{0}_{j}) +$$

$$+\frac{\sqrt{N}}{2}\sum_{\alpha\beta}\sum_{\gamma\lambda}\sum_{j}\Phi_{(\alpha\beta)(\gamma\lambda)}(^{-\ -\ 0}_{\ \ \ \ j})u_{\alpha\beta}u_{\gamma\lambda}\,Q(^{0}_{j}) +$$

$$+\tfrac{1}{2}\sum_{y}\sum_{j}\omega^{2}(^{y}_{j})Q(^{y}_{j})Q(^{-y}_{j}) +$$

$$+\tfrac{1}{2}\sum_{\alpha\beta}\sum_{y}\sum_{jj'}\Phi_{(\alpha\beta)}(^{-\ y\ -y}_{\ \ j\ j'})u_{\alpha\beta}\,Q(^{y}_{j})Q(^{-y}_{j'}) +$$

$$+\tfrac{1}{4}\sum_{\alpha\beta}\sum_{\gamma\lambda}\sum_{y}\sum_{jj'}\Phi_{(\alpha\beta)(\gamma\lambda)}(^{-\ -\ y\ -y}_{\ \ \ \ j\ j'})u_{\alpha\beta}u_{\gamma\lambda}\,Q(^{y}_{j})Q(^{-y}_{j'})+\ldots.$$
$$(40.4)$$

The systematic notation for the coefficients points directly to their relations to the derivatives of Φ; for example

$$\Phi_{(\alpha\beta)(\gamma\lambda)}(^{-\ -\ y\ -y}_{\ \ \ \ j\ j'}) = \sum_{lk}\sum_{l'k'}\sum_{k\mu}\sum_{k'\nu}\Phi_{\alpha\gamma\mu\nu}(^{l\ l'\ 0\ l'}_{k\ k'\ k\ k'})x_{\beta}(^{l}_{k})x_{\lambda}(^{l'}_{k'})\times$$

$$\times\frac{1}{(m_{k}m_{k'})^{\frac{1}{2}}}e_{\mu}(k|^{y}_{j})e_{\nu}(k'|^{-y}_{j'})e^{-2\pi iy\cdot x(l')}. \quad (40.5)$$

In order to obtain the total energy we have to add to (40.4) the energy in the homogeneous configuration described by $u_{\alpha\beta}$. As we have already emphasized in the last chapter (§ 25), the method of direct expansion of Φ does not lead to an unambiguous value for the energy density due to a homogeneous deformation. This difficulty arises from the fact that the nuclear displacements in an external strain increase indefinitely with the distance of the nuclei from the origin and do not fulfil the periodic condition; thus the normalization procedure which has been used in connexion with the displacements (40.1) is no longer applicable.

The energy density due to an external strain may, however, be obtained with the help of the elastic constants (about the static equilibrium configuration X^0) derived in the last chapter with the method of long waves. From the elastic constants we can construct the strain energy function

$$\tfrac{1}{2}\sum_{\alpha\gamma}\sum_{\beta\lambda}c_{\alpha\gamma,\beta\lambda}u_{\alpha\gamma}u_{\beta\lambda}. \tag{40.6}$$

This expression includes not only the energy due to the external strain described by $u_{\alpha\beta}$ but also the energy (negative!) of the internal strain induced by the external strain. The contribution due to the internal strain can be eliminated as follows. In view of the physical interpretation of the first-order equation (26.18), one sees readily that in the expression for the elastic constants (see (27.26)),

$$c_{\alpha\gamma,\beta\lambda} = [\alpha\beta,\gamma\lambda]+[\beta\gamma,\alpha\lambda]-[\beta\lambda,\alpha\gamma]+(\alpha\gamma,\beta\lambda), \tag{40.7}$$

only the term $(\alpha\gamma,\beta\lambda)$ is due to the induced internal strain. If the internal strain is suppressed, this term drops out and the square brackets are left unaffected. Hence upon leaving out $(\alpha\gamma,\beta\lambda)$ in (40.7) and substituting the latter in (40.6), we obtain the energy density due to the external strain in the following form:

$$\tfrac{1}{2}\sum_{\alpha\gamma}\sum_{\beta\lambda}\{[\alpha\beta,\gamma\lambda]+[\beta\gamma,\alpha\lambda]-[\beta\lambda,\alpha\gamma]\}u_{\alpha\gamma}u_{\beta\lambda}. \tag{40.8}$$

Denoting the energy per cell of the undeformed lattice by ϕ_0, we find by adding (40.4), (40.8) and $N\phi_0$ the total potential energy of N cells:

$$N\phi_0+\tfrac{1}{2}Nv_a\sum_{\alpha\gamma}\sum_{\beta\lambda}\{[\alpha\beta,\gamma\lambda]+[\beta\gamma,\alpha\lambda]-[\beta\lambda,\alpha\gamma]\}u_{\alpha\gamma}u_{\beta\lambda}+$$

$$+\sqrt{N}\sum_{\alpha\beta}\sum_{j}\Phi_{(\alpha\beta)}(_{\,-}^{\,0}{}_j)u_{\alpha\beta}Q(_j^0)+\frac{\sqrt{N}}{2}\sum_{\alpha\beta}\sum_{\gamma\lambda}\sum_{j}\Phi_{(\alpha\beta)(\gamma\lambda)}(_{\,-}^{\,-}{}_j^0)u_{\alpha\beta}u_{\gamma\lambda}Q(_j^0)+$$

$$+\tfrac{1}{2}\sum_{y}^{N}\sum_{j}\omega^2(_j^y)Q(_j^y)Q(_{\,-j}^{\,-y})+$$

$$+\tfrac{1}{2}\sum_{\alpha\beta}\sum_{y}^{'N}\sum_{jj'}\Phi_{(\alpha\beta)}(_{\,-j}^{\,-y}{}_{j'}^{\,-y})u_{\alpha\beta}Q(_j^y)Q(_{j'}^{\,-y})+$$

$$+\tfrac{1}{4}\sum_{\alpha\beta}\sum_{\gamma\lambda}\sum_{y}^{N}\sum_{jj'}\Phi_{(\alpha\beta)(\gamma\lambda)}(_{\,-\,-j}^{\,-\,-y}{}_{j'}^{\,-y})u_{\alpha\beta}u_{\gamma\lambda}Q(_j^y)Q(_{j'}^{\,-y})+\ldots. \tag{40.9}$$

We may recall that the square brackets are defined in terms of the second derivatives of Φ by (26.5) and (26.32); combining these formulae, we have

$$[\alpha\beta,\gamma\lambda] = \frac{-1}{2v_a}\sum_{l}\sum_{kk'}\Phi_{\alpha\beta}(_{kk'}^{\,l})x_\gamma(_{kk'}^{\,l})x_\lambda(_{kk'}^{\,k}). \tag{40.10}$$

If an electric field is present, we have to include in the Hamiltonian an effective interaction term between the nuclear motion and the field. The use of Placzek's approximation (§ 20) shows immediately that in order to reproduce the eigenvalues (18.11) correctly for the vibrational states (in the electronic ground state), the interaction term must be of the form

$$- \sum_\alpha M_\alpha(X) E_\alpha - \tfrac{1}{2} \sum_{\alpha\beta} P_{\alpha\beta}(0, X) E_\alpha E_\beta. \qquad (40.11)$$

In expressing the interaction term explicitly as a series in $u_{\alpha\beta}$, $\mathbf{E}$, and $Q(^y_j)$, we can consider it as the sum of two contributions, namely, the value of the interaction term in the homogeneous configuration described by $u_{\alpha\beta}$ and the additional contribution due to the displacements (40.1) from the homogeneous configuration. The latter can be written down directly by the procedure used in obtaining (40.4):

$$-\sqrt{N} \sum_\alpha M_\alpha(^0_j) E_\alpha Q(^0_j) - \frac{\sqrt{N}}{2} \sum_{\alpha\beta} P_{\alpha\beta}(^0_j) E_\alpha E_\beta Q(^0_j) -$$

$$-\sqrt{N} \sum_\alpha \sum_{\beta\gamma} M_{\alpha,(\beta\gamma)}(^-\ ^0_j) E_\alpha u_{\beta\gamma} Q(^0_j) -$$

$$-\tfrac{1}{2} \sum_y^N \sum_{jj'} \sum_\alpha M_\alpha(^y_j \ ^{-y}_{j'}) E_\alpha Q(^y_j) Q(^{-y}_{j'}) -$$

$$-\tfrac{1}{4} \sum_y^N \sum_{jj'} \sum_{\alpha\beta} P_{\alpha\beta}(^y_j \ ^{-y}_{j'}) E_\alpha E_\beta Q(^y_j) Q(^{-y}_{j'}) -$$

$$-\tfrac{1}{2} \sum_y^N \sum_{jj'} \sum_\alpha \sum_{\beta\gamma} M_{\alpha,(\beta\gamma)}(^-\ ^y_j \ ^{-y}_{j'}) E_\alpha u_{\beta\gamma} Q(^y_j) Q(^{-y}_{j'}) + ..., \qquad (40.12)$$

where the coefficients are transforms of the derivatives of the moment and the polarizability defined in the same way as in the case of the potential function; for example,

$$M_{\alpha,(\beta\gamma)}(^-\ ^y_j \ ^{-y}_{j'})$$

$$= \sum_{lk} \sum_{k\mu} \sum_{l'k'\nu} M_{\alpha,\beta\mu\nu}(^l_k \ ^0_k \ ^{l'}_{k'}) x_\gamma(^l_k) \frac{1}{(m_k m_{k'})^{\frac{1}{2}}} e_\mu(k|^y_j) e_\nu(k'|^{-y}_{j'}) e^{-2\pi i y \cdot x(l')}. \qquad (40.13)$$

Since the components of $\mathbf{E}$ are macroscopic parameters like $u_{\alpha\beta}$, to be consistent with the desired accuracy we have only to consider terms in $\mathbf{M}(X)$ up to the first order in $u_{\alpha\beta}$, and in $P_{\alpha\beta}(0, X)$ we can ignore all terms dependent on $u_{\alpha\beta}$.

In the interaction term for the homogeneous configuration described by $u_{\alpha\beta}$, we write the part independent of $u_{\alpha\beta}$ as follows:

$$-N \sum_\alpha \mu_\alpha^0 E_\alpha - \frac{N}{2} \sum_{\alpha\beta} p_{\alpha\beta}^0 E_\alpha E_\beta, \qquad (40.14)$$

where μ_α^0 and $p_{\alpha\beta}^0$ are respectively the moment and electronic polarizability per cell in the undeformed configuration X^0. Besides (40.14), we have to consider only the terms of $\mathbf{M}(X)$ linear in $u_{\alpha\beta}$, all other terms being of the third and higher orders in the macroscopic parameters. Using the displacements (36.4), we find that the terms of $\mathbf{M}(X)$ linear in $u_{\alpha\beta}$ are

$$\sum_{lk\beta} M_{\alpha,\beta}\binom{l}{k} \sum_\gamma u_{\beta\gamma} x_\gamma\binom{l}{k} = \sum_{l\beta\gamma} u_{\beta\gamma} x_\gamma(l) \sum_k M_{\alpha,\beta}(k) + \sum_l \sum_{\beta\gamma} u_{\beta\gamma} \sum_k M_{\alpha,\beta}(k) x_\gamma(k). \qquad (40.15)$$

The first term on the right-hand side vanishes identically owing to the invariance relation (23.24); the second term, on the other hand, shows that every cell contributes to the moment the amount

$$\sum_{\beta\gamma} u_{\beta\gamma} \sum_k M_{\alpha,\beta}(k) x_\gamma(k). \qquad (40.16)$$

The corresponding energy for N cells is

$$-N \sum_{\alpha\beta\gamma} \left\{ \sum_k M_{\alpha,\beta}(k) x_\gamma(k) \right\} E_\alpha u_{\beta\gamma}. \qquad (40.17)$$

Adding (40.12), (40.14), and (40.17), we obtain for the interaction term, normalized to N lattice cells, the following expression:

$$-N \sum_\alpha \mu_\alpha^0 E_\alpha - \frac{N}{2} \sum_{\alpha\beta} p_{\alpha\beta}^0 E_\alpha E_\beta - N \sum_{\alpha\beta\gamma} \left\{ \sum_k M_{\alpha,\beta}(k) x_\gamma(k) \right\} E_\alpha u_{\beta\gamma} -$$

$$- \sqrt{N} \sum_\alpha M_\alpha\binom{0}{j} E_\alpha Q\binom{0}{j} - \frac{\sqrt{N}}{2} \sum_{\alpha\beta} P_{\alpha\beta}\binom{0}{j} E_\alpha E_\beta Q\binom{0}{j} -$$

$$- \sqrt{N} \sum_\alpha \sum_{\beta\gamma} M_{\alpha,(\beta\gamma)}\binom{-\ 0}{\ j} E_\alpha u_{\beta\gamma} Q\binom{0}{j} -$$

$$- \tfrac{1}{2} \sum_\gamma^N \sum_{jj'} \sum_\alpha M_\alpha\binom{y\ \ \bar{y}}{j\ \ \bar{j}} E_\alpha Q\binom{y}{j} Q\binom{\bar{y}}{\bar{j}} -$$

$$- \tfrac{1}{4} \sum_\gamma^N \sum_{jj'} \sum_{\alpha\beta} P_{\alpha\beta}\binom{y\ \ \bar{y}}{j\ \ \bar{j}} E_\alpha E_\beta Q\binom{y}{j} Q\binom{\bar{y}}{\bar{j}} -$$

$$- \tfrac{1}{2} \sum_\gamma^N \sum_{jj'} \sum_\alpha \sum_{\beta\gamma} M_{\alpha,(\beta\gamma)}\binom{-\ y\ \ \bar{y}}{\ j\ \ \bar{j}} E_\alpha u_{\beta\gamma} Q\binom{y}{j} Q\binom{\bar{y}}{\bar{j}}. \qquad (40.18)$$

The sum of the kinetic energy (38.29), the potential energy (40.9), and the interaction energy (40.18) represents the normalized Hamiltonian for a homogeneously strained lattice in the presence of an electric field.

41. The free energy

Subdividing the Hamiltonian as in (17.2), we can write in the present case

$$H_0 = N\phi_0 + \tfrac{1}{2}\sum_{y}^{\downarrow N}\sum_{j}\sum_{\lambda}\{P_\lambda^2(\tbinom{y}{j}) + \omega^2(\tbinom{y}{j})q_\lambda^2(\tbinom{y}{j})\},$$

$$H_1 = \sum_{\alpha\beta} g^{(\alpha\beta)}(q)u_{\alpha\beta} + \sum_{\alpha} g^{\alpha}(q)E_\alpha,$$

$$H_2 = \tfrac{1}{2}\sum_{\alpha\beta} g^{\alpha\beta}(q)E_\alpha E_\beta + \sum_{\alpha}\sum_{\beta\gamma} g^{\alpha(\beta\gamma)}(q)E_\alpha u_{\beta\gamma} + \tfrac{1}{2}\sum_{\alpha\beta}\sum_{\gamma\lambda} g^{(\alpha\beta)(\gamma\lambda)}(q)u_{\alpha\beta}u_{\gamma\lambda},$$

$$(41.1)$$

where, as in § 17,

$$g^{\alpha\beta} = g^{\beta\alpha}, \qquad g^{(\alpha\beta)(\gamma\lambda)} = g^{(\gamma\lambda)(\alpha\beta)}. \qquad (41.2)$$

All the g-coefficients are series of the form

$$g = g_0 + \sum_{j} g_j Q(\tbinom{0}{j}) + \tfrac{1}{2}\sum_{y}^{N}\sum_{jj'} g(\tbinom{y}{j}\tbinom{-y}{j'})Q(\tbinom{y}{j})Q(\tbinom{-y}{j'}) + \dots. \qquad (41.3)$$

The coefficients in the series can be directly read off from (40.9) and (40.18); they are listed below for various cases:

$$g_0^\alpha = -N\mu_\alpha^0, \qquad g^{(\alpha\beta)} = 0,$$

$$g_j^\alpha = -\sqrt{N}\, M_\alpha(\tbinom{0}{j}), \qquad g_j^{(\alpha\beta)} = \sqrt{N}\,\Phi_{(\alpha\beta)}(\tbinom{-\ 0}{\ j}),$$

$$g^\alpha(\tbinom{y}{j}\tbinom{-y}{j'}) = -M_\alpha(\tbinom{y}{j}\tbinom{-y}{j'}), \qquad g^{(\alpha\beta)}(\tbinom{y}{j}\tbinom{-y}{j'}) = \Phi_{(\alpha\beta)}(\tbinom{-\ y}{\ j}\tbinom{-y}{j'}),$$

$$g_0^{\alpha\beta} = -Np_{\alpha\beta}^0, \qquad g_0^{\alpha(\beta\gamma)} = -N\sum_k M_{\alpha,\beta}(k)x_\gamma(k),$$

$$g_0^{(\alpha\beta)(\gamma\lambda)} = \{[\alpha\gamma,\beta\lambda] + [\beta\gamma,\alpha\lambda] - [\gamma\lambda,\alpha\beta]\}Nv_a = g_0^{(\gamma\lambda)(\alpha\beta)}\ \text{(see (27.25))},$$

$$g_j^{\alpha\beta} = -\sqrt{N}\,P_{\alpha\beta}(\tbinom{0}{j}), \qquad g_j^{\alpha(\beta\gamma)} = -\sqrt{N}\,M_{\alpha,(\beta\gamma)}(\tbinom{-\ 0}{\ j}),$$

$$g_j^{(\alpha\beta)(\gamma\lambda)} = \sqrt{N}\,\Phi_{(\alpha\beta)(\gamma\lambda)}(\tbinom{-\ -\ 0}{\ \ j}), \qquad g^{\alpha\beta}(\tbinom{y}{j}\tbinom{-y}{j'}) = -P_{\alpha\beta}(\tbinom{y}{j}\tbinom{-y}{j'}),$$

$$g^{\alpha(\beta\gamma)} = -M_{\alpha,(\beta\gamma)}(\tbinom{-\ y}{\ j}\tbinom{-y}{j'}), \qquad g^{(\alpha\beta)(\gamma\lambda)}(\tbinom{y}{j}\tbinom{-y}{j'}) = \Phi_{(\alpha\beta)(\gamma\lambda)}(\tbinom{-\ -\ y}{\ \ j}\tbinom{-y}{j'}).$$

$$(41.4)$$

All the second-order coefficients satisfy the relations

$$g(\tbinom{y}{j}\tbinom{-y}{j'}) = g(\tbinom{-y}{j'}\tbinom{y}{j}), \qquad (41.5)$$

$$g^*(\tbinom{y}{j}\tbinom{-y}{j'}) = g(\tbinom{-y}{j'}\tbinom{y}{j}), \qquad (41.6)$$

which can be verified individually by arguments similar to those used in § 39. More generally, as a moment's consideration will show, these relations arise as the direct consequence of the relations

$$\Phi^{\mathrm{def}}(\tbinom{y}{j}\tbinom{-y}{j'}) = \Phi^{\mathrm{def}}(\tbinom{-y}{j'}\tbinom{y}{j}) = \{\Phi^{\mathrm{def}}(\tbinom{-y}{j'}\tbinom{y}{j})\}^*, \qquad (41.7)$$

which are valid for any homogeneously deformed configurations such as described by $u_{\alpha\beta}$, and the similar relations for the derivatives of

$M(X)$ and $P_{\alpha\beta}(0,X)$; for the second-order g-coefficients are essentially the expansion coefficients of $\Phi^{\text{def}}(\begin{smallmatrix}y\\j\end{smallmatrix}\begin{smallmatrix}-y\\j'\end{smallmatrix})$, etc., expressed as series in the parameters $u_{\alpha\beta}$.

When expressed in terms of the real coordinates $Q(\begin{smallmatrix}0\\j\end{smallmatrix})$ and $q_\lambda(\begin{smallmatrix}y\\j\end{smallmatrix})$ ($y \neq 0$, see (38.33), (38.34)), (41.3) clearly assumes the form

$$g = g_0 + \sum_j g_j Q(\begin{smallmatrix}0\\j\end{smallmatrix}) + \tfrac{1}{2} \sum_y^{\frac{1}{2}N} \sum_{jj'} \sum_{\lambda\lambda'} g_{\lambda\lambda'}(\begin{smallmatrix}y\\jj'\end{smallmatrix}) q_\lambda(\begin{smallmatrix}y\\j\end{smallmatrix}) q_{\lambda'}(\begin{smallmatrix}y\\j'\end{smallmatrix}), \qquad (41.8)$$

i.e. the linear terms depend only on $Q(\begin{smallmatrix}0\\j\end{smallmatrix})$ and the quadratic terms are products of $q_\lambda(\begin{smallmatrix}y\\j\end{smallmatrix})$ (or $Q(\begin{smallmatrix}0\\j\end{smallmatrix})$, not explicitly indicated in (41.8)) belonging to the wave-number y. $Q(\begin{smallmatrix}0\\j\end{smallmatrix})$ and $q_\lambda(\begin{smallmatrix}y\\j\end{smallmatrix})$ represent in the present case the internal coordinates denoted by q_j in §17; (41.8) is thus the exact equivalent of the expansions (17.3). With (17.3) replaced by (41.8), we can accordingly rewrite the formulae (17.26) for the coefficients in the free energy as follows:

$$A^0 = N\phi_0 + 2kT \sum_y^{N/2} \sum_j \ln\{2\sinh\tfrac{1}{2}\beta(\begin{smallmatrix}y\\j\end{smallmatrix})\},$$

$$A^s = g_0^s + \tfrac{1}{2} \sum_y^{N/2} \sum_j \overline{q^2(\begin{smallmatrix}y\\j\end{smallmatrix})} \sum_\lambda g_{\lambda\lambda}^s(\begin{smallmatrix}y\\jj\end{smallmatrix}),$$

$$A^{st} = g_0^{st} - \sum_j [\omega(\begin{smallmatrix}0\\j\end{smallmatrix})]^{-2} g_j^s g_j^t +$$
$$+ \sum_y^{\frac{1}{2}N} \sum_j \left\{ \tfrac{1}{2}\overline{q^2(\begin{smallmatrix}y\\j\end{smallmatrix})} \sum_\lambda g_{\lambda\lambda}^{st}(\begin{smallmatrix}y\\jj\end{smallmatrix}) - \frac{\overline{q^2(\begin{smallmatrix}y\\j\end{smallmatrix})}}{4\omega^2(\begin{smallmatrix}y\\j\end{smallmatrix})} \sum_\lambda g_{\lambda\lambda}^s(\begin{smallmatrix}y\\jj\end{smallmatrix}) g_{\lambda\lambda}^t(\begin{smallmatrix}y\\jj\end{smallmatrix}) \right\} +$$
$$+ \frac{1}{2} \sum_y^{\frac{1}{2}N} \sum_{j\lambda} {\sum_{j'\lambda'}}' \frac{\overline{q^2(\begin{smallmatrix}y\\j\end{smallmatrix})}}{\omega^2(\begin{smallmatrix}y\\j\end{smallmatrix}) - \omega^2(\begin{smallmatrix}y\\j'\end{smallmatrix})} g_{\lambda\lambda'}^s(\begin{smallmatrix}y\\jj'\end{smallmatrix}) g_{\lambda\lambda'}^t(\begin{smallmatrix}y\\jj'\end{smallmatrix}) -$$
$$- \frac{1}{4kT} \sum_y^{\frac{1}{2}N} \sum_j \left\{ [\overline{q^2(\begin{smallmatrix}y\\j\end{smallmatrix})}]^2 - \frac{\hbar^2}{4\omega^2(\begin{smallmatrix}y\\j\end{smallmatrix})} \right\} \sum_\lambda g_{\lambda\lambda}^s(\begin{smallmatrix}y\\jj\end{smallmatrix}) g_{\lambda\lambda}^t(\begin{smallmatrix}y\\jj\end{smallmatrix}), \qquad (41.9)$$

where the indices s and t stand for either a single index or an index-pair depending on whether they refer to the field components or the deformation parameters. In writing down (41.9) we have taken account of the fact that the frequency $\omega(\begin{smallmatrix}y\\j\end{smallmatrix})$ is independent of the λ-index and hence are also $\beta(\begin{smallmatrix}y\\j\end{smallmatrix})$ and $\overline{q^2(\begin{smallmatrix}y\\j\end{smallmatrix})}$:

$$\beta(\begin{smallmatrix}y\\j\end{smallmatrix}) = \hbar\omega(\begin{smallmatrix}y\\j\end{smallmatrix})/kT, \qquad \overline{q^2(\begin{smallmatrix}y\\j\end{smallmatrix})} = \frac{\hbar}{2\omega(\begin{smallmatrix}y\\j\end{smallmatrix})} \coth\left\{\frac{\hbar\omega(\begin{smallmatrix}y\\j\end{smallmatrix})}{2kT}\right\}. \qquad (41.10)$$

In order to work out A^0, A^s, A^{st} explicitly with the help of (41.4), we have still to express the second-order coefficients in (41.9) in terms of the second-order coefficients $g(\begin{smallmatrix}y\\j\end{smallmatrix}\begin{smallmatrix}-y\\j'\end{smallmatrix})$. A comparison of (41.8) with (41.3) leads to the identities

$$\sum_{\lambda\lambda'} g_{\lambda\lambda'}(\begin{smallmatrix}y\\jj'\end{smallmatrix}) q_\lambda(\begin{smallmatrix}y\\j\end{smallmatrix}) q_{\lambda'}(\begin{smallmatrix}y\\j'\end{smallmatrix}) = g(\begin{smallmatrix}y\\j\end{smallmatrix}\begin{smallmatrix}-y\\j'\end{smallmatrix}) Q(\begin{smallmatrix}y\\j\end{smallmatrix}) Q(\begin{smallmatrix}-y\\j'\end{smallmatrix}) + g(\begin{smallmatrix}-y\\j'\end{smallmatrix}\begin{smallmatrix}y\\j\end{smallmatrix}) Q(\begin{smallmatrix}-y\\j'\end{smallmatrix}) Q(\begin{smallmatrix}y\\j\end{smallmatrix}). \qquad (41.11)$$

Expressing the complex coordinates on the right-hand side in terms of the real coordinates $q_\lambda(\substack{y \\ j})$ with the help of (38.33) and (38.34) and afterwards equating the corresponding coefficients on the two sides of (41.11), we find the desired relations between the two types of second-order coefficients, which may be written

$$g_{\lambda\lambda'}(\substack{y \\ jj'}) = a_{\lambda\lambda'} \cdot g(\substack{y \ -y \\ j \ \ j}) + a_{\lambda\lambda'}^* \cdot g(\substack{-y \ y \\ \ j \ j}), \qquad (41.12)$$

where $\qquad a_{11} = \dfrac{1}{2}, \qquad a_{12} = \dfrac{-i}{2}, \qquad a_{21} = \dfrac{i}{2}, \qquad a_{22} = \dfrac{1}{2}.$ $\qquad (41.13)$

Except in the terms

$$\frac{1}{2} \sum_y^{\frac{1}{2}N} \sum_{j\lambda} \sideset{}{'}\sum_{j'\lambda'} \frac{\overline{q^2(\substack{y \\ j})}}{\omega^2(\substack{y \\ j}) - \omega^2(\substack{y \\ j'})} g_{\lambda\lambda'}^s(\substack{y \\ jj'}) g_{\lambda\lambda'}^t(\substack{y \\ jj'}), \qquad (41.14)$$

the second-order coefficients appear in (41.9) only in the form $g_{\lambda\lambda'}(\substack{y \\ jj})$. Upon putting $j' = j$ in (41.5) and (41.6), we observe that $g(\substack{y \ -y \\ j \ \ j})$ is real, and furthermore the g-coefficients in the two terms on the right-hand side of (41.12) are equal, so that (41.12) can be written as

$$g_{\lambda\lambda'}(\substack{y \\ jj}) = g(\substack{y \ -y \\ j \ \ j})\{a_{\lambda\lambda'} + a_{\lambda\lambda'}^*\}. \qquad (41.15)$$

Thus, using the values (41.13) for $a_{\lambda\lambda'}$, we find that

$$g_{\lambda\lambda}(\substack{y \\ jj}) = g(\substack{y \ -y \\ j \ \ j}), \qquad (41.16)$$

and $\qquad g_{\lambda\lambda'}(\substack{y \\ jj}) = 0 \quad$ if $\lambda' \neq \lambda$. $\qquad (41.17)$

Hence, in connexion with expressing (41.9) in terms of the coefficients listed in (41.4), only the terms (41.14) require special mention. The prime over the summation in (41.14) excludes only the terms with both $j' = j$ and $\lambda' = \lambda$; there are thus a number of terms with a vanishing denominator, namely, the terms with $j' = j$ but $\lambda' \neq \lambda$. However, as shown by (41.17), the g-coefficients in these terms are equal to zero. We recall that the terms (41.14) arise originally (see § 17) from a second-order perturbation calculation; the vanishing g-coefficients indicate that the matrix elements concerned are equal to zero. The terms $j' = j$, $\lambda' \neq \lambda$ are thus to be omitted from (41.14), leaving

$$\frac{1}{2} \sum_y^{\frac{1}{2}N} \sum_j \sideset{}{'}\sum_{j'} \frac{\overline{q^2(\substack{y \\ j})}}{\omega^2(\substack{y \\ j}) - \omega^2(\substack{y \\ j'})} \left\{ \sum_{\lambda\lambda'} g_{\lambda\lambda'}^s(\substack{y \\ jj'}) g_{\lambda\lambda'}^t(\substack{y \\ jj'}) \right\}, \qquad (41.18)$$

where the prime over the summation now excludes all terms with $j' = j$.

With the help of (41.12) and (41.13), we find that

$$\sum_{\lambda\lambda'} g^s_{\lambda\lambda'}(^{\mathbf{y}}_{jj'})g^t_{\lambda\lambda'}(^{\mathbf{y}}_{jj'}) = \left\{\sum_{\lambda\lambda'} a^2_{\lambda\lambda'}\right\}\{g^s(^{\mathbf{y}}_j {}^{-\mathbf{y}}_{j'})g^t(^{\mathbf{y}}_j {}^{-\mathbf{y}}_{j'})\}+$$

$$+\left\{\sum_{\lambda\lambda'} a_{\lambda\lambda'}\,a^*_{\lambda\lambda'}\right\}\{g^s(^{\mathbf{y}}_j {}^{-\mathbf{y}}_{j'})g^t(^{-\mathbf{y}}_j {}^{\mathbf{y}}_{j'})+g^s(^{-\mathbf{y}}_j {}^{\mathbf{y}}_{j'})g^t(^{\mathbf{y}}_j {}^{-\mathbf{y}}_{j'})\}+$$

$$+\left\{\sum_{\lambda\lambda'} (a^*_{\lambda\lambda'})^2\right\}\{g^s(^{-\mathbf{y}}_j {}^{\mathbf{y}}_{j'})g^t(^{-\mathbf{y}}_j {}^{\mathbf{y}}_{j'})\}$$

$$= g^s(^{\mathbf{y}}_j {}^{-\mathbf{y}}_{j'})g^t(^{-\mathbf{y}}_j {}^{\mathbf{y}}_{j'})+g^s(^{-\mathbf{y}}_j {}^{\mathbf{y}}_{j'})g^t(^{\mathbf{y}}_j {}^{-\mathbf{y}}_{j'}).$$

Thus (41.18) reduces to

$$\frac{1}{2}\sum_{\mathbf{y}}^{\tfrac{1}{2}N}\sum_{j}\sum_{j'}{}' \frac{\overline{q^2(^{\mathbf{y}}_j)}}{\omega^2(^{\mathbf{y}}_j)-\omega^2(^{\mathbf{y}}_{j'})} \{g^s(^{\mathbf{y}}_j {}^{-\mathbf{y}}_{j'})g^t(^{-\mathbf{y}}_j {}^{\mathbf{y}}_{j'})+g^s(^{-\mathbf{y}}_j {}^{\mathbf{y}}_{j'})g^t(^{\mathbf{y}}_j {}^{-\mathbf{y}}_{j'})\}. \quad (41.19)$$

In view of (41.16) and (41.19), (41.9) can be readily written in terms of the coefficients given in (41.4) as follows:

$$A^0 = N\phi_0+2kT\sum_{\mathbf{y}}^{\tfrac{1}{2}N}\sum_{j}\ln\{2\sinh\tfrac{1}{2}\beta(^{\mathbf{y}}_j)\},$$

$$A^s = g^s_0 + \sum_{\mathbf{y}}^{\tfrac{1}{2}N}\sum_{j}g^s(^{\mathbf{y}}_j {}^{-\mathbf{y}}_j)q^2(^{\mathbf{y}}_j),$$

$$A^{st} = g^{st}_0 - \sum_{j}[\omega(^0_j)]^{-2}g^s_j g^t_j+$$

$$+\sum_{\mathbf{y}}^{\tfrac{1}{2}N}\sum_{j}\left\{g^{st}(^{\mathbf{y}}_j {}^{-\mathbf{y}}_j)-\frac{1}{2\omega^2(^{\mathbf{y}}_j)}g^s(^{\mathbf{y}}_j {}^{-\mathbf{y}}_j)g^t(^{\mathbf{y}}_j {}^{-\mathbf{y}}_j)\right\}\overline{q^2(^{\mathbf{y}}_j)}+$$

$$+\tfrac{1}{2}\sum_{\mathbf{y}}^{\tfrac{1}{2}N}\sum_{j}\sum_{j'}{}' \{g^s(^{\mathbf{y}}_j {}^{-\mathbf{y}}_{j'})g^t(^{-\mathbf{y}}_j {}^{\mathbf{y}}_{j'})+g^s(^{-\mathbf{y}}_j {}^{\mathbf{y}}_{j'})g^t(^{\mathbf{y}}_j {}^{-\mathbf{y}}_{j'})\}\frac{\overline{q^2(^{\mathbf{y}}_j)}}{\omega^2(^{\mathbf{y}}_j)-\omega^2(^{\mathbf{y}}_{j'})}-$$

$$-\frac{1}{2kT}\sum_{\mathbf{y}}^{\tfrac{1}{2}N}\sum_{j}g^s(^{\mathbf{y}}_j {}^{-\mathbf{y}}_j)g^t(^{\mathbf{y}}_j {}^{-\mathbf{y}}_j)\left\{[\overline{q^2(^{\mathbf{y}}_j)}]^2-\frac{\hbar^2}{4\omega^2(^{\mathbf{y}}_j)}\right\}. \quad (41.20)$$

Written explicitly in terms of the macroscopic parameters, the free energy (17.25) becomes in the present case

$$F = A^0 + \sum_{\alpha}A^\alpha E_\alpha + \sum_{\alpha\beta}A^{(\alpha\beta)}u_{\alpha\beta}+\tfrac{1}{2}\sum_{\alpha\beta}A^{\alpha\beta}E_\alpha E_\beta+$$

$$+ \sum_{\alpha}\sum_{\beta\gamma}A^{\alpha(\beta\gamma)}E_\alpha u_{\beta\gamma}+\tfrac{1}{2}\sum_{\alpha\beta}\sum_{\gamma\lambda}A^{(\alpha\beta)(\gamma\lambda)}u_{\alpha\beta}u_{\gamma\lambda}. \quad (41.21)$$

We shall now derive certain invariance relations between the coefficients which will enable us to express the free energy in terms of the strain parameters $\bar{u}_{\alpha\beta} \equiv \bar{u}_{\beta\alpha}$ and $\mathbf{E}$ (see § 37). Consider a simultaneous infinitesimal rotation of the specimen and the field. Let us describe the rotation to the first order by the antisymmetric matrix

$$\omega_{\mu\nu} = -\omega_{\nu\mu}. \quad (41.22)$$

It follows from (37.19) that the rotation of the specimen changes the deformation parameters $u_{\mu\nu}$ into

$$u_{\mu\nu}+\omega_{\mu\nu} + \sum_\sigma \omega_{\mu\sigma} u_{\sigma\nu}; \tag{41.23}$$

the field components become, on the other hand,

$$E_\mu + \sum_\nu \omega_{\mu\nu} E_\nu. \tag{41.24}$$

Replacing the deformation parameters and the field components in (41.21) by (41.23) and (41.24) respectively, we obtain up to the first-order terms in $\omega_{\mu\nu}$ the following:

$$\begin{aligned}
F = A^0 &+ \sum_\alpha A^\alpha E_\alpha + \sum_{\alpha\beta} A^\alpha \omega_{\alpha\beta} E_\beta + \sum_{\alpha\beta} A^{(\alpha\beta)} u_{\alpha\beta}+ \\
&+ \sum_{\alpha\beta} A^{(\alpha\beta)}\omega_{\alpha\beta} + \sum_{\alpha\beta\gamma} A^{(\alpha\beta)}\omega_{\alpha\gamma} u_{\gamma\beta}+ \\
&+ \sum_{\alpha\beta\gamma} A^{\alpha(\beta\gamma)} E_\alpha \omega_{\beta\gamma} + \sum_{\alpha\beta}\sum_{\gamma\lambda} A^{(\alpha\beta)(\gamma\lambda)} u_{\alpha\beta}\,\omega_{\gamma\lambda}+\ldots \quad (41.25)
\end{aligned}$$

where terms of the second order in the macroscopic parameters are not written explicitly and use has been made of the fact that

$$A^{\alpha\beta} = A^{\beta\alpha}, \qquad A^{(\alpha\beta)(\gamma\lambda)} = A^{(\gamma\lambda)(\alpha\beta)}. \tag{41.26}$$

Since the rotation cannot affect the value of the free energy, (41.25) must be independent of the parameters $\omega_{\mu\nu}$ to the first order. Thus we may set equal to zero the first derivatives of (41.25) with respect to $\omega_{\mu\nu} = -\omega_{\nu\mu}$, obtaining

$$\begin{aligned}
I = A^\mu E_\nu &- A^\nu E_\mu + A^{(\mu\nu)} - A^{(\nu\mu)} + \sum_\beta \{A^{(\mu\beta)}u_{\nu\beta} - A^{(\nu\beta)}u_{\mu\beta}\}+ \\
&+ \sum_\alpha \{A^{\alpha(\mu\nu)} - A^{\alpha(\nu\mu)}\}E_\alpha + \sum_{\alpha\beta} \{A^{(\alpha\beta)(\mu\nu)} - A^{(\alpha\beta)(\nu\mu)}\}u_{\alpha\beta}+\ldots \\
&\equiv 0. \quad (41.27)
\end{aligned}$$

This relation must be fulfilled identically for any values of $u_{\alpha\beta}$ and **E**. It follows that the constant term and all derivatives of the expression on the left-hand side (denoted by I) with respect to the parameters must vanish:

$$A^{(\mu\nu)} - A^{(\nu\mu)} = 0, \tag{41.28}$$

$$\frac{\partial I}{\partial E_\gamma} = A^\mu \delta_{\gamma\nu} - A^\nu \delta_{\gamma\mu} + A^{\gamma(\mu\nu)} - A^{\gamma(\nu\mu)} = 0, \tag{41.29}$$

$$\frac{\partial I}{\partial u_{\gamma\lambda}} = A^{(\mu\lambda)}\delta_{\gamma\nu} - A^{(\nu\lambda)}\delta_{\gamma\mu} + A^{(\gamma\lambda)(\mu\nu)} - A^{(\gamma\lambda)(\nu\mu)} = 0, \text{ etc.} \tag{41.30}$$

In view of (41.26), (41.30) can also be written alternatively as follows:

$$A^{(\mu\lambda)}\delta_{\gamma\nu} - A^{(\nu\lambda)}\delta_{\gamma\mu} + A^{(\mu\nu)(\gamma\lambda)} - A^{(\nu\mu)(\gamma\lambda)} = 0. \tag{41.31}$$

The invariance relations (41.28)–(41.31) are sufficient to enable us to write the free energy in terms of $\bar{u}_{\alpha\beta} \equiv \bar{u}_{\beta\alpha}$ and $\mathbf{E}$ up to terms of the second order in these parameters. Let us rewrite the two linear terms in (41.21) as follows:

$$\sum_\alpha A^\alpha E_\alpha = \sum_\gamma A^\gamma E_\gamma = \sum_\gamma A^\gamma \Big(E_\gamma + \sum_\beta u_{\beta\gamma} E_\beta \Big) - \sum_{\beta\gamma} A^\gamma u_{\beta\gamma} E_\beta$$

$$= \sum_\gamma A^\gamma \bar{E}_\gamma - \sum_{\alpha\beta\gamma} A^\gamma \delta_{\alpha\beta} E_\alpha u_{\beta\gamma}, \tag{41.32}$$

$$\sum_{\alpha\beta} A^{(\alpha\beta)} u_{\alpha\beta} = \sum_{\beta\lambda} A^{(\beta\lambda)} u_{\beta\lambda} = \sum_{\beta\lambda} A^{(\beta\lambda)} \{ \tfrac{1}{2}(u_{\beta\lambda} + u_{\lambda\beta}) \}$$

$$= \sum_{\beta\lambda} A^{(\beta\lambda)} \Big\{ \tfrac{1}{2}(u_{\beta\lambda} + u_{\lambda\beta}) + \tfrac{1}{2} \sum_\alpha u_{\alpha\beta} u_{\alpha\lambda} \Big\} - \tfrac{1}{2} \sum_{\alpha\beta\lambda} A^{(\beta\lambda)} u_{\alpha\beta} u_{\alpha\lambda}$$

$$= \sum_{\beta\lambda} A^{(\beta\lambda)} \bar{u}_{\beta\lambda} - \tfrac{1}{2} \sum_{\alpha\beta\gamma\lambda} A^{(\beta\lambda)} \delta_{\alpha\gamma} u_{\alpha\beta} u_{\gamma\lambda}, \tag{41.33}$$

where, we notice, the replacement of $u_{\beta\lambda}$ by the symmetrized expression $(u_{\beta\lambda} + u_{\lambda\beta})/2$ is permissible owing to (41.28); thus afterwards dividing (41.21) by Nv_a, we can write the free energy per unit volume (of the undeformed lattice) in the form

$$F = F^0 + \sum_\alpha F^\alpha \bar{E}_\alpha + \sum_{\alpha\beta} F^{(\alpha\beta)} \bar{u}_{\alpha\beta} + \tfrac{1}{2} \sum_{\alpha\beta} F^{\alpha\beta} E_\alpha E_\beta +$$

$$+ \sum_{\alpha\beta\gamma} F^{\alpha(\beta\gamma)} E_\alpha u_{\beta\gamma} + \tfrac{1}{2} \sum_{\alpha\beta} \sum_{\gamma\lambda} F^{(\alpha\beta)(\gamma\lambda)} u_{\alpha\beta} u_{\gamma\lambda}, \tag{41.34}$$

where

$$F^0 = \frac{A^0}{Nv_a}, \qquad F^\alpha = \frac{A^\alpha}{Nv_a}, \qquad F^{(\alpha\beta)} = \frac{A^{(\alpha\beta)}}{Nv_a} = F^{(\beta\alpha)},$$

$$F^{\alpha\beta} = \frac{A^{\alpha\beta}}{Nv_a} = F^{\beta\alpha}, \qquad F^{\alpha(\beta\gamma)} = \frac{1}{Nv_a} \{ A^{\alpha(\beta\gamma)} - A^\gamma \delta_{\alpha\beta} \} = F^{\alpha(\gamma\beta)},$$

$$F^{(\alpha\beta)(\gamma\lambda)} = \frac{1}{Nv_a} \{ A^{(\alpha\beta)(\gamma\lambda)} - A^{(\beta\lambda)} \delta_{\alpha\gamma} \} = F^{(\beta\alpha)(\gamma\lambda)} = F^{(\gamma\lambda)(\alpha\beta)}. \tag{41.35}$$

The symmetry relations indicated can be readily verified with the help of (41.26) and the invariance relations (41.28)–(41.31). Owing to these symmetry relations, the deformation parameters in the second-order terms may be replaced by the corresponding symmetrized expressions as follows:

$$F = F^0 + \sum_\alpha F^\alpha \bar{E}_\alpha + \sum_{\alpha\beta} F^{(\alpha\beta)} \bar{u}_{\alpha\beta} + \tfrac{1}{2} \sum_{\alpha\beta} F^{\alpha\beta} E_\alpha E_\beta +$$

$$+ \sum_{\alpha\beta\gamma} F^{\alpha(\beta\gamma)} E_\alpha \{ \tfrac{1}{2}(u_{\beta\gamma} + u_{\gamma\beta}) \} +$$

$$+ \tfrac{1}{2} \sum_{\alpha\beta\gamma\lambda} F^{(\alpha\beta)(\gamma\lambda)} \{ \tfrac{1}{2}(u_{\alpha\beta} + u_{\beta\alpha}) \} \{ \tfrac{1}{2}(u_{\gamma\lambda} + u_{\lambda\gamma}) \}.$$

By addition of terms of the third and higher orders in the macroscopic parameters, we can rewrite the free energy finally as a function of the parameters $\bar{u}_{\alpha\beta} \equiv \bar{u}_{\beta\alpha}$ and $\mathbf{E}$ in the form

$$F = F^0 + \sum_\alpha F^\alpha \bar{E}_\alpha + \sum_{\alpha\beta} F^{(\alpha\beta)}\bar{u}_{\alpha\beta} + \tfrac{1}{2}\sum_{\alpha\beta} F^{\alpha\beta}\bar{E}_\alpha \bar{E}_\beta +$$

$$+ \sum_{\alpha\beta\gamma} F^{\alpha(\beta\gamma)}\bar{E}_\alpha \bar{u}_{\beta\gamma} + \tfrac{1}{2}\sum_{\alpha\beta\gamma\lambda} F^{(\alpha\beta)(\gamma\lambda)}\bar{u}_{\alpha\beta}\bar{u}_{\gamma\lambda}. \quad (41.36)$$

By the use of (41.35), (41.20) and the g-coefficients given in (41.4), it is now straightforward to write down the explicit expressions for the coefficients in the free energy (41.36). It is convenient to convert all the summations over the wave-number $\mathbf{y}$ into integrals as follows:

$$\sum_{\mathbf{y}}^{\frac{1}{2}N} \to Nv_a \int d\mathbf{y}. \quad (41.37)$$

This is possible because, for large N, the representative points of the permitted wave-numbers in the reciprocal space are very dense and the density of the points is equal to $V = Nv_a$ (see § 38).

The explicit expressions for the coefficients in the free energy are listed below (see (41.4)):

$$F^0 = \frac{\phi_0}{v_a} + 2kT \sum_j \int \ln[2\sinh \tfrac{1}{2}\beta(\substack{\mathbf{y}\\j})]\, d\mathbf{y},$$

$$F^\alpha = -\frac{\mu_\alpha^0}{v_a} - \sum_j \int M_\alpha(\substack{\mathbf{y}\\j}\,\substack{-\mathbf{y}\\j})\overline{q^2(\substack{\mathbf{y}\\j})}\, d\mathbf{y},$$

$$F^{(\alpha\beta)} = \sum_j \int \Phi_{(\alpha\beta)}(\substack{-\mathbf{y}\\j}\,\substack{-\mathbf{y}\\j})\overline{q^2(\substack{\mathbf{y}\\j})}\, d\mathbf{y},$$

$$F^{\alpha\beta} = -\frac{p_{\alpha\beta}^0}{v_a} - \frac{1}{v_a}\sum_j \frac{M_\alpha(\substack{0\\j})M_\beta(\substack{0\\j})}{\omega^2(\substack{0\\j})} -$$

$$- \sum_j \int \left\{ P_{\alpha\beta}(\substack{\mathbf{y}\\j}\,\substack{-\mathbf{y}\\j}) + \frac{1}{2\omega^2(\substack{\mathbf{y}\\j})} M_\alpha(\substack{\mathbf{y}\\j}\,\substack{-\mathbf{y}\\j})M_\beta(\substack{\mathbf{y}\\j}\,\substack{-\mathbf{y}\\j}) \right\} \overline{q^2(\substack{\mathbf{y}\\j})}\, d\mathbf{y} +$$

$$+ \tfrac{1}{2}\sum_j \sum_{j'}{}' \int \{ M_\alpha(\substack{\mathbf{y}\\j}\,\substack{-\mathbf{y}\\j'})M_\beta(\substack{-\mathbf{y}\\j}\,\substack{\mathbf{y}\\j'}) + M_\alpha(\substack{-\mathbf{y}\\j}\,\substack{\mathbf{y}\\j'})M_\beta(\substack{\mathbf{y}\\j}\,\substack{-\mathbf{y}\\j'}) \} \frac{\overline{q^2(\substack{\mathbf{y}\\j})}\, d\mathbf{y}}{\omega^2(\substack{\mathbf{y}\\j}) - \omega^2(\substack{\mathbf{y}\\j'})} -$$

$$- \frac{1}{2kT}\sum_j \int M_\alpha(\substack{\mathbf{y}\\j}\,\substack{-\mathbf{y}\\j})M_\beta(\substack{\mathbf{y}\\j}\,\substack{-\mathbf{y}\\j}) \left\{ [\overline{q^2(\substack{\mathbf{y}\\j})}]^2 - \frac{\hbar^2}{4\omega^2(\substack{\mathbf{y}\\j})} \right\}\, d\mathbf{y},$$

$$F^{\alpha(\beta\gamma)} = -\frac{1}{v_a}\sum_k M_{\alpha,\beta}(k)x_\gamma(k) + \frac{1}{v_a}\sum_j \frac{M_\alpha\binom{0}{j}\Phi_{(\beta\gamma)}\binom{-0}{j}}{\omega^2\binom{0}{j}} -$$

$$-\sum_j \int \left\{ M_{\alpha,(\beta\gamma)}\left(\begin{smallmatrix} -y & -y \\ j & j \end{smallmatrix}\right) - \frac{1}{2\omega^2\binom{y}{j}} M_\alpha\left(\begin{smallmatrix} y & -y \\ j & j \end{smallmatrix}\right)\Phi_{(\beta\gamma)}\left(\begin{smallmatrix} -y & -y \\ j & j \end{smallmatrix}\right) \right\} \overline{q^2\binom{y}{j}}\, d\mathbf{y} -$$

$$-\tfrac{1}{2}\sum_j \sum_{j'} {}' \int \left\{ M_\alpha\left(\begin{smallmatrix} y & -y \\ j & j' \end{smallmatrix}\right)\Phi_{(\beta\gamma)}\left(\begin{smallmatrix} - & -y & y \\ & j' & j \end{smallmatrix}\right) + M_\alpha\left(\begin{smallmatrix} - & y & y \\ & j' & j \end{smallmatrix}\right)\Phi_{(\beta\gamma)}\left(\begin{smallmatrix} - & y & -y \\ & j & j' \end{smallmatrix}\right) \right\} \frac{\overline{q^2\binom{y}{j}}\, d\mathbf{y}}{\omega^2\binom{y}{j} - \omega^2\binom{y}{j'}} +$$

$$+\frac{1}{2kT}\sum_j \int M_\alpha\left(\begin{smallmatrix} y & -y \\ j & j \end{smallmatrix}\right)\Phi_{(\beta\gamma)}\left(\begin{smallmatrix} - & y & -y \\ & j & j \end{smallmatrix}\right)\left\{ [\overline{q^2\binom{y}{j}}]^2 - \frac{\hbar^2}{4\omega^2\binom{y}{j}} \right\} d\mathbf{y} +$$

$$+\delta_{\alpha\beta}\frac{\mu_\gamma^0}{v_a} + \delta_{\alpha\beta}\sum_j \int M_\gamma\left(\begin{smallmatrix} y & -y \\ j & j \end{smallmatrix}\right)\overline{q^2\binom{y}{j}}\, d\mathbf{y},$$

$$F^{(\alpha\beta)(\gamma\lambda)} = \{[\alpha\gamma,\beta\lambda] + [\beta\gamma,\alpha\lambda] - [\gamma\lambda,\alpha\beta]\} - \frac{1}{v_a}\sum_j \frac{\Phi_{(\alpha\beta)}\binom{-0}{j}\Phi_{(\gamma\lambda)}\binom{-0}{j}}{\omega^2\binom{0}{j}} +$$

$$+\sum_j \int \left\{ \Phi_{(\alpha\beta)(\gamma\lambda)}\left(\begin{smallmatrix} - & - & y & -y \\ & & j & j \end{smallmatrix}\right) - \frac{1}{2\omega^2\binom{y}{j}}\Phi_{(\alpha\beta)}\left(\begin{smallmatrix} - & -y & y \\ & j & j \end{smallmatrix}\right)\Phi_{(\gamma\lambda)}\left(\begin{smallmatrix} - & y & -y \\ & j & j \end{smallmatrix}\right) \right\} \overline{q^2\binom{y}{j}}\, d\mathbf{y} +$$

$$+\tfrac{1}{2}\sum_j \sum_{j'} {}' \int \left\{ \Phi_{(\alpha\beta)}\left(\begin{smallmatrix} - & y & -y \\ & j & j' \end{smallmatrix}\right)\Phi_{(\gamma\lambda)}\left(\begin{smallmatrix} - & -y & y \\ & j' & j \end{smallmatrix}\right) + \right.$$

$$\left. + \Phi_{(\alpha\beta)}\left(\begin{smallmatrix} - & -y & y \\ & j' & j \end{smallmatrix}\right)\Phi_{(\gamma\lambda)}\left(\begin{smallmatrix} - & y & -y \\ & j & j' \end{smallmatrix}\right) \right\} \frac{\overline{q^2\binom{y}{j}}\, d\mathbf{y}}{\omega^2\binom{y}{j} - \omega^2\binom{y}{j'}} -$$

$$-\frac{1}{2kT}\sum_j \int \Phi_{(\alpha\beta)}\left(\begin{smallmatrix} - & y & -y \\ & j & j \end{smallmatrix}\right)\Phi_{(\gamma\lambda)}\left(\begin{smallmatrix} - & y & -y \\ & j & j \end{smallmatrix}\right)\left\{ [\overline{q^2\binom{y}{j}}]^2 - \frac{\hbar^2}{4\omega^2\binom{y}{j}} \right\} d\mathbf{y} -$$

$$-\delta_{\alpha\gamma}\int \Phi_{(\beta\lambda)}\left(\begin{smallmatrix} - & y & -y \\ & j & j \end{smallmatrix}\right)\overline{q^2\binom{y}{j}}\, d\mathbf{y}. \tag{41.38}$$

42. The static (non-vibrating) lattice

In (41.38) the terms involving integrals over $\mathbf{y}$ represent the effects due to nuclear vibrations. When such dynamical terms are omitted, the free energy reduces to the energy density for a static lattice. We note that, owing to the persistence of the zero-point vibrations, a real lattice does not reduce to the static model even at the absolute zero of temperature. The static case is, however, of interest, since the mechanical and electrical properties following from the corresponding energy density can be directly compared with the results obtained by the long-wave method, which makes no allowance for the internal nuclear vibrations. In particular, it is instructive to see how in the present treatment the internal strain induced by an elastic strain is taken into account.

For the static case, we find upon putting

$$u_{\alpha\beta}^T = 0, \qquad \dot{u}_{\alpha\beta}^T = 0$$

in (37.38) that

$$c_{\alpha\gamma,\beta\lambda} = c_{\beta\lambda,\alpha\gamma} = \frac{1}{4}\left\{\frac{\partial^2 F}{\partial \bar{u}_{\beta\lambda}\,\partial \bar{u}_{\alpha\gamma}} + \frac{\partial^2 F}{\partial \bar{u}_{\beta\lambda}\,\partial \bar{u}_{\gamma\alpha}} + \frac{\partial^2 F}{\partial \bar{u}_{\lambda\beta}\,\partial \bar{u}_{\alpha\gamma}} + \frac{\partial^2 F}{\partial \bar{u}_{\lambda\beta}\,\partial \bar{u}_{\gamma\alpha}}\right\}_{\text{static}}.$$

(42.1)

Using the free energy (41.36) and remembering the symmetry relations (41.35), we obtain the following:

$$c_{\alpha\gamma,\beta\lambda} = \{F^{(\alpha\gamma)(\beta\lambda)}\}_{\text{static}}$$

$$= [\alpha\beta,\gamma\lambda]+[\beta\gamma,\alpha\lambda]-[\beta\lambda,\alpha\gamma]-\frac{1}{v_a}\sum_j \frac{\Phi_{(\alpha\gamma)}(-\,^0_j)\Phi_{(\beta\lambda)}(-\,^0_j)}{\omega^2(^0_j)}, \quad (42.2)$$

where $\{F^{(\alpha\gamma)(\beta\lambda)}\}_{\text{static}}$ is obtained from (41.38) by leaving out the dynamical terms. It will now be shown that the last term in (42.2) is equal to $-(\alpha\gamma,\beta\lambda)$ (see (26.33)). The arguments which we shall follow are essentially similar to the discussion given in § 34.

By definition (§ 40)

$$\Phi_{(\alpha\gamma)}(-\,^0_j) = \sum_{lk}\sum_{k'\mu} \Phi_{\alpha\mu}(^l_k\,^0_{k'})x_\gamma(^l_k)\frac{1}{\sqrt{m_k}}e_\mu(k'|^0_j). \quad (42.3)$$

In view of the invariance relation (23.16), we can replace $x_\gamma(^l_k)$ by $x_\gamma(^l_{kk'})$ $(= x_\gamma(^l_k)-x_\gamma(k'))$ and write (42.3) in terms of the coefficients $C^{(1)}_{\alpha\beta,\gamma}(kk')$ defined in (26.4) as follows:

$$\Phi_{\alpha\gamma}(-\,^0_j) = -\frac{1}{2\pi}\sum_{k'\mu}\left\{\sum_k \sqrt{m_k}\,C^{(1)}_{\alpha\mu,\gamma}(kk')\right\}e_\mu(k'|^0_j). \quad (42.4)$$

It follows that the last term in (42.2) can be written alternatively as

$$-\frac{1}{4\pi^2 v_a}\sum_{k\mu}\sum_{k'\nu}\left\{\sum_{k''}\sqrt{m_{k''}}\,C^{(1)}_{\alpha\mu,\gamma}(kk'')\right\}\left\{\sum_{k'''}\sqrt{m_{k'''}}\,C^{(1)}_{\beta\nu,\lambda}(k'k''')\right\}\times$$

$$\times\left\{\sum_j \frac{e_\mu(k|^0_j)e_\nu(k'|^0_j)}{\omega^2(^0_j)}\right\}. \quad (42.5)$$

Putting $\mathbf{y}=0$ in (38.24) and remembering (38.5) and (38.21), we find that the polarization vectors $\mathbf{e}(k|^0_j)$ satisfy the equations

$$\sum_{lk'\beta}\frac{1}{(m_k m_{k'})^{\frac12}}\Phi_{\alpha\beta}(^l_{kk'})e_\beta(k'|^0_j) = \omega^2(^0_j)e_\alpha(k|^0_j), \quad (42.6)$$

which, expressed in terms of the coefficients $C^{(0)}_{\alpha\beta}(kk')$ defined in (26.3), become

$$\sum_{k'\beta}C^{(0)}_{\alpha\beta}(kk')e_\beta(k'|^0_j) = \omega^2(^0_j)e_\alpha(k|^0_j). \quad (42.7)$$

Regarding the right-hand side formally as the inhomogeneous part of the equations, we find with the help of the Γ-matrix introduced in (26.25) and (26.27) that

$$e_\mu(k|{}^0_j) = \omega^2({}^0_j) \sum_{k'\sigma} \Gamma_{\mu\sigma}(kk'')e_\sigma(k''|{}^0_j). \qquad (42.8)$$

After (42.8) is multiplied by

$$\frac{e_\nu(k'|{}^0_j)}{\omega^2({}^0_j)}$$

and summed over j, the use of the orthogonality relations (note that $e_\nu^*(k'|{}^0_j) = e_\nu(k'|{}^0_j)$) leads immediately to the relation

$$\sum_j \frac{e_\mu(k|{}^0_j)e_\nu(k'|{}^0_j)}{\omega^2({}^0_j)} = \Gamma_{\mu\nu}(kk'). \qquad (42.9)$$

Using (42.9) in (42.5) we see that the last term in (42.2) is equal to the round-bracket expression (26.33), in view of the symmetry relation given in (26.4). Thus the elastic constants given in (42.2) are in complete agreement with the expression (27.26) obtained previously by the long-wave method. It is evident from the above discussion that in the present treatment the effect of the internal strain is reproduced through the normal coordinates $Q({}^0_j)$; this is exactly as to be expected, for we have observed (§ 38) that these coordinates describe the relative shifts between the constituent Bravais lattices, or, in other words, the internal strains.

We have also remarked in § 38 that if the lattice is ionic the coordinates $Q({}^0_j)$, as introduced there, are necessarily ambiguous. In the consideration of the free energy, we have taken account of the macroscopic field explicitly by an additional term in the Hamiltonian. It is thus clear that in this connexion the coordinates $Q({}^0_j)$ are to be defined by eliminating first the contribution due to the macroscopic field. In order to do so explicitly, it is necessary to consider a definite model, at least in so far as the Coulomb part of the atomic forces is concerned. Thus for the rigid ion model (cf. § 31), the coordinates $Q({}^0_j)$ are to be determined by (42.8), wherein we have to use the Γ-matrix (31.35) defined in terms of $\bar{C}^{(0)}_{\alpha\beta}(kk')$; the latter, we remember, are the coefficients $C^{(0)}_{\alpha\beta}(kk')$ with the contribution due to the macroscopic field eliminated. It is then readily verified that the static parts of the piezoelectric constants and the dielectric tensor as given by (37.39), (37.46), (41.36), and (41.38) in the present treatment are in complete agreement with the corresponding results (32.12) and (32.18) obtained by the long-wave method. The proof required is straightforward; we need only note that in the rigid ion model

$$P^0_{\alpha\beta} = 0,$$
$$\sum_k M_{\alpha,\beta}(k)x_\gamma(k) = \delta_{\alpha\beta} \sum_k e_k x_\gamma(k) = \delta_{\alpha\beta}\mu^0_\gamma, \qquad (42.10)$$

and it is implied in the long-wave method that $\mu^0 = 0$, for otherwise there would be an ambiguous macroscopic field in addition to the field (31.14) produced by the ion displacements.

43. The T^4-law

As shown in § 37, all the normal mechanical and electrical properties of a crystal can be described in terms of its free energy. The explicit formulae governing these properties are obtained by substituting the free energy given by (41.36) and (41.38) in the general results derived in § 37. The temperature dependence of these formulae is in general very complex; a general discussion is possible only in the limit of very low temperatures.

Let us examine how the coefficients (41.38) of the free energy vary with the temperature at the low-temperature limit. Subtracting from the coefficients their respective limiting values at $T = 0$, we obtain the following:

$$F^\alpha(T) - F^\alpha(0) = -\sum_j \int M_\alpha\binom{y}{j}{}^{-y} \left\{ \overline{q^2\binom{y}{j}} - \frac{\hbar}{2\omega\binom{y}{j}} \right\} dy,$$

$$F^{(\alpha\beta)}(T) - F^{(\alpha\beta)}(0) = \sum_j \int \Phi_{(\alpha\beta)}\binom{-y}{j}{}^{-y} \left\{ \overline{q^2\binom{y}{j}} - \frac{\hbar}{2\omega\binom{y}{j}} \right\} dy,$$

$$F^{\alpha\beta}(T) - F^{\alpha\beta}(0) =$$

$$-\sum_j \int \left\{ P_{\alpha\beta}\binom{y}{j}{}^{-y} + \frac{1}{2\omega^2\binom{y}{j}} M_\alpha\binom{y}{j}{}^{-y} M_\beta\binom{y}{j}{}^{-y} \right\} \left\{ \overline{q^2\binom{y}{j}} - \frac{\hbar}{2\omega\binom{y}{j}} \right\} dy +$$

$$+\tfrac{1}{2}\sum_j\sum_{j'}{}' \int \{ M_\alpha\binom{y}{j}{}^{-y}M_\beta\binom{-y}{j'}{}^{y} + M_\alpha\binom{-y}{j'}{}^{y}M_\beta\binom{y}{j}{}^{-y} \} \left\{ \frac{\overline{q^2\binom{y}{j}} - (\hbar/2\omega\binom{y}{j})}{\omega^2\binom{y}{j} - \omega^2\binom{y}{j'}} \right\} dy -$$

$$-\frac{1}{2kT}\sum_j \int M_\alpha\binom{y}{j}{}^{-y}M_\beta\binom{y}{j}{}^{-y} \left\{ [\overline{q^2\binom{y}{j}}]^2 - \frac{\hbar^2}{4\omega^2\binom{y}{j}} \right\} dy,$$

$$F^{\alpha(\beta\gamma)}(T) - F^{\alpha(\beta\gamma)}(0) =$$

$$-\sum_j \int \left\{ M_{\alpha,(\beta\gamma)}\binom{-y}{j}{}^{-y} - \frac{1}{2\omega^2\binom{y}{j}} M_\alpha\binom{y}{j}{}^{-y}\Phi_{(\beta\gamma)}\binom{-y}{j}{}^{-y} \right\} \left\{ \overline{q^2\binom{y}{j}} - \frac{\hbar}{2\omega\binom{y}{j}} \right\} dy -$$

$$-\tfrac{1}{2}\sum_j\sum_{j'}{}' \int \{ M_\alpha\binom{y}{j}{}^{-y}\Phi_{(\beta\gamma)}\binom{-y}{j'}{}^{y} + M_\alpha\binom{-y}{j'}{}^{y}\Phi_{(\beta\gamma)}\binom{-y}{j}{}^{-y} \} \times$$

$$\times \left\{ \frac{\overline{q^2\binom{y}{j}} - (\hbar/2\omega\binom{y}{j})}{\omega^2\binom{y}{j} - \omega^2\binom{y}{j'}} \right\} dy +$$

$$+\frac{1}{2kT}\sum_j \int M_\alpha\binom{y}{j}{}^{-y}\Phi_{(\beta\gamma)}\binom{-y}{j}{}^{-y} \left\{ [\overline{q^2\binom{y}{j}}]^2 - \frac{\hbar^2}{4\omega^2\binom{y}{j}} \right\} dy -$$

$$-\delta_{\alpha\beta}\{ F^\gamma(T) - F^\gamma(0) \},$$

$$F^{(\alpha\beta\chi\gamma\lambda)}(T) - F^{(\alpha\beta\chi\gamma\lambda)}(0) =$$

$$\sum_{\bar{j}} \int \left\{ \Phi_{(\alpha\beta\chi\gamma\lambda)}(\begin{smallmatrix}-&-&-^{\mathbf{y}}&\mathbf{y}\\ &&\bar{j}&\bar{j}\end{smallmatrix}) - \frac{1}{2\omega^2(\begin{smallmatrix}\mathbf{y}\\\bar{j}\end{smallmatrix})} \Phi_{(\alpha\beta)}(\begin{smallmatrix}-&\mathbf{y}&-^{\mathbf{y}}\\ &\bar{j}&\bar{j}\end{smallmatrix}) \Phi_{(\gamma\lambda)}(\begin{smallmatrix}-&\mathbf{y}&-^{\mathbf{y}}\\ &\bar{j}&\bar{j}\end{smallmatrix}) \right\} \times$$

$$\times \left\{ \overline{q^2(\begin{smallmatrix}\mathbf{y}\\\bar{j}\end{smallmatrix})} - \frac{\hbar}{2\omega(\begin{smallmatrix}\mathbf{y}\\\bar{j}\end{smallmatrix})} \right\} d\mathbf{y} +$$

$$+ \tfrac{1}{2} \sum_{\bar{j}}\sum_{\bar{j}'}{}' \int \left\{ \Phi_{(\alpha\beta)}(\begin{smallmatrix}-&\mathbf{y}&-^{\mathbf{y}}\\ &\bar{j}&\bar{j}'\end{smallmatrix}) \Phi_{(\gamma\lambda)}(\begin{smallmatrix}-&-^{\mathbf{y}}&\mathbf{y}\\ &\bar{j}'&\bar{j}\end{smallmatrix}) + \Phi_{(\alpha\beta)}(\begin{smallmatrix}-&-^{\mathbf{y}}&\mathbf{y}\\ &\bar{j}'&\bar{j}\end{smallmatrix}) \Phi_{(\gamma\lambda)}(\begin{smallmatrix}-&\mathbf{y}&-^{\mathbf{y}}\\ &\bar{j}&\bar{j}'\end{smallmatrix}) \right\} \times$$

$$\times \left\{ \frac{\overline{q^2(\begin{smallmatrix}\mathbf{y}\\\bar{j}\end{smallmatrix})} - (\hbar/2\omega(\begin{smallmatrix}\mathbf{y}\\\bar{j}\end{smallmatrix}))}{\omega^2(\begin{smallmatrix}\mathbf{y}\\\bar{j}\end{smallmatrix}) - \omega^2(\begin{smallmatrix}\mathbf{y}\\\bar{j}'\end{smallmatrix})} \right\} d\mathbf{y} -$$

$$- \frac{1}{2kT} \sum_{\bar{j}} \int \Phi_{(\alpha\beta)}(\begin{smallmatrix}-&\mathbf{y}&-^{\mathbf{y}}\\ &\bar{j}&\bar{j}\end{smallmatrix}) \Phi_{(\gamma\lambda)}(\begin{smallmatrix}-&\mathbf{y}&-^{\mathbf{y}}\\ &\bar{j}&\bar{j}\end{smallmatrix}) \left\{ [\overline{q^2(\begin{smallmatrix}\mathbf{y}\\\bar{j}\end{smallmatrix})}]^2 - \frac{\hbar^2}{4\omega^2(\begin{smallmatrix}\mathbf{y}\\\bar{j}\end{smallmatrix})} \right\} d\mathbf{y} -$$

$$- \delta_{\alpha\gamma}\{ F^{(\beta\lambda)}(T) - F^{(\beta\lambda)}(0) \}, \tag{43.1}$$

where we have used the fact (evident from the following discussion) that the integrals involving the factor

$$[\overline{q^2(\begin{smallmatrix}\mathbf{y}\\\bar{j}\end{smallmatrix})}]^2 - \frac{\hbar^2}{4\omega^2(\begin{smallmatrix}\mathbf{y}\\\bar{j}\end{smallmatrix})}$$

vanish faster than T at the low-temperature limit. The differences given in (43.1) represent the temperature-dependent parts of the coefficients and are seen to depend only on the dynamical terms. The integrals in (43.1) contain one of the following temperature-dependent factors:

$$\overline{q^2} - \frac{\hbar}{2\omega} = \frac{\hbar}{2\omega} \frac{2e^{-\hbar\omega/2kT}}{e^{\hbar\omega/2kT} - e^{-\hbar\omega/2kT}}, \tag{43.2}$$

$$[\overline{q^2}]^2 - \frac{\hbar^2}{4\omega^2} = \frac{\hbar^2}{4\omega^2} \frac{4}{[e^{\hbar\omega/2kT} - e^{-\hbar\omega/2kT}]^2}. \tag{43.3}$$

For $\hbar\omega$ large compared with kT both factors fall rapidly with increasing frequency, essentially in an exponential way. Since the frequencies $\omega(\begin{smallmatrix}\mathbf{y}\\\bar{j}\end{smallmatrix})$ for all except the acoustic branches $j = 1, 2, 3$ have finite lower limits, for sufficiently low temperatures we need consider only the contributions due to the low-frequency portions of the acoustic branches, in other words, the contributions due to the long acoustic vibrations. For these vibrations we can write

$$\omega(\begin{smallmatrix}\mathbf{y}\\\bar{j}\end{smallmatrix}) = 2\pi c_j(\theta\phi)y, \tag{43.4}$$

where (y, θ, ϕ) are the polar coordinates of $\mathbf{y}$ and $c_j(\theta\phi)$ is the velocity for elastic waves on the branch j and travelling in the direction $(\theta\phi)$. Hence

all the integrals in (43.1) which are relevant at the low-temperature limit
can be written in one of the following forms:

$$\int_0^\pi d\phi \int_0^\pi \sin\theta \, d\theta \int_0^\infty y^2 \, dy \, K(\mathbf{y}) \frac{1}{yc(\theta\phi)} \times$$

$$\times \left\{ \frac{\exp[-yhc(\theta\phi)/2kT]}{\exp[yhc(\theta\phi)/2kT] - \exp[-yhc(\theta\phi)/2kT]} \right\}, \quad (43.5)$$

$$\frac{1}{T} \int_0^\pi d\phi \int_0^\pi \sin\theta \, d\theta \int_0^\infty y^2 \, dy \, K(\mathbf{y}) \frac{1}{y^2 c^2(\theta\phi)} \times$$

$$\times \left\{ \frac{1}{\exp[yhc(\theta\phi)/2kT] - \exp[-yhc(\theta\phi)/2kT]} \right\}^2, \quad (43.6)$$

where $K(\mathbf{y})$ stands for an arbitrary function of $\mathbf{y}$. The upper limit for
the integration over y is not clearly definable; but, whenever the temperature is sufficiently low to permit the rejection of all but the long
acoustic vibrations, the exact value of the upper limit is immaterial
and we can take it as infinite (cf. the discussion of the T^3-law in § 6).
Let us represent $K(\mathbf{y})$ as a series in y as follows:

$$K(\mathbf{y}) = K_0(\theta\phi) + K_1(\theta\phi)y + \tfrac{1}{2}K_2(\theta\phi)y^2 + ..., \quad (43.7)$$

where the coefficients are functions of the polar angles (note that the
series is of the type used in the perturbation method in § 26, with ϵ
and $\mathbf{y}$ effectively replaced by y and $\mathbf{y}/y$). Upon substituting (43.7) in
(43.5) and (43.6) and introducing

$$\xi = yhc(\theta\phi)/2kT$$

as the integration variable in place of y, we find that (43.5) and (43.6) become respectively

$$\sum_n \left(\frac{2kT}{h}\right)^{n+2} \left\{ \int_0^\infty \frac{\xi^{n+1}e^{-\xi} \, d\xi}{e^\xi - e^{-\xi}} \right\} \int_0^\pi d\phi \int_0^\pi \sin\theta \, d\theta \left\{ \frac{K_n(\theta\phi)}{[c(\theta\phi)]^{n+3}} \right\}, \quad (43.8)$$

$$\sum_n \left(\frac{2k}{h}\right)^{n+1} T^n \left\{ \int_0^\infty \frac{\xi^n \, d\xi}{(e^\xi - e^{-\xi})^2} \right\} \int_0^\pi d\phi \int_0^\pi \sin\theta \, d\theta \left\{ \frac{K_n(\theta\phi)}{[c(\theta\phi)]^{n+3}} \right\}. \quad (43.9)$$

Thus if the lowest non-vanishing term in $K(\mathbf{y})$ is of the nth order in y
the integral (43.5) approaches the low-temperature limit as T^{n+2} whereas
the integral (43.6) approaches the limit more slowly as T^n.

In all cases, the function $K(\mathbf{y})$ is a combination of the second-order
coefficients:

$$M_\alpha(\tfrac{y}{j} \tfrac{-y}{j}), \quad \Phi_{(\alpha\beta)}(-\tfrac{y}{j} \tfrac{-y}{j}), \quad M_{\alpha,(\beta\gamma)}(-\tfrac{y}{j} \tfrac{-y}{j}), \quad \text{etc.}$$

These coefficients are essentially expansion coefficients of the following quantities (see § 39):

$$M_\alpha^{\text{def}}(\begin{smallmatrix}y\\j\end{smallmatrix}\begin{smallmatrix}-y\\j\end{smallmatrix}), \quad \Phi^{\text{def}}(\begin{smallmatrix}y\\j\end{smallmatrix}\begin{smallmatrix}-y\\j\end{smallmatrix}), \quad P_{\alpha\beta}^{\text{def}}(\begin{smallmatrix}y\\j\end{smallmatrix}\begin{smallmatrix}-y\\j\end{smallmatrix}), \tag{43.10}$$

with respect to the parameters $u_{\alpha\beta}$ which define the homogeneously deformed configuration. We shall now show that, when expressed as series in the manner of (43.7), the quantities (43.10) have no zero-order terms if $j, j' = 1, 2, 3$ (acoustic branches); and, moreover, the first-order terms also vanish if $j = j'$. Consider, for instance, $\Phi^{\text{def}}(\begin{smallmatrix}y\\j\end{smallmatrix}\begin{smallmatrix}-y\\j\end{smallmatrix})$; by definition,

$$\Phi^{\text{def}}(\begin{smallmatrix}y\\j\end{smallmatrix}\begin{smallmatrix}-y\\j\end{smallmatrix}) = \sum_{k\alpha} \sum_{l'k'\beta} \Phi_{\alpha\beta}^{\text{def}}(\begin{smallmatrix}0\\k\end{smallmatrix}\begin{smallmatrix}l'\\k'\end{smallmatrix}) \frac{1}{(m_k m_{k'})^{\frac{1}{2}}} e_\alpha(k|\begin{smallmatrix}y\\j\end{smallmatrix})e_\beta(k'|\begin{smallmatrix}-y\\j\end{smallmatrix})\exp[-2\pi i y . \mathbf{x}(l')]. \tag{43.11}$$

As we have remarked in § 38, the polarization vectors $\mathbf{e}(k|\begin{smallmatrix}y\\j\end{smallmatrix})$ differ from the solutions $\mathbf{w}(k|\begin{smallmatrix}y\\j\end{smallmatrix})$ considered in the last chapter (see § 26) essentially only by a phase factor $\exp[2\pi i y . \mathbf{x}(k)]$. Thus if we assume the solutions $\mathbf{w}(k|\begin{smallmatrix}y\\j\end{smallmatrix})$ to be suitably normalized, we can write (43.11) as

$$\Phi^{\text{def}}(\begin{smallmatrix}y\\j\end{smallmatrix}\begin{smallmatrix}-y\\j\end{smallmatrix})$$

$$= \sum_{k\alpha} \sum_{l'k'\beta} \Phi_{\alpha\beta}^{\text{def}}(\begin{smallmatrix}0\\k\end{smallmatrix}\begin{smallmatrix}l'\\k'\end{smallmatrix}) \frac{1}{(m_k m_{k'})^{\frac{1}{2}}} w_\alpha(k|\begin{smallmatrix}y\\j\end{smallmatrix})w_\beta(k'|\begin{smallmatrix}-y\\j\end{smallmatrix})\exp\{2\pi i y . [\mathbf{x}(k)-\mathbf{x}(\begin{smallmatrix}l'\\k'\end{smallmatrix})]\}. \tag{43.12}$$

For $j, j' = 1, 2, 3$ it follows from (26.14) that as $\mathbf{y} \to 0$ along a fixed direction $(\theta\phi)$,

$$\mathbf{w}(k|\begin{smallmatrix}y\\j\end{smallmatrix}) \to \mathbf{w}^{(0)}(k|\begin{smallmatrix}y\\j\end{smallmatrix}) = \sqrt{m_k}\,\mathbf{u}_j(\theta\phi), \tag{43.13}$$

$\mathbf{u}_j(\theta\phi)$ being a vector independent of the base index k. Hence

$$\lim_{y\to 0} \Phi^{\text{def}}(\begin{smallmatrix}y\\j\end{smallmatrix}\begin{smallmatrix}-y\\j\end{smallmatrix}) = \sum_{\alpha\beta} u_j(\theta\phi)u_{j'}(\theta\phi)\Big\{\sum_{k}\sum_{l'k'} \Phi_{\alpha\beta}^{\text{def}}(\begin{smallmatrix}0\\k\end{smallmatrix}\begin{smallmatrix}l'\\k'\end{smallmatrix})\Big\}. \tag{43.14}$$

Since the deformed configuration concerned is a perfect lattice, the invariance relation (23.16) obtained by considering translational invariance is valid also for the deformed configuration; i.e.

$$\sum_{l'k'} \Phi_{\alpha\beta}^{\text{def}}(\begin{smallmatrix}0\\k\end{smallmatrix}\begin{smallmatrix}l'\\k'\end{smallmatrix}) = 0. \tag{43.15}$$

It follows that

$$\lim_{y\to 0} \Phi^{\text{def}}(\begin{smallmatrix}y\\j\end{smallmatrix}\begin{smallmatrix}-y\\j\end{smallmatrix}) = 0. \tag{43.16}$$

In fact, as the index k is not summed in (43.15) it is clear that (43.16) is true if only one of j or j' is an acoustic branch.

When the solution $\mathbf{w}(k|\begin{smallmatrix}y\\j\end{smallmatrix})$ for an acoustic branch is expanded in the manner of (43.7) all the odd terms are imaginary and all the even terms are real. This is immediately evident from the equivalent expansion (26.10), for the perturbation equations are all real. The same is evidently true for the exponential factor in (43.12), and hence also for $\Phi^{\text{def}}(\begin{smallmatrix}y\\j\end{smallmatrix}\begin{smallmatrix}-y\\j\end{smallmatrix})$;

i.e. when expanded in the manner of (43.7), $\Phi^{\text{def}}\!\left(\begin{smallmatrix}y & -y\\ j & j'\end{smallmatrix}\right)$ has imaginary odd terms and real even terms. On the other hand, we have observed in § 39 that

$$\Phi^{\text{def}}\!\left(\begin{smallmatrix}y & -y\\ j & j'\end{smallmatrix}\right) = \Phi^{\text{def}}\!\left(\begin{smallmatrix}-y & y\\ j & j'\end{smallmatrix}\right) = \left[\Phi^{\text{def}}\!\left(\begin{smallmatrix}-y & y\\ j & j'\end{smallmatrix}\right)\right]^{*}. \tag{43.17}$$

Putting $j = j'$ in (43.17), we see that $\Phi^{\text{def}}\!\left(\begin{smallmatrix}y & -y\\ j & j'\end{smallmatrix}\right)$ must always be real. This being the case for all values of y, it follows that the odd terms in the expansion of $\Phi^{\text{def}}\!\left(\begin{smallmatrix}y & -y\\ j & j'\end{smallmatrix}\right)$ must vanish.

Since there are relations exactly analogous to (43.15) and (43.17) for $M_\alpha^{\text{def}}\!\left(\begin{smallmatrix}y & -y\\ j & j'\end{smallmatrix}\right)$ and $P_{\alpha\beta}^{\text{def}}\!\left(\begin{smallmatrix}y & -y\\ j & j'\end{smallmatrix}\right)$, similar conclusions to the above hold also in the case of $M_\alpha^{\text{def}}\!\left(\begin{smallmatrix}y & -y\\ j & j'\end{smallmatrix}\right)$ and $P_{\alpha\beta}^{\text{def}}\!\left(\begin{smallmatrix}y & -y\\ j & j'\end{smallmatrix}\right)$. Moreover, as the conclusions are valid for any homogeneously deformed configuration, they can be immediately extended to the second-order coefficients $M_\alpha\!\left(\begin{smallmatrix}y & -y\\ j & j'\end{smallmatrix}\right)$, $\Phi_{(\alpha\beta)}\!\left(\begin{smallmatrix}- & y & -y\\ & j & j'\end{smallmatrix}\right)$, etc., which, as pointed out above, are the expansion coefficients of $M_\alpha^{\text{def}}\!\left(\begin{smallmatrix}y & -y\\ j & j'\end{smallmatrix}\right)$, $\Phi^{\text{def}}\!\left(\begin{smallmatrix}y & -y\\ j & j'\end{smallmatrix}\right)$, etc., with respect to the deformation parameters $u_{\alpha\beta}$. Therefore, when expanded in the manner of (43.7), the second-order coefficients

$$M_\alpha\!\left(\begin{smallmatrix}y & -y\\ j & j\end{smallmatrix}\right), \quad \Phi_{(\alpha\beta)}\!\left(\begin{smallmatrix}- & y & -y\\ & j & j\end{smallmatrix}\right), \quad M_{\alpha,(\beta\gamma)}\!\left(\begin{smallmatrix}- & y & -y\\ & j & j\end{smallmatrix}\right), \quad \text{etc.,}$$

begin in general with terms of the second order in y, whereas the coefficients with $j \neq j'$,

$$M_\alpha\!\left(\begin{smallmatrix}y & -y\\ j & j'\end{smallmatrix}\right), \quad \Phi_{(\alpha\beta)}\!\left(\begin{smallmatrix}- & y & -y\\ & j & j'\end{smallmatrix}\right), \quad M_{\alpha,(\beta\gamma)}\!\left(\begin{smallmatrix}- & y & -y\\ & j & j'\end{smallmatrix}\right), \quad \text{etc.,}$$

begin with terms of the first order in y.

In (43.1) we notice that $F^\alpha(T) - F^\alpha(0)$ and $F^{(\alpha\beta)}(T) - F^{(\alpha\beta)}(0)$ are each equal to an integral of the form (43.5), in which the function $K(\mathbf{y})$ begins with terms of the second order in y. Hence both vary as T^4 at very low temperatures. Apart from terms of this kind, $F^{\alpha\beta}(T) - F^{\alpha\beta}(0)$, $F^{\alpha(\beta\gamma)}(T) - F^{\alpha(\beta\gamma)}(0)$, and $F^{(\alpha\beta)(\gamma\lambda)}(T) - F^{(\alpha\beta)(\gamma\lambda)}(0)$ each contain three further integrals. Two of the integrals are of the form (43.5); in both, the function $K(\mathbf{y})$ begins with terms of the second order in y. The remaining integral has the form (43.6), but the $K(\mathbf{y})$ function begins with terms of the fourth order in y. Therefore all these integrals vary as T^4 at very low temperatures.

Thus despite the very different structures of the various coefficients in the free energy, their temperature-dependent parts all vary as the fourth power of the absolute temperature at very low temperatures.

Putting $\mathbf{E} = 0$ in (41.36) and minimizing F with respect to the strain parameters, we get the relations

$$F^{(\alpha\beta)}(T) + \sum_{\gamma\lambda} F^{(\alpha\beta)(\gamma\lambda)}(T)\bar{u}_{\gamma\lambda}^{T} = 0, \tag{43.18}$$

which determine the thermal strain components $\bar{u}_{\alpha\beta}^{T}$. Subtracting from

(43.18) the same relations for $T = 0$, we can write the difference in the form

$$\sum_{\gamma\lambda} F^{(\alpha\beta\chi\gamma\lambda)}(T)(\bar{u}^T_{\gamma\lambda} - \bar{u}^0_{\gamma\lambda}) = -\{F^{(\alpha\beta)}(T) - F^{(\alpha\beta)}(0)\} -$$
$$- \sum_{\gamma\lambda} \{F^{(\alpha\beta\chi\gamma\lambda)}(T) - F^{(\alpha\beta\chi\gamma\lambda)}(0)\}\bar{u}^0_{\gamma\lambda}. \quad (43.19)$$

Since the right-hand side approaches $T = 0$ as T^4 whereas the coefficients $F^{(\alpha\beta\chi\gamma\lambda)}(T)$ on the left approach finite limits, $\bar{u}^T_{\gamma\lambda} - u^0_{\gamma\lambda}$ must vary as T^4 at very low temperatures.

In view of this fact, when the free energy (41.36) is used in (37.32), (37.38), (37.39), and (37.46), it is immediately seen that the temperature-dependent parts of $\mathbf{P}^{\text{pyro}}$, $c_{\alpha\gamma,\beta\lambda}$, $e_{\beta,\alpha\gamma}$, and $a_{\alpha\beta}$ all vary as T^4 at very low temperatures. As easily verified, the same is true of all the other alternative coefficients for describing the mechanical and electrical properties of a crystal (see (37.52), (37.53), and (37.54)).

THE OPTICAL EFFECTS

44. The microscopic theory of dispersion

THE theories of dispersion discussed in previous chapters (see §§ 7–9 and §§ 33–35) are partly based on phenomenological considerations. We have taken the macroscopic Maxwell equations for granted and used microscopic models only in so far as they are necessary for the derivation of the dispersion formula. We shall now consider a microscopic theory, which gives a self-contained account of the optical waves without the aid of the macroscopic Maxwell equations for dielectric media.

In the phenomenological theory given in § 8 it was found that the consideration of lattice vibrations led naturally to the optical waves and their dispersion law. In Chapter V we have discussed the vibrations of ionic lattices on the assumption that the ions interact with the un-retarded Coulomb forces. We can build up the microscopic theory of dispersion by taking account of the retardation of the electric inter-action. The following treatment is thus essentially a revision of the theory of lattice vibrations given in Chapter V. For the sake of sim-plicity, we shall use the rigid ion model as in § 31.

In a microscopic theory the macroscopic Maxwell equations are replaced by the Maxwell–Lorentz equations:

$$\nabla . \mathbf{E} = 4\pi\rho, \tag{44.1}$$

$$\nabla . \mathbf{H} = 0, \tag{44.2}$$

$$\nabla \wedge \mathbf{H} = \frac{1}{c}(\dot{\mathbf{E}} + 4\pi\rho\mathbf{v}), \tag{44.3}$$

$$\nabla \wedge \mathbf{E} = -\frac{1}{c}\dot{\mathbf{H}}. \tag{44.4}$$

The fields, we note, are no longer macroscopic quantities, but represent values defined at every point in space. In order to realize clearly the connexion between the present discussion and the theory of lattice vibrations in Chapter V, let us decompose the vector functions **E**, **H**, and $(\rho\mathbf{v})$ into their irrotational and solenoidal parts as follows:

$$\mathbf{E} = \mathbf{E}^{\parallel} + \mathbf{E}^{\perp}, \tag{44.5}$$

$$\mathbf{H} = \mathbf{H}^{\parallel} + \mathbf{H}^{\perp}, \tag{44.6}$$

$$\rho\mathbf{v} = (\rho\mathbf{v})^{\parallel} + (\rho\mathbf{v})^{\perp}, \tag{44.7}$$

where
$$\nabla \wedge \mathbf{E}^{\parallel} = \nabla \wedge \mathbf{H}^{\parallel} = \nabla \wedge (\rho \mathbf{v})^{\parallel} = 0, \qquad (44.8)$$

$$\nabla . \mathbf{E}^{\perp} = \nabla . \mathbf{H}^{\perp} = \nabla . (\rho \mathbf{v})^{\perp} = 0. \qquad (44.9)$$

If an irrotational vector is resolved into plane waves, the waves are all longitudinal; the plane-wave components of a solenoidal vector are on the other hand all transverse. Therefore the irrotational and solenoidal parts of the electromagnetic field are known respectively as the longitudinal and transverse fields. We shall first show that the use of the unretarded Coulomb interaction is equivalent to neglecting the transverse electromagnetic field.

It follows from (44.2) that the magnetic field is solenoidal; in other words:
$$\mathbf{H}^{\parallel} = 0. \qquad (44.10)$$

Therefore (44.1) and (44.4) reduce to
$$\nabla . \mathbf{E}^{\parallel} = 4\pi \rho, \qquad (44.11)$$

$$\nabla \wedge \mathbf{E}^{\perp} = -\frac{1}{c} \dot{\mathbf{H}}^{\perp}. \qquad (44.12)$$

Moreover, since the curl of any vector is solenoidal, we find upon splitting (44.3) into its irrotational and solenoidal parts that
$$0 = \dot{\mathbf{E}}^{\parallel} + 4\pi (\rho \mathbf{v})^{\parallel}, \qquad (44.13)$$

$$\nabla \wedge \mathbf{H} = \frac{1}{c} (\dot{\mathbf{E}}^{\perp} + 4\pi (\rho \mathbf{v})^{\perp}). \qquad (44.14)$$

Taking account of the irrotational character of $\mathbf{E}^{\parallel}$, we observe that (44.11) is identical with the equation encountered in electrostatics. It follows from a well-known result in potential theory that
$$\mathbf{E}^{\parallel} = \nabla \int \frac{\rho(\mathbf{x}', t)}{|\mathbf{x} - \mathbf{x}'|} \, d\mathbf{x}'. \qquad (44.15)$$

Thus $\mathbf{E}^{\parallel}$ represents the instantaneous (or unretarded) Coulomb field of the charge distribution.

The equation (44.13) is essentially redundant if we take account of the equation of continuity
$$-\dot{\rho} = \nabla . (\rho \mathbf{v}) = \nabla . (\rho \mathbf{v})^{\parallel}. \qquad (44.16)$$

For, owing to the irrotational character of $(\rho \mathbf{v})^{\parallel}$, we can once more use the result of potential theory quoted above and rewrite (44.16) as
$$(\rho \mathbf{v})^{\parallel} = -\frac{1}{4\pi} \nabla \int \frac{\dot{\rho}(\mathbf{x}', t)}{|\mathbf{x}' - \mathbf{x}|} \, d\mathbf{x}'. \qquad (44.17)$$

In view of this relation, we can obtain (44.13) from (44.15) by differentiating with respect to time.

The remaining equations (44.12) and (44.14) determine the transverse electromagnetic field and do not affect the longitudinal field. Therefore, if we ignore the transverse field, the Maxwell–Lorentz equations lead simply to the unretarded Coulomb electric field. This is the approximation underlying the theory of the lattice vibrations given previously in Chapter V. Apart from the special case of strictly longitudinal vibrations, the vibrations of an ionic lattice are associated with a non-vanishing transverse current density $(\rho v)^{\perp}$. The latter produces a transverse field, which is determined by (44.12) and (44.14). Once the transverse field is taken into account, the vibrational modes describe what we observe experimentally as optical waves.

The electric forces transmitted respectively by the longitudinal and transverse fields differ in one essential respect; namely, the former can be described with the help of a potential function (the Coulomb potential) whereas the latter cannot. Accordingly, we write the equations of motion for an ionic lattice (in the harmonic approximation) as follows:

$$m_k \ddot{u}_\alpha(^l_k) = - \sum_{l'k'} \Phi_{\alpha\beta}(^{ll'}_{kk'}) u_\beta(^{l'}_{k'}) +$$

$$+ e_k \lim_{\mathbf{x} \to \mathbf{x}(^l_k)} \{E^{\perp}_\alpha(\mathbf{x}) - (\text{trans. field at } \mathbf{x} \text{ due to } (^l_k))_\alpha\}, \quad (44.18)$$

where Φ includes the Coulomb energy between the ions. Without the last term, (44.18) would be identical with the equation of motion underlying the discussion of the vibrations of ionic lattices given in § 31. The last term represents the force exerted on (^l_k) by the transverse field due to the other ions. The magnetic field $\mathbf{H}^{\perp}$ exerts on (^l_k) a force

$$\frac{e_k}{c} [\dot{\mathbf{u}}(^l_k) \wedge \mathbf{H}^{\perp}(\mathbf{x}(^l_k))].$$

Since $\mathbf{H}^{\perp}$ is produced by the motion of the ions, the force is of the second order in the ion displacements and is thus to be neglected in the harmonic approximation.

We have to consider the above equation of motion together with (44.12) and (44.14), which determine the transverse field. We shall see that these equations are still satisfied by solutions of the following form (see (24.5)):

$$\mathbf{u}(^l_k) = \frac{1}{\sqrt{m_k}} \mathbf{w}(k) \exp\{2\pi i \mathbf{y} . \mathbf{x}(^l_k) - i\omega t\}. \quad (44.19)$$

Consider first the transverse field produced by the ions moving in accordance with (44.19). The current density can be expressed with the help of Dirac's δ-function as

$$\rho\mathbf{v} = \sum_{lk} e_k \dot{\mathbf{u}}(^l_k)\delta(\mathbf{x}-\mathbf{x}(^l_k)) = -i\omega \sum_{lk} e_k \mathbf{u}(^l_k)\delta(\mathbf{x}-\mathbf{x}(^l_k))$$

$$= \left\{ -i\omega \sum_{lk} \frac{e_k}{\sqrt{m_k}} \mathbf{w}(k)\delta(\mathbf{x}-\mathbf{x}(^l_k))\exp[2\pi i\mathbf{y}.(\mathbf{x}(^l_k)-\mathbf{x})] \right\}\exp[2\pi i\mathbf{y}.\mathbf{x}-i\omega t].$$

$$(44.20)$$

The expression given in the curly brackets is clearly a periodic function in the lattice. We can thus represent it by a Fourier series

$$\sum_h \mathbf{J}(h)e^{2\pi i\mathbf{y}(h).\mathbf{x}}, \qquad (44.21)$$

where the coefficients are given by (see (22.22))

$$\mathbf{J}(h) = \frac{-i\omega}{v_a} \int_{\text{zero-cell}} \sum_{lk} \frac{e_k}{\sqrt{m_k}} \mathbf{w}(k)\delta(\mathbf{x}-\mathbf{x}(^l_k)) \times$$

$$\times \exp\{2\pi i\mathbf{y}.(\mathbf{x}(^l_k)-\mathbf{x})-2\pi i\mathbf{y}(h).\mathbf{x}\}\,d\mathbf{x}$$

$$= \frac{-i\omega}{v_a} \sum_k \frac{e_k}{\sqrt{m_k}} \mathbf{w}(k)\exp\{-2\pi i\mathbf{y}(h).\mathbf{x}(k)\}. \qquad (44.22)$$

Let us split $\mathbf{J}(h)$ into two component parts $\mathbf{J}^{\parallel}(h)$ and $\mathbf{J}^{\perp}(h)$ respectively parallel and perpendicular to $\mathbf{y}+\mathbf{y}(h)$. Then we can write

$$(\rho\mathbf{v})^{\parallel} = \sum_h \mathbf{J}^{\parallel}(h)\exp\{2\pi i(\mathbf{y}+\mathbf{y}(h)).\mathbf{x}-i\omega t\}, \qquad (44.23)$$

$$(\rho\mathbf{v})^{\perp} = \sum_h \mathbf{J}^{\perp}(h)\exp\{2\pi i(\mathbf{y}+\mathbf{y}(h)).\mathbf{x}-i\omega t\}; \qquad (44.24)$$

for it is easily verified that (44.23), (44.24) are respectively irrotational and solenoidal. The transverse electromagnetic field can be represented by similar series:

$$\mathbf{E}^{\perp} = \sum_h \mathbf{E}^{\perp}(h)\exp\{2\pi i(\mathbf{y}+\mathbf{y}(h)).\mathbf{x}-i\omega t\}, \qquad (44.25)$$

$$\mathbf{H}^{\perp} = \sum_h \mathbf{H}^{\perp}(h)\exp\{2\pi i(\mathbf{y}+\mathbf{y}(h)).\mathbf{x}-i\omega t\}, \qquad (44.26)$$

where the coefficients are vectors perpendicular to $\mathbf{y}+\mathbf{y}(h)$ so that the expressions are solenoidal. We can determine the coefficients by substituting the series in (44.12) and (44.14):

$$2\pi \sum_h (\mathbf{y}+\mathbf{y}(h)) \wedge \mathbf{E}^{\perp}(h)\exp\{2\pi i(\mathbf{y}+\mathbf{y}(h)).\mathbf{x}-i\omega t\}$$

$$= \frac{\omega}{c} \sum_h \mathbf{H}^{\perp}(h)\exp\{2\pi i(\mathbf{y}+\mathbf{y}(h)).\mathbf{x}-i\omega t\}, \quad (44.27)$$

$$2\pi \sum_h (\mathbf{y}+\mathbf{y}(h)) \wedge \mathbf{H}^{\perp}(h)\exp\{2\pi i(\mathbf{y}+\mathbf{y}(h)).\mathbf{x}-i\omega t\}$$

$$= \frac{1}{c} \sum_h (-\omega\mathbf{E}^{\perp}(h)-i4\pi\mathbf{J}^{\perp}(h))\exp\{2\pi i(\mathbf{y}+\mathbf{y}(h)).\mathbf{x}-i\omega t\}. \quad (44.28)$$

The different plane waves being independent, it follows that

$$(\mathbf{y}+\mathbf{y}(h))\wedge \mathbf{E}^{\perp}(h) = \frac{\omega}{2\pi c}\mathbf{H}^{\perp}(h), \qquad (44.29)$$

$$(\mathbf{y}+\mathbf{y}(h))\wedge \mathbf{H}^{\perp}(h) = \frac{\omega}{2\pi c}\left(-\mathbf{E}^{\perp}(h)-\frac{4\pi i}{\omega}\mathbf{J}^{\perp}(h)\right). \qquad (44.30)$$

Forming the vector product of $(\mathbf{y}+\mathbf{y}(h))$ and (44.29), and eliminating the magnetic field with (44.30), we get

$$-(\mathbf{y}+\mathbf{y}(h))^2\mathbf{E}^{\perp}(h) = \left(\frac{\omega}{2\pi c}\right)^2\left(-\mathbf{E}^{\perp}(h)-\frac{4\pi i}{\omega}\mathbf{J}^{\perp}(h)\right), \qquad (44.31)$$

$$\mathbf{E}^{\perp}(h) = \left\{\frac{-\dfrac{4\pi i}{\omega}}{1-\dfrac{4\pi^2 c^2(\mathbf{y}+\mathbf{y}(h))^2}{\omega^2}}\right\}\mathbf{J}^{\perp}(h). \qquad (44.32)$$

The first term of either series (44.25) or (44.26) has the phase factor

$$\exp\{2\pi i\mathbf{y}.\mathbf{x}-i\omega t\}$$

and represents the macroscopic value of the corresponding field. We shall signify the constant amplitude of the macroscopic field by an overhead bar. Thus we write the macroscopic electric field as

$$\bar{\mathbf{E}}^{\perp}\exp\{2\pi i\mathbf{y}.\mathbf{x}-i\omega t\}, \qquad (44.33)$$

where, as follows from (44.32),

$$\bar{\mathbf{E}}^{\perp} = \mathbf{E}^{\perp}(0) = \frac{-\dfrac{4\pi i}{\omega}}{1-\dfrac{4\pi^2 c^2 y^2}{\omega^2}}\mathbf{J}^{\perp}(0). \qquad (44.34)$$

The dielectric polarization due to the motion of the ions is given by

$$\bar{\mathbf{P}}\exp\{2\pi i\mathbf{y}.\mathbf{x}-i\omega t\} \qquad (44.35)$$

with the amplitude

$$\bar{\mathbf{P}} = \frac{1}{v_a}\sum_k \frac{e_k}{\sqrt{m_k}}\mathbf{w}(k). \qquad (44.36)$$

Comparing with (44.22), we find that

$$\mathbf{J}(0) = -i\omega\bar{\mathbf{P}}, \qquad (44.37)$$

from which we obtain, after subtracting the part parallel to $\mathbf{y}$,

$$\mathbf{J}^{\perp}(0) = -i\omega\left\{\bar{\mathbf{P}}-\frac{\mathbf{y}}{|\mathbf{y}|}\left(\frac{\mathbf{y}}{|\mathbf{y}|}.\bar{\mathbf{P}}\right)\right\}. \qquad (44.38)$$

Therefore (44.34) can also be written

$$\mathbf{E}^{\perp} = 4\pi\left(\frac{4\pi^2 c^2 y^2}{\omega^2} - 1\right)^{-1}\left\{\mathbf{P} - \frac{\mathbf{y}}{|\mathbf{y}|}\left(\frac{\mathbf{y}}{|\mathbf{y}|}.\mathbf{P}\right)\right\}. \tag{44.39}$$

The macroscopic value of the longitudinal field is identical with the macroscopic field considered in Chapter V. Therefore it follows from (30.26) that

$$\mathbf{E}^{\parallel} = -4\pi\frac{\mathbf{y}}{|\mathbf{y}|}\left(\frac{\mathbf{y}}{|\mathbf{y}|}.\mathbf{P}\right). \tag{44.40}$$

Adding (44.39) and (44.40), we obtain for the amplitude of the macroscopic electric field

$$\mathbf{E} = \mathbf{E}^{\perp} + \mathbf{E}^{\parallel} = \frac{4\pi}{(n^2-1)}\left\{\mathbf{P} - n^2\frac{\mathbf{y}}{|\mathbf{y}|}\left(\frac{\mathbf{y}}{|\mathbf{y}|}.\mathbf{P}\right)\right\}, \tag{44.41}$$

where

$$n = \frac{2\pi c|\mathbf{y}|}{\omega}$$

is clearly the refractive index of the optical wave under consideration.

(44.41) is the basic equation of crystal optics (cf. Appendix VIII). For the discussion of the plane optical waves we can replace the macroscopic Maxwell equations in dielectric media by (44.41). It can, in fact, be easily verified that for plane waves the Maxwell equations reduce to the equation (44.41).

We have so far only considered the field produced by the ions moving in accordance with (44.19). The equation (44.18), which determines the motion of the ions, will, as we shall see, lead to the dispersion formula. It is shown in Appendix IX that the dispersion formula, together with the basic equation (44.41), gives a self-contained description of the optical waves. Thus we shall obtain a theory for the optical waves which is entirely free of phenomenological considerations.

Let us substitute (44.19) in the equations of motion (44.18). Since, apart from the last term of (44.18), the equations are identical with the corresponding equations underlying the discussion given in § 31, the equations multiplied by

$$-\frac{1}{\sqrt{m_k}}\exp\{-2\pi i \mathbf{y}.\mathbf{x}(_k^l) + i\omega t\}$$

are directly comparable with (31.19) and can thus be written as

$$\omega^2 w_\alpha(k) = \sum_{k'\beta}\bar{C}_{\alpha\beta}(_{kk'}^{\,\mathbf{y}})w_\beta(k') - \frac{e_k}{\sqrt{m_k}}\bar{E}_\alpha^{\parallel} - \frac{e_k}{\sqrt{m_k}}\exp\{-2\pi i \mathbf{y}.\mathbf{x}(_k^l) + i\omega t\}\times$$

$$\times \lim_{\mathbf{x}\to\mathbf{x}(_k^l)}\{E_\alpha^{\perp}(\mathbf{x}) - (\text{trans. field at } \mathbf{x} \text{ due to } (_k^l))_\alpha\}. \tag{44.42}$$

After writing $\mathbf{E}^{\perp}(\mathbf{x})$ as a series (see (44.25)) and detaching the first term, we can rewrite (44.42) as

$$\omega^2 w_{\alpha}(k) = \sum_{k'\beta} \bar{C}_{\alpha\beta}(\substack{y\\kk'}) w_{\beta}(k') - \frac{e_k}{\sqrt{m_k}} \bar{E}_{\alpha}^{\parallel} - \frac{e_k}{\sqrt{m_k}} \exp\{-2\pi i \mathbf{y} \cdot \mathbf{x}(\substack{l\\k}) + i\omega t\} \times$$
$$\times \lim_{\mathbf{x} \to \mathbf{x}(\substack{l\\k})} \left\{ \sum_{h \neq 0} E_{\alpha}^{\perp}(h) \exp[2\pi i (\mathbf{y} + \mathbf{y}(h)) \cdot \mathbf{x} - i\omega t] - \right.$$
$$\left. - (\text{trans. field at } \mathbf{x} \text{ due to } (\substack{l\\k}))_{\alpha} \right\}. \quad (44.43)$$

The last term of (44.43) gives the effect of the transverse part of the inner field. The inner field which has been discussed in Chapter V, and which is absorbed in the coefficients $\bar{C}_{\alpha\beta}(\substack{y\\kk'})$, actually represents only the longitudinal part, or the value of the inner field in the electrostatic approximation. The last term is the correction due to the retardation effect. The inner field, we remember, is that part of the electric field which is determined by local conditions; in other words, it gives the part of the electric interaction which has effectively a limited range. So long as the time required to traverse such a distance with the light velocity c is inappreciable compared with the period of vibration concerned, the last term of (44.43) is negligible. For a distance of the order 10^{-8} to 10^{-7} cm., the time required is of the order 10^{-17} sec., whereas the vibrational period of an optical wave in the infra-red region is roughly of the order 10^{-13} sec. Therefore for our purpose the last term in (44.43) can be ignored and the equation may be written as

$$\omega^2 w_{\alpha}(k) = \sum_{k'\beta} \bar{C}_{\alpha\beta}(\substack{y\\kk'}) w_{\beta}(k') - \frac{e_k}{\sqrt{m_k}} \bar{E}_{\alpha}. \quad (44.44)$$

The equation leads to a dispersion formula essentially similar to that obtained previously in §§ 33, 34 with the help of a phenomenological energy density. Thus upon substituting (33.3), (34.7), (34.8), (34.9) in the equation (33.22) of the phenomenological theory, we find that that equation can be written as

$$\omega^2 \left(\sqrt{v_a} \, \bar{w}_{\alpha}(k) \right) = \sum_{k'\beta} \bar{C}_{\alpha\beta}^{(0)}(kk') \left(\sqrt{v_a} \, \bar{w}_{\beta}(k') \right) - \frac{e_k}{\sqrt{m_k}} \bar{E}_{\alpha}. \quad (44.45)$$

We observe that (44.45) and (44.44) are almost identical: in particular $\sqrt{v_a} \, \bar{\mathbf{w}}(k)$ has the same meaning as $\mathbf{w}(k)$ (see (33.3) and (33.21)); both represent the reduced displacements of the ions. The only difference between the equations is that the coefficients $\bar{C}_{\alpha\beta}(\substack{y\\kk'})$ in (44.44) are replaced in (44.45) by the long wave limits $\bar{C}_{\alpha\beta}^{(0)}(kk')$ $(= \lim_{y \to 0} \bar{C}_{\alpha\beta}(\substack{y\\kk'}))$. (44.45), hence also the dispersion formula (33.37) derived from it, is independent of the wave-number $\mathbf{y}$. This is an inherent weakness of the phenomenological theory, for the use of the energy density (33.1) directly implies complete microscopic homogeneity, which is of course only approximately true in an actual optical wave. Thus, strictly speaking, the dispersion formula discussed in §§ 33, 34 is correct only in the limit of infinitely long waves.

For optical waves in the infra-red region, the wave-lengths are so large ($\sim 10^5$ times the lattice constant) that the difference between the coefficients $\bar{C}_{\alpha\beta}(^{y}_{kk'})$ and $\bar{C}^{(0)}_{\alpha\beta}(kk')$ is very small. Owing, however, to this difference, (44.44) leads to a dispersion formula which is capable of a distinct type of optical effect, namely, the optical rotation (see Appendix IX). By regarding $\bar{E}_\alpha$ in (44.44) as an externally imposed field (cf. § 31) we can write down the dispersion formula from (44.44) by following the procedure of § 33. In analogy with (33.14) we consider the linear homogeneous equations obtained from (44.44) by leaving out the last term. These homogeneous equations have $3n$ independent solutions for each y. The corresponding frequencies are the solutions of the secular equation

$$|\omega^2 \delta_{kk'}\,\delta_{\alpha\beta} - \bar{C}_{\alpha\beta}(^{y}_{kk'})| = 0. \qquad (44.46)$$

We denote these frequencies (for notation see § 24) by $\bar{\omega}(^{y}_{j}), j = 1, 2,..., 3n$, where the overhead bar has been introduced to distinguish the solutions of the homogeneous equations from those of (44.44). For a given frequency $\bar{\omega}(^{y}_{j})$ the $3n$ quantities $\bar{w}_\alpha(k|^{y}_{j}), k = 0, 1,..., n-1$, satisfy identically the equations†

$$\bar{\omega}^2(^{y}_{j})\bar{w}_\alpha(k|^{y}_{j}) = \sum_{k'\beta} \bar{C}_{\alpha\beta}(^{y}_{kk'})\bar{w}_\beta(k'|^{y}_{j}). \qquad (44.47)$$

It follows from (31.25) that the $\bar{C}_{\alpha\beta}(^{y}_{kk'})$ form a $3n \times 3n$ Hermitian matrix. Hence all the frequencies are real and we can choose the solutions so that (cf. § 38)

$$\sum_{k\alpha} \bar{w}_\alpha(k|^{y}_{j})\bar{w}^*_\alpha(k|^{y}_{j'}) = \delta_{jj'}, \qquad \sum_{j} \bar{w}_\alpha(k|^{y}_{j})\bar{w}^*_\beta(k'|^{y}_{j}) = \delta_{kk'}\,\delta_{\alpha\beta}, \quad (44.48)$$

and
$$\bar{w}^*(k|^{y}_{j}) = \bar{w}(k|^{-y}_{-j}). \qquad (44.49)$$

Following the method of § 33, we write the solution of the inhomogeneous equation (44.44) as a linear combination of the $3n$ solutions of the homogeneous equations

$$\mathbf{w}(k) = \sum_{j} a_j \bar{\mathbf{w}}(k|^{y}_{j}). \qquad (44.50)$$

---

† $\bar{w}_\alpha(k|^{y}_{j})$ should not be confused with the $\bar{w}_\alpha(k)$ occurring in (44.45).

Substituting (44.50) in (44.44) and using (44.47), we get

$$\omega^2 \sum_j a_j \bar{w}_\alpha(k|\tfrac{y}{j}) = \sum_{j'} a_{j'} \bar{\omega}^2(\tfrac{y}{j'})\bar{w}_\alpha(k|\tfrac{y}{j'}) - \frac{e_k}{\sqrt{m_k}}\bar{E}_\alpha. \qquad (44.51)$$

Multiplying the equation by $\bar{w}_\alpha^*(k|\tfrac{y}{j})$ and summing over k, α we find upon using (44.48) that

$$a_j = \frac{1}{\bar{\omega}^2(\tfrac{y}{j}) - \omega^2}\bigg\{ \sum_{k\alpha} \frac{e_k \bar{w}_\alpha^*(k|\tfrac{y}{j})}{\sqrt{m_k}}\bigg\}\bar{E}_\alpha. \qquad (44.52)$$

The dielectric polarization can be obtained readily from (44.50), (44.52) in accordance with (44.36); thus we find that

$$\bar{P}_\alpha = \frac{1}{v_a}\sum_k \frac{e_k w_\alpha(k)}{\sqrt{m_k}} = \sum_\beta a_{\alpha\beta}(\mathbf{y},\omega)\bar{E}_\beta, \qquad (44.53)$$

where

$$a_{\alpha\beta}(\mathbf{y},\omega) = \frac{1}{v_a}\sum_j (\bar{\omega}^2(\tfrac{y}{j}) - \bar{\omega}^2)^{-1}\bigg\{ \sum_{k'} \frac{e_{k'}\bar{w}_\beta^*(k'|\tfrac{y}{j})}{\sqrt{m_{k'}}}\bigg\}\bigg\{ \sum_k \frac{e_k \bar{w}_\alpha(k|\tfrac{y}{j})}{\sqrt{m_k}}\bigg\} \qquad (44.54)$$

is the dielectric susceptibility tensor; this, we note, depends on the wave-number $\mathbf{y}$ as well as on the frequency.

It is evident from (44.54) that the susceptibility tensor is Hermitian:

$$a_{\alpha\beta}(\mathbf{y},\omega) = a_{\beta\alpha}^*(\mathbf{y},\omega). \qquad (44.55)$$

Moreover, owing to (44.49),

$$a_{\alpha\beta}(-\mathbf{y},\omega) = a_{\alpha\beta}^*(\mathbf{y},\omega). \qquad (44.56)$$

Therefore if we expand $a_{\alpha\beta}(\mathbf{y},\omega)$ as a series in $\mathbf{y}$, the even and odd terms must be respectively real and imaginary. Splitting the tensor into its real and imaginary parts, we thus have

$$a_{\alpha\beta}(\mathbf{y},\omega) = a_{\alpha\beta}^R(\mathbf{y},\omega) + i a_{\alpha\beta}^I(\mathbf{y},\omega), \qquad (44.57)$$

where, expressed as series,

$$a_{\alpha\beta}^R(\mathbf{y},\omega) = a_{\alpha\beta}(\omega) + \tfrac{1}{2}\sum_{\gamma\lambda} a_{\alpha\beta,\gamma\lambda}(\omega)y_\gamma y_\lambda + \cdots, \qquad (44.58)$$

$$a_{\alpha\beta}^I(\mathbf{y},\omega) = \sum_\gamma a_{\alpha\beta,\gamma}(\omega)y_\gamma + \cdots. \qquad (44.59)$$

As the leading term of $a_{\alpha\beta}^I(\mathbf{y},\omega)$ is linear in the wave-number, for optical waves in the infra-red region the imaginary part is very small compared with the real part. It follows from (44.55) that $a_{\alpha\beta}^R$ is a symmetric tensor and $a_{\alpha\beta}^I$ is an antisymmetric tensor. Therefore we can rewrite (44.53) as

$$D_\alpha = E_\alpha + 4\pi\bar{P}_\alpha = \sum_\beta \epsilon_{\alpha\beta}E_\beta + i(\mathbf{E}\wedge\mathbf{G})_\alpha, \qquad (44.60)$$

where $$\epsilon_{\alpha\beta} = \delta_{\alpha\beta} + 4\pi a_{\alpha\beta}^{R}(\mathbf{y}, \omega) = \epsilon_{\beta\alpha}, \tag{44.61}$$

and $$G_1 = 4\pi a_{23}^{I}(\mathbf{y}, \omega) = -4\pi a_{32}^{I}(\mathbf{y}, \omega), \text{ etc.} \tag{44.62}$$

G is known as the gyration vector. It is shown in Appendix IX that the real dielectric tensor $\epsilon_{\alpha\beta}$ and the gyration vector **G** are respectively responsible for the phenomena of double refraction and optical rotation.

The series (44.58) and (44.59) can be given explicitly if we have $\bar{\omega}^2(\mathbf{y}_j)$ and $\bar{w}(k|\mathbf{y}_j)$ in the form of series in y. After writing (see (31.20))

$$\bar{C}_{\alpha\beta}(\mathbf{y}_{kk'}) = \bar{C}_{\alpha\beta}^{(0)}(kk') + i\sum_{\gamma} \bar{C}_{\alpha\beta,\gamma}^{(1)}(kk')y_\gamma + \dots \tag{44.63}$$

we can obtain the series for $\bar{\omega}^2(\mathbf{y}_j)$ and $\bar{w}(k|\mathbf{y}_j)$ from the homogeneous equations (44.47) by the perturbation method. Thus if $\bar{\omega}^2(j)$, $\bar{w}(k|j)$ represent the solutions of (44.47) in the limit $\mathbf{y} = 0$, i.e. if

$$\bar{\omega}^2(j)\bar{w}_\alpha(k|j) = \sum_{k'\beta} \bar{C}_{\alpha\beta}^{(0)}(kk')\bar{w}_\beta(k'|j), \tag{44.64}$$

it is readily found that the perturbation method gives

$$\bar{\omega}^2(\mathbf{y}_j) = \bar{\omega}^2(j) + i\sum_{\gamma}\left[\sum_{kk'}\sum_{\alpha\beta} \bar{C}_{\alpha\beta,\gamma}^{(1)}(kk')\bar{w}_\alpha(k|j)\bar{w}_\beta(k'|j)\right]y_\gamma + \dots, \tag{44.65}$$

$$\bar{w}_\alpha(k|\mathbf{y}_j) = \bar{w}_\alpha(k|j) +$$

$$+ i\sum_{\gamma}\left\{\sum_{j'} \frac{\sum_{k'k''}\sum_{\mu\nu} \bar{C}_{\mu\nu,\gamma}^{(1)}(k'k'')\bar{w}_\mu(k'|j')\bar{w}_\nu(k''|j)}{\bar{\omega}^2(j) - \bar{\omega}^2(j')} \bar{w}_\alpha(k|j')\right\}y_\gamma + \dots. \tag{44.66}$$

We note that in the above formulae $\bar{\omega}^2(j)$, $\bar{w}(k|j)$, $\bar{C}_{\alpha\beta}^{(0)}(kk')$, $\bar{C}_{\alpha\beta,\gamma}^{(1)}(kk'),\dots$ are all real quantities (see § 31).

For the discussion of optical waves in the infra-red region we need retain only the leading term in (44.58) and (44.59). In this approximation, $a_{\alpha\beta}^R$ is obtained directly by replacing $\bar{\omega}^2(\mathbf{y}_j)$ and $\bar{w}(k|\mathbf{y}_j)$ in (44.54) by $\bar{\omega}^2(j)$ and $\bar{w}(k|j)$; in other words $\epsilon_{\alpha\beta}$ has the value corresponding to the long-wave limit $\mathbf{y} = 0$ and is thus identical with the dispersion formula obtained previously in §§ 33, 34. The gyration vector **G**, on the other hand, is obtained by expanding (44.54) in a series with the help of the expansions (44.65), (44.66) and picking out the linear terms. With the help of the relation (see (26.4) and § 31)

$$\bar{C}_{\alpha\beta,\gamma}^{(1)}(kk') = -\bar{C}_{\beta\alpha,\gamma}^{(1)}(k'k), \tag{44.67}$$

we find after some simplification that

$$G_1 = -(4\pi/v_a) \sum_\gamma y_\gamma \sum_j \sum_{j'} \frac{1}{(\bar{\omega}^2(j)-\omega^2)(\bar{\omega}^2(j')-\omega^2)} \times$$

$$\times \left\{ \sum_k \frac{e_k}{\sqrt{m_k}} \bar{w}_2(k|j) \right\} \left\{ \sum_{k'} \frac{e_{k'}}{\sqrt{m_{k'}}} \bar{w}_3(k'|j') \right\} \times$$

$$\times \left\{ \sum_{k''\mu} \sum_{k'''\nu} \bar{w}_\mu(k''|j) \bar{C}^{(1)}_{\mu\nu,\gamma}(k''k''') \bar{w}_\nu(k'''|j') \right\}, \quad (44.68)$$

and the other components can be obtained by cyclic permutation of the indices 2, 3 on the right-hand side. Owing to (44.67) the expression on the right-hand side reverses its sign, if we interchange the indices 2 and 3. Hence we can also express (44.68) in the vector form:

$$\mathbf{G} = (-2\pi/v_a) \sum_\gamma y_\gamma \sum_j \sum_{j'} \frac{1}{(\bar{\omega}^2(j)-\omega^2)(\bar{\omega}^2(j')-\omega^2)} \times$$

$$\times \left\{ \sum_k \sum_{k'} \frac{e_k e_{k'}}{(m_k m_{k'})^{\frac{1}{2}}} \bar{\mathbf{w}}(k|j) \wedge \bar{\mathbf{w}}(k'|j') \right\} \times$$

$$\times \left\{ \sum_{k''\mu} \sum_{k'''\nu} \bar{w}_\mu(k''|j) \bar{C}^{(1)}_{\mu\nu,\gamma}(k''k''') \bar{w}_\nu(k'''|j') \right\}. \quad (44.69)$$

As we have observed, if the transverse electromagnetic field is ignored, the dynamics of the system can be described by a potential function. The corresponding vibrational modes, which we have discussed in Chapter V, describe essentially the free oscillations of the normal coordinates corresponding to the potential function Φ. In this approximation the use of quantum instead of classical mechanics affects the modes only in so far as the normal coordinates should be treated as quantum mechanical variables. The above treatment takes account in a classical way of the fact that such a vibrational mode emits a transverse electromagnetic field and is modified by the latter; with this modification incorporated, the vibrational mode represents an optical wave. It can be shown† that if we assume the same transverse electromagnetic field as in the classical treatment acting on the normal coordinates now treated quantum mechanically, the quantum mechanical currents induced at the ions are identical with the values given by the classical treatment. It thus follows that a quantum mechanical treatment of the motion of the ions leaves the results of the above classical theory completely unaffected.

† See K. Huang, *Proc. Camb. Phil. Soc.* **45**, 452 (1949).

45. The local treatment of optical effects

The same formula for the dielectric tensor $\epsilon_{\alpha\beta}(\omega)$ as given by the microscopic theory can also be obtained by applying the results established in Chapter IV. We obtained there certain general formulae relating to the optical effects of a molecular system which is subject to an electromagnetic wave, the wave-length of the latter being large compared with the dimensions in the molecular system. Now the dispersion formula for a crystal is essentially a local relation, i.e. it relates the macroscopic field to the dielectric polarization at the same point (in the macroscopic sense). The general formulae of Chapter IV, therefore, can be used directly to establish the dispersion formula for a crystal, if an arbitrary local portion of a crystal behaves virtually as an independent system and if upon the local portion the macroscopic field has virtually the effect of an externally imposed field. This is indeed the case, as we have observed in connexion with the interpretation of the first-order equation (31.34); thus we have found that once we interpret the macroscopic field as a field externally imposed on a local portion (small compared with the wave-length), the latter, dynamically speaking, is virtually decoupled from the rest of the crystal. Furthermore, we recall that, owing to the separation of the macroscopic field, the restoring forces arising from the locally homogeneous internal strain are determined by the coefficients $\bar{C}^{(0)}_{\alpha\beta}(kk')$ rather than $C^{(0)}_{\alpha\beta}(kk')$, the latter being in fact divergent for ionic lattices. If we introduce for such a local portion of a crystal the normal coordinates as explained in § 38, the internal strain is described by the coordinates $Q\binom{0}{j}$ for zero wave-number. Thus, in defining the coordinates $Q\binom{0}{j}$, we have only to replace the equation (38.24) for the case $\mathbf{y} = 0$ by

$$\sum_{k'\beta} \bar{C}^{(0)}_{\alpha\beta}(kk')e_{\beta}(k'\,|^0_j) = \omega^2\binom{0}{j}e_{\alpha}(k\,|^0_j).$$

(45.1)

In this connexion, we need only remember the relation between lattice waves and normal coordinates as discussed in § 38; note in particular that the coefficients $D_{\alpha\beta}(^{\mathbf{y}}_{kk'})$ and $C_{\alpha\beta}(^{\mathbf{y}}_{kk'})$ differ only by a phase factor which reduces to unity for $\mathbf{y} = 0$). A moment's consideration will show that this definition for $Q\binom{0}{j}$ agrees exactly with that already introduced in § 42 in connexion with the static phenomena.

Let us now apply the results of Chapter IV to a local portion of N cells in an ionic lattice using the rigid ion model as in the last section. The total electric moment in this case,

$$\mathbf{M} = \sum_k e_k \mathbf{u}(^l_k),$$

(45.2)

is linear in the nuclear displacements. It follows that the series expansion (39.11) for the electric moment of N cells reduces to

$$\delta M_\alpha = \sqrt{N} \sum_j M_\alpha({}^0_j) Q({}^0_j). \tag{45.3}$$

Using (45.2), we find that (for notation see (23.3), (23.5))

$$M_{\alpha,\beta}(k) = M_{\alpha,\beta}({}^l_k) = e_k \delta_{\alpha\beta}; \tag{45.4}$$

thus it follows from (39.13) that the coefficients in (45.3) are given by

$$M_\alpha({}^0_j) = \sum_k \frac{e_k}{\sqrt{m_k}} e_\alpha(k|{}^0_j). \tag{45.5}$$

Since (45.3) has only linear terms, we have only first-order dispersion in this case (see § 21), the transition frequencies being $\pm\omega({}^0_j)$. Therefore the polarizability is given directly by the formula (21.12):

$$\begin{aligned}
\{P^{vv}_{\alpha\beta}(\omega)\}_{Av} &= N \sum_j \frac{M_\alpha({}^0_j) M_\beta({}^0_j)}{\omega^2({}^0_j) - \omega^2} \\
&= N \sum_j \frac{1}{\omega^2({}^0_j) - \omega^2} \left\{ \sum_k \frac{e_k e_\alpha(k|{}^0_j)}{\sqrt{m_k}} \right\} \left\{ \sum_{k'} \frac{e_{k'} e_\beta(k'|{}^0_j)}{\sqrt{m_{k'}}} \right\},
\end{aligned} \tag{45.6}$$

where the term in (21.12) describing electronic polarization vanishes owing to the assumption of rigid ions. Dividing (45.6) by the volume $N v_a$, we obtain for the dielectric susceptibility:

$$a_{\alpha\beta}(\omega) = \frac{1}{v_a} \sum_j \frac{1}{\omega^2({}^0_j) - \omega^2} \left\{ \sum_k \frac{e_k}{\sqrt{m_k}} e_\alpha(k|{}^0_j) \right\} \left\{ \sum_{k'} \frac{e_{k'}}{\sqrt{m_{k'}}} e_\beta(k'|{}^0_j) \right\}. \tag{45.7}$$

The formula is completely equivalent to the dielectric tensor obtained in the last section and the results of §§ 33, 34. Thus upon substituting (34.7), (34.8), (34.9), and (33.3) in (33.14) and (33.18), we see immediately that the dispersion formula obtained there is identical with (45.7).

As compared with the method of local treatment, the microscopic treatment gives a deeper understanding of the detailed mechanism of refraction and is, moreover, capable of accounting for effects depending essentially on the finite wave-length of an optical wave, such as optical rotation. However, the microscopic method lacks flexibility and is very involved. The method requires, in the first place, a precisely defined model (e.g. rigid ions, polarizable ions), and owing to its complexity it does not lend itself readily to the discussion of more involved problems. In the following sections we shall discuss the more complex optical effects on the basis of the method of local treatment as exemplified above.

46. The effect of the anharmonic potential on dispersion

The dispersion formula (45.7) for general lattice structures is similar to the dispersion formula (7.5) for isotropic diatomic crystals, the only difference being in the appearance of, in general, more than one dispersion frequency in the former. It has been seen in § 10 that a formula of this type gives a correct description of optical dispersion only for frequencies not too near the dispersion frequencies. Supplementary to (45.7) we have, furthermore, the anti-Hermitian dielectric susceptibility which is readily obtained by using (45.3) and (45.5) in (21.13):

$$\{R^{vv}_{\alpha\beta}(\omega)\}_{Av} = \frac{i\pi}{2v_a} \sum_j \frac{1}{\omega\binom{0}{j}} \left\{ \sum_k \frac{e_k}{\sqrt{m_k}} e_\alpha(k|\substack{0\\j}) \right\} \times$$

$$\times \left\{ \sum_{k'} \frac{e_{k'}}{\sqrt{m_{k'}}} e_\beta(k'|\substack{0\\j}) \right\} \{\delta(\omega+\omega(\substack{0\\j})) - \delta(\omega-\omega(\substack{0\\j}))\}. \quad (46.1)$$

The formula gives infinitely sharp absorption lines at the dispersion frequencies $\omega(\substack{0\\j})$, again at variance with the experimental results which we have reviewed in § 10.

It has been remarked in § 10 that the failure of the theoretical formulae in the neighbourhood of the dispersion frequencies is a consequence of the harmonic approximation. In the harmonic approximation we neglect in the potential function all terms of the third or higher orders, with the result that the motion of a lattice can be resolved into independent modes of vibration. In actual fact, the higher-order terms in the potential function, known usually as the anharmonic terms, provide a coupling between the various vibrational modes. The effect of the coupling on the optical waves, as we have explained in § 10, is expected to be particularly strong in the neighbourhood of the dispersion frequencies. Broadly speaking, the experimental evidence points to a broadening of the absorption lines at the dispersion frequencies $\omega(\substack{0\\j})$. The analogous broadening of spectral lines of gaseous molecules is well known to be due to certain means by which a molecule can exchange energy with its surroundings (e.g. spontaneous emission, intermolecular collisions). Similarly in a crystal the anharmonic potential provides a channel for the exchange of energy between the dispersion oscillators and other vibrational modes.

The effect of the anharmonic potential on dispersion has been investigated by Pauli;[†] Born and Blackman;[‡] Blackman;[§] Barnes, Brattain,

---

[†] W. Pauli, *Verh. d. D. Phys. Ges.* (3) **6**, 10 (1925).
[‡] M. Born and M. Blackman, *Zeit. f. Phys.* **82**, 551 (1933).
[§] M. Blackman, ibid. **86**, 421 (1933).

and Seitz.† Pauli considered the problem before the secondary structure in dispersion (§ 10) was experimentally established. Using a linear chain of alternate positive and negative ions of the same mass, Pauli obtained a dispersion formula identical with the *ad hoc* formula (10.6). As the formula cannot account for the appearance of the observed secondary structure, still using a linear model Born and Blackman extended Pauli's consideration to the case of ions of unequal masses. Their result remains similar to (10.6) with, however, the important difference that the damping constant is a function of the frequency ω. Generally speaking, their result is consistent with the appearance of the secondary structure. Blackman's further extension of the theory to a three-dimensional model leads to roughly the same conclusion. Whereas these authors used the method of classical mechanics, Barnes, Brattain, and Seitz developed a quantum mechanical theory. Their treatment demonstrates that there should be superimposed upon the sharp absorption lines a continuous absorption. In other words, their result is consistent with the presence of absorption at frequencies other than the dispersion frequencies $\omega\binom{0}{j}$ but still presumes the infinitely sharp absorption lines at the dispersion frequencies. This latter weakness of their theory is due to the application of the ordinary second-order perturbation method to a case where the energy spectrum is practically continuous. In the following, we shall attempt to give a more satisfactory quantum mechanical treatment of the problem on the basis of a method due originally to Weisskopf and Wigner.‡

Let us consider a finite portion of a crystal consisting of N lattice cells and subject to a macroscopic field

$$\mathbf{E}(t) = \mathbf{E}e^{-i\omega t} + \mathbf{E}^*e^{i\omega t}. \qquad (46.2)$$

According to the method of local treatment, $\mathbf{E}(t)$ is to be considered as an externally imposed field; thus, in the presence of $\mathbf{E}(t)$, we have to add to the Hamiltonian of the N cells the interaction term

$$\mathbf{E}(t).\left\{\sqrt{N}\sum_{j}\mathbf{M}\binom{0}{j}Q\binom{0}{j}\right\}. \qquad (46.3)$$

We have represented the electric moment by the expansion (39.11) and ignored all terms of the second and higher orders. (The approximation implied thereby includes the rigid ion model as a special case.)

---

† R. B. Barnes, R. R. Brattain, and F. Seitz, *Phys. Rev.* **48**, 582 (1935).
‡ V. Weisskopf and E. Wigner, *Zeit. f. Phys.* **63**, 54 and **65**, 18 (1930).

From (39.2) we can write down directly the anharmonic potential for N cells:

$$\frac{1}{6}\frac{1}{\sqrt{N}}\sum_{yj}^{N}\sum_{y'j'}^{N}\sum_{y''j''}^{N}\Delta(\mathbf{y}+\mathbf{y}'+\mathbf{y}'')\Phi(^{y\ y'\ y''}_{j\ j'\ j''})Q(^{y}_{j})Q(^{y'}_{j'})Q(^{y''}_{j''})+$$

$$+\frac{1}{24}\frac{1}{N}\sum_{yj}^{N}\sum_{y'j'}^{N}\sum_{y''j''}^{N}\sum_{y'''j'''}^{N}\Delta(\mathbf{y}+\mathbf{y}'+\mathbf{y}''+\mathbf{y}''')\times$$

$$\times\Phi(^{y\ y'\ y''\ y'''}_{j\ j'\ j''\ j'''})Q(^{y}_{j})Q(^{y'}_{j'})Q(^{y''}_{j''})Q(^{y'''}_{j'''}),\qquad(46.4)$$

where the coefficients are defined in (39.9). We observe that in the approximation expressed in (46.3) only the modes $(^{0}_{j})$ interact directly with the field and determine the induced polarization. Therefore, in this connexion, the other oscillators (vibrational modes) function essentially as a viscous medium dissipating the energy of the dispersion oscillators $(^{0}_{j})$ which are set in motion by the field. The effect of the terms in (46.4) not containing $Q(^{0}_{j})$ is thus to modify somewhat the nature of these dissipating oscillators; for a first approximation, we can ignore these terms. Of the remaining terms in (46.4), only those linear in $Q(^{0}_{j})$ need be taken into account, for it can be shown that the rest of the terms will affect the results only by terms proportional to $1/\sqrt{N}$ or its higher powers. Moreover, the different dispersion oscillators can influence one another only indirectly through the dissipating oscillators; owing to the large number of the latter, such effects can be ignored. Hence we can consider the dielectric polarization due to each of the dispersion oscillators separately. For the discussion of the contribution by an oscillator we can thus use the anharmonic potential

$$\Phi_{j}^{A}=\frac{1}{2}\frac{1}{\sqrt{N}}\,Q(^{0}_{j})\sum_{yj'}^{N}\sum_{y'j''}^{N}\Delta(\mathbf{y}+\mathbf{y}')\Phi(^{0\ y\ y'}_{j\ j'\ j''})Q(^{y}_{j'})Q(^{y'}_{j''})+$$

$$+\frac{1}{6}\frac{1}{\sqrt{N}}\,Q(^{0}_{j})\sum_{yj'}^{N}\sum_{y'j''}^{N}\sum_{y''j'''}^{N}\Delta(\mathbf{y}+\mathbf{y}'+\mathbf{y}'')\times$$

$$\times\Phi(^{0\ y\ y'\ y''}_{j\ j'\ j''\ j'''})Q(^{y}_{j'})Q(^{y'}_{j''})Q(^{y''}_{j'''})+\dots,\qquad(46.5)$$

where we have used the fact that $\Delta(\mathbf{y}+\mathbf{y}'+\dots)\Phi(^{y\ y'}_{j\ j'}\dots)$ is completely symmetric in the indices $(^{y}_{j})$, $(^{y'}_{j'})$,... (see § 39).

In the following we shall consider the joint effect of the anharmonic potential (46.5) and the electric interaction (46.3) by a perturbation method similar to that used by Weisskopf and Wigner. This method will now be explained.

In general, if $\quad\quad \psi_n = e^{-\omega_n t}\phi_n \quad (\hbar\omega_n = E_n)$

represents the normalized time-dependent solutions of the Schrödinger equation

$$H^0\psi_n = i\hbar\,\frac{\partial}{\partial t}\,\psi_n,$$

the solutions of the perturbed wave equation

$$(H^0 + H')\Phi = i\hbar\,\frac{\partial}{\partial t}\,\Phi$$

can be expressed in terms of the functions ψ_n as follows:

$$\Phi = \sum_n a_n \psi_n. \tag{46.6}$$

It is well known that the coefficients a_n are functions of time satisfying the equations

$$i\hbar\,\frac{da_n}{dt} = \sum_{n'} \langle n|H'|n'\rangle a_{n'}\exp\{i(\omega_n-\omega_{n'})t\}, \tag{46.7}$$

where $\langle n|H'|n'\rangle$ represents the matrix element of H' formed with ϕ_n^* and $\phi_{n'}$. It will save trouble if we first explain the method of Weisskopf and Wigner for calculating the lifetime of a state ψ_0 in the presence of a perturbation H'. Representing the state in the manner of (46.6) we have the initial conditions

$$a_0 = 1, \quad\quad a_n = 0 \quad (n \neq 0) \quad\quad \text{at } t = 0, \tag{46.8}$$

i.e. we know that, at $t = 0$, the system is in the state ψ_0. Following Weisskopf and Wigner, we shall try to solve the equations (46.7) by putting

$$a_0 = e^{-\Gamma t}, \tag{46.9}$$

where $1/2\Gamma$ is the time of probability decay to the fraction $1/e$ of its value at $t = 0$, or the *lifetime*. Moreover, in the equations for $n \neq 0$ we retain on the right-hand side only the term containing a_0 (first-order perturbation). Thus the equations for $n \neq 0$ become

$$i\hbar\,\frac{da_n}{dt} = \langle n|H'|0\rangle\exp\{-\Gamma t + i(\omega_n-\omega_0)t\} \quad (n \neq 0). \tag{46.10}$$

In view of the initial conditions (46.8) we integrate (46.10) with respect to t from $t = 0$ and obtain

$$a_n = -\frac{i}{\hbar}\,\frac{\exp[-\Gamma t + i(\omega_n-\omega_0)t] - 1}{-\Gamma + i(\omega_n-\omega_0)}\,\langle n|H'|0\rangle \quad (n \neq 0). \tag{46.11}$$

The equation (46.7) for $n = 0$ still has to be considered. We find that when (46.9) and (46.11) are substituted, the equation reduces to the following:

$$\Gamma = \frac{1}{\hbar^2} \sum_{n'} \frac{1 - \exp[\Gamma t - i(\omega_{n'} - \omega_0)t]}{i(\omega_{n'} - \omega_0) - \Gamma} |\langle n'|H'|0\rangle|^2. \qquad (46.12)$$

Evidently the equation can be fulfilled only if the expression on the right-hand side is a constant independent of t. This is true only under circumstances which we shall presently describe.

The Weisskopf–Wigner solution is designed for cases where the states n' cover practically a continuous energy spectrum. For such states we can order the states n' according to the transition frequency

$$\omega = \omega_0 - \omega_{n'}, \qquad (46.13)$$

. and define a function of ω as follows:

$$\gamma_0(\omega) = \frac{\pi}{\hbar^2} \lim_{\Delta\omega \to 0} \frac{1}{\Delta\omega} \sum_{(n')}^{\omega} |\langle n'|H'|0\rangle|^2, \qquad (46.14)$$

where the symbol $\sum_{(n')}^{\omega}$ means that the summation is over the states n' which have transition frequencies in the interval from ω to $\omega + \Delta\omega$. The subscript of γ refers to the initial state with respect to which the transition frequency is defined.† With the help of this function, we can evidently rewrite the equation (46.12) in the form

$$\Gamma = \frac{1}{\pi} \int \left\{ \frac{1 - \exp[\Gamma t + i\omega t]}{-i\omega - \Gamma} \right\} \gamma_0(\omega)\, d\omega. \qquad (46.15)$$

It is usually assumed that the integral on the right-hand side of (46.15) has the value

$$\int \left\{ \frac{1 - \exp[\Gamma t + i\omega t]}{-i\omega - \Gamma} \right\} \gamma_0(\omega)\, d\omega = \pi\gamma_0(0). \qquad (46.16)$$

This formula, as it stands without qualification is, however, far from being correct. In the first place, there is in general an imaginary part to the integral; this part is usually ignored because, for the type of problem (e.g. damping) to which the formula is applied, the real part is much the more significant. Furthermore, the real part as given by (46.16) is true only for real values of Γ and, even then, only for a restricted

† A function of this type can be defined with reference to any state provided the states n' cover a continuous energy spectrum.

range of the parameter t. This range will be the wider, the more uniform the function $\gamma_0(\omega)$ is near $\omega = 0$; in order that the solution $a_0 = \exp[-\Gamma t]$ should be valid for the greater part of its lifetime, $\gamma_0(\omega)$ must be constant over a range of frequencies comparable with Γ. For the present we shall assume that (46.16) is valid; at the end of this section, we shall derive the formula and discuss its various qualifications.

It follows from (46.16) that

$$\int \left\{ \frac{1-\exp[\Gamma t + i(\omega-c)t]}{-i(\omega-c)-\Gamma} \right\} \gamma_0(\omega)\, d\omega = \pi\gamma_0(c). \qquad (46.17)$$

This formula will be useful for later discussions.

In view of (46.16), (46.15) reduces to

$$\Gamma = \gamma_0(0) \qquad (46.18)$$

which, on the one hand, shows that (46.9) is an appropriate solution and, on the other hand, determines the value of the damping constant Γ. Since $|a_0|^2 = \exp[-2\Gamma t]$ is the probability of the system being in the state 0 at the time t, the mean lifetime of this state is

$$\int_0^\infty e^{-2\Gamma t} t\, dt \Big/ \int_0^\infty e^{-2\Gamma t}\, dt = \frac{1}{2\Gamma}. \qquad (46.19)$$

As we have explained, we shall treat the electric interaction (46.3) and the anharmonic potential (46.5) jointly as the perturbation. The wave functions of the unperturbed system are products of simple harmonic oscillator wave functions with real normal coordinates as arguments; for the latter, we shall use the real normal coordinates of the second kind (§ 38). Thus a stationary state of the system in the absence of the perturbation is specified by the quantum numbers $v\binom{0}{1}$, $v\binom{0}{2},..., v\binom{y}{j}$, which describe the states of the individual oscillators corresponding to the real normal coordinates $q\binom{y}{j}$.

Let us suppose that at a time $t = t_0$ the system is in a state 0 with the quantum numbers

$$v^0\binom{0}{1},\ v^0\binom{0}{2},..., v^0\binom{y}{j},... \quad \text{(state 0)}. \qquad (46.20)$$

The states obtained by changing the quantum number of one of the dispersion oscillators by ± 1, namely the states with the quantum numbers

$$v^0\binom{0}{1},\ v^0\binom{0}{2},..., v^0\binom{0}{j}\pm 1,..., v^0\binom{y}{j},... \quad \text{(states } \pm j) \qquad (46.21)$$

will be referred to as the states $\pm j$. All other states we shall designate

generally by the letter s. We note that from the state 0 the electric
interaction (46.3) has non-vanishing matrix elements only for transitions
to the states $\pm j$ $(j = 1,..., 3n)$.

In the perturbation scheme discussed earlier the perturbation H' now
stands for the sum of (46.3) and (46.5), and the expansion (46.6) is made
with respect to the unperturbed states 0, $\pm j$, and s. According to our
supposition, the expansion coefficients satisfy the following initial
conditions:

$$a_0 = 1, \text{ and all } a_{\pm j}, a_s = 0 \quad \text{at } t = t_0. \tag{46.22}$$

If we write down the equation (46.7) for the state 0, we have on the right-
hand side two groups of terms due respectively to the anharmonic
potential and the electric interaction. For our purpose we can neglect
the latter as compared with the former, for in discussing the phenomenon
of dispersion in so far as it is independent of the intensity of the optical
wave, we can consider the electric field as infinitesimal. The problem
of determining the coefficient a_0 thus reduces to the Weisskopf–Wigner
problem, the perturbation being the anharmonic potential Φ_j^A (46.5).
Therefore we have for a_0

$$a_0 = e^{-\Gamma_0\tau}, \qquad \Gamma_0 = \gamma_0(0), \tag{46.23}$$

where
$$\gamma_0(0) = \frac{\pi}{\hbar^2} \lim_{\Delta\omega \to 0} \frac{1}{\Delta\omega} \sum_{(n')}^{0} |\langle n'|\Phi_j^A|0\rangle|^2, \tag{46.24}$$

and τ is the time counted from t_0,

$$\tau = t - t_0. \tag{46.25}$$

The equation (46.7) for the states $\pm j$ can be written as follows:

$$i\hbar \frac{da_{\pm j}}{dt} = \sqrt{N}\, \mathbf{M}\binom{0}{j} \times$$
$$\times \{\mathbf{E} \exp[i(\pm\omega\binom{0}{j}-\omega)t] + \mathbf{E}^* \exp[i(\pm\omega\binom{0}{j}+\omega)t]\}\langle\pm j|Q\binom{0}{j}|0\rangle e^{-\Gamma_0\tau} +$$
$$+ \sum_s a_s \langle\pm j|\Phi_j^A|s\rangle \exp[i(\omega_{\pm j}-\omega_s)t]. \tag{46.26}$$

The first term on the right-hand side describes the transitions from the
state 0 owing to the electric interaction. The sum over the states s
describes the damping of the states $\pm j$ because of the anharmonic
potential. The coupling to the states s owing to the electric interaction
has been ignored for the same reason as in deriving the coefficient a_0.
The coefficients a_s in the damping term can be eliminated as in the

Weisskopf–Wigner problem with the help of the equations (46.7) for a_s:

$$i\hbar\frac{da_s}{dt} = a_{\pm j}\langle s|\Phi_j^A|\pm j\rangle\exp[i(\omega_s-\omega_{\pm j})t].$$ (46.27)

(46.26) and (46.27) are to be considered together with either the upper or the lower signs in both. The use of (46.27) effectively ignores the mutual interference between the damping of the various states $\pm j$; this is clearly a secondary effect.

In order to solve the equations (46.26) and (46.27) we require a trial solution for the coefficients $a_{\pm j}$. Let us put

$$a_{\pm j} = A_{\pm j}\exp[i(\pm\omega_j^{(0)}-\omega)t_0]\{\exp[i(\pm\omega_j^{(0)}-\omega)\tau-\Gamma_0\tau]-e^{-\Gamma_{\pm j}\tau}\}+$$
$$+B_{\pm j}\exp[i(\pm\omega_j^{(0)}+\omega)t_0]\{\exp[i(\pm\omega_j^{(0)}+\omega)\tau-\Gamma_0\tau]-e^{-\Gamma_{\pm j}\tau}\}.$$ (46.28)

The expression evidently fulfils the initial conditions (46.22). The factors containing t_0 could, of course, have been absorbed in the as yet undetermined constants $A_{\pm j}$, $B_{\pm j}$; however, as the expression stands, the coefficients $A_{\pm j}$, $B_{\pm j}$ will turn out to be independent of t_0. Now substituting (46.28) in (46.27) and integrating with respect to τ from 0, we get

$$a_s = -\frac{i}{\hbar}\Bigg\{A_{\pm j}\bigg(\frac{\exp[i(\omega_s-\omega_{\pm j}\pm\omega_j^{(0)}-\omega)\tau-\Gamma_0\tau]-1}{i(\omega_s-\omega_{\pm j}\pm\omega_j^{(0)}-\omega)-\Gamma_0} -$$
$$-\frac{\exp[i(\omega_s-\omega_{\pm j})\tau-\Gamma_{\pm j}\tau]-1}{i(\omega_s-\omega_{\pm j})-\Gamma_{\pm j}}\bigg)\exp[i(\omega_s-\omega_{\pm j}\pm\omega_j^{(0)}-\omega)t_0]+$$
$$+B_{\pm j}\bigg(\frac{\exp[i(\omega_s-\omega_{\pm j}\pm\omega_j^{(0)}+\omega)t-\Gamma_0\tau]-1}{i(\omega_s-\omega_{\pm j}\pm\omega_j^{(0)}+\omega)-\Gamma_0} -$$
$$-\frac{\exp[i(\omega_s-\omega_{\pm j})\tau-\Gamma_{\pm j}\tau]-1}{i(\omega_s-\omega_{\pm j})-\Gamma_{\pm j}}\bigg)\exp[i(\omega_s-\omega_{\pm j}\pm\omega_j^{(0)}+\omega)t_0]\Bigg\}\langle s|\Phi_j^A|\pm j\rangle.$$ (46.29)

After substituting (46.29) in (46.26), we can convert the damping terms into integrals with respect to the transition frequencies $\omega_{\pm j}-\omega_s$ from $\pm j$ to s:

$$\omega_\pm = \omega_{\pm j}-\omega_s$$ (46.30)

by introducing the following functions in analogy with (46.14):

$$\gamma_{\pm j}(\omega_\pm) = \frac{\pi}{\hbar^2}\lim_{\Delta\omega_\pm\to 0}\frac{1}{\Delta\omega_\pm}\sum_{(s)}|\langle s|\Phi_j^A|\pm j\rangle|^2.$$ (46.31)

Thus we find that (46.26) can be written as follows:

$$A_{\pm j}[i(\pm\omega(^0_j)-\omega)-\Gamma_0]\exp[i(\pm\omega(^0_j)-\omega)t-\Gamma_0\tau]+$$
$$+A_{\pm j}\Gamma_{\pm j}\exp[i(\pm\omega(^0_j)-\omega)t_0-\Gamma_{\pm j}\tau]+B_{\pm j}[i(\pm\omega(^0_j)+\omega)-\Gamma_0]\times$$
$$\times\exp[i(\pm\omega(^0_j)+\omega)t-\Gamma_0\tau]+B_{\pm j}\Gamma_{\pm j}\exp[i(\pm\omega(^0_j)+\omega)t_0-\Gamma_{\pm j}\tau]$$
$$=-(i/\hbar)\sqrt{N}\{\mathbf{E}.\mathbf{M}(^0_j)\exp[i(\pm\omega(^0_j)-\omega)t-\Gamma_0\tau]+$$
$$+\mathbf{E}^*.\mathbf{M}(^0_j)\exp[i(\pm\omega(^0_j)+\omega)t-\Gamma_0\tau]\}\langle\pm j|Q(^0_j)|0\rangle-$$
$$-\frac{1}{\pi}A_{\pm j}\exp[i(\pm\omega(^0_j)-\omega)t-\Gamma_0\tau]\times$$
$$\times\int\gamma_{\pm j}(\omega_\pm)\left\{\frac{1-\exp[i(\omega_\pm\mp\omega(^0_j)+\omega)\tau+\Gamma_0\tau]}{-i(\omega_\pm\mp\omega(^0_j)+\omega)-\Gamma_0}\right\}d\omega_\pm+$$
$$+\frac{1}{\pi}A_{\pm j}\exp[i(\pm\omega(^0_j)-\omega)t_0-\Gamma_{\pm j}\tau]\times$$
$$\times\int\gamma_{\pm j}(\omega_\pm)\left\{\frac{1-\exp[i\omega_\pm\tau+\Gamma_{\pm j}\tau]}{-i\omega_\pm-\Gamma_{\pm j}}\right\}d\omega_\pm-$$
$$-\frac{1}{\pi}B_{\pm j}\exp[i(\pm\omega(^0_j)+\omega)t-\Gamma_0\tau]\times$$
$$\times\int\gamma_{\pm j}(\omega_\pm)\left\{\frac{1-\exp[i(\omega_\pm\mp\omega(^0_j)-\omega)\tau+\Gamma_0\tau]}{-i(\omega_\pm\mp\omega(^0_j)-\omega)-\Gamma_0}\right\}d\omega_\pm+$$
$$+\frac{1}{\pi}B_{\pm j}\exp[i(\pm\omega(^0_j)+\omega)t_0-\Gamma_{\pm j}\tau]\times$$
$$\times\int\gamma_{\pm j}(\omega_\pm)\left\{\frac{1-\exp[i\omega_\pm\tau+\Gamma_{\pm j}\tau]}{-i\omega_\pm-\Gamma_{\pm j}}\right\}d\omega_\pm.\quad(46.32)$$

Using the values for the integrals as given by (46.17), we get

$$A_{\pm j}[i(\pm\omega(^0_j)-\omega)-\Gamma_0]\exp[i(\pm\omega(^0_j)-\omega)t-\Gamma_0\tau]+$$
$$+A_{\pm j}\Gamma_{\pm j}\exp[i(\pm\omega(^0_j)-\omega)t_0-\Gamma_{\pm j}\tau]+$$
$$+B_{\pm j}[i(\pm\omega(^0_j)+\omega)-\Gamma_0]\exp[i(\pm\omega(^0_j)+\omega)t-\Gamma_0\tau]+$$
$$+B_{\pm j}\Gamma_{\pm j}\exp[i(\pm\omega(^0_j)+\omega)t_0-\Gamma_{\pm j}\tau]$$
$$=-\frac{i}{\hbar}\sqrt{N}\langle\pm j|Q(^0_j)|0\rangle\{\mathbf{E}.\mathbf{M}(^0_j)\exp[i(\pm\omega(^0_j)-\omega)t-\Gamma_0\tau]+$$
$$+\mathbf{E}^*.\mathbf{M}(^0_j)\exp[i(\pm\omega(^0_j)+\omega)t-\Gamma_0\tau]\}-$$
$$-A_{\pm j}\exp[i(\pm\omega(^0_j)-\omega)t-\Gamma_0\tau]\gamma_{\pm j}(\pm\omega(^0_j)-\omega)+$$
$$+A_{\pm j}\exp[i(\pm\omega(^0_j)-\omega)t_0-\Gamma_{\pm j}\tau]\gamma_{\pm j}(0)-$$
$$-B_{\pm j}\exp[i(\pm\omega(^0_j)+\omega)t-\Gamma_0\tau]\gamma_{\pm j}(\pm\omega(^0_j)+\omega)+$$
$$+B_{\pm j}\exp[i(\pm\omega(^0_j)+\omega)t_0-\Gamma_{\pm j}\tau]\gamma_{\pm j}(0).\quad(46.33)$$

In the equation appear time-dependent factors of the following types:

$$\exp[i(\pm\omega(^0_j)-\omega)t-\Gamma_0\tau], \quad \exp[i(\pm\omega(^0_j)+\omega)t-\Gamma_0\tau], \quad \exp[-\Gamma_{\pm j}\tau].$$

The equation can be satisfied by collecting together terms with the same time-dependent factor and equating the corresponding coefficients to zero. Thus we obtain the following values for the various constants:

$$A_{\pm j} = -\frac{1}{\hbar}\frac{\sqrt{N}\,\mathbf{M}(^0_j).\mathbf{E}\langle\pm j|Q(^0_j)|0\rangle}{\pm\omega(^0_j)-\omega+i[\Gamma_0-\gamma_{\pm j}(\pm\omega(^0_j)-\omega)]}, \qquad (46.34)$$

$$B_{\pm j} = -\frac{1}{\hbar}\frac{\sqrt{N}\,\mathbf{M}(^0_j).\mathbf{E}^*\langle\pm j|Q(^0_j)|0\rangle}{\pm\omega(^0_j)+\omega+i[\Gamma_0-\gamma_{\pm j}(\pm\omega(^0_j)+\omega)]}, \qquad (46.35)$$

and
$$\Gamma_{\pm j} = \gamma_{\pm j}(0). \qquad (46.36)$$

These constants completely determine the expression (46.28) for the coefficients $a_{\pm j}$. $A_{\pm j}$, $B_{\pm j}$, hence also the coefficients $a_{\pm j}$, are linear in the electric field.

Now the induced electric moment due to the dispersion oscillator (^0_j) is the expectation value of the electric moment operator

$$\sqrt{N}\,\mathbf{M}(^0_j)Q(^0_j).$$

Forming the expectation value of this operator using the perturbed wave function $\Phi = \sum a_n\psi_n$ and retaining the non-vanishing terms of the lowest order only, we obtain for the induced electric moment due to the dispersion oscillator (^0_j) the expression

$$\sqrt{N}\,\mathbf{M}(^0_j)\{\langle 0|Q(^0_j)|+j\rangle a_0^* a_{+j}e^{-i\omega(^0_j)t}+\langle 0|Q(^0_j)|-j\rangle a_0^* a_{-j}e^{i\omega(^0_j)t}+$$
$$+\text{complex conjugate}\}. \quad (46.37)$$

Dividing the expression by the volume Nv_a and using the expressions (46.23), (46.28) for a_0 and $a_{\pm j}$ respectively, we get the dielectric polarization, which may be written as follows:

$$\frac{1}{v_a\sqrt{N}}\mathbf{M}(^0_j)\{A_{+j}\langle 0|Q(^0_j)|+j\rangle e^{-i\omega t}[\exp(-2\Gamma_0\tau)-$$
$$-\exp(-i(\omega(^0_j)-\omega)\tau-(\Gamma_0+\Gamma_{+j})\tau)]+$$
$$+B_{+j}\langle 0|Q(^0_j)|+j\rangle e^{i\omega t}[\exp(-2\Gamma_0\tau)-\exp(-i(\omega(^0_j)+\omega)\tau-(\Gamma_0+\Gamma_{+j})\tau)]+$$
$$+A_{-j}\langle 0|Q(^0_j)|-j\rangle e^{-i\omega t}[\exp(-2\Gamma_0\tau)-\exp(+i(\omega(^0_j)+\omega)\tau-(\Gamma_0+\Gamma_{-j})\tau)]+$$
$$+B_{-j}\langle 0|Q(^0_j)|-j\rangle e^{i\omega t}[\exp(-2\Gamma_0\tau)-\exp(+i(\omega(^0_j)-\omega)\tau-(\Gamma_0+\Gamma_{-j})\tau)]+$$
$$+\text{complex conjugate}\}, \quad (46.38)$$

where we have expressed t_0 as $t-\tau$. Owing to the factors given in the square brackets, the above expression, in its time dependence, is out of phase with the electric field (46.2). This is indeed what we should expect, for we have assumed an initial state which is, so to speak, created at an arbitrary time t_0. In fact (46.38) is of the nature of a pulse starting at $\tau = 0$ (i.e. $t = t_0$) and afterwards decaying with increasing values of τ. This is due to the fact that the initial state has only a lifetime of $1/(2\Gamma_0)$ (see 46.19). In order, therefore, to obtain the actual dielectric polarization at time t, we must integrate (46.38) with respect to t_0 from $-\infty$ to t and divide by

$$\int_{-\infty}^{t} e^{-2\Gamma_0(t-t_0)}\, dt_0.$$

This can be interpreted by assuming that the initial state is being continuously recreated $2\Gamma_0\, dt_0$ times in the interval from t_0 to t_0+dt_0. In other words, in order to obtain the actual dielectric polarization, we have to multiply (46.38) by $2\Gamma_0\, dt_0$ and integrate with respect to t_0 from $-\infty$ to the time of observation t. Hence taking τ as the integration variable and carrying out the integration from $\tau = 0$ to ∞, we obtain for the dielectric polarization $\mathbf{P}^j(t)$ due to the oscillator $\binom{0}{j}$ the following formula:

$$\mathbf{P}^j(t) = \frac{1}{v_a\sqrt{N}}\, \mathbf{M}\binom{0}{j}\Bigg\{ A_{+j}\langle 0|Q\binom{0}{j}|+j\rangle e^{-i\omega t}\left[1 + \frac{2\Gamma_0}{-i(\omega\binom{0}{j}-\omega)-(\Gamma_0+\Gamma_{+j})} \right] +$$

$$+ B_{+j}\langle 0|Q\binom{0}{j}|+j\rangle e^{i\omega t}\left[1 + \frac{2\Gamma_0}{-i(\omega\binom{0}{j}+\omega)-(\Gamma_0+\Gamma_{+j})} \right] +$$

$$+ A_{-j}\langle 0|Q\binom{0}{j}|-j\rangle e^{-i\omega t}\left[1 + \frac{2\Gamma_0}{i(\omega\binom{0}{j}+\omega)-(\Gamma_0+\Gamma_{-j})} \right] +$$

$$+ B_{-j}\langle 0|Q\binom{0}{j}|-j\rangle e^{i\omega t}\left[1 + \frac{2\Gamma_0}{i(\omega\binom{0}{j}-\omega)-(\Gamma_0+\Gamma_{-j})} \right] \Bigg\} +$$

$$+\text{complex conjugate.} \qquad (46.39)$$

When the values for the constants $A_{\pm j}$, $B_{\pm j}$, Γ_0, $\Gamma_{\pm j}$ given respectively by (46.34), (46.35), (46.23), (46.36) are substituted, we find after grouping the terms according to the time factors $\exp(\pm i\omega t)$ that (46.39) can be written

$$P_\alpha^j(t) = \sum_\beta a_{\alpha\beta}^j(\omega)E_\beta\, e^{-i\omega t}+\text{complex conjugate,} \qquad (46.40)$$

where

$$a^j_{\alpha\beta}(\omega) = -\frac{1}{v_a} M_\alpha(^0_j) M_\beta(^0_j) \frac{1}{2\omega(^0_j)} \left\{ \left[\frac{v^0(^0_j)+1}{\omega(^0_j)-\omega+i[\gamma_0(0)-\gamma_{+j}(\omega(^0_j)-\omega)]} \right] \times \right.$$

$$\times \left[1 + \frac{2\gamma_0(0)}{-i(\omega(^0_j)-\omega)-(\gamma_0(0)+\gamma_{+j}(0))} \right] +$$

$$+ \left[\frac{v^0(^0_j)}{-\omega(^0_j)-\omega+i[\gamma_0(0)-\gamma_{-j}(-\omega(^0_j)-\omega)]} \right] \times$$

$$\times \left[1 + \frac{2\gamma_0(0)}{i(\omega(^0_j)+\omega)-(\gamma_0(0)+\gamma_{-j}(0))} \right] +$$

$$+ \left[\frac{v^0(^0_j)+1}{\omega(^0_j)+\omega-i[\gamma_0(0)-\gamma_{+j}(\omega(^0_j)+\omega)]} \right] \times$$

$$\times \left[1 + \frac{2\gamma_0(0)}{i(\omega(^0_j)+\omega)-(\gamma_0(0)+\gamma_{+j}(0))} \right] +$$

$$+ \left[\frac{v^0(^0_j)}{-\omega(^0_j)+\omega-i[\gamma_0(0)-\gamma_{j}(-\omega(^0_j)+\omega)]} \right] \times$$

$$\times \left. \left[1 + \frac{2\gamma_0(0)}{-i(\omega(^0_j)-\omega)-(\gamma_0(0)+\gamma_{-j}(0))} \right] \right\}, \qquad (46.41)$$

where we have used the following values of the matrix elements of $Q(^0_j)$:

$$|\langle \pm j | Q(^0_j) | 0 \rangle|^2 = \left. \begin{matrix} v^0(^0_j)+1 \\ v^0(^0_j) \end{matrix} \right\} \left(\frac{\hbar}{2\omega(^0_j)} \right).$$

$a^j_{\alpha\beta}(\omega)$ represents the partial dielectric susceptibility due to the dispersion oscillator (^0_j); the total susceptibility tensor is obtained by summing $a^j_{\alpha\beta}(\omega)$ over all dispersion oscillators.

A discussion of the susceptibility tensor will be given in the next section. We shall now return to consider the extensively used formula (46.16). Let us suppose that the constant Γ in the integral is in general complex,

$$\Gamma = \gamma + i\delta \quad (\gamma > 0). \qquad (46.42)$$

The integral

$$\int \left\{ \frac{1-\exp[\Gamma t + i\omega t]}{-i\omega - \Gamma} \right\} \gamma_0(\omega) \, d\omega = \int \left\{ \frac{1-\exp[\gamma t + i(\omega+\delta)t]}{-i(\omega+\delta)-\gamma} \right\} \gamma_0(\omega) \, d\omega$$

$$(46.43)$$

can be written as the sum of the following three integrals:

(i) $\quad i \int \frac{(\omega+\delta)}{(\omega+\delta)^2+\gamma^2} \gamma_0(\omega) \, d\omega, \qquad$ (ii) $\quad - \int \frac{\gamma}{(\omega+\delta)^2+\gamma^2} \gamma_0(\omega) \, d\omega,$

(iii) $\quad \int \frac{\exp[\gamma t + i(\omega+\delta)t]}{i(\omega+\delta)+\gamma} \gamma_0(\omega) \, d\omega.$

Our investigation of the integral (46.43) has two aspects: (a) how far and under what circumstances is the integral independent of the parameter t? (b) under these circumstances, what is the value of the integral? The answer to (a) will give the time limits within which a Weisskopf–Wigner type of solution exists; the answer to (b) determines the damping constant in the solution.

As the integrals (i) and (ii) are independent of t, we need only consider the integral (iii) in connexion with the question (a). Let us suppose that within frequency intervals of the order $2\epsilon_0$ the variation of $\gamma_0(\omega)$ is inappreciable and consider the integral

$$I(\epsilon) = \int_{-\delta-\epsilon}^{-\delta+\epsilon} \frac{\exp[\gamma t + i(\omega+\delta)t]}{i(\omega+\delta)+\gamma} \, d\omega$$

$$= \int_{-\epsilon}^{\epsilon} \frac{\exp[\gamma t + i\omega't]}{i\omega'+\gamma} \, d\omega' \quad (\omega' = \omega+\delta). \quad (46.44)$$

If we should find that, for certain values of t, $I(\epsilon)$ has practically the same value for all values of $\epsilon > \epsilon_0$, we can conclude that the main contribution to (iii) is due to the frequency interval from $\omega = -\delta-\epsilon_0$ to $-\delta+\epsilon_0$. In that case, we have for (iii)

$$\int \frac{\exp[\gamma t + i(\omega+\delta)t]}{i(\omega+\delta)+\gamma} \gamma_0(\omega) \, d\omega \cong \gamma_0(-\delta)I(\epsilon_0). \quad (46.45)$$

In order to see for what values of t the above situation is realized, let us make the substitution

$$\frac{\exp[\gamma t + i\omega't]}{i\omega'+\gamma} = \int_{-\infty}^{t} \exp[\gamma t' + i\omega't'] \, dt' \quad (46.46)$$

in (46.44). After carrying out the integration over ω', we get

$$I(\epsilon) = 2 \int_{-\infty}^{t} e^{\gamma t'} \left(\frac{\sin \epsilon t'}{t'}\right) dt'. \quad (46.47)$$

If the factor $e^{\gamma t'}$ is ignored, the main contribution to the integral comes from a range of t' comparable with $2\pi/\epsilon$; i.e. the integral is then largely independent of t for all values $\epsilon > \epsilon_0$, if

$$t \gg \frac{2\pi}{\epsilon_0}. \quad (46.48)$$

Moreover a loop of the function $(\sin \epsilon t')/t'$ contributes roughly $2\pi/\epsilon t$ as much as the central loop at $t' = 0$; therefore the factor $e^{\gamma t'}$ in (46.47) can no longer be ignored if $e^{\gamma t}(2\pi/\epsilon t)$ is comparable with unity. Thus, in order that $I(\epsilon)$ should be independent of $\epsilon > \epsilon_0$, we have an upper limit for t given by

$$e^{\gamma t} \ll (\epsilon_0 t/2\pi). \tag{46.49}$$

It follows from (46.48) and (46.49) that the larger ϵ_0, the wider is the range of t for which $I(\epsilon)$ is largely independent of values of $\epsilon > \epsilon_0$. Within this range we can write (46.47) approximately as

$$2 \int_{-\infty}^{\infty} \frac{\sin \epsilon t'}{t'} dt' = 2\pi \tag{46.50}$$

(ignore $e^{\gamma t'}$ and let $t \to \infty$), so that the integral (iii) becomes (see (46.45))

$$2\pi\gamma_0(-\delta). \tag{46.51}$$

A Weisskopf–Wigner type of solution is physically significant only if the solution holds for the greater part of the lifetime. That means that the above limits for t must allow $e^{\gamma t}$ to be considerably larger than unity. It then follows from (46.49) that ϵ_0 must be considerably larger than γ. Under such circumstances it is clear from the integral

$$\int_{-\delta-\epsilon_0}^{-\delta+\epsilon_0} \frac{\gamma}{\gamma^2+(\omega+\delta)^2} d\omega = 2\tan^{-1}\frac{\epsilon_0}{\gamma} \cong \pi \tag{46.52}$$

that the integral (ii) is mainly determined by the range of ω from $-\delta-\epsilon_0$ to $-\delta+\epsilon_0$; within this range $\gamma_0(\omega)$, by assumption, does not vary appreciably. Therefore we have for (ii)

$$-\int \frac{\gamma}{\gamma^2+(\omega+\delta)^2} \gamma_0(\omega)\, d\omega \cong -\gamma_0(-\delta) \int_{-\delta-\epsilon_0}^{-\delta+\epsilon_0} \frac{\gamma}{\gamma^2+(\omega+\delta)^2}\, d\omega$$

$$\cong -\pi\gamma_0(-\delta). \tag{46.53}$$

The imaginary part (i) presents a different situation. Since the function

$$\frac{\omega+\delta}{(\omega+\delta)^2+\gamma^2}$$

is odd about the point $\omega = -\delta$, the symmetric interval $-\delta-\epsilon_0$ to $-\delta+\epsilon_0$ contributes practically nothing to the integral. Because, however, of the slower convergence of the above factor, the more remote part of the frequency range may give an appreciable contribution and no simple formula for the integral can be given in this case.

Collecting together the above results, we have thus

$$\int \frac{1-\exp[\gamma t+i(\omega+\delta)t]}{-i(\omega+\delta)-\gamma}\gamma_0(\omega)\,d\omega = \pi\gamma_0(-\delta)+i\int\left\{\frac{\omega+\delta}{(\omega+\delta)^2+\gamma^2}\right\}\gamma_0(\omega)\,d\omega.$$

$$(46.54)$$

As we have seen, this formula should be approximately valid if $\epsilon_0 \gg \gamma$ and t is within the limits given by (46.48) and (46.49); ϵ_0, we remember, is a measure of the maximum frequency interval within which a variation of $\gamma_0(\omega)$ is still negligible.

If the formula (46.54) had been used in place of the formula (46.16) in discussing the Weisskopf–Wigner problem, we should have found the complex value for the damping constant:

$$\Gamma = \gamma+i\delta, \qquad (46.55)$$

where

$$\gamma = \gamma_0(-\delta)$$

and δ is the solution of the equation

$$\delta = \frac{1}{\pi}\int\left\{\frac{\omega+\delta}{(\omega+\delta)^2+\gamma^2}\right\}\gamma_0(\omega)\,d\omega.$$

The solution (46.18) which we obtained earlier with the formula (46.16) is thus equivalent to ignoring δ. If we did not put δ equal to zero, the more important modifications of the formula (46.41) for the dielectric susceptibility would be the presence of certain shifts in the frequencies appearing in the various denominators and in the arguments of the damping constants $\gamma_0(\omega)$, $\gamma_\pm(\omega)$.

Mathematically speaking, Born and Blackman's classical theory[†] bears a close resemblance to the above treatment. One simplification made in their work is equivalent to putting $a_0 = 1$ in integrating the equation (46.7) for $n \neq 0$. A solution obtained in this way holds for a much smaller part of the lifetime than a Weisskopf–Wigner type of solution. The effect of this simplification is roughly equivalent to putting Γ_0, $\Gamma_{\pm j}$ equal to zero; this makes a significant difference in the result only at comparatively high temperatures, where Γ_0, $\Gamma_{\pm j}$ are appreciable and cannot be ignored (see also § 47).

47. The dispersion formula with damping

Let us first consider for a moment the nature of the functions $\gamma_0(\omega)$, $\gamma_{\pm j}(\omega)$, assuming for this purpose a third-order anharmonic potential. Retaining only the third-order terms in (46.5) and expressing the complex normal coordinates in terms of the auxiliary variables a_+, a_- (see

† M. Born and M. Blackman, loc. cit.

(38.38)), we have

$$\Phi_j^A = \frac{1}{2\sqrt{N}}\left(a_+\binom{0}{j}+a_-\binom{0}{j}\right)\sum_{y'}^{N}\sum_{j'}\sum_{j''}\Phi\left(\begin{smallmatrix}0&y'&-y'\\j&j'&j''\end{smallmatrix}\right)\left(a_+\binom{-y'}{j'}+a_-\binom{y'}{j'}\right)\left(a_+\binom{y'}{j''}+a_-\binom{-y'}{j''}\right)$$

$$= \frac{1}{2\sqrt{N}}\left(a_+\binom{0}{j}+a_-\binom{0}{j}\right)\sum_{y'}^{N}\sum_{j'}\sum_{j''}\Phi\left(\begin{smallmatrix}0&y'&-y'\\j&j'&j''\end{smallmatrix}\right)\times$$

$$\times\{a_+\binom{-y'}{j'}a_+\binom{y'}{j''}+a_-\binom{y'}{j'}a_+\binom{y'}{j''}+a_+\binom{-y'}{j'}a_-\binom{-y'}{j''}+a_-\binom{y'}{j'}a_-\binom{-y'}{j''}\},\quad(47.1)$$

where, owing to the factor $\Delta(y'+y'')$ (see (38.13)) in (46.5), we have only a single summation over y'.

In view of the definition of the auxiliary variables,

$$a_+ = \frac{1}{2}\left\{\frac{\dot{q}}{\omega}+iq\right\} = \frac{1}{2}\left\{\frac{p}{\omega}+iq\right\}, \qquad a_- = \frac{1}{2}\left\{\frac{\dot{q}}{\omega}-iq\right\} = \frac{1}{2}\left\{\frac{p}{\omega}-iq\right\},$$
$$(47.2)$$

we find that their only non-vanishing matrix elements [with respect to the oscillator wave functions of the real normal coordinates $q\binom{y}{j}$] are the following:

$$\langle v+1|a_+|v\rangle = i\left(\frac{\hbar}{2\omega}\right)^{\frac{1}{2}}(v+1)^{\frac{1}{2}}, \qquad \langle v-1|a_-|v\rangle = -i\left(\frac{\hbar}{2\omega}\right)^{\frac{1}{2}}v^{\frac{1}{2}}. \quad(47.3)$$

Before discussing the matrix elements of the anharmonic potential, we note that the terms in (47.1) are not all different from one another. In fact, owing to the symmetry relation (§ 39)

$$\Phi\left(\begin{smallmatrix}0&y'&-y'\\j&j'&j''\end{smallmatrix}\right) = \Phi\left(\begin{smallmatrix}0&-y'&y'\\j&j''&j'\end{smallmatrix}\right), \qquad (47.4)$$

practically all terms (except a few terms with $y' = 0$ and $j' = j''$) occur in pairs. Further, if we replace y' by $-y'$ and interchange j' and j'', the first and last terms in the curly brackets remain unaltered, while the second and third terms are changed into one another. Hence, by combining such pairs, we can rewrite (47.1) as follows:

$$\Phi_j^A = \frac{1}{\sqrt{N}}\left(a_+\binom{0}{j}+a_-\binom{0}{j}\right)\Bigg\{\sum_{j'>j''}\sum_{j''}\sum_{y'}^{N}\left[a_+\binom{-y'}{j'}a_+\binom{y'}{j''}+a_-\binom{y'}{j'}a_-\binom{-y'}{j''}\right]+$$

$$+ \sum_{j''}\sum_{y'}^{\frac{1}{2}N}\left[a_+\binom{-y'}{j'}a_+\binom{y'}{j'}+a_-\binom{y'}{j'}a_-\binom{-y'}{j'}\right]+$$

$$+ \sum_{j'}\sum_{j''}\sum_{y'}^{N}a_-\binom{y'}{j'}a_+\binom{y'}{j''}\Bigg\}\Phi\left(\begin{smallmatrix}0&y'&-y'\\j&j'&j''\end{smallmatrix}\right), \quad(47.5)$$

where the sum with $\frac{1}{2}N$ over the summation signifies, as before, that the summation selects only one of a pair of wave-numbers y' and $-y'$. In (47.5) all terms are distinct, each giving rise to a distinct transition involving three oscillators, one of these being the dispersion oscillator $\binom{0}{j}$

(with the exception of the terms with $j' = j''$ in the last sum; these terms lead to first-order transitions only and are irrelevant for anharmonic damping). The various types of transitions and the squares of the corresponding matrix elements of Φ_j^A, obtained readily with the help of (47.3), are

Quantum numbers occurring in the squares of matrix elements of Φ_j^A			Transition frequencies
$\left(v\binom{0}{j}+1\right)$	$\left(v^0\binom{-y'}{j'}+1\right)$	$\left(v^0\binom{y'}{j\cdot}+1\right)$	$-\omega\binom{0}{j} - \omega\binom{y'}{j'} - \omega\binom{y'}{j\cdot}$
$\left(v\binom{0}{j}+1\right)$	$v^0\binom{y'}{j\cdot}$	$\left(v^0\binom{y'}{j\cdot}+1\right)$	$-\omega\binom{0}{j} + \omega\binom{y'}{j'} - \omega\binom{y'}{j\cdot}$
$\left(v\binom{0}{j}+1\right)$	$v^0\binom{y'}{j\cdot}$	$v^0\binom{-y'}{j}$	$-\omega\binom{0}{j} + \omega\binom{y'}{j'} + \omega\binom{y'}{j\cdot}$
$v\binom{0}{j}$	$\left(v^0\binom{-y'}{j}+1\right)$	$\left(v^0\binom{y'}{j\cdot}+1\right)$	$\omega\binom{0}{j} - \omega\binom{y'}{j'} - \omega\binom{y'}{j\cdot}$
$v\binom{0}{j}$	$v^0\binom{y'}{j\cdot}$	$\left(v^0\binom{y'}{j\cdot}+1\right)$	$\omega\binom{0}{j} + \omega\binom{y'}{j'} - \omega\binom{y'}{j\cdot}$
$v\binom{0}{j}$	$v^0\binom{y'}{j\cdot}$	$v^0\binom{-y'}{j}$	$\omega\binom{0}{j} + \omega\binom{y'}{j'} + \omega\binom{y'}{j\cdot}$

$$(47.6)$$

where the second and fifth types describe genuine third-order transitions for $j' \neq j$. The squares of the matrix elements can be used directly to calculate the functions $\gamma_0(\omega)$, $\gamma_{\pm j}(\omega)$, where for the three cases, the quantum number $v\binom{0}{j}$ in (47.6) should be given the values $v^0\binom{0}{j}$, $v^0\binom{0}{j}+1$, and $v^0\binom{0}{j}-1$ respectively.

We observe that, despite the appearance of N in the matrix elements, the functions $\gamma_0(\omega)$, $\gamma_{\pm j}(\omega)$ are independent of N. In constructing these functions, we can consider the contributions due to the six types of transition separately. In order, for instance, to calculate the contribution due to the first type of transition, we have to consider the sum of

$$\frac{\hbar^3 |\Phi\binom{0\ y'\ -y'}{j\ j'\ j\cdot}|^2}{8N\omega\binom{0}{j}\omega\binom{y'}{j'}\omega\binom{y'}{j\cdot}} \left(v\binom{0}{j}+1\right)\left(v^0\binom{-y'}{j}+1\right)\left(v^0\binom{y'}{j\cdot}+1\right) \qquad (47.7)$$

over all values of the wave-number $\mathbf{y'}$ such that $-\omega\binom{0}{j} - \omega\binom{y'}{j'} - \omega\binom{y'}{j\cdot}$ lies between ω and $\omega+\Delta\omega$. Since the density of allowed values of $\mathbf{y'}$ is proportional to N we find upon forming the sum a result independent of N. This conclusion, as can be easily verified, holds equally for anharmonic terms of any order.

Let us discuss the dispersion formula (46.41) for a few special cases:

(i) *The low-temperature limit* $T = 0°$

For this case, $v^0\binom{0}{j} = 0$; thus the second and the fourth terms in (46.41) drop out. Moreover, the damping constants $\gamma_0(0)$ and $\gamma_{\pm j}(\omega+\omega\binom{0}{j})$ are equal to zero. We can see this readily as follows: $\gamma_0(0)$ is determined by transitions with practically vanishing transition frequencies, i.e. transitions involving practically no energy change. If such transitions were possible, as the oscillator $\binom{0}{j}$, being in its lowest state, must necessarily go up, at least one other oscillator would have to go down in energy. This being impossible (all oscillators being already in their lowest state), it follows that no such transition can occur and $\gamma_0(0)$ in consequence vanishes; clearly the value of $\gamma_0(0)$ will only become appreciable when the temperature is high enough to raise a considerable fraction of the oscillators to their excited states. A similar argument applies to the case of $\gamma_{+j}(\omega+\omega\binom{0}{j})$ which is due to transitions leading to states with an energy $\hbar(\omega+\omega\binom{0}{j})$ below the state $+j$, i.e. an energy $\hbar\omega$ below the initial state. Since the initial state is the lowest state of the system in the present case, such transitions do not exist and we have thus to put $\gamma_{+j}(\omega+\omega\binom{0}{j})$ equal to zero [$\gamma_{-j}(\omega+\omega\binom{0}{j})$ is obviously zero].

Putting $v^0\binom{0}{j}$, $\gamma_0(0)$, and $\gamma_{+j}(\omega+\omega\binom{0}{j})$ equal to zero in (46.41), we find that in this case the dispersion formula reduces to

$$a^j_{\alpha\beta} = -\frac{1}{v_a} M_\alpha\binom{0}{j} M_\beta\binom{0}{j} \frac{1}{2\omega\binom{0}{j}} \left\{ \frac{1}{\omega\binom{0}{j}-\omega-i\gamma_{+j}(\omega\binom{0}{j}-\omega)} + \frac{1}{\omega\binom{0}{j}+\omega} \right\}.$$

$$(47.8)$$

The damping constants are in general small compared with $\omega\binom{0}{j}$ (j refers to an optical branch; note that $\mathbf{M}\binom{0}{j} = 0$ for the acoustic branches). In considering the absorption region, where the damping constant is comparable with $|\omega\binom{0}{j}-\omega|$ or larger, we can ignore the second term in (47.8). Moreover, in this region $2\omega\binom{0}{j} \sim \omega\binom{0}{j}+\omega \sim 2\omega$; we can thus write (47.8) approximately as follows:

$$a^j_{\alpha\beta}(\omega) \cong -\frac{M_\alpha\binom{0}{j} M_\beta\binom{0}{j}}{v_a} \left\{ \frac{1}{\omega^2\binom{0}{j}-\omega^2-i\omega\binom{0}{j}[2\gamma_{+j}(\omega\binom{0}{j}-\omega)]} \right\}.$$

We observe that the formula is similar in form to the elementary formula (10.6) with, however, the important difference that the constant γ in (10.6) is here replaced by $2\gamma_{+j}(\omega\binom{0}{j}-\omega)$, which is a function of the frequency ω. Outside the absorption region we can neglect $\gamma_{+j}(\omega\binom{0}{j}-\omega)$ in (47.8) and the damping constant γ in the elementary formula (10.6); the two formulae then become identical. Our conclusion is thus similar to that reached by Born and Blackman, namely, that the only important difference of the theoretical formula from the formula (10.6) is the replacement of the constant γ by a function of frequency.

If we assume that only the third-order anharmonic terms are significant, there must then exist an upper limit for ω, above which the function

$\gamma_{+j}(\omega(^0_j)-\omega)$ vanishes. For if ω_{max} represents the highest lattice vibrational frequency, it is obvious from (47.6) that the algebraically lowest transition frequency is

$$-\omega(^0_j)-2\omega_{max}.$$

Therefore $$\gamma_{+j}(\omega(^0_j)-\omega) = 0$$

if ω is so high that $\omega(^0_j)-\omega$ falls below $-\omega(^0_j)-2\omega_{max}$, i.e. if

$$\omega > 2(\omega(^0_j)+\omega_{max}). \qquad (47.9)$$

One thus expects that, starting from the dispersion frequency $\omega(^0_j)$, as ω is raised, the absorption should show a conspicuous drop as the limiting value $2(\omega(^0_j)+\omega_{max})$ is reached. In the case of NaCl, ω_{max} is equal to ω_l, the frequency of the long longitudinal vibrations in the optical branch. Using the values $\omega_l = 4\cdot9 \times 10^{13}$/sec., $\omega(^0_j) = 3\cdot1 \times 10^{13}$/sec. (see § 7), we find that the limiting frequency corresponds to a wave-length about $12\,\mu$. Czerny† has actually observed that the absorption in NaCl begins to drop conspicuously from about $20\,\mu$ downwards, ending in negligible values in the neighbourhood of $12\,\mu$.

(ii) *Wings of the absorption region $(\omega-\omega(^0_j) >$ the damping constants)*

Except at very low temperatures, the various damping constants in the dispersion formula (46.41) should be of roughly the same order of magnitude. The value of the damping constant in the formula (10.6) has been estimated by Czerny to be 1/20 of the dispersion frequency in the case of NaCl. Therefore there is a frequency region where the absorption is still appreciable, yet terms of higher powers in $\gamma/[\omega-\omega(^0_j)]$ can be approximately ignored. In this approximation, one finds that the dispersion formula (46.41) can be written as follows:

$$a^j_{\alpha\beta}(\omega) \cong -\frac{1}{v_a} M_\alpha(^0_j)M_\beta(^0_j)\Big\{\frac{1}{\omega^2(^0_j)-\omega^2}+$$

$$+i\Big[\frac{2\omega\gamma_0(0)}{(\omega^2(^0_j)-\omega^2)}+\frac{\gamma_{+j}(\omega(^0_j)-\omega)}{(\omega(^0_j)-\omega)^2}-\frac{\gamma_{+j}(\omega(^0_j)+\omega)}{(\omega(^0_j)+\omega)^2}\Big]+$$

$$+iv^0(^0_j)\Big[\frac{1}{(\omega(^0_j)-\omega)^2}\Big(\gamma_{+j}(\omega(^0_j)-\omega)-\gamma_{-j}(-\omega(^0_j)+\omega)\Big)+$$

$$+\frac{1}{(\omega(^0_j)+\omega)^2}\Big(\gamma_{-j}(-\omega(^0_j)-\omega)-\gamma_{+j}(\omega(^0_j)+\omega)\Big)\Big]\Big\}. \qquad (47.10)$$

† loc. cit., p. 118.

The real part of the expression is the same as the dispersion formula based upon the harmonic approximation and is independent of the temperature. The imaginary terms are temperature-dependent, owing to the damping constants. If one considers only the third-order anharmonic terms, the damping constants are built up from terms, each containing the product of three oscillator quantum numbers (see (47.6)). At high temperatures, the thermal average of an oscillator quantum number is roughly proportional to T. Therefore the damping constants are approximately proportional to T^3. It follows that at high temperatures the first group of imaginary terms vary roughly as T^3. The remaining imaginary terms in (47.10) contain an extra factor $v^0(^0_j)$; it is probable, however, that these terms, taken together, do not vary any more rapidly than T^3, for the differences between two damping constants probably do not vary with temperature so rapidly as the damping constants (see below).

(iii) *Centre of the dispersion region*

Upon putting $\omega = \omega(^0_j)$ in (46.41) and neglecting powers of the ratios of the damping constants to $\omega(^0_j)$ we find, after some simplification, that

$$a^j_{\alpha\beta}(\omega(^0_j)) = \frac{1}{v_a} M_\alpha(^0_j) M_\beta(^0_j) \frac{1}{2\omega(^0_j)} \left\{ \frac{1}{2\omega(^0_j)} + \right.$$

$$\left. +i\left[\frac{1}{\gamma_0(0)+\gamma_{+j}(0)} + \frac{v^0(^0_j)(\gamma_{-j}(0)-\gamma_{+j}(0))}{(\gamma_0(0)+\gamma_{+j}(0))(\gamma_0(0)+\gamma_{-j}(0))} \right] \right\}. \quad (47.11)$$

It follows from the same argument as used above that the first term given in the square brackets should decrease with rising temperature as T^{-3}. The second term in the square brackets contains $\gamma_{-j}(0)-\gamma_{+j}(0)$ as a factor. One readily sees that the summations involved in constructing $\gamma_{-j}(0)$ and $\gamma_{+j}(0)$ are identical, the only difference being the different values given to the quantum number $v(^0_j)$ in the matrix elements (see (47.6)). Thus if we represent separately the contributions due respectively to the first three and the last three types of transitions listed in (47.6), we can write $\gamma_{-j}(0)$ and $\gamma_{+j}(0)$

$$\gamma_{-j}(0) = v^0(^0_j)A + (v^0(^0_j)-1)B, \qquad \gamma_{+j}(0) = (v^0(^0_j)+2)A + (v^0(^0_j)+1)B,$$

$$\quad (47.12)$$

where A and B are sums of terms, each containing a product of two oscillator quantum numbers. The difference between the two damping constants

$$\gamma_{-j}(0)-\gamma_{+j}(0) = -2(A+B) \qquad (47.13)$$

clearly varies as T^2 at high temperatures. Therefore the last term in (47.11), like the preceding term, varies with temperature as T^{-3}.

The cases (ii) and (iii) taken together thus indicate that the absorption region broadens with rising temperature, with the wings growing at the expense of the centre.

No detailed comparison between the theoretical formula and the experimental results is as yet available. In the following, we shall attempt to obtain a very crude estimate of the order of magnitude of the damping constants.

Let us first fix our attention on one of the six types of transitions described in (47.6). To obtain the corresponding contribution to a particular damping constant (e.g. $\gamma_0(\omega)$, $\gamma_{\pm j}(\omega)$) at a specific value of its argument ω, one should, strictly speaking, consider all transitions with transition frequencies between ω and $\omega + \Delta\omega$ and calculate the ratio of the sum of the squares of the transition matrix elements of Φ_j^A to $\Delta\omega$. To obtain a rough estimate, without regard to the argument of the damping constant, we shall sum the squares of the matrix elements over all possible transitions of the given type (i.e. over all permissible values of the wave-number $\mathbf{y}'$) and afterwards divide the sum by the total range of transition frequency thus traversed. This latter range should be roughly of the order of the mean lattice vibration frequency which we denote by ω. Following this procedure, we find, after summing over all possible types of transitions, that γ can be written in the form

$$\gamma = C\left(\frac{\pi}{\hbar^2}\right)\frac{1}{\omega}\left(\frac{\hbar^3\bar{v}^3}{8N\bar{\omega}^3}\right)\sum_{j'}\sum_{j''}\sum_{\mathbf{y}'}^{N}|\Phi(\begin{smallmatrix}0\\j\end{smallmatrix}\begin{smallmatrix}\mathbf{y}'\\j'\end{smallmatrix}\ \begin{smallmatrix}-\mathbf{y}'\\j''\end{smallmatrix})|^2, \qquad (47.14)$$

where we have replaced the frequencies $\omega(\begin{smallmatrix}0\\j\end{smallmatrix})$, $\omega(\begin{smallmatrix}\mathbf{y}'\\j'\end{smallmatrix})$, $\omega(\begin{smallmatrix}\mathbf{y}'\\j''\end{smallmatrix})$ by $\bar{\omega}$ and the quantum numbers by a single mean quantum number $\bar{v}$. In (47.14) C is a constant of the order of magnitude unity (perhaps 1 to 2). Although, in arriving at this formula, we have not taken account of details it should, nevertheless, reproduce roughly the order of magnitude of the damping constants. Expression (47.14) refers to the damping constants related to a particular dispersion oscillator $(\begin{smallmatrix}0\\j\end{smallmatrix})$. We shall take a further average by summing (47.14) over j and dividing the result by $3n$, the total number of dispersion oscillators. Thus we obtain finally (putting $\omega \sim \bar{\omega}$)

$$\gamma = \frac{Ch}{48Nn}\frac{\bar{v}^3}{\bar{\omega}^4}\sum_{j}\sum_{j'}\sum_{j''}\sum_{\mathbf{y}'}^{N}|\Phi(\begin{smallmatrix}0\\j\end{smallmatrix}\begin{smallmatrix}\mathbf{y}'\\j'\end{smallmatrix}\ \begin{smallmatrix}-\mathbf{y}'\\j''\end{smallmatrix})|^2. \qquad (47.15)$$

Let us express the expansion coefficients of Φ in terms of the derivative of Φ in two steps (cf. § 39):

$$\Phi\begin{pmatrix} 0 & y' & \bar{}y' \\ j & j & j \end{pmatrix} = \sum_{k\alpha} \sum_{k'\beta} \sum_{k'\gamma} e_\alpha(k|\begin{smallmatrix} 0 \\ j \end{smallmatrix}) e_\beta(k'|\begin{smallmatrix} y' \\ j \end{smallmatrix}) e_\gamma(k''|\begin{smallmatrix} \bar{}y' \\ j \end{smallmatrix}) \Phi_{\alpha\beta\gamma}\begin{pmatrix} 0 & y' & \bar{}y' \\ k & k' & k'' \end{pmatrix}, \quad (47.16)$$

where

$$\Phi_{\alpha\beta\gamma}\begin{pmatrix} 0 & y' & \bar{}y' \\ k & k' & k'' \end{pmatrix} = \frac{1}{(m_k m_{k'} m_{k''})^{\frac{1}{2}}} \sum_{l'l''} \Phi_{\alpha\beta\gamma}\begin{pmatrix} 0 & l' & l'' \\ k & k' & k'' \end{pmatrix} \exp\{2\pi i y' . (\mathbf{x}(l') - \mathbf{x}(l''))\}.$$
$$(47.17)$$

Upon substituting (47.16) in (47.15) and using the orthonormal relations (38.25), we get

$$\gamma = \frac{Ch}{48Nn} \frac{\bar{v}^3}{\bar{\omega}^4} \sum_{k\alpha} \sum_{k'\beta} \sum_{k'\gamma} \sum_{y'}^{N} |\Phi_{\alpha\beta\gamma}\begin{pmatrix} 0 & y' & \bar{}y' \\ k & k' & k'' \end{pmatrix}|^2. \quad (47.18)$$

By substituting for $\Phi_{\alpha\beta\gamma}\begin{pmatrix} 0 & y' & \bar{}y' \\ k & k' & k'' \end{pmatrix}$ in (47.18) from (47.17) and using the same argument as in discussing the function Δy in § 38 (see particularly pp. 294–6), we can reduce (47.18) to the form

$$\gamma = \frac{Ch}{48Nn} \frac{\bar{v}^3}{\bar{\omega}^4} \sum_{k\alpha} \sum_{k'\beta} \sum_{k'\gamma} \frac{1}{(m_k m_{k'} m_{k''})} \sum_{l'l''} \sum_{\substack{l'l''' \\ (l-l'=l''-l''')}} \Phi_{\alpha\beta\gamma}\begin{pmatrix} 0 & l' & l'' \\ k & k' & k'' \end{pmatrix} \Phi_{\alpha\beta\gamma}\begin{pmatrix} 0 & l & l' \\ k & k' & k'' \end{pmatrix},$$
$$(47.19)$$

where, as indicated, the summations over l, l', l'', l''' are restricted to the values fulfilling the condition that $l-l' = l''-l'''$.

Let us obtain a rough numerical estimate for the NaCl lattice by considering only the overlap energy between nearest neighbours. If the overlap energy is represented as a function $\psi(r^2)$ of the square of the distance between neighbours, one finds after some tedious work that (47.19) reduces in this case to the form

$$\gamma = C \frac{2r_0^2 h \bar{v}^3}{\bar{\omega}^4} \left(\frac{1}{M_+^2 M_-} + \frac{1}{M_-^2 M_+} \right) \{6[\psi''(r_0^2)]^2 + [2r_0^2 \psi'''(r_0^2) + 3\psi''(r_0^2)]^2\},$$
$$(47.20)$$

where M_+ and M_- are respectively the masses of the Na$^+$ and Cl$^-$ ions and r_0 is the nearest neighbour distance. For ψ we shall use the exponential form

$$\psi(r^2) = \phi(r) = \lambda_{+-} \exp[-r/\rho], \quad (47.21)$$

where the values of the constants are given in Table 9 of § 3. In the case of NaCl, the dispersion frequency $3 \cdot 09 \times 10^{13}$/sec. should give a fair estimate of the mean lattice frequency. Thus, using this value for $\bar{\omega}$, we find as the result of the numerical calculation

$$\frac{\gamma}{\omega} = C \times 0.043\bar{v}^3 \qquad (47.22)$$

as compared with the value 0.045 for γ/ω, estimated by Czerny by adapting the elementary formula (10.6) to the experimental observations. Since at room temperatures $\bar{v}$ is of the order unity the agreement between the two values is satisfactory. (That C turns out to be almost exactly unity is accidental.)

48. The effect of the second-order electric moment

The consideration of the dispersion formula has thus far been based upon the assumption that the electric moment of a crystal is a linear function of the nuclear displacements. In general, however, the electric moment must be considered as a general Taylor's series in the nuclear displacements consisting of terms of all orders. In an ionic crystal the ionic charges contribute directly only to the linear terms (§ 21); the higher-order terms are due to the distortion of the electronic cloud by the displacements of the nuclei. We shall see that in a crystal the higher-order terms give rise to effective transitions to states which cover a continuous range of energy values. Such transitions should evidently lead to continuous absorption in contrast to the line absorption due to the linear electric moment. However, as in ionic crystals the linear terms will be abnormally large compared with the higher-order terms, we expect that the effect of the higher-order moments could be observable only in regions away from the primary dispersion frequencies $\omega(^0_j)$. In other words, the higher-order electric moments cannot be primarily responsible for the observed absorption in the close neighbourhood of the dispersion frequencies $\omega(^0_j)$. Moreover, unlike the anharmonic potential, the higher-order moments do not remove the infinite sharpness of the absorption lines at $\omega(^0_j)$.

A significant instance bearing on the problem has been pointed out by Burstein, Oberley, and Pyler.† They observed that the MgO crystal, as indicated by its various properties, deviates more markedly from the ideal ionic structure than the alkali halides. For instance, one expects in MgO a more extensive overlap between neighbouring ions, or a more pronounced tendency for the electrons on the negative ions to form homopolar bonds. Consequently, there should be a stronger distortion of the electronic cloud when the nuclei are displaced. The above authors

† E. Burstein, J. J. Oberley, and E. K. Pyler, *Proc. Ind. Acad. Sci.* **28**, 388 (1948).

have actually found that, as compared with the alkali halides, the MgO crystal gives a more extended absorption region. They have further pointed out that MgO has a higher Debye temperature and a lower coefficient of expansion; both features are such as to indicate a relatively small anharmonic effect. From these observations, it would appear that the higher-order electric moments are responsible for the wider extensions of the absorption region in the MgO crystal. It thus seems plausible to suppose that, in general, the anharmonic effect is responsible for the damping of the absorption lines at the frequencies $\omega(^0_j)$ and the absorption regions close to these frequencies, whereas the higher-order electric moments account for the observed continuous absorption further removed from the frequencies $\omega(^0_j)$. In the following, we shall use the method of local treatment to discuss the effect of the second-order electric moment. Since the consequent absorption is continuous, the effect of the anharmonic potential is less drastic. It will presumably modify to some extent the actual continuous distribution of absorption; for the sake of simplicity, however, we shall ignore the anharmonic effect in the following discussion.

To represent a local region, we consider a block of N cells as in § 38. Adding the general formulae (20.18) and (20.19) for the Hermitian and anti-Hermitian parts of the polarizability tensor respectively, and dividing by the volume Nv_a, we obtain for the complex dielectric susceptibility tensor the following expression:

$$a_{\alpha\beta}(\omega) = \frac{1}{Nv_a\hbar} \sum_{v''\neq v} \left\{ \frac{\langle v|M_\alpha|v''\rangle\langle v''|M_\beta|v\rangle}{\omega_{v''v}+\omega} + \frac{\langle v|M_\beta|v''\rangle\langle v''|M_\alpha|v\rangle}{\omega_{v''v}-\omega} + \right.$$
$$+ i\pi[\langle v|M_\alpha|v''\rangle\langle v''|M_\beta|v\rangle\delta(\omega+\omega_{v''v}) -$$
$$\left. - \langle v|M_\beta|v''\rangle\langle v''|M_\alpha|v\rangle\delta(\omega-\omega_{v''v})] \right\}. \quad (48.1)$$

For the discussion of the effect of the second-order electric moment, we can replace **M** in the formula by its second-order part (see (39.11))

$$M_\alpha^{(2)} = \tfrac{1}{2} \sum_y^N \sum_j \sum_{j'} M_\alpha(^y_j \,^{-y}_{j'})Q(^y_j)Q(^{-y}_{j'}), \quad (48.2)$$

where the coefficients are as defined in (39.15).

We shall describe the system by the real normal coordinates of the first kind, for, as we shall find, the matrix elements are then real. Anticipating this fact, we may thus put

$$\langle v|\mathbf{M}^{(2)}|v''\rangle = \langle v|\mathbf{M}^{(2)}|v''\rangle^* = \langle v''|\mathbf{M}^{(2)}|v\rangle. \quad (48.3)$$

Upon subdividing the summation over y in (48.2) into two groups according to whether y lies on one side or the other of an arbitrary plane through the origin of y-space, we find that $M_\alpha^{(2)}$ can be expressed in terms of the real normal coordinates of the first kind in the following form:

$$M_\alpha^{(2)} = \tfrac{1}{4} \sum_{y}^{\frac{1}{2}N} \sum_{jj'} M_\alpha(^y_j \, ^{-y}_{j'})[q_1(^y_j)+iq_2(^y_j)][q_1(^y_{j'})-iq_2(^y_{j'})] +$$

$$+ \tfrac{1}{4} \sum_{y}^{\frac{1}{2}N} \sum_{jj'} M_\alpha(^{-y}_j \, ^y_{j'})[q_1(^y_j)-iq_2(^y_j)][q_1(^y_{j'})+iq_2(^y_{j'})], \quad (48.4)$$

where both summations over y are over values on the same side of the plane. Rearranging the terms, we get

$$M_\alpha^{(2)} = \tfrac{1}{4} \sum_{y}^{\frac{1}{2}N} \sum_{jj'} [M_\alpha(^y_j \, ^{-y}_{j'}) + M_\alpha(^{-y}_j \, ^y_{j'})][q_1(^y_j)q_1(^y_{j'}) + q_2(^y_j)q_2(^y_{j'})] -$$

$$- \frac{i}{4} \sum_{y}^{\frac{1}{2}N} \sum_{jj'} [M_\alpha(^y_j \, ^{-y}_{j'}) - M_\alpha(^{-y}_j \, ^y_{j'})][q_1(^y_j)q_2(^y_{j'}) - q_2(^y_j)q_1(^y_{j'})]. \quad (48.5)$$

It follows from the relation (see (39.17))

$$M_\alpha(^y_j \, ^{-y}_{j'}) = M_\alpha(^{-y}_{j'} \, ^y_j)$$

that all cross-product terms in (48.5) occur exactly twice with the same coefficient. This is not only true for the terms $q_1(^y_j)q_1(^y_{j'})$ and $q_2(^y_j)q_2(^y_{j'})$ $(j \neq j')$ but also true for the cross-product terms $q_1(^y_j)q_2(^y_{j'})$ and $q_2(^y_j)q_1(^y_{j'})$. The coefficients

$$\tfrac{1}{2}[M_\alpha(^y_j \, ^{-y}_{j'}) + M_\alpha(^{-y}_j \, ^y_{j'})] \quad \text{and} \quad -\tfrac{1}{2}i[M_\alpha(^y_j \, ^{-y}_{j'}) - M_\alpha(^{-y}_j \, ^y_{j'})]$$

are thus directly comparable with the second-order expansion coefficients $A_{jj'}$ and $B_{jj'}$ in the general formula (21.5). Hence (21.10) can be used directly to work out the explicit expression for (48.1). In the first place, (21.10) shows that there are six types of transitions $v \to v''$ corresponding in the present case to the transition frequencies $\omega_{vv''} = \pm 2\omega(^y_j)$, $\pm(\omega(^y_j)+\omega(^y_{j'}))$, $\pm(\omega(^y_j)-\omega(^y_{j'}))$. The summation in (48.1) is simply a summation over all possible transitions; thus it resolves in the first place to six partial sums corresponding to the six types of transitions, each partial sum being a summation over all choices for the oscillator or pair of oscillators which can participate in a transition of the specified type. (21.10), moreover, gives the thermal average values of the relevant products of matrix elements for all six types of transitions. Thus with the help of (21.10) we find after some simplification that the thermal average of the susceptibility tensor (48.1) is given explicitly by the formula

$$a_{\alpha\beta}(\omega) = -\frac{1}{Nv_a\hbar} \sum_{y}^{\frac{1}{2}N} \sum_{j} M_\alpha(\tbinom{y}{j}\tbinom{-y}{j}) M_\beta(\tbinom{y}{j}\tbinom{-y}{j}) C^2(\tbinom{y}{j})[1-e^{-2\beta(\tbinom{y}{j})}] \times$$

$$\times \left\{ \frac{4\omega(\tbinom{y}{j})}{4\omega^2(\tbinom{y}{j})-\omega^2} + i\pi[\delta(\omega+2\omega(\tbinom{y}{j}))-\delta(\omega-2\omega(\tbinom{y}{j}))] \right\} +$$

$$+ \frac{1}{Nv_a\hbar} \sum_{y}^{\frac{1}{2}N} \sum_{j>j'} \tfrac{1}{2}\{[M_\alpha(\tbinom{y}{j}\tbinom{-y}{j'})+M_\alpha(\tbinom{-y}{j}\tbinom{y}{j'})][M_\beta(\tbinom{y}{j}\tbinom{-y}{j'})+M_\beta(\tbinom{-y}{j}\tbinom{y}{j'})] -$$

$$- [M_\alpha(\tbinom{y}{j}\tbinom{-y}{j'})-M_\alpha(\tbinom{-y}{j}\tbinom{y}{j'})][M_\beta(\tbinom{y}{j}\tbinom{-y}{j'})-M_\beta(\tbinom{-y}{j}\tbinom{y}{j'})]\}C(\tbinom{y}{j})C(\tbinom{y}{j'}) \times$$

$$\times \left\{ [1-e^{-\beta(\tbinom{y}{j})-\beta(\tbinom{y}{j'})}]\left[\frac{2(\omega(\tbinom{y}{j})+\omega(\tbinom{y}{j'}))}{(\omega(\tbinom{y}{j})+\omega(\tbinom{y}{j'}))^2-\omega^2} + \right. \right.$$

$$\left. + i\pi\{\delta(\omega+\omega(\tbinom{y}{j})+\omega(\tbinom{y}{j'}))-\delta(\omega-\omega(\tbinom{y}{j})-\omega(\tbinom{y}{j'}))\}\right] +$$

$$+ [e^{-\beta(\tbinom{y}{j'})}-e^{-\beta(\tbinom{y}{j})}]\left[\frac{2(\omega(\tbinom{y}{j})-\omega(\tbinom{y}{j'}))}{(\omega(\tbinom{y}{j})-\omega(\tbinom{y}{j'}))^2-\omega^2} + \right.$$

$$\left. \left. + i\pi\{\delta(\omega+\omega(\tbinom{y}{j})-\omega(\tbinom{y}{j'}))-\delta(\omega-\omega(\tbinom{y}{j})+\omega(\tbinom{y}{j'}))\}\right]\right\} \right\}, \quad (48.6)$$

where $\quad C(\tbinom{y}{j}) = \dfrac{\hbar/2\omega(\tbinom{y}{j})}{1-\exp[-\hbar\omega(\tbinom{y}{j})/kT]}, \qquad \beta(\tbinom{y}{j}) = \hbar\omega(\tbinom{y}{j})/kT. \quad (48.7)$

In writing down (48.6) one is assisted by certain simplifying circumstances. Thus one readily sees that the contributions due respectively to the terms $q_1(\tbinom{y}{j})q_1(\tbinom{y}{j'})$ and $q_2(\tbinom{y}{j})q_2(\tbinom{y}{j'})$ in (48.5) are identical; the contributions due to the terms $q_1(\tbinom{y}{j})q_2(\tbinom{y}{j'})$ and $q_2(\tbinom{y}{j})q_1(\tbinom{y}{j'})$ are also identical. Moreover the six types of transitions are in three pairs of Stokes and anti-Stokes transitions; one finds that the Stokes and anti-Stokes contributions can be readily combined in all three cases.

Finally we replace the summations over y by integrals as follows:

$$\sum_{y}^{\frac{1}{2}N} \rightarrow Nv_a \iiint dy$$

(remember that the density of permitted wave-numbers is equal to the volume Nv_a) and introduce the two types of functions

$$+M_{\alpha\beta}^{jj'}(\omega) = \lim_{\Delta\omega\to 0} \frac{1}{\Delta\omega} \iiint_{\omega<\omega(\tbinom{y}{j})+\omega(\tbinom{y}{j'})<\omega+\Delta\omega} [M_\alpha(\tbinom{-y}{j}\tbinom{y}{j'})M_\beta(\tbinom{y}{j}\tbinom{-y}{j'})+$$

$$+ M_\alpha(\tbinom{y}{j}\tbinom{-y}{j'})M_\beta(\tbinom{-y}{j}\tbinom{y}{j'})]C(\tbinom{y}{j})C(\tbinom{y}{j'})[1-e^{-\beta(\tbinom{y}{j})-\beta(\tbinom{y}{j'})}]\,dy, \quad (48.8)$$

$$-M_{\alpha\beta}^{jj'}(\omega) = \lim_{\Delta\omega\to 0} \frac{1}{\Delta\omega} \iiint_{\omega<(\omega(\tbinom{y}{j})-\omega(\tbinom{y}{j'}))<\omega+\Delta\omega} [M_\alpha(\tbinom{-y}{j}\tbinom{y}{j'})M_\beta(\tbinom{y}{j}\tbinom{-y}{j'})+$$

$$+ M_\alpha(\tbinom{y}{j}\tbinom{-y}{j'})M_\beta(\tbinom{-y}{j}\tbinom{y}{j'})]C(\tbinom{y}{j})C(\tbinom{y}{j'})[e^{-\beta(\tbinom{y}{j'})}-e^{-\beta(\tbinom{y}{j})}]\,dy. \quad (48.9)$$

It is then found that (48.6) can be written

$$a_{\alpha\beta}(\omega) = -\frac{1}{2\hbar}\Bigg\{ \sum_j \sum_{j'} \int \big[{}^+M_{\alpha\beta}^{jj'}(\omega') + {}^-M_{\alpha\beta}^{jj'}(\omega') \big] \frac{2\omega'}{(\omega')^2 - \omega^2}\, d\omega' +$$

$$+ i\pi \Big[\sum_j \sum_{j'} \big({}^+M_{\alpha\beta}^{jj'}(-\omega) + {}^-M_{\alpha\beta}^{jj'}(-\omega) - {}^+M_{\alpha\beta}^{jj'}(\omega) - {}^-M_{\alpha\beta}^{jj'}(\omega) \big) \Big] \Bigg\}.$$

(48.10)

The formula shows essentially that the function

$$\sum_j \sum_{j'} \{ {}^+M_{\alpha\beta}^{jj'}(\omega) + {}^-M_{\alpha\beta}^{jj'}(\omega) \}$$

describes the distribution of the strength of continuous absorption.

Unlike the first-order dispersion, the second-order dispersion is not independent of temperature. As $T \to 0$,

$$e^{-\beta\binom{y}{j}} = e^{-\beta\binom{y}{j'}} \to 0.$$

It follows that the functions ${}^-M_{\alpha\beta}^{jj'}(\omega)$ vanish at the absolute zero of temperature. If these functions are responsible for certain observed maxima and minima, these should tend to lie on the low-frequency end (optical long-wave-length end) of the spectrum and they should diminish with decreasing temperature and finally disappear at very low temperatures. At high temperatures we have the expansions

$$C\binom{y}{j}C\binom{y}{j'}[1 - e^{-\beta\binom{y}{j} - \beta\binom{y}{j'}}] = kT\left\{ \frac{\hbar\{\omega\binom{y}{j} + \omega\binom{y}{j'}\}}{4\omega^2\binom{y}{j}\omega^2\binom{y}{j'}} + O\left(\frac{1}{T}\right) + \cdots \right\},$$

$$C\binom{y}{j}C\binom{y}{j'}[e^{-\beta\binom{y}{j'}} - e^{-\beta\binom{y}{j}}] = kT\left\{ \frac{\hbar\{\omega\binom{y}{j} - \omega\binom{y}{j'}\}}{4\omega^2\binom{y}{j}\omega^2\binom{y}{j'}} + O\left(\frac{1}{T}\right) + \cdots \right\}.$$

Thus, at sufficiently high temperatures, the M-functions vary linearly with temperature. The variation with temperature is therefore much less rapid than that of the anharmonic effect, and the measurement of temperature-dependence should provide useful clues for distinguishing between the two types of effects.

49. The first- and second-order Raman effects

At practically the same time as Raman and Krishnan† discovered the Raman effect in liquids and gaseous vapours, Landsberg and Mandel-stam‡ detected a frequency change in the radiation scattered by a quartz crystal and correctly interpreted the effect as due to the excitation of an infra-red vibration of the lattice. The very much weaker second-order Raman effect was later observed in the scattering from NaCl by

† C. V. Raman and K. S. Krishnan, *Nature*, **121**, 501 (1928).
‡ G. Landsberg and L. Mandelstam, *Naturwiss.* **16**, 557 (1928).

Fermi and Rasetti,[†] who have also given a qualitative explanation of its continuous frequency spectrum. The first-order Raman scattering is often known as the thermal scattering by the crystal; it is characterized by minute frequency changes which are, moreover, strongly dependent on the angle of scattering. For the discussion of the thermal scattering it is essential to take account of the finite wave-length of the incident and scattered radiation; we shall discuss this effect separately in the next section. A classical theory of the first-order Raman effect has been given by Mandelstam, Landsberg, and Leontowitsch[‡]; a quantum mechanical theory due to Tamm[§] led to the same intensity formula as in the classical theory of the first-order Raman effect, apart from a factor describing the temperature dependence. These authors have taken formal account of the finite wave-length of the radiation, which plays, however, no essential role in the phenomena to be considered in this section. For the following discussion we shall thus follow the method of Born and Bradburn,[||] which, while neglecting the effect of the finite wave-length, is adequate for the discussion of Raman effects of any order. This is essentially an application of the method of local treatment, wherein the general formulae developed in §§ 20, 21 can be directly utilized.

According to (20.15), the intensity of Raman scattering per unit solid angle due to a transition from a vibrational state v to another state v' can be written

$$I = \frac{\omega_0^4}{2\pi c^2} \sum_{k=1,2} \sum_{\alpha\beta} \sum_{\gamma\lambda} n_\alpha^k n_\beta^k i_{\alpha\gamma,\beta\lambda} E_\gamma^- E_\lambda^+, \qquad (49.1)$$

where we have ignored the small frequency change in the factor in front and $i_{\alpha\gamma,\beta\lambda}$ denotes the product of matrix elements of the electronic polarizability:

$$i_{\alpha\gamma,\beta\lambda} = \{\langle v'|P_{\alpha\gamma}^*|v\rangle\langle v|P_{\beta\lambda}|v'\rangle\}_{Av}. \qquad (49.2)$$

The vectors $\mathbf{n}^1$, $\mathbf{n}^2$, we remember, are two mutually perpendicular unit vectors both perpendicular to the direction of scattering. If the frequency spectrum of the scattered radiation is continuous, as in the case of the second- and higher-order scatterings, the states v' for which $i_{\alpha\gamma,\beta\lambda}$ has a non-vanishing value cover a continuous energy spectrum. It is then convenient to introduce instead of $i_{\alpha\gamma,\beta\lambda}$, a function of frequency

$$i_{\alpha\gamma,\beta\lambda}(\omega) = \lim_{\Delta\omega\to 0} \sum_{v'}^{\omega<\omega_0+\omega_{vv'}<\omega+\Delta\omega} \{\langle v'|P_{\alpha\gamma}^*|v\rangle\langle v|P_{\beta\lambda}^*|v'\rangle\}_{Av}, \qquad (49.3)$$

† E. Fermi and F. Rasetti, *Zeit. f. Phys.* **71**, 689 (1931).
‡ L. Mandelstam, G. Landsberg, and M. Leontowitsch, *Zeit. f. Phys.* **60**, 334 (1930).
§ Ig. Tamm, *Zeit. f. Phys.* **60**, 345 (1930).
|| M. Born and Mary Bradburn, *Proc. Roy. Soc.* A, **188**, 161 (1947).

so that

$$I(\omega)\,d\omega = \frac{\omega_0^4}{2\pi c^2} \sum_{k=1,2} \sum_{\alpha\gamma} \sum_{\beta\lambda} n_\alpha^k n_\beta^k i_{\alpha\gamma,\beta\lambda}(\omega) E_\gamma^- E_\lambda^+\,d\omega \qquad (49.4)$$

gives the intensity of scattered radiation in the frequency range from ω to $\omega+d\omega$.

We have seen in § 39 that the electronic polarizability for N cells depends on the complex normal coordinates as follows:

$$\delta P_{\alpha\beta} = \sqrt{N} \sum_j P_{\alpha\beta}\binom{0}{j} Q\binom{0}{j} + \tfrac{1}{2} \sum_y^N \sum_j \sum_{j'} P_{\alpha\beta}\binom{y\ \ -y}{j\ \ j'} Q\binom{y}{j} Q\binom{-y}{j'} + \dots, \quad (49.5)$$

where the coefficients are as defined in (39.14) and (39.16). The first-order transitions are due to the linear terms in the expansion, and the corresponding transition frequencies are clearly the dispersion frequencies $\omega\binom{0}{j}$, $j = 4,\dots, 3n$ (remember that we are only concerned with the optical branches). The first-order Raman scattering then consists of spectral lines with the Stokes and anti-Stokes frequencies

$$\omega_0 \mp \omega\binom{0}{j} \quad (j = 4, 5,\dots, 3n).$$

Using (49.5), we find immediately with the help of formula (21.8) for the thermal averages of products of first-order transition matrix elements that

$$i_{\alpha\gamma,\beta\lambda} = N P_{\alpha\gamma}\binom{0}{j} P_{\beta\lambda}\binom{0}{j} C\binom{0}{j} \begin{cases} 1 & \text{(Stokes line)} \\ e^{-\beta\binom{0}{j}} & \text{(anti-Stokes line)}, \end{cases} \qquad (49.6)$$

where we may note that $P_{\alpha\gamma}\binom{0}{j}$ is real. The temperature-dependence given by the formula has been experimentally demonstrated by Landsberg and Mandelstam† by careful experiments on quartz.

The second-order Raman scattering is due to the second-order terms in the expansion (49.5), and these can be expressed in terms of the real normal coordinates of the first kind exactly as in the case of the second-order electric moment,

$$P_{\alpha\beta}^{(2)} = \frac{1}{4} \sum_y^{\frac{1}{2}N} \sum_j \sum_{j'} \left[P_{\alpha\beta}\binom{y\ \ -y}{j\ \ j'} + P_{\alpha\beta}\binom{-y\ \ y}{j\ \ j'} \right]\left[q_1\binom{y}{j} q_1\binom{y}{j'} + q_2\binom{y}{j} q_2\binom{y}{j'} \right] -$$

$$- \frac{i}{4} \sum_y^{\frac{1}{2}N} \sum_j \sum_{j'} \left[P_{\alpha\beta}\binom{y\ \ -y}{j\ \ j'} - P_{\alpha\beta}\binom{-y\ \ y}{j\ \ j'} \right]\left[q_1\binom{y}{j} q_2\binom{y}{j'} - q_2\binom{y}{j} q_1\binom{y}{j'} \right].$$

$$(49.7)$$

---

† G. Landsberg and S. L. Mandelstam, *Zeit. f. Phys.* **73**, 502 (1931).

Clearly we have the same second-order transitions as in the last section. The Raman scattering due to the six different types of transition will be given the designation numbers 1, 2,..., 6 as indicated below:

Designation

$$
\begin{array}{ll}
1 & \omega_{jj}^{(1)}(\mathbf{y}) = \omega_0 - 2\omega(\tfrac{\mathbf{y}}{j}) \\[4pt]
2 & \omega_{jj}^{(2)}(\mathbf{y}) = \omega_0 + 2\omega(\tfrac{\mathbf{y}}{j})
\end{array}
\left.\begin{array}{c}\\ \text{overtone}\\ \\ \end{array}\right\}
\left\{\begin{array}{l}\text{Stokes}\\ \\ \text{anti-Stokes}\end{array}\right.
\left.\begin{array}{c}\\ \end{array}\right\} q_1^2(\tfrac{\mathbf{y}}{j}),\ q_2^2(\tfrac{\mathbf{y}}{j})
$$

$$
\begin{array}{ll}
3 & \omega_{jj}^{(3)}(\mathbf{y}) = \omega_0 - [\omega(\tfrac{\mathbf{y}}{j}) + \omega(\tfrac{\mathbf{y}}{j'})] \\[4pt]
4 & \omega_{jj}^{(4)}(\mathbf{y}) = \omega_0 + [\omega(\tfrac{\mathbf{y}}{j}) + \omega(\tfrac{\mathbf{y}}{j'})] \\[4pt]
5 & \omega_{jj}^{(5)}(\mathbf{y}) = \omega_0 - [\omega(\tfrac{\mathbf{y}}{j}) - \omega(\tfrac{\mathbf{y}}{j'})] \\[4pt]
6 & \omega_{jj}^{(6)}(\mathbf{y}) = \omega_0 + [\omega(\tfrac{\mathbf{y}}{j}) - \omega(\tfrac{\mathbf{y}}{j'})]
\end{array}
$$

summation / difference, Stokes / anti-Stokes

$$q_1(\tfrac{\mathbf{y}}{j})q_1(\tfrac{\mathbf{y}}{j'}),\ q_2(\tfrac{\mathbf{y}}{j})q_2(\tfrac{\mathbf{y}}{j'}),$$
$$q_1(\tfrac{\mathbf{y}}{j})q_2(\tfrac{\mathbf{y}}{j'}),\ q_2(\tfrac{\mathbf{y}}{j})q_1(\tfrac{\mathbf{y}}{j'}),$$

$$(49.8)$$

where, for the last four types to be distinct, we assume that $j > j'$ (or $\omega(\tfrac{\mathbf{y}}{j}) > \omega(\tfrac{\mathbf{y}}{j'})$). In the last column of (49.8), we have given the terms in (49.7) responsible for the scattering in the various cases. The Raman frequency of each type varies continuously with the wave-number $\mathbf{y}$; thus a given branch of vibrations j gives rise to two continuous Raman spectra designated above by the numbers 1 and 2, and a given pair of branches jj' gives rise to four distinct continuous spectra designated by the numbers 3, 4, 5, 6. The experimentally observed spectrum is, of course, a superposition of all the various spectra. Accordingly we subdivide the function introduced in (49.3),

$$
i_{\alpha\gamma,\beta\lambda} = \sum_{j \geqslant j'} \sum_{s=1}^{6} i_{\alpha\gamma,\beta\lambda}^{(s)}(\omega)_{jj'},
\qquad (49.9)
$$

where

$$
i_{\alpha\gamma,\beta\lambda} = 0 \quad \text{if } j \neq j' \text{ for } s = 1,\, 2
$$

$$
= 0 \quad \text{if } j = j' \text{ for } s = 3,\, 4,\, 5,\, 6.
$$

The meaning of (49.9) is fairly evident: every term on the right-hand side of (49.9) describes a separate continuous spectrum with s giving its designation number and the indices jj or jj' describing the branch or branches of lattice vibrations involved.

The explicit expressions for the functions on the right-hand side of (49.9) can be derived with the help of considerations entirely analogous to those leading to (48.6) of the last section. Thus, using the general formula given in (21.10) for the thermal averages of products of second-order transition matrix elements, we find after some simplification that

$$\left.\begin{matrix} i^{(1)}_{\alpha\gamma,\beta\lambda}(\omega)_{jj'} \\ i^{(2)}_{\alpha\gamma,\beta\lambda}(\omega)_{jj'} \end{matrix}\right\rangle = Nv_a \lim_{\Delta\omega\to 0} \frac{1}{\Delta\omega} \iiint_{\omega < \omega_0 \mp 2\omega({y \atop j}) < \omega+\Delta\omega} dy\, P_{\alpha\gamma}({y \atop j}{}^{jy})P_{\beta\lambda}({y \atop j}{}^{jy})C^2({y \atop j})\left\langle{1 \atop e^{-2\beta({y \atop j})},}\right.$$

$$\left.\begin{matrix} i^{(3)}_{\alpha\gamma,\beta\lambda}(\omega)_{jj'} \\ i^{(4)}_{\alpha\gamma,\beta\lambda}(\omega)_{jj'} \end{matrix}\right\rangle = Nv_a \lim_{\Delta\omega\to 0} \frac{1}{\Delta\omega} \iiint_{\omega < \omega_0 \mp [\omega({y \atop j})+\omega({y \atop j'})] < \omega+\Delta\omega} dy\, [P_{\alpha\gamma}(^{jy}{}_j^y)P_{\beta\lambda}({y \atop j}{}^{jy})+$$

$$+P_{\alpha\gamma}({y \atop j}{}^{jy})P_{\beta\lambda}(^{jy}{}_{j'}^y)]C({y \atop j})C({y \atop j'})\left\langle{1 \atop e^{-\beta({y \atop j})-\beta({y \atop j'})},}\right.$$

$$\left.\begin{matrix} i^{(5)}_{\alpha\gamma,\beta\lambda}(\omega)_{jj'} \\ i^{(6)}_{\alpha\gamma,\beta\lambda}(\omega)_{jj'} \end{matrix}\right\rangle = Nv_a \lim_{\Delta\omega\to 0} \frac{1}{\Delta\omega} \iiint_{\omega < \omega_0 \mp [\omega({y \atop j})-\omega({y \atop j'})] < \omega+\Delta\omega} dy\, [P_{\alpha\gamma}(^{jy}{}_{j'}^y)P_{\beta\lambda}({y \atop j}{}^{jy})+$$

$$+P_{\alpha\gamma}({y \atop j}{}^{jy})P_{\beta\lambda}(^{jy}{}_{j'}^y)]C({y \atop j})C({y \atop j'})\left\langle{e^{-\beta({y \atop j'})} \atop e^{-\beta({y \atop j})}.}\right.$$

$$(49.10)$$

On the basis of the above theory the Raman spectra of the NaCl and diamond lattices have been discussed in some detail by Born and Bradburn[†] and by Helen Smith[‡] respectively. The NaCl lattice does not give any first-order scattering, the reason being the following. In the NaCl lattice a coordinate $Q({0 \atop j})$ (for the optical branches) describes a shift of the Na$^+$ ions as a whole relative to the Cl$^-$ ions; $-Q({0 \atop j})$ describes the same configuration after an inversion. Since the polarizability, being a second-rank tensor, is invariant under an inversion operation, we have

$$\sqrt{N} \sum_j P_{\alpha\beta}({0 \atop j})Q({0 \atop j})+\text{etc.} = -\sqrt{N} \sum_j P_{\alpha\beta}({0 \atop j})Q({0 \atop j})+\text{etc.}$$

It follows that $$P_{\alpha\beta}({0 \atop j}) = 0; \qquad\qquad (49.11)$$

therefore, there are no linear terms in the expansion of the polarizability tensor.

For the discussion of the second-order scattering, Born and Bradburn considered the integrands in (49.10) as constants which are given approximately by the values of the integrands at the point

$$\mathbf{y} = \tfrac{1}{2}(\mathbf{b}^1+\mathbf{b}^2+\mathbf{b}^3).$$

The component spectra described by $i^{(s)}_{\alpha\gamma,\beta\lambda}(\omega)_{jj'}$ are thus taken to vary with the frequency as the density of transitions:

$$z^{(s)}_{jj'}(\omega) = \lim_{\Delta\omega\to 0} \frac{1}{\Delta\omega} \iiint_{\omega < \omega^{(s)}_{jj'}(\mathbf{y}) < \omega+\Delta\omega} d\mathbf{y},$$

† M. Born and Mary Bradburn, *Proc. Roy. Soc.* A, **188**, 161 (1947).
‡ Helen Smith, *Phil. Trans. Roy. Soc.* A, **241**, 105 (1948).

where $\omega_{jj}^{(s)}(\mathbf{y})$ stands for the appropriate Raman frequency as given in (49.8). They have shown that if only the mutual influence between the nearest neighbours is taken into account, only three independent constants are required to describe the values of the integrands. They have calculated the functions $z_{jj}^{(s)}(\omega)$ from the frequency distributions for NaCl as given by Kellermann† and found that by suitable choice of the

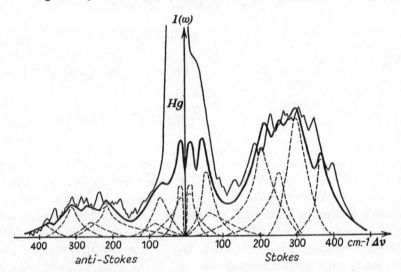

FIG. 26. Fine line represents Krishnan's microphotometric record of Raman spectrum of rock-salt (Krishnan‡). Broken lines represent theoretical contributions to intensity of pairs of branches of frequency. Thick line represents theoretical intensity obtained from superposition of these.

independent constants the Raman spectrum observed by Krishnan can be reproduced fairly well. The theoretical and experimental curves in Fig. 26 are reproduced from the paper·by Born and Bradburn.

The frequencies $\omega({0 \atop j})$ for all three optical branches are identical for the diamond lattice (cf. § 6). The first-order spectrum should thus show two lines with the frequencies $\omega \mp \omega({0 \atop j})$. These lines as found by experiments show that $\omega({0 \atop j}) = 2 \cdot 56 \times 10^{14}/\text{sec}$. It follows that the intensity ratio between the two lines should be (see (49.6))

$$\exp[\hbar\omega({0 \atop j})/kT] = 586 \quad (T = 300° \text{ K.}).$$

The experimental value given by Krishnan§ is 575.

The second-order spectrum of diamond has been considered by Helen

† E. W. Kellermann, *Phil. Trans. Roy. Soc.* A, **238**, 513 (1940).
‡ R. S. Krishnan, *Nature*, **156**, 267 (1945).
§ R. S. Krishnan, *Proc. Ind. Acad. Sci.* **24**, 25 (1946).

Smith, who used the same method as Born and Bradburn to treat the
NaCl spectrum. The comparison of her theoretical results and the experimental intensity distribution given by Krishnan[†] is reproduced in Fig. 27
from her paper.

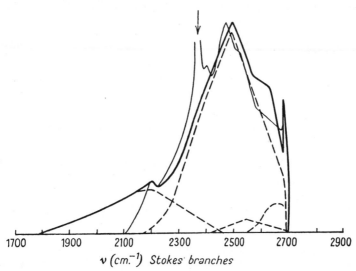

$\nu\ (cm.^{-1})$ Stokes branches

Fig. 27. Second-order Raman spectrum of diamond. Fine line represents Krishnan's
microphotometric record of the Raman spectrum of diamond (Krishnan 1946). Broken
lines represent theoretical contributions to the intensity of frequency density functions
$z_{jj'}(\omega)$ after multiplication by appropriate factors. Thick line represents the superposition
of these functions, i.e. the total theoretical intensity.

50. The Brillouin components of the thermal scattering of light

When light passes through an optical medium, a certain amount of
scattering can be detected even if the medium is entirely free of structural
inhomogeneities (e.g. impurities, cluster formations in liquids, residual
strain in solids). The scattering is due to the fluctuating dielectric
inhomogeneities associated with the thermal agitation of the medium.
From the classical viewpoint, the scattering can be interpreted as the
result of reflections from the thermal elastic waves. Owing to the effect
of interference, for scattering in a given direction, only the elastic waves
of a certain wave-number vector are operative. Moreover, as a result of
the Doppler effect, the light scattered by elastic waves of the frequency ν
suffers the frequency changes $\pm\nu$. Thus Brillouin[‡] predicted a long time
ago that the thermally scattered light should show certain frequency
changes $\pm\nu$ which are functions of the scattering angle. The existence

† R. S. Krishnan, *Proc. Ind. Acad. Sci.* **24**, 25 (1946).
‡ L. Brillouin, *Ann. d. Phys.*, Paris, **17**, 88 (1922).

of the Brillouin components has since been confirmed experimentally both in liquids and in solids.[†]

Leontowitsch and Mandelstam[‡] have given a general theory of the thermal scattering by crystals from the classical standpoint. They obtained an intensity formula depending only on the elastic and elasto-optical constants of the scattering crystal. The order of magnitude of the effect as given by their theory has been confirmed by recent experiments carried out by Krishnan[§] on diamond crystals.

From the standpoint of the quantum theory, the thermal scattering is simply the first-order Raman scattering associated with a transition in the vibrational states of the acoustic vibrations. Tamm[‖] has developed a theory where both the crystal and the radiation are treated quantum mechanically. In his treatment, he considers, however, only the dielectric effects due to density fluctuations. Whereas the above authors considered the medium macroscopically as an elastic medium, Theimer[††] has more recently given a microscopic theory on the basis of the lattice model. However, in order to obtain numerical estimates, he has eventually to express the atomic constants in terms of the elastic and elasto-optical constants. The results thus obtained would be expected to coincide with those of Leontowitsch and Mandelstam, because (as we shall see presently) the vibrational quanta involved in thermal scattering are so small that the quantum treatment should give the same results as the classical treatment. Actually, there is a disagreement between the results of these two approaches, which is due to an error in Theimer's paper. In expressing the atomic constants in terms of the elasto-optical constants he has overlooked that the former refer to a given number of particles, the latter to a given volume; an elastic deformation changes the number of particles per volume.[‡‡]

In the following discussion we shall consider the dielectric polarization phenomenologically with the help of the elasto-optical constants in order that the intensity formula can readily give numerical estimates for concrete examples. As elsewhere in the book, we shall treat the radiation classically and the lattice quantum mechanically. Since, except at the

---

[†] E. Gross, *Nature*, **126**, 201, 400, 603 (1930); *Zeit. f. Phys.* **63**, 685 (1930); W. Ramm, *Phys. Zeit.* **35**, 111, 756 (1934); R. S. Krishnan, *Nature*, **159**, 740 (1947).

[‡] M. Leontowitsch and S. Mandelstam, jr., *Phys. Zeit. Sowjet.* **1**, 317 (1931); *Zeit. f. Phys.* **75**, 350 (1932).

[§] R. S. Krishnan, *Proc. Ind. Acad. Sci.* A, **26**, 399 (1947).

[‖] Ig. Tamm, *Zeit. f. Phys.* **60**, 345 (1930).

[††] O. Theimer, *Proc. Phys. Soc.* **64**, 1012 (1951); ibid. **65**, 38 (1952).

[‡‡] In Theimer's paper, loc. cit. (1952), it is the formula (3.11) which introduces the error.

very lowest temperatures, the quantum effects are negligible, the following treatment is in substance the same as the theory of Leontowitsch and Mandelstam.

Let us consider a crystal which is in an arbitrary state of elastic deformation (in general inhomogeneous) and denote the corresponding change of the dielectric tensor at a point $\mathbf{x}$ by $\delta\epsilon_{\alpha\beta}(\mathbf{x})$. When subject to the electric field

$$\mathbf{E}^-e^{-i\omega t+2\pi i \mathbf{s}_0 \cdot \mathbf{x}/\lambda}+\mathbf{E}^+e^{i\omega t-2\pi i \mathbf{s}_0 \cdot \mathbf{x}/\lambda} \quad (\mathbf{E}^- = (\mathbf{E}^+)^*) \qquad (50.1)$$

of an electromagnetic wave travelling in the direction of the unit vector $\mathbf{s}_0$, the dielectric inhomogeneities act as scattering centres. Thus a volume element $d\tau$ emits the radiation (see § 19) associated with the excess electric moment

$$\mathbf{m}^-(\mathbf{x})e^{-i\omega t+2\pi i \mathbf{s}_0 \cdot \mathbf{x}/\lambda}+\mathbf{m}^+(\mathbf{x})e^{i\omega t-2\pi i \mathbf{s}_0 \cdot \mathbf{x}/\lambda}, \qquad (50.2)$$

where $$m_\alpha^-(\mathbf{x}) = (m_\alpha^+(\mathbf{x}))^* = \frac{1}{4\pi}\sum_\beta \delta\epsilon_{\alpha\beta}(\mathbf{x})E_\beta^-. \qquad (50.3)$$

The electric field in the scattered light can be calculated with the help of (19.3). Thus, taking the origin $\mathbf{x} = 0$ at an arbitrary point within the crystal, we find that the electric field at a distant point $\mathbf{R} = \mathbf{s}R$ due to the dipole at $\mathbf{x}$ is given by

$$\mathbf{E}(t) = -\frac{\omega^2 d\tau}{|\mathbf{R}-\mathbf{x}|^3 c^2}\Big[(\mathbf{R}-\mathbf{x})\wedge[(\mathbf{R}-\mathbf{x})\wedge\{\mathbf{m}^-(\mathbf{x})^{-i\omega t+2\pi i(\mathbf{s}_0 \cdot \mathbf{x}+|\mathbf{R}-\mathbf{x}|)/\lambda}+$$
$$+\mathbf{m}^+(\mathbf{x})e^{i\omega t-2\pi i(\mathbf{s}_0 \cdot \mathbf{x}+|\mathbf{R}-\mathbf{x}|)/\lambda}\}]\Big]. \quad (50.4)$$

For $\mathbf{R}$ sufficiently large compared with $\mathbf{x}$ we can ignore the latter in the above expression except in the exponential factors; in the latter one has the expansion

$$|\mathbf{R}-\mathbf{x}| = \{R^2+|\mathbf{x}|^2-2R\mathbf{s}\cdot\mathbf{x}\}^{\frac{1}{2}} = R-\mathbf{s}\cdot\mathbf{x}+|\mathbf{x}|\Big\{0\Big(\frac{|\mathbf{x}|}{R}\Big)+...\Big\},$$

and can replace $|\mathbf{R}-\mathbf{x}|$ by $R-\mathbf{s}\cdot\mathbf{x}$, as long as $|\mathbf{x}|^2 \ll \lambda R$. Therefore, if $(R\lambda)^{\frac{1}{2}}$ is large compared with the dimensions of the crystal, (50.4) can be written

$$\mathbf{E}(t) = -\frac{\omega^2 d\tau}{R^3 c^2}\Big[\mathbf{R}\wedge[\mathbf{R}\wedge\{\mathbf{m}^-(\mathbf{x})e^{-i\omega t+2\pi i(\mathbf{s}_0-\mathbf{s})\cdot \mathbf{x}/\lambda+2\pi iR/\lambda}+$$
$$+\mathbf{m}^+(\mathbf{x})e^{i\omega t-2\pi i(\mathbf{s}_0-\mathbf{s})\cdot \mathbf{x}/\lambda-2\pi iR/\lambda}\}]\Big],$$

or

$$\mathbf{E}(t+R/c) = -\frac{\omega^2 d\tau}{R^3 c^2}\Big[\mathbf{R}\wedge[\mathbf{R}\wedge\{\mathbf{m}^-(\mathbf{x})e^{-i\omega t+2\pi i(\mathbf{s}_0-\mathbf{s})\cdot \mathbf{x}/\lambda}+$$
$$+\mathbf{m}^+(\mathbf{x})e^{i\omega t-2\pi i(\mathbf{s}_0-\mathbf{s})\cdot \mathbf{x}/\lambda}\}]\Big]. \quad (50.5)$$

Upon integrating (50.5) over the whole crystal, we find that the total electric field at $\mathbf{R}$ can be written in the same form as (19.3), in which the expression (19.1) for $\mathbf{m}(t)$ is to be used with the amplitude

$$m_\alpha^- = (m_\alpha^+)^* = \int_V m_\alpha^-(\mathbf{x})e^{2\pi i(\mathbf{s}_0-\mathbf{s})\cdot\mathbf{x}/\lambda}\,d\mathbf{x}. \tag{50.6}$$

It is convenient for the following discussion to write

$$\frac{(\mathbf{s}_0-\mathbf{s})}{\lambda} = -\frac{\mathbf{s}_\theta}{\lambda_\theta}, \tag{50.7}$$

where $\mathbf{s}_\theta$ is a unit vector parallel to $(\mathbf{s}_0-\mathbf{s})$ and λ_θ is related to the scattering angle θ as follows:

$$\lambda_\theta = \frac{\lambda}{2\sin\frac{1}{2}\theta}. \tag{50.8}$$

Substituting (50.3) and (50.7) in (50.6), we get

$$m_\alpha^- = (m_\alpha^+)^* = \sum_\beta \frac{E_\beta^-}{4\pi} \int_V \delta\epsilon_{\alpha\beta}(\mathbf{x})e^{-2\pi i \mathbf{s}_\theta\cdot\mathbf{x}/\lambda_\theta}\,d\tau. \tag{50.9}$$

Classically, the radiation scattered by the dielectric inhomogeneities can be obtained directly by using (50.9) in (19.7) or (19.8).

To the first order of approximation, $\delta\epsilon_{\alpha\beta}(\mathbf{x})$ can be taken as a linear function of the elastic strain components at $\mathbf{x}$. It follows that (50.9) has no diagonal matrix elements with respect to the vibrational states of the crystal. Therefore, from the quantum mechanical viewpoint, the corresponding scattering must appear as Raman scattering. The intensity of the Raman scattering can be obtained exactly as in the case of the classical scattering, the only difference being the replacement of $\mathbf{m}^-$ and $\mathbf{m}^+$ by their transition matrix elements.

The effect of an elastic strain on the dielectric tensor is usually expressed as follows:

$$\delta(\epsilon^{-1})_{\alpha\beta} = (\epsilon^{-1})_{\alpha\beta}-(\epsilon_0^{-1})_{\alpha\beta} = \sum_{\gamma\eta} p_{\alpha\beta,\gamma\eta}\frac{\partial u_\gamma}{\partial x_\eta}, \tag{50.10}$$

where ϵ^{-1} denotes the inverse of the dielectric tensor and $\mathbf{u}$ represents the elastic displacement; the coefficients

$$p_{\alpha\beta,\gamma\eta} = p_{\beta\alpha,\gamma\eta} = p_{\alpha\beta,\eta\gamma} \tag{50.11}$$

are Pockel's *elasto-optical constants* in tensor notation. The elasto-optical constants can be derived from the general theory by considering the simultaneous effect of an elastic strain and an external field which is not static but represents a light wave. The procedure is similar to that used in § 37; the difference is that the field is periodic (not constant) and that the deformation produced by it is superposed over an elastic strain (not a thermal one, as in (37.46)). The dielectric susceptibility tensor components $a_{\alpha\beta}$ are still connected to the $\epsilon_{\alpha\beta}$ by (44.61).

Differentiating the tensor (matrix) identity

$$\epsilon^{-1}\epsilon = 1 \qquad (50.12)$$

and multiplying by ϵ_0 from the left, we get to the first order of approximation

$$\delta\epsilon = -\epsilon_0\,\delta\epsilon^{-1}\epsilon_0. \qquad (50.13)$$

Hence we can express (50.10) alternatively as

$$\delta\epsilon_{\alpha\beta} = \sum_{\gamma\eta} k_{\alpha\beta,\gamma\eta}\frac{\partial u_\gamma}{\partial x_\eta}, \qquad (50.14)$$

where

$$k_{\alpha\beta,\gamma\eta} = -\sum_{\mu\nu} (\epsilon_0)_{\alpha\mu}\,p_{\mu\nu,\gamma\eta}(\epsilon_0)_{\nu\beta}. \qquad (50.15)$$

Since we shall be concerned with elastic deformations only, we can express **u** in terms of the complex normal coordinates $Q(\substack{\mathbf{y}\\j})$ $(j = 1, 2, 3)$ of the acoustic vibrations

$$\mathbf{u(x)} = \Big\{N\sum_k m_k\Big\}^{-\frac{1}{2}} \sum_{\mathbf{y}}^{N}\sum_{j=1}^{3} \mathbf{e}(\substack{\mathbf{y}\\j})Q(\substack{\mathbf{y}\\j})e^{2\pi i\mathbf{y}\cdot\mathbf{x}}, \qquad (50.16)$$

where $\mathbf{e}(\substack{\mathbf{y}\\j})$ represents the unit polarization vector for elastic waves belonging to the branch j and in the direction of **y**. (Remember that the elastic waves are identical with the zero-order acoustic vibrations. Thus the polarization vectors $\mathbf{e}(k|\substack{\mathbf{y}\\j})$ in § 38 reduce to

$$m_k\Big(\sum_k m_k\Big)^{-\frac{1}{2}}\mathbf{e}(\substack{\mathbf{y}\\j}),$$

the mass factor being required by the orthonormal relations (38.25).) For the convenience of the following discussion, we write (50.16) in the integral form

$$\mathbf{u(x)} = \Big\{N\sum_k m_k\Big\}^{-\frac{1}{2}}\sum_{j=1}^{3} V\int \mathbf{e}(\substack{\mathbf{y}\\j})Q(\substack{\mathbf{y}\\j})e^{2\pi i\mathbf{y}\cdot\mathbf{x}}\,d\mathbf{y}$$

$$= \Big(\frac{V}{\rho}\Big)^{\frac{1}{2}}\sum_{j=1}^{3}\int \mathbf{e}(\substack{\mathbf{y}\\j})Q(\substack{\mathbf{y}\\j})e^{2\pi i\mathbf{y}\cdot\mathbf{x}}\,d\mathbf{y}, \qquad (50.17)$$

where ρ is the mass density. Substituting (50.17) in (50.14) and afterwards using the latter in (50.9), we get

$$m_\alpha^- = (m_\alpha^+)^* = \frac{i}{2}\Big(\frac{V}{\rho}\Big)^{\frac{1}{2}}\sum_{\beta\gamma\eta} E_\beta^-\,k_{\alpha\beta,\gamma\eta}\times$$

$$\times \sum_{j=1}^{3}\int\Big\{e_\gamma(\substack{\mathbf{y}\\j})y_\eta\,Q(\substack{\mathbf{y}\\j})\int_V \exp\Big[2\pi i\Big(\frac{-\mathbf{s}_\theta}{\lambda_\theta}+\mathbf{y}\Big)\cdot\mathbf{x}\Big]\,d\mathbf{x}\Big\}\,d\mathbf{y}. \qquad (50.18)$$

Upon replacing the integral over $\mathbf{x}$ by Dirac's δ-function $\delta(\mathbf{y}-\mathbf{s}_\theta/\lambda_\theta)$, we finally obtain the relation

$$m_\alpha^- = (m_\alpha^+)^* = \frac{i}{2\lambda_\theta}\left(\frac{V}{\rho}\right)^{\frac{1}{2}} \sum_{\beta\gamma\eta} E_\beta^- \, k_{\alpha\beta,\gamma\eta} \sum_{j=1}^{3} e_\gamma(^{\mathbf{s}_\theta/\lambda_\theta}_j)s_{\theta\eta}\, Q(^{\mathbf{s}_\theta/\lambda_\theta}_j),$$

$$(50.19)$$

where $s_{\theta\eta}$, $\eta = 1$, 2, 3, denotes the components of $\mathbf{s}_\theta$.

The above replacements of the summation over $\mathbf{y}$ by an integral in (50.17) and of the integral over $\mathbf{x}$ by a δ-function in (50.18) are, strictly speaking, permissible only if the crystal is infinitely large. For a finite crystal, the summation over $\mathbf{y}$ will restrict the scattering to certain discrete directions and the finite integral over $\mathbf{x}$ leads to a broadening of each of these scattering directions. So long as such fine features of the scattering (angle $\sim \lambda$/dimension of crystal) are unresolved, the above results can be used.

Expressing the complex coordinates in terms of the real normal coordinates of the first kind, we can write (50.19) as

$$m_\alpha^- = (m_\alpha^+)^* = \frac{1}{\lambda_\theta}\left(\frac{V}{\rho}\right)^{\frac{1}{2}} \sum_{j=1}^{3} \sum_{\beta} g_{\alpha\beta}(j)\{iq_1(^{\mathbf{s}_\theta/\lambda_\theta}_j)-q_2(^{\mathbf{s}_\theta/\lambda_\theta}_j)\}E_\beta^-,$$

$$(50.20)$$

where

$$g_{\alpha\beta}(j) = \frac{1}{2^{\frac{1}{2}}} \sum_{\gamma\eta} k_{\alpha\beta,\gamma\eta}\, e_\gamma(^{\mathbf{s}_\theta/\lambda_\theta}_j)s_{\theta\eta}. \qquad (50.21)$$

If $\mathbf{n}^1$, $\mathbf{n}^2$ are two mutually perpendicular unit vectors both perpendicular to $\mathbf{s}$, we find that the rate of Raman scattering per unit solid angle is given by (see (19.8))

$$\frac{\omega^4}{2\pi c^3} \sum_{i=1}^{2} \sum_{\alpha\beta} n_\alpha^i n_\beta^i \langle v|m_\alpha^+|v'\rangle\langle v'|m_\beta^-|v\rangle. \qquad (50.22)$$

Since (50.20) is linear in the normal coordinates there are only first-order transition matrix elements and (50.22) accordingly gives only first-order Raman scattering. For scattering in the direction of $\mathbf{s}$ the transition frequencies are clearly $\pm\omega(^{\mathbf{s}_\theta/\lambda_\theta}_j)$ ($j = 1$, 2, 3). In other words, the scattered light should be composed of six lines with the following circular frequencies

$$\omega\pm\omega(^{\mathbf{s}_\theta/\lambda_\theta}_j) = \omega\pm\frac{2\pi}{\lambda_\theta}c_j(\mathbf{s}_\theta)$$

$$= \omega\left\{1\pm\frac{2c_j(\mathbf{s}_\theta)}{c}\sin\theta/2\right\}, \qquad (50.23)$$

where we have introduced the phase velocities $c_j(\mathbf{s}_\theta)$ ($j = 1$, 2, 3) for the elastic waves travelling in the direction of $\mathbf{s}_\theta$. The fractional change of frequency is thus very small, being of the order of the ratio of the elastic wave velocities to the light velocity. We note that each of these lines is due to two distinct transitions corresponding to the two coordinates $q_1(^{s\theta|\lambda\theta}_j)$ and $q_2(^{s\theta|\lambda\theta}_j)$.

Using (21.8), we find directly that the thermal average of the product of the relevant transition matrix elements can be given as follows:

$$\{\langle v|m_\alpha^+|v'\rangle\langle v'|m_{\bar\beta}^-|v\rangle\}_{Av} \qquad\qquad \omega_{vv'}$$

$$\left(\frac{V}{\rho}\right)\frac{1}{\lambda_\theta^2}\sum_{\gamma\eta} g_{\alpha\gamma}(j)g_{\beta\eta}(j)E_\gamma^+\,E_\eta^- \begin{cases} C(^{s\theta|\lambda\theta}_j)e^{-\beta(^{s\theta|\lambda\theta}_j)} & \omega(^{s\theta|\lambda\theta}_j) \\ C(^{s\theta|\lambda\theta}_j) & -\omega(^{s\theta|\lambda\theta}_j) \end{cases} \qquad (50.24)$$

Since the wave-length λ_θ is of the order 10^{-4} to 10^{-3} cm. and the elastic wave velocities are of the order 10^5 to 10^6 cm./sec., the vibrational quanta concerned, namely $\hbar\omega(^{s\theta|\lambda\theta}_j)$, are very small compared with kT at practically any temperature. Thus we have approximately

$$C(^{s\theta|\lambda\theta}_j)e^{-\beta(^{s\theta|\lambda\theta}_j)} \simeq C(^{s\theta|\lambda\theta}_j) \simeq \frac{kT}{2\omega^2(^{s\theta|\lambda\theta})} = \frac{\lambda_\theta^2\,kT}{8\pi^2c_j^2(\mathbf{s}_\theta)}. \qquad (50.25)$$

Upon using (50.24) and (50.25) in (50.22) and remembering that every transition is actually double (due respectively to q_1 and q_2 in (50.20)) we find that the intensity of scattering per solid angle for any particular Brillouin component is given by the formula

$$I_j = \left(\frac{V}{\rho}\right)\frac{kT\omega^4}{8\pi^3c^3c_j^2(\mathbf{s}_\theta)}\sum_i\sum_{\alpha\beta\gamma\eta} n_\alpha^i\,n_\beta^i\,g_{\alpha\gamma}(j)g_{\beta\eta}(j)E_\gamma^+\,E_\eta^-\,, \qquad (50.26)$$

where j is the particular branch of acoustic vibrations responsible for the Brillouin component (either the Stokes or the anti-Stokes component) under consideration.

Let us consider the special case of a cubic crystal, with the incoming radiation along one cubic axis (X-axis) and the scattered radiation observed along another (Y-axis). For a cubic crystal, the dielectric tensor is equivalent to a scalar ϵ, therefore (50.15) reduces to

$$k_{\alpha\beta,\gamma\eta} = -\epsilon^2 p_{\alpha\beta,\gamma\eta}. \qquad (50.27)$$

Moreover, the only non-vanishing elasto-optical constants are the following:

$$p_{11,11} = p_{22,22} = p_{33,33} = p_{11},$$

$$p_{11,22} = p_{22,11} = p_{22,33} = p_{33,22} = p_{33,11} = p_{11,33} = p_{12},$$

$$p_{12,12} = p_{21,21} = p_{12,21} = p_{21,12} = p_{23,23} = p_{23,32}$$
$$= p_{32,23} = p_{32,32} = p_{31,31} = p_{31,13} = p_{13,31} = p_{13,13} = p_{44}.$$

(50.28)

Since $\mathbf{s}_0$ and $\mathbf{s}$ are respectively parallel to the X and Y axes, $\mathbf{s}_\theta = \mathbf{s} - \mathbf{s}_0$ is in the direction of an XY diagonal. Putting

$$\mathbf{y} = \mathbf{s}_\theta/\lambda_\theta = (-1, 1, 0)/(\lambda_\theta \sqrt{2})$$

in the equation (27.10) for the elastic waves and remembering that the only non-vanishing elastic constants of a cubic crystal are

$$c_{11,11} = c_{22,22} = c_{33,33} = c_{11},$$

$$c_{11,22} = c_{22,33} = c_{22,11} = c_{33,11} = c_{33,22} = c_{11,33} = c_{12},$$

$$c_{12,12} = c_{21,12} = c_{21,21} = c_{12,21} = c_{23,23} = c_{23,32}$$

$$= c_{32,32} = c_{32,23} = c_{31,31} = c_{31,13} = c_{13,31} = c_{13,13} = c_{44}$$

one readily finds that the three elastic vibrations can be described as follows:

j	$c_j^2(\mathbf{s}_\theta)$	$e_1\binom{\mathbf{s}_\theta}{j}$	$e_2\binom{\mathbf{s}_\theta}{j}$	$e_3\binom{\mathbf{s}_\theta}{j}$
1	c_{44}/ρ	0	0	1
2	$(c_{11}-c_{12})/2\rho$	$1/\sqrt{2}$	$1/\sqrt{2}$	0
3	$(c_{11}+c_{12}+2c_{44})/2\rho$	$-1/\sqrt{2}$	$1/\sqrt{2}$	0

(50.29)

With the help of (50.28), (50.29), and (50.21), we obtain for the g-coefficients the following values:

j	$g_{11}(j)$	$g_{22}(j)$	$g_{33}(j)$	$g_{23}(j)=g_{32}(j)$	$g_{31}(j)=g_{13}(j)$	$g_{12}(j)=g_{21}(j)$
1	0	0	0	$-\tfrac{1}{4}\epsilon^2 p_{44}$	$\tfrac{1}{4}\epsilon^2 p_{44}$	0
2	$\dfrac{\epsilon^2(p_{11}-p_{12})}{2^{5/2}}$	$\dfrac{-\epsilon^2(p_{11}-p_{12})}{2^{5/2}}$	0	0	0	0
3	$\dfrac{-\epsilon^2(p_{11}+p_{12})}{2^{5/2}}$	$\dfrac{-\epsilon^2(p_{11}+p_{12})}{2^{5/2}}$	$\dfrac{-\epsilon^2 p_{12}}{2^{3/2}}$	0	0	$\dfrac{\epsilon^2 p_{44}}{2^{3/2}}$

(50.30)

For unpolarized incident radiation we can rewrite (50.26) as follows:

$$I_j = \left(\frac{V}{\rho}\right)\frac{I_0 kT\omega^4}{8\pi^2 c^4 c_j^2(\mathbf{s}_\theta)} \sum_{i=1,2} \sum_{k=1,2} \sum_{\alpha\beta\gamma\eta} n_\alpha^i n_\beta^i g_{\alpha\gamma}(j) g_{\beta\eta}(j) \bar{n}_\gamma^k \bar{n}_\eta^k, \quad (50.31)$$

where I_0 is the intensity of the incident radiation and $\bar{\mathbf{n}}^k$ ($k = 1, 2$) are two mutually perpendicular unit vectors both perpendicular to the direction of incidence. Now choosing the unit vectors as follows:

$$\mathbf{n}^1 = (1, 0, 0), \qquad \mathbf{n}^2 = (0, 0, 1), \qquad \bar{\mathbf{n}}^1 = (0, 1, 0), \qquad \bar{\mathbf{n}}^2 = (0, 0, 1),$$

we find with the help of (50.29) and (50.30) that the intensity of scattering for the various Brillouin components is

$$I_1 = \frac{VkT\omega^4\epsilon^4}{64\pi^2c^4}\left\{\frac{p_{44}^2}{c_{44}}\right\}I_0 \qquad (j = 1),$$

$$I_2 = \text{zero} \qquad\qquad\qquad (j = 2), \qquad (50.32)$$

$$I_3 = \frac{VkT\omega^4\epsilon^4}{64\pi^2c^4}\left\{\frac{2(p_{44}^2+p_{12}^2)}{c_{11}+c_{12}+2c_{44}}\right\}I_0 \quad (j = 3).$$

By considering $i = 1, 2$; $k = 1, 2$ in (50.31) separately, we can discuss the polarization of the scattered radiation due to an incident beam linearly polarized along a cubic axis. Suppose that the XY plane is horizontal (i.e. both the direction of incidence and the direction of observation are in the horizontal plane), we find that the condition of polarization can be described as follows:

j	electric field of incident radiation	electric field of scattered radiation	intensity of scattering
1	horizontal	vertical	$\dfrac{VkT\omega^4\epsilon^4}{64\pi^2c^4}\left\{\dfrac{p_{44}^2}{c_{44}}\right\}I_0$
	vertical	horizontal	$\dfrac{VkT\omega^4\epsilon^4}{64\pi^2c^4}\left\{\dfrac{p_{44}^2}{c_{44}}\right\}I_0$
3	horizontal	horizontal	$\dfrac{VkT\omega^4\epsilon^4}{32\pi^2c^4}\left\{\dfrac{p_{44}^2}{c_{11}+c_{12}+2c_{44}}\right\}I_0$
	vertical	vertical	$\dfrac{VkT\omega^4\epsilon^4}{32\pi^2c^4}\left\{\dfrac{p_{12}^2}{c_{11}+c_{12}+2c_{44}}\right\}I_0$

Krishnan discovered in experiments on liquids that for transverse observation the vertical component of the scattered radiation from a horizontally polarized beam is equal in intensity to the horizontal component of the radiation scattered from a vertically polarized beam. On the basis of the results of Leontowitsch and Mandelstam, Mueller[†] pointed out that the same reciprocity relation holds also in the case of cubic crystals. It is clear from the above table that the scattered radiation involved in such observations is due to the transverse elastic waves vibrating in the vertical direction.

[†] H. Mueller, *Proc. Roy. Soc. A,* **166**, 425 (1938).

APPENDIXES

I

SOME COMMON LATTICE STRUCTURES (p. 1)

In Fig. 28 (a), (b), (c) are shown the three cubic Bravais lattices, where a possible set of basic vectors and an elementary cell are indicated in each case. The CsCl structure on the one hand and the NaCl, diamond, and ZnS structures on the other hand are composite structures derived respectively from the simple cubic and the face-centred lattices; the corresponding basic vectors and elementary cells are the same as in the underlying Bravais lattices.

The dimensions of the structures can be specified by the edge length a of the unit cube. With the exception of that for the body-centred lattice, each figure given represents one unit cube of the structure; in the case of the body-centred lattice we have shown in the figure a layer of four unit cubes. For convenience of reference some data relating to the structures illustrated are given in the following table:

	$s.c.$	CsCl	$b.c.c.$	$f.c.c.$	NaCl	$diamond$	ZnS
Coordination no. M	6	8	8	12	6	4	4
Neighbour dist. r	a	$\sqrt3\,a/2$	$\sqrt3\,a/2$	$\sqrt2\,a/2$	$\sqrt3\,a/2$	$\sqrt3\,a/4$	$\sqrt3\,a/4$
Particles per cell	1	2	1	1	2	2	2
Base vectors $\mathbf{x}(1)-\mathbf{x}(0)$	..	$\left(\dfrac{a}{2},\dfrac{a}{2},\dfrac{a}{2}\right)$	..	..	$\left(\dfrac{a}{2},\dfrac{a}{2},\dfrac{a}{2}\right)$	$\left(\dfrac{a}{4},\dfrac{a}{4},\dfrac{a}{4}\right)$	$\left(\dfrac{a}{4},\dfrac{a}{4},\dfrac{a}{4}\right)$
Cell vol. v_a	a^3		$a^3/2$	$a^3/4$			

		$s.c.$, CsCl	$b.c.c.$	$f.c.c.$, NaCl, $diamond$, and ZnS
Basic vectors	$\mathbf{a_1}$	$(a, 0, 0)$	$\left(\dfrac{a}{2},\dfrac{a}{2},\dfrac{a}{2}\right)$	$\left(0,\dfrac{a}{2},\dfrac{a}{2}\right)$
	$\mathbf{a_2}$	$(0, a, 0)$	$\left(\dfrac{a}{2},\dfrac{a}{2},\dfrac{-a}{2}\right)$	$\left(\dfrac{a}{2}, 0,\dfrac{a}{2}\right)$
	$\mathbf{a_3}$	$(0, 0, a)$	$\left(\dfrac{-a}{2},\dfrac{a}{2},\dfrac{a}{2}\right)$	$\left(\dfrac{a}{2},\dfrac{a}{2}, 0\right)$
Reciprocal vectors	$\mathbf{b^1}$	$\left(\dfrac{1}{a}, 0, 0\right)$	$\left(\dfrac{1}{a}, 0,\dfrac{1}{a}\right)$	$\left(\dfrac{-1}{a},\dfrac{1}{a},\dfrac{1}{a}\right)$
	$\mathbf{b^2}$	$\left(0,\dfrac{1}{a}, 0\right)$	$\left(0,\dfrac{1}{a},\dfrac{-1}{a}\right)$	$\left(\dfrac{1}{a},\dfrac{-1}{a},\dfrac{1}{a}\right)$
	$\mathbf{b^3}$	$\left(0, 0,\dfrac{1}{a}\right)$	$\left(\dfrac{-1}{a},\dfrac{1}{a}, 0\right)$	$\left(\dfrac{1}{a},\dfrac{1}{a},\dfrac{-1}{a}\right)$
Reciprocal lattice		$s.c.$	$f.c.c.$	$b.c.c.$

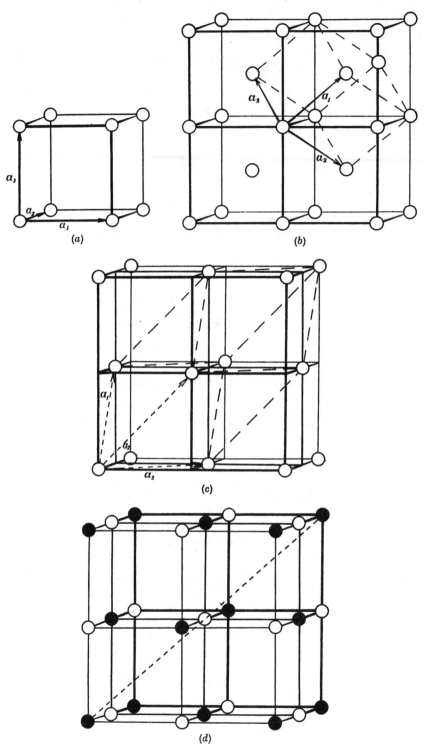

FIG. 28. (a) Simple cubic (s.c.); (b) body-centred cubic (b.c.c.);
(c) face-centred cubic (f.c.c.); (d) NaCl.

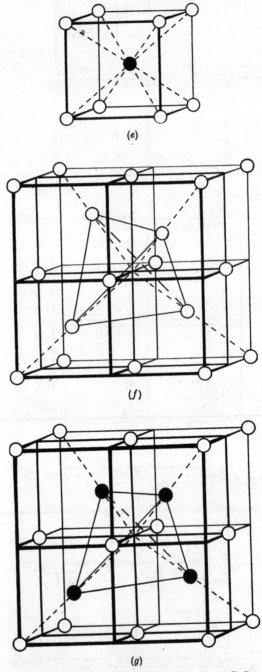

(e)

(f)

(g)

Fig. 28. (cont.). (e) CsCl; (f) diamond; (g) ZnS.

II

MADELUNG'S ENERGY (p. 3)

CONSIDER a general lattice structure, where the lattice sites are occupied by point charges. Let the charges on the sites k, measured in terms of the smallest charge ze in the lattice as unit, be ξ_k. For the cells to be electrically neutral we must have

$$\sum_k \xi_k = 0. \tag{II.1}$$

The electrostatic energy per cell can be written as follows:

$$\frac{\alpha'(ze)^2}{r}, \tag{II.2}$$

where r is the distance between the nearest sites and the Madelung constant α' is given by the lattice sum

$$\alpha' = \frac{1}{2} \sum_k \sum_{l'k'}' \frac{r\xi_k\xi_{k'}}{|\mathbf{x}(^{l'}_{k'}) - \mathbf{x}(k)|}. \tag{II.3}$$

As indicated by the prime over the summation sign, the sum does not include the terms with $(^{l'}_{k'}) = (^0_k)$.

. The sum can be converted into quickly convergent series by the method of theta-function transformation, which we have discussed in § 30. Let us consider the function

$$F(\mathbf{x}) = \sum_{l'k'} \frac{\xi_{k'}}{|\mathbf{x}(^{l'}_{k'}) - \mathbf{x}|}. \tag{II.4}$$

In terms of this function we can write the Madelung constant

$$\alpha' = \frac{1}{2} \sum_k r\xi_k \lim_{\mathbf{x} \to \mathbf{x}(k)} \left\{ F(\mathbf{x}) - \frac{\xi_k}{|\mathbf{x}(k) - \mathbf{x}|} \right\}. \tag{II.5}$$

The function $F(\mathbf{x})$ can be expressed in terms of integrals (see (30.11)):[†]

$$F(\mathbf{x}) = \int_0^\infty \left\{ \frac{2}{\sqrt{\pi}} \sum_{l'k'} \xi_{k'} \exp[-|\mathbf{x}(^{l'}_{k'}) - \mathbf{x}|^2\rho^2] \right\} d\rho. \tag{II.6}$$

In the theta-function transformation formula (30.16) we put $\mathbf{y} = 0$ and replace $\mathbf{x}$ by $\mathbf{x} - \mathbf{x}(k')$. Afterwards multiplying the formula by $\xi_{k'}$ and summing over k', we get

$$\frac{2}{\sqrt{\pi}} \sum_{l'k'} \xi_{k'} \exp\{-|\mathbf{x}(^{l'}_{k'}) - \mathbf{x}|^2\rho^2\}$$

$$= \frac{2\pi}{v_a} \sum_{k'} \xi_{k'} \sum_h' \frac{1}{\rho^3} \exp\left\{ -\frac{\pi^2}{\rho^2} |\mathbf{y}(h)|^2 + 2\pi i \mathbf{y}(h) \cdot [\mathbf{x} - \mathbf{x}(k')] \right\}, \tag{II.7}$$

where the term $h = 0$ may be omitted (indicated by a prime) in view of (II.1). Subdividing the integral in (II.6) and using the two alternative expressions for the

[†] Concerning the interchange of the order of summation and integration, see P. P. Ewald, *Ann. d. Phys.* **54**, 519 (1917); **64**, 253 (1921).

integrand given respectively by the two sides of (II.7), we obtain $F(\mathbf{x})$ in the form

$$F(\mathbf{x}) = \int\limits_{R}^{\infty} \left\{ \frac{2}{\sqrt{\pi}} \sum_{l'k'} \xi_{k'} \exp[-|\mathbf{x}(_{k'}^{l'}) - \mathbf{x}|^2 \rho^2] \right\} d\rho +$$

$$+ \int\limits_{0}^{R} \left\{ \frac{2\pi}{v_a} \sum_{k'} \xi_{k'} \sum_{h}' \frac{1}{\rho^3} \exp\left[-\frac{\pi^2}{\rho^2} |\mathbf{y}(h)|^2 + 2\pi i \mathbf{y}(h) \cdot [\mathbf{x} - \mathbf{x}(k')] \right] \right\} d\rho.$$

$$(\text{II}.8)$$

When (II.8) is substituted in (II.5) the Madelung constant becomes (see (30.18))

$$\alpha' = \tfrac{1}{2} \sum_{k} r\xi_k \lim_{\mathbf{x} \to \mathbf{x}(k)} \left\{ R \sum_{l'k'} \xi_{k'} H(R|\mathbf{x}(_{k'}^{l'}) - \mathbf{x}|) + \right.$$

$$\left. + \frac{\pi}{v_a R^2} \sum_{k'} \xi_{k'} \sum_{h}' G(\pi^2 |\mathbf{y}(h)|^2 / R^2) \exp[2\pi i \mathbf{y}(h) \cdot (\mathbf{x} - \mathbf{x}(k'))] - \frac{\xi_k}{|\mathbf{x}(k) - \mathbf{x}|} \right\}.$$

$$(\text{II}.9)$$

Apart from the last term and the term $(_{k'}^{l'}) = (_{k}^{0})$, all terms in (II.9) are regular functions of $\mathbf{x}$ in the neighbourhood of $\mathbf{x}(k)$. In the latter terms we can directly put $\mathbf{x}$ equal to $\mathbf{x}(k)$ and obtain

$$\alpha' = \frac{R}{2} \sum_{l'k'k}' r\xi_k \xi_{k'} H(R|\mathbf{x}(_{k'k}^{l'})|) +$$

$$+ \frac{\pi}{2v_a R^2} \sum_{kk'} \sum_{h}' r\xi_k \xi_{k'} G(\pi^2 |\mathbf{y}(h)|^2 / R^2) \exp[2\pi i \mathbf{y}(h) \cdot (\mathbf{x}(k) - \mathbf{x}(k'))] +$$

$$+ \frac{1}{2} \sum_{k} r\xi_k^2 \lim_{\mathbf{x} \to \mathbf{x}(k)} \left\{ RH(R|\mathbf{x}(k) - \mathbf{x}|) - \frac{1}{|\mathbf{x}(k) - \mathbf{x}|} \right\}. \quad (\text{II}.10)$$

The value of the last term is easily found as follows:

$$\lim_{\zeta \to 0} \left\{ RH(R\zeta) - \frac{1}{\zeta} \right\} = \lim_{\zeta \to 0} \frac{2}{\sqrt{\pi}} \frac{1}{\zeta} \left\{ \int\limits_{R\zeta}^{\infty} e^{-x^2} dx - \frac{\sqrt{\pi}}{2} \right\}$$

$$= \lim_{\zeta \to 0} \frac{2}{\sqrt{\pi}} \frac{1}{\zeta} \left\{ - \int\limits_{0}^{R\zeta} e^{-x^2} dx \right\}$$

$$= \lim_{\zeta \to 0} \frac{2}{\sqrt{\pi}} \frac{1}{\zeta} \left\{ - \int\limits_{0}^{R\zeta} \left[1 - x^2 + \frac{x^4}{2} - \ldots \right] dx \right\}$$

$$= \frac{-2R}{\sqrt{\pi}}. \quad (\text{II}.11)$$

Therefore (II.10) reduces to

$$\alpha' = \frac{R}{2} \sum_{l'k'k}' r\xi_k \xi_{k'} H(R|\mathbf{x}(_{k'k}^{l'})|) + \frac{\pi}{2v_a R^2} \sum_{kk'} r\xi_k \xi_{k'} \sum_{h}' G(\pi^2 |\mathbf{y}(h)|^2 / R^2) \times$$

$$\times \exp[2\pi i \mathbf{y}(h) \cdot (\mathbf{x}(k) - \mathbf{x}(k'))] - \frac{R}{\sqrt{\pi}} \sum_{k} r\xi_k^2. \quad (\text{II}.12)$$

The arbitrary parameter R has the dimension of an inverse length. Therefore we may write

$$R = \frac{c}{r}, \tag{II.13}$$

where c is a pure number. For quick convergence of both series in (II.12), c is usually chosen as a small number of the order unity. For a given structure the cell volume v_a is proportional to the third power of r:

$$v_a = sr^3, \tag{II.14}$$

where s is a dimensionless constant. When R and v_a are expressed in terms of r (II.12) becomes

$$\alpha' = \frac{c}{2} \sum_{l'k'k}' \xi_k \xi_{k'} H(c|\mathbf{x}(_{k'k}^{l'})|/r) + \frac{\pi}{2sc^2} \sum_{kk'} \xi_k \xi_{k'} \sum_h' G(\pi^2|\mathbf{y}(h)|^2 r^2/c^2) \times$$
$$\times \exp\{2\pi i \mathbf{y}(h) . (\mathbf{x}(k) - \mathbf{x}(k'))\} - \frac{c}{\sqrt{\pi}} \sum_k \xi_k^2. \tag{II.15}$$

The expression is now completely independent of the absolute dimension of the lattice; for a suitable choice of c it may be evaluated without excessive labour.

For the method originally used by Madelung[†] and other general methods for evaluating the electrostatic energy, one may refer to the accounts given by Born in his book *Atomtheorie des festen Zustandes* and by Born and Mayer in their article[‡] in *Handbuch der Physik*.[§]

We quote below the Madelung constants for some common ionic structures. The constants are given for cubic lattices in the two alternative ways as explained in § 1. Thus if r and d denote respectively the nearest ion–ion distance and the lattice constant, the electrostatic energy per cell is

$$\frac{\alpha'(ze)^2}{r} = \frac{\alpha''(ze)^2}{d}. \tag{II.16}$$

	CsCl	NaCl	ZnS (*Zinc Blende*)	CaF$_2$ (*Fluorite*)	TiO$_2$ (*Rutile*)	Cu$_2$O (*Cuprite*)
α'	1·7627	1·7476	1·6381	5·0387	4·82	4·1155
α''	2·0354	3·4951	3·7829	11·6365	(not cubic)	9·5044

[†] E. Madelung, *Phys. Zeit.* **19**, 524 (1918).
[‡] M. Born and M. Göppert-Mayer, *Handb. d. Physik*, **24**, 2nd part, pp. 708–14 (1933).
[§] Further publications are quoted in a paper by O. Emersleben, *Math. Nachr.* **9**, 221 (1953).

EVALUATION OF SIMPLE LATTICE SUMS (p. 23)

A SIMPLE procedure for evaluating a sum

$$S_n = \sum_l{}' \frac{1}{|\mathbf{x}(l)|^n} \tag{III.1}$$

over a simple lattice is to obtain by direct summation the contributions by the lattice points $\mathbf{x}(l)$ within a certain radius R and to replace the summation over the points beyond this radius by an integral.

As will be seen presently, it is, however, not desirable to carry out the integration simply over the space beyond the radius R. Suppose that there are N points (counting also the point $\mathbf{x}(l) = 0$) within the radius R, we should carry out the integration beyond a radius ρ to be chosen as follows. If v_a is the cell volume, the average density of points is $1/v_a$; ρ is chosen equal to the radius of the sphere which contains exactly N points. That is, ρ is to be determined from the relation

$$\frac{4\pi\rho^3}{3v_a} = N. \tag{III.2}$$

Using this radius for the integration, we find that approximately

$$S_n = \sum_{|\mathbf{x}(l)| < R}{}' \frac{1}{|\mathbf{x}(l)|^n} + \frac{4\pi}{v_a} \int_\rho^\infty \frac{dr}{r^{n-2}} = \sum_{|\mathbf{x}(l)| < R}{}' \frac{1}{|\mathbf{x}(l)|^n} + \frac{1}{(3N)^{\frac{1}{3}n-1}(n-3)} \left(\frac{4\pi}{v_a}\right)^{\frac{1}{3}n}. \tag{III.3}$$

The disadvantage of using R directly as the lower limit for the integral is fairly evident. The points on a lattice can usually be segregated into a series of spherical shells. For two values of R chosen immediately within and outside such a shell, the corresponding integrals do not differ appreciably, if they are evaluated directly from R outwards, although the approximate values obtained in the two cases should differ by the total contribution of the points lying on the shell concerned. In the method suggested above, the difference in the number of points included in the direct summation is suitably compensated in the choice of the integration limit ρ.

The accuracy of (III.3) evidently increases with the radius R (or the number N). An idea of the error may be obtained by observing the variation of the calculated value with the value used for R.[†]

This simple method is obviously useful for large n. Another method applicable to all n down to $n = 4$ is based on the ϑ-transformation formula, but in a somewhat different way from that used by Ewald (see Appendix II) for the evaluation of electrostatic lattice sums.[‡] We explain the method for the lattice sum (III.1) extended over a simple cubic lattice.

Using the well-known formula

$$r^{-\frac{1}{2}n} = \frac{1}{\Gamma(\frac{1}{2}n)} \int_0^\infty e^{-r^2 u} u^{\frac{1}{2}n-1} \, du, \tag{III.4}$$

[†] R. Dh. Misra, *Proc. Camb. Phil. Soc.* **36**, 173 (1940), has given explicit upper and lower limits for the error committed by choosing a definite radius R.

[‡] This method, due to A. Erdélyi and M. Born, has been published in the paper by Misra quoted above.

we can represent the lattice sum (III.1) by an integral

$$S_n = \sum_l{}' \frac{1}{(l_1^2+l_2^2+l_3^2)^{\frac{1}{2}n}} = \frac{1}{\Gamma(\frac{1}{2}n)} \int_0^\infty u^{\frac{1}{2}n-1}\sigma(u)\, du, \tag{III.5}$$

where

$$\sigma(u) = \sum_l{}' \exp\{-(l_1^2+l_2^2+l_3^2)u\} = \left\{\sum_{l=-\infty}^{+\infty} \exp(-l^2u)\right\}^3 - 1. \tag{III.6}$$

Introducing one of the ϑ-functions well known from the theory of elliptic functions, namely $\vartheta_3(0,q) = \vartheta_3(q)$, one has

$$\sigma(u) = [\vartheta_3(e^{-u})]^3 - 1. \tag{III.7}$$

This function can be represented by rapidly converging power series,

$$\vartheta_3(q) = 1+2q+2q^4+2q^9+\dots \tag{III.8}$$

and satisfies the transformation law

$$\vartheta_3(e^{-\pi\beta}) = \frac{1}{\sqrt{\beta}}\vartheta_3(e^{-\pi/\beta}). \tag{III.9}$$

Now (III.5) can be written

$$S_n = \frac{\pi^{m+1}}{\Gamma(m+1)}s_m, \qquad m = \tfrac{1}{2}n-1, \tag{III.10}$$

with

$$s_m = \int_0^\infty \beta^m\{(\vartheta_3(e^{-\pi\beta}))^3 - 1\}\, d\beta. \tag{III.11}$$

In order to obtain quick convergence we split the integral into two parts, the first extended from 0 to 1, the second from 1 to ∞. In the first part we use the ϑ-transformation (III.9) and replace β by $1/\beta$; then the interval becomes 1 to ∞, and we have

$$s_m = \int_1^\infty \beta^m\{[\vartheta_3(e^{-\pi\beta})]^3 - 1\}\, d\beta + \int_1^\infty \frac{[\vartheta_3(e^{-\pi\beta})]^3}{\beta^{m+\frac{3}{2}}}\, d\beta - \frac{1}{m+1}. \tag{III.12}$$

From (III.8) we obtain the power series

$$[\vartheta_3(q)]^3 = 1+6q+12q^2+8q^3+6q^4+\dots. \tag{III.13}$$

We define the function

$$\phi_m(x) = \int_1^\infty \beta^m e^{-\beta x}\, d\beta \tag{III.14}$$

and obtain by substituting (III.13) into (III.12) and integrating term by term

$$s_m = 6\phi_m(\pi)+12\phi_m(2\pi)+8\phi_m(3\pi)+6\phi_m(4\pi)+\dots$$
$$+\frac{1}{m-\frac{1}{2}}+6\phi_{-m-\frac{1}{2}}(\pi)+12\phi_{-m-\frac{1}{2}}(2\pi)+8\phi_{-m-\frac{1}{2}}(3\pi)+$$
$$+6\phi_{-m-\frac{1}{2}}(4\pi)+\dots-\frac{1}{m+1}. \tag{III.15}$$

The functions $\phi_m(x)$ satisfy the recurrence formula

$$\phi_m(x) = \frac{e^{-x}}{x}+\frac{m}{x}\phi_{m-1}(x) \tag{III.16}$$

and reduce for $m = 0$, $-\frac{1}{2}$, -1 to the well-known tabulated functions

$$\left.\begin{aligned}
\phi_0(x) &= \frac{e^{-x}}{x} \\
\phi_{-\frac{1}{2}}(x) &= \sqrt{\left(\frac{\pi}{x}\right)} \left(1 - \Phi(x)\right) \\
\phi_{-1}(x) &= -\mathrm{Ei}(-x)
\end{aligned}\right\} , \qquad (III.17)$$

where $\Phi(x)$ is Gauss's error function and $\mathrm{Ei}(x)$ the logarithmic integral. From these three functions (III.17) one can obtain all the others for $m = 0$, $\pm\frac{1}{2}$, ± 1,..., with the help of (III.16).

A table of the $\phi_m(x)$ for the arguments needed in (III.15) is given by Misra (loc. cit.).

The method can easily be extended to other lattice sums, e.g. of the form

$$\sum_{l}{}' \frac{l_1^{n_1} l_2^{n_2} l_3^{n_3}}{(l_1^2 + l_2^2 + l_3^2)^{\frac{1}{2}n}},$$

where n, n_1, n_2, n_3 are integers, and to other types of lattices. For the face-centred and body-centred cubic lattice one can use combinations of the ϑ-functions ϑ_2 and ϑ_3 and their derivatives.

A still more general type of lattice sum is

$$\sum_{l}{}' \frac{l_1^{n_1} l_2^{n_2} l_3^{n_3}}{(l_1^2 + l_2^2 + l_3^2)^{\frac{1}{2}n}}\, e^{-2\pi i(l_1 v_1 + l_2 v_2 + l_3 v_3)},$$

where y_1, y_2, y_3 is a point of the reciprocal space. This has been treated by Born and Bradburn† with the ϑ-function method, and tables have been computed for some simple cases.

† Max Born and Mary Bradburn, *Proc. Camb. Phil. Soc.* **39**, 104 (1942).

THE APPROXIMATION TO THE VIBRATIONAL SPECTRUM WITH THE HELP OF THE CYCLIC BOUNDARY CONDITION (p. 45)

THE theorem that the distribution of the frequencies of a large crystal lattice of arbitrary shape is approximately the same as that of a hypothetical lattice with cyclic boundary conditions has been subject to attacks by Sir C. V. Raman and his school. Stressing the fact that the number of the observed absorption and Raman lines of crystals is small they reject the current theory which leads to a practically continuous vibrational spectrum and explains the small number of the optically observable frequencies by selection rules based on the actual physical circumstances. Although Raman's theory contradicts the laws of classical mechanics and quantum mechanics as well, it seems to be not superfluous to show that his main objection to the current theory, directed against the validity of the cyclic boundary condition, is not correct.

A rigorous and general proof of this theorem has been given by Ledermann (loc. cit., p. 45), who reduced it to an algebraic problem about the roots of the characteristic determinant (latent roots). The following algebraic theorem is proved:

If in a Hermitian matrix the elements of r rows and their corresponding columns are modified in any way whatever, provided only that the matrix remains Hermitian, then the number of latent roots which lie in a given interval cannot increase or decrease by more than $2r$.

The application of this theorem to the crystal problem is obvious. The frequencies are the latent roots of a Hermitian, in fact a symmetric, matrix (see (24.4), p. 223). If the boundary is modified the matrix elements representing the interaction of the boundary with interior atoms are changed; as each coordinate of a particle corresponds to one line and one column of the matrix, a finite number r of lines and corresponding columns are affected, and this number r is relatively small for a large crystal. Thus the change of the distribution ($2r$ in any interval of frequency) is negligible.

This consideration is rigorous but is not quite simple. Therefore Peierls has offered another simpler proof † for the most important part of the statement, namely: consider a piece of the crystal similar in shape to the single cell and having a free surface; its frequency distribution is practically the same as that of a hypothetical lattice of the same shape and with cyclic boundary conditions.

It is, of course, not claimed that the normal modes in both cases are identical, but merely that the distribution of the frequencies, i.e. their average number in an interval $\Delta\omega$, large enough to contain many frequencies, is the same in the two cases up to the leading order in L, where L is the linear dimension of the crystal in terms of the lattice spacing.

For the sake of simplicity it is further assumed that the lattice has cubic symmetry and that the shape of the cell and of the whole crystal is cubic. The

† R. E. Peierls, *Proc. National Institute of Science of India*, **20**, 121 (1954). Very similar reasoning was used in an earlier paper by the same writer in *Monthly Notices R.A.S.* **96**, 780 (1936) concerned with justifying the cyclic boundary condition in the derivation of the equation of state of a relativistic gas.

argument could easily be freed from these restrictions, at the expense of some slight complexity in notation.

The idea of the proof is to show a connexion between the frequency distribution and the propagation of disturbances through the crystal. In particular, from the knowledge of the propagation of disturbances over times up to a certain time τ, we can uniquely determine the frequency distribution except for its fine structure which concerns frequency intervals less than $1/\tau$. Now a disturbance starting from a point at a distance d from the nearest surface will propagate in the same manner as in an infinite crystal for a time less than d/c, where c is the maximum velocity of sound, which is of the order 10^5 cm./sec. Hence for a crystal dimension of 1 cm. d/c will be of the order of 10^{-5} sec. and certainly larger than 10^{-10} sec. for practically the whole crystal. It follows, therefore, that over such times the propagation of disturbances is the same in the real and fictitious crystal. Hence also the frequency distribution is the same except for oscillations in frequency intervals of the order of 10^{10} sec.$^{-1}$ (or in spectroscopic terminology about $0\cdot3$ cm.$^{-1}$).

The number of cells along a side of the cubic crystal considered may be called L, so that the cell indices $l(l_1, l_2, l_3)$ are restricted by

$$1 \leqslant l_\alpha \leqslant L \quad (\alpha = 1, 2, 3). \tag{IV.1}$$

The number of particles in a cell may be denoted, as before, by n, so that

$$1 \leqslant k \leqslant n. \tag{IV.2}$$

There exists† a set of normal coordinates $q_j(t)$ connected with the displacements $u_\alpha\binom{l}{k}t)$ by linear relations

$$u_\alpha\binom{l}{k}t) = \sum_j e_\alpha\binom{l}{k}j)q_j(t), \tag{IV.3}$$

where the $e_\alpha\binom{l}{k}j)$ satisfy the orthogonality conditions

$$\sum_{lk\alpha} e_\alpha\binom{l}{k}j)e_\alpha\binom{l}{k}j')m_k = \delta_{jj'}, \tag{IV.4}$$

and the index j is restricted by

$$1 \leqslant j < 3nL^3. \tag{IV.5}$$

The $q_j(t)$ satisfy the second-order differential equation of an oscillator

$$\ddot{q}_j + \omega_j^2 q_j = 0,$$

which has the two independent solutions

$$q_j = \begin{cases} A_j \cos \omega_j t, \\ B_j \sin \omega_j t, \end{cases} \tag{IV.6}$$

from which one obtains the displacement $u_\alpha\binom{l}{k}t)$ by substituting in (IV. 3).

The vectors $e_\alpha\binom{l}{k}j)$ are found, together with the frequencies ω_j, from the equations of motion, and the constants A_j, B_j from the initial conditions.

In order to represent the frequency distribution we define the function

$$D(x) = \begin{cases} 1 & \text{for } |x| < \tfrac{1}{2}\Delta, \\ 0 & \text{,, } |x| > \tfrac{1}{2}\Delta. \end{cases} \tag{IV.7}$$

Then the number of frequencies in the interval $\omega + \tfrac{1}{2}\Delta$, $\omega - \tfrac{1}{2}\Delta$ is given by

$$F(\omega) = \sum_j D(\omega_j - \omega). \tag{IV.8}$$

† The more elaborate notation of § 38 is not necessary here.

As we are not interested in the exact values of the frequencies we replace this discontinuous function by a smooth function, defined with the help of a continuous weight function, also denoted by $D(x)$ and defined by the following properties:

$$\int_{-\infty}^{\infty} D(x)\,dx = \Delta, \qquad D(x) \ll 1, \quad \text{if } x \gg \Delta. \qquad (IV.9)$$

In addition we assume that $D(x)$ is so smooth that its Fourier transform converges rapidly. We write

$$g(\tau) = \frac{1}{2\pi} \int_{-\infty}^{\infty} e^{-i\tau\omega} D(\omega)\,d\omega, \qquad (IV.10)$$

$$D(\omega) = \int_{-\infty}^{\infty} e^{i\tau\omega} g(\tau)\,d\tau; \qquad (IV.11)$$

then the condition $\qquad\qquad g(\tau) \ll \Delta \quad \text{if } \tau \gg 1/\Delta \qquad (IV.12)$

is compatible with (IV.9). For instance, a possible choice of D is

$$D(\omega) = \frac{1}{\sqrt{(2\pi)}}\, e^{-\omega^2/2\Delta^2}, \qquad g(\tau) = \frac{\Delta}{2\pi}\, e^{-\Delta^2\tau^2/2}.$$

It is clear that, with this definition of D, the expression (IV.8) represents the spectrum of vibrations except for the fine structure, provided Δ is small compared with the maximum frequency ω_{max}.

It will be convenient to replace $F(\omega)$ by

$$G(\omega) = F(\omega) + F(-\omega)$$
$$= \sum_j \{D(\omega_j - \omega) + D(\omega_j + \omega)\}. \qquad (IV.13)$$

Here the extra term $D(\omega_j + \omega)$ is practically negligible for all but the lowest frequencies of the spectrum, where $\omega \sim \Delta$; for otherwise, as $\omega_j > 0$, $\omega_j + \omega$ will appreciably exceed Δ, hence according to (IV.9) $D(\omega_j + \omega)$ will be small.

Now we express $G(\omega)$ in terms of the Fourier transform $g(\tau)$ of $D(\omega)$ by inserting (IV.11):

$$G(\omega) = \sum_j \int_{-\infty}^{\infty} d\tau\, g(\tau)[e^{i\omega_j\tau}e^{-i\omega\tau} + e^{i\omega_j\tau}e^{i\omega\tau}]$$

$$= \int_{-\infty}^{\infty} d\tau\, 2g(\tau)\cos\omega\tau \sum_j e^{i\omega_j\tau}. \qquad (IV.14)$$

If $D(x)$ is an even function, $g(\tau)$ will also be even, hence

$$G(\omega) = 4\int_0^{\infty} d\tau\, g(\tau)\cos\omega\tau \sum_j \cos\omega_j\tau. \qquad (IV.15)$$

However, because of (IV.12) it is sufficient to include Fourier components belonging to values of τ not much greater than $1/\Delta$, say γ/Δ, where γ is a numerical constant of order 1. Therefore we can write

$$G(\omega) = 4\int_0^{\gamma/\Delta} d\tau\, g(\tau)\cos\omega\tau\, \Phi(\tau), \qquad (IV.16)$$

where $\qquad\qquad \Phi(\tau) = \sum_j \cos\omega_j\tau, \quad \tau < \gamma/\Delta. \qquad (IV.17)$

Now it can be shown that this $\Phi(\tau)$ has a simple physical significance. Using the orthogonality relation (IV.4) we can write

$$\Phi(\tau) = \sum_j \sum_{lk\alpha} m_k e_\alpha^2(^l_k j) \cos \omega_j \tau$$

$$= \sum_{lk\alpha} \Phi_\alpha(^l_k \tau),$$ (IV.18)

where $$\Phi_\alpha(^l_k \tau) = m_k \sum_j e_\alpha^2(^l_k j) \cos \omega_j \tau.$$ (IV.19)

This quantity has the following meaning. At $t = 0$ let all atoms be in their equilibrium positions except that the atom of kind k in the cell l is displaced by an infinitesimal amount u in the α-direction and is there released from rest. Then the value of the α-coordinate of this atom at time τ is exactly $u\Phi_\alpha(^l_k \tau)$.

To see this, remember that the general solution of the equation of motion is given by $$u_\alpha(^l_k t) = \sum_j e_\alpha(^l_k j)[A_j \cos \omega_j t + B_j \sin \omega_j t],$$ (IV.20)

where the A_j and B_j are arbitrary constants. These can be expressed in terms of the initial positions and velocities of all atoms; one has

$$u_\alpha(^l_k 0) = \sum_j A_j e_\alpha(^l_k j),$$

$$\dot{u}_\alpha(^l_k 0) = \sum_j \omega_j B_j e_\alpha(^l_k j),$$ (IV.21)

and these equations can be solved with the help of the orthogonality relation (IV.4): $$A_j = \sum_{kl\alpha} m_k u_\alpha(^l_k 0) e_\alpha(^l_k j),$$

$$\omega_j B_j = \sum_{kl\alpha} m_k \dot{u}_\alpha(^l_k 0) e_\alpha(^l_k j).$$ (IV.22)

Substituting this in (IV.20) we get

$$u_\alpha(^l_k t) = \sum_j e_\alpha(^l_k j) \sum_{k'l'\beta} e_\beta(^{l'}_{k'} j) m_{k'} \left\{ u_\beta(^{l'}_{k'} 0) \cos \omega_j t + \frac{1}{\omega_j} \dot{u}_\beta(^{l'}_{k'} 0) \sin \omega_j t \right\}.$$ (IV.23)

In the special case in which at $t = 0$ the atom l_0, k_0 is pushed out in the α_0-direction from its equilibrium position by the amount u and released without impulse, while all the other atoms are kept at rest in their equilibrium position, one has

$$u_\alpha(^l_k 0) = u\delta_{ll_0}\delta_{kk_0}\delta_{\alpha\alpha_0}, \qquad \dot{u}_\alpha(^l_k 0) = 0.$$ (IV.24)

Then (IV.23) reduces to

$$u_\alpha(^l_k t) = u m_{k_0} \sum_j e_{\alpha_0}(^{l_0}_{k_0} j) e_\alpha(^l_k j) \cos \omega_j t.$$ (IV.25)

The result announced follows now by taking $l_0 = l$, $k_0 = k$, $\alpha_0 = \alpha$, $t = \tau$:

$$u_\alpha(^l_k \tau) = u\Phi_\alpha(^l_k \tau).$$ (IV.26)

Now the frequency distribution $F(\omega)$, which is practically equivalent to the function $G(\omega)$, is determined by $\Phi(\tau)$; $g(\tau)\Phi(\tau)$ is the Fourier transform of $G(\omega)$, and $\Phi(\tau)$ consists, according to (IV.18), of the additive contributions $\Phi_\alpha(^l_k \tau)$ of the single particles. The theorem to be proved is therefore reduced to the following two statements. (I) Each function $\Phi_\alpha(^l_k \tau)$, which contributes essentially to the sum (IV.18), is practically identical for the finite lattice with that for the infinite lattice $\Phi_\alpha^\infty(^l_k \tau)$. (II) The sum (IV.18) $\Phi(\tau)$ of these contributions for the finite lattice is practically identical with the corresponding sum $\Phi^\infty(\tau)$ for the infinite lattice.

To demonstrate (I), let $d(l)$ denote the distance of the cell l from the nearest boundary of the finite crystal. According to (IV.26) $\Phi_\alpha(^l_k\tau)$ represents the action of the cell on itself; this action will be the discontinuous analogue of a spherical wave with the centre at l and reflected at the boundary, and this will be almost independent of the boundary for a time $t < 2d(l)/c$, as then the wave will not have returned to the source. Here c represents the maximum of the group velocity for all waves, which is of the same order of magnitude as the ordinary velocity of sound. Now according to (IV.17) τ is restricted, for a given frequency interval Δ, by $\tau < \gamma/\Delta$, where $\gamma \sim 1$. Hence only the waves from those cells for which $d(l) < c/2\Delta$ are strongly influenced by the wall. Or in other words, the functions $\Phi_\alpha(^l_k\tau)$ for which $d(l) > c/2\Delta$ are practically the same as the functions $\Phi_\alpha^\infty(^l_k\tau)$ for the infinite crystal.

Now consider the statement (II). We divide the crystal into an interior and a surface part, by a surface at the distance $d = c/2\Delta$ from the boundary, where Δ is a fixed frequency interval which determines the accuracy of frequency measurement. Let us assume that Δ is a certain fraction β of the maximum frequency ω_{max} of the crystal and d a certain fraction δ of the side S of the cube,

$$\Delta = \beta\omega_{max}, \qquad d = \delta S; \qquad (IV.27)$$

then

$$\delta\beta = \frac{c}{2\omega_{max}S} = \frac{\lambda_{min}}{4\pi S},$$

where $\lambda_{min} = 2\pi c/\omega_{max}$ is the corresponding minimum wave-length. Now this latter is of the order of the lattice constant a, $\lambda_{min} \sim 2\pi a$, and $S = La$, hence

$$\delta\beta \sim \frac{1}{2L}. \qquad (IV.28)$$

Thus the two fractions δ and β are reciprocal; the smaller the frequency interval Δ the thicker the boundary layer. (This is quite similar to Heisenberg's uncertainty relations.) The ratio of the number of atoms in the boundary layer to those in the interior is obviously given by

$$\epsilon = \frac{N_{surface}}{N_{interior}} = \frac{6dS^2}{S^3} \sim 6\delta;$$

hence from (IV.28)

$$\epsilon\beta \sim \frac{3}{L}. \qquad (IV.29)$$

Now for a crystal of 1 cm. length one has $L \sim 10^8$, and therefore ϵ can be small even if β is a small fraction; e.g. for $\beta \sim 10^{-4}$ one has $\epsilon \sim 3 \times 10^{-4}$. Thus the contribution of the surface layer to the sum (IV.18) is negligible, and the statement (II) is proved.

The formulae (IV.28) and (IV.29) show at the same time the limitations of this statement. If a very high accuracy of frequency measurement is demanded, i.e. if β is very small, then ϵ cannot be very small, hence the surface atoms contribute to the sum (IV.18) and the sums $\Phi(\tau)$ and $\Phi^\infty(\tau)$ differ appreciably.

The proof shows actually that the spectral distribution is independent of the boundary and therefore establishes the practical identity of the distribution for the finite, the infinite, or the cyclic lattice. The latter has the advantage that the spectrum can be determined by elementary methods, as shown in the text of the book.

V

ENERGY DENSITY IN IONIC CRYSTALS (p. 83)

WE shall show that the relation

$$b_{12} = b_{21} \tag{V.1}$$

in the phenomenological theory given in § 7 follows from the principle of energy conservation.

Let us place a point charge e at the origin, while keeping the ions in the configuration $\mathbf{w}(\mathbf{x}) \equiv 0$, and consider the following cycle:

(a) Keeping the charge fixed, displace the ions into the 'irrotational' configuration:

$$\mathbf{w}(\mathbf{x}) = \nabla\psi(\mathbf{x})$$

continuously, according to

$$\mathbf{w}(\mathbf{x}) = \xi\nabla\psi(\mathbf{x}), \tag{V.2}$$

by raising ξ from 0 to 1, the function ψ being arbitrary.

(b) Keeping the ions at $\mathbf{w}(\mathbf{x}) = \nabla\psi(\mathbf{x})$, move the charge to $\Delta\mathbf{x}$.

(c) With the charge fixed at $\Delta\mathbf{x}$, reverse the process (a), namely, by reducing ξ from 1 to 0 in accordance with (V.2).

(d) Finally complete the cycle by restoring the charge to the origin. The total work required to complete the cycle must clearly vanish.

The right-hand side of the equation (7.1), namely,

$$\mathbf{f} = b_{11}\mathbf{w} + b_{12}\mathbf{E}, \tag{V.3}$$

represents the force conjugate to $\mathbf{w}$ (refer, for instance, to the microscopic treatment in § 9); in other words, $-\mathbf{f}.\Delta\mathbf{w}$ is the work per unit volume required to change $\mathbf{w}$ to $\mathbf{w} + \Delta\mathbf{w}$. Thus generally to move the ions from a configuration $\mathbf{w}(\mathbf{x})$ to $\mathbf{w}(\mathbf{x}) + \Delta\mathbf{w}(\mathbf{x})$, the total work expended on the system is given by the integral

$$-\int \mathbf{f}(\mathbf{x}).\Delta\mathbf{w}(\mathbf{x})\,d\mathbf{x} = -\int \{b_{11}\mathbf{w}(\mathbf{x}) + b_{12}\mathbf{E}(\mathbf{x})\}.\Delta\mathbf{w}(\mathbf{x})\,d\mathbf{x}. \tag{V.4}$$

Using the first part of (7.23) (for its derivation, no use has been made of the relation $b_{12} = b_{21}$) we find that during the process (a) the electric field is given by

$$\mathbf{E}(\mathbf{x}) = \frac{-4\pi b_{21}}{1 + 4\pi b_{22}}\xi\nabla\psi(\mathbf{x}) - \frac{1}{1 + 4\pi b_{22}}\nabla\phi(\mathbf{x}), \tag{V.5}$$

where the potential of the point charge *in vacuo*

$$\phi(\mathbf{x}) = \frac{e}{|\mathbf{x}|}$$

satisfies the Poisson equation

$$\nabla^2\phi(\mathbf{x}) = -4\pi e\delta(\mathbf{x}), \tag{V.6}$$

$\delta(\mathbf{x})$ being the Dirac δ-function. The infinitesimal ion displacements during (a) are (see (V.2))

$$\Delta\mathbf{w}(\mathbf{x}) = \{\nabla\psi(\mathbf{x})\}\Delta\xi. \tag{V.7}$$

Substituting (V.2), (V.5), (V.7) in (V.4) and integrating ξ from 0 to 1, we obtain for the work expended during (a):

$$\int \left\{\frac{1}{2}\left[\frac{4\pi b_{12}b_{21}}{1 + 4\pi b_{22}} - b_{11}\right](\nabla\psi(\mathbf{x})).(\nabla\psi(\mathbf{x})) + \frac{b_{12}}{1 + 4\pi b_{22}}(\nabla\phi(\mathbf{x})).(\nabla\psi(\mathbf{x}))\right\}d\mathbf{x}. \tag{V.8}$$

The second term on the right-hand side of (V.5) represents the spherically symmetric field due to the point charge and exerts no force on the latter. The field acting on the charge during the process (b) is thus obtained by putting $\xi = 1$ and $x = 0$ (to the first order of accuracy) in the first term on the right-hand side of (V.5), namely,

$$\frac{-4\pi b_{21}}{1+4\pi b_{22}}\nabla\psi(0).$$

Therefore the work expended in moving the charge to $\Delta\mathbf{x}$ is given by

$$\frac{4\pi b_{21}e}{1+4\pi b_{22}}(\nabla\psi(0)).\Delta\mathbf{x}. \tag{V.9}$$

The process (c) is the exact reverse of (a), apart from the altered position of the charge. Thus, upon reversing the sign of (V.8) and replacing $\phi(\mathbf{x})$ by $\phi(\mathbf{x}-\Delta\mathbf{x})$, we obtain the work done during (c):

$$-\int\left\{\frac{1}{2}\left[\frac{4\pi b_{12}b_{21}}{1+4\pi b_{22}}-b_{11}\right](\nabla\psi(\mathbf{x})).(\nabla\psi(\mathbf{x}))+ \right.$$
$$\left. +\frac{b_{12}}{1+4\pi b_{22}}(\nabla\phi(\mathbf{x}-\Delta\mathbf{x})).(\nabla\psi(\mathbf{x}))\right\}d\mathbf{x}. \tag{V.10}$$

In the last step, $\mathbf{w}(\mathbf{x}) \equiv 0$ and no field acts on the charge. Therefore no work is required to move the charge back to the origin.

Putting the sum of (V.8), (V.9), and (V.10) equal to zero, we get

$$\frac{b_{12}}{1+4\pi b_{22}}\left\{\int(\nabla\phi(\mathbf{x})).(\nabla\psi(\mathbf{x}))\,d\mathbf{x} - \int(\nabla\phi(\mathbf{x}-\Delta\mathbf{x})).(\nabla\psi(\mathbf{x}))\,d\mathbf{x}\right\}+$$
$$+\frac{4\pi b_{21}e}{1+4\pi b_{22}}(\nabla\psi(0)).\Delta\mathbf{x} = 0. \tag{V.11}$$

With the help of Green's theorem, we find that

$$\int(\nabla\phi(\mathbf{x})).(\nabla\psi(\mathbf{x}))\,d\mathbf{x} = -\int\nabla^2\phi(\mathbf{x})\psi(\mathbf{x})\,d\mathbf{x} = 4\pi e\psi(0),$$

where use has been made of (V.6). Similarly

$$\int(\nabla\phi(\mathbf{x}-\Delta\mathbf{x})).(\nabla\psi(\mathbf{x}))\,d\mathbf{x} = -\int\nabla^2\phi(\mathbf{x}-\Delta\mathbf{x})\psi(\mathbf{x})\,d\mathbf{x} = 4\pi e\psi(\Delta\mathbf{x}).$$

Hence

$$\int(\nabla\phi(\mathbf{x})).(\nabla\psi(\mathbf{x}))\,d\mathbf{x} - \int(\nabla\phi(\mathbf{x}-\Delta\mathbf{x})).(\nabla\psi(\mathbf{x}))\,d\mathbf{x}$$
$$= 4\pi e\{\psi(0)-\psi(\Delta\mathbf{x})\} = -4\pi e(\nabla\psi(0)).\Delta\mathbf{x}. \tag{V.12}$$

Taking account of this relation, we find immediately that (V.11) reduces to the relation (V.1), which we set out to prove.

One may observe that the relation (V.1) makes it possible to deduce the phenomenological equations from an energy density as a function of $\mathbf{w}$ and $\mathbf{E}$:

$$u = -\tfrac{1}{2}\{b_{11}\mathbf{w}^2+2b_{12}\mathbf{w}.\mathbf{E}+b_{22}\mathbf{E}^2\}. \tag{V.13}$$

Thus
$$\ddot{w}_\alpha = f_\alpha = -\frac{\partial u}{\partial w_\alpha}, \qquad P_\alpha = -\frac{\partial u}{\partial E_\alpha}. \tag{V.14}$$

That an energy density of this form should exist for an ionic crystal is by no means *a priori* obvious. For, as already emphasized in the text, the properties of a volume element cut off from an ionic lattice are essentially shape-dependent and the field $\mathbf{E}$ cannot be considered as an externally imposed field in a simple sense.

An energy density (or free energy density) of this nature, we remember, has also been postulated in the general discussions given in § 33 and § 37.

THE INNER FIELD IN UNIFORMLY POLARIZED
CRYSTALS WITH TETRAHEDRAL SYMMETRY
(THE LORENTZ FIELD) (p. 104)

WE shall first establish a simple theorem on tensors which are invariant under the tetrahedral group of operations.

Operating on any vector $\mathbf{X}(X_1, X_2, X_3)$, a Cartesian tensor $T_{\alpha\beta}$ transforms the vector into another vector $\mathbf{Y}$ with the components

$$Y_\alpha = \sum_\beta T_{\alpha\beta} X_\beta. \tag{VI.1}$$

Suppose that a certain rotation sends $\mathbf{X}$ and $\mathbf{Y}$ into $\mathbf{X}'$ and $\mathbf{Y}'$ respectively. The tensor $T_{\alpha\beta}$ is said to be invariant under the rotation, if, for any arbitrary vector $\mathbf{X}$,

$$Y'_\alpha = \sum_\beta T_{\alpha\beta} X'_\beta; \tag{VI.2}$$

in other words, if a vector $\mathbf{X}$ is subject to the rotation, the transformed vector $\mathbf{Y}$ is similarly rotated. We shall show that if a tensor $T_{\alpha\beta}$ is invariant under the tetrahedral group of rotations, namely, rotations by π about three mutually perpendicular axes (2-fold axes) and rotations by $\pm 2\pi/3$ about the corresponding diagonal directions (3-fold axes), the tensor is equivalent to a scalar; i.e.

$$T_{\alpha\beta} = T\delta_{\alpha\beta}. \tag{VI.3}$$

Let $\mathbf{X}$ and $\mathbf{Z}$ be any two arbitrary vectors. If $T_{\alpha\beta}$ is invariant under a rotation which sends $\mathbf{X}, \mathbf{Z}$, and $\mathbf{Y}$ (the transformed vector of $\mathbf{X}$) into $\mathbf{X}', \mathbf{Z}'$, and $\mathbf{Y}'$, we have evidently the relation

$$\sum_{\alpha\beta} Z_\alpha T_{\alpha\beta} X_\beta = \sum_{\alpha\beta} Z'_\alpha T_{\alpha\beta} X'_\beta, \tag{VI.4}$$

in view of the fact that the scalar product between two vectors is invariant under any rotation, and so
$$\sum_\alpha Z_\alpha Y_\alpha = \sum_\alpha Z'_\alpha Y'_\alpha.$$

Let the Cartesian axes be chosen in the directions of the 2-fold axes. Then, for the rotation about the first 2-fold axis,

$$X'_1 = X_1, \qquad X'_2 = -X_2, \qquad X'_3 = -X_3,$$
$$Z'_1 = Z_1, \qquad Z'_2 = -Z_2, \qquad Z'_3 = -Z_3. \tag{VI.5}$$

Accordingly, (VI.4) reduces to

$$T_{12} Z_1 X_2 + T_{21} Z_2 X_1 + T_{13} Z_1 X_3 + T_{31} Z_3 X_1$$
$$= -T_{12} Z_1 X_2 - T_{21} Z_2 X_1 - T_{13} Z_1 X_3 - T_{31} Z_3 X_1. \tag{VI.6}$$

$\mathbf{Z}$ and $\mathbf{X}$ being arbitrary, it follows that

$$T_{12} = T_{21} = T_{13} = T_{31} = 0. \tag{VI.7}$$

Similar considerations of the rotations about the other 2-fold axes lead to relations which can be obtained from (VI.6) by cyclic permutation of the indices. Therefore we find further that
$$T_{23} = T_{32} = 0. \tag{VI.8}$$

Thus all the non-diagonal components of $T_{\alpha\beta}$ are equal to zero, and (VI.4) reduces to

$$\sum_\alpha T_{\alpha\alpha} Z_\alpha X_\alpha = \sum_\alpha T_{\alpha\alpha} Z'_\alpha X'_\alpha. \qquad (VI.9)$$

Further consideration of a rotation about a 3-fold axis, for instance

$$X'_2 = X_1, \qquad X'_3 = X_2, \qquad X'_1 = X_3,$$
$$Z'_2 = Z_1, \qquad Z'_3 = Z_2, \qquad Z'_1 = Z_3, \qquad (VI.10)$$

shows immediately that all the diagonal components $T_{\alpha\alpha}$ ($\alpha = 1, 2, 3$) are equal.

In § 30 it has been shown that the electric field at a lattice point (the exciting field) in a dipole lattice (see p. 253) is the sum of the macroscopic field at the point and an inner field; and a unique limit exists for the latter for large wave-lengths of the dipolar polarization. Put more physically, when conditions are uniform over a large number of cells so that macroscopic considerations are possible, the inner field is completely determined by the local conditions (in the macroscopic sense). In such a microscopically homogeneous neighbourhood, the inner field is given by the second term in (30.30) in the limit $\mathbf{y} = 0$, namely,

$$\sum_{k'\beta} Q_{\alpha\beta}(\begin{smallmatrix}0\\kk'\end{smallmatrix}) p_\beta(k'), \qquad (VI.11)$$

where, according to (30.31),

$$Q_{\alpha\beta}(\begin{smallmatrix}0\\kk'\end{smallmatrix}) = R^3 \sum_{l'} H_{\alpha\beta}(R\mathbf{x}(\begin{smallmatrix}l'\\k'k\end{smallmatrix})) -$$

$$-\frac{4\pi^3}{R^2 v_a} \sum_h{}' y_\alpha(h) y_\beta(h) G(\pi^2 |\mathbf{y}(h)|^2/R^2) \exp\{2\pi i \mathbf{y}(h) \cdot (\mathbf{x}(k) - \mathbf{x}(k'))\}, \quad (VI.12)$$

$\mathbf{p}(k')$ being the dipoles on the different types of sites.

Consider a lattice with tetrahedral symmetry such that under an operation of the tetrahedral group each constituent Bravais lattice is separately invariant. Then $Q_{\alpha\beta}(\begin{smallmatrix}0\\kk'\end{smallmatrix})$, considered as a Cartesian tensor (indices α, β), is invariant under the tetrahedral group of operations. In order to see that this is the case, let us consider the following two alternative processes:

(a) We rotate all the dipoles $\mathbf{p}(k')$ on the sites k' each on its own site.

(b) Alternatively we may imagine the dipoles completely fixed in the lattice structure and rotate the lattice as a whole.

If the k' lattice is invariant under the rotation concerned, the two alternative processes clearly lead to exactly the same final arrangement of dipoles. After the process (a) the inner field at a site k due to dipoles on sites k' is given by

$$\sum_\beta Q_{\alpha\beta}(\begin{smallmatrix}0\\kk'\end{smallmatrix}) p'_\beta(k'),$$

where $\mathbf{p}'(k')$ represents the dipoles after rotation; whereas the process (b) simply rotates the original inner field (due to the dipoles on sites k'),

$$E_\alpha = \sum_\beta Q_{\alpha\beta}(\begin{smallmatrix}0\\kk'\end{smallmatrix}) p_\beta(k'), \qquad (VI.13)$$

with the lattice, sending it into $\mathbf{E}'$. Since the inner field must be the same in both cases, we have

$$E'_\alpha = \sum_\beta Q_{\alpha\beta}(\begin{smallmatrix}0\\kk'\end{smallmatrix}) p'_\beta(k'). \qquad (VI.14)$$

(VI.13) and (VI.14), we note, are identical with (VI.1) and (VI.2), showing that $Q_{\alpha\beta}(\begin{smallmatrix}0\\kk'\end{smallmatrix})$ is an invariant tensor under the rotation.

Thus, for structures with the symmetry stipulated above, $Q_{\alpha\beta}(^{0}_{kk'})$ is equivalent to a scalar, which we denote by $Q(^{0}_{kk'})$; and we have the relation

$$Q(^{0}_{kk'}) = Q_{11}(^{0}_{kk'}) = Q_{22}(^{0}_{kk'}) = Q_{33}(^{0}_{kk'}) = \tfrac{1}{3}\sum_{\alpha} Q_{\alpha\alpha}(^{0}_{kk'}). \qquad (VI.15)$$

Let us use (VI.12) in the expression on the right-hand side and take for the arbitrary parameter R the limiting value $R \to 0$. Thus the sum over h drops out, leaving

$$Q(^{0}_{kk'}) = \tfrac{1}{3}\lim_{R\to 0} R^3 \sum_{l'}\sum_{\alpha} H_{\alpha\alpha}(R\mathbf{x}(^{l'}_{k'k}))$$

$$= \frac{2}{3\sqrt{\pi}}\lim_{R\to 0} R^3 \sum_{l'}\left\{ \sum_{\alpha} \frac{\partial^2}{\partial x_{\alpha}^2}\left[\frac{1}{|\mathbf{x}|}\int_{|\mathbf{x}|}^{\infty} e^{-x^2}\,dx\right]\right\}_{\mathbf{x}=R\mathbf{x}(^{l'}_{k'k})}. \qquad (VI.16)$$

Upon expressing $\mathbf{x}$ and the Laplacian operator

$$\sum_{\alpha} \frac{\partial^2}{\partial x_{\alpha}^2}$$

in polar coordinates, we find that

$$Q(^{0}_{kk'}) = \frac{2}{3\sqrt{\pi}}\lim_{R\to 0} R^3 \sum_{l'}\left\{\frac{1}{r^2}\frac{d}{dr}r^2\frac{d}{dr}\left[\frac{1}{r}\int_{r}^{\infty} e^{-x^2}\,dx\right]\right\}_{r=R|\mathbf{x}(^{l'}_{k'k})|}$$

$$= \frac{4}{3\sqrt{\pi}}\lim_{R\to 0} R^3 \sum_{l'} \exp\{-R^2|\mathbf{x}(^{l'}_{k'k})|^2\}. \qquad (VI.17)$$

We observe that as R is progressively reduced, the predominant contributions to the lattice sum in (VI.17) come from increasingly more distant lattice points (i.e. points with large $|\mathbf{x}(^{l'}_{k'k})|$). Hence in the limit $R \to 0$ we can rigorously replace the lattice sum by an integral. Thus remembering that there are $1/v_a$ points per unit volume, we find upon introducing

$$R\mathbf{x}(^{l'}_{k'k})$$

as the integration variable that

$$Q(^{0}_{kk'}) = \frac{4}{3\sqrt{\pi}\,v_a}\int_{0}^{\infty} e^{-|\mathbf{x}|^2}\,d\mathbf{x}. \qquad (VI.18)$$

The integral is easily evaluated with polar coordinates, giving

$$Q(^{0}_{kk'}) = \frac{16\sqrt{\pi}}{3v_a}\int_{0}^{\infty} e^{-r^2}r^2\,dr = \frac{4\pi}{3v_a}. \qquad (VI.19)$$

The formula is in complete agreement with the result obtained in § 9, for the inner field (VI.11) is now given by

$$\sum_{k'\beta} Q_{\alpha\beta}(^{0}_{kk'})p_{\beta}(k') = \sum_{k'\beta}\delta_{\alpha\beta}\,Q(^{0}_{kk'})p_{\beta}(k')$$

$$= \frac{4\pi}{3}\left\{\frac{1}{v_a}\sum_{k'} p_{\alpha}(k')\right\}, \qquad (VI.20)$$

the expression given in the curly brackets being clearly the α-component of the dielectric polarization.

For crystals with tetrahedral symmetry in general, the different constituent lattices are not necessarily separately invariant under the operations of the tetrahedral group. Thus such an operation may send the k-sites into the positions originally occupied by the k'-sites. For such crystals the different constituent lattices may be classified into groups; the members of each group can be sent into one another by certain of the operations of the tetrahedral group. In this general case we have to stipulate that the constituent lattices belonging to the same group must be occupied by identical dipoles. Under this stipulation it is easily seen that the same result as above follows. The proof is practically the same; the only difference is that instead of $Q_{\alpha\beta}(^0_{kk'})$, the sum over constituent lattices of the same g-group,

$$\sum_{k'}^{(\text{group})} Q_{\alpha\beta}(^0_{kk'}),$$

is an invariant tensor; and this sum replaces $Q_{\alpha\beta}(^0_{kk'})$ in every step of the proof given above for the simple case.

THE ADIABATIC APPROXIMATION (p. 170)

THE perturbation calculation in § 14 will now be carried on to the fourth-order terms.

The perturbation equations of the third and fourth orders for the electronic motion (see (14.14)) are

$$(H_0^{(0)} - \Phi_n^{(0)})\phi_n^{(3)} = -H_0^{(1)}\phi_n^{(2)} - (H_0^{(2)} - \Phi_n^{(2)})\phi_n^{(1)} - (H_0^{(3)} - \Phi_n^{(3)})\phi_n^{(0)}, \quad \text{(VII.1)}$$

$$(H_0^{(0)} - \Phi_n^{(0)})\phi_n^{(4)} = -H_0^{(1)}\phi_n^{(3)} - (H_0^{(2)} - \Phi_n^{(2)})\phi_n^{(2)} - (H_0^{(3)} - \Phi_n^{(3)})\phi_n^{(1)} - (H_0^{(4)} - \Phi_n^{(4)})\phi_n^{(0)}; \quad \text{(VII.2)}$$

the similar equations for the rigorous molecular motion (see (14.19)), on the other hand, are given by

$$(H_0^{(0)} - \Phi_n^{(0)})\psi_n^{(3)} = -H_0^{(1)}\psi_n^{(2)} - (H_0^{(2)} + H_1^{(2)} - E_n^{(2)})\psi_n^{(1)} - (H_0^{(3)} - E_n^{(3)})\psi_n^{(0)}, \quad \text{(VII.3)}$$

$$(H_0^{(0)} - \Phi_n^{(0)})\psi_n^{(4)} = -H_0^{(1)}\psi_n^{(3)} - (H_0^{(2)} + H_1^{(2)} - E_n^{(2)})\psi_n^{(2)} - (H_0^{(3)} - E_n^{(3)})\psi_n^{(1)} - (H_0^{(4)} - E_n^{(4)})\psi_n^{(0)}. \quad \text{(VII.4)}$$

In writing down the above equations we have taken account of the fact that $E_n^{(1)} = \Phi_n^{(1)} \equiv 0$ (see (14.23), (14.25)).

In view of (14.29), the second-order equation (14.28) can be written

$$(H_0^{(0)} - \Phi_n^{(0)})\{\psi_n^{(2)} - \chi^{(0)}\phi_n^{(2)} - \chi^{(1)}\phi_n^{(1)}\} = 0. \quad \text{(VII.5)}$$

The corresponding general solution is

$$\psi_n^{(2)} - \chi^{(0)}\phi_n^{(2)} - \chi^{(1)}\phi_n^{(1)} = \chi^{(2)}\phi_n^{(0)},$$

where $\chi^{(0)}$ stands for a function of u, as yet completely arbitrary; the solution may be written in the alternative form

$$\psi_n^{(2)} = \chi^{(0)}\phi_n^{(2)} + \chi^{(1)}\phi_n^{(1)} + \chi^{(2)}\phi_n^{(0)}. \quad \text{(VII.6)}$$

We substitute the explicit expressions for $\psi_n^{(0)}$, $\psi_n^{(1)}$, and $\psi_n^{(2)}$ given by (14.20),(14.26), and (VII.6) in the third-order equation (VII.3), thus obtaining

$$(H_0^{(0)} - \Phi_n^{(0)})\psi_n^{(3)} = -H_0^{(1)}(\chi^{(0)}\phi_n^{(2)} + \chi^{(1)}\phi_n^{(1)} + \chi^{(2)}\phi_n^{(0)}) - (H_0^{(2)} + H_1^{(2)} - E_n^{(2)})(\chi^{(0)}\phi_n^{(1)} + \chi^{(1)}\phi_n^{(0)}) - (H_0^{(3)} - E_n^{(3)})\chi^{(0)}\phi_n^{(0)}. \quad \text{(VII.7)}$$

From this equation we subtract $\chi^{(0)}$ times (VII.1), $\chi^{(1)}$ times (14.14) (c), and $\chi^{(2)}$ times (14.14) (b). Remembering that the operators $H_0^{(r)}$ do not operate on the functions of u: $\chi^{(0)}$, $\chi^{(1)}$, $\chi^{(2)}$, we obtain the equation

$$(H_0^{(0)} - \Phi_n^{(0)})(\psi_n^{(3)} - \chi^{(0)}\phi_n^{(3)} - \chi^{(1)}\phi_n^{(2)} - \chi^{(2)}\phi_n^{(1)})$$
$$= -(H_1^{(2)} + \Phi_n^{(2)} - E_n^{(2)})(\chi^{(0)}\phi_n^{(1)} + \chi^{(1)}\phi_n^{(0)}) - (\Phi_n^{(3)} - E_n^{(3)})\chi^{(0)}\phi_n^{(0)}. \quad \text{(VII.8)}$$

Using (14.29) and remembering that $\phi_n^{(1)}$ is linear in u, we find from (14.16) that

$$(H_1^{(2)} + \Phi_n^{(2)} - E_n^{(2)})\chi^{(0)}\phi_n^{(1)} = \phi_n^{(1)}(H_1^{(2)} + \Phi_n^{(2)} - E_n^{(2)})\chi^{(0)} - \frac{\hbar^2}{m}\sum\left(\frac{M_0}{M}\right)\left(\frac{\partial}{\partial u}\chi^{(0)}\right)\left(\frac{\partial}{\partial u}\phi_n^{(1)}\right)$$
$$= -\frac{\hbar^2}{m}\sum\left(\frac{M_0}{M}\right)\left(\frac{\partial}{\partial u}\chi^{(0)}\right)\left(\frac{\partial}{\partial u}\phi_n^{(1)}\right). \quad \text{(VII.9)}$$

The product of this expression with $\phi_n^{(0)}$, when integrated over the electronic coordinates, vanishes if the electronic wave functions $\phi_n(x, X)$ are always chosen to be real (possible in the absence of magnetic field). This conclusion follows from the normalization condition

$$1 = \int [\phi_n(x, X)]^2 \, dx = \int [\phi_n^{(0)}(x)]^2 \, dx + 2\kappa \int \phi_n^{(0)}(x)\phi_n^{(1)}(x, u) \, dx +$$

$$+ \kappa^2 \Big\{ 2 \int \phi_n^{(0)}(x)\phi_n^{(2)}(x, u) \, dx + \int [\phi_n^{(1)}(x, u)]^2 \, dx \Big\} + \dots, \quad \text{(VII.10)}$$

which is to be fulfilled for all values of κ; thus we have

$$\int [\phi_n^{(0)}(x)]^2 \, dx = 1, \quad \text{(VII.11)}$$

$$\int \phi_n^{(0)}(x)\phi_n^{(1)}(x, u) \, dx = 0, \quad \text{(VII.12)}$$

$$2 \int \phi_n^{(0)}(x)\phi_n^{(2)}(x, u) \, dx + \int [\phi_n^{(1)}(x, u)]^2 \, dx = 0. \quad \text{(VII.13)}$$

The above conclusion is readily verified by differentiating (VII.12) with respect to the nuclear coordinates, denoted here symbolically by u.

The condition for solubility of (VII.8) is obtained by multiplying the right-hand side of the equation by $\phi_n^{(0)}$, integrating over the electronic coordinates, and equating the result to zero. Making use of the fact proved in the last paragraph, and remembering that $\phi_n^{(0)}$ does not depend on u, we find that the solubility condition can be written

$$\{H_1^{(2)} + \Phi_n^{(2)} - E_n^{(2)}\}\chi^{(1)} = -(\Phi_n^{(3)} - E_n^{(3)})\chi^{(0)}. \quad \text{(VII.14)}$$

When this condition is fulfilled the third-order equation (VII.8) becomes (see (VII.9))

$$(H_0^{(0)} - \Phi_n^{(0)})(\psi_n^{(3)} - \chi^{(0)}\phi_n^{(3)} - \chi^{(1)}\phi_n^{(2)} - \chi^{(2)}\phi_n^{(1)}) = \frac{\hbar^2}{m} \sum \Big(\frac{M_0}{M}\Big)\Big(\frac{\partial}{\partial u} \phi_n^{(1)}\Big)\Big(\frac{\partial}{\partial u} \chi^{(0)}\Big). \quad \text{(VII.15)}$$

Hence we can write

$$\psi_n^{(3)} = \chi^{(0)}\phi_n^{(3)} + \chi^{(1)}\phi_n^{(2)} + \chi^{(2)}\phi_n^{(1)} + \chi^{(3)}\phi_n^{(0)} + F(x, u), \quad \text{(VII.16)}$$

where $F(x, u)$, a solution of the inhomogeneous equation

$$(H_0^{(0)} - \Phi_n^{(0)})F(x, u) = \frac{\hbar^2}{m} \sum \Big(\frac{M_0}{M}\Big)\Big(\frac{\partial}{\partial u} \phi_n^{(1)}\Big)\Big(\frac{\partial}{\partial u} \chi^{(0)}\Big), \quad \text{(VII.17)}$$

is an involved function of u as well as x; the term $\chi^{(3)}\phi_n^{(0)}$ in (VII.16) represents, on the other hand, the general solution of the homogeneous equation, $\chi^{(3)}$ being an arbitrary function of u.

After substituting the expressions for $\psi_n^{(0)}$, $\psi_n^{(1)}$, $\psi_n^{(2)}$, and $\psi_n^{(3)}$ given by (14.20), (14.26), (VII.6), and (VII.16) in the fourth-order equation (VII.4), we can simplify the equation by following a similar procedure as in the case of the second- and third-order equations, namely, by subtracting from the equation $\chi^{(0)}$ times (VII.2), $\chi^{(1)}$ times (VII.1), $\chi^{(2)}$ times (14.14) (c), and $\chi^{(3)}$ times (14.14) (b). Thus we obtain the fourth-order equation in the form

$$(H_0^{(0)} - \Phi_n^{(0)})(\psi_n^{(4)} - \chi^0\phi_n^{(4)} - \chi^{(1)}\phi_n^{(3)} - \chi^{(2)}\phi_n^{(2)} - \chi^{(3)}\phi_n^{(1)})$$

$$= -H_0^{(1)}F(x, u) - (H_1^{(2)} + \Phi_n^{(2)} - E_n^{(2)})(\chi^{(0)}\phi_n^{(2)} + \chi^{(1)}\phi_n^{(1)} + \chi^{(2)}\phi_n^{(0)}) -$$

$$- (\Phi_n^{(3)} - E_n^{(3)})(\chi^{(0)}\phi_n^{(1)} + \chi^{(1)}\phi_n^{(0)}) - (\Phi_n^{(4)} - E_n^{(4)})\chi^{(0)}\phi_n^{(0)}. \quad \text{(VII.18)}$$

The corresponding solubility condition is

$$(H_1^{(2)} + \Phi_n^{(2)} - E_n^{(2)})\chi^{(2)} + (\Phi_n^{(3)} - E_n^{(3)})\chi^{(1)} + (\Phi_n^{(4)} - E_n^{(4)})\chi^{(0)}$$

$$= -\int \phi_n^{(0)} H_n^{(1)} F(x,u)\, dx - \int \phi_n^{(0)}(H_1^{(2)} + \Phi_n^{(2)} - E_n^{(2)})\chi^{(0)}\phi_n^{(2)}\, dx -$$

$$- \int \phi_n^{(0)}(H_1^{(2)} + \Phi_n^{(2)} - E_n^{(2)})\chi^{(1)}\phi_n^{(1)}\, dx, \quad \text{(VII.19)}$$

where we have taken account of the orthogonality between $\phi_n^{(0)}$ and $\phi_n^{(1)}$ (see (VII.12)). We shall show that the right-hand side of (VII.19) is equal to $\chi^{(0)}$ multiplied by a constant.

Consider first the third term on the right-hand side of (VII.19), namely,

$$- \int \phi_n^{(0)}(H_1^{(2)} + \Phi_n^{(2)} - E_n^{(2)})\chi^{(1)}\phi_n^{(1)}\, dx.$$

Using the explicit expression (14.16) for $H_1^{(2)}$ and remembering that $\phi_n^{(1)}$ is a linear function in u, we can write the term as

$$- \int \phi_n^{(0)}\phi_n^{(1)}(H_1^{(2)} + \Phi_n^{(2)} - E_n^{(2)})\chi^{(1)}\, dx + \frac{\hbar^2}{m}\sum \left(\frac{M}{M_0}\right)\left(\frac{\partial \chi^{(1)}}{\partial u}\right)\frac{\partial}{\partial u}\int \phi_n^{(0)}\phi_n^{(1)}\, dx.$$

Further use of (VII.12) shows immediately that the term vanishes:

$$- \int \phi_n^{(0)}(H_1^{(2)} + \Phi_n^{(2)} - E_n^{(2)})\chi^{(1)}\phi_n^{(1)}\, dx = 0. \quad \text{(VII.20)}$$

On the other hand the second term on the right-hand side of (VII.19) can be transformed as follows:

$$- \int \phi_n^{(0)}(H_1^{(2)} + \Phi_n^{(2)} - E_n^{(2)})\chi^{(0)}\phi_n^{(2)}\, dx = -\int \phi_n^{(0)}\phi_n^{(2)}\, dx\, \{(H_1^{(2)} + \Phi_n^{(2)} - E_n^{(2)})\chi^{(0)}\} +$$

$$+ \frac{\hbar^2}{m}\sum \left(\frac{M_0}{M}\right)\left(\frac{\partial}{\partial u}\chi^{(0)}\right)\int \phi_n^{(0)}\frac{\partial}{\partial u}\phi_n^{(2)}\, dx + \frac{\hbar^2}{2m}\sum \left(\frac{M_0}{M}\right)\chi^{(0)}\int \phi_n^{(0)}\frac{\partial^2}{\partial u^2}\phi_n^{(2)}\, dx$$

$$= \frac{\hbar^2}{m}\sum \left(\frac{M_0}{M}\right)\left(\frac{\partial}{\partial u}\chi^{(0)}\right)\int \phi_n^{(0)}\frac{\partial}{\partial u}\phi_n^{(2)}\, dx + \frac{\hbar^2}{2m}\chi^{(0)}\sum \left(\frac{M_0}{M}\right)\int \phi_n^{(0)}\frac{\partial^2}{\partial u^2}\phi_n^{(2)}\, dx,$$

$$\text{(VII.21)}$$

where use has been made of the equation (14.29).

Lastly, as $H_0^{(r)}$ are Hermitian and the functions $\phi_n^{(r)}$ have been chosen real, we can transform the first term on the right-hand side of (VII.19) with the help of (14.14)(b):

$$- \int \phi_n^{(0)} H_0^{(1)} F(x,u)\, dx = -\int F(x,u)(H_0^{(1)}\phi_n^{(0)})\, dx$$

$$= \int F(x,u)(H_0^{(0)} - \Phi_n^{(0)})\phi_n^{(1)}\, dx$$

$$= \int \phi_n^{(1)}(H_0^{(0)} - \Phi_n^{(0)})F(x,u)\, dx.$$

Eliminating $F(x,u)$ with (VII.17), we get

$$- \int \phi_n^{(0)} H_0^{(1)} F(x,u)\, dx = \frac{\hbar^2}{m}\sum \left(\frac{M_0}{M}\right)\left(\frac{\partial \chi^{(0)}}{\partial u}\right)\int \phi_n^{(1)}\frac{\partial}{\partial u}\phi_n^{(1)}\, dx$$

$$= -\frac{\hbar^2}{m}\sum \left(\frac{M_0}{M}\right)\left(\frac{\partial \chi^{(0)}}{\partial u}\right)\int \phi_n^{(0)}\frac{\partial}{\partial u}\phi_n^{(2)}\, dx, \quad \text{(VII.22)}$$

where we have used the following relation, obtained by differentiating (VII.13) with respect to the nuclear coordinates:

$$\int \phi_n^{(0)}\frac{\partial}{\partial u}\phi_n^{(2)}\, dx + \int \phi_n^{(1)}\frac{\partial}{\partial u}\phi_n^{(1)}\, dx = 0. \quad \text{(VII.23)}$$

Adding (VII.20), (VII.21), and (VII.22), we find that the right-hand side of (VII.19) becomes

$$- C\chi^{(0)},$$

where $$C = -\frac{\hbar^2}{2m} \sum \left(\frac{M_n}{M}\right) \int \phi_n^{(0)} \frac{\partial^2}{\partial u^2} \phi_n^{(2)} \, dx \qquad \text{(VII.24)}$$

is clearly a constant, for $\phi_n^{(2)}$ is a quadratic function of the nuclear coordinates. When the right-hand side of (VII.19) is written as $- C\chi^{(0)}$ we arrive at the equation quoted in (14.32) of the text.

The presence of the function $F(x, u)$ in $\psi^{(3)}$ indicates that the adiabatic interpretation breaks down when we proceed beyond the second-order terms in the wave function.

VIII

ELIMINATION OF THE ELECTRONIC MOTION (p. 172)

In the text (see § 14) and in Appendix VII, the laws of motion for the nuclei are obtained by a systematic expansion in powers of the parameter $\kappa = (m/M_0)^{\frac{1}{4}}$ (14.5), and it is shown that the adiabatic approximation, in which the electronic motion is calculated as if the nuclei were at rest, is valid up to terms of the fourth order in κ. The averaged electronic energy (eigenvalue) in a given state plays, to this approximation, the part of the potential energy of the nuclei.

It has, however, been found, in particular by studying molecular vibrations, that the adiabatic model has a wider application than predicted by this theory. There exists in fact another method which contains this practical result, with the only modification that the potential energy of the nuclei is not the energy eigenvalue of the electronic state considered, but a slightly different quantity. This method has the further advantage that it leads to a system of simultaneous equations for all electronic states which represent the coupling of electronic and nuclear motion in a rigorous way.

In the notation of IV, § 14, the total Hamiltonian is

$$H = T_E + T_N + U(x, X), \tag{VIII.1}$$

and that corresponding to fixed nuclei

$$H^0 = T_E + U(x, X). \tag{VIII.2}$$

It is assumed, as in IV, § 14, that the latter problem is solved; in the equation

$$(H^0 - \Phi_n(X))\phi_n(x, X) = 0 \tag{VIII.3}$$

the functions $\Phi_n(X)$ and $\phi_n(x, X)$, which represent the energy and the wave function of the electrons in the state n for a fixed nuclear configuration X, are regarded as known. The actual wave equation is

$$(H - E)\Psi(x, X) = 0; \tag{VIII.4}$$

we try to solve it by an expansion

$$\Psi(x, X) = \sum_n \psi_n(X)\phi_n(x, X). \tag{VIII.5}$$

Substituting this in (VIII.4), multiplying the result by $\phi_n^*(x, X)$ and integrating over x we obtain, with $T_N = \frac{1}{2} \sum_k P_k^2/M_k$, (14.1):

$$(T_N + \Phi_n(X) - E)\psi_n(X) + \sum_{n'} C_{nn'}(X, P)\psi_{n'}(X) = 0, \tag{VIII.6}$$

where

$$C_{nn'} = \sum_k \frac{1}{M_k}(A_{nn'}^{(k)} P_k + B_{nn'}^{(k)}), \tag{VIII.7}$$

and

$$A_{nn'}^{(k)}(X) = \int \phi_n^*(x, X) P_k \phi_{n'}(x, X)\, dx,$$

$$B_{nn'}^{(k)}(X) = \frac{1}{2} \int \phi_n^*(x, X) P_k^2 \phi_{n'}(x, X)\, dx. \tag{VIII.8}$$

Consider the diagonal elements of these matrices. For stationary states the $\phi_n(x, X)$ can be chosen as real functions; then

$$A_{nn}^{(k)}(X) = -\frac{i\hbar}{2} \frac{\partial}{\partial X_k} \int \phi_n^2(x, X)\, dx = 0, \tag{VIII.9}$$

since the ϕ_n can be supposed to be normalized to the same constant value for all values of X_k. Hence C_{nn} is independent of the differential operator P; it is a multiplication operator, a function of X.

We can now write (VIII.6) in the form

$$(T_N + U_n(X) - E)\psi_n(X) + \sum_{n'}' C_{nn'}(X, P)\psi_{n'}(X) = 0, \qquad \text{(VIII.10)}$$

where the dash at the summation symbol indicates that the term $n' = n$ has to be omitted. Here $U_n(X)$ is defined by

$$U_n(X) = \Phi_n(X) + \sum_k \frac{1}{M_k} B_{nn}^{(k)}. \qquad \text{(VIII.11)}$$

It is this quantity, and not $\Phi_n(X)$, which plays the part of the potential energy of the nuclei, provided the coupling of different electronic states, represented by the sum in (VIII.10) with the coefficients $C_{nn'}$ ($n' \neq n$), can be neglected. The difference between $U_n(X)$ and $\Phi_n(X)$ as given by (VIII.11) can be determined when the electronic eigenfunctions for fixed nuclei are known.†

The equation of the nuclear motion under the assumption of negligible coupling is

$$(T_N + U_n(X) - E)\psi_n(X) = 0. \qquad \text{(VIII.12)}$$

The question, under what conditions the coupling parameters $C_{nn'}$ will be small, cannot be answered in general. Even if they are not very small their influence will be negligible if the electronic state n is separated from all others by a large gap, as can be seen from well-known perturbation formulae. This will be the case for the ground state of many molecules and non-conducting crystals; then the zero approximation is a non-harmonic nuclear vibration with the potential energy $U_0(X)$, and the coupling with higher electronic states can be calculated from (VIII.10) by perturbation methods. For metals, however, where the electronic states form a quasi-continuum, the sum in (VIII.10) cannot be regarded as a small perturbation; it will go over into an integral, and the equation (VIII.10) will become an integro-differential equation, which expresses the coupling of electronic and nuclear motion in a rigorous way.

† The method outlined here has been published by M. Born, *Gött. Nachr. math. phys. Kl.* (1951), 1.

DOUBLE REFRACTION AND OPTICAL ROTATION (p. 333)

THE basic equation (see (44.41))

$$E = \frac{4\pi}{n^2-1}\{P - n^2 s(s.P)\} \quad (s = y/|y|) \tag{IX.1}$$

and the dispersion formula (44.60), p. 336,

$$D_\alpha = \sum_\beta \epsilon_{\alpha\beta} E_\beta + i(E \wedge G)_\alpha \tag{IX.2}$$

together provide a self-contained description of crystal optics. We shall indicate this by a brief discussion of the phenomena of double refraction and optical rotation.

Let us transform (IX.1) as follows. Forming the scalar product between s and (IX.1), we get

$$s.E = -4\pi s.P. \tag{IX.3}$$

Eliminating (s.P) from (IX.1) and (IX.3), we obtain

$$4\pi P = (n^2-1)E - n^2 s(s.E). \tag{IX.4}$$

Hence

$$D = E + 4\pi P = n^2\{E - s(s.E)\}. \tag{IX.5}$$

Further, it follows from (IX.5) that

$$D.s = 0. \tag{IX.6}$$

Consider first the case when the gyration vector G vanishes. With the Cartesian axes chosen along the principal axes of the symmetric tensor $\epsilon_{\alpha\beta}$, (IX.2) becomes in this case

$$D_\alpha = \epsilon_\alpha E_\alpha, \tag{IX.7}$$

where ϵ_1, ϵ_2, ϵ_3 denote the principal dielectric constants. Dividing (IX.7) and the α-component of (IX.5) by ϵ_α and n^2 respectively and taking their difference, we get

$$D_\alpha\left(\frac{1}{n^2} - \frac{1}{\epsilon_\alpha}\right) = -s_\alpha(s.E), \tag{IX.8}$$

or

$$D_\alpha = \frac{-s_\alpha(s.E)}{(1/n^2 - 1/\epsilon_\alpha)}. \tag{IX.9}$$

Substituted in (IX.6), (IX.9) gives

$$\left\{\frac{s_1^2}{1/n^2 - 1/\epsilon_1} + \frac{s_2^2}{1/n^2 - 1/\epsilon_2} + \frac{s_3^2}{1/n^2 - 1/\epsilon_3}\right\}(s.E) = 0. \tag{IX.10}$$

It follows from (IX.5) that if s.E vanishes, D must be parallel to E. Such is the case only when s is along one of the principal axes. Apart from these few special cases, we have the Fresnel formula

$$\left\{\frac{s_1^2}{1/n^2 - 1/\epsilon_1} + \frac{s_2^2}{1/n^2 - 1/\epsilon_2} + \frac{s_3^2}{1/n^2 - 1/\epsilon_3}\right\} = 0. \tag{IX.11}$$

For a given direction of propagation s (IX.11) is an equation of the second degree in n and gives in general two distinct values for the refractive index. In other words, for a given direction of propagation, there are in general optical waves of two different phase velocities (double refraction).

A crystal with a non-vanishing gyration vector G exhibits the phenomenon of optical rotation. We shall discuss the phenomenon in the simple case when the real dielectric tensor $\epsilon_{\alpha\beta}$ is isotropic (e.g. in a crystal with tetrahedral symmetry). For this case (IX.2) reduces to

$$\mathbf{D} = \epsilon\mathbf{E}+i(\mathbf{E}\wedge\mathbf{G}). \tag{IX.12}$$

Equating (IX.5) and (IX.12), we get

$$(n^2-\epsilon)\mathbf{E} = n^2\mathbf{s}(\mathbf{s}.\mathbf{E})+i(\mathbf{E}\wedge\mathbf{G}). \tag{IX.13}$$

Forming the scalar product between (IX.13) and $\mathbf{s}$, and the vector and scalar products between (IX.13) and the gyration vector $\mathbf{G}$, we obtain the three relations

$$-\epsilon(\mathbf{s}.\mathbf{E}) = i\mathbf{s}.(\mathbf{E}\wedge\mathbf{G}), \tag{IX.14}$$

$$(n^2-\epsilon)\mathbf{G}\wedge\mathbf{E} = n^2(\mathbf{s}.\mathbf{E})(\mathbf{G}\wedge\mathbf{s})+i\{G^2\mathbf{E}-(\mathbf{G}.\mathbf{E})\mathbf{G}\}, \tag{IX.15}$$

$$(n^2-\epsilon)(\mathbf{G}.\mathbf{E}) = n^2(\mathbf{G}.\mathbf{s})(\mathbf{s}.\mathbf{E}). \tag{IX.16}$$

After eliminating $(\mathbf{G}.\mathbf{E})$ from (IX.15) with (IX.16), we form the scalar product between (IX.15) and $\mathbf{s}$, obtaining

$$(n^2-\epsilon)\mathbf{s}.[\mathbf{G}\wedge\mathbf{E}] = i(\mathbf{s}.\mathbf{E})\Big\{G^2-\frac{n^2(\mathbf{G}.\mathbf{s})^2}{n^2-\epsilon}\Big\}. \tag{IX.17}$$

Comparing (IX.14) and (IX.17), we find that

$$\epsilon(n^2-\epsilon) = \frac{n^2(\mathbf{G}.\mathbf{s})^2}{n^2-\epsilon}-G^2, \tag{IX.18}$$

or alternatively $\qquad (n^2-\epsilon)^2 = \dfrac{1}{\epsilon}\{n^2(\mathbf{G}.\mathbf{s})^2-(n^2-\epsilon)G^2\}.$ $\qquad$ (IX.19)

As we have mentioned in § 44, for optical waves in the infra-red region $\mathbf{G}$ is very small compared with ϵ. (IX.19) shows that $n^2-\epsilon$ is of the order of $\mathbf{G}$. Hence we can approximately put n^2 equal to ϵ on the right-hand side of (IX.19), getting

$$n^2-\epsilon = \pm(\mathbf{G}.\mathbf{s}). \tag{IX.20}$$

Thus owing to the gyration vector, a crystal, which is otherwise free from double refraction, transmits for a given direction $\mathbf{s}$ optical waves of two slightly different velocities, namely,

$$\frac{c}{n} = \frac{c}{\{\epsilon\pm(\mathbf{G}.\mathbf{s})\}^{\frac{1}{2}}} \simeq \frac{c}{\sqrt{\epsilon}}\Big\{1\mp\frac{1}{2\epsilon}(\mathbf{G}.\mathbf{s})\Big\}. \tag{IX.21}$$

Consider now the light vector $\mathbf{D}$ of an optical wave travelling in the Z-direction. It follows from (IX.5) that

$$D_1 = n^2E_1, \qquad D_2 = n^2E_2, \qquad D_3 = 0. \tag{IX.22}$$

Thus upon taking the scalar product between $\mathbf{G}$ and (IX.12), we get

$$(G_1D_1+G_2D_2)\Big(1-\frac{\epsilon}{n^2}\Big) = \epsilon E_3G_3. \tag{IX.23}$$

Since $\mathbf{s}$ is a unit vector in the Z-direction, the third component of (IX.13) reduces to

$$-\epsilon E_3 = i(E_1G_2-E_2G_1). \tag{IX.24}$$

After eliminating E_3 from (IX.23) with (IX.24), we find with the help of (IX.22) that

$$(G_1D_1+G_2D_2)(n^2-\epsilon) = -iG_3(D_1G_2-D_2G_1). \tag{IX.25}$$

Since $\mathbf{G}.\mathbf{s} = G_3$, (IX.20) used in (IX.25) gives

$$\mp(G_1D_1+G_2D_2) = i(D_1G_2-D_2G_1) \tag{IX.26}$$

so long as $G_3 \neq 0$. From (IX.26) we readily obtain the ratio

$$\frac{D_2}{D_1} = \frac{\pm G_1 + i G_2}{\mp G_2 + i G_1} = \mp i. \qquad (IX.27)$$

If we write out the phase factor of the displacement vector explicitly and take the real part, we see readily that the two alternative signs in (IX.27) describe the two alternative circular polarizations. These have the different phase velocities given by (IX.21).

X

RECENT PUBLICATIONS

WHILE this volume was in preparation a number of papers have appeared which deal with related subjects. The following list contains some of these publications with summaries of their contents and references to the corresponding sections of the book.

E. BAUER. 'The vibrational spectrum and specific heat of sodium', *Phys. Rev.* **92**, 58 (1953).

The method of Houston (see p. 71), with a modification due to T. Nakamura (*Progr. Theor. Phys.* **5**, 213 (1950)) is applied to Na which, though cubic, is elastically extremely anisotropic. The distribution of the vibrations is calculated and compared with Debye's formula and with an improvement of it where the longitudinal and transverse vibrations are cut off separately. (This method is ascribed to Brillouin: in fact it is much older; see M. Born, *Atomtheorie des festen Zustandes*.) Then the specific heat is determined and its accuracy judged by plotting θ_D (Debye temperature) against T for the theoretical and experimental C_v-curves (see p. 80). (II, § 6.)

M. BORN. 'Die Gültigkeitsgrenze der Theorie der idealen Kristalle und ihre Überwindung', *Festschr. Gött. Akad. 1951, math. phys. Kl.* p. 1.
—— 'Kopplung der Elektronen- und Kernbewegung in Molekülen und Kristallen', *Göttinger Nachr. math.-phys. Kl.* 1951, p. 1.
—— 'Un nouveau point de vue sur la théorie de la matière condensée', *C.R.* $2^{i\grave{e}me}$ *Réun. Chim. Phys.* Paris, 1952, p. 334.

A new thermodynamical method is developed which makes no use of the assumption that the amplitudes of the vibrations about the static equilibrium configuration are small. An attempt is made to deduce the existence of irregularities (block-structure) as a consequence of the anharmonic forces. The most important application is that to solid helium where the zero-point vibrational energy is large compared with the static energy; this case has been worked out by D. J. Hooton (publication in preparation). (V, § 23.)

O. EMERSLEBEN. 'Die elektrostatische Gitterenergie endlicher Stücke heteropolarer Kristalle', *Z. f. phys. Chem.* **199**, 170 (1952).
—— 'Eine Darstellung des Einflusses der Kristallbegrenzung auf die Gitterenergie endlicher Ionenkristalle', *Z. f. Elektrochemie*, **56**, 305 (1952).
—— 'Über das Restglied der Gitterenergieentwicklung neutraler Ionengitter', *Math. Nachr.* **9**, 221 (1953).

The methods for calculating the electrostatic lattice energy are perfected in such a way that the influence of the finite size (surface) can be estimated. (I, § 1.)

A. HERPIN. 'Les forces de polarisabilité dans les cristaux', *J. Phys. Radium*, **14**, 611 (1953).

The forces between polarizable particles are systematically derived from the Coulomb interaction between the electronic clouds with the help of quantum-mechanical perturbation theory in the adiabatic approximation (see p. 171). They are ordered according to the power of the distance, into dipole, quadrupole, etc., coupling. It is shown that in lattices where each ion is a centre of symmetry the dipole coupling contributes nothing to the elastic constants. The quadrupole forces, however, do. They are calculated for crystals of the NaCl type and can be expressed in terms of the ionization energies and the polarizabilities of the ions. The deviation from the Cauchy relation, $c_{12} - c_{44}$, can thus be explained, in fair agreement with the observations, for NaCl, KCl, NaBr, and KBr. (I, § 1.)

T. A. HOFFMANN and A. KÓNYA. 'Linear atomic chain and the metallic state', *J. Chem. Phys.* **16**, 1172 (1948).

—— 'Some investigations in the field of the theory of solids, I to V', *Acta Phys. Hung.* **1**, 5 (1951), 175 (1951); **2**, 97 (1952), 101 (1952), 195 (1952).

Detailed studies of models of crystals, in particular linear chains, from the standpoint of the electronic theory; the main purpose is the understanding of the metallic state, of alloys, etc. (I, § 1.)

J. HOVE and J. A. KRUMHANSL. 'The evaluation of lattice sums for cubic crystals', *Phys. Rev.* **92**, 569 (1953).

A modified Ewald method for determining lattice sums with a better efficiency in convergence. (I, § 1.)

W. KROLL. 'On the determination of the elastic spectra of solids from specific heats', *Prog. Theor. Phys.* **8**, 457 (1952).

The specific heat can be expressed as an integral over the density of the vibrations. Conversely, the distribution of the vibrations can be found from a given specific heat curve by a reciprocal integration process. (II, § 6.)

G. LEIBFRIED und W. BRENIG. 'Zur spezifischen Wärme fester Körper', *Z. f. Phys.* **134**, 451 (1953).

Two kinds of forces are assumed, central and volume forces (like those due to the Fermi distribution of free electrons), and the atomic parameters (coefficients of the second order terms in the potential energy) determined from the elastic constants. The specific heat is represented by a Debye term and an Einstein term. The numerical results are in good agreement with observations and other (more rigorous) calculations; in particular some anomalies of the alkali metals are explained.

—— 'Schwingungsspektren fester Körper', *Fortschritte der Physik*, **1**, 187 (1953).

A comprehensive report of the present situation. (II, §§ 4, 5, 6.)

J. A. POPLE. 'Dielectric polarization of a dipolar lattice', *Phil. Mag.* (VII), **44**, 1276 (1953).

This paper belongs to a branch of crystal theory which is not considered in this book. To each lattice point of a rigid lattice some variable quantity is attached and it is assumed that there is an interaction between these quantities at neighbouring lattice points (e.g. two kinds of atoms are distributed over the lattice points, the energy depending on whether neighbours are alike or different. This leads to a thermodynamical theory of alloys, mixed crystals in the state of order or degrees of disorder.) In the present case rigid, freely rotating dipoles are distributed, and their thermodynamical behaviour calculated with the help of a method due to L. Onsager (*J. Amer. Chem. Soc.* **58**, 1486 (1936)). (Preface.)

Yosio Sakamoto. 'Calculation of Madelung's coefficient of NaCl', *J. Sci. Hiroshima Univ.* A, **16**, 569 (1953).

—— 'Calculation of Madelung's coefficient of the rhombic $CaCO_3$ (aragonite)', ibid. A, **16**, 573 (1953).

—— 'Calculation of Madelung's coefficient of $PbSO_4$, $BaSO_4$, $SrSO_4$, and $CaSO_4$', ibid. A, **16**, 581 (1953).

—— 'Electrostatic lattice energy of $PbSO_4$, $BaSO_4$, $SrSO_4$, and $CaSO_4$', *Bull. Chem. Soc. Japan*, **26**, 283 (1953).

Using Ewald's method the Madelung constant is recalculated for several lattices; first, for checking purposes, with exaggerated accuracy (16 decimals!) that of NaCl which is found in good agreement with the result of Emersleben (5 decimals in 1923, 14 in 1953); then of rhombic crystals of the sulphates of the alkaline earth metals. (I, § 1; V, § 30; Appendix II.)

J. C. Slater. 'Electronic structure of solids, I and II', *Publ. of the Mass. Inst. of Techn.*

A systematic and comprehensive representation of the electronic theory of crystal structures. (I, § 1; III, § 13.)

R. Stratton. 'A surface contribution to the Debye specific heat', *Phil. Mag.* (VII), **44**, 519 (1953).

The frequency distribution of the elastic waves in a finite rectangular block of an isotropic substance is calculated; the effect of the boundary results in a term proportional to k^2 (k is the wave-number) in addition to the ordinary Debye term, proportional to k^3. The Debye formula for the specific heat has to be corrected by a corresponding surface term which explains the difference of the specific heat of activated charcoal and graphite, observed by F. Simon and R. C. Swain (*Z. f. phys. Chem.* B, **28**, 189 (1935)). (II, § 6.)

J. H. C. Thompson. 'Instability and melting of the alkali halides', *Phil. Mag.* (VII), **44**, 131 (1953).

Several attempts have been made to interpret melting as mechanical instability of the lattice (see, for instance, M. Born, *J. Chem. Phys.* **7**, 591 (1939)). They were not successful (see R. Fürth, *Proc. Camb. Phil. Soc.* **37**, 34 (1941)) because melting is actually a question of coexistence of two thermodynamical phases (liquid—solid). The author investigates the stability of the vibrations of an elastically extended lattice of the NaCl type and finds that instability occurs for expansions of 6 to 7 per cent. This is of the same order but considerably larger than the observed thermal expansions at the melting-point. (III, § 12.)

K. S. Viswanathan. 'The characteristic vibrations of a rectangular lattice', *Proc. Ind. Ac. Sci.* A, **36**, 306 (1952).

—— 'The characteristic vibrations of crystal lattices', ibid. A, **37**, 424, 435 (1953).

Raman rejects the standard theory of crystal vibrations because it leads to a quasi-continuous spectrum while only a few lines are observed in absorption and scattering experiments. The author tries to justify Raman's own theory which maintains that only a finite (small) number of vibrations exists in a lattice. He starts from the standard theory and shows that Raman's vibrations can be characterized by a simple property (vanishing of the group velocity). He shows that any disturbance can be written asymptotically (for large time) as a super-position of Raman's vibrations but with time-dependent amplitudes. This does not mean that eventually only these frequencies are excited; the initial energy distribution in the frequency spectrum is unaltered. (II, §§ 5, 6; Appendix IV.)

L. I. VIDRO and B. I. STEPANOV. 'Die Intensitätsverteilung in den Schwingungs-spektren linearer Ketten', *Doklady Ak. Nauk USSR*, Ser. **82,** 557 (1952).

Infra-red and Raman spectra are discussed with the model of a linear chain. (II, §§ 8, 9; VII, §§ 47, 49.)

S. YANAGAWA. 'Theory of the normal modes of vibrations in crystals', *Prog. Theor. Phys.* **10,** 83 (1953).

The normal modes of the vibrations in a lattice are classified by group-theoretical methods and the secular equation giving the frequencies for special wave vectors is reduced to a set of equations of lower degree. By using Houston's approximation (see p. 71) the method is numerically applied to the NaCl and diamond lattices. (II, §§ 5, 6; V, § 24.)

INDEX

absorption, 120, 194, 197, 208.

—, continuous, 359, 363, 367.

—, dependence on T, 361.

— lines, 341.

acoustic frequency branches, 57 ff.

— waves, 231, 277.

— — in ionic lattices, 255.

— —, long, 153, 226, 229, 237, 298, 323.

additive radii, 15 ff.

adiabatic approximation, 171, 172, 204, 218, 402, 406, 411.

alkali halides, 2, 26, 28, 31, 32, 162.

anharmonic damping, 355.

— potential, 171, 341, 343, 347, 355 ff.

anisotropic medium, 62.

— stresses, 227, 245, 247.

anti-Stokes lines, 203, 366, 369, 379.

atom refractions, 107.

atomic forces in crystals, 1, 13.

— polarizability, 101, 104 ff., 115.

Avogadro's number, 41.

Barnes, R., 123 ff, 341, 342.

base index, 213.

basic vectors, 1, 23, 68, 213 ff., 278, 280.

basis of lattice, 213.

Bauer, E., 411.

Blackman, M., 61 ff., 71 ff., 81, 341, 342, 355, 358.

block-structure, 411.

Boltzmann factor, 178.

Born, M., 21, 29, 30, 33, 35, 36, 45, 49, 55, 61, 107, 111, 117, 140, 144, 154, 163, 166, 229, 341, 342, 355, 358, 368, 371 ff., 387, 388, 390, 407, 411, 413.

Born–Haber cycle, 35.

Bradburn, Mary, 368, 371, 372, 373, 390.

Brattain, R., 126, 341, 342.

Bravais lattice, 1, 129, 132, 134, 140, 142, 144, 150, 152, 213 ff., 233, 248, 302, 321, 382, 399.

Brenig, W., 74, 412.

Bridgman, P. W., 160.

Brillouin, L., 373.

— scattering, 373 ff., 379.

— — in cubic lattice, 379, 381.

broadening of absorption lines, 341.

Brody, E., 29.

Buckingham, R. A., 22.

Burstein, E., 363.

caesium salts, structure of, 159.

Cartwright, C. H., 126.

Cauchy relations, 136 ff., 142, 145, 149, 412.

cell, lattice, 1, 23, 67.

— indices, 213.

— parameters, 132.

central forces, 19 ff., 226, 245.

centre of symmetry, 136.

characteristic temperature, see Debye; Einstein.

— vibrations of lattice, see lattice.

charges, motion in ionic crystal, 88.

clamped dielectric constants, 293.

Clausius–Mosotti law, 107.

Clusius, K., 81.

coefficient of thermal expansion, 52.

cohesive energy, 7, 8, 13, 24 ff., 34, 52, 53.

— — in alkali halides, 36.

complex normal coordinates, see normal coordinates.

compressibility, 25, 50 ff., 110 ff.

continuous absorption, see absorption.

contravariant components of basis, 278, 280, 282.

— field components, 282.

coordination number, 24, 155, 382.

Coulomb energy, 3 ff., 166, 330.

— field, unretarded, 328 ff.

— — in dipole lattice, 248 ff., 261.

— — in ionic lattice, 255 ff., 273, 275, 321, 330.

— static interaction, 9, 12, 115, 136, 240, 255, 411.

coupling of electronic and nuclear motion, 170, 171, 407.

crystal plane, 216.

— radii, 19, 27.

crystals, finite, 217, 218.

—, ionic, 1, 2, 12, 14, 24, 136, 227.

—, valency, 2, 14, 15.

—, van der Waals, 1, 8, 12, 13, 14.

cubic lattice structure, 140, 382.

— symmetry, 137, 142.

current density in lattice, 331.

cyclic boundary condition, 45, 46, 391.

Czerny, M., 118, 123, 125, 126, 127, 359, 363.

damping constants, in dispersion, 346, 353, 355, 357 ff.

Debye model, 41, 42 ff., 52, 60, 74, 79, 413.

— temperature, 43, 44, 48, 65, 66, 79, 411.

deformation, arbitrary, 152, 154.

—, energy of, see energy.

—, homogeneous, see homogeneous.

Dehlinger, W., 111.

A
LIFE
OF TWO HALVES

FOOTBALL, FINANCE AND FAITH
– THE FULL STORY

RT REVD DR DAVID CARR OSL

RIVER
PUBLISHING

River Publishing & Media Ltd
Barham Court
Teston
Maidstone
Kent
ME18 5BZ
United Kingdom

info@river-publishing.co.uk

ISBN 978-1-908393-39-5

Printed in the United Kingdom
Cover design by www.SpiffingCovers.com
Cover photograph by Ashley Carr; www.ashley-carr.co.uk

Contents

Dedication

I dedicate this book to Molly, my dear wife.
Throughout forty-four years of marriage she has selflessly
given me over to working for Christ and His Church ...
...and she still refuses to let me retire!
Love you.

Foreword by Cyrille Regis MBE

I had the pleasure of meeting Dave Carr in the late seventies whilst I was playing football and he was in financial services. Dave is an "upfront" man – what you see is what you get – and he held a reputation in the footballing world as a man who genuinely cared for his clients. Dave consistently displayed the character and values of a godly man, whether dealing with professional footballers or ministering to the brokenhearted.

This book peels back the layers of the person I know as "Dave" and provides a refreshing insight into the journey he and his family have travelled to this point. It is easy to applaud a man when he appears to have reached the heights of success and ignore the challenges, the sacrifices, the mistakes, the ups and the downs that he would have encountered along the way.

Reading about his early years in particular, struggling with dyslexia and acute shyness, seems a far cry from the man I see and know today. God's hand was clearly on his life and the transformation from then to now is astonishing. Indeed, it is not where we start in life but where we finish that is important and Dave's story is a living testimony to us all that God has a plan and purpose for each one of us.

Dave talks of a loving family and a caring father. He also speaks candidly about his role as a father to others, and how these two experiences have helped to mould him into the father that he is today, both naturally and spiritually. His transparency is refreshing and only serves to endear him to me more as a role model.

Dave has certainly lived up to his philosophy: "love people for who they are and not what they do". Stories about his work as a tailor and bouncer are interlaced with heartwarming "Dave" humour. Even back then you can see his love for people and how God was shaping his character and saturating his heart with love and compassion.

Dave's relationship with God is at the heart of who he is. His obedience to God is a great example to us all. In August 1972 Dave felt God was calling him to start a church. For one year there were only 12 people attending. Through much turmoil and hardship, Dave and Molly stayed the course. Dave offers a candid and inspiring historical account of the journey of the church and shares examples of good and faithful relationships and also of relationships that caused him pain.

Honesty and integrity shine through his career as a financial advisor. Just one act of love and compassion opened the door for him to manage the finances of hundreds of professional footballers. Dave lived what he taught. He practiced what he preached and, most importantly, he put God first. It is not surprising that his legacy in the football world has followed him, so that today so many professional sports people are drawn to his leadership and mentorship, including myself.

In reading this book, I see a man who is not afraid to reveal who he is. If you want to know what it takes to build and lead a church of the size and reach of Renewal Christian Centre then this book is an excellent starting point. You will learn about a man who worked hard to develop himself as a husband, father, colleague and pastor. You will read about the highs and lows in his life, both personally and professionally. Dave draws back the curtains so you and I can have a glimpse into the journey of his life. He teaches us about love, forgiveness, obedience and determination. It is, simply put, a compelling story of God's unmerited grace and favour.

Cyrille Regis MBE

Foreword by Darren Moore

One of the things I love about Pastor Dave Carr is the love and empathy that oozes out of him for others. While he naturally has this for his family, his love and appreciation extends to countless people outside. Yet through this, he is still able to bring spiritual discipline and offer the right, honest advice to people in their particular circumstances.

I met Pastor Dave in the season of 1990-91, when I was on the journey to becoming a professional footballer. At the time I was beginning my apprenticeship at Torquay United Football Club. Pastor Dave advised me and a fellow apprentice, Duane Darby, to sign up and start a pension with the Professional Footballers Association. I took to him instantly. I remember him being warm, but speaking with authority, as well as plenty of humour.

Having first met him in a footballing context, I saw him again and witnessed his dry humour again when he conducted the marriage of my great friend Arvel Lowe and his wife, Anne, in March 2000. Anne was running a little late and Pastor Dave walked down the aisle to the front, declaring out loud, "Don't worry, I'm not the bride!" I found it very funny and remembered him (and his lovely white hair!) from my time in Torquay.

I joined Renewal Christian Centre in 2001 when I signed for West Bromwich Albion Football Club and ever since the day I first met him until now he has been a great help with his advice regarding footballing matters and his pastoral care throughout my years both in and outside of the game. In 2003 I had a very serious anterior cruciate knee injury in a premier league match against Chelsea and the medical team at WBA were not giving me encouraging messagess about my future! I got my wife to drive me straight over to Pastor Dave and Molly's house. It was 8:00 in the evening, but Pastor Dave came straight out of the house and prayed for me whilst I lay in the car. This is typical Pastor Dave. He has been there for me throughout my life in the football industry – looking after me, along with so many other players – and now in the church, serving God with complete love and dedication.

It is an honour to write this foreword for this remarkable man. I pray that the Lord will give him the desires of his heart.

Darren Moore,

Coach Educator and Equalities Executive,

Professional Footballers Association

Introduction

I never realised until now the difficulty inherent in writing a story about yourself! What do you include and what do you leave out? This is a story of a frightened young boy who felt disappointed with life. When you are a child, full of phobic fears and very low self esteem, then what happens to you when you are born during a time when many families lived through obvious trauma deeper than my fears? What effect does this have on later life? And is it possible to break through all of that and become someone else?

The title of this book is a parody on the football saying, "A game of two halves". I chose it because the two major parts of my life have been divided by working in professional football and being the leader of a Christian community. For almost eighteen years these two very different worlds overlapped until I basically walked away from football after recovering from heart trouble. Life can be seen through the philosophy of sport. We all desire to win and not suffer defeat. Often we settle for a draw, because something is better than nothing. But the truth is, the prize only goes to those who win more than they lose.

You could say that the title is also descriptive of the change in my life from being completely illiterate as a young adult to being able to hold my own with most. Our early years are crucial in the development of our character and academic formation. Back in the 40s people like me, who today are identified as dyslexic, were dismissed as being either thick or lazy. This made growing up very tough. The effects of other's disdain were no less damaging than if they'd tattooed the word "failure" across my forehead. Feeling grossly insecure affects every part of life. The normal dreams a man has – getting a satisfying job, finding a wife, having children – seem like outlandish fantasies to those who see no value in themselves.

However ... despite such bleak beginnings, I hope to demonstrate to those who read this book, whose backgrounds are littered with similar obstacles and hardships, that there is always hope. A vibrant faith coupled with courageous action can change your situation and the final outcome of your life. From the little council house in Birmingham, England, where I grew up, I have travelled to many exotic destinations, staying at world renowned hotels, castles and other amazing places. God, in His grace, is the Master of the turnaround.

Fears are not always unjustified. Nelson Mandela said, "I learned that courage was not the absence of fear, but the triumph over it. The brave man is not he who does not feel afraid, but he who conquers the fear." So true. My Christian faith didn't remove all my fears and phobias – it gave me the strength and courage to face each one and triumph over them.

In life I have seldom achieved the desired results first time around. I learn by my mistakes. This has led to many unusual circumstances. I remember the first ever baptism I conducted. There was an elderly lady who was, let's say, slightly larger than her doctor would have liked. So large, in fact, that one baptismal gown wasn't going to cover it. She required two gowns, pinned together. This was fine, but she neglected to wear a swimsuit beneath the gowns, as was the norm. I ducked her beneath the water and, to my horror, she ascended from the deep like a certain animal rising from the watering hole, as naked as the day she was born. Because she was "portly" all the pins had popped out under the pressure!

The first wedding I conducted was a similar disaster. I noticed that the Bride looked breathtakingly beautiful and made a mental note to mention later that she looked "as pretty as a picture". After the service, my horrified wife dragged me to one side to inform me that what I'd actually said was, "I think you'll all agree with me, she's no oil painting"!

Yes, I've made a few mistakes over the years, but my heart has always been to serve God and to love people. When you have faced personal loneliness and a deep sense of unworthiness yourself, you can identify with the fears and sorrows of others – those who would normally feel so worthless that they'd be ashamed to darken the doorstep of their local church.

Be it a football club, a church, a hospital or a supermarket, I have always endeavoured to simply be "Dave". Titles have been bestowed upon me over the years that have dictated a certain wardrobe to go with them. But we must never forget who we really are, whether we're meeting a football star, an Archbishop, or visiting a frightened patient in AE.

There are many precious people that I have not mentioned in this book. These omissions aren't intentional. It's just that I have lived a life with so many stories that I can't recall all of them. If you are missing from the narrative, I trust you'll forgive me for not referring to your impact on my life. Write down any stories you think would be useful and send them to me. Maybe they will go into a sequel!

My wife, Molly, has been so faithful in sharing me with so many others for almost forty-four years – often sacrificing her own personal time and that of a young family. I am blessed to have a great family – our children still bring joy to us in their individual ways; our grandchildren are very precious; and there are the dogs! One that we've had from a puppy is in need of psychological assessment. The other we inherited after his owner died is angelic. It might give you an indication of the stability of the owners!

A few "thank you"s are in order. Firstly, thank you to the leadership of Renewal for their consistent love and support over the years, especially my PA's who have had to put up with my unpredictability.

Thank you to Mandy Cooper who edited this book and encouraged me to publish it, and also to her husband, Dave. You are both really good friends and an important part of our spiritual family.

To the publisher of all my books, Tim Pettingale. You have worked sacrificially over the years to make me look good. I think it might have worked!

Thank you to my son Ashley who produced the front cover picture that reinforces my split personality!

Thanks to my two brothers, both of whom have served me at Renewal.

To Melvyn Carman who has travelled with me for many years and witnessed many of the miracles.

To Dr John Andrews who became Senior Associate Pastor to me at Renewal in 2012. I know that when I become Founding Pastor you and I will will take Renewal to fantastic heights!

If you have paid for this book, know that the proceeds will go to the blessing of others. I always give it back to the Church. Alternatively, if this was a present, you can always exchange it!

Enjoy...

Dave.

FIRST HALF...

Chapter 1
Pig Bins and Hand-Me-Downs

Within five days of my birth both Mussolini and Hitler were dead. Within three months, victory had been achieved in the Second World War. I don't think I can take any credit for this!

I was born in Birmingham in the April of 1945. I was a twin, but my sister was not fully formed and didn't survive. This loss still resonates with me today. It continues to echo out of the overwhelming sense of community that existed at that time – people still reeling from the scars of a war in which, during seventy seven air raids, 2,241 people were killed and over 12,000 homes destroyed. Although the war was over, a sense of sombre mourning prevailed over the city.

But, we were a resilient lot! During the late 40s and early 50s, Jewish children, orphaned by German concentration camps, came to Birmingham and were adopted by local Jewish families. Ex-servicemen from the Caribbean, India and Poland settled in the city. Peace was embraced, roads were revitalised, communities were repaired and new housing started to spring up. Many of these prefabricated bungalows were described as "temporary" houses, yet they but remained in situ until the late 1980s.

Our own home at 76 Pineapple Road, Stirchley, was a council house with an outside toilet. Bathing involved a large tin bath, filled with water and placed before the open coal fire.

My dad, Albert, was a charge hand and skilled leather worker at Cadbury's – the famous chocolate makers. He worked there for 46 years, man and boy. I will never forget the powerful aroma of leather and chocolate combined that so epitomised my father whenever he entered the room. My mother, Winifred, also worked for Cadbury's, but only until she was married. In those days no married woman could work for this strict Quaker family and, upon their impending nuptials, all female employees received a rose, a Bible and the sack! I also have two brothers. Anthony (Tony) who is 12 years older and Michael, 11 years older than me.

As a child I obviously had little understanding of the effect upon the country of six years of war. Food was rationed until I was 9 years old. I now understand (although it was confusing at the time) why my parents would leave me at the table with the words, "You'll stay there until you eat it. There isn't anything else!"

Buying a car in the 40s would set you back just over £600; a house just over £3,000. Numbers like that were astronomical and well beyond the reach of most of us, who commonly wore the "hand-me-down" clothes of our siblings. Our rationed clothes had little circles with a section cut out as a label – a utility mark. This was the aftermath of a Nation virtually reduced to its knees.

Outside on the pavement was a network of dustbins, or as we knew them, "pig bins". After every preparation or completion of a meal, any leftovers or potato peel was carried out and placed in the bin. I'm sure many of us who grew up in these times find ourselves smiling as we are now encouraged to "recycle" everything. After all, we invented it! Once a week the wagon would come and empty the bins. The contents were used to feed pigs. It was the philosophy of "waste not, want not" and we all understood how it worked.

Only the bin collectors and the coal man had motor wagons. Mom would say to me, "David, keep an eye on them to see if they deliver the correct number of coal sacks." The coal was stored outside in the coal house. Bread and Milk were delivered by horse-drawn wagons. If the wagon brakes were not properly applied, the horses would wander down to the pig bins, knock the lid off with their noses and start eating the waste food. Yes, we also invented home shopping deliveries! When the wagons moved on, the horses would inevitably leave something behind on the road. Mom would call to us, "Go out and shovel up the manure for your dad's garden." We really didn't waste anything!

This was a time of change for transport in the city though. Whilst horses drinking from troughs was still a common sight in the towns, so were trams running on rails through the city and the single and double decker buses. Then we had two main railway stations: Snow Hill Great Western and New Street London Midland Scottish.

These were the days when you didn't lock your front door and us kids would often be sent to, "borrow a cup of sugar from next door and tell her I'll give it back later." We played in the streets and the parks till the sun diminished, with no thought of child abductions. Our toys would have held little for appeal for a 21st century child, as our own imaginations created guns. Forts were made from odd bits of wood and cardboard boxes.

I was envied among all the children in our neighbourhood because my dad, with his leather and wood turning skills, could make me guns and holsters that real. So real, in fact, that one day a policeman stopped me because he thought the gun was genuine! I was a soldier one day, then a cowboy, a Canadian mounted policeman the next and even an Indian chief ... but that's another story for later in the book.

I started school in 1950, aged 5. Due to my father's profession,

I qualified to attend Bournville School, situated on The Green – a "model" village set in the midst of the industrial City of Birmingham.

This enclave of unique Quaker ideology had provided all the infrastructure needed for its employees: doctors, dentists, schools, housing, sports, social activities, clubs and even an operatic society. Yet, no pub or bar. No! Bournville would not have alcohol for many years.

The school stood with its tall tower holding a carillon of 48 bells. It was first constructed in 1906 and held 22 bells back then. These bells are not attached to ropes but a keyboard. The ringer had leather straps on his hands to protect him as he struck the stick-like keys. My dad used to make these straps.

My fears and phobias were firmly grounded right here, during my first ten years of education. I was a complex, fearful child; acutely shy and developing a number of additional irrational phobias. To fear almost everything at such a young age was horrific.

My main issue involved the dark. I could not be in a dark room or situation without hyper ventilating and until my teenage years I never slept without a light on. I felt that the dark was restricting my space, suffocating me, and I would panic. Years later I would be forced to confront this fear head on when I worked in security for the Bullring shopping centre – a vast area that at night time was patrolled by two men, one of which was me. We couldn't keep our torches on for fear that someone hiding would throw a piece of scaffolding at us. So we had to walk this area in the pitch black, guided by touch, feeling our way through.

Since I still couldn't openly admit to my fear, I had to trust in my faith in Christ. Because we were frequently understaffed, I often had to conduct the patrol totally on my own – with not a crack of daylight to be seen and yet with all the noises of a large area. I came out of that experience some nights dripping with sweat, but after surviving that I could face anything. Sometimes, your faith doesn't remove the fear, it supports you in it. And victory comes from walking through the shadow.

I think some of my fears had undoubtedly transferred from my mother, who permanently lived with fear. I would get hysterical if I was near water. I couldn't even take a shower in case the water came near my face. I was afraid of animals, especially dogs and cats. As for dentists, well, I didn't have a serious appointment until I was almost thirty. Add to that doctors and pretty much the whole medical profession and you can understand that I was a quite seriously damaged person.

My acute shyness rendered me unable or unwilling to interact with many people. The only thing that I could share anything with was my teddy bear. I would cry over it and talk to it (although I never thought to give it an actual name!). It was the only thing, apart from my parents, that never laughed at me. My childhood wish would have been that my teddy came to life and became my best friend. He was my therapy, my confidante, my companion. I felt safe when I had him with me. Many, many years later, when clearing out the loft, I would look at this battered, torn, grubby little scrap of fabric and filling and still feel the emotion welling within me as I discarded him once and for all.

I was constantly bullied and ridiculed because I was timid and because I couldn't read or write – and this was by both teachers and pupils alike. It was almost as if they could sense the fear within me and preyed upon it. I would often be kept off from school to keep my mother company, since she was also living in fear. She managed to justify this by pretending it was to protect me from being picked on.

This pattern of bullying and intense emotional pain continued until one day I snapped. My dad had always told me to stand up for myself, but I was never brave enough to do it – until one particular day. A certain lad was bullying me as usual. But this time, instead of cowering and taking it, in a rage I picked him up and threw him straight through a hedge. No one picked on me after that. I was still terrified of them all, but somehow everything had changed.

I visited my old school recently. I had remembered seeing a big rocking horse on my first day there and sixty years later, he was still there. The only difference being that he appeared so much smaller than he had before. This was such a powerful metaphor for the fear I had experienced as a child, which in adulthood had been settled by the grace of God.

I understood much later that my inability to read and write was as a consequence of Dyslexia. But in the 50s and 60s we were classed as either stupid, lazy or insolent. This was so damaging that only the Lord's miraculous intervention some years later was able to fully restore my confidence and written communication abilities.

My relationship with my brothers was a little blurred. They almost weren't a part of my life, being so much older than me. Their experiences were always ahead of mine – I was small when they were teens. I can't recollect any "play time" with them, although I'm sure there must have been. Despite the fact that I had two brothers, I wasn't fully part of their lives and felt very much that I suffered the loneliness of being an only child. My brother Tony would later reveal to me that he had felt he didn't really know me. After that we set about building many decades as brothers and friends and to this day, we have never exchanged a cross word.

Tony became a male nurse when I was just 6. This career would eventually take him all over the country, ending up in Newcastle upon Tyne as the Chief Nursing Officer for the Region. Michael took over the Sunday School at Pineapple Junior school from Colin Merryman, the former incumbent, which I attended from the age of 7. It was only about 2 years ago that I attended the funeral of Pearl, one of my first Sunday School teachers. She was loving and kind and although I probably didn't appreciate this at the time, I later realised that it's not always the dynamic stuff that changes people. It's the consistency of turning up, week after week, building structure into a life that counts. What a blessing it was to look back at her faithfulness and see that she was part of my spiritual destiny. Few children's workers ever imagine that their weekly preparation for a mixed group of fractious kids will produce not only transformed lives, but lead to these children becoming transformers themselves. Similarly, I remember a man at an Elim church who prepared the stage and refreshed the water, consistently, week after week. This is what he did and he did it to the best of his ability. I would see him week in and week out. There was a safety and security in this consistency. I learned, from these two and many others that often, this is all that people need.

It was Michael, who later became an architect, who became my first influence towards the Christian faith. Although I'd prayed a prayer at the age of seven, I feel now that there was little maturity in it. Listening to Michael, I learned most of my theology. I still hold him as one of the finest Bible teachers I ever heard. I didn't always understand, or need to understand, but sitting beneath this teaching for 18 years, 3 times a week, there was, of course, an impartation. At 18 I was filled with the Spirit and suddenly all of this teaching came into strong, active application. I started stepping into roles within the church – youth ministry, cleaning, stewarding, leading worship – and I was trained in every aspect of church life by virtue of my willingness to serve and fill any available gap. I served Michael's vision for his newly established church, Calvary Chapel, until I was 25 years old.

Chapter 2
Milk Monitors and Misfits

The 50s were largely a decade of recovery from the disasters of the 40s. But despite this we were still overshadowed by the Korean War, in which the lives of British young men were being taken again.

The highlight of the decade for us children was 2nd June 1953 – the Coronation of Queen Elizabeth II. What a day! Over 8,000 guests inside and outside Westminster Abbey London saw the new Queen being crowned by the Archbishop of Canterbury Dr Geoffrey Fisher. Everything stood still for street parties, mugs, races and games in the street. For us it was orange juice and a few sandwiches. I can recall running in a race in Wellington boots, wearing a red, white and blue sash over my shirt and shorts. We only saw the black and white recording of it much later when visiting the "Pictures" and watching the Pathe News.

I remember other good experiences around that time. Because of the link with Cadbury's, we would visit their sports' ground on

May Day and dance round a large Maypole with different coloured ribbons. We were also able to attend the annual pantomime in the theatre at Christmas. But apart from being punctuated by the occasional bright moment, my childhood was largely a mixture of fear and loneliness. During my time in infant school, lads from the junior playground "raided us" and, during one such incident I was thrown against a wall, breaking my nose for the first time. It caused me to develop such severe sinusitis that I was admitted to Moseley Hall Children's Hospital for surgery. Over the years I would suffer acutely with this complaint; with swollen eyes and the need to protect myself from bright lights. This situation wasn't helped by my second breakage – however that's not a story for now.

This debilitating condition remained until I was about 23. I was a singer in a Gospel band at the time and one night I was feeling so ill that my band mates had to transport me to the gig lying down in the van. I got up to sing and afterwards I was profusely sick! I cried to The Lord, "I can't go on like this, please heal me!" He did and from that point on I never suffered with the condition again. Praise God!

As a child my family lived in a suburb of Birmingham that had little to offer. It was on the south side of the city and anyone who lived there was automatically initiated as a Birmingham City fan. In those days, football affiliation was mainly geographical. The north of the city was Villa land.

Stirchley was located mainly along the double sided Pershore Road, littered with frail shops that closed one by one. It was the next district (or village as it was then) by the name of Kings Heath that presented the larger and more lively shopping area with Wrensons, Masons, Woolworths and the Kingsway Cinema – stores and companies that are no longer household names. It was here that I spent my early years, sitting in the window waiting for my dad to come home from work. At Cadbury's the bull (a loud horn) would sound proclaiming both the start and finish of the working day. It could be heard from up to a mile away.

Mom would say, "Put your toys away, your dad's coming," and he would arrive, riding his bike, a small man of 5'7", stocky and with beautiful white hair covered by a flat cap. I never knew him with dark hair. He worked in Fitting Shop B, maintaining the conveyor belts that moved the chocolate from one department to another. Sometimes working up to 60 feet high, his job was often dangerous and always hard work. Often, he was so hot that Mom needed to peel his shirt off his back when he arrived home.

As a treat, dad would reach into his bag and pull out a bag of "waste chocolate", distorted in the manufacturing process. It was packaged in a fawn bag and only available to employees from the works' shop. But that strange mixture, the aroma of chocolate and leather that was so uniquely my dad, was always a strong reminder of him. Before I lost my sense of smell, those aromas always flooded back memories of him.

Washing day was Monday. In those days there was no concept of what is now termed "white goods". Rather than some sleek kitchen appliance there was a metal "dolly tub" that my mother would fill up with boiling water from the kettle. The washing was submerged in it and any stains received a vigorous scrubbing with a large bar of soap and a brush. Then there was the dolly, a long pole with a round wooden base with sections cut out. This would be pounded up and down into the tub to create a similar movement to that which the future washing machine would do so much faster. Then, I would help her to transfer the washing to the mangle, a large cast iron machine with two rollers rotated by a large wheel with a handle.

As a child, to see the water squeezed out of our clothes was very exciting! Then, we'd cart the washing out to the garden to be pegged on the line. In the winter, the sheets would freeze so solidly that they could barely be wrenched from the line. Mom would get frustrated and I would laugh at the comedy of the sight. It was great fun. Once the washing was dry (and defrosted in some cases!) came the ironing. The large flat iron, heated in front of the fire, was the only way this could be done. To this day, I wonder how she did all of this: cleaned the house, ferried me back and forth to school on the number 27 bus and still had dad's meal on the table when he finished work.

Dad worked hard for so many hours so that, each year, we could visit Kewstoke at Weston Super Mare – a small private caravan and chalet park run by Mr and Mrs Needham. The holiday journey would begin by catching a bus to Digbeth Coach Station, followed by a Midland Red Coach to Cheltenham. We'd then transfer to a Black and White Coach to Weston Coach Station. Parked outside the coach station would be Mr Needham, sitting in his Triumph Mayflower. We'd then travel, crammed together in the car with our luggage, through the woods, past the toll barrier and right to our accommodation next to the Parish Church.

Our holiday home was a wooded bungalow and from the window you could just see the sea at Sand Bay. There was a black cat that lived nearby and he would visit us each day and cheekily flip the door knocker – his way of demanding a saucer of milk. These holidays held fondly in my memory. We didn't have much money, but we had great times. I would always climb onto dad's shoulders for the walk from the main beach, through the woods, to our little holiday home. When money did allow, there might be a trip to Cheddar Gorge or to Wells Cathedral where the swans would ring a bell in order to be given food.

At the appropriate age, I joined the Cubs – 1st Grove Kings Heath. With my grey pullover and neckerchief I promised, "to do my best and do my duty to God and the Queen, to keep the laws of the wolf cub pack, and do a good turn to somebody every day." This was my parents' endeavouring to integrate me into some form of social activity. But because of my illiteracy, and therefore my painful shyness, my integration was always limited. I even found it difficult to learn the cub scout oath. It took me nine months to get my badge, purely because it took that long for me to learn it and be able to recite it in front of everyone.

Neighbours played a big part in community life in those days. Often, in a childish temper, I would pack my toys in a shopping bag and leave home, journeying all the way to the house next door to tell them that I was not loved and that I was never going back. I would be given some orange squash and a biscuit and then, sometime around tea time, I'd decide to go home!

I left Bournville School in 1956 as there were no senior facilities at the school. Despite Dame Elizabeth Cadbury having left money for a new school to be built "for the continuance of the pupils" I was disappointed to learn that I was one of the few who fell into the new concept of' "catchment areas". I was outside the catchment for this new school. That left me on my own to choose between Queensbridge Road and Wheelers Lane Boys School – both located in Kings Heath. Wheelers Lane was a Secondary Modern Boys and, I have to say, it could definitely be classed as secondary! And it certainly wasn't modern. All I can say with any accuracy was that it was a boys' school. Needless to say, I wasn't anticipating a great education here.

It suited me to be at a boys' school, however, since I was very embarrassed around girls. This matter hadn't been helped by the especially provocative nature of one girl in particular. She was a tall girl with dark hair, who bullied me incessantly. Her name was Susan. She made my life so miserable that for years, the very name "Susan" stirred up a whirlpool of negative emotions. This continued until, in due course, circumstances conspired to surround me with so many nice Susans that I was forced to re-evaluate the name and its association!

Wheelers Lane continued my disastrous education by disowning me as an abject failure and a societal misfit. For one year, our class – a collection of other misfits – was located in the minor hall at the local Kings Heath Baptist Church. There was no proper playground, so we'd play outside in the mud and go back into class filthy and soaked. We had no contact whatsoever with the main school, but I teamed up with Tony George who only lived in the next road to me, in Fordhouse Lane. We were 14 at the time. Tony had lost his mom and we shared a dependency upon one another.

Our family had recently moved to a new council house in a private part of Pineapple Road. We were now at No. 230 – the increase in house number signifying a commensurate move up in the world! It was if we'd died and gone to the promised land. There were only six of these semi-detached, three bedroom houses in our road, with a through lounge and a long garden without an air raid shelter. We also had a proper kitchen and a proper bathroom. We suddenly thought ourselves terribly posh.

At the bottom of our garden, about 60 feet beneath us, was the Hazelwell railway line. I loved to go down the garden to see the steam trains rushing past. The dad of one of my friends was the signalman. He'd be there with his waistcoat and pocket watch, sleeves rolled up and we would visit him and hear the bells ringing to tell him to pull the massive levers to either change the track or work the signals. I remember the smell of oil and his tin mug full of fresh tea on his desk. He had a doctor's style bag filled with railway tools and flags. It was all thrilling to behold.

Back at school, Tony George and I were considered unteachable. I remember after one test I was ranked twelfth in the fourth set. I was unbelievably proud not to have come bottom of the class for once! My exuberance was short lived, however, as my teacher told me, "Carr, you would have done better if you hadn't turned up."

I hated virtually every moment of school life. Despite being generally regarded as useless, the one job I was given was that of milk monitor. Both my mate Tony and I were made "sub-prefects in charge of milk delivery". During the fifties (and for a couple of decades thereafter) each school child would receive a small bottle of milk every day. Tony and I set up our office in the boiler house. It was like the start of quite a business career! We numbered each class, placing bottles in each crate and then releasing them to the individual classroom monitors. When the crates were returned, we would see if there was a consistent number not taken, as some children didn't like milk. We then removed that number on the next delivery and, if there was no complaint, we'd do a deal with the caretaker who knew how to "move them on". Not bad initiative for an illiterate, "unteachable" thirteen year old!

Through my contact with the Baptist Church I discovered the Boys Brigade – the 6th Birmingham with Captain Ken Bushell. My time in Boys Brigade became the military service that I had missed by two years. The brigade officers had been serving in the war less than a decade previously, so we had drill every Friday night. Even now, at 68, I could still do that drill. The discipline and the emphasis on impeccable uniforms was incredibly character building.

On a Thursday we would have to be certain that our uniform was spotless. Our company had a full uniform of blue short-sleeved shirt with epaulettes, white lanyard, blue tie, blue shorts and a brown leather belt with a brass buckle. Then we had a white pouch on a shoulder strap, blue socks with two white stripes at the top and a forage cap on our heads.

Church parade on a Sunday saw us assemble by All Saints Parish Church and, with a full band accompaniment, march down the High Street to the shout of the officer, "Left, left, left-right left." Those days have sadly passed away.

My phobia of animals (transferred to me from mom) only began to abate when I was allowed to own two goldfish. Billy and Sammy they were called. One was gold and black, the other all gold. They used to know exactly when it was feeding time and would zoom round the bowl at turbo speed. They lived about 12 years – an ancient age for domestic fish – and then died within one week of each other.

Living in fear wasn't helped by the fact that my brothers would listen to the wireless and such programs as "Journey into Space" with over 60 episodes made during the 50s. Captain Andrew "Jet" Morgan, Lemmy Barnet and Doc Matthews' adventures were so frightening for a young child that I would put my hands over my ears. By today's standards, the stories would be laughable, but our imagination is always more terrible than reality.

Once we'd moved to our spacious new house, I was soon to become the last child left at home. Tony had moved away with his nursing, had married Alice and they had already had Alison who was exactly the same difference in age to me as I was to her dad. Michael had married Pat and they had bought a house in Cornhill Grove, about two roads away. Much later, Dad bought that house when Michael built himself a house in Moseley.

Our new next door neighbour was Mrs Green, who always seemed sad. When I was older, my dad told me the reason why. During the war she had been notified, on the same day, that both her sons were missing. One had died on HMS Hood, blown from stem to stern by the German Battleship Bismarck on 24 May 1941. The other son, Mr and Mrs Green were advised, was missing and presumed dead in the desert. Mr Green, a police officer, died of a heart attack on hearing the news. A few weeks later, the missing son was found alive. As a child you are largely blissfully unaware of tragedies such as these that can change the direction of some people's lives forever.

Chapter 3
Fashion Icons and Electric Shocks

My father was an only child. My grandma and grandpa lived in May Lane, Alcester Lanes End, Birmingham. My grandpa suffered from Parkinson's disease and in those days, without medication, he trembled all over. I would sit on his lap and it would be like sitting on a ride at the fair! He was a tall, bent man, wearing a grandad shirt, waistcoat and flat cap. He had a special cup with a "lip" section on which to rest his bushy moustache. I would help to hold the cup so that it wouldn't spill and then I would hold his hand and try to drag him up the garden. I loved my grandpa.

The day he died there was a knock at our front door. As Mom opened it, there stood a Telegraph Boy, a young postal worker who would ride a BSA Bantam motorbike. The presence of a telegraph boy meant either good or bad news, delivered in a short economic sentence, similar to posting a tweet.

After a few moments of staring down at the piece of paper that had been delivered, Mom walked slowly back to the living room and said, "David, Grandpa has gone to heaven." I ran from the house, down the road, screaming, "I want my Grandpa. I want my Grandpa!" This was my first taste of death. I was 7.

My paternal grandma was strange woman. She had accumulated savings that in today's money would amount to around £500,000. But this ex-teacher always bought out of date food and second hand clothes and furniture. When my dad was young, his uncle, who had no children, wanted to send him to university and make him his prodigy and heir. Grandma refused, saying that my dad had to go to work like everyone else. So she made him go to Cadbury's – even though his uncle was worth at least £10 million in modern terms. Despite all this, Dad never said a bad word about her; even though she showed no affection towards me at all.

My mother's parents died well before my birth, but I did have uncles, the favourite being Fred Lloyd. He played drums in the Cadbury's band and always gave me a lot of attention. I desperately wanted to be like him – and especially a drummer. Although I don't play any instrument, I always look for the beat in everything. A legacy from Uncle Fred. My cousins were part of Lloyds Chemists, a well known chain in the UK, and Carr's Paints, an automotive paint brand.

I was expected to go to Sunday school each week. Mom went to church (The Full Gospel Tabernacle) each Sunday evening and dad would babysit. The moniker "Baby" or "Bab" would stick with me well into my teens. That's the problem with having older brothers! When I was small, Dad used to sit me on his lap and tickle my arm. His hands were rough and leathery, but they conveyed to me a great sense of strength and safety. I would tuck my head deep into his coat, listening to the beat of his heart, breathe in the essence of chocolate and leather and listen to him singing the Thomas Dekker poem over me, which later became a Beatle's song:

"Golden slumbers fill your eyes, smiles await you when you rise, sleep pretty darling, do not cry, and I will sing a lullaby."

I think that the security associated with being tickled on my arm started with Dad. I still crave that today, even from my grown children!

After the singing, and as the musical strains of Max Jaffa and the Palm Court Orchestra – a long running series that aired for 30 years on Sunday evenings – started to play over the gramophone, it was a piggy back ride "up the wooden hill" to bed. I loved my Dad dearly and can feel the tears forming now, just at the memory of his love for me. Then he would put me into bed, press all the bed clothes round me and say, "You're as snug as a bug in a rug." Then I would constantly call down, "Goodnight" through fear, checking that I wasn't alone. My dad would patiently call back in reply, "Goodnight, son." The light would always be left on until I fell asleep.

My Dad provided the only real security I ever had as a child. I can't explain it, but he was my final hope of safety in a world full of fear. He would be up at the crack of dawn and by the time I arrived downstairs, my breakfast was on the table and my shoes fully polished. Dad was so gifted that, sadly, he did everything for me, rather than with me. So, when I lost him, he hadn't transferred many of his skills to me. His passion for clean shoes, however, remains today. Like my dad, I would go on to be the family shoe cleaner. This meant that, later on in my life, my own children had the cleanest "spit and polished" shoes in the school. But even the best of plans can go astray. On one particular occasion, my eldest daughter, Melanie, had a new pair of school shoes, but I couldn't understand why they appeared so dull. I spent a long time working on those shoes, until I was almost shattered, but eventually I had a shine on them. The next morning I heard one of Melanie's dramatic wails echoing around the house. Apparently the shoes were meant to have a matt finish!

Despite my loving family, those childhood school years fashioned my life through their fears and apprehensions. Feeling as though I had achieved little if anything, I limped into teenage frustration. As an illiterate teenager with little hope of an ongoing education, April 1960 became the date prescribed for me to leave school. At 15 years old I was the only student not offered the opportunity to sit a GCE. In the words of my teacher, "Carr, it would only be a waste of paper." They gave me two weeks off during the exams instead.

A new teacher, Mr Cunningham, who arrived a few months before I left school, gave me a glimmer of hope. He looked at me one day and said, "You're not thick, Carr. I don't know what is wrong, and I don't have long enough to make a difference, but I'll teach you to write your name and address." Many years later I would return to that same school as a successful businessman – invited as the guest of honour at the annual prize giving.

Having left school I was immediately faced with the dilemma of what to do. I wanted to be a hairdresser at the local Co-Op and applied, but they forgot to call me for the interview. Then I was offered a job as a trainee window dresser at C&A in the city centre, but I turned it down. I figured that I didn't wear my trousers that tight, even though pink was always my favourite colour! I eventually chose Kenneth Hirst – a private tailor and men's outfitters on the High Street in Kings Heath.

Working there was like being in a TV show. The owner, Kenneth, and his wife were very, "Just going to the tennis club, dahling..." Ron Foster was the manager, Gordon Sutter the assistant, and I was the junior. My job was to clean the windows, sweep the floor with "dust mow" (damp sand), parcel up goods and take them to the post office, tidy the suits and learn to serve the customers.

My first embarrassment occurred on day one. I was asked by the boss to write a note for the till that said, "Paid Joyce the cleaner 10 shillings". I stood in front of them all, unable to spell or write any of these words. I was forced to confess that I couldn't even read properly, much less write. It was a miracle that I wasn't sacked. Mr Hirst just gasped, shook his head in disbelief, then did it himself.

I liked the store. Especially the regimental section, with blazer badges of all the British Regiments and their ties. It was only fifteen years since the end of the war, so there were many differing badges, including the air force squadrons and naval ships. For many years after I could describe the design and colours of each. I had only been at work about three weeks when Gordon said I should go with him to buy my first serious record. He obviously thought I needed a bit of culture in my life! I still have it in my office, 53 years on: Grieg's Peer Gynt, Suites 1 and 2 from the Nordwestdeutsche Philharmonic Orchestra. This has remained a fond piece of music amongst my other favourites consisting mainly of '60s rock!

I worked for Hirst's for about two years and during that time I also joined the St John Ambulance Nelson Transport Division. This particular team covered at Birmingham City football matches, The Tulip Festival and other major activities that the City celebrated. It was at one of these events that I ended up in Dudley Road Hospital. On entering our medical unit I had reached above the door and accidentally touched a main cable that the council had laid on for our lighting. I felt an extreme power surge pulsate through my body. I could hear myself scream, and yet it seemed detached from my body. Then a nurse touched me, thinking I was having a fit. The poor girl earthed me, throwing her across the room and me out through the door to the wet ground beneath. I came round in the back of the ambulance, on the way to the hospital. There was no lasting damage. I was however bruised all over and suffered a deep nervousness towards anything electrical for many months to come!

I used to help cover the emergency calls at Bristol Road Station, by the University, on a Friday night. I was the assistant to Clive, who was the driver. He, in is early twenties, developed what he thought was back strain. His widowed mother informed us that he actually had terminal cancer. She didn't want him to know, however, since he would have been traumatised at the thought of leaving her alone. This situation brought great conflict to me. I felt that, because he didn't know the limited time he had left, I had not given him an opportunity to hear about eternal life. I couldn't talk to him about eternity because I didn't have a context in which to do so. As far as he was concerned, he was a young man with his whole life to live.

One day our Superintendent, Dick Sanderson, knocked my door and said that he had just left the hospital. Clive had passed away crying, "Don't let me die!" We both found this totally heart breaking. The guilt was terrible and it was a pivotal moment in my Christian faith. Never again would I miss an opportunity to witness to someone.

On occasion I used to sing at the Full Gospel Tabernacle Bible class with a chap called Graham Homer. I took the opportunity of visiting their youth club one night with a couple of friends. When I walked in, I saw a person wearing a sloppy pullover and a pair of jeans. I say "person" because at the time I wondered, "Is that a boy or a girl?"

It turned out to be Molly, who was later to become my wife.

She told me later that her first impression of me was not very satisfactory either! Molly lived near Kings Heath with her mother. She had a half brother who was about 15 years her senior and he was one of the Elders of the church that ran our youth club. Molly was a hairdressing apprentice, working on the Stratford Road. She was rather modern for my aging parents because, as was the fashion, she was attired with a mini skirt, cape, long leather boots and a bow in the back of her hair. It gets me excited just thinking about it!

If I was on duty with the St John Ambulance she would say, "Don't meet me with that uniform on!" It just wasn't the right image for this modern dresser. My Dad would often say to her, "You should have a vest on wench, you'll get a cold!" And Mom ventured, "She's a bit worldly, isn't she?" Worldly is a term that Christians often use when they don't like something!

Molly persuaded me to buy a pink shirt. In those days, whether for work or special occasions, you would wear a white shirt with a stiff, detachable collar which you could send off to be cleaned. Imagine my mom's horror when I arrived home with a pink shirt. "She is leading you astray, mark my words!"

Little did we know that this relationship would continue for another nine years before we eventually got engaged. Before I left Hirst's, I would wave to Molly as she waited for the bus outside the shop, on her way to spend her day off each week at her brother's house in Hollywood. Molly was full fashion. Christians at this time tended to be somewhat "bland" and lacked any sense of design.

It's funny what you discover about people. For a long time Molly never told me that she had bad eyesight. This led to various funny situations. I discovered her problem when we went to see Birmingham City play one weekend. Molly thought that the pitch finished at the half way line! Often, when shopping, she would leave me standing in some store with the words, "Don't move until I get back." I, of course, would always move about thirty feet away and stand there watching her trying to find me! Once she came out of the ladies toilets and put her arm through what she thought was mine. The strange man said to her, "Who are you?!" However, I didn't encourage her to get some glasses until after we married. I thought it best she keep a more romantic view of what she was getting!

I moved from this small, privately owned tailor's business to a multi outlet chain called John Collier. Some will remember its strapline: the window to watch. I worked at both Birmingham Corporation Street and then a new store at the opening of the Bull Ring Centre in 1964. I was 19 and becoming frustrated by my lack of education and the diminishing hope of any real, progressive career.

Outside of work my passion was cycling and I had saved for many months to buy a Claude Butler racing bike from Tommy Godwin, the Olympic Medallist, at his shop on Silver Street in Kings Heath. This bike was metallic green (quite unusual at that time) and I would ride about 200 miles per week. I actually travelled everywhere with it.

My friend at the time, Peter Dyer, had a Dawes, and on his first day out with it we tracked a car on its journey, using the moving vehicle as a wind break. We were travelling at a good speed. Peter, however, kept talking to me with his head turned back, looking behind him. Suddenly, I shouted, "PETER!" as I saw the car in front come to a halt. Peter said, "What?" but the question was unanswered as a second later he hit the now stationery bumper and flew, still on his bike, over the roof, bounced off the bonnet and finished, still on his bike, upside down in the road. Peter was okay. His sparkling new bike, however, was so distorted it had to be carried home.

Peter only loved one girl, Wendy, and for months she refused to go out with him. We would cycle past her house most days and he would ask her out, but she would always say no. One day, however, he walked her up the hill to her home and, after 30 minutes, I cycled home alone. They have now been married for 48 years. I met up with them again when I was 68. Wendy had recovered from a brain tumour and Peter is bravely fighting cancer. How the passage of time continues.

Chapter 4
From Bouncing to Bible Studies

My working life took a bit of a different direction when I was offered a job with what was actually a private police force at the Bull Ring Centre. Being privately owned, this vast new concept of shopping had no one to police its malls, banks, pubs and night life. Chief Superintendent Jack Love, recently retired from the Birmingham City force, became its leader, with Inspector Eric Eastow as his assistant. Then there was Sgt George Davis, perpetually with a pipe in mouth and "Alright my son?" as his greeting.

This gave me the opportunity I was seeking to escape from the dreariness of tailoring. I was uniformed and carried a truncheon, whistle, notebook and first aid pouch. We worked three shifts, had "parade" before we started and also had use of a cell for any detainees. The mixture of backgrounds was interesting: former Palestinian police, RAF regiments, Guards Division, Fire Service, the Police and then there was me!

The duties could be quite diverse: bank escorts, patrolling, locking up and unlocking, clearing out the nightclub and pubs, breaking up the many gangs of drug and alcohol-related thugs and ex-convicts. Every now and again there would be a running battle between us and groups of very unsavoury characters. It made us as hard as nails. We were more like bouncers than private police.

One cold winter's night we discovered a little girl sitting in the roadway. She was dressed in a summer dress, wellingtons and had long matted hair. The smell was repugnant. We called a woman officer who took this little mite into the ladies' wash room, where she stripped her of these flimsy garments only to find out she was actually a boy! This was the tragedy of inner city life: a young boy of eight found eating a stolen sandwich like a dog on the floor. Then there was the grammar school girl who collapsed in the ladies room with a Heroin overdose; and the 15 year old found aborting in the toilets. This was the life I lived for two years.

One day, I passed Gino's Pizzeria to hear panicked voices. The chef had sliced his arm from his elbow to the muscle at the top. There was blood everywhere. I secured the wound, stopped the bleeding and contributed to the saving of his arm. For that, I was granted free pizzas!

As a team we had great comradeship. My best friend was John Young and we worked together on two different occasions. But he was hit with an iron bar one day and, for me, that started the thoughtful process of just how dangerous life could be. We lived on the edge of problems. Once, I was confronted by four drug addicts who became violent towards me. We struggled and one of them head butted me, breaking my nose … again! I managed to hold on to the one, but the other three ran off. This second breakage reduced my ability to smell to the absolute minimum. Much later in life, I would find that I could just about smell the sulpher in the geysers of New Zealand, but not much else.

As a result of these brushes with thugs I lived on my nerves. One day, a girl from one of the shops crept up behind me and jabbed me in the ribs, playfully. Without thinking, my automatic reaction was to whirl around and punch whoever was assaulting me, convinced it was an attack. I laid the poor girl out cold.

* * *

I had been attending church since the age of seven – a legacy from my brother Michael – and I had a faith of sorts. But in many ways it was lacking in reality. I had helped out at church by doing some cleaning on occasion and assisting with the children's work, but I was still an emotional mess, struggling with my phobias. I also felt like someone living a double life, because it seemed to me that my job and my faith had little in common. I remember telling the church leadership that I needed to step down from church duties, so that I could give some thought to my life. I assumed this would be sympathetically received. I was somewhat surprised to be treated with great disdain. They looked at me as if I had let the side down. The implied attitude was, if I wasn't going to help them, they certainly had no time to stand and talk to me. At the time, church was very much activity driven rather than relational. It was more about what you did than who you were. This experience helped to shape the philosophy that I maintain to this day – if you are valued only for what you do, who will you be if you lose that position? I love people for who they are, not for what they can do for me.

The turning point in my spiritual life came when two young boys from our youth group invited me to go with them to Hockley Mission, a Pentecostal Church in Birmingham that was nationally known for its extreme and powerful presentation of the Gospel. The pastors/ founders were two ladies, Miss Reeve and Miss Fisher. They were originally from the Brethren, but were somewhat too much for them to handle, so they founded their own church. They held a service every Saturday night. Mostly they had no idea who would be speaking, but without fail, led by God, many internationally known speakers would just turn up. Hockley Mission had people from a diverse mix of spiritual backgrounds – Catholics, monks, Pentecostals, you name it, they were there. Miss Fisher would say, "All are welcome. Our brothers and sisters from the Catholic church are here to be filled with Spirit."

I remember one Saturday night, Miss Fisher said, "If anyone on the first two rows doesn't want a blessing, please move now." The speaker approached the platform and the first two rows of people were knocked down by the power of the Spirit. It is very difficult to describe the experience that awaited anyone who ventured into this mission hall situated in the midst of this challenging area. The music was loud, addictive and produced a genuine fervour. The two ladies moved up and down the aisles seeking those who the Holy Spirit was touching. When they had that look in their eyes it was advisable to duck!

This particular night was deeply challenging to me. The lads I'd gone with had been praying that I would be filled with the Holy Spirit. For those who are not familiar with this expression, this began when the Holy Spirit descended on the disciples on the day of Pentecost. They began to speak in tongues (that is, a known language previously unknown to the speaker).

The preacher that night had a personality I didn't particularly warm to. His message wasn't connecting with me, or so I thought. The Holy Spirit, however, had no such personality hang-ups. A sudden rush of His power hit me, threw me to the floor, and there I stayed for about forty minutes. Throughout this time I felt that I was living in a world of bliss. It was like I was living on another planet. This one night changed the course of my history. It was the day my destiny was confirmed.

I remembered that, as a child of about 9, I had been prayed over by a man called J. Nelson Parr, the pastor of the well known Bethshan Tabernacle, Manchester. He was a small, firey man with a church of a thousand people, who preached mainly on hell. Then, at age 11 I was prayed for by William F. P. Burton, who died in 1971 and was known as the Apostle of Faith. He worked in South Africa and founded many churches. Once, when he had been diagnosed with cancer, he was visited by a young couple. When they arrived he said, "Pray for me, because the Lord told me that the next two visitors would pray a prayer of healing for me." They prayed and the cancer disappeared.

From this night at Hockley Mission onwards my personality started to be refined. My previously uncontrollable temper started to diminish; my compassion for others started to increase. Everything in my life started to come into focus. I felt unable to continue in my job in the private security business. I had lost the hard edged toughness that was needed to show trouble makers who was the boss.

Change of direction

I had little in the way of spiritual gifts or training, but since experiencing this baptism in the Holy Spirit I had developed an immense hunger for reading the Bible. I would sit and read it over and over again. It was the American Standard version, so even today I still occasionally lapse into American spelling. Through reading the Bible God gradually healed my dyslexia, so that within about two years I could read virtually anything that was put in front of me.

Some time after this I travelled to Weston Super Mare for a week's holiday on my own and had an amazing experience. I visited a Baptist Church in Upper Bristol Road and felt deeply convicted by the message that was preached. That evening, returning to my bed and breakfast accommodation, I lay on my bed and had either a vision or a dream, I'm not sure which, of myself dressed in a Nurse's uniform, holding a black baby.

I heard God speak to me saying, "Will you nurse my babies?"

It made such a powerful impression on me that I initially thought

God was calling me to nurse children in Africa. Wanting to cooperate with Him I decided to approach Selly Oak Hospital and do some pre-nursing training, to see if it was for me. As I mentioned, I had been part of the St John Ambulance, so I felt I was capable of dealing with the sick.

The hospital kitted me out with a pair of white trousers and a white coat, fastened at the shoulders. My first days there coincided with the 1966 World Cup final. I was assigned to Ward E2 and the fierce ward Sister advised the patients that they were allowed to gather around the TV at the top of the ward to watch the matches – as long as they didn't "get too excited". Her warning proved to be oddly prophetic as England kicked off and one of the patients promptly had a heart attack and died! I missed most of the match while I was assisting in laying him out – although I did manage to get back for, "They think its all over, it is now!"

I mentioned in Chapter 1 that my brother, Tony, was a male nurse and had risen through the ranks to be chief nursing officer for a large region of the country. He was one of the first men to have made it big in nursing and of all places, I had to be at the hospital where he had trained. I was quickly identified as "Anthony Carr's brother" and, still struggling as I was with great personal insecurity, this was hard to bear. In fact, the pressure of having a "celebrity" brother in nursing was intolerable. Every day, a Matron, Deputy or Assistant would come and visit the ward to see if I was getting on alright. At that time the structure of the NHS made them like senior officers in the Army, with the Sisters being the equivalent of Sergeant Majors. One day the Sister said to me, "Why did you come here? You are a pain, I can't run my ward with a visit from a Matron or Deputy every day!"

One day I had an interview with the Matron. She sat looking at me in her dark blue uniform with its white frilly cap and brought out a silver tea pot to serve us both tea. Of course, she had to mention Tony.

"We are very proud of Mr Carr. What does your brother think of you considering nursing?"

I looked at her and said, "I wouldn't know Matron, I haven't told him. I'm David Carr and it's my decision."

I realised at that moment that full time nursing would never be for me. If I did well, people would say, "Well, he is Mr Carr's brother!" If I did badly, they'd say, "He isn't as good as his brother, is he?" I didn't want to live my life under that shadow.

I prayed about it and asked God for clarity. I heard the answer come back, "Don't train." This left me deeply confused. Did I see that vision or not? It would be 25 years before I discovered its meaning.

Having walked away from nursing, I had no option but to return to tailoring. A Midlands based, high quality department store, Beatties of Wolverhampton opened a new store in Solihull, near Birmingham, and I decided to look for a job there. Solihull is only eight miles from Birmingham, but I had never been there! To us city dwellers Solihull was like the Promised Land! The Borough of Solihull is in the middle of England. It has a population of 200,000 and is divided into the south (a high quality area with beautiful housing, a lavish shopping centre and a number of private schools) and the north – a more deprived area with 40,000 people historically transferred from the Council clearance list of Birmingham.

I couldn't know that this journey to be interviewed at Beatties would one day determine my future life, home and love.

Chapter 5
Tape Measure and Wedding Plans

I started in the Menswear Department. It wasn't easy getting used to a customer service industry again when, only months earlier, I was throwing people out of nightclubs. Because of my background, I was also appointed to the roles of store nurse and backup to the store security. I remember one day Mr Gerrard, the General Manager, calling me into his office and saying, "If I pass out and end up in Solihull Hospital, don't bother to come back to work again!" It didn't have a very good reputation at the time.

Beatties was like a little community all on its own. Eventually I became the Chair of the Staff Board and went down in the record books as the only person in the history of the company to call a strike. The staff restaurant received no subsidy. It was very expensive and you weren't allowed to bring your own sandwiches. Since the management refused to negotiate, I called an embargo on the catering facilities. For one whole week, the catering staff just stood there each day with not one person to serve. At the end of the week, I was summoned to Wolverhampton HQ to meet James Beattie himself!

I entered the plush office and saw, behind a large desk, a little man with white hair, a dark blue business suit and gold rimmed glasses. He was a man of few words and got right to the point:

"Sit down Mr Carr. But why? But why, Mr Carr?"

They had never had a strike before. I told him the restaurant was too expensive for the staff and that we should not be forced to eat there. I returned with a discounted menu and permission to eat sandwiches in the staff area. The General Manager, however, called me into his office and said, "David, you may have won the battle, but you have lost the war. You will never be promoted as long as you are with us."

Molly and I had been going out with each other, on and off, since I was seventeen and she was sixteen. She had qualified as a hair stylist in the city and counted many of the wives of leading Jewish businessmen as her clients, along with a fair few of Central TV's personalities of the time. She was earning more than I was at that time and the gifts she received from well wishing clients at our engagement were expensive.

Molly will tell you that I am loyal. Yes. But romantic I am not! I don't do the candlelight and dimmed lights. I'm too matter of fact and life to me is too much yes or no, can or can't, do or don't. We had been going out with each other for nearly a decade when she asked me, sometime near Christmas, if we were going to be married. I asked her if I could have a week to consider it. She offered me 24 hours! Her mother had told her, "If he has kept you waiting this long, have the best ring in the shop." Molly doesn't need telling twice! She had a uniquely made, 13 diamond ring, shaped like a bow. Sadly, it was stolen seven years ago during a house burglary.

Molly grew up living in the same house that she and her mother were born in. Gertrude Sale was my future mother-in-law, a lovely woman who suffered considerably in life. Having become pregnant in the early '30s, her fiancé ran off leaving her with the stigma of being a single mother. She lived with her parents and worked hard – first as a cook at the local Moseley Golf Club, then at the Billesley Arms – to bring up her son Keith, putting him through university and seeing him develop into a successful businessman and an a Elder of The Full Gospel Tabernacle.

When Keith was in his teens, Gerty met a soldier who was stationed nearby on the Anti-aircraft guns of the Royal Artillery. They married and she became Mrs Gosling. Molly was born and, when she was less than a year old, Gerty became pregnant again. Just before the birth of what was to be Molly's brother, Brian, there was a knock at the front door and Gerty opened it to find a woman standing there. "Is this where George Gosling lives?" she demanded. "He is my husband!"

It appeared that George was a bigamist. This shattered my poor mother-in-law and caused her to have a complete breakdown. At this time in our society there was both the pain and humiliation of moral judgement, coupled with a lack of social concern.

George and his first wife wanted to take Molly to live with them in Doncaster, but Gerty was inseparable from her. With no husband, money or self-confidence she surrendered Brian to his father, at birth, since she had no means of keeping him. A further sad chapter of this story is that by the age of 9, Brian was placed in an orphanage. At 16 he went into the army and spent 23 years there. But there is a happy ending: Molly met up with Brian about 20 years later.

Gerty returned to being Mrs Sale, in keeping with the name of her fifteen year old son, and Molly remained Molly Gosling. Thinking she herself was adopted, she gradually began using the surname Sale, so that all of her friends thought that was her real name. It was only when Molly and I got engaged and began planning our wedding that the fully story came out. Even then, there were still chunks of Molly's history missing. She had discovered a letter when she was about ten from wife number one explaining that baby Brian was doing well. This had stayed in her mind. It all caused deep confusion! We had Molly's name changed by deed poll just before our wedding. Within three weeks she changed her name from Molly Gosling to Molly Sale and to Molly Carr – that must be some kind of record!

During our engagement, and realising that my promotion prospects had all but evaporated due to my rebellious strike action, I was approached by one of the agency concessions in the store. Hector Powe of Regents Street, London, was a highly respected tailor to the those who paid more for their suits than I earned in a month! I moved to a new job with them and trained at Bishops Gate, London. It was like a blast from the past. Suits with waistcoats, jackets off, tailor's triangular chalk in our pockets, tape measures round our necks – it was like something out of Charles Dickens.

I stayed in a bed and breakfast on Ealing Common for the six weeks of my training, then returned home as "the personal representative to the Managing Director, Solihull Branch". I received a 60% discount on all suits and two per year, at cost, for business purposes.

We had a unique fitting system that was like a coat of many tape measures. It took about fifty minutes to be measured, so we only saw customers by appointment. We then brought the client back in when the suit was still in its early stages in case it needed to be altered. We needed a fancy job title to stop some of the richer businessmen treating us with contempt. Since they spent hundreds of pounds (at 1969 prices) they expected to be attended to by the boss. We even had one consultant, Bert Trillow, who would travel to their offices or an airport to measure them up! So I did eventually become a manager in Beatties, just not for them.

I had some good friends there. I appreciated Peter Haywood, Peter Thewlis and Mr Lucas the shoe department manager. We would play each other at table tennis during lunch hours and I also played cricket for the store team.

In this upper class, "I'm free" environment, reminiscent of the old British department store sitcom "Are You Being Served?", there would occasionally be a call over the store tannoy: "Will the duty Manager kindly go to the restaurant." As acting security officer, this was basically polite code for, "Dave, get up there now, there's trouble!" It was normally either a drunk or punk who was shouting their mouth off. I would quietly lean against the bar and whisper a few words in their ear. Invariably they would quietly leave the store with me.

One slightly more "gentle" male manager said to me once, "Mr Carr, I don't know what you say, but you are very persuasive!" I told him, "I just ask them to choose between walking down the escalators nicely or being thrown down them!" I also said that if I needed to be a little more persuasive, I would point to my misshapen nose and say, "I didn't get this picking strawberries!"

Meanwhile, Molly and I started planning our wedding. We had arranged to buy a little house in the area of our church, Calvary Chapel, so that we could continue to attend. With just weeks before the wedding, my new brother-in-law, Keith, found in the planning section of the Birmingham Mail that a new road in the area was imminent. It followed a route that would effectively wipe out most of our front garden-to-be. It was a close thing. Who would have expected Keith to scan the applications and discover that?

In another strange turn of events, the very next week I found myself in trouble with our church leadership. It was the church's Annual General Meeting and I was verbally attacked by our pastor's wife for the heinous crime of praying for her serious illness! I had prayed for her several times and each time she convulsed involuntarily. Oddly, it was me who was accused of being demonically influenced. It became a highly contentious issue and, to save a split in the church, I volunteered to leave – with Molly weeping and a great argument still continuing amongst the members.

The trauma we had at Calvary changed the entire course of my Christian life and, at the time, I felt sad about that. But it also taught me a valuable lesson: to forgive and not seek to get my own back. I could have been vindictive, but I wasn't. Years later I would embrace them again. This wasn't the end of the story.

However, I had been in that church from the age of seven until I was twenty five. During that time I had assisted with most ministries. Now, in the space of a week or two, I was suddenly faced with having no new home to go to, no church to attend, and possibly no wedding venue! That evening Molly and I stood outside in the dark night air, not knowing what God was doing with us. The next day was Sunday and we decided we would go and visit another church. It felt strange. I had never visited another church on a Sunday before, as I was completely loyal to my own. We chose Selly Oak Elim church. It was a much larger church than our old one. After our first visit, on our way out of the building the pastor, Alan Caple, stopped us and asked, "Which church do you belong to?"

I said, "We don't."

He said, "Well, come here!"

That was great, but no one knew us. We didn't feel we could get married there. Later, Keith reminded us: "Molly used to go to my church. You visited the Bible class and both your brothers were married there."

I thought, "Why not?"

So, on August 29th, 1970, we were married at the Full Gospel Tabernacle, Yardley Wood Road, Billesley, Birmingham by Rev Ron Bailey. Tony, my eldest brother, was my best man, and both Tony and Keith's daughters were bridesmaids. Keith gave Molly away.

It was a beautiful hot day. Molly looked stunning in a white dress with a fantastic train and a wonderful bouquet of yellow roses. The girls wore blue and white, a petrol type colour. I, much slimmer then, had a light fawn suit with a thin, peach coloured stripe, a peach shirt and tie, brown shoes and a yellow rose as a button hole. We held the reception at the Red Lion, Vicarage Road, Kings Heath. Because Molly's mother had no money we had saved up and paid for everything ourselves.

In the end our wedding had gone ahead as planned (same date, same time, just a different venue!) and the highlight was our following one-week honeymoon in Bournemouth. We spent our first night en route in the Ship Motel B&B, Gloucester. It wasn't a salubrious venue, to say the least, and a bed and breakfast isn't what most young couples would expect today, but to my generation it was as good as it got!

The rest of the honeymoon included a night watching Bournemouth FC play Peterborough! We arrived ten minutes late, Bournemouth had already scored, and that was it. On the Sunday morning we visited the famous Lansdowne Baptist Church with the well known preacher Francis W. Dixon. We arrived fifteen minutes before the service started, but it was so packed we were still forced to sit in the overflow room. In the evening we went to hear Jack Osman, the Elim Pastor in Bournemouth.

Around this time I was singing and preaching in a band called "Second Advent". We travelled throughout the UK for a number of years, playing in different venues. We would frequently bump into similar groups of the 1970s, such as Stuart Bell's "The Advocates" (Stuart is a great friend and now leads the Ground Level network of churches). They sounded a bit like the Dave Clark Five. Then there was Dave Pope, Len Magee, Phil & John, John Pantry, Parchment, Kevin Gould...

At the time I had a light grey Ford Anglia van as gig transport. There were no back seats, so anyone additional passengers had to sit on the collection of pillows and bed covers I'd piled in there for people to sit on. I'd had to sit my test three times – much to the delight of my wife who passed hers first time. The first time I was apparently too slow – those who know me today will appreciate the irony of this. The second time I was too hesitant. I got there eventually.

The band's musical style was similar to The Seekers. Gaynor, our lead singer, was a "Judy Durham" – what a voice. Then there was Alan on the bass, John from Trowbridge – a fantastic guitarist who could play any style; be it classical, Spanish or rock, he was your man. Derek was on keys and guitar. Phil was on guitar and vocals. I was tasked with vocals and preaching. We had a ministry that impacted the lives of many young people. Only eternity will record the extent of it. Sadly, the former band members have little or no contact now. Two got married and then withdrew into a navel gazing group of no-laughter Christians, condemning the rest of us for being superficial. Two divorced and then remarried. One survived and continued to play wonderful guitar for his local Baptist Church. And then there is me!

Chapter 6
Small Steps of Obedience

During this time, Molly and I moved, by some miracle, into a lovely semi-detached house in Hollywood, on the borders of Worcestershire. There was no way we could afford the mortgage, but at the recommendation of my brother-in law (in the days when loans were decided by people not computers) the manager of the Halifax Building Society in Solihull gave us a mortgage to buy the house for £4,300. At that time, that was a massive amount of money!

We'd taken a leap of faith to go for it. Soon the first mortgage payment was due and although we'd been given a number of lovely gifts to start us off – a carpet, sofa, second hand table and chairs, even a white TV from Spectra – we had no actual money. We prayed, "Lord, why would you give us this house when we can't even pay the first mortgage payment? We have another 24 years and 11 months to go!" I came downstairs the next morning to make breakfast and found a letter on the doormat. It was from the solicitor who had arranged the house purchase. The letter said he felt as though he should repay us his fee. I could hardly believe it. It was the exact amount we needed for the first payment. For the next 25 years we never missed a payment.

Spiritually speaking, I was learning so much. The presence of the Holy Spirit was educating me in all kinds of ways. But I had no "formal" education that society might recognise and, though I couldn't face the future as a tailor (which I found soul destroying), I couldn't think what else to do. Molly had an idea. "Why don't you do the same as my brother, Keith?" I wasn't greatly impressed by that advice. Keith had been educated at Reading University and had gone on to be very successful working for Sun Life of Canada.

I prayed about it fearfully, however, and secured an interview with a company called National Mutual of Australasia. I wasn't surprised when I didn't get the job! I was surprised, however, when two weeks later I received a phone call from them. The person they had appointed had left his wife and they now didn't trust him to be working on his own. They asked me if I would still like to join them. I agreed and once and for all my tailoring career was finished.

So began the start of a long stint in the finance business. Much of this was spent in the arena of professional football, but I'm saving those stories for later. Initially, I moved about a bit. To say that I struggled in my first finance job is a massive understatement. I had a lot to learn. I soon moved Dominion Lincoln of Canada. That was also challenging, but didn't offer any real career progression opportunity. It was with my third company, Crown Life of Canada, that I eventually became established in the business community.

I completed the training and my manager, Terry Marsden (who was the double of Basil Fawlty) told me that I was the worst student they'd ever had come through the company! In fact, he said that the only reason he hadn't sacked me was because he was frightened of me! He gave me £10 and told me to take Molly out for a meal to talk and decide whether this was really meant to be what I did with my life. I felt I had to continue though, as Molly was pregnant with Ashley, our first born, and that created added financial pressure.

So I proceeded to be the worst financial salesperson in the history of the industry – for a while, at least. My utter lack of confidence resulted in a complete lack of new business. I would say goodbye to Molly and leave the house in the morning, only to drive around the corner, park up and stay there because I had no appointments! Or, I would drive to Stoke-on-Trent where our head office was, be physically sick with fear outside, then return home, having been unable to enter the office building.

So far, the only advantage of the new job was the replacement of the old Ford Anglia Van. My first company car was a Morris Minor, brown with red seats. It was more of a mini tank than a car. Later I had a one-year old Ford Escort. Wow! This was posh. I had to pick it up in Manchester and drive it home. When I got to St Chad's Island in Birmingham I paused and, a second later, was promptly catapulted across the island. A lorry had run into the back of me. Me and the car limped home with the boot roughly wired together. Molly was all excited when I arrived home and asked, "Well, what's it like?" I said, "It's a two seater convertible!"

Our son, Ashley, was born on the 28 February. Molly suffered from post natal depression and life was difficult. When we married we had come from two different worlds. Molly had never known her dad and was totally committed to her mother. I, however, came from a home where dad was my hero. Molly was acutely jealous and I was hot tempered – not ingredients that generally make for a quiet life!

One day, not long after we married, we had a massive argument. Molly packed her bags and returned to mother. Some hours later I received a frosty phone call: "Are you going to pick me up?" Me – "Mr hard man" – retorted, "If you found your way there, you can find your way home" and promptly put the phone down! It was

God's grace that changed us both in the end. Molly would never have survived the life that was about to unfold for us – imagine being jealous when your husband is dealing with football stars and their hangers on all the time and then running a church as a pastor? And what good would I have been – aggressive; always willing to face off with anyone. We are called to be strong and bold, but never hard.

Two experiences I had around this time had a profound affected on me. One altered the way in which I thought about and conducted my business. The other set me on a new journey to establish a church. God used them both to shape and change my character.

Since committing my life to Jesus I had made a simple promise: "Lord, I will always put you first." So if I felt God wanted me to do something, I did it. One such example was accepting an invitation to go and speak at a youth camp for deprived youngsters on the south coast. There was no real payment, but it was "for the Lord", so I agreed to go.

A few days before leaving for the youth camp, I was given a business lead in Stoke-on-Trent. A gentleman named Mr Fox was interested in taking out a small, company pension scheme. If he went ahead and bought the policy I would receive a commission which was equivalent to a full month's wages for me. But there was a problem. Mr Fox could only see me during the week of the youth camp. I tried hard to find another date that would suit him, but no convenient alternative date could be agreed upon. I could have simply cancelled speaking at the youth event, but I'd agreed to go and I didn't want to renege on my promise. Mr Fox understood and said, "I'll leave it and phone you when you get back." I knew that meant I wouldn't hear from him again.

I went to youth camp as promised, but on my return I felt desolate. I left for work the following Monday, leaving Molly and Ashley at home, and all that was on my mind was the prospect of no real finances for the coming month. I had no appointments, no place to go, so I did my usual routine of driving round the corner and pulling up at the kerbside. I pleaded with the Lord and complained, "Why, when I put You first, am I penniless?"

A response came back immediately. An inner voice said, "Drive into the city and look for Essex Street." I had gotten used to hearing this voice, as many Christians do. The Bible calls it the "still, small voice"; God speaking into our mind. If ever I heard God speak to me, I was accustomed to doing whatever He told me, unfailingly and without question. So I set off and drove into Birmingham. I drove round and round but I couldn't find such a street. I needed to stop and ask someone. As I was driving up Hurst Street, I saw a man walking along the pavement. I pulled over, got out of the car and spoke to him.

"Excuse me, do you know were Essex Street is?"

He replied, "No, sorry. I'm not from round here, I'm from Stoke."

"That's funny," I said. "I should have been in Stoke last week to see someone."

The stranger said, "Well, that's interesting, because someone from here should have come to see me last week in Stoke."

I couldn't quite believe this, but after a pause I ventured, "Mr Fox?"

"Yes!" he replied, startled.

On the basis of this meeting we went on to strike the original deal and I finished that month having both served at youth camp AND secured this new client and the commission. Oh, and by the way, there was a Essex Street on the other side of the city, but that wasn't really important!

Not long after this, I received a life-changing phone call to the office. That morning, Molly, having been more than conscious of the debilitating effect of our lack of finances, had said to me: "Dave, if you are to fail, go out fighting." I left for work with her advice ringing in my ears. I was the only consultant in the office that day, so when the phone rang I answered. It was the first team Coach at Walsall FC. It was a conversation that would open up a whole new horizon of my career in finance. I'll share that story later in the book. For now, suffice to say that through that one phone call I would eventually go on to help manage the finances of around 700 professional footballers. It is amazing what God can do when you decide to put Him first.

* * *

Parallel to my desperation to succeed in business God manoeuvred me into starting a church! I say manoeuvred because although I obeyed His instruction, I was reluctant to do so and it happened with not a little squirming and protesting on my part.

Since my first visit to Solihull and early days working at Beatties, I was now living only one road away from the Solihull boundary in a place called Solihull Lodge. One day, driving down into Solihull Lodge, I heard that same inner voice suddenly command me to "Turn left." I did so and found myself outside a building that looked like a church. I got out and had a look at it. On closer inspection, I realised it was a community centre. I thought to myself, "That place is perfect. Someone should start a church there." Then the inner voice, that I know to be the Holy Spirit, said, "Yes, you do it."

I responded and explained that I had no intention of becoming a pastor and made an attempt at arguing. Finally, feeling that I wasn't being convincing, I decided to test the Lord and said, "Lord, I only know one Christian in Solihull. If you truly mean for me to start a church here, let me meet her ... now!"

The Lord fell silent and I smiled to myself, thinking, "That's got You!" I got back into the car. As I waited at the junction to turn out of the road onto the main carriageway, there was a knock on my window. To my surprise (and horror) there was the Christian lady I knew, frantically waving and smiling at me. I slowly wound down my window. My face must have been ashen.

"Hello!" she exclaimed. "Fancy seeing you here! Have you seen that building round the corner? You should start a church there!"

The Lord had given me the specific confirmation I'd asked for. But I was still a bit confused. I had no theological training, no desire to do the work of a pastor, no financial security that I could depend on. Plus, the band was at its height, so I was busy with that. Why would I want to start a church?

However, I felt compelled by God to make some investigations, so reluctantly I made some enquiries and ended up phoning the Chair of the Residents' Committee that operated the building, a Mr Alex Goulding. I asked Mr Goulding if they would be willing to hire the room out on a Sunday for use as a church. I was delighted when he replied that a number of other churches had already expressed an interest. He added that if I left my details with him he would perhaps get back to me. I felt extremely relieved at this. I had no denomination behind me, no money, no people, no charity! In fact, I was such an unlikely candidate in unlikely circumstances that I decided the Lord had merely been testing me – like Abraham. He'd never intended me to actually start a church, He just wanted to make sure that I would do as I was told. I'd done it! I'd passed the test! The Lord had tested me and I had passed. I was elated.

But then, the voice came again: "Start the church in that Community Centre."

And I realised He meant it.

I just couldn't rest, so I asked for another confirmation – one which, to me, seemed highly unlikely, if not impossible. "Lord, if I'm to start a church there, the Committee will call me before 1.00pm tomorrow and offer me the building!" I immediately felt more at ease. I hadn't said "no" to God, I'd just created what I thought were impossible circumstances.

The next day at work was fairly uneventful and I arrived home at about 6.00pm. I was a little apprehensive as I opened the front door, remembering the deadline I'd presented to God, but I had no real expectation of anything. I walked in, greeted Molly and then, nonchalantly asked, "Has anyone called today?"

"No," she said, busy in the kitchen making dinner.

My sigh of relief was only half exhaled when she continued, "Oh, no wait ... yes ... there was a call. It was the Chair of that Residents' Association. He wants you to ring him back."

I gulped. "What time did he call?"

"I don't know," she said, "I was busy." And then, "Oh, yes, I remember ... when I put the phone down, the pips went on the radio for the One O'Clock news."

Not for the first time in a few days the blood drained from my face and I was transfixed to the spot. After a few minutes, I gathered myself together and went to the phone, with more than a little trepidation.

"Yes, hello David," Alex greeted me, "thanks for calling back. Well, good news, there were about a dozen churches on the list ahead of you, but the Assemblies of God, who were at the top of the list, phoned this morning to withdraw. We went down the rest of the list and, to be honest, we thought they were all either a bit too confusing or too specific for the local community. We thought you'd probably be the most neutral, so you can start your church in August ... David? Are you still there?"

Chapter 7
Mentors and Monks

To say that I was unimpressed was putting it mildly. I had never seen myself as a pastor. My passion and gifting was as an evangelist. With the band we were travelling the country, seeing literally hundreds and hundreds of people find faith in Christ. My heartbeat was for those who didn't know Christ.

But my mind suddenly flew back to the time when, aged eighteen, I was sitting in the pews of Calvary Chapel. The inner voice of the Spirit had spoken to me. I look back now and acknowledge it as the leading of the Spirit, telling me to read Ezekiel 33. I remember that as I turned to this scripture, my brother's sermon had dimmed into the background and these words of scripture started ringing in my heart and mind.

"I have set thee as a watchman to your Nation."

The call on my life was to sound a warning – to warn God's people about what the enemy of God would bring to this beloved country of ours. I didn't know exactly how that word might be fulfilled, but I understood that a watchman held the responsibilities of protection and warning, rather than decision-making or policy.

At the time when the word came, I sensed that if I didn't accept God's mandate and challenge I could easily slip into apathetic religion. So despite my reluctance to become the pastor of a church, I knew that God had bigger plans and, like it or not, I had to cooperate.

Over the years, it has been this call that has anchored me in my ministry in England. I would travel to many other countries and see the ministry impact churches in other nations, but my call was to England – the heart of England and to Solihull, the place that I have remained and served ever since. When opportunities came over the years to accept ministerial positions in some beautiful countries and some significant churches, it was this scripture that helped me to turn them down before I could be tempted!

Through the band, we had ministered across many denominational boundaries – all kinds of church streams. I think this was foundational in fashioning a desire for unity in my heart. In the future I would be passionate about the issues of "oneness" – not of some religious structure, but of individual Christians.

An early mentor

As I wrestled with my calling and the establishment of a church I met a Gloucester born Evangelist by the name of Eddie Smith. It's difficult to describe Eddie, but he was essentially an "old school" preacher. "Methinks" was a favourite saying of his, as he imagined the glory of The Lord. He was a one off. He didn't fit any mould. He and his wife, Rhoda, became great friends. For me, as young man with the desire to move in the gifts of the Spirit, Eddie was a gift from God.

We went with him one day to a church in Lane End, High Wycombe. Eddie was preaching and the service started at 11am on the Sunday morning. After we had sung the first hymn he asked us to just stand in the presence of God with our hands lifted in the air. A deep, heavy peace that came into that building. You could hardly even hear the sound of anyone breathing.

After what seemed like about ten minutes I tried to put my arms down. But they just wouldn't move! We stayed in this state of total bliss for a while until Eddie asked us to sit down. Then he said, "Please don't look at your watches, but what time do you think it is?"

We had started at 11am, sung one hymn and stood for a short while with our arms in the air, so the answers ranged from 11.15 to 11.30am. A sense of awesome fear gripped us when we realised that we had stood, timeless, until 1.00pm with our arms in the air. Not one of us had been able to take them down for almost two hours!

It was the only time in my life when I witnessed what seemed to be a suspension of the natural, temporal laws. Have you ever held your arms in the air for two hours? I couldn't imagine ever being able to do that again! The service ended there and then. There was nothing else to say. We left having been in the presence of the Almighty in a way I have never again experienced.

Eddie had a sharpness in the prophetic that was frightening. I was with him one night at the church when, in the congregation, there was a man who that same day had seen the roof of his house fall in. Eddie told him, "Because you have been faithful in coming to the house of God, the correct amount of money to repair your roof will be waiting for you, in an envelope on the door mat, when you return home." It was awesome! The man wept, went home and there was the money, waiting on the mat.

I sat with Eddie for two years, watching, listening and learning how to hear from God. On the way home from meetings I would ask him lots of questions. As we progressed, I would tell him that before he called certain people out, that I had also been given a prophetic word for them. After two years he said to me, "I've taught you all you need to know. It's time to release you into the purposes of God." He was not holding me to his ministry, but commissioning me to my destiny.

Moving in the anointing is not always exciting. Over the years it has been "by faith" – taking a risk to step out and obey. My departure from my first church had been linked to using the gifts of the Spirit, so I realised that I had to become accountable for the administration of those gifts. Ever since I have always made myself accountable to the leaders of whatever church I happen to be ministering in.

Here is how my (I felt) untimely departure from my first church played out. With this anointing growing in my life, I had asked my pastor if I could pray for people at the Tuesday night prayer and Bible study meeting. I felt God prompting me to pray for a lady called Mrs Wells and the pastor's own wife. I didn't know the nature of Mrs Wells' illness, but I knew the pastor's wife was suffering from Myasthenia Gravis, a neuromuscular disease.

The pastor said I could pray and suddenly my hands started to burn. I laid my hands on Mrs Wells and, for some reason, felt led to pray for her throat. I then prayed for the pastor's wife – not physically, because she was actually at home in bed, just out loud.

About two weeks later at church, I again felt the prompting of the Spirit to pray for the pastor's wife. Once again I asked the pastor for permission. This time he was a little hesitant, but he still said yes, so I stood up and publicly prayed that the sickness would be broken and then sat down again. On the following Sunday morning, one of the Elders stopped me as I was leaving and asked if I could meet with the Elders after the evening service for ten minutes. I said that would be fine. I was just on my way to see them when I bumped into Mrs Wells, who stopped me to say, "Thank you so much!" She went on to explain that she had a lump in her throat and that the doctor's had thought it was cancer, but now it had now gone! Praise The Lord. I didn't know that this testimony would help to keep me focussed on the Lord in what was to happen next.

As I entered the room, the elders and the pastor sat at a long table. It looked and felt like a court martial. They relayed the information to me that on the two occasions I had prayed for the pastor's wife as she lay in bed at home, she had immediately began to convulse. They asked me to give an explanation as to why this was. I didn't know the answer. I hadn't got a clue! I was new to all this. I had, however, seen similar things happen with Eddie.

"It almost sounds spiritual or demonic," I suggested, unwittingly mentioning the very thing they were thinking – except not in the same way as me! They immediately leaped on this statement: "So, you admit it then? You are demonic!"

They disciplined me and told me I wasn't allowed to pray for anybody in the church again. I left that meeting agreeing to obey them, but totally confused. All I could think about was what had happened to Mrs Wells. Surely, that wasn't demonic? Sadly, the matter wasn't allowed to rest there. The following week I was given an ultimatum – leave the church or I would be responsible for the pastor resigning his post and potentially causing a church split. I left. Sometimes moving in the gifts does have a heavy cost to pay.

* * *

Molly and I had been married for about two years and I was trying to establish myself in the insurance industry. I had made the jump into the world of football and was beginning to manage the finances of a number of footballers. One day I was out and about and due to head over to Coventry City when I suddenly noticed I needed petrol and realised that I'd left my wallet at home. I was more than a little angry with myself, annoyed that I would have to make a trip home first to rescue the situation. But as I put my key in the front door, the house phone was ringing. It was Mom.

Apparently Dad had been mowing the lawn on his day off and had come in for a cup of tea. Mom was worried as he appeared to be in pain. She said to me, "Please come over and see him. He will listen to you." That was Mom's way of passing the problem over to someone else! I now had no chance of getting back to Coventry before the players left for the day, so I dropped what I was doing and went over.

My parents had moved when I was in my teens, two roads away from our original home and had bought my brother's house from him. For the ten years or so I was living with them I became the unofficial First Aid person of Cornhill Grove. Often the doorbell would ring and a neighbour's child would be there: "Excuse me, Mrs Carr, is David in?"

I arrived at the house to find dad looking very much as if he'd had a cardiac attack or was in the process of having one. I phoned the family doctor. In those days they had a relationship with the families they cared for. I explained dad's condition. The doctor said, "You have been in the St John Ambulance, what do you think?" I said, "Dad needs hospitalisation". The ambulance was sent for and, before my dad could object, he was admitted.

It was shortly after this that his heart arrested. If he hadn't been admitted, there's every possibility he would have died that day. As it was, he lived for a further 18 years thereafter. And it was through this serious incident that he became a Christian.

* * *

With all this chaos swirling around, I was on the brink of planting a church! Me, an evangelist, pioneering and pastoring a church. It was almost unthinkable. I had no team, no equipment, no money ... nothing. But I did have a start date: 13th August 1972!

Well, for the first week, I supplied the band. We needed a name for the church, so I called it Trinity Free Church, which seemed to be a good start. I arranged for some black and white leaflets to be printed, emblazoned with the words:

Trinity Free Church
Grafton Road Community Centre
Solihull Lodge at 10.30 am

My friends at Selly Oak Elim church helped to push the leaflets through all the houses in this border district, 500 yards from my native Birmingham. You had to have hymn books in those days and we used the Redemption Hymnal; those good old red covered books. Dr Stephen Brown, who would become such a kingpin in Stirchley Elim, kindly offered to play the piano for me till I could arrange for a regular pianist.

During the week before we started, I placed an advert in the Solihull News, the local free newspaper for the area. It caused something of a riot! The press asked the other churches in the area what they thought about us and we bore the brunt of their insecurities. They were not happy at our arrival in "their" town.

I had been brought up a Protestant and believed that Catholics lived under deception, so imagine my reaction when, upon answering a knock at my front door, I saw before me a full blown monk; bearded, wearing a brown habit, sandals and a rope around his waist. He apologised for his unscheduled arrival and then also apologised for the reaction we had suffered from the local churches. He simply said, "We need many churches to declare the kingdom of God."

He then asked for permission to pray for me and, of course, I agreed. His prayer was that I would be blessed with much fruit and, upon completing this, he made the sign of the cross and bade me goodbye. When I stopped him and asked how far he had to journey, he told me that he had walked four miles and would now be returning the same way. Of course, I offered him a lift home, but he refused with a smile, saying, "The walk gives me time to meditate upon the Lord."

I never saw him again but that visit has stayed with me throughout my ministry. He was a man of peace and reconciliation and, although I had problems with some of his theology, I knew his lifestyle was authentic and I could not deny the reality of his Christian walk.

They say there is no such thing as bad publicity. Sitting in Mell Square, the town's most central shopping area at the time, reading the local paper were Geoff and Althea Greenaway, two young, married teachers who had been instructed by the Lord to move to the area from their home in Slough. God had not yet, however, given them the reason for it.

After leaving Solihull, they drove for a little while around Lapworth and, after a short while, Geoff stopped at a phone box (no mobile phones in those times) and dialled the telephone number that I had added to my advert. I remember the conversation as if it was yesterday.

"David Carr?"

"Yes," I replied.

"Will you tell me what you believe?"

"Yes," I replied again, before launching into the basic theological stance we would take.

When I finished, Geoff's final words were, "We will see you on Sunday."

That was almost 41 years ago. That Sunday, many came to have a look at what we were doing, yet never returned! But Althea is still with us and Geoff was with us right up until the Lord called him home, three years ago.

So we started our church with four people. I figured that I had doubled our congregation in the first week, so that was a good start! It was with these humble beginnings that we turned to face the future.

Chapter 8
New Names and New Places

Personality-wise, there tends to be a big difference between an evangelist and a pastor. Evangelists are often charismatic by nature and their gifting draws people in. By contrast, Pastors tend to be more measured, more suited to the week-in-week-out slog of maintaining the local church. I was an evangelist, not a pastor. This led me to make some youthful statements I lived to regret. One classic was this: "Any pastor that cannot raise a church of 150 within one year isn't fit for ministry!"

The Lord reminded me of this statement a years after I'd made it when, after beginning with 4 people, we still only had 12 in our church. Molly was in hospital having Ashley. I sat with her, reading A. W. Pink's book on "The Life of Elijah". The text was recalling that, after a successful confrontation with the King, Elijah found himself in the wilderness, depending on water from a stream, meat from the mouths of ravens and food from a starving widow. The Lord said to me, "David, I've put you in this wilderness to teach you humility."

He then continued, "You said that if you hadn't 150 in the church you would not be fit for purpose. Well?" I said deep within my heart, "Lord, forgive me. If you wish me to spend my life looking after just these twelve, I will." Right there and then, on 28 February 1973, the Lord gave me the vision that we still work to, 41 years on. If we are willing to be faithful in little, He will trust us with much. Today when people today ask me why we have a congregation that is so significant, I tell them it's because I was prepared to spend my life caring for just twelve. This foundational lesson was painful, but it provided the firm footings of future greatness.

Meanwhile, it was not easy trying to spin so many plates – family, church, business. Other than my faith, nothing in my life seemed firmly established. Molly was suffering with post natal depression, the business was struggling, and the church was in its infancy. But in the 1970s there was little known of mentoring, life coaching, marriage enrichment or financial planning. It was very much a case of "just get on with it"!

Growing responsibilities meant that it impossible to keep the band going, so sadly it was time to finish touring and singing and pursue our individual ministries. I had our new church, so former band members Gaynor and Alan came to us for a season before moving on with Dick Maund to a small community of self-regulating Christians. We don't see them now. Derek moved back to the Black Country and established a Christian music ministry. John got married and established himself in the local Baptist church. Phil became my assistant at the church for many years until, sadly, he ended his marriage and left the ministry.

I think that during that time there was a serious lesson to learn. We all stood equal under the anointing of the Lord; we all bathed in the presence of the Holy Spirit; we all celebrated the birth of new Christians – yet our future paths left little in common between us. The disciples found the same disparaging issues, yet the Holy Spirit can and did change all of that on the day of Pentecost.

I had only intended to stay in Solihull for 18 months, but God challenged me to stay. My thought was to build the church and then hand it over to someone else at some point, though this might prove difficult since we weren't affiliated to any church stream or denomination. The truth is, most of the pastors I knew were just looking on and giving me 18 months to fail!

I decided to contact a few denominations to ask for information, including the Free Methodist Church of North America. I had read an article on their history and found their Wesleyan theology, holiness and receptiveness to the Holy Spirit attractive. We had recently changed our name to Trinity Pentecostal Church and it seemed obvious to contact the two main Pentecostal Groups: Elim Pentecostal Church and the Assemblies of God.

The results of those enquiries left me a little limited. The Free Methodists thanked me for my enquiry, but said they didn't operate in the UK. The Assemblies of God arranged to visit the church, but didn't turn up! Then there was Elim, who did come along to see us! On that basis our small group became affiliated to the Elim Church.

I was initially appointed as the "honorary" pastor, since I had received no formal training. I had considered entering theological training, but the Bible College Principal I spoke to had the wisdom to tell me not to give up the unique opportunity of working with professional footballers. Therefore, I studied every Wednesday for five years, submitting papers on the doctrine of the Holy Spirit and Salvation. I also spent two years studying comparative religions, namely the Latter Day Saints (Mormons) and Jehovah's Witnesses.

I remember looking for a secretary to type up my dissertation. One of my client offered his PA, so I gratefully gave her the manuscript to type.

Apparently, as she was typing, she started to manifest, ripped off her headset, ran from the office screaming and locked herself in her home for a week. After that, I tested it with a few others until I found someone who was fascinated rather than terrified!

At that time, only churches with the word "temple" after their name had permission to have a name that included a reserved designation, so "Trinity" became replaced by Solihull Elim Pentecostal Church. That was our third name change. The journey with Elim lasted for the next 28 years, during which time I trained with them and was ordained at their Clacton on sea Conference in 1979.

I served them as Presbytery Youth Director and Superintendent, overseeing 32 churches and became a member of the Elim Church Incorporated Executive. I split this oversight into two areas: the mainly Birmingham and South of the City area became known as the Heart of England Presbytery and the rest of the region became the Black Country Presbytery, to which I appointed another Superintendent. I travelled extensively for Elim, preaching at a great number of churches and conferences, including twice representing them at the Elim Church, New Zealand.

We moved out of Grafton Road Community Centre, Solihull Lodge in 1976. It was interesting that being in this "outpost" of Solihull, we had little relationship with the life of the wider community. I didn't really know what to do and decided to test the waters of possibility. This was one of those times that I had NOT received a "Thus saith the Lord!" proclamation, so I embarked upon a six day mission – three days in Solihull Lodge, two days in Knowle Village Hall and two days in Solihull town centre's flagship Civic Hall.

The big faith step came, unsurprisingly, at the Civic Hall. There were two rooms to choose from: a 1,000-seater auditorium and the smaller 300-seater auditorium. At this time, we'd grown to around 40 people, so we felt the 300-seater was a faith statement in itself! We booked it for Sunday 13 November 1976 at 8pm.

My friend, David Woodfield, was the preacher for our mission. The Renewal movement had started to spread within the established churches. It was their version of Pentecostalism. I had moved in the authority of the Spirit for several years, yet never felt completely comfortable with the practices of traditional Pentecostalism.

The obvious name for our outreach was "Renewal '76". We put leaflets everywhere. Mostly black and white again. However, this time the title was in red! We travelled the local streets with a loud speaker, canvasing support. The whole church of forty signed up to help. Wow! Solihull Lodge and Knowle attracted virtually no one. It was very disappointing. But we all held on to the prospect of the Civic Hall. The day arrived and suddenly I thought, "What on earth am I thinking? People don't do church on a winter's night at 8.00pm!"

I was pacing up and down whilst our teams started to take their places on the doors and the PA desk. The worship team, led by Tony Burchell, stood ready. Tony was an incredible keyboard player. The church was only six weeks old when, one Sunday morning, his wife, Diane, walked into our service. We were singing without music at the time, as we'd sent Stephen back to his own church and his replacement left a little to be desired in the "tuning" department. So, there we were, singing a cappella! Diane asked if we might like the help of her husband. She explained that he was a professional musician, currently accompanying the likes of Peters and Lee and playing on the TV programme "Sale of the Century". So, the next week, along came Tony. He wasn't a Christian and didn't know any of our songs. He just said, "You sing it, I'll play it!" He was the most incredible musician I had ever heard – and he was playing at our church. In the midst of the music, there was an altar call and Tony, without missing a beat, put up his hand and gave his life, there and then. That was 41 years ago and he's still on our worship team to this day.

Back to our wintery Sunday night. Everything was ready. The only thing missing was the people. I stood behind the curtain ready for the 8pm start. I walked through the curtains, unsure what to expect, and to my complete surprise the room was filled with 300 people! The service was excellent and the next night was equally well attended. This was the sign we needed. Things were starting to move.

Then we met the Towlers, a family who had previously been mission workers in Nigeria. They came from a Methodist/Anglican background and felt the conviction of the Holy Spirit to be baptised. I sent them to their local priest to gain permission as I believe in maintaining good relationships between churches. The priest reluctantly agreed and they were one family in an increasingly large group of people who started to join us. We booked the swimming baths for the baptism, since we had no actual church building, and we advertised the service in the Solihull News.

Then things quickly began happening all at once. We heard from the BBC who wanted to come and film the baptisms. I was interviewed by their well known reporter, Tom Coin. The Solihull News featured us in a centre spread and, more traumatically, the Towler family had to stand before the priest the following Sunday, in front of the whole congregation, while they were publicly excommunicated from their church for their association with me.

I was noted in a pew slip, that was sent to the five local Anglican churches, stating that I was a heretic and that no one should fellowship with me. It was only last year that I fully appreciated that a few hundred years ago, I would have been burned at the stake in the middle of the town!

This beautiful Towler family have served the vision with us in an incredible way. All the family, at some time, have been involved in ministry. At the time of writing, dad Chris is a trustee and lead in the Camera Department. Mum, Anthea has been involved in drama, poetic readings and proof reading of church literature. Their daughter, Heidi, is a major part of the women's "Blessed" ministry and is married to one of our elders. Son, Justyn is our Associate Pastor of Children and Family Ministries. Now, the grandchildren are joining in too, with Chloe accompanying many worship songs through beautiful interpretive dance. All of the children of this family have made valuable contributions to the church.

Although we'd changed names many times in a short period, this new era seemed to demand a fresh, motivational name change – one that would reflect our emerging identity. We had seen the impact of the word "Renewal" in our mission to Solihull, so I informed Elim that we needed to identify with the mission and its success. In December 1976 we became The Renewal Christian Centre or "Renewal" for short.

The problem was, our home base was still in Solihull Lodge and we really needed to be in the centre of Solihull. Our numbers had considerably increased since the mission. I then had a call from Terry Law Ministries, from the USA, asking if I wanted a fantastic band and singers, 35 in all, to put on a festival of Christian Worship. This included a young man who one day would become one of the world's leading worship leaders: Don Moen. They had been booked at the largest Elim Church in Birmingham, but they couldn't find enough members to put everyone up for the weekend. I thought to myself, "If they can't, what chance have we got?"

But I asked our congregation and a room was pledged for every single one of the 35 musicians immediately. Wow! The Library Theatre had just opened in Solihull, so I secured one of the first bookings. The performance was full and the music and presence of God was awesome.

After the service, I was standing in the lobby, speaking with one of our leaders. We were discussing the difficulties of finding a building suitable for our Sunday morning. At that moment, the Library Theatre manager, Derek Johns, said to me, "Excuse me, if you can put on a show like that, we need you in this building. Would you like the exhibition room?"

What a fantastic facility it was! Every Sunday morning we met, right in the centre of the town. We then booked the theatre for the first Saturday night of each month, too, for what we called

"Cross-Word Rally". The logo was a cross, flowing into the Word. We did, on occasion, get a crossword boffin turn up, thinking it was something entirely different, but with this extra night we had really started to build the church.

We didn't have the facilities to put on an evening service, but then I was told of a United Reformed Church just on the Solihull bypass, called St Ninian's. They allowed the Jews to meet on a Saturday till their synagogue was built. The local congregation was very small and met on Sunday morning and once a month at 6pm for one hour.

I had asked Phil Saleh, from the band, to join me as my Associate, but he hadn't been convinced. However, he then had a dream in which he saw a building with a wooden floor, marked out for sport. A gush of wind came and the room was filled with people praising God with their hands in the air. The Lord said, "I'm going to visit Renewal, go join."

I phoned the minister of St Ninian's, the Rev John Forsyth, and he said there was a good possibility that we could be accommodated. "Come and have a look at it" he said. We met at the front door, but John said, "No, let's go through the back door." As we entered the building I could see through the doors into the main sanctuary and yes, it had a wooden floor marked out for badminton. I said, "Yes! We'll take it!"

John looked surprised and said, "I haven't told you how much, yet!" It didn't matter. At that moment, all I knew was that this was God's designated place. So within weeks of "Renewal '76" we had two new venues, a new name and had tripled in number. Later in the story, I'll describe the incredible spiritual experiences that manifested themselves during the eight years that we lodged in this "tent".

Chapter 9
Faithful Friends and
Supernatural Happenings

Through the mission we connected with Dick Maund. Dick was an RE teacher at what was then Malvern Hall Girls Grammar School. It later became a comprehensive and is now the home of St Martin's Girls Public School. At that time, the school was situated in Station Road in Solihull. Dick invited me into the school to run their Christian Union. What happened then, between the two of us, was breath taking.

The Holy Spirit would tangibly descend whilst Dick was strumming his guitar. Then I, the transitioning evangelist, would share a very simple word and the girls wept their way to Christ. A mini-revival developed so quickly that we had to move the Christian Union meeting time to after school, because the girls would either be face down or have their arms raised, weeping or laughing in the presence of the Holy Spirit.

God so impacted the social lives of these girls that it became worrying to both secular and religious parents alike! Rather than partying, dancing and carrying on with other "normal" teenage activities, all these girls wanted to do was to read the Bible, pray, listen to Christian worship music and attend SNYF (pronounced "SNIFF"): Saint Ninian's Youth Fellowship. In due course it had spread to St Martins School, Solihull Boys' School and individuals from other schools started coming too.

We built a room in the roof of St Ninian's so that, on a Friday night, the youth could gather. To behold it was beyond your greatest dreams come true. All these wonderful teen aged girls and guys were just totally sold out in worship. Wow! Dick, myself and our friend, Phil ran the youth group and we were merely three men acting as facilitators for the Holy Spirit.

A number of parents became concerned, however. They couldn't understand what was going on and some decided to come to the school to sit in on the Christian Union. Not a few were obviously determined to expose the "cult" that their children had been drawn into! I could perfectly understand that. It was nothing like normality – even to religious families who were faithfully attending church, but perhaps not taking the lifestyle so seriously. Some contacted local clergy and asked, "What is this 'born again' teaching?" Some of the clerics responded with confusion. I remember one rector coming out with the statement of the century: "I wouldn't worry. It's a youthful expression. I did that once but, like me, they will grow out of it "!

My God, may that never happen! Grow into it, Yes! Thankfully, however, many parents who came to find out what was going on at the school or to our evening service at St Ninians, immediately received Christ themselves. Many of these parents became our future leaders.

Faithful friends

Through all these exciting, chaotic times, Geoff and Althea

Greenaway stood by us, with a faithfulness and loyalty not seen in many. In many ways, Geoff was a complex man. He was always totally focused on the things that mattered to him. To Geoff, everything was black and white: his faith, his wife, his family, Renewal, his relationship with me and steam trains! It may not have been in that order, however. I may be a little presumptuous putting myself ahead of his passion for steam trains!

From when they joined the church that very first week, they lived in Kings Heath – less than a mile from where I was born and 8 miles from the church. It seemed strange that as a new, emerging church we would think nothing of traveling to Geoff and Althea's at 82 Avenue Road for our Bible study and prayer meeting. This large pre-war house would be so full that Geoff would, on occasion, remove the door in order to get more people in! Then there was always Ringo, the large ginger cat, who would simply walk over people since there was no floor left for him!

I don't know how I could have continued in so many challenging and sometimes disappointing situations without the faithfulness of these dear friends.

Geoff started as our treasurer. He then held every office in the church until, in due course, he would surrender it, at personal emotional cost, to others who he felt could do the job better. I don't know how he had the moral fibre to do this time and time again. He was a founding member, yet he never sought or asked for any recognition. Who better then, to be appointed a trustee?

Althea had changed from teaching home economics to teaching those with severe disabilities – something she continued to do for years to come – demonstrating the great care and compassion she has. Geoff took early retirement and also worked with vulnerable people, driving a bus for those with disabilities.

Geoff, who was a perfect driver and also the kind of person who could recall events in great detail, arrived late to our house group one night. Geoff was never late! On arriving, he laughed and said that he had forgotten how to get there and had hit the kerb on the way. This was so out of character. For my brother, Tony, now retired from being Chief Nursing Officer, this sounded a warning bell and he made Geoff promise to see his doctor.

Sadly, a brain tumour was diagnosed. Geoff endured an operation and the ensuing treatment impeccably, but it came back and he started to become immobile. Despite this, he never complained. He was eventually admitted into a hospice.

Geoff was independent and in many ways a very private man, without in any way seeming distant. It was heart breaking when, towards the end, he couldn't even feed himself. One day when I went to visit him, without thinking I simply picked up his food and began to feed him. He gave no resistance.

One day, he just fell asleep and two days later he just gently stopped breathing. Althea said to me, "When I saw you feeding him and him willingly letting you, I saw the depth of your friendship for one another." I'm crying as I write this. It brinks back the friendship that kept me going for the thirty seven years that I knew him. I have given him this portion of the book to show you that my story is not all about me, but about the incredibly special people who made me what I am today.

It's difficult to have to take the funeral of such a friend. I held it together until the crematorium. Then, just before the committal, during my favourite last hymn, I wept. I lost it. Althea stood up, stepped out of the front row and put her arms around me while I continued the committal.

Some time after this, I was speaking to our church leaders about this and said how I felt my emotional display had let the family down. Dr Chris Oyede, one of our pastors who is also a consultant psychiatrist, disagreed.

He said, "Firstly, you showed that pastors can have friends. Secondly, you showed how much you loved him. Thirdly, you deflected the most painful moment for his wife, her thoughts being directed towards you. Actually, pastor, I thought you acted perfectly."

Althea now has moved to live on the same road as the church and runs a ministry for widows and ladies suffering from loss. She travels extensively overseas, ministering to those who have suffered loss through bereavement or divorce.

Their two children, Pauline and Philip, grew to see me as their "Uncle Dave" and that has continued into adulthood. Pauline, is married and lives in Scotland, having studied theology at Aberdeen. Philip, their son, was made a trustee, taking his dad's place. At his appointment he cried and said, "I always wanted to be like my dad." He is, and even more! His two daughters affectionately call me their "Dr Dave".

Abigail is Phil and Jessica's eldest daughter. When she was dedicated, I lifted her up and prophesied, "This child will grow to worship and dance for The Lord!" Everyone said, "Amen." However, a few months later they discovered, after a scan, that a part of her brain had never developed. They were told that she would probably never be able to function in any normal way, since the defective area was one of the most important in communications and movement. What of the prophecy? I went before the Lord saying, "Lord, what is going on? Did I get it wrong? How do I explain this to the parents and the congregation?"

A still voice just said, "It's true."

I do a lot of crying as a Christian. This Easter, 2013, there she was – standing in front of the children, hands in the air, fervently leading worship in front of 1,700 people over two services. Yes! What a fantastic and passionate worshipper. She is a little paralysed on one side, but destined to be an incredible woman of God and part of the third generation of Greenaways to impact Renewal. And just in case I forgot to mention it, she also dances before the Lord!

During the move of the Holy Spirit many future leaders came into the Church. Geoff and Dorrie Banks sold their shop to serve the vision by Geoff acting as one of our first administrators and the couple opened a coffee lounge that became a relational base for Christian and non- Christian alike. Then came Ken and Kath Osborn. Ken became a trustee. What a gentleman! Both Geoff and Ken are now with the Lord.

Many more faithful helpers followed these incredible men of God. During an influx of boys from the public school, one of their masters, Niall Cluley, attended. I can picture him now in his sheepskin coat! At the final prayer he walked to the front and surrendered his life to Christ, right in front of his pupils! Not an easy task. We used to place these new contacts on a "Foundations of Faith" 16-week training programme. Niall came and completed the course three times! He claimed that it wasn't until the third time that he fully understood it. He now oversees the Bible Training Centre and is an elder at Renewal.

Supernatural happenings

Those days had an ignorant bliss attached to them. We really didn't have a clue what to do to build the church. Although I guess, at that time, no one did! Molly was being thrust into situations she never believed possible.

1973 saw the release of the horror film "The Exorcist". In total ignorance of the demonic, I gave out leaflets to those leaving the cinema in Solihull. In the centre of Birmingham, such an act would have been quickly lost in the ether. But this was Solihull – renowned for its conservative attitudes and politics. The people had never seen what they perceived as such a strident protest from a churchman! I received a phone call from Tom Coin of the BBC and the cameras duly arrived to film me giving out leaflets.

At the time, Geoff Greenaway commented, "After this, we will have to put our money where our mouth is!" The next day dawned and, little did I know, I would be catapulted into a spiritual experience, the like of which I had never before encountered.

The BBC phoned to say that a hotel owner had been watching the programme and needed to speak with me urgently. It was a place in Coventry where, apparently, increasing levels of "spiritual" activity was disturbing the guests. I didn't really know what to do, so I solicited the help of the Elim Pastor at the Coventry church, the Rev Jack Newman. Jack was a solid, experienced minister and I was confident that he would be able to help me.

I travelled to the hotel and met Jack outside. I quickly established that he was good at New Testament Greek, but not very experienced with the demonic! In short, he had no more idea than me. However, we went in, doing our best to look very knowledgeable. Apparently, voices could be heard in the corridors in the middle of the night. When the guests opened their doors to see what was going on, the voices continued but there were no people there. Then there were reports of furniture being moved around the wall of another bedroom and of a man being attacked by a pair of scissors that suddenly flew across the room. It was not a nice situation.

I started to pray in the main lounge, in front of a group of guests. As I did, the room temperature plummeted dramatically – as the hair on the back of my neck rose dramatically! Suddenly, a blood curdling scream filled the lounge. I stopped immediately and looked up to see a young soldier lying face down, shuddering vigorously and stuttering the name, "JESUS". When he had regained his composure, he told me and Jack that, as I was praying, he felt the room temperature drop immediately. He thought someone had opened a window and opened his eyes to look around. This young man then started weeping. His face shining. After a further delay, to compose himself again, he told us that standing next to me was Jesus. He was standing with His arms wide open, right next to me. This young man then gave his life to Christ immediately. What a start!

This inspired me to continue, so we visited every bedroom in the hotel and prayed for the peace of God to cleanse it and for the Holy

Spirit to bring liberty. We entered one bedroom, however, where the guest became verbally resistant to our presence and began to argue with Jack over the validity of the Scriptures. As they argued, I prayed in the room and then left.

The only part of the hotel not yet prayed over was the owner's private living accommodation. I noticed that a number of crucifixes were hung on his walls, so I asked if he and his wife had Catholic connections. "No," came the reply. But the owner told me he had invited a local priest to come and do an exorcism, but nothing happened. There and then the Holy Spirit revealed to me that the problem lay with the owner's wife. I asked if they had received help from anybody else and he replied yes, a spiritualist had laid his hands on his wife and she had received an electric shock that went through her body. The Holy Spirit instructed me to immediately place a chair in front of the door so that this woman could not leave the room.

I started to read Isaiah 61. The owner's wife started to fidget and then asked me to stop reading about, "You know who." I continued to read and she then started to swear at me and curse me. By this time, the husband was beside himself with embarrassment and confusion. "My wife has never used language like this," he gasped. I told him that this was not his wife speaking, but a demonic influence. I commanded it to leave her and, as if a cork had been popped from a bottle, it left.

As this was taking place, a member of staff came to say that the guest who had been resistant to us praying had specifically asked for me to go immediately to his room. I said, "No," because I was busy with the present situation and asked Jack to go. Jack was away for about ten minutes. Oddly, on his return, he said that he had become sidetracked and hadn't had an opportunity to visit this guest. However, we had the pleasure of leading both the owner and his wife to faith in Christ and Jack baptised them both some weeks later.

We left at about 2.00am after six hours of intense ministry! Jack said to me, "David, did this really happen?" I said, "Yes, Jack, but no one is going to believe us!"

The next day the hotel owner phoned me to ask if he should phone the police. Early that morning, the guest who had sent for me checked out. He was agitated. He told the owner that he was a satanist and that he had been sent to the hotel to keep things going there. He opened his case and the receptionist noticed a gun. The guest told her that he had sent for me the night before to kill me, but I hadn't visited him.

The BBC wanted to do a reconstructed documentary, but neither I nor Jack wanted anything to do with that. I was retelling the story at a ministers' conference many years later and explained that Jack must be with the Lord by now, when suddenly a shout went up from the audience. It came from a frail, retired minister, clinging to a walking stick. Jack stood up and confirmed to the people that the story was completely true.

The downside to all this was that the Christian Church tended, as it does, to major on the minors. I was subsequently inundated with pastors bringing people or situations to me that might be demonic. Some of them would have made horror films that defied all logic!

Chapter 10
Faith, Finance and
Furry Friends

During the next few years I seemed to be confronted with a considerable amount of demonic attention. Sometimes, I would enter a building – anything from churches to old people's retirement homes – only to have some person begin to manifest, just by virtue of me being there. I remember one particular night when I was preaching at St Ninian's and felt the presence of evil enter the room. I just paused and said, "Be gone" and then continued preaching. At the end of the service one of our members, a teacher, told me that as I'd paused, he looked to his right and saw a disfigured shape pointing at me. As I spoke, it vanished. Nevertheless, I never taught on the demonic, nor became obsessed with the subject. I'm a firm believer in moderation in all doctrine and teaching. Biblical balance keeps one from excess or error.

St Ninian's was not a large building. It held around 220 people. But we continued to have a steady growth in numbers. We had been meeting in the Library Theatre Complex on a Sunday morning and attracted about 100 people. I believed that the Lord had promised that when 120 of us were regularly sharing communion together, we would have our own building. But try as we might, we could never manufacture that number.

The evening service was a different story. It would be filled with fifteen minutes to spare before it was due to start. We saw salvation, healings, miracles and, on one occasion, the congregation saw a blue haze around we whilst I was ministering. We had nothing in the way of mentoring or guidance. We tended to face either isolation from other Pentecostals or opposition from the establishment.

The magnitude of the Holy Spirit's presence, however, was awesome. One night I prayed for the sick and a tall soldier who, at a distance appeared to be black, came slowly forward. This young man had come for prayer for his little baby who was at home with his wife. The child was facing being admitted to hospital if her condition didn't change by the morning. It was then that I discovered that his skin colour was caused by a severe burning to the face he'd received after the engine of a truck he was driving had exploded.

I prayed that his child would be healed on his return home. Then I prayed that his skin would be healed and become like that of a baby's. He thanked me and left. I found out later that when the soldier returned home, on entering the house, he was greeted by perfect silence. His wife told him that the fever had left their daughter at 9.00pm – the exact time at which I had prayed.

The next day he went to his local church to thank God for His mercy and to take communion. As he knelt at the altar rail, he felt what he later described as two large, firm hands holding his face, pulling back, stretching his damaged skin. He put his hands up to his face in panic and dead skin came away in his hands. Within a few moments he lifted his face towards the minister and he had a soft pink face with two rose coloured cheeks – just like his baby girl's.

So we seemed to be functioning at two different venues and two different levels of effectiveness. I believe, and always have, that any move of God or "outpouring" should enhance the ministry of the local Church, not take it over. And the local church continued to grow.

Growing family

Our family was growing too, with the addition of Melanie. Ashley was 4 and it was left to me to take him to see his little sister at the hospital. She had hair so long at birth that Molly had to cut it as soon as she came home. I couldn't understand why, during our walk from the car park, Ashley kept falling over. However Molly, after 36 years, still proudly tells everyone who will listen that when she saw him, she found his shoes on the wrong feet and the buttons of his clothes fastened to the wrong holes. My lawyer still contests such slander!

My business was starting to succeed and we found ourselves being invited to conferences in many exotic places. This was totally outside my social upbringing. Until the age of 25, the furthest I had travelled was to Cornwall. A work colleague, Gerry Smith and his wife Sheila, suggested that we all went to Paris for the weekend. Wow! The thought of this panicked my mother. Why would we want to fly and go to France? To her, this was a near crisis. But despite my mother's fears for our safety, we went. It was exciting to see the historic sights and it was a great experience. However, as the time came to return to the airport, I was frantically encouraging everyone to hurry up, worrying that we might be late. I had an obsession with always wanting to arrive early that was compelling me to get a move on. Gerry, however, who explained that he was a "more experienced traveller", would not be rushed, stating that we had "plenty of time". We missed the plane. Having no more money between us, we slept on the floor in the departure lounge until the first flight home the next day!

Molly and I had moved from our first house in Arundel Road, Hollywood, Birmingham to Robin Hood Lane, Hall Green. A journey of just 5 miles, but one which took us well into the south of the city. This large, three bedroom plus box room house was about fifty years old and in total need of refurbishment. The previous owner, an old lady, had died in the house. We have bought two houses under such provenance!

The woodworm in this place was so bad that when we picked up a brush that was lying around, it virtually disintegrated. It had all been eaten away! But over time we worked on it and gradually restored it to a beautiful house. We lived next to a dear, elderly lady who suffered terribly with arthritis, and who would regularly call over the fence, "Molly, my fingers have locked again. Can you put them back in?"

Molly had grown up always having dogs in the family. Conversely, I had not, since animals had been banned from our home due to Mom being frightened of anything that made a noise or had the tendency to chew. I wasn't keen on having a dog, not readily accepting the prospect of moulting hair everywhere in the house. But, in response to the "dripping tap" of constant request, I finally gave in. On Molly's birthday, I walked into the house with a ball of fluff in my hand – a toy poodle that a delighted Molly subsequently named Joshua.

One night when I was out at church, Molly, with Ashley as a toddler and Melanie in her arms, opened the front door and Joshua ran out. When I came home, Molly was frantic that he had not returned. I started to search for him and was stopped by a policeman who had received a report that two youths in a car had deliberately run him over as he stood in the road frightened. Poor Joshua. The policeman suggested I bury him in a deep hole in the park so that the foxes couldn't get him. I, Mr Strongman, couldn't do this on my own so late at night, so dear Dr Ken Carter, our Sunday School leader, came out and helped me.

Joshua was the first of four poodles. Next we owned a manic, king sized airedale who would eat everything from soap to, well, anything that was available really. He would chase joggers and once pushed a fisherman into a pond. He had the ability to urinate consistently on anything he felt he owned. During one Ecumenical event, I had the town church leaders visit my home. As I sat on a stool chatting with these distinguished guests, I leaned back against the French windows. Unfortunately, they were unlocked and overbalanced and fell right through them onto the patio. Before I could move, Amos the Airedale bolted the length of the garden and relieved himself profusely over my head! Dragging myself off the ground and wiping my face, I uttered to the shocked dignitaries the first thing that came into my mind: "They say it's good for rheumatism!"

We also had two lovely miniature schnauzers, Penny and Max. Max would go with Ashley to University College and sit with him as he painted during his fine art degree. Ashley came and held him when, through frailty, we had to have him put down. Then we had the most beautiful Giant Schnauzer, Fleur. She was so loving. Although she had received no "official" training, she was the most wonderfully obedient dog and a faithful friend to Molly.

Unfortunately, one dark night Molly didn't put the light on when she went downstairs and fell over the dog, breaking both her wrists and thus ending her forty year hairdressing career.

We lost Fleur through old age just six months ago. We now have Elsa, a standard schnauzer. We had been warned that this middle of the three sizes are the most difficult to train. Yes, we had been told. Yes, we still bought her. At the time of writing, I am convinced that we have the dog that inherited the demons that left the swine when they ran into the sea! We love her, by our will and also by virtue of the fact that we paid £750 for her! If anything happens, I will tell you before we finish the book.

We loved our house in Hall Green. We had the sweet old lady

on one side of us and on the other, a henpecked husband, his rather robust, dominant wife and their three boys. The night that Melanie was born, I retuned home from visiting just as dusk was settling. As I put the key in the door I felt someone standing by my side. I turned slightly and had just a second to register it was Mrs "Robust", who shouted, "Your dog has been barking!" before punching me in the face! The rest of this episode is testimony to the grace of God. Since I'd had my nose broken in earlier fights, anyone who touched my face would end up being flattened. Now, however, I turned to this obnoxious woman and said, "If that's what turns you on, you can hit the other side." I then added, "However, having turned the other cheek, I would recommend that you quietly go inside, because the third time is my opportunity and then you will need hospital attention. Go home!" She quickly left.

Ashley was a blonde haired, slightly built boy. Molly is of similar build. Melanie, however, was like me. She was seven and a half pounds born, but quickly became very tall and had long black hair. Ashley's and Melanie's personalities were completely different. Ashley would wind her up and she, a real actress, was always so dramatic.

Melanie was four years younger than Ashley and I nicknamed her "Mel Rose" after a recently commissioned ship. Ashley was born just before I went on a mission. In those days, the fathers were asked to leave the hospital once the baby was born. The same happened with Mel. However, I can still picture them as new born babies. Ashley with his blonde hair and beautifully sculptured features and Melanie with amazing locks of black hair and those incredible big button eyes.

While Ashley, and later our younger daughter, Lucy, would draw much of their DNA from Molly, Mel was saddled with her dad's personality. I was amazed as she instinctively relived my relationship with my dad. She would hide underneath my coat, just as I had done so many years earlier.

That house at Robin Hood Lane became an extension of the church. Having no church building of our own we had no place to see people, so they headed to our house. On some occasions it was like a doctor's surgery waiting room as they queued down the hall and sat on the stairs. Sadly, it was our children who suffered, having to be restricted to their bedrooms.

The ability to reflect means that I can now say to any young minister, "Yes, to pastor is sacrificial, however never reduce your own family to the waiting list." Nevertheless, many incredible occurrences happened in that house: salvations, healings and deliverances. I believe that the hand of God was protecting my children upstairs, whilst genuine demonic situations manifested in our front room!

Sadly, I didn't spend much time with my two older children in their early years but, by the time Lucy arrived 7 years later, I was beginning to learn.

Ashley has memories of an "Airfix kit" of a Messerschmitt ME 109e that we built together over many weeks, when he was 6 years old. It was this that triggered the start of his love for model making, moving from planes to tanks. The days of model making are long since ceased, yet we both share a love for war history.

Children see you as you have never seen yourself! Ashley remembers our holidays and me, apparently, daily checking the money to make sure that it lasted the entire holiday. School reports would always say, "Ashley is easily distracted; he could do better" so being a budding artist, he decided to copy my signature. As I was always so busy, I never thought to ask why I hadn't received this year's report. Ashley would sign it and return it himself. He then taught Mel to do the same!

One day, the school phoned me. It was the Head. "Excuse me, Mr Carr, would I be correct in thinking you have a very successful career dealing with footballers?"

"Yes," I said "you are correct."

She continued, "Is it also correct that you had no formal education?"

"Yes," I said "that is true also."

She continued to tell me that Ashley had convinced himself that he didn't need to work, so he could just be like his Dad! She asked if I would I please have a word with him. I certainly did! Ashley was a beautiful boy, yet frustrating. If anybody said he was good at anything he immediately lost interest: swimming, long distance running; he dropped both.

Even during his fine art degree, when he was superb at photography, and I mean superb, he didn't want to do it, preferring to be a fine artist. Ashley studied at Solihull College, which was an outpost of Coventry University and gained a 2.1 in Fine Art. He then went on to work at the renowned "Custard Factory" in Birmingham, where he met his future wife, Fiona. They later married and today Ashley is slowly establishing a high quality commercial photography company, dealing with leading companies. They have two children, Camille and Sebastian.

I've already mentioned that Melanie was dramatic. Ashley would wind her up and then vanish. If you moved towards Mel to discipline her she would throw herself on the floor shouting, in the most plaintive way, "Don't hit me, oh, please don't hit me!" You would be at least ten feet away from her!

Melanie struggled at her first school, leaving with a reading age four years behind most pupils. It would be true to say that the children had a mild form of their father's dyslexia. But she always had a great voice, going on to sing at Ronnie Scott's Jazz Clubs, both in Birmingham and London. On leaving school, she went to Coventry University to study for a degree in drama. However, after the first year, the course had developed into the history of acting rather than the applied aspects of it, so she left and pursued a career in estate agency. Mel has not enjoyed the best of health in her life, but continues to work hard selling like her Dad – excellently! She continues to sing both solo and in a group within the Worship Team at Renewal. She married Dave Metcalfe who works at Rover. He is a cameraman at the church, works in the children's department and is a good man.

Lucy was always our little girl in so many ways. She would actually grow to have the same logic as her Mom! Her shyness was channelled into dancing and slowly she became confident and started winning many medals and grades in contemporary dance. She left with just one more exam to qualify as a teacher. Like her mother, Lucy trained to be a hair dresser, but never pursued the career after graduating from the Birmingham College. She continued her studies in beauty, but has also started working in the retail trade.

I mentioned earlier that she married a boy from the church. Together they had a little girl called Poppy. Unfortunately the marriage didn't last. This was a very difficult time for the whole family. Lucy and Poppy moved into our new bungalow and lived with us for about 18 months. It was strange having a baby, and later, of course, a toddler in the house. She was a wonderful mother, but assumed she would not marry again until a wonderful young man joined the church. Treating her with respect, he married her and willingly embraced the care of Poppy as if she was his own daughter. Lucy and Richard now have two children, Poppy and Hattie.

Chapter 11
Dreams and Protests

My business was progressing well and I had made connections with accountants and other professional organisations. But the most purpose driven area of my work, my connection with professional football, was really blossoming.

Alongside this, with the church expanding and growing in influence, it seemed a necessity to move house and actually live in Solihull itself. We didn't really have any desire to do so since, apart from the church, we didn't like the town much and the house prices were at least 30% higher. But it was practical to do so.

It was around this time that I lost my Dad. After his heart attack at the age of 60, he had retired from Cadbury's and, basically just pottered around at home, mainly doing things for Mom. He loved travelling around in his Reliant Robin – the infamous three-wheeler car that, stupidly, the law permitted to be driven on a motorcycle licence. He was lethal in it!

The Cotswold's was his favourite place. He loved photography and his world radio. He would listen to Christian radio from all over the world, taping the programmes on his cassette deck. He also wrote poetry. He had become a Christian following his heart attack and both he and Mom attended Stirchley Elim Church. He always had a pocket full of Fox's Glacier mints; a tradition carried on by Mom after his death.

It was 10th May 1985. Dad was 78 and he and Mom had been gardening at the front of the house when Mom suggested they go inside for a cup of tea. Dad sat on the wall facing the footpath and, when Mom turned round, he had vanished. He had suffered a full-on heart attack and had fallen onto the public footpath.

I was sitting in my office in the centre of Birmingham when my secretary, Maureen, put a call through from a neighbour. Apparently, Dad had been rushed to hospital. I immediately rushed out of the office and then checked myself. Hang on, which hospital. So I went back inside and phoned Mom. She wept on the phone. "Your dad isn't in hospital. He's laid out in the front room, dead."

That dull day became blurred by the tears that streamed down my face. I put on a pair of sunglasses to hide my reddened eyes from the rest of the office staff as I left to drive the 5 miles to my parents' home. On arrival, I found an ambulance and a police car outside the house. Mom was in denial, so I had to go in and identify him, my friend and hero, lying lifeless by the fireplace. The ambulance crew then removed him and left me to contact my brothers.

Some time later, Mom asked me to clear his tapes from the table as she couldn't bear to do it. As I was doing so, I noticed one new one, standing proud of the others in the box. I looked at it and noticed that Dad had written on the label, "To Winnie". I took it into the front room and played it. It was Dad's farewell to Mom.

I don't know how, but he must have known that his time was coming to an end. On the recording, he told her that he loved her more now than when he first saw her long blond hair. He continued to say that he had never been able to give her much of this world's goods, but he was leaving a legacy of three preaching sons. He then said his goodbyes and told her he would be waiting for her in glory.

On the day of the funeral, Ashley looked at the coffin and said to me, "My Granddad isn't in there, is he?"

I said, "No, son, he isn't. He's with Jesus."

Mom, who was totally dependent on Dad, lived another 18 years, dying after a short illness on the 19th July 2003 aged 96. In her later years, her neighbours became her carers. We transferred the allowances and benefits to them to thank them for their care. They had known her from when they were young married couples, all moving into the Grove at the same time. At her funeral, her three sons stood by her grave and made the committal together.

* * *

The church was steadily growing and the need for our own building was increasing. We looked at numerous possible sites, but never had the opportunity to buy anything. During our eight year joint tenancy of the United Reformed Church building, we had a small leadership team. Laurie Pedler was our first Elder – a man of the deepest spiritual and ethical integrity. He joined us from Sparkbrook Elim Church were he had served on their leadership team for 25 years. Prior to this he had served the respected Rev Percy Brewster at Cardiff City Temple. His wife, Marjorie, was a solid support for him and they had one daughter, Jean, who was dedicated by the well known Evangelist, Alex Tee. Jean went on to marry one of our present elders, Niall Cluley.

On his appointment, Laurie asked, "What do you want me to do?"

I replied, "Nothing. I want you to be." What I meant by that was simply that elders should be able to teach, be the husband of one wife, rule their home with credibility and so on – according to the qualifications outlined in Scripture. Then I said to him that, if he wanted an official title or position, he could become my Dad, as I had lost mine. Laurie took that to heart and played the role superbly.

Not long after, a strange thing happened. My brother, Michael, had resigned from ministry at his church, due to the on going illness of his wife, Pat, and they decided to move to live in Solihull. I had forgiven them both for a situation that had happened in the past and when he requested permission to attend the church, for the sake of his youngest child, I readily agreed. After a few months, the leadership team, noting that I was not full time and that the church was growing, suggested I appoint Michael as my assistant.

This was interesting for a number of reasons. When our family had fallen into disagreement about certain issues, and my brother had fallen out with both me and Mom and Dad, Dad had a dream. This was significant because Dad wasn't an "I've had a dream" type of man! It was very unlike him.

Historically, I had felt very much "the runt of the litter" in our family, compared to my brothers – especially academically. But Dad said to me, "Son, I had a dream last night about Joseph and his brothers. God wants you to know that both your brothers will serve you one day."

If it had been anybody but Dad I would have laughed, but not at him. He had never shared a dream like that before and he never did again. I hid this deep in my heart and, unlike Joseph in his immaturity, I told no one. However, to hear our leaders proposing such a thing freaked me out. Michael was not the type of person who could work for anybody, especially not his little brother. I loved my brother and recognised his teaching ability as one of the finest in the UK. Molly was more reticent, based largely on her knowledge of our traumatic past. But I talked with him and, to my amazement, he immediately agreed to join the team as my assistant.

The future developed into Michael taking the majority of the Sunday morning teaching services and me taking the evening service which was a much larger, revival type of outreach. But, after a few years Michael came to me one day and suggested that, as we had differing styles of leadership, he should take the morning congregation and me the evening congregation and that we become two separate churches. This idea of splitting the church into two was not a road I was willing to travel. Shortly after Michael move to London to work with Wynne Lewis who was, at that time, pastor of Kensington Temple.

Not long after this we moved from St Ninian's.

The Sunday mornings at the Library Theatre were not ideal. The Lord had clearly told me that when we had 120 people regularly taking communion together, we would have our own building. Try as we might, we still couldn't consistently gather that number. Then one day Geoff said to me, "Dave, we've had 120 attending each week for the last six weeks." It was then that things rapidly began to happen.

I made such a nuisance of myself with Solihull Council that, one day, I received a phone call inviting me to visit the Education Department. The Council offered me a building, The Moat Infant School in Frankton Close. Only the Head Teacher had any knowledge of its impending closure. It had the typical facilities you might expect at any school – a number of large class rooms, a full kitchen, an assembly hall, an office, staff room, and children's and staff toilets.

My business expertise came into play during the negotiations. The Council wanted more than we could afford and also wanted to restrict any future sale to charitable purchasers. We settled on a reduced figure with no restrictions on the resale and a 20% payment to the Council on any profit.

So we had our own place! The school was situated in a beautiful garden and the playground was ideal for car parking. We used the hall as the sanctuary, but soon needed to knock down two classrooms to extend it on two levels to hold about 220 people.

Three other classrooms became used alternately by Children's work, Youth groups and for Bible teaching and mid-week activities. The last room was turned into a coffee lounge with its own counter, sink, tables and chairs. This was faithfully established and run by Geoff and Dorrie Banks.

We built an extension that housed our offices and administration and could be used as an additional meeting facility. The staff room became my office and the old head teacher's office was made ready for a pastoral leader. We left St Ninian's after eight spectacular years, most of them fantastic and highly rewarding.

I have only one regret regarding the time. It was around then that, due to listening to others, I stepped out of the mighty outpouring that had been happening. A few prominent leaders with hidden motives suggested that the crowds only came to our meetings because of the miracles and spectacle. Listening to this information eventually caused me to miss out on future anointing. These people suggested that I should stand back and let others teach. They told me that then I would then see the real reason that people came – because I healed them and for no other reason.

I complained to the Lord about this and, of course, He corrected me. "So ... you heal them, do you?" I floundered and fell back into my old insecurities. The Holy Spirit was then limited by my insecurity and the miracles stopped. Ironically, the first people to leave were those who thought I should step back and let others teach. That action of foolishness robbed me of two years of unstoppable blessing.

Meanwhile, the new building was an incredible blessing to us. We knew that it wouldn't be our final resting place, but it was a great provision on our journey. I was still working in finance, closely linked to the footballing profession and I gave every Wednesday to the church. However, we desperately needed a full time Pastor.

I prayed, "Lord, is it time for me to go full time?"

The answer came, "NO!"

I was drawn to the story of Paul. He chose not to be salaried and supplemented any gifts he received by tent making. As elders we prayed together and a few months later appointed our first paid leader. Youth pastor Derek McDonald and his wife Linda were fresh from the Elim Bible college. We found them a nice rented maisonette in Hall Green to live in. Derek was from the South of England and Linda was from Scotland. It was strange having someone on the payroll of the church, yet we had always had a significant youth work and this investment was what was needed.

Around this time we also met Mike and Anna Webster from New Zealand. Mike would later become on of our pastors. They had been mission workers in Botswana, sent out from Queen Street Assemblies of God in Auckland. Mike was a softly spoken preacher who had trained as a probation officer.

We had some beautiful times in this ex-school complex, but our first two years found us being monitored by the Department of the Environment for making excessive noise! Our immediate neighbour was a deeply bitter lady who had a nephew at the Embassy in Stockholm who managed to make waves for us. These pensioners basically formed a lynch mob and, on occasions, invaded the services! Once, an elderly neighbour refused to leave the foyer until she had spoken to me. I was busy preaching at the time so Niall, one of our elders – who stands 6'4" tall – spoke to her. He was a little taken aback when she looked up at him and said, "I have come to speak to the organ grinder, not the monkey!" We could not pacify these people. Regardless of what we did to try and calm the situation, they just didn't want us there. We prayed for a breakthrough and then two separate things started to happen.

Firstly, it came to our attention that a bedridden lady who lived in the retirement bungalows was dying and on her own. Two of our church partners, qualified nurses, decided to take it in turns to visit her, wash her and take her to the toilet her and prepare her for the night hours. Good news, like negative gossip, eventually gets out. Maybe positive things take a little longer, but the neighbours began to talk about this.

As this was happening, one of church members, George Gardner, visited one of our neighbours and started mowing the lawn of the most antagonistic neighbour. George, Gardner by name and gardener by gifting, maintained the most beautiful flower beds in front of the church. The neighbour would shake her fist at him through the window, but George maintained his ministry to her, continuing by planting beautiful flower borders. All this brought her to the door shouting, "Don't think I'm paying for all this! I didn't ask you to do it!" Then, as the weeks progressed, she started bringing him a cup of tea out. Then, later, she invited him in to look at photos of when she was a young model, meeting actors such as Gregory Peck. George was no preacher, he was a gardener, but his literal "groundwork" opened the way for someone to visit her just before her death and lead her to Christ. Then, not only did the opposition amongst the local residents cease immediately, but they would also strongly defend us against anyone who opposed what we were doing.

Our first wedding in the building was that of future elder Niall Cluley and Jean, the daughter of our first elder, Laurie Pedler. During this period we lost our first leader. Dick Vane and his wife, Sheila, gave oversight to a district just over the borders in Birmingham – Sheldon, near the Airport. Dick came from the motor trade background. No messing about, straight to the point. He became famous for a favourite saying in meetings. After I had explained our future projects, if he had an opinion, he would often say, "Excuse me, Dave, is this written in stone or is there room for a comment?"

I would say, "Why, Dick?"

Then, without fail, he would reply, "If it's in stone, I'll get on with it. If not, I have a question." Dick, like most of my pastoral workers, shared my heart for caring for the flock. He would visit people constantly, encouraging the needy.

One day, I received a phone call. Dick was visiting Heartlands Hospital when he had a massive heart attack. As he was dying, his thoughts were only for the person he was supposed to be visiting and his last words were, "Don't forget to visit the man on ward..."

I rushed to his home to comfort Sheila. I went in, sat down next to her and burst into tears. She put her arms round me and comforted me! Afterwards she said that me crying showed her the depth of love I had for him. She was right. Sheila still remains with us today, one of our prestigious Platinum Partners having been part of the vision for over thirty years. Dick was one of our historic heroes who laid the foundations for the church we have today. Today's church only has this book as a memory of Renewal's history makers.

Chapter 12
Pain and Preparation

We developed a friendship with a small fellowship in the area. The lead elder was a local solicitor, an evangelical with Brethren connections. Over some months, it was decided that this church ought to merge into Renewal. They brought about 35 people with them and, to show good will, I did a very foolish thing. I appointed their leader as one of our elders. Why foolish? For a start I didn't know him that well. Yes, he had said that he believed everything we believed and that it would be foolish not to join us. But I didn't know him.

At this time, my original assistant pastor, friend and visionary strayed from the faith. The pain was incredible. I was the Senior Pastor and the Overseer for Elim in that region, so it fell to me to confront him both as line manager and pastor. He showed little concern or remorse at all. The Elders stood by me wonderfully. I had to remove him from the church, ministry and all duties. I said to him, "The day you repent, I will travel to the ends of the earth to be with you." I have never seen him again. I lost a friend, a pastor and a part of the vision. He had helped us to put on the musical "The Witness" which was so good that we performed it at the largest Pentecostal Church in the UK. Sadly, he never aspired to what God had prepared for him.

We lived in times when sin was called "sin" and not explained away as "problems", "difficulties", "challenges", "slip ups" or "mistakes". In those days, pastors who strayed were committing sin, repented and received forgiveness from God – and faced the consequences of moral failure. Many ministries today can violate their covenant, even more than once, and be back in ministry within months, teaching on the valuable lessons they "learned" through their sinful actions – and, of course, producing a best selling book in the process!

We soon outgrew this lovely facility and needed a larger Sunday venue. Geoff Wright was Head of Education for the local Government and part of Renewal. He told us that the largest assembly hall was actually in a school only a mile from our church, at Lyndon School. This school had a hall with a balcony that could hold at least another 120 people with a large car park behind the building. So we moved our Sunday services there and put a bus on to take the children back to our church for Sunday School.

The first Sunday, as two of our lady stewards stood on the door, a man covered in blood staggered across the road, having been bludgeoned by a hammer. He died in their arms. Our vehicles started to be vandalised, so we mustered the full congregation, both morning and night and, in the darkness, circled the school ground, praying. The attacks ceased immediately. I actually didn't want to be the pastor. I was an evangelist and longed to be back on the road. So I would give the emerging team every opportunity to speak.

Two of our people had journeyed to South Africa to become

practical mission workers. They attended Rhema Church, which was a huge mega-church. They sent me a letter saying, "Pastor Dave, we have continued in your vision. Come out and see." It was nice to be invited, but I felt I needed some kind of confirmation before jetting off to South Africa. Not long after I was speaking at a conference when a church leader came up to me with a young man, with long hair down his back. He told me an incredible story as he introduced his companion and explained, "This young man was being controlled by the demonic and entered a church service. He started to manifest and fell to the ground. The church was not open to the things of the Spirit and therefore assumed he had fitted and called an ambulance."

He continued to tell me that, when in hospital, a South African doctor had enquired as to whether the young man was South African, since he was speaking in his tribal dialect, "...cursing a man called David Carr and forbidding him to go to South Africa!" The man was eventually discharged and taken to this Pentecostal pastor who released him from demonic control and brought him to faith in Christ. This young man had never heard of me. The pastor, knowing I was at that conference, brought him to meet me and to tell the story. That was a spectacular confirmation, I think! So I decided to go.

A businessmen I had led to the Lord asked if he could take me to the airport. Just before I left church, the youth pastor told me that the Lord would reveal my future when in Africa. Both Molly and I felt that this visit would be a watershed moment. I actually believed that God would release me from pastoral ministry. Apparently (and unknown to me at the time) some of the team believed the same.

South Africa is one of the most beautiful countries in the world and they don't speak English – at least not the Queen's English! When asking directions, I was confronted with a man who kept on saying, "Ya" and told me to, "turn left at the Robots." What is a robot? I eventually realised he was referring to the traffic lights.

From here things got progressively worse. I was asked if I took supper. For someone from Birmingham, that is what you have just before you go to bed. I politely declined. In South Africa, it's your main evening meal, so I didn't eat that night! The next day when I was asked if I liked a Braai, I just put my hands up and confessed that I didn't understand a word of their language! Apparently a Braii was a barbeque. The food there was incredible.

We overcame our cultural differences and started to appreciate the magnificent ministry that Rhema Church had. The associate pastor had arranged various preaching engagements, one of which was in Soweto. I tended to think of it as a large town, however the population of 1.3 million made it bigger than Birmingham, the second largest city in the UK. Entering this vast area was dramatic. Roads lined with poor quality properties were punctuated with a sprinkling of the plush buildings of the successful. Bonfires were burning everywhere.

We arrived at the church which was full of families squeezed into a space created to hold half the amount of people. I was introduced to two different interpreters and suddenly, there I was, ready to go. It didn't take long for me to realise that they didn't relate to me at all. The Lord spoke directly to my heart and halted my preaching. He said, "Stop and ask the pastor if you can bless the babies and the children." I did as I was instructed and the pastor looked surprised. He announced what I had asked and nothing happened. After a short pause, he then led his own children to me to be blessed. As I picked them up, cuddled them, kissed them and blessed them, the Hoy Spirit fell and people began running to the front, pushing their babies into my arms.

You'll recall that, after breaking my nose for the second time, many years previously whilst dealing with four drug addicts in the Bull Ring Centre, I had lost my sense of smell. At this point, the Lord used this to my great advantage. Some of these poor little ones touched my heart. They were dirty and unkempt, with running noses and matted hair. I could only imagine what the smell would have been like and thanked God that he had rendered me able to minister to them without flinching. Many of the parents responded to salvation and healing after that. Demonic manifestations appeared and people were delivered.

On leaving that evening, the leadership presented me with a clock depicting Africa. How could I receive such a gift from those so poor? Yet I knew I must honour their generosity, for the sake of the Kingdom. Sitting in the car, driving away with the haze of bonfire smoke hovering all around, I closed my eyes and had a flashback to that day in Weston Super Mare when the Lord had called me to nurse His babies. I opened my eyes and looked down. I was wearing the white shirt with epaulettes, white trousers and white shoes. The same Voice spoke to me again that day, after 25 years. "Today," He said, "you nursed my babies." I had imagined the original prophecy to be medically related. Instead it was practical and spiritual. I now understood the reality of God's prophetic word over my life.

I was then taken to a restaurant and, sitting at a table nearby, was Ray McCauley, the Pastor of Rhema Bible Church. He called me over and, even though he didn't know me at the time, literally hauled me over the table and said, "Stop running away from what the Lord has called you to do! Go back!" That really was a "now" word.

My phobic fears of the dark had been overcome, yet the test came when our mission workers, Ray and Angie, together with their children, booked us all into a survival camp consisting of three tents on a mountain, in a vast game park. A protective guard had a little tin hut a way down the path and there was a toilet: a hole with a bamboo petition, together with half an oil drum as a sink. The camp fire was always kept burning at night to keep the wild animals away.

We found that, in Africa, when the sun went down it became so dark that you couldn't see your hand in front of you. As the night progressed, it was decided that the family would have their three tents arranged together, circling the fire, and I would have one just down the track, in the dark on my own! I had a Davy lamp, just like a miner, and that was it. To be in a tent, listening to all the animals walking about and making the most horrific noises, wasn't the most comfortable way to try and drift off to sleep. As the sun started to rise, my courage flooded back and I felt I was now something of a seasoned explorer!

But nothing could have prepared me for what happened next. I was sitting in my tent when the Lord clearly spoke to me, saying, "If you stay, these will be the people who will leave you." He then flashed a sequence of faces before my eyes, people that I would never have imagined would be part of such a list. This was distressing to say the least. On returning home, I called the elders to a day retreat at a Monastic Centre.

In this place of contemplation I told them about Pastor Ray's prophetic word, telling me to "go back". I announced that I would be staying and not giving up the leadership of the church. Immediately, the Youth Pastor and the recently appointed elder of the church who had joined us, stood up and said, "If you're staying, we're leaving." This brought the meeting into total meltdown. I offered my resignation, but the rest of the elders refused. Over the course of the next month, about 120 people, 22 leaders, a pastor and an elder had left the church! Even families were split by this action – and it was all about aspiring to position.

The deep pain all this caused distorted my emotions for many months. I couldn't open a letter marked PRIVATE AND CONFIDENTIAL without experiencing a sick feeling, deep inside, fearing more bitterness and antagonism. The disruption to the church was immense. One Sunday our worship leader, just 15 minutes before the service was due to start, came into my office and said, "Excuse me Pastor Dave, the Lord has told me it's time to move on."

I said, "When?" He said, "Now!" and walked out of my office and into the car park. I left my office to find the entire worship team in tears. I had to lead the worship that day, as well as preach.

The next week, a friend of mine from Newcastle came to preach. He looked into the congregation, singled out a young man that I didn't know, and told everybody, "He is your new worship leader!" I thought, "Lord, that's just what I don't need: a stranger I don't know standing up and trying to sing!" Well, 27 years and numerous songs written later, Phil Jones has just moved to our Stratford church to help develop our church plant there. What a faithful servant he became. Tragically, shortly after the wife of the worship leader who walked out left him and he was later found beaten up and lying in the road.

During this time of upset and devastation, I was still working in football and had to travel to Scarborough. The team played in the second division. It was a stormy, windy day with the rain was lashing against anyone foolish enough to be out in it. I was pleading with God to just let me vanish into the distance and not to return to Solihull. I felt The Lord leading me to the top of a hill where the War Memorial was standing proudly against the storm. Standing in front of this celebration of the lives of local fallen soldiers, I heard the Holy Spirit say to me, "Read their names out loud." This seemed pointless as I was the only one there and the wind was so strong that no one could hear anyway. But with tears being blown from my face by the gale, I started to shout their names out. When I had finished, the Lord said to me, "These men gave their lives for their country. Now, I want you to do something."

I said, "What's that Lord?"

He said, "Go back to Solihull and live for Me."

Still crying, I got back into the car and turned to drive back to a place that, at that moment, made my stomach churn with all the painful memories and the chaos that was still playing out. I told the Lord that if I met these two leaders again I would forgive them for the devastation they had caused. Within two weeks, both were sat in my office, asking for forgiveness. The youth pastor then vanished and I haven't seen him since.

It was a desolate time. After some months, I was reflecting on this painful event and said to the Lord, "How could 120 people get it so wrong? Did I make any mistakes? I was like a big brother to them. I gave them anything they wanted."

The Holy Spirit said to me, "That's your problem. I called you to be a father not a big brother; change!"

From that moment on I sought to become a spiritual father figure and asked God to help me.

SECOND HALF...

Chapter 13
Footballing Legends

I think I've alluded to it enough in previous chapters, so now for some of those stories from my time in professional football!

As a young man suffering from Dyslexia, not being able to read or write, I used to collect cigarette cards of football teams. I had a great love for the game of football. I didn't really play myself until I was in my teens, as I was living in permanent fear of everything and had the confidence of a "dead rat" as we would say in Brum!

I learned the names of most of the football teams in the four divisions. I knew the names of all the grounds, the colours of the first team and reserve kits and all the players nicknames. I subscribed to two monthly magazines, even though I was not able to fully read the contents. But despite all this, I had never met a footballer!

When I joined the St John Ambulance, however, part of my work was based at the Birmingham City football ground and there I could watch such players such as Harry Hooper, Gill Merrick, Trevor Smith,

Ron Wylie, Dave and Bob Latchford, Terry Hennessey and others.

However, my favourite team in those early 60's was Newcastle United. Players such as Ivor Allchurch, Jimmy Scouler, Bob Stokoe, Bobby Mitchell, Len White and Ronnie Simpson played in a silk-look kit with white edging round the hems of the shorts that didn't rest below the knee!

On becoming a Spirit-filled Christian, I was convicted about the amount of time I spent on my football "hobby", so I cancelled my subscription to the magazines and burned the hundreds of copies I already had. I didn't follow football again until, at the age of 27, I walked into the office of the assurance company where I worked and answered that phone call I referred to earlier, from someone who needed help.

It was Jimmy McEwan, the former Villa player who was now a coach at Walsall FC. He needed advice, so I visited him at Fellows Park. After we concluded our business he thanked me and asked which other footballers I had helped. I admitted that he was the only one I had ever met! So he introduced me to the new manager, John Smith the former Spurs player, who allowed me to meet the players.

I became friends with George Andrews who, would you believe it, scored the winning goal against Newcastle United in the FA Cup Third Round in 1975!

I suddenly found that I could help these lads plan for their uncertain futures. John suggested I visit his ex-colleague Dave Mackay, who was the manager of Nottingham Forest. Dave had been a legend at Spurs and later, after leaving Forest, moved to Derby County after the shock resignations of Brian Clough and Peter Taylor who went to Brighton.

Dave sent me to Don Howe at West Bromwich Albion and Vic Crowe at the Villa, however, on arriving at the training ground at Villa I was met by Ron Wylie, who put me through the tenth degree and warned me that if I, "step't out of line" I would be for the high jump!

I think my real break came through the sadness of a tragedy. I visited Oxford United on the 19th of March 1977 to see Peter Housman. He had joined Oxford from Chelsea, playing 269 times for them. Peter had played 64 times for Oxford and was due to play against Palace the next day.

Peter had had difficulties in his previous financial dealings, not having had the correct advice in certain areas. This had left him vulnerable and he needed some security as he had a young family and his wife, Sally, to care for. I don't know why, but I felt a compulsion to advise him to take out the cheapest Term Assurance to protect the family immediately, which could be changed to something else when his needs diminished. He readily agreed. I sat there with a cup of tea and filled out the application form with him. When we'd finished, I wished him well and left to travel back home.

On Sunday, I finished at church and returned home for lunch. As I entered the house, Molly was flustered. She said, "What was the name of the player you visited on Friday?" I replied that it was Peter Houseman. She looked at me and said, "He and his wife and two of their friends are dead." Apparently, they had gone out for something to eat after the match and had been involved in a road traffic accident.

The news hit me like a tidal wave. What could I do? The application hadn't even been submitted; it was still in my bag. All I could think about were those young children, now orphaned and with no future security.

I phoned the company, Crown Life, first thing in the morning. There was no application, no premium and, in all probability, no chance. I explained the situation. After a pause, the underwriter said that to show goodwill, they would pay out the sizeable amount, in full, to his little children. Wow! What a miracle. They didn't have any liability to do so.

It was surreal. By Friday, I was sitting in the same lounge where I'd been the week before, chatting with Peter and Sally. But now, the meeting was with Mick Brown, the manager, going through the couple's financial affairs.

The insurance paid up enough money to buy the house outright. A close relative moved in to be guardian and custodian of the children. At a time when they had lost their parents, I was moved by the fact that they didn't also have to lose their home. It was little comfort, but I held on to it, and also to the fact that this house would help to provide some stability for their future.

Mick Brown said to me, "I will endorse you any time for what you have done for this family."

Mick moved to West Bromwich Albion to work with Ron Atkinson and, following this, they ended up at Manchester United. When "Big Ron" went to the Albion, he called me in and said to me, "The players are on wages lower than some of my previous players at Cambridge! Can you help some of the longstanding players?"

People like Tony Brown had been faithful servants of the club for many years. "Bomber", as he was known, had never had the opportunity to take out a pension, yet he had years of service. During the summer, he signed for the New England Tea Men in the USA league. This enabled him to fund a reasonable pension before finally transferring to Torquay United. It was there that I met Bryan Robson and, with Tony Price and Brian Hassell, transferred Bryan to Manchester United.

My first trip to Luton Town, who at that time played in the equivalent of the Premier League, was disastrous. I had booked to see the manager, Harry Haslam, at 12.30pm and I arrived in good time and sat in reception. A young footballer came and sat by me, waiting to see the boss. We talked and when the assistant came in, he spoke to the lad next to me, who then left. By 2.30pm, there was still no Harry, so I asked the club secretary who told me that he had left for the day! I was not very happy to say the least and I journeyed home, woeful. The next day, Harry phoned me laughing.

"Alright lad? A funny thing happened yesterday. I was just coming to see you when the assistant manager walked into my office. After few words, I told him that Dave Carr was waiting to see me and he said, 'Don't worry, boss, I've sorted it' so I left. It was only this morning that I realised the footballer sat by the side of you was also called David Carr!"

Harry was one of the last of the old breed of managers. I would arrive to see him after I had met with the players and he would offer me a cup of tea and a packet of pork scratchings! Harry knew everybody who had ever played or managed and his stories would leave me enthralled.

It was there that I met the captain, Alan West, the former Burnley player. He and Cathy asked me to oversee their financial management just before they spent the summer in the USA with Alan playing for Minnesota Kicks. It was the Millennium Celebrations and for a month American soccer was suspended. Cathy had relatives in Australasia who had decided to divorce, so Alan and Cathy felt that this would be the last opportunity to see the couple together. They travelled to meet them and were confused to find about forty people there to welcome them at the arrivals gate. The relatives had become Christians just before they split up and it had transformed them. All the people had come from the local Pentecostal church.

During their stay, these people seemed to expose a deep loneliness inside both Alan and Cathy, who both wanted what these people had. So, independently of one another and in different parts of the house, they both became committed Christians and were later baptised in a swimming pool. On returning to the USA it seemed normal for them to attend church, bur what about his return to Luton Town?

I had been tipped off by someone in Christians in Sport, so I phoned Cathy on their return and asked if anything special had happened during the trip? After a few hesitant moments she said they had become Christians.

I said, "Praise God!"

She paused and then shouted to Alan: "Dave's a Christian too!"

He didn't know how to tell his fellow players. It's not easy when you have been "one of the lads" and you are their captain. After pre-season training, sitting in the communal bath, a player asked him, "Hey, Westy, what's different about you? You haven't been swearing since you've come back!"

Alan and Cathy have Pastored the Elim Church in Luton now for many years. They are dear people.

Football is full of colourful characters and none was more so than Brian Clough. To meet him was both a challenge and an experience! I visited him at Nottingham Forest. He had a team of successful players who were household names. When I first laid eyes on him he quizzed me, "Who are you lad?" After I explained, he continued, "There are only two problems in football: football and money. I'll deal with the one, you deal with the other. Be here next Tuesday at 11.50am to see them about their futures – and don't be late!"

Tuesday came. I was at the City Ground at 11.20am. At 11.50am, he arrived.

"Who are you, lad?" he asked again as if he didn't know me. This was his way of keeping people under his thumb.

"I'm Dave Carr "

"Correct!" he says. "Why have you come?"

I explained that he had arranged for me to see the players about their futures.

"So I did!" was his reply. He assembled the players, most with international status, and said to them: "This gentleman has come to speak to you about your future. I saw you play on Saturday, you don't have a future!"

With that he called the meeting off and walked out. That was Brian Clough at his best!

I was outside his office one day. The door was open and he sat with his feet on the desk, wearing tracksuit bottoms and a green goalkeeper's top, his trusted dog curled up on the floor nearby. In came an international player, complaining that he had been left out of the team.

After a few minutes, Brian said, "I love a man who speaks his mind. Well done lad. I promise you will not be playing in the reserves next week."

"Thanks Boss," said the player as he left.

Brian kept his word. The following week, he played him in the youth team!

Then, Cloughy was trying to sign a player from Notts County. Sitting in a hotel, eating a meal, he asked the player to sign.

"What about the loyalty bonus?" the player asked.

A loyalty bonus was a payment made for each year that the player maintained his contract.

Brian stood up, tapped the table with his fork and announced to the restaurant at large, "Excuse me ladies and gentlemen, I have a player who plays for the oldest team in the country and we are European Champions. He wants a loyalty bonus. Would anybody agree that he hasn't been loyal to me yet?"

With that, he sat down and the player meekly signed!

When Martin Keown was thinking of leaving Villa in 1989, three clubs showed interest: Nottingham Forest, Everton and Manchester United. I travelled with Martin and warned him about Cloughy. We arrived and Brian asked the same question of me that he asked every time we met: "Who are you, lad?" After that procedure was over he addressed Martin. "Stand up lad, you're a big one." Then he asked, "Can you head a high ball or a low one?" This was somewhat confusing and, during the conversation, Martin started referring to Cloughy as "Brian". He was quickly stopped in his tracks. "Lad, no one calls me Brian unless they are family or friends."

Then Cloughy asked Martin to sign for him – and to let him write what he was worth into the contract afterwards! Martin said he couldn't do that. Brian said, "Don't blame you, lad. I don't trust me either!" Brian, who was a master of research, asked Martin if he was married. Martin said he was due to be married that coming week. Flowers and chocolates appeared almost at a click of his fingers. Brian said, "Do you want children?" Martin replied that he did. Brian said, "If you join us, West Bridgford is a lovely place to live; nice schools. If you go to United Bowden, that's a good place to live. If you go to Liverpool, you must be brain dead!"

He then told Martin to go and see what the other two clubs offered and get me to phone him and let him know. No football talk was exchanged, just "Can you head a high ball or a low one?"

It was Everton who appealed to Martin. I phoned Brian to tell him the offer and he responded by saying, "Good God! I wouldn't pay the whole team that. Good bye!"

You never really knew where you stood with him, it was always a game between him and you. Brian was complex in his ways, yet he motivated players of average ability and made quality players out of them.

Chapter 14
More of Football's Great Characters

One day I was sitting at home when Graham Taylor phoned me from Villa to ask if I could accompany Mark Walters to both Everton and Glasgow Rangers. On arriving at Everton, the manager did everything possible to prevent us from going on to visit Graeme Souness, the Glasgow Rangers manager. They wanted Mark and they made a really good offer. However, we had promised to meet Graeme and wanted to honour our commitment, so we left for the airport, expecting to be back later that today to accept the offer and sign the contract.

We arrived at Ibrox Stadium where we were met by a distinguished member of the Core of Commissioners who promptly confiscated a green and white umbrella Mark happened to be carrying with the words, "We don't bring that colour in here sir!" I didn't realise the fierce traditions of such a club.

We met with Graeme and the conversation was not about

football. They had already decided that they wanted Mark. But Mark was black and Graeme was more concerned about discussing how he might be received by the club's supporters.

"I don't know what the reaction will be up here to you being black," he said. "We don't have much experience."

Mark said that he had coped with it in England. Then the next question was, "You're not a Catholic are you? A black Catholic may be a challenge!"

Mark laughed and assured him he wasn't. Then Graeme said that he had booked us all in at a local Italian restaurant and didn't want us to be late, so he got onto the finances.

"I would imagine that Everton offered you this ... I will double it!"

Mark and I looked at each other, he with eyes glazed and a little bulging!

"Any questions?" asked Graeme.

Mark, recovering slightly, said, "Is there a car?"

The answer was yes, a SAAB .

The deal was done. It was decided that Mark would stay overnight for medicals and the press. He was wearing a designer track suit, but this was a fashion statement that was totally unacceptable to Rangers. A fist full of bank notes appeared and someone was sent back to the airport shops to buy a suit, shoes, ties and a couple of shirts. "We expect our players to arrive smart and then change at the ground," we were told. Mark was well received in Glasgow. He encountered very little abuse compared to being in England. In fact, any fans found to be guilty of such behaviour faced lifetime bans.

Soon after starting to deal with professional footballers I joined forces with Brian Hassell, an ex-teacher from Hednesford in Staffordshire. Over the years we formed "Professional Promotions Financial and Marketing Partnership" through which, at its height, we managed the finances of almost 700 players.

We visited the majority of clubs during the season and had clients at most. We received the PFA (Professional Footballers

Association) endorsement as an ethical company. We transferred many household names, travelling thousands of miles in the process. There are many stories of emotional situations that are so personal it would not be right to share, even after all these years. There are other tales that typify the courage and commitment of some of the "old school" characters in the game – such as one particular manager who sadly lost his wife on the morning of a match day, but still turned up at the ground, telling no one what had happened, and performed his many managerial duties before asking the Chairman for permission to return to his grieving family.

We dealt with players of all levels of ability and from a wealth of different backgrounds, but in general, I think it is fair to say, the majority of them were troubled by some degree of insecurity. If one player was seen to be getting a better deal than the others then, invariably, there would be trauma and uproar.

In my time I had my fair share of embarrassing moments – such as the time I was speaking to a player at the Cliff in Broughton, where Manchester United trained and unwittingly put him down by not recognising him.

Bryan Robson was standing by my table when this player asked to see me. "Great to see you," I said. "Now let me tell you, I don't just deal with the famous players. You might never be as successful as Bryan here but, nevertheless, I will help you. Now, what's your name?"

"Frank Stapleton," came the reply.

Frank had played 225 times for Arsenal. He would play 223 times for United and 71 times for his country! Bryan had tears rolling down his face.

There were many times during my career in football that touched me, such as when I had been off for many months following my heart attack. One of the Villa players needed to see me and I suggested meeting at his house so that I could see his wife and family. I duly arrived for the appointment, went through the business and left. A few weeks later I was speaking to someone when this same player's name came up.

"It is tragic that their marriage has broken down, isn't it?" said this person, much to my surprise.

"When?" I asked.

He shared that it had been three to four months ago. I thought this couldn't be true; I had been there only a few weeks before. So, I phoned him. It was true. I asked the obvious question: "What was that all about at your house if you are living apart?" He explained that since they both respected me so much, they hadn't wanted me to be upset, so they pretended they were still together.

Not long afterwards, Villa phoned me again and asked me to accompany Tony Dorigo, the Australian footballer. The year was 1987 and by now I was known as the "Rev" to the football fraternity. We arrived at Stamford Bridge and were shown directly into the Chairman's Office. The white haired Ken Bates looked at me and said, "I gather I can't swear as Billy Graham is conducting the transfer?" I asked why he had said that and he replied that Doug Ellis, the Villa Chairman, had warned him!

Villa again requested that I travel with a player, this time Paul Elliott who was going out to talk with a club in Pisa, Italy. We travelled in a private jet and received a very warm welcome from the President of the club, a very successful business man. We stayed in top class hotel overnight and entered into negotiations the next day, prior to returning home on a commercial flight, carrying a Pisa football bag each filled with heavy cans of virgin olive oil!

That same President contacted us some years later to see if we had any contacts in the oil industry (crude, not olive oil!) as he was seeking to sell some barrels. Brian Hassell had some contacts so he spent what I felt was excessive amount of time on the deal – bearing in mind we stood to make the princely sum of 6p a barrel! After two months of hard work we pulled if off and the contracts were ready to sign. Then, a certain Saddam Hussein invaded Kuwait. This, of course, was where the oil was. This untimely excursion cost us a mere £250,000!

We never did have much success in pulling off those "special" deals. Harry Haslam had asked us to partner with him in bringing Diego Maradona to a leading London Club. We spent about £5,000 sending Harry to Argentina and, for a while, it all looked very good. We would go to the World Cup in Spain and then travel to the UK to complete the signing. However, as the political situation between Britain and Argentina reached the point of no return, and before we had a chance to conclude the contract, Mrs Thatcher gave orders to attack the General Belgrano, an Argentine light cruiser. The submarine HMS Conqueror duly sank the vessel at the tragic cost of some 323 lives. Although somewhat trivial by comparison, this act of war meant an end to the deal and the loss of our investment. Looking back, I am of the opinion that we must have had a special "war curse" put on our activities. Who could have anticipated two wars in the midst of our transactions? We lost out to the tune of about £300,000 each.

I met a number of interesting people during my time in the game. Bert McGee, the Chairman of Sheffield Wednesday FC, was a stereotypically successful businessman. I first met him when he phoned my office one day and said,

"Mr Carr, I need to meet you, 3.30pm"

Looking at my diary I asked, "What day?"

He replied, "Today! Don't be late, you can park in my wife's space."

He wouldn't take no for an answer. Bert was the Chairman of the Football Club as well as being the Chair at Presto Tools and Southfields Health. You didn't say no to Bert! I left all my paperwork and drove to Sheffield and actually parked in his wife's parking space! I was shown into the Board Room, traditionally panelled, with a large oak table that dominated the space. Bert was sitting at the head of the table, immaculately dressed in a blue pin stripe suit and white shirt with cuffs showing off a beautiful pair of cuff links. His pure white hair was carefully groomed and the gold in his signet ring and fob watch glinted in the light. He sat with his head in his hands. His secretary brought me a cup of tea in a beautiful china cup. I was thanked for coming at short notice and asked the following:

"Mr Carr, I have a manager who has a wife who is dying. Can I do the following three things for his pension so that he can retire?"

I replied "Yes" to the first request, "no" to the second, and "I've never heard of this before, I'll have to check my books" to the third.

Mr McGee simply said, "Thank you, you may go now."

I couldn't believe it. I'd driven all that way for five minutes! I thanked him and drove all the way back to Birmingham. The next morning I received a phone call from Graham McCrail, the Chief Executive at Sheffield Wednesday. He said,

"I hear you met Mr McGee for five minutes yesterday?"

I told him that I was not impressed at going all that way when he could have asked me those questions on the phone. Graham said, "It was three questions that you answered 'Yes', 'No' and 'Don't know' wasn't it?"

I said, "Yes, how do you know?"

He went on to tell me that it had been a test to see a) if I would turn up at a moment's notice, b) if I knew my subject and c) to determine if I was honest. Graham told me, "You see, David, Mr McGee already knew the answers, yes and no. The third was a test. The question had no answer because it was made up! You told the truth, saying you had never heard of such a thing. The outcome? You can now quote not only for Sheffield Wednesday, but for all of his companies."

That was to be the most productive five minutes of my entire business life! I would have lunch with him three or four times during the season. We would eat at the Omega Suite and I learned so much from him in the field of business acumen. He would tell me things like, "If you can't put your argument down on one side of A5 paper, then you don't have one!" In other words, top men only have minimal time on their hands. One-liners will explain it to them. Technical manuals are for the managers to interpret. Bert was a delight to be with.

On April 15, 1989, the national news showed us the horrific sight of the disaster that occurred at the semi-final between Liverpool and Nottingham Forest, played at Hillsborough Stadium, the home of Sheffield Wednesday. This tragic event left 766 injured and 96 dead. During the following week the staff at Wednesday showed signs of considerable stress through being exposed to such carnage. They had the offer of professional agencies to assist with their grief, but the staff said, "We want Dave Carr." So Gordon Taylor seconded me to them for ten days.

I arrived at what was a club in total shock. The staff had been exposed to the dead being lined up in the gym as a temporary mortuary and, as a result, many people there walked round the office, one minute blank and expressionless and the next minute bursting into tears. In the days that followed we had coach loads of what we thought were survivors coming back from Liverpool to receive counselling. However, after the second lot of "victims" I refused to see any more people. Most of them hadn't even been at the match! The local radio station had announced that buses had been laid on if supporters wanted to visit the ground, the scene of the disaster. There were so many who witnessed and survived the tragedy that I had no time for those who came merely out of curiosity.

The very nature of our work meant that Brian and myself covered a ridiculously large number of miles throughout the year. Brian lived at Hood Farm in Upper Longdon, north of Lichfield, and we would share the driving if the two of us needed to travel to an appointment. Any club north of Lichfield I would travel to him, leave my car there and we would travel together. Any club South of

Lichfield, I would do the driving. We would take it in turns for the longer distances.

Between us, we helped shape the early pension provisions for professional footballers. They could retire at 35 with a pension for life. Then they had the opportunity to secure a new career.

Chapter 15
Inside the "Beautiful Game"

The job did had its perks. We were often having Christmas dinner with the players and staff at Manchester United or watching a key match from the Director's Box. But I frequently found myself dealing with slightly "off the wall" situations too – such as the English international player who kept ducks costing £300 a time, which he kept losing because no one had told him he needed to remove their flight feathers! Or the Birmingham City player who drove the Club's mower into a ditch so that it would be difficult to cut the grass, because he didn't like the manager!

We offered a service that, if used, gave players all that agents could offer and more. The PFA didn't charge percentage commissions like agents did, just a set fee. To the players, this meant the difference between parting with forty thousand pounds (the agent's 10% fee of a £400,000 contract) and £2,000 (ten of our hours at £200 per hour). Even with this vast difference of £38,000 for what amounted to pretty much the same service, many players would still choose to go with an agent!

Some managers were easy to work whilst others, by contrast, were supremely difficult. Lawrie McMenemy, when he was at Sunderland, was very aggressive and confrontational. I had a disagreement with Kenny Dalglish, who kept interrupting me during my presentation. I told him that, as the manager of Blackburn Rovers, he had the authority to refuse to have the PFA in his club, however it gave him no authority to be offensive. To my surprise he left the changing room and went to his office.

I was invited to players' weddings, I dedicated babies, and buried Paul Birch the ex-Villa and Wolves star. When Ashley was a young boy he would go to school and say,"My dad had tea with Andy Gray yesterday." No one believed him, so he stopped telling them.

The first time we went to Manchester United "Big Ron" Atkinson, looked at the players training and said, "Something's wrong, why don't we have any black players?" Apparently, at that time most of the youth players coming into the squad came from either Scotland or Northern or Southern Ireland. Few, if any, black footballers could be found at the time. Ron went on to sign several of the emerging black players. At Albion he had three sensational players in Cyrille Regis, Brendon Batson and Laurie Cunningham, who tragically lost his life in a car crash at the age of 33. To hear recently of Ron being accused in the media of racism for a remark he made off-mic (which obviously was unacceptable) surprised me. He is not a racist, but comes from a footballing background and time that had zero political correctness. The language, the banter and the direct discussions that went on in his day left little to the imagination. As a Christian I managed to see through this superficial, linguistic license to the true heart of the person. In all of my dealings with Ron, I have to say I would find it difficult to see him as a racist – it's just that his "old school" style is no longer acceptable in modern Britain.

I once received a golden cockerel pin from the directors of

Tottenham Hotspur for successfully transferring Danny Thomas to them. I left it in my bedside draw for some thirty years. On the wedding day of my daughter, Lucy, however, it became a present to my future son-in-law who is an avid Spurs fan. I knew it would come in useful eventually!

The time I spent travelling was incredible. With 92 football league clubs I could be anywhere in a given week: Newcastle Monday, Charlton Tuesday, in the office Wednesday, Plymouth Thursday and so on! But I redeemed the hours spent driving by educating myself with knowledge on many different subjects, all of which helped to equip me to be a Pastor with street credibility! Initially I had intended to learn languages during the many hours on the road, but I listened to programmes on many subjects from medicine, to politics, to world affairs and to historic and social reform. After 21 years, there were very few subjects that I didn't have some at least some knowledge of.

Finally I left football to become the full time minister of the church that Molly and I had started 21 years before. I will describe this somewhat traumatic transition in the following chapter, but my time in the beautiful game had taught me many things. I came from a simple working class background, born in a council house, yet I came be sitting in the boardrooms of the biggest clubs in the UK. I was part of a team negotiating contracts providing more money in a week than I received in a year! It set me in good stead for the future where, as a pastor, I would not become overwhelmed by the size or importance of certain projects, or indeed the pressure of making things work.

So my pastoral apprenticeship was not honed in a theological college, but in the dug outs of football clubs, standing in the rain talking to injured players who needed motivating, learning to resolve conflicts and not allowing myself to be intimidated by chairmen or managers. All that I had learned in football would combine to give me a unique insight into human behaviour. Just as a footballer would live a normal person's working life of fifty or so years in the space of seventeen to twenty, I would experience the challenges of every stage of life concentrated into a relatively short space of time.

I could begin to believe that one day I could have a congregation of 7,000 people – after all, that is only just above the size of a crowd attending the average Division One club! And to have to purchase a church building for hundreds of thousands of Pounds – well, that just the equivalent of one average Championship player! All this stopped me being overwhelmed by the opportunities that God was opening up. I had in the natural seen bigger things and touched the impossible. Now the Spirit would translate my football experiences into prophetic realities.

Many of the young players I encountered never reached the level of success that they had dreamed of. Life is not always fair. But then, many of the young men had come from backgrounds totally unable to prepare them for fame and wealth. They often could not discern between genuine friendships and the "hangers on" who either wanted to be seen with them or help them spend their money.

Footballers often married at a young age and many, of course, lived miles from their family since they moved with the job. Isolation and loneliness was a problem. Getting married gave them the family they were missing. I met some really good players' wives who tried hard to fashion a lifestyle for their husband to reflect a near-normal family. But there was the ever present temptation for these young guys to be seduced by young women seeking attention for themselves.

Then there was the financial controversies...

To say there wasn't a culture of "bungs" in the game would not be true. I remember one manager, known for not being without his side "interests", being asked by a fellow manager, "What are you doing dealing with Dave Carr, don't you know he is straight down the line?" He answered, "I do all my 'normal' business through him because it looks good!"

I received a visit one day at the office from a special investigations team who asked to see and take away certain files of a number of high profile managers. These guys from the Revenue had a remarkable amount of power.

After I retired from the PFA, I was visiting the Merry Hill shopping complex one day when I heard a voice call, "Are you David Carr?" It was one of the special investigators. I asked him how the cases had worked out and he told me that, after years of intensified investigation, he was told he must "let it go". The number of prosecutions that could have been acted on would have caused such a crisis that the national game would suffer. The impact of this intervention had caused him to have a breakdown and he told me that he no longer worked for them.

I think most people in those days understood what was going on. I think we often missed out on some lucrative contracts because we would not consider any "special arrangements". Football is high on insecurity. You can understand this in light of the player or manager who receives a vote of confidence from the Board and then is sacked or released shortly after! Many clubs excluded players from the "inner circle" if they were not currently in the first team squad through injury or form. It could be a lonely place to be.

I would honestly say that, in my experience, football is very much like a faith community. It has a church building (the ground) and even the smallest clubs in the league have larger congregations than most churches in their vicinity. Not one of their "churches" has a roof or heating and still people turn up in the snow! They have compulsory "tithing" (admission) that can range anywhere from £16-£60 per week and some times two services per week. Those who tithe in advance get to choose their seat in the church (ironically the Church of England used to do the same).

They have different churches that belong to the equivalent of

"Churches Together" or the "Free Church Council" (Premier League or Championship!). Football has worship, yet no leader or band. They have eleven priests (players) who are allowed to minister at any given time, a High Priest (the manager), cell groups (supporters groups) and a Bible (the match programme). The Congregation (supporters) are not ashamed to testify (wearing team shirts, hats and scarves). There are a number of mega-churches (top Premiership teams) who attract partners from miles away (up to 50,000 attending from all over the country and abroad). They have vision (League title, cup win, European trophies, etc) and they firmly believe in team building and strive to appeal to the entire family.

There is no separate children's ministry – parents bring their children of all ages and make them stay in the main service (the match). Parents name their children after their "leaders" and so it goes on. The "churches" have the latest technology – big screen TVs and some even have their own TV and radio stations. The "congregations" sing, prophecy, put their hands in the air, hug and kiss each other and get married at times that don't conflict with their "services". They even have their ashes scattered at their "church"!

You can see why football is such a driving force throughout the world. It has a passion that is infectious. Humanity was made to worship and if it is not God, then it will find another outlet. I was a missionary to this twentieth century religion!

So much more could be said. I have many stories left untold. Maybe they should remain that way. This is but a sample of the many diverse situations I encountered in the parallel universe known as football.

From that one phone call from Walsall FC, a series of events was launched that changed my life. Through football I had the privilege of travelling to many parts of the world, stopping in the most famous hotels and speaking in the leading conference centres.

Molly and I were honoured guests on the Orient Express and we visited palaces, castles and other luxurious venues most can only dream about, meeting all manner of famous people along the way.

From Bermuda to Monte Carlo, to Paris, Vienna, Athens, Venice and cities all across the USA, this frightened little boy from a council house in Birmingham who couldn't even read travelled by plane, train, boat and limousine to see some of the wonders of the world. I became Fellow of a number of Institutes and Associations and spoke at many conferences in world famous auditoriums. From the humble beginnings of Carr & Co Pension Consultants came Carr & Price Consultancy Ltd and then Professional Promotions before being integrated into PFA Financial Management Ltd.

I left school a failure but returned to give out prizes as a "success story". All this was training for reigning. Building a business and growing it in the world of football was undoubtedly my template for church building. My training may have been very different, but it proved to be very effective. Thirty-three years ago I was rewarded for writing over six million pounds of life business in one year. I had never been a winner at anything until the Lord transformed my life. Now I have boxes full of awards with no walls to put them on, they are already full!

Chapter 16
"Full Time" Whistle

I had become a leader in business, but God had called me to be a father. The difference between a leader and a father is considerable. A leader surrounds himself with those who can help them get to where they want to be. A father surrounds himself with children, so that he can encourage them to go where he has never been. The first is threatened by the success of others; the latter is proud of the success of others.

During this period I looked after the youth, led worship and carried out my pastoral duties. We invited the youth to have supper with us every Sunday night, to rebuild the shattered confidence they had suffered during the split.

One Sunday, returning from the Evening services, I couldn't find my keys so had to ring the doorbell. Someone let me in and as I came through the door, Molly shouted from the kitchen, "Who is it?" From the lounge, which was already filling up with the evening's visitors, Mike White, one of our youth workers, called back, "It's only Dad." This marked the beginning of my new role.

More leaders

Around this time my eldest brother Tony, who lived in Tynemouth near Newcastle upon Tyne, was driving through Birmingham on his way home with his wife Alice. Suddenly, the Lord spoke to him and said, "Tony, come to Solihull and serve your younger brother as they did with Joseph." He turned to Alice and told her that God had just spoken to him. Alice, being very perceptive in God, said, "I know, we're coming back to the Midlands."

Tony and Alice have been so supportive of me and the ministry I've been called to. He is a brother who has gladly served in many different aspects of the Lord's work, sitting with me on the Executive Presbytery of the Elim Church Incorporated and becoming a Bishop alongside me with the Communion of Evangelical Episcopal Churches. He is now 82 and retiring from local Pastoral ministry, yet being acknowledged by Renewal as a Pastor Emeritus. To me he will always remain my big brother who I love dearly.

Then there was a young lady called Anne who suffered from a muscular disease. She became actively involved with the choir and the worship. Her parents, John and Vi, lived in a mobile home on a country estate in Warwickshire. John had been in the Parachute Regiment during the war, but had returned to his profession as a cabinet maker in the shop fitting industry. John had no time for God until he noted the change in his disabled daughter after she found faith in Christ Jesus.

They lived next door to a historic church in their village, but felt a compulsion to come and visit Renewal, some 9 miles away. They both committed their lives to the Lord and John became a house group leader. John and Vi had the gift of offering an outstanding quality of hospitality to others – and on that basis alone their ministry blossomed into four house groups! I then appointed

John as one of our area leaders. During the church split his area lost nobody. He was a pastor personified. I then put him in charge of pastoral care and, even though he had no formal training, he was the most naturally gifted pastor I have ever met. I taught him how to conduct funerals and he took most of them for the church. In his latter years he gave oversight to the "Golden Partners", our ministry to the over 60's and he trained his successor, Rob Jones, who has built on this fine man's gifting. People can tell who has genuine empathy and compassion. It reveals itself when you are not expected to give it or deliver it, but you just do it! It's about going the extra mile without having to ask the question. John was one of those pastors.

* * *

We had started to have our Sunday morning services at Lyndon School. Despite having a two-tier auditorium, we wouldn't let anybody sit in the balcony until we had filled the ground floor.

One Sunday, a young lady walked into the church and asked me all about my beliefs. I wondered what on earth she was on about. Little did I know that she would become my first PA. She had been serving Bryn Jones, a well known Charismatic leader, who had shaken the UK by his teaching on Kingdom living and submission. At that moment, the Lord spoke to Sue, telling her that her long and faithful commitment to Harvestime Ministries was coming to an end and that she would serve me in the future.

Only a short time later, the eldership decided together that it was time for me to leave the world of football and go full time.

Sue now had the opportunity to test that prophetic word and applied to be my P.A. She was offered the position immediately.

Transition to full time ministry
Earlier I hinted that my journey into full time ministry was traumatic.

Here is how it unfolded: I was a partner in a flourishing financial and marketing company, dealing with a third of all professional footballers as well as overseeing 32 churches for Elim and running Renewal, which attracted about 400 people to each service at that time. As well as all these activities, I was a contributor on commercial TV and attempting to be a husband and a father!

Eventually all this intense activity caught up with me. On 2 September 1987, I received a call to go immediately to Tottenham Hotspur Football Club for a meeting. Ignoring the lift, I bolted up three flights of stairs and ran to my car. I was not exactly fit and this left me gasping for breath as I drove from the car park towards the motorway link. Journeying down the M6 motorway, I reached the junction with the M42 when I had a severe pain in my chest, throat and arms. I was driving at 70 miles per hour and beginning to pass out. Instinctively, I started to sing in tongues – the spiritual language identified with the day of Pentecost and recorded in Acts 2. My head cleared and I was able to pull over onto the hard shoulder.

I slumped in my seat and the pain in my chest felt like the weight of two house bricks. It was decision time: should I continue to London for the important meeting or turn off the motorway and go for medical assistance? Thankfully, I understood I needed medical attention and logic won the day. Turning off the motorway, I headed back to Solihull to my GP. As I arrived, the only doctor available was a new partner. She came back into her consulting room, put me on an ECG and, having seen the results, I was immediately transferred to the intensive coronary unit at Solihull Hospital.

I had to phone Molly, as she thought I was in London. I tried to play things down and more or less inferred that I was having a superior check up! However, later, as I lay wired up with monitors and drips, I heard the distinctive click, click of her high heeled shoes walking briskly – a sound that always told me she was stressed. On entering the ward she launched into blaming this on me rushing around the country all the time, being addicted to cheese, and anything else she could cram into a single breath! My monitor went ballistic. I told her that, if she loved me, she should go home before she killed me!

There were only four of us in the unit. The guy to my left died the first night I was there and the one opposite me a couple of nights later. It was very thought provoking; a test of one's faith and beliefs. When facing uncertain outcomes, you find out the reality of your faith. I asked the Lord for a sign regarding my future. The Bible at my bedside table was the Living Translation and the Holy Spirit led me to Revelation 2:10:

"Do not fear any of those things which you are about to suffer. Indeed the devil is about to throw some of you into prison, that you might be tested, and you will have tribulation ten days. Be faithful until death, and I will give you the crown of life."

This scripture was like a spotlight in the midst of a dark room. The Holy Spirit told me,

1. Not to be frightened by what was happening

2. That this was being used by the enemy to try and destroy me

3. That imprisonment for ten days was the mandatory stay in hospital!

4. The comment about the Crown referred to my employer, Crown Life of Canada

5. I was to expect a glorious future

I smiled, turned my head to one side and fell asleep in the peace of Christ. The following months could have been the most frustrating time of my life, however they became the foundation for the rest of my life.

Some years earlier, we'd had the precious gift of a second daughter – Lucy, seven years younger than Melanie and eleven years younger than Ashley. Lucy was the only one born in Solihull and the only one whose birth I was present for. So here I was, in hospital, acutely aware that I had responsibility for a wife and three children, the youngest only 4 years old.

I was sitting up in my bed reading when a massive display of flowers arrived for me, then another and another. One of the other patients laughed and said, "Are you a ballet dancer or something?" The other guys joined in the laughter. Then, as they looked at the cards, they gasped. They were messages from the managers and players of Manchester United, Sheffield Wednesday and Aston Villa. The laughter turned to admiration as they discovered that I was involved with the football stars. I became a bit of a celebrity on the ward!

I was taken to other hospitals for tests until Dr Gordon Murray released me home to three months of recovery. My business partner, Brian Hassall, had to take the strain as I developed the lifestyle of a retired gentleman at the age of 42!

On my first day home, my daily walk was to the next house. Then, after a rest, I would make the return journey! The heart is a complex organ. It slows you down to protect itself. My initial goal was to walk three miles per day, but I couldn't make more than 50 yards.

As I worked out my routine, I surprisingly found myself living the dream. I would arise, wash and get breakfast for the children. Once I was able to drive, I would take Melanie to school and return home for a rest and my daily reading of the Bible. Then, after helping make the beds, I would drive to the church with instructions to park my car and not go in! The area included a large park with a lake and, at the top of the hill, the parish church. I would often sit in the grave yard, weeping at the thought of God's grace in saving me from death. I would look at the grave stones and become aware of the lives of those who had died younger than me. He had saved me for service. I would then return home for a sleep, before maybe making a trip with Molly to Solihull for a cup of tea. Then it was a further rest until it was time to pick Melanie up from school and to watch for Ashley walking home. Lucy probably saw more of me than either of the other two. However she wouldn't have understood that.

I became so relaxed and thin. I lost over four stone and it left me looking uniquely not me! I was allowed to attend church, but only if I arrived 10 minutes after the service had started and left 10 minutes before it finished. The bliss of this lifestyle was beyond description. If the Lord had asked me to give up preaching and remain like this forever, I would gladly have accepted.

Don't get me wrong. I have always loved preaching. But now it was not my driving force and focus. The Lord Jesus Christ was. And to just spend time with Him and my family was so precious.

When I was discharged from full medical supervision, I was allowed to phase my return to work: two days, then three and then five. I was not allowed to drive in cities nor for long distances, so I used trains.

One day, during my recovery, the door bell rang. Maureen, Molly's friend, opened it to find Ron Saunders, the Manager of Birmingham City, standing there. His sleek, gold-coloured sports car as the backdrop. Maureen was dumb struck by her hero! Ron arranged to take me with him to his golf club in Harboune, to walk the course with him. I received so many similar, thoughtful gestures from the football fraternity.

It was at this time, appreciating the vulnerability of a two-man business, that the Professional Footballers Association (PFA) approached our company, Professional Promotions, to take us over and make us directors of the Union's financial management company. The advantage of merging our 700 players with this company was the respect and influence that the PFA commanded within the footballing profession. Then, there was the long term aspect. If we'd continued with them until retirement, we would have been very secure.

Gordon Taylor, the Chief Executive, made Brendon Batson, his Deputy, our Chairman; with Brian Hassall as Sales Director and myself as Marketing Director. Our head office became the PFA's head office in Bishops Gate, Manchester and we opened a Regional

Office in Water Street, Birmingham in a restored lock keepers house alongside the major canal link in the city centre. We managed as much for our players as they desired: contracts, book deals, advertising, boot endorsements, house purchases, accountancy, investments, transfers and media.

The church continued to grow and, at my six monthly check up, Dr Murray said to me, "Well, Mr Carr, I think you will only need to see me one more time."

I said to him, "Great, that's good news."

He just looked at me and said, "Not really, because you will probably be dead."

I looked shocked, as he continued, "I see I have your attention."

He continued to explain that my lifestyle was equivalent to the output of three men. He said that if I wanted to continue to live, I needed to readdress my priorities. That was a loud warning bell going off in my head. What he was saying was absolutely true. But, I needed a sign from God. The next week, the Elders suggested that, with the congregation continuing to grow, they needed me to consider going full time.

Strategically, this was not a good idea. I had merged my business with the PFA on the premise that, at retirement, I would be given my share of its overall value. I was now 42 and with a heart complaint, so I had no opportunity to gain insurance and faced a future without any pension or security.

I was about to retire on medical grounds when my partner, Brian Hassall, left the company suddenly. I had to maintain my own integrity, so I stayed on for a further year before my heart consultant signed me off for three months, and then retired me on medical grounds during the Easter of1994.

To this day, Gordon Taylor has never said goodbye. I remember a board meeting at the PFA head office in Manchester. The offer they made me was so unacceptable that I left the meeting, with Brendon chasing after me. My partner Brian had left with their mutual agreement and had received a package greater than I was offered. The reason? They explained, "Because we don't want you to go." I managed to negotiate a consultancy agreement at a minimal retainer for five years and the purchase of my old company car that had 80,000 miles on the clock. This however was the faith step. Was the PFA or the Lord going to be my support?

There was no going back. My leaving meant that I had to guarantee I would not return to my field of expertise for a number of years. I remember saying goodbye to my staff. What a great team we had. They filled up, I filled up. Brendon gave me a great tribute. He wrote that he always knew that football would give way to God.

Chapter 17
Building to Building

The following Monday I became a pastor, but at a fraction of my previous income. No company car, clothing allowance, phone, meal expenses and free seats in the executive boxes at football matches. I was now the full time Senior Pastor at Renewal Christian Centre with my own PA, Sue Byrne from Southend: "Wanna cup of tea, luv?"

I'd had three assistants up until that point, but Sue was a new breed. She had strong values, gained through her previous employment, that it was her role to honour and serve her boss with generosity of spirit and commitment. Sue has been a valued member of the team and a personal friend to the family for many years. She was so committed to serving me that, for the sake of her own future in the event that I should die, I released her to the National Health Service, so that she would have a progressive career should and when I transfer to the eternal presence. Sue, however, still acts as my personal host for any visiting speakers or dignitaries. She has been given the difficult job of preparing me for Tuesday night meetings and Sunday services, with notes, Bible, instructions, water etc. She shows a spiritual loyalty seldom seen today.

We still had our beautiful school, which was to small for Sunday meetings and were hiring the larger Lyndon School. We only had one of everything, however. One worship band with Phil leading, one PA team run by Simon Nall and Neil Deakin who, with a small team, laboured endlessly. What a gift they were.

The frailty of some of my faithful gems of the seventies was beginning to show. Ken Osborn and his dear wife Kath had joined us during the 70's youth revival. Their daughter Jane was one of the school children who had attracted their parents to come. Ken, who had always had difficulty breathing as he had only one lung, passed away.

Until now, for no other reason than our demography, our entire congregation had been white. Then one day the Lord spoke to me clearly and said, "Would you like a multi-coloured church including many nations?" That was something new to me, but I said, "Yes, Lord." Not long after Jackie walked in and has been with us ever since.

As the church started to filter more and more into the balcony we needed a breakthrough. The largest building in the town was the civic Hall: at 800 seats, it was the pride of Solihull. We had no opportunity to hire it on a one off basis, however, there was the expectation of multiple bookings.

I spoke to the council. For Sunday mornings and evenings we had an expected cost of £1,000 per week! This was about 1995. My salary, this rental, the upkeep of our building ... what was I thinking even contemplating it? Well, we received the message, "No way will we let you have this building every week."

However, after much prayer, they finally allowed us to take the building each week, apart from the dates of the long standing model train exhibition that was held every year. On that day we would move to a number of venues, including the Motorcycle Museum and The National Exhibition Centre. At that time, that was a huge leap of faith.

God demonstrated His presence in the Civic Hall. The miracles we saw there – wow! We had many leading preachers from all over the world: Casey Treat, David Wilkinson, John Bevere, Dr Edwin Louis Cole, Ray McCauley, Rick Godwin from Eagles Nest, and many more. We remained estranged from most of the churches in town. We believed that the rest had rejected us and, in a way, they had.

On reflection I realised that both their lack of spiritual understanding and our brashness (that could been seen as superiority) were to blame. We must remember that Solihull had never had a designated Pentecostal Church in over 800 years!

The numbers grew in this central complex. We would have about 450 people during our morning service and people would be healed and saved regularly. Our evening service was more geared towards outreach, yet the Holy Spirit started to fall on the services and we soon started a third service, eight till late. The same team worked constantly to supply the requirements needed to maintain a flow of blessing. The PA team started every Sunday at 7.00am and didn't finish till almost midnight – and they did it every single week!

Occupying the Civic Hall was a kind of test for us, to see whether, when the time came, we would have the collective faith to step out and purchase our own, larger facility – the building that is the church's home today. We had identified a possible location for us to occupy.

But it was indeed a big step of faith. We would need to sell our existing building to help finance any new development. Our church at the time in Frankton Close stood on 1.4 acres of land and was valued at about £250-275k. We had become locked into negotiations with a building developer who owned a vast amount of green belt land in the prime district of Solihull. They offered us a fantastic package that would enable us to build a 2,000-seater church with all the additional buildings needed for community and designated ministry. All of this, of course, on the condition that we could gain planning authority.

It was suggested that we employ a Christian architect. I didn't see why a Christian would automatically be the best, but agreed that we would interview two Christian companies and one secular. They all received the brief and were asked to present concepts for us to consider. The first Christian brought nothing. He said, "Fellowship as Christians is the most important thing." He was dismissed! The second brought a sterile concept of an "A" shaped church with traditional pews. We thanked him and quickly dismissed him.

The third, a Canadian, admitted to not having a faith, but had secretly sat in the congregation for one month to gain the ethos of the church – wow!

His analysis of us was interesting. He told us that he saw...

1. We loved worship, fellowship, the Word, prayer and the local community
2. We loved the traditional and the contemporary equally
3. We wanted to be relevant, without being outrageous

He designed a concept that will remain in my heart till my passing. It was a circular sanctuary, built into the hillside, to hold 2000 people. In the summer, the back wall opened up to an amphitheatre built into the hill, so that in nice weather we could project the services to the hillside. There would have been a separate youth/children's block and administration block, all covered with a dome that provided a fellowship area, very much like a hotel reception, serving coffee and snacks. The whole complex was then surrounded by cloisters looking out into the fields, set every few yards for prayer. The woods were to be changed into a community park with fishing and BBQ facilities. It was absolutely fantastic!

Sadly, however, when the local residents' committee got wind of our plans, they protested strongly, parading with banners and boards, campaigning to hold onto the vast amounts of greenbelt land. We lost our bid for planning permission.

Shortly after, however, we heard that another building was for sale. It was Wilsdon's Body Works – a leading truck and fire tender builder. As soon as I saw it, I felt the prompting of the Holy Spirit that in fact this would be our church home. It stood on the main road that included the Rover Motor Company, the local hospital and a 1,000 pupil school directly opposite. In many ways it was perfect. We would have main road exposure, which at that time was unusual, since most modern churches found themselves stuck down side streets.

We offered £550,000 for it, but lost out to a national leisure company. I was so incensed that I prayed and dialled the telephone number for the building owners, British Federated Foods and, without anybody ever understanding why, I was given the personal number of Mr Weston the Chairman! I spoke to him and we immediately disagreed with each other. I said that he should have offered it to us. He told me I should have offered a reasonable price. He finally said to me, "What would be your top offer?" I didn't have the money anyway, so I offered £720,000. He told me that only put us in third place.

I put the phone down and said to Sue, "Fax him that offer. I don't care what he said, that's our new church." At six o'clock, a fax came back saying: "You have it." I know, like those before you, you may be thinking, "Well he gained almost a quarter of a million out of you." Well, two months later the person who offered more than us offered me one million pounds to sell it on to them! I could have made £280,000 profit in two months, but it was God's provision for us. We received other tempting offers over the years to sell, yet to relocate all that we have now would cost about £15-20million. The most we have been offered is £6million.

In order to fund the purchase we needed to sell our present building. Ideally, we needed a whopping £400,000 for it – despite the fact that the maximum valuation was about £275,000. After talking to a sceptical agent, I dismissed the idea of going the traditional sales route and began contacting builders, asking each one to tender their best offer. With one week to go about five major builders submitted offers for developing houses on the site. The highest figure was only £245,000. Then, an engineering building company that wanted to enter the housing market submitted an offer one day before the deadline. I opened it and it was for £400,000. I gasped. Only God could have caused such a increase.

Our next hurdle was to persuade Solihull Council to allow a change of use for the building. The recommendation from their planning department was to reject the request. One day before the planning meeting, the Mayor stated in the local paper that the planning meeting would be considering our application. He said that, in spite of the planners' advice, he would be voting for us. We won on a six/four majority. Praise God!

So, we had our building, but the factory was a total mess, with holes and pits everywhere. It would, with hindsight, have been easier to have pulled it down and started again. The purchase of this historic factory seemed, to many people, doomed to fail. The whole building was basically falling down!

On top of that, there was an agreement that insisted a portion of the factory still remained operational as a manufacturing outlet. As a church, we had neither the expertise or desire to enter the business world in manufacturing. I thought, if I could find a firm who would buy one of the units from me for £150k, with the guarantee of first refusal should they wish to sell in the future, this would enable me to finish the refurbishment of the premises and turn it into a church.

Well, we found them! A company who built display stands for the National Exhibition Centre moved in and stayed until we needed the extra space and the agreement had been removed. It did cost us, however, considerably more to get it back!

We now have it, but the project to refurbish it will take about five years. The total cost for purchasing it back and refurbishing it will total £5 million!

Chapter 18
A New Movement

Renewal had previously been part of the Elim Pentecostal movement. Having joined in 1974 I had been Ordained at their Clacton Conference in 1979, then served them as District Youth Director and Superintendent overseeing 32 Churches. All whilst holding down a full time job! I had become a Member of the Elim Church Incorporated Executive alongside my brother Tony. My assistant Sue had moved to the NHS and Yvonne East had taken over as my PA. Yvonne was a lovely, caring person who sadly faced considerable pressures in her life. She overcame them with great courage and only left me when it came time for her to get married and move to a Greek island to live for six months of each year. What people will do to leave me!

Next, Barbara joined as my PA. I met her and her husband Richard when they were pastors serving in Leicestershire. She served me for six years before moving, with Richard, to team up together once again, caring for business professionals in an international rehab clinic.

I was asked to visit the Elim HQ in Cheltenham to meet with Wynne Lewis, the Leader of the Movement. Standing in the gentlemen's toilet he said to me, "Do you see yourself taking over from me?" I must be truthful, my shoes became a little wet!

"No," I said. "I'm at Solihull till the Lord takes me."

Wynne was not impressed with that and asked me to pray about it. I replied, "I did before you offered it to me."

I asked who else he considered suitable. He mentioned John Glass. What a wonderful choice and he has proven to be the right man for the job. His consistent high quality leadership has established Elim as a credible denomination.

As I journeyed back up the M5, I felt the Lord affirming my decision. He asked me to pull off the motorway and read Genesis 26. It was Isaac who saw the wells of his father all filled in by the Philistines. He desired to reopen them, yet retain their original names. He found considerable opposition and negative behaviour, yet he continued until the Lord found him space. The Lord spoke to me clearly: "Leave Elim and open my traditional wells. The first well is this: become Methodist."

I found that farcical. Me, a Methodist? A raving Pentecostal? Yet I had been christened at Hazelwell Methodist Church over fifty years before. All those years before, at Grafton Road, Solihull Lodge, I had contacted the Assemblies of God, Elim and the Free Methodist Church to see who we could be affiliated to. We needed to be identified with someone. You'll recall that the Assemblies of God didn't turn up and the Free Methodist's replied saying that they didn't operate in the UK, so by default we met with Elim and we joined!

The Lord said, "The Free Methodists."

By now I had a young associate pastor, Richard Taylor, working with me. On arriving back at the church I relayed the story to

Richard, thoroughly confused by what the Lord had said. "You wouldn't have heard of the Free Methodist Church," I told him, "they don't operate in the UK."

Richard replied, "They do. I know a couple of friends from Bible collage who are part of them."

Apparently they had formed a base in the UK about two years after we had joined Elim. This was interesting. Now I looked for God to confirm the validity of such a "now" word. I was due to speak in Devon at a Conference and I travelled with my long standing colleague Melvyn Carman. Melvyn had been an elder of the fellowship that merged with Renewal years before and our elder, knowing that I was being invited to travel more and more, wanted someone of spiritual maturity to travel with me. Melvyn stepped forward and has been a fantastic companion in the many, many parts of the country to which we have travelled. What a wonderful man of spiritual integrity. Thank you to his wife, Marion, for letting him serve me in The Lord.

So, we travelled to the mission in Devon. They gave me the address of our accommodation. I must confess, when I saw where we were staying I was incensed. We were being put up in a smoke infested pub with dirty bedrooms and bedding that I wouldn't have let my dog sleep in! I admit that I totally lost the plot, demanding that Melvyn, "Phone their office and tell them I'm off back home. Tell them I haven't come this far to be insulted!"

Mel walked with me up the High Street and gently reasoned with me, talked me down and then got us transferred to a beautiful country hotel that actually cost less! In the end I was so glad I hadn't returned home. The next day, after my session, the other speaker who was from Jack Hayford's Church on the Way in California, asked me a simple question: "Are you Pentecostal?"

"Yes," I said

Then he said, "The Lord has given me a message for you. You will leave your denomination and will open an old well and there will be a move of God." This was the first confirmation.

I moved from that mission on to Cefn Lea Bible Week to speak. As I was due to leave, the next speaker, who was due to arrive the day I left, arrived from South Africa a day early. He was the leader of the Assemblies of God RSA. He met me after the final night's meeting and asked if we could have breakfast the following morning. Over breakfast he asked me the same thing. It was a total re-run of the conversation in Devon.

I arrived back in time for the Sunday morning service at Renewal. As I was sitting on the front row, a man came and sat by me. His name was Andrew and he was from the NFI Church. He had previously been a monastic brother. He looked at me and simply said, "I have been praying and fasting for you this week. The Lord says that He has given you two prophetic words. He will give you just this one more."

He then suggested I read John 4 – the story of Jesus sitting by an old well and giving a woman living fresh new waters! Well that was it. I couldn't say that I didn't understand.

We had three meetings with Elim and they obviously didn't understand why we wanted to go. It was Colin Dye from Kensington Temple who stopped the questioning by saying to the Executive, "Listen, he isn't leaving Elim, he is leaving Pentecostalism." That was true. Yes I was still very much Pentecostal, but no longer part of the denominational expression that bears that tradition.

It was, sadly, a frosty goodbye. Twenty eight years service was now basically wiped from the ledgers of acceptability. I received just three phone calls over the next twelve months out of the hundreds who I had helped and encouraged. That silent time continued until 2012 when we had just finished a Free Methodist Ministers' Conference at the Elim Complex at Malvern, Worcestershire. John Glass had left a message at reception for me to see him. We had a wonderful time together, restoring the years that had been lost.

The Free Methodist Church are a Holiness Movement. They try

to impact lifestyle rather than promote platform stardom. As a team, we attended the Nazarene College in Manchester for three days of lectures on Wesleyan Theology. It was outstanding. I was invited to visit the Free Methodist Conference for Ministers in San Antonio USA. It was my first encounter with their pastors.

The Bishop asked for reports from the pastors. One had just recovered from a serious car accident that had left him immobilised for almost six months. As the pastor shared, he started to cry. The Bishop called me forward to pray for him. At this point I had no idea that, whilst speaking in tongues was acceptable in the movement, it was never done publicly. I began praying and speaking in tongues over this man whilst embracing him. When I finished, I kissed him and told him that all would be revealed to him on a specific date.

The room fell silent. The Bishop eventually spoke. "After seeing this, we can't carry on. Can anybody tell us what we have just witnessed?" A voice said, "He spoke in tongues in public!" "Yes," said the Bishop, "but it seemed acceptable." Then another voice said, "He kissed him!" The Bishop responded, "It was sacred though, wasn't it?" Yet another voice: "He told him the date it would be resolved." The Bishop said, "That's prophetic!"

By now the pastor I'd prayed for had now stopped crying. He stood up and said, "You don't understand. My late father was French Canadian. He would always put his arms around me and quote a scripture over me. David, speaking in French Canadian, just quoted that scripture." I don't speak French. I was simply speaking in tongues.

The Bishop said, "What has God given us?" Someone shouted, "A father in the faith." That was the beginning of my Free Methodist Ministry.

Over the last eleven years I have helped to restructure the leadership of the movement. The UK church was a group of Methodist ministers who had left the Methodist Union in the 60's because of a feeling of humanistic tendencies creeping into the nation. Appreciating the Wesleyan theological viewpoint, they found in the Free Methodist Church a strong Evangelical theology, with accommodation for the working of the Holy Spirit. I was a founding member of the National Leadership Team which replaced a full time Superintendent Minister.

Chapter 19
Wells are Reopened

In response to the call to "reopen old wells" we began to reach out beyond our own comfort zone. A new Bishop had just been appointed in Birmingham, the Rt Revd Dr John Tucker Mugabi Sentamu. I was invited to meet him at his house in Bishop's Croft, Harbourne. The only problem was that there was no sign to identify the house. I am phobic about being late, but arrived five minutes after the appointed time because I couldn't find it. He was sitting at the top of a long table with other senior clergy scattered around it.

I apologised and mentioned that it was difficult to find with no sign. He contradicted me and said, "There is a sign." We then exchanged a forceful conversation until his chaplain, Andrew, came in and confirmed that someone had actually stolen the sign. After this slightly tense start, however, the meeting became very relational. It was the start of a great friendship that would eventually reopen for us the "Anglican well".

Some time later Bishop John acknowledged me as an "Honorary Bishop" and insisted that I wore a Bishop's shirt when ministering in the Church of England.

One day, before a breakfast meeting with him, the Lord told me that John would be the next Archbishop of York. That is the second highest office in the Church of England. I told him this and he was somewhat surprised, since he had only just become a full Bishop. We prayed over this and then had breakfast. A few weeks later John called me to come and see him and told me that the next day the Prime Minister would be announcing his appointment to York!

One evening, when taking a leaders meeting, my mobile phone rang. That doesn't normally happen as I usually turn it off. It was John telling me (John doesn't ask) that I was going to be at his consecration in London. That day was unique. Only the Bishops and a few invited guests attend. It was held at St Mary le Bow Church in London, followed by a reception at the Mansion House. To see the Church governance with the judges in full regalia and the parchment with a massive Queen's Seal was spectacular. The Archbishop of Canterbury was the officiating minister, installing John on behalf of the Queen, who is the head of Church of England.

The installation into York Minster was a battle against the cold. This ancient city is dominated by the incredible church. It has been a diocese since 314AD and its full title is "The Cathedral and Metropolitan Church of St Peter in York", but such a building has virtually no heating! The service lasted two hours and was classical Anglican mixed with African. I was asked to pray and give a prophetic word over John. As I was escorted to the steps, I was surrounded by the leaders of the other expressions of Christendom: Catholic, Orthodox and a few I couldn't recognise. They stood about twenty feet away and just extended their hands towards him. John looked at me and mouthed, "Come and lay hands on me." As I moved towards him, they came with me, thankfully. I laid hands on him and started to speak in tongues. I think the others thought it was a variation of Latin! John started to fill up and I handed him my handkerchief. The service was watched by an estimated five million people. That was my best yet!

I was once having breakfast with John and, as I was leaving, he stopped me. His sister-in-law had arrived from Uganda and had been diagnosed with cancer. Could I pray for her? "Of course," I replied. John knocked on her bedroom door to see if we could enter. She sat on the end of the bed looking shocked. She had literally just suffered a terribly frightening panic attack that had paralysed her with fear and prevented her from calling John or his wife, Margaret, for help. All she could do was pray inwardly, "I don't know how, Lord, but please send a man of God to me." There I was. Over the months, I visited her at John's house weekly, until she recovered enough to return home.

I also had the pleasure of travelling to York in the midst of winter to pray for Grace, his daughter, who had been afflicted with a form of M.E. I returned in another blizzard to see her married at the Minister, with John both conducting the service and cooking the meal for about 250 guests! I sat next to my god friend J. John and his wife, Killy. Thankfully, we were next to a fantastic heating unit!

Perhaps my greatest accomplishment, however, was to teach John to speak with a Birmingham accent. Now that he is established in York, we don't hear from each other as much as we would like, but we know that we are there for each other, if ever we are needed.

* * *

Over the years, our team grew and changed, with different members coming and going. In the Youth Department we had Derek McDonald who developed into a leader with New Frontiers, Richard Ellerington who now ministers in France, Matt Spina who continues with us at Renewal, but is now the Ethos Principal at a

Christian Academy in the North of Solihull, Jim Wilkinson who is now pastoring a church in Bridgend, Wales, and our current Youth Pastor, Richard Louis, who trained in Sweden.

In the central church we have had Mike Webster who returned to New Zealand and revisited his old profession in the probation service. Michael Carr who is Senior Pastor of a church in Harrow; Richard Taylor, who after nine years returned to his native Wales and is planting significant churches in his homeland, and Peter Jenkins, who moved to Chaplaincy and Missions for his friend Richard Taylor.

The new Senior Associate Pastor is Dr John Andrews, who could easily be the man to follow me in time. Pastoral Care is overseen by Justin Marsh, who is a son of the house, as is Justyn Towler who oversees Children and Community at the Family Centre.

Alongside these gifted men are the elders. Our first was Laurie Pedler, who had become surrogate father to me, confiscated my diary after I had my heart attack, and cancelled all my preaching engagements without my permission! He became my point of accountability and was a gentleman of the highest level. Our other elders included his son-in-law, Niall, who gave his life to the Lord at the height of our encounter with God during the late 70's; Phil Lane who was in the Youth during our incredible explosion of young people being saved; David Russell who managed the massive move from Frankton Close to Lode Lane; Arvel Lowe, who came into the church from the world of professional sport and is now a funeral director; and Rob Williams, who came to us from Elim. He is a National leader with the Free Methodist Church and a man of the highest integrity.

With the completion of the building as we had designated it, the central area, against my preference, was lengthways, so creating a back row that tapered into the distance, rather than side on, enabling all to be reasonably situated. We created a massive foyer, toilets and eventually additional rooms that could be used for conferences.

Over the years the complex has grown extensively. We recently set out the designs for a new Sanctuary to seat 1,300 people and purchased the building adjacent to our campus. This laundry and fitness club had originally belonged to Wilsdon's, but had been sold off before we acquired the site. The design of what was to be called the "Praise Cathedral" was superb. We arranged for a DVD and full model to be produced and a loan was offered to build it. Then the Lord said to me one day, "Don't build it."

What did I hear Him say?

We had basically committed ourselves to it. "Don't build it" came the same message. "There is going to be a dreadful recession and you will go bankrupt!"

Let me take you back to the time when I had just given up my position with the PFA to become the full time Minister. We arranged to take about 50 men to Cliff College, the Methodist Training Centre at Calver in Derbyshire, for a weekend. I need to give you the background so that you will fully appreciate what happened that night. We were staying in a first floor unit, which was double glazed and held about 70 people. Phil Jones was leading worship with Tony Burchell playing the keyboard. During the singing of the second song, the presence of God descended to such a degree that I felt myself being sucked out of my body into a throne room, void of any music or singing.

I was standing behind the throne of Christ and I could only see the back of Him. Without opening His mouth He spoke to me. He called me to look over a shelf-like edge. As I looked, I saw many knights in their armour, leaning on shields with a variety of crests displayed. They had goblets in their hands and were drinking and laughing. The Lord again spoke to me, "This is My Church in your country. Differing crests, drinking and making merry, yet not unified in battle."

He then told me what would start to happen in the next ten years, both to the nation and to the corporate Church. He told me that the nation would reach a financial crisis and to advise churches not to take on unsustainable debt. He told me that persecution would hit the Church regarding the changing of laws to permit Gay marriage. This would lead to Christian leaders being prosecuted for being classed as homophobic and that many churches would actually have their wedding licences revoked.

The Church of England would become disestablished and the next King would not be expected to be Head of the Church. Churches that did not comply with humanistic rules would lose both the rebate of tax on their offerings and, in some cases, even their Trust status. There would also be an expectation for Christian charities to prove that they actually benefited the community by the activities they provided. Then the registration of leaders and ministers would take place. Any who had not received mandatory training and been credentialed by an authorised body would be refused permission to either preach publicly or rent buildings.

God told me not to reveal this to anyone. Those who needed to know would ask and that it was to them that I was being sent. I then felt myself being sucked back into my body. All around me I saw men draped over chairs, face down on the floor, crying and calling on the Lord. Fear like I have never felt before gripped me. Not the kind of fear associated with harm or danger. This was an awesome hold fear. I jumped to my feet and hurried to my bedroom. I was shaking with the awesomeness of the occasion. About half an hour later, the elders came into my room. One cradled me in his arms and just reassured me.

The next morning, the men still showed signs of shock from the intensity of the presence of God's Spirit. I was walking across the campus for breakfast when one of the managers of the complex approached me and said, "Revd Carr, last night was strange." I asked him why and he explained that they'd also a concert the previous night with really loud music, but the "male voice choir" we'd had in our meeting room had been like a noise blanket covering the whole campus. I told him that this could not be, because firstly we had no choir and secondly, as the accommodation had double glazing, they should not have heard any noise. It was truly an angelic visitation. Altogether it was a memorable encounter. God had said people would come to me. I just didn't know how.

About two weeks later the office put a call through to me from Pastor Wynne Lewis. He simply said, "David, I'm sorry to disturb you, but has the Lord given you anything for us?"

"Yes" I replied.

He then arranged for me to visit Cheltenham and address the Executive Council. I sat at the table with them and, as I relayed my experience, the Holy Spirit started to hit them with most of them slipping off their chairs onto the floor! I can assure you that, knowing the personalities of these men, few could be accused of emotionalism. I remember that Gordon Hills, the Overseer of the Ministerial Board, just stood with his back to the wall. I pointed to him and said, "Your time will come shortly." That came true. He visited Toronto and ended up walking everywhere doubled up in hysterics. To have seen this traditional Northern Irish preacher walking through a hotel lobby, doubled up giggling would have been worth all the embarrassment of walking beside him!

God then sent to the Elim Bible College. The Lord took them out!

Then came a call from the Assemblies of God Executive. I arrived, bumped into Ken Gott and gave him a prophetic word that referred to a discussion he had just had. God did the same with them and their Theological College.

A few months passed and I received what was a mixture between an invitation and a command from Bob Gordon to go down to the Kings Coach HQ to meet his team. That visit sparked another friendship. Bob was a very clever man who craved the truth of the Holy Spirit. One of his favourite sayings was, "The Spirit and the Word." During one of my visits, I actually prophesied his death, in the middle of my teaching. I turned to him and said, "Bob, you will not be leading this work after September."

After the session, we went flying in his plane, over the beautiful sunshine over the coastal villages of Norfolk and the former home of Lord Horatio Nelson, First Viscount Nelson KB Vice Admiral – one of our mutual historic heroes.

Bob was fascinated by the prophetic word. He tried to tease the meaning out of me. I didn't know other than that his elders would determine the future of the church. He anticipated change and had a compulsion to finish his new book before September! That was the last time I saw him alive. September came and Bob fell ill with what turned out to be Pancreatitis. He rapidly lost ground and he went to be with his Maker.

Bryn Jones of Harvestime Ministries, the well known "House Church" movement invited me to Nettle Hill Conference Centre near Ansty, just outside Coventry. This revelation was a massive door opener for me. Did they all fully take on board the warning? I would say no. Like most of us, it faded into history to replaced by the "now" word that tended to come in every meeting. It has, however, been a directional tool for us at Renewal as we have walked through the last two decades.

* * *

After God had spoken, instead of progressing plans to build the Praise Cathedral, we changed our plans to develop the site by retaining the units and converting them into a Family Centre – a facility which would meet the needs of the local community. Those who had worked on the cathedral project obviously felt some frustration, but bankruptcy was not an option I was prepared to embrace. I shared the new project with our congregation and asked them to stand with us.

The next day I received a call from the school opposite the church. There was a situation with the mother of a pupil and they asked if our youth pastor would accompany the Deputy Head in a house visit. It was our youth pastor's day off, so I offered to go. On arriving, we found a tragic situation. The mother was in a traumatised state. There was no food in the house and the place was a mess – even her bed was broken. Plus she needed urgent medical treatment.

I phoned the family doctor and arranged for him to visit. Then I phoned the church and arranged for food and a new bed to be delivered. I held the lady's hand as she wept. I felt a deep compassion for her. We left the house and sat in my car. Before I could start the vehicle, the Lord clearly spoke to me: "Do you want to be a TV mega-church or a large, significant community church?" That was unexpected. I hesitated for a moment before I said, "The second." Then I drove back to the church.

When I entered the offices, Richard met me, emotional. He told me that a disabled man had offered 25% of his pension to help build the Family Centre. It had happened at the precise moment I had agreed with the Lord that we should be a community-focused church.

In retrospect, the Family Centre has been the most significant project we've ever undertaken to reach our community. In it there is a specially designed sensory room for children with Autism, multi-functional rooms, a changing and shower room for the disabled, a coffee shop and restaurant, a playground adapted for disabled children (including the only wheelchair swing in the region), a dance studio and provision for private consultations on issues such as family, health and financial problems.

The local community and health service have adopted the facilities to engage with the community. The school opposite the church – now an academy – has contributed to our philosophy of care and, with over one thousand pupils, has given us the opportunity to extend our daily pastoral oversight for our community. I am Vice Chair of Governors and Chair of the Board of Discipline. I also have

Governor oversight for RE and Music.

I have always seen schools as having a vital connection to the reality of the Church's mission. We have a responsibility to impact the next generation of leaders with a positive example of true faith in action, amid growing social needs. Our church has supplied mentors, governors and even teaching materials to assist the staff, especially in the RE department. The Head, John Burton, is a visionary like me. We spark off each other and together delight in seeing the obvious, which sadly often seems to be beyond the sight of the majority! He is retiring at Christmas, but will remain as a consultant in a partnership initiative that should propel Renewal into being a major player within the educational strategy of future generations.

We prayed energetically for Margaret, his wife, who had a difficult diagnosis with cancer. I visited her at the Queen Elizabeth Medical Centre. We prayed as a church and she is now enjoying total remission. Mark Wilson, his Deputy, has been appointed as his replacement. He is a good man and will continue to keep the school as a cutting edge establishment.

Chapter 20
More on the Miraculous

People often ask me to tell them about the amazing miracles I've seen God perform. I don't talk about these things a lot, but it would be remiss of me to leave out some of these stories because, ultimately, all the glory goes to God. It is He who moves, by His grace, according to His will.

From those early days of the Second Advent band, I had an expectation that after delivering God's word, we would see miracles happen. Since then the miracles, particularly those to do with healing and wholeness, have been incredible. I have witnessed the grace of God in wonderful ways. Here are some of the stories:

Praying in Tongues

One year at the Elim Conference, I was the last person to leave the stage when a pastor's wife stopped me and asked, "Please give me a moment of your time." A young "gothic" looking lady stood next to her, dressed head to toe in black with black makeup. I was a bit pressured for time, but I agreed to pray for her. As I prayed I began speaking in tongues. What happened then was amazing.

I continued in what sounded very much to me like French and it became almost conversational in style. The young lady not only began answering back, but also became deeply emotional. Then I, with great authority, waved my finger at her. She screamed and vomited over my shoes, then fell to the floor and broke down and wept. I was still in the flow, speaking in tongues. I continued to speak over her and she then began praising God. I then left (having first wiped my shoes!)

Before the evening meeting, the pastor's wife found me. She was still in shock at what had happened. She asked me a question, the answer to which was going to blow her away.

"Do you speak French?"

"No," was my reply.

Then she told me the girl's story. They had found her when they were travelling to the conference. She had been trying to hitch a lift. She had been part of a demonic cult and was planning to commit suicide the day they found her. The pastor's wife was a French teacher and was going to interpret for me. But I started to not only speak in French, but in her regional dialect! I had led her to faith in Christ and saw her filled with the Holy Spirit without understanding a single word I had said! This has happened on a number of occasions since that day.

I think perhaps God used me in this way to help me overcome an irrational dislike of the French I harboured for a number of years. I used to have a profound dislike for the French with no real reason for it. Once, an American asked me, "Why were the British at war with France for one hundred years?" I replied, "We ran out of ammunition."

God started my rehabilitation when I was a specialist advisor at a conference at the National Exhibition Centre. People from all over the world attended and many committed their lives to The Lord. I was given an interpreter and sent to the French! That experience changed my opinion forever.

I have spoken in ancient French, Canadian French and Guernsey French – all as a witness to those who knew I had never had any formal education. I have appreciated what is called "speaking in tongues". Paul calls it the language of Angels and of men. To all believing Christians, it is obvious that God is not English and that this is not the uniform language of the Spirit! I know that this is a shock to some.

There are natural languages that came into being at the building of the Tower of Babel. Until the fall of Adam, we can only assume that man, who communed with God daily, spoke the language of the Spirit.

I was at an Assemblies of God conference when I prayed for missions' workers, prophesying over them in the language of their adopted nation! At a Brighton conference of New Frontiers I prayed for healing of cancer in Mandarin. I prayed one Tuesday night at Renewal Solihull in Italian, and a prostitute from Italy came into the service and was challenged by what she heard me say in her own language. I witnessed to an Albanian Muslim in his own language and prayed the prayer that a French Canadian pastor often prayed over his son.

But the most dramatic would have to be in Native American. A married couple, both bank managers visited Renewal for one of our Tuesday evening prayer meetings. The lady, Wendy, was terminally ill with cancer and was receiving palliative care. I found out later that Wendy had decided that they should attend so that her husband, Bob, might be born again and, after her death, might also be able to find a lovely Christian wife. When you know Wendy, you will understand that her heart is such that even in her own fear, she was focussed on the wellbeing of her family.

I didn't know them, but I felt led to call Bob and Wendy out to

A LIFE OF TWO HALVES

the front and spoke in a language that frightened even me. It was deep and sounded like Sitting Bull or Crazy Horse! As I was speaking I saw a tepee with a medicine man fixing his eyes on me from the flap. I kept looking at him until he went inside and lowered the flap door. The room went frighteningly quiet. I had just quoted the verse from Revelation 2:17 and said that God would give them a stone with a new name written on it. The couple had both fallen to the floor, out in the Spirit.

This was so frightening, that I was deeply disturbed within my spirit. I had never experienced anything like this in my life. One of the leaders said to me afterwards, "Pastor Dave, we need to meet." I was expecting it. If I hadn't experienced it for myself, I would probably have believed it to be demonic.

Bob and Wendy, however, had no idea that this wasn't a "normal" night at church. After the service, they rushed to her parents' house to try and share that Pastor Dave spoke over them as War Chief in a Native American language and told them Christ had given them a stone with a new name on it. Rather than the mother reacting as if they had joined a cult, she walked across the room to a cupboard, retrieved a stone and placed it on the table.

She told a story that freaked us all out. Wendy's brother lived in America and, on hearing of his sister's terminal illness, had visited a reservation and been given a healing stone by a Medicine man, exactly as I saw in my experience. Wendy said to her mum that she didn't need this one as God had already given her a new stone! We had broken that charm during our encounter.

That was many years ago now. Wendy was totally healed. Bob became a Christian and is on the church staff as our finance manager. Wendy is in the worship team and a part of our creative design ministry and, by the way, even though the doctor predicted she wouldn't see her children grow up, she has attended both her son's university graduations and her daughter's wedding.

The native American tongue lasted about three more months. I had just returned from South Africa when Steve Davis, one of our youth ministry team, complained of pains in his joints. We laughed and said he had premature aging! But he visited his doctor who decided, just for safety, to send him for blood tests. The results came back showing that he had Leukaemia. I remember it well. He came into the service late. I called him to the front and said, "What did the Doctor say, then?"

He replied, "He told me to go home and enjoy my family as I have about a year to live." I then said an amazingly stupid thing: "Well, Steve, that's great. The doctor promised you a year and God only promises you a day at a time!"

During the worship, the Holy Spirit told me to ask Steve to lie down, then told me to lie down head to tail with him. I then started to speak in this same, deep throated native voice. It frightened the daylights out of me. As I was praying, an electric shock went from me to him and I saw clearly a tepee with his wife and children huddled in it, with a warrior about to kill them with a spear. Then, out of the blue, a Chief with a pure white war bonnet arrived on a horse and broke the spear and banished the warrior.

Once again, I was so disturbed by all this that I asked another leader to finish the service and left to go home. The next Sunday, I approached Steve to say that I had no answer for what happened. I didn't want to discuss it until he said, "You spoke in Native American. Did you see that Chief in the white war bonnet?"

He had actually seen the same thing. He went into remission and, over time, he and his family moved on from Renewal. He was clear for 17 years, but was recently was diagnosed with cancer once more. I offered to pray again for him, but he didn't feel it was necessary for him at that time and he has since passed into the presence of the Lord.

Healed of cancer

Another couple, Lesley and Andy, both had serious illnesses and visited the church. Lesley had terminal cancer that began in her stomach and it was feared that it had reached her brain. They both received prayer. Over the next few days, Andy was wonderfully released from a deep viral infection and at a later service, Lesley came running down the aisle of the church waving two scans in the air.

I called two of our leaders over, both consultant doctors. They looked at the scans, discussed them and started laughing. The first scan showed the stomach totally overtaken by the cancer. The second showed a perfectly normal stomach with scars showing the aftermath of a perfect operation. The only problem was, she hadn't had an operation! I reckon that if Jesus is the great Physician, then the Holy Spirit is the great Surgeon, able to "divide bone from marrow". The Word is sharper than a two edged sword (scalpel)!

A Consultant calls

On another occasion, I was sitting in my office when I received a phone call from a Consultant at Heartlands, our area general hospital. He introduced himself as an agnostic with no real faith, however he wanted to speak to me regarding one of his patients who had given him permission to phone. The story goes like this:

The patient had scans that revealed a cluster of abscesses that, normally, if not submitted to surgery, would take considerable time to treat. He gave her a prescription until long term procedures could take place. This was on the Friday. On Sunday, this lady came forward at the morning service for prayer and I simply laid hands in her and prayed a short prayer for healing. She tells me that immediately the pain went and she knew that she was healed. The next miracle was that she returned to the hospital on the Monday with no appointment, telling the story to the nurse and was told there would be no chance of a consultation. The nurse, however, spoke to him and it happened that his next patient had not turned up. So, curious, he sent her for further tests and found, to the surprise of everyone, that she was completely whole!

The doctor said to me, "Reverend Carr, I have these two scans in front of me. This cannot happen. Is this what you would call a divine intervention?"

I said to him that we would be happy to call it that. He asked if he could bring his wife and family to the Sunday service to experience it for themselves. He came and I thanked him for being so open as to accommodate something beyond medical science.

A family touched

In the family of Sarah, one of our worship leaders, we have seen two touches of healing. Maxine, her sister-in-law, was part of our church with her husband, Andy Bromley, just before they relocated to a church in the USA to take up the position of worship leader. She discovered white lesions in her mouth and after hospital investigations was referred to a cancer consultant.

A few days before the appointment I prayed for her and, on the day of her appointment, the Consultant was confused to find her mouth totally void of any form of disease. Sarah's sister, only a few weeks before writing this, visited the church from her home in Barnsley. Ali is a senior nurse and found a bleeding lesion on her back. She asked Matt, her husband, to photograph it and immediately he didn't like what she saw. The GP confirmed her suspicion and she was referred to hospital. After their diagnosis, it was arranged for her to be admitted for two days for it to be removed and a biopsy taken. They came to the Tuesday night meeting and, for some reason, as I passed down the prayer line, I revisited her three times. On the third time, the power of God hit her. When she arrived home she felt for the lesion but couldn't find it. She called Matt who confirmed it had totally disappeared! Ali visited the GP who was amazed at the fact that it had vanished. When he was told it was due to prayer, his reply was, "Some prayer!" The hospital appointment was cancelled.

A little girl's finger

When I was a speaker at the Grapevine conference at the Lincoln Showground, I called for the sick to come forward and a little girl came out, wearing a beautiful pink coat. A pastor's wife brought her to me. She had been born without a knuckle in one finger, so the finger was straight and would not bend. The surgeons were planning to operate and implant a plastic knuckle. As I prayed for her I was not looking at her hand, but suddenly there was a gasp from the pastor's wife. The knuckle had grown, instantly and miraculously, and the little girl was now bending her finger.

A sport's coach healed

A young sports coach attended the same conference. He was restricted to a wheel chair by virtue of a syndrome that could not be treated. He was committed to a bed in the front room and so frail that he could not even hold his young child. There was no known cure. I just looked at him and said, "Tonight is your night!" He shot straight out of the chair and ran. This young man's doctor was in the meeting and was shocked to see him running. On returning home, the local daily newspaper did a front page article on the healing.

Deafness healed

Then there was Christine who was deaf. She used to sit on the front row at church and watch me intently. I just thought she was a really committed member of the congregation until I found out later that she couldn't hear and was trying to lip read. On the evening after we had baptised her, she experienced a pain in her ears and a loud bang. The next morning she came to see me and told me that she could now hear perfectly.

Awareness of our spiritual battle

Since Renewal began, we have seen so many incredible manifestations of God's grace and on more than one occasion God has allowed me to see the reality of the spiritual war being waged all around us. Sitting in my office one morning, I received a call from reception to say that a smartly dressed man had asked to see me. I agreed and he thanked me for my time and said to me, "My master has asked me to give you a message."

I said, "Who is your master?"

He looked surprised, as if he thought I already knew. "Satan, of course," he said.

I asked him about his background and how he had become involved with Satanism. He told me that he had sold his soul to Satan in return for status, wealth and long life. He told me he had double degrees in psychology from a red brick university and told me that everything he touched turned to gold. He was very prosperous, naturally speaking.

He told me that his "master" had said that he didn't mind our large church and good music, but that he wanted me to stop preaching the Word and to stop praying for the sick. We talked for some time. He told me that he had been instructed to become a "plant" in a Christian church and that his purpose was to slowly bring confusion amongst the people. He had even become an Alpha Course leader in one church!

I asked him if he had considered asking God into his life. He told me that once, an elderly preacher at a church in Bristol had looked at him and cried out "Diablo!" and began to pray. As he did, the Satanist's spine began to twist and creak. He was in agony and stopped the man from praying. He told me that he wanted to stay where he was – that Satan had vowed to kill him if he ever went seeking God.

I was regularly invited to appear on Central Television's "Central Weekend" programme with Nicky Campbell. I was always wheeled in to offer the Church's stance on various matters. One particular week they filled the studio with witches and pagans. After the show, the other guests were in the green room and began to manifest, screeching that they wanted to kill me. The producers, who had taken the whole subject very lightly and were more interested in the entertainment value, were horrified. The crew, in terror, quickly ushered me out of the building with offers of a security escort home.

Why are some not healed?

Many people ask me, "Why are so many not healed?" or say to me, "They were such faithful Christians, they had faith and they still died."

There is no formula for healing. I've heard them all, yet the truth is we don't know. We do know that we are eternal beings clothed in flesh. Yet, not to pray for people on the basis that not all will be healed would be a big mistake.

Sometimes I have prayed for people over many months and still seen them slip away. I think of Rosie, a team vicar's wife. She was a teacher and worship leader. Discovering she had a terminal brain tumour, I visited her almost every Wednesday for a year. We would laugh, pray and share faith and friendship. As she deteriorated, I would sit by her and hold her hand and just talk to her about the Lord.

She became bed bound. I only knew her with deformity, but radiant beauty shone from that broken body. The last time I saw Rosie was just before her death when she whispered to me, "I want to go home now." I told her about Heaven and the beauty awaiting her. We wept. I kissed her and later that week she stepped into eternity.

A number of those I lost to sickness left messages of thanks for bringing them into a deeper relationship with Christ. They left with a smile of peace on their faces.

To record all the miracles that God has done over the years would need a book just of its own, but this is a flavour of the goodness of God.

Chapter 21
Ministry Further Afield

I have had the pleasure of preaching in many parts of the world. The first time I visited New Zealand it was to represent the Elim Pentecostal Movement. Wynne Lewis recommended Garuda International. I had never heard of them.

Wynne warned me, "You need to understand that you'll be lucky to cover your costs. They aren't fully up to blessing people over there."

That had never been a problem to me, so I contacted them. I told them that I would be bringing Molly, as we would be celebrating our 25th Wedding Anniversary during the visit, but I made sure that they understood that I would be paying for her as the invitation didn't include her.

As the time grew nearer, I emailed them and gave them my card details and asked them to book a nice hotel for our anniversary. I couldn't believe the deal that they found for us. For only £300 more per person than the other airlines, we could travel Business Class. Fantastic! We left Heathrow with no hitches. The plane was a bit old but the service was good and we landed at our first stop in Malaysia. That was where things started to go wrong.

After changing planes, we were positioned at the end of the runway, ready for take off. Molly, who is not the best flyer, said, "I don't like the sound of the engine." I thought to myself, "As if she would know!" Well, as we thundered down the runway the pilot suddenly aborted the take off, telling us, "I don't like the sound of the engine!" So we returned to the terminal and, after a 10-hour wait, there was still no replacement plane. When I asked about it I was told, "We are looking for one." I had visions of them going through hangers, lifting dust sheets to see if they had a spare.

When we finally boarded the plane again, we saw an engineer in overalls walk past us on the tarmac with spanners and screwdrivers sticking out of his pockets! A thunderous banging started and Molly, in a deeply concerned state, asked a hostess, "We are not flying in this plane, are we? I have children back home!"

At which the retort came, "Don't worry, Madam. We don't want to die either!"

Molly, now in full panic mode, said, "Die? Who mentioned dying?!"

It was like something out of a film! Then Molly looked and saw a pilot, fast asleep in the next row with lumps of ice on his eye lids. Molly, ready to run for her life, said, "Tell me he isn't the pilot?" I said, confidently, "No, don't be silly, he's just having a lift after his own flight." In due course, the noise of the maintenance man banging woke him up. He smiled at me and I said to him, "Finished for the day, are you?" He replied, "I will be, when I've flown this plane to Auckland."

Our prayer life improved considerably during that flight. When we arrived, I apologised for the delay to our host who was waiting for us at the airport. He told us not to worry, saying that firstly, they never turned up on time to collect anyone from the Garuda flight because it was always delayed, and secondly that they knew I must be a man of faith by flying with them in the first place! I gather they have bought new planes since then.

We had a brilliant stay there. It started with a journey on a coach from Auckland to Wellington. Ash from Mount Tongariru was hanging over many places, as it had erupted recently and was still going when we arrived. Wellington was dark with a storm brewing as we arrived. By this point Molly just wanted to go home.

I was due to speak within forty minutes. The place was full and I met, for the first time, some Samoan guys. Some were professional rugby players and they made me look positively anorexic! Over my subsequent visits they became close friends.

We had incredibly powerful meetings and, against all the warnings and low expectations, Molly and I were incredibly blessed by these wonderful people. They gave us air tickets to go to see the Molly's aunt in Napier (a fantastic visit as Molly had previously never met her, having only discovered each other recently). I also received tickets to visit my nephew, Gresham, who was a doctor in Christchurch. They had booked us a beautiful hotel for our 25th Anniversary. When I came to pay the bill, the receptionist told me that the Elim Pentecostal Church had already paid it.

Towards the end of our trip we returned to Wellington. Pastor Mike Knott stopped the car outside the most expensive hotel on the harbour front and walked in carrying our luggage. He told us that his congregation had been so moved by our ministry that they had paid for us to have a meal and overnight stay in a executive room before we ministered on our final night.

If this wasn't the pinnacle to the most memorable of trips, then this was to come when we arrived at Auckland airport for our homeward journey. The Elim secretary handed me a cheque to cover the cost of Molly's ticket. I lovingly protested and told them that I had covered the cost as the invite was only for me. She said the reason that I had received so many blessings was because I was one of the few that had never talked about money to them. Secondly, I was one of only a few who stopped and prayed for every person needing prayer, even it ran into masses.

So we left a beautiful country and some beautiful people. I would return in the future, however not with Molly. On that second visit she remained at home with a young lady who was like family to us that had lost a baby. I decided to take her husband with me to help him in his grief. I prophesied that they would soon have another child, a girl, and she would be called Promise. It wasn't too long before this prophecy was realised.

During a visit to Dunedin we spoke at a University outreach night and as I stood up to preach, I found myself delivering the whole sermon in rhyme. It blew them away!

Having represented Elim in New Zealand, I was invited to Northern Ireland to the Hollywood Bible Week. It was my first visit to that beautiful place. The miracles that I have seen over there have been incredible. On that first visit, a lady crippled with arthritis was seen going up and down the aisles on her knees to prove that the pain had immediately left her limbs and she was no longer deformed! At that very moment an Elim leader had a bright beam of light hit him. He had been born colour blind, but could now see all colours perfectly.

We saw much impact during the next three years of visits there. A police training officer had stress fractures in both legs, but after prayer he immediately went on a three mile run! At a men's weekend the Holy Spirit fell at the communion service and every man in the building fell in the presence of God for over an hour. We had a mission at the Waterfront theatre in Derry and a deaf lady was totally healed during the sermon. At the same time the town drunk was saved and vowed never to touch drink again. We booked the famous Kings Hall in Belfast and saw traffic congestion on the way there because of all the people who were trying to get in! I was on the BBC and they received over 21,000 phone calls for information.

Then there was the graffiti artist – man who was very well known and who had painted many of the images that had become so well known on the buildings in the North. He had become a born again Christian and, through his friend and mine, Chris Orr, I was introduced to him. It was a miracle that we ever became friends, since he had lost two cousins, both shot dead by British troops. His acceptance of anything English and, as he would think, Protestant would have been very limited. To quote his words some time later, when he introduced me at a service, "If you had ever told me that I would be not only introducing an English man, but also calling him my friend, I would have spat in your face."

He said to me one day, "If I could arrange a meeting with some people, although I couldn't guarantee your safety, would you agree to come?" I took a little while to respond to that. In middle England we are not often asked to put our life on the line for our faith. I prayed for a week and, against the advice of other Christian leaders, I agreed to go to the meeting. Two car rides later and in and out of several houses, I actually ended up in the IRA headquarters.

About a dozen of them, with a few wives, were sat on chairs looking at me. There was an armed guard on the door. I remember thinking how ordinary they all looked. To me, they were just normal men and women, but this room had seen the planning of many killings.

I spoke for 8 minutes – my shortest sermon ever! Then I sat down and the Lord said to me, "Now stand up and speak over their lives." I stood, in genuine fear, but I was obedient. I had come too far not to be. First, I told one of them what happened to them at the age of 16. I continued, one by one, speaking over them and, as I worked through the number, they fell off their chairs by the power of the Holy Spirit. I left, with them all spaced out on the floor!

The next day, the artist found a black ribbon on his front door.

He said it was a warning not to bring me again. I prophesied over him that when the violence was over, his wall paintings would go around the world. I also added that, "From little acorns come great oaks."

When peace was secured, the Derry city authorities asked him to transfer his "war art" to canvas, as they wanted to send him to Australia and the USA to open exhibitions amongst the Irish communities. They also included his work in their official welcome brochure. As he sat in the airport lounge waiting for his first flight, he picked up the daily news paper and on the front page were the words, "From little acorns come great oaks". It was talking about Derry. I know now that the word Derry means "the place of great oaks"!

Back to Chris Orr. He and I met when Chris's wife, Hilda, had brought their son to the Hollywood Bible week. Hilda had stayed for the service and was so convinced of my "trickery" that she had stayed behind to try to expose it as such. Chris, a TV cameraman, was also a fairly staid evangelical, so Hilda phoned him to come up, which he did. She was convinced that I was controlling people's minds, but that night I "read their mail". It blew them away and was the start of a beautiful friendship between us.

Chris became the producer of my film ministry and we had a tremendous breakthrough when a twelve week series on Ireland with cultural music, history and interviews was shown on a Republican TV station over the twelve days of Christmas. It brought a great response. It was those programmes that led to the mission on the Waterfront.

The Roman Catholic Bishop invited the Catholic community to attend my services! I was warned by a prominent Christian that the political infighting amongst the denomination that I belonged to would close the door to this very successful ministry.

Living in England, I didn't believe it. However, it happened and I was slowly pushed back across the Irish sea or, to be more accurate, over the border into the Republic.

After speaking at a conference in the Republic one day, a lady called Jackie came out with her sister. I asked her if she wanted prayer. She said she did not, but wanted to ask me a question. She said, "Where do children go when they die?" She told me that she had lost her young daughter, Emily, only seven days before.

I asked her to come back that night to the next service. I went back to my room at the hotel and prayed for guidance. I wrote a poem on death for her and changed my sermon to accommodate any who had suffered loss. At the end she cried and I said to her, "If you lived in England, I would be your dad."

Gary, her husband, was later standing outside. He was suffering from the same genetic condition that had claimed the life of their daughter and he was outside using a nebuliser. He came up to me and said, "Did you mean that?" He told me that Jackie's father had passed away and that he was now facing death himself and would really appreciate knowing there was someone looking out for Jackie.

That meeting started a relationship that saw them eventually build a church in their back garden, a beautiful building that holds about eighty people, in County Roscommon. Gary lived to see the vision realised and we celebrated each stage of the building project together, from foundation stone to consecration. Gary passed into the presence of the Lord and we celebrated his life in the same church.

The connection with the Free Methodist church grew and, having been elected to their three-man leadership team, I received a call from a missions worker in Bulgaria. Mike Long, from the USA FM, had met a group of 100 Churches who had been dramatically affected by a merger with a Pentecostal denomination from the States.

Many of the churches had lost their buildings and about 10,000 people had been lost to this group. They didn't trust Mike as he was from the States and, as Pentecostals, they couldn't consider affiliation with Methodists! Mike, however, continued to love and care for them in their vulnerable state. He mentioned me and Renewal as a possible relationship. The British didn't need visas, we were Pentecostal so they understood us, and we were also Methodist, just to confuse them a bit!

We paid for three of their leadership to come to the UK and talk. This resulted in Renewal funding a mission in Sophia, the capital city, in the former Communist conference hall, embracing four thousand people.

Their worship team was equal to Hillsong and was led by the pianist of the Sophia Philharmonic Orchestra. Richard Taylor and Geoff Cartwright managed the teams from Renewal and Bulgaria in a fantastic way. On the Sunday morning, we served communion to the leadership of four denominations who had been in broken relationships.

This process of building relationship has taken seven years to develop into a formal agreement between the Church of God, Bulgaria and the Free Methodist Church. They have about 100 churches at this time. Al Mellinger from the USA has become the link man between the USA and the UK. What a fantastic worker he is.

Having preached in many places, however, the call to return to Solihull has always been there. It's my Bethlehem, my "house of bread", and I always to return to it.

I've recently heard that I am now to be appointed European Team Member for the Free Methodist Church and Overseer for Bulgaria and Archbishop of the Province of Europe [UK] for the Communion of Evangelical Episcopal Churches ... not bad for 68!

Chapter 22
Grateful for Grace

My personal life, at the age of sixty eight and amidst this manic activity, is satisfying. I have already touched on the pressures of family life during those early years, but to enjoy the love of grandchildren is a precious bonus. The eldest, Poppy who is eight at the time of writing, had a difficult beginning with her father leaving home when she was only months old. We gave her a trip to Euro Disney when she was seven. The joy of seeing her face was the only reward necessary. It was her Nanna that helped lead her to the Lord.

Hattie, who is two, calls out to me, "GRANDA!" She can't fully call me granddad yet! Then there is Camille, just a little mite, but one who loves her Granddad and Nanna. She is a bundle of uncontrollable energy and loves dancing – a lovely little four year old. Then there is Sebastian (Seb), a little boy with a mop of hair. He is a sweetie.

I'm a blessed man, granted mercies beyond my worthiness. Having survived birth as a twin, with my sister not making it, I have since been saved from serious injury or death on a number of occasions. Here are a few of them:

Once, I was driving home from a preaching engagement in Ipswich on a dark, foggy night. It was about 11.30pm and the fog was really heavy. Conditions closed in to the extent that I decided to leave the M6 to travel on the M45, a little used motorway that ends to the south of Coventry. I slowly turned to enter the M45 when the inner voice of the Spirit said to me, "The enemy of your soul requires your life tonight. Commit yourself to Me now!"

I thought, how will they identify me? Do I have my driving licence on me? Then I prayed, "Into Your hands I commit myself."

Suddenly, out of the thick blanket of fog, bright lights penetrated the gloom. The headlights were on my side of the carriageway! Time stood still and event took place in slow motion. I swerved and the approaching car hit the side of my car and vanished into the fog-infested night. I limped to a phone and reported the accident to the Police. They didn't seem to believe me at first, yet took my details and I drove slowly home. At 6.00am the next morning, the Northamptonshire Police phoned to say they had closed the M6 and apprehended two men. The driver was so drunk that he had not realised that he had driven down the wrong side of the motorway.

Then I had survived the situation with the Satanist at the hotel in Coventry, when he was waiting in his room to shoot me. The Lord told me not to go in.

There was the time I had my heart attack on the motorway at 70 miles per hour, where singing in the Holy Spirit saved me from passing out.

You may recall I was electrocuted when standing in a medical unit. Holding onto a mains cable I was thrown through the door and finished up in hospital, but I only sustained heavy bruising and a burn to the palm of my hand.

On another occasion, I was returning home from Rotherham on the M1 at about 10.00pm one Saturday night. There was nothing on the roads at all and I was driving at 70 miles per hour when the inner voice again spoke: "Reduce your speed to 50 and pull over immediately into the slow lane."

There seemed to be no reason to do this, however I have learned to obey the Holy Spirit. On entering the slow lane, seconds later a car shot past me at high speed and hit the central reservation. It literally spun over the top of my car and vanished over the bank that rose away from the motorway. On clambering up the embankment, I saw a shattered car with a young lady driver strapped in and trapped in the wreck. She turned out to be a backslidden Christian.

Then there was the time when I was travelling down the motorway and my car left the carriageway. Actually, it's not accurate to say that it "left" the carriageway – it was literally "picked up" and place in the inside lane. My passenger, Tony, was in total shock and couldn't speak for a long time. Eventually he managed to relay to me that he had seen an angel, about 9' tall, who had picked the car up, laid it back down and smiled before disappearing. Ahead, in the lane we had been travelling in, lay a pile of rubble into which we would have crashed.

On another occasion, an ex-member of the Paratroop Regiment had determined to do me harm. As he approached the church one dark night, he saw what he supposed to be a spotlight above the front door. But as this man approached the door he could see a being, about 9' tall, leaning on a sword. The angel looked down at him and simply said, "I don't think you want to cross me do you?" With that, the man ran at speed down the road!

Then a demonically controlled lady came to church one Tuesday night with the express purpose of plunging a pair of decorating shears into my chest. I wasn't there that night. She was not happy, made a scene, and had to be escorted from the building. Later, she attacked one of the police officers who worships at our church with the same weapon. She was arrested and eventually sentenced. This woman could talk in demonic tongues and knew that the officer was a Christian without natural knowledge.

Another time, Molly and her prayer group felt the urgent need to pray for me on a Monday morning. As they did, I arrived at an office block and received a call on my mobile. I briskly walked through the door only to hear a terrific explosion. I had walked straight though what I thought was an open door, but it was actually a full, plate glass window. It shattered with glass everywhere and large jagged shards all around me. Staff stood and stared in amazement as I escaped without so much as a scratch.

Alongside this, I have been attacked by four drug addicts and had my nose broken. I have had Satanists try to strangle me and do me physical harm. I've wrestled with a demonically influenced man on the floor, trying to disarm him as he endeavoured to kill me with a knife! And so it goes on. I guess I should just call these occupational hazards! In all of this, I am grateful for God's amazing grace.

Chapter 23
The Order of St Leonard
and Passing the Baton

OSL

Since the turn of the millennium both I and Renewal have "boldly gone to places never seen before". The reopening of "old wells" had embraced the Methodist connection, yet there was more to come. Renewal Solihull is housed in a factory – beautifully refurbished, but it's still a factory. Some of our young couples desired the traditional style of church building for their marriages, so I asked the team to look for a redundant traditional building.

David Tompkins, one of our Pastoral team, lived in a Warwickshire village near Wroxall Abbey. This establishment dated back to 1141 when the Priory of St Leonard was formed. It has an incredible history. The Shakespeare family worshipped there. One of William's relatives was even a nun in the Priory but, after the reformation, Cromwell's men destroyed the Priory, leaving behind only the sister chapel. Sir Christopher Wren purchased the entire site some time later and many of Wren's relatives are buried there. The sister chapel became known as St Leonard's Church of England. An Elizabethan house was rebuilt on the site, containing a spectacular stained glass window depicting the unique history of Wroxall.

Dating back to 1141, the story is told that Hugh De Hatton, a local dignitary, had left to fight in the crusades. He was captured by Saracen warriors and chained to a wall for seven years. St Leonard miraculously appeared to him, freed him from his chains, and he was able to return to his native Warwickshire home. So many years at war had, however, taken their toll. He was unkempt, with long hair and a beard and his wife did not recognise him. Romantically, however, the couple had broken a wedding band before he had left, and each retained a separate piece. The couple fitted the two pieces of the ring together and were reunited. Hugh's joy was such that he dedicated the lands at Wroxall to St Leonard and built the Priory on the site. An order of Black nuns were consecrated and the Prioress was commissioned by the Pope. The site has a powerful female influence, which has been reiterated throughout its history, including the fact that the site also went on to become a girls' school.

When we arrived, the site was now in the hands of an exclusive hotel group who had turned the old house into a wonderful hotel. They were very amenable to us reopening the church for services and marriages. As I walked into this ancient house of God, the doors to which had been closed for only seven years since 1141, I felt the Holy Spirit distinctly say, "Open up this well." So we commissioned it once more with a new name: "Wren's Chapel, Wroxall Abbey".

A small village congregation was formed, very different from the style of our normal services, yet very meaningful. It was then that I received a call from Bishop Charles Travis of the Communion of Evangelical Episcopal Churches to see if I could be a link with the Church of England through my friendship with Archbishop John Sentamu. After much discussion, they asked if I would consider being appointed a Bishop with them so that I could represent them in the UK.

This was a strange request. I was a Overseer with the Free Methodist Church. However, as a unique exception, the Bishops of the Free Methodist Church decided, as a one off, that they would let me receive consecration and still maintain my role with them. They said that I was a Kingdom man, and shouldn't be contained by any one group.

So the CEEC came and firstly appointed me a Presbyter and then returned two months later and consecrated me as the Bishop of Wroxall Abbey. Because of my calling to "open up old wells" they suggested that I open an Order. I didn't fully appreciate what an Order was, so I researched it. I found that it was basically a group of people committed by lifestyle, faith and service to the community they serve. I realised, after much prayer, that this could be the way forward. It had been estimated that about thirty three thousand denominational groups had been formed in the world. The world really doesn't need any more denominations. John 17 is the prayer of Christ for oneness, yet we as a body of believers have slowly become fractured by doctrine over the two centuries of our existence.

On looking into the history of St Leonard, it was obvious that this was the biggest well yet to open. He lived in the year five hundred as a Frank. This preceded what we would call France today. He left financial security to become a cleric – so did I. He was ordained the same day as his brother – so was I. He had a healing ministry – so have I. He preached a strong message of repentance – so do I. He had a effective community program – so do we.

This was now taking me deeper into history, yet I was still that Pentecostal fire brand from Birmingham!

At the time we had been working for about 8 years with a couple who were media consultants. They'd helped us with a variety of marketing materials for the church, so I commissioned them to create an identity and materials to reflect the new Order of St Leonard. Dave and Mandy Cooper could not be mistaken for your average Christians. In fact, at that time, they weren't saved at all! Dave very much personified his 25 year history in Rock 'n' Roll – long hair, tattoos and clothing that wouldn't confuse him with a bank manager. Mandy was very creative, yet came from a background of spiritual darkness. They were the perfect choice to help determine our future ministry!

Mandy was tasked with researching the history and putting together timelines and genealogies to uncover the foundations of this old well. The first thing that she did was visit the site. She went alone, with a camera, and walked around the ruins of the old abbey, taking photographs. On her return to her desk that afternoon, she was looking through the pictures and found something that surprised her. At the centre of one of the photographs was a bright white cross that looked as if it lay on the driveway next to the chapel. On checking the location, she found that although it was a driveway today, this was very much in the region of where the high altar of the old priory would have stood. She sent me the photograph, which I found to be most significant. Mandy, however, unsaved and unimpressed decided that, although it was "quite cool" it was probably just a trick of the light.

A little while later, I asked Mandy if she had looked closely at the stained glass windows around the site. She told me that she had. Indeed, she had lots of photographs of them. I just smiled at her and suggested that she hadn't looked properly and perhaps she should go and have a look at the largest stained glass window, now sited in the main house.

What she saw in that window was the catalyst to her salvation shortly afterwards. There, in the final three panes of glass, the story of a novice nun was portrayed. The nun shared Mandy's initials. This lady had received a divine visitation and instructions as to where to build the chapel. The instruction was accompanied by the image of a white cross, laid across the floor in front of the high altar.

I asked Mandy what she thought of this now. She replied, "It's a sign for you to start the Order."

Since then, my friends J. John and Chris Bowater have joined the Order, together with about eighty other leaders in the UK at the time of writing. Unbelievable growth, however, has come from Africa. We now operate in ten countries covering over five thousand churches and one hundred denominations. We have also opened in Bulgaria working with the Pentecostal, Evangelical, Catholic and Orthodox Churches. And then recently we have been to Pakistan. Altogether, a total of three million people!

The story of this alone will be continued in a further volume, but I must note the work and support of David Insull, a close friend who has worked tirelessly in helping formulate the basis of the Order's credibility. He and his wife, Jackie, are like gold dust.

And Dave Cooper? Well Dave and Mandy are now saved. Ex-rocker Dave has just finished his Masters Degree in Applied Theology and has started touring with me. Mandy still works with me on the development of OSL and is Director of a group of companies.

Passing the Baton

With age creeping up on me, it was sensible to start considering a future replacement. Richard Taylor, who had served with youthful inspiration as my number two, came to see me and emotionally informed me that he felt a deep call to return to Wales, his home country, and become Director of Victory Outreach – the drug rehab ministry under which he himself was saved and nurtured.

Since then, he has established a vibrant church with a number of church plants and they are experiencing a great outpouring at this time. That left me light of a number two, so my dear friend Peter Jenkins, who was our Missions Director, stepped up as a temporary Associate until the Lord found us an Associate Pastor. Peter did a fantastic job for the next three years but, as we are both the same age, we could both see that this arrangement wasn't a permanent solution to our dilemma.

Around that time I was attending a conference in Hawaii and met a man called Ian Green. Ian only lives about eleven miles from me, yet having not met for a number of years we end up on the same table half way round the world! Ian is a true Evangelist, licensed by the Assemblies of God. He asked me if I had found a future replacement. I said that I hadn't and explained that the good ones already had churches that they had developed and that they wouldn't wish to serve another man's vision. Ian simply said, "Just list all the people who could, regardless of whether or not you think they would."

As I went down the list I mentioned John Andrews. John was the pastor of the Hub in Rotherham. Not a very large church, but he was a fine teacher of the Word and on the National Leadership Team of the Assemblies of God. On returning from Hawaii I received a phone call from John asking to meet for a chat. I asked him if he had been talking to Ian Green and he replied that he had not.

I took him out for a meal and during our conversation I asked him if he could ever see himself pastoring a large church. He said, "Until recently no. However, I feel I could in the future." I then mentioned Renewal and he said, "No, you are a legend in your own time. No one could follow you." I said that actually I wasn't going to retire, but that I would eventually change my role to be Founding Pastor and not Senior Pastor.

John went very pale! On his return home, he greeted his wife Dawn with, "You'll never guess what happened." Dawn replied, "Pastor Dave has asked you to join him." She could see it all over him. After much prayer and meetings with our Elders, John felt it was of the Lord to accept. At the time of writing he has been with us about one year. I have delegated many of my administration duties to him and I have become the overseer, endorsing or questioning to help equip him, should he become the Senior Pastor.

I have been asked by the Elders to decide the future role I wish to play in the vision of the house till my departure to the Paradise of God. It seems to be working very well at this moment. Long may it continue.

Chapter 24
Leaving a Legacy

If asked to consider what legacy I wish to leave behind for future generations, there are two major things. One is my considerable passion of pastoring. The other is my natural relationship with the supernatural.

True pastoring

As Christians we should, by virtue of our belief, develop a lifestyle of integrity that far outweighs our previous life, outside of Christ. But I find it common to encounter many moral weaknesses in dealing with Christians, whether leaders or followers. Sadly, Christians are often unable to receive a truthful assessment of their character without taking offence and leaving their place of worship.

Some years ago, five young men visited me and asked me to mentor them and to tell them if I could see areas of their lives that needed attention. I gladly agreed and, with thought and prayer, addressed the strengths and weaknesses of each. It was then that I realised that many Christians are actually seeking endorsement not mentorship. Three took offence and left the church. One I never saw again. Only one still remains today!

Those seeking endorsement feel they should have free access to your time, thought and attention and don't appreciate there are many more needy of care and attention. There are people who have been part of our church for years who would never think of troubling me and others that if I haven't phoned them to talk about every problem they have, decide that I obviously don't care.

Dealing with people is seldom taught in preparation for ministry, but is a skill that is worth its weight in gold. More people leave churches over offence than bad doctrine. I appreciate the need for learning, degrees in theology, divinity, mission and all other related subjects – each of which can be exceptionally useful in determining directional paths. It must be said, however, that few ever receive the "hands on" training needed for ministry. Pastoral work is not a rota for visitation, but a heart for those in the kingdom of God, their wellbeing and the effect that situations can have on these individuals, their families and loved ones.

The question I would ask is similar to that set by Mother Teresa of Calcutta. She wouldn't accept a sister into the Order until they had nursed a wretched homeless person unto death. Often these pathetic individuals had maggots feeding off their open wounds that had turned to puss. She would tell her workers that, in the natural, they would be repulsed at the sight. However, as in Matthew 25, they needed to see Jesus as the recipient of their care.

This might seem a little severe in our culture, but pastoral care, like hospital care, must never be reduced to a file on a computer. The holding of the hand and personal prayer is still the way of Christ.

Sadly, many think that compassion is one of the gifts of the Spirit, or is for those who are exceptional pastors. If my reading of Scripture is correct, the Lord Jesus Christ, modelled it to all His disciples. In other words, He expected it to be a prerequisite of ministry.

Name me one of the Apostolic gifts to the Church that you can honestly say doesn't need compassion? Would you want to work with an Apostle who had not entered into the pain of those he seeks to establish and build? The Teacher who doesn't think of the consequences of his style or delivery and the emotional damage they could bring? The Prophet who declares, "Stuff you everybody, thus says The Lord!"

The worship song says it all: "Everyone needs Compassion."

Pastoring is not the gift of administration that checks procedures and ticks boxes. It is a gift of compassion that binds up the broken and afflicted.

Embracing the supernatural

Legacy is a new concept at Renewal that has replaced our Ministry of Power meetings. It takes place each Tuesday evening at Renewal Solihull at 7.30pm. It is an equipping for ministry all who truly believe and want to know the reality of God's supernatural power at work in their lives.

Many queried me becoming a Methodist, feeling that I had somehow disowned my heritage as a Pentecostal, but I have never been at ease with stamping my relationship with the Holy Spirit with a denominational label. Although I faithfully served the Elim Church for 28 years and enjoyed a deep relationship with the Assembles of God for even longer, I am not a Pentecostalist, I'm a Pentecostal. The reality of this hit home to me on night in 2002 when, as a powerful Tuesday night meeting I drove our visiting FM Bishop Richard back to his accommodation. He commented that Wesley was also Pentecostal, yet did not adopt the style of Pentecostalism. This simple revelation revealed who I really was. A true "Pentecostal" can belong to almost any denomination.

If that wasn't enough, a few weeks later I was addressing the Pastors' Conference at Bedworth for the Assemblies of God. My good friend Paul Weaver, who was the General Superintendent at that time, announced that I would be speaking that night and the next morning. Ken Gott and one of his team had been discussing only the week before the lack of Pentecostals about today. His team member queried that statement, as all the team had been filled with the Spirit. Ken described the categories as Pentecostalists, Charismatics and Pentecostals. He then gave a description of what a Pentecostal was: someone who was open to the Holy Spirit, unpredictable, could not be boxed in and maybe a bit frightening! He said that such a description fitted Smith Wigglesworth perfectly. Then he went on to tell his team, "Dave Carr is as near as you will get. Listen to him and then we'll have breakfast with him in the morning."

That night, I prophesied over the mission workers in the language of their calling. I also "read some mail". The next morning, one young man admitted that it had frightened him silly, but that he now knew the difference. After over fifty years, so did I!

I would like to think that as so many Christians are now familiar with the Person and personality of the Holy Spirit, that we are seeing more supernatural encounters. However, what I often see is a higher level of professionalism rather than prophetic dependency.

Traditional Pentecostalism has, over the years, placed what I believe to be over emphasis on speaking in tongues, rather than a personal relationship with the Giver.

We live in a day when discernment is of vital importance and liberty has, in some cases, turned into licence. Professional qualifications are a great benefit. I have some myself. However, they must never be a substitute for the actions of the Spirit. One word of knowledge can achieve what weeks of professional therapy can't do. If you have both, working hand in hand, we gain the best of both worlds – the Kingdom of Heaven and the Kingdom of this World.

The pure Gospel

If I can leave one further thought as a legacy to the next generation it is this: you may change your style of delivery, but the message must remain the same. Sin is still sin. Righteousness has not changed its name. Salvation is not a good idea, but a God idea. And Christianity is not to be rebranded by the world to make it more acceptable.

Our message has never been popular. But it never will be and it's not meant to be! Change the wrapping paper if you wish, change the box for a different shape, that fine – but the product itself doesn't belong to you, so leave it alone!

* * *

Since Dr John Andrews has joined us, it has enabled me to meet with people more. As a Pastor, that is the desire of my heart. We all think that we are mature and can acclimatise to change, but few can. Change is called for when frustration is pending, yet the time to change is when you don't need to. In football, the good teams change before it becomes critical. That is why they win more than others.

At Renewal we are changing whilst still successful, not for the sake of change but for the glorious future. I'm not ready to climb the mountain or get into a chariot of fire and I believe I still have a great deal to achieve, yet we must train the Elishas and the Joshuas ready for the day of their calling.

We have defined our mission in three simple yet inextricable strands. A three-strand rope is not easily broken. We must Reach, Care and Equip.

As a senior leader, I have spent many hours involved in leadership meetings and, by virtue of the office, away from non-Christian people. I cried out to the Lord that I wanted to spend time with the people.

Heidi Lane is the wife of one of our Elders, Phil. Heidi is a ward sister for Cardiology at our local General Hospital. One day, the Matron asked her how she seemed to manage when so many others faced crisis. Heidi explained that her faith, family and pastor supported her in this. The Matron asked to see me and asked if I would mentor her staff over their three-hospital group.

This was the opportunity for me to meet people who mostly wouldn't know me as either Pastor Dave or Bishop David. I agreed to trial it on Wednesdays, my day off, to see if it would work. It was not good to be independent from the Christian Community so I contacted the hospital chaplaincy and suggested we meet.

Pam, who was head of Diversity and acting head of Chaplaincy, was very pleased to see me. Chaplaincy can either be a formal job or the opportunity to minister to vulnerable people with exceptional compassion. I have found many in the first category, but few in the second. I have now been there some 18 months and have secured three wards as their chosen spiritual advisor. Many staff have been helped and some have found Christ. I was initially there to see Cardiology patients, however other departments have seen the results and asked for the same opportunities.

We minister to about ten thousand people including patients and staff. It was threatening to the Chaplains to have a Bishop working in the team, however my good friend Roy Boffey, who I've known for forty years, was the key to me finally breaking into the team.

The Management appreciate my attendance and actually bring it up in Regional meetings. It was a great surprise to me when Lord Hunt, Chair of the Trust, awarded me the title Honorary Chief Chaplain.

As I walk the corridors I am called Father, Padre, Rev, Vicar, Bishop and Pastor. Underneath all the titles, I'm Dave. Unless I'm in trouble and then I'm David!

Thank You

I could have rambled on for many more chapters. Many fascinating and worthy people have entered and left my life. They have not been ignored; they are for future telling. You may be part of my history and your story has not been told. If so, please send me your memories for the next instalment in our journey. I have left out more than I have put in of my life and the impact that many people have had on me. No life of success can ever manifest itself in isolation.

What about the man outside the pulpit? You could ask my wife, but hopefully you won't!

Unlike my public persona, I'm very switched off. I like to be on my own for periods of time and I don't speak unless spoken to. I love TV programmes on medicine, the countryside, animals and history – and I like war films. I enjoy most kinds of music except jazz, heavy opera and heavy metal! I enjoy football (even though I follow Oxford United!). I play bowls and I am the Chaplain for the Birmingham Bowls Association. I love visiting historic buildings and cathedrals.

My heroes are Billy Graham, John Wesley, Mother Teresa and Nelson Mandela – together with respected ministers of the old school, Eldin Corsie and David Shearman.

I love watches (so I'm fortunate that Ashley deals in them). I drive a 12 year old BMW and I live in a Bungalow in Solihull.

I still just about have Elsa, our pet Standard Schnauzer. My wife has decided that Elsa belongs to me. I don't know what I've done to upset her! Elsa now has a friend: an extremely well behaved puppy called Oliver, who exhibits such serenity that he is almost monastic. I'm hopeful that this good nature might somehow influence Elsa.

I want to thank Molly for sacrificially giving me away to thousands of people over the last forty three years. Yes, it was her gift to God, yet I know that it has been very lonely at times for both of us. Having grown up in poverty as a child, crying to God in the darkest of nights when her only friend, her mum, was gasping for breath, it's not easy to give away those you love. But the Lord restored another part of her family. Her brother, who was taken to live with her father at just a few weeks old, was located with a revelation from God. With just a letter from 1948 and after 20 years of searching in the natural, it took just one phone call after the Lord told her He would restore the secrets of generations past.

We found Brian, who had spent his career as a staff sergeant in the army. We learned that after being taken from the family home by his father he had, shortly afterwards, been sent to a children's home. He believed his step-mother was his real mother and hadn't wanted him. He never knew he had a sister. When these two met it was like something from a film, both running to hug and kiss one another. It's strange to listen to him talk of "Our Moll" in his broad Yorkshire accent, while Molly talks about "Our Brian" in broad Brummie! We were even able to reunite him and introduce him to his mother just before she died.

Having many nations in Renewal, me and Molly have become "Father and Mother" of the house. I have given away young ladies at weddings when they didn't have a dad. I think of Rose who married in Antigua on the beach. That was a good one!

I don't feel 68. Inside I'm about 35! I don't really know what has happened to the time, but we are but frail children of dust. I don't think about what it is like to be in eternity. Eye has not seen nor ear heard, or even thought ever contemplated the things that God has prepared for us.

Finally, I do have deep concern that because Eschatology has now fragmented the present day Church, we are one of the first generation of Christians who rarely teach on the Second Coming of Jesus Christ. We have a Christian Community that is driven by desires rather than destiny, where the gospel is one of temporal gratification rather than eternal gratitude.

Many are living for today rather than planning for tomorrow. We are not truly looking for His appearing, however He is only coming for those who are. So preachers, you don't have to teach dispensationalism to acknowledge His Coming, but over twice as many prophesies declare the Second Advent than the First. The world knows much about the First Advent and the Church knows little about the Second Advent.

This must change.

That's my rant over!

Thank you

Thank you my wonderful children: Ashley and his wife Fiona; Melanie and her husband David; Lucy and her husband Richard. I'm very proud of you all. Stay in the faith of Christ. I would love to see you on the other side.

To my fantastic grandchildren: Poppy, Hattie, Camille and Sebastian – and any future ones! When you are older and read this book, remember the God of your Granddad and follow Him.

To all our "spiritual" adopted children and grand children, we love you all. I know that if Dan and Matthew are not mentioned they will tell their Nan!

To my wife, Molly, thank you for loving and supporting me for fifty years of knowing each other. I love you.

To friends and fellow National leaders – thank you.

To my PA's Susan, Yvonne, Barbara and Lyn – it will get better!

To team Renewal past and present and its many partners. It could not have happened without you.

And to our Renewal Platinum Partners, who have journeyed with us for thirty years or more. You are fantastic!

To those quality partners who have arrived in Paradise. I'm getting nearer each year to seeing you all ... one day!

And to all I have disappointed, I am sorry.

To Mandy, who has translated this into English, thank you.

And to the most psychologically disturbed dog in the world, Elsa. Remember, I'm the only one who loves you! Don't chew this book!

* * *

So that's the story folks. It's been good telling it to you. Maybe we will talk again in the future – *Dave*.